Reproductive Physiology of Vertebrates

Reproductive

SECOND EDITION

COMSTOCK PUBLISHING ASSOCIATES, a division of

Physiology of Vertebrates

Ari van Tienhoven, Ph.D.
Professor of Animal Physiology
Department of Poultry and Avian Sciences
   and Division of Biological Sciences
Cornell University

CORNELL UNIVERSITY PRESS Ithaca and London

First published 1983 by Cornell University Press.
Published in the United Kingdom by Cornell University Press Ltd.,
Ely House, 37 Dover Street, London W1X 4HQ.

International Standard Book Number 0-8014-1281-1
Library of Congress Catalog Card Number 82-71595

Printed in the United States of America

*Librarians: Library of Congress cataloging information appears
on the last page of the book.*

*The paper in this book is acid-free and meets the guidelines
for permanence and durability of the Committee on Production
Guidelines for Book Longevity of the Council on Library Resources.*

KM
3-14-83

To Ans

Preface

The second edition of this book, like the first, is based on lectures given in a course on reproductive physiology of vertebrates. It was originally written to fill the need for a textbook that compared the reproductive physiology of different classes of vertebrates, from cyclostomes to humans, and it continues to serve that purpose. It is intended both as a textbook for advanced students familiar with concepts of neuroendocrinology and reproductive physiology, and as a reference book for researchers.

About ten years after the first edition was published, two of my colleagues, Dr. Neena B. Schwartz and Dr. Terry G. Baker, stated that they wished a newer edition were available. Their encouragement led me to undertake the writing of this second edition, in which I incorporated much of the new information about hormonal fluctuations gained through radioimmunoassay techniques. I rearranged the chapters so that Chapter 1, "Sex and Its Determination," is now followed by a chapter on sexual development through the prenatal period. A chapter on intersexes is included because of its importance, especially in teleost fishes. A new chapter, "Puberty," precedes "Anatomy of the Reproductive System." The chapters on the physiology of the testis and the ovary lay the foundation for the understanding of the reproductive cycles. The chapter on insemination and fertilization, followed by those on care of the embryo and fetus and the expulsion of the oocyte, embryo, or fetus, give a chronological and logical sequence of events. Chapter 12, "Reproduction and Immunology," and especially Chapters 13 and 14, "Endocrinology of Reproductive Behavior" and "Environment and Reproduction," integrate the information given in previous chapters.

In the revision, the chapters "The Nongonadal Endocrine Glands," "The Hypothalamus," and "Effects of Nutrition on Reproduction" are omitted, partly because the readers of this book are assumed to be familiar with the neuroendocrinology of reproduction and partly because the information is incorporated in other chapters to provide a better and more logical sequence. The choice of topics and their treatment reflect the bias of my education and interests. The reader will find little in the present edition about the biochemical or molecular biological aspects of reproductive physiology.

Even in a much-studied species such as the laboratory rat, much basic information is lacking, and information about nonlaboratory animals is often fragmentary. I hope that in bringing some diverse information together I will encourage others to do some much-needed research and provide the context that will aid in the interpretation of this information.

It is a pleasure to acknowledge the help and cooperation of many people. First and foremost I thank my wife, Ans, for her continuous confidence, her patience, and her contributions in prodding me to rewrite and rethink. She also proofread the final version of the manuscript and the page proofs. Thanks are due to Dr. Neena B. Schwartz and Dr. Terry G. Baker for their encouragement to rewrite this book. The typists of the Department of Poultry and Avian Sciences at Cornell University, Olympia McFall, Mary Ellen McParland, Theresa Rinkcas, Barbara Smagner, Diane Wittner, and Helen DeGraf were most helpful and patient; Teresa Ferrara retyped the entire final version of the manuscript. The personnel of Cornell University's Mann Library Reference and Loan departments have given frequent and valuable help and were always willing to go an extra mile.

My friend, critic, and delightful devil's advocate, Dr. Howard E. Evans, has contributed in many more ways than he knows to this book, and I will always be in debt for many kindnesses he has extended to me.

The editor of the manuscript, Antoinette M. Wilkinson, deserves a special acknowledgment of my gratitude for her many questions, for her patient explanations, and for ignoring my sometimes caustic comments; she has improved the book substantially. I am also indebted to Allison Dodge of Cornell University Press for her patience and for letting me argue my case.

I gratefully acknowledge authors and publishers who gave their permission to use tables, graphs, and figures from their publications. In these tables and legends small changes made by me are placed in brackets.

Last but not least, I thank Cornell University and the head of my department, Robert Baker, for letting me write without questioning or interfering.

ARI VAN TIENHOVEN

Ithaca, New York

Contents

9 Insemination and Fertilization 250

10 Care of the Embryo and Fetus 270

Abbreviations

The following abbreviations have been used in the text of this book. Abbreviations used in tables and figures are given in their legends.

ABP	Androgen-binding protein	DNA	Deoxyribonucleic acid
A-Ca	Activator calcium	E_1	Estrone
ACTH	Adrenocorticotrophic hormone	E_2	Estradiol
ADP	Androstanediol propionate	E_3	Estriol
AFP	Alpha feto-protein	EB	Estradiol benzoate
AH	Anterior hypothalamus	FGP	First growth phase
AH-POA	Anterior hypothalamic preoptic area	FM	Fluoxymesterone
AP	Anterior pituitary	FMP	Forward-mobility protein
A̶P̶	Hypophysectomy	FSH	Follicle-stimulating hormone
ATD	1,4,6-androstradien-3,17-dione	FTP	6α-fluoro-testosterone propionate
ATP	Adenosinetriphosphate	GSI	Gonadosomatic index
AVT	Arginine vasotocin	GTH	Gonadotrophic hormone
CA^{2+}	Calcium ion	GVBD	Germinal vesicle breakdown
cAMP	Cyclic $3',5'$adenosine monophosphate	h	Hour
°C	Degrees Celsius	hCG	Human chorionic gonadotrophic hormone
C	Corticosterone	HPL	Human placental lactogen
CG	Chorionic gonadotrophin	3β-HSD	3β-hydroxysteroid dehydrogenase
CL	Corpus luteum or corpora lutea	HVc	Hyperstriatum ventralis
COMT	Catechol-O-methyl-transferase	I	Iodine
d	Days	IC	Infundibular complex
DAS	Delayed anovulatory syndrome	ICo	Nucleus intercollicularis
DD	Continuous darkness	irLH	Immunoreactive LH
DES	Diethylstilbestrol	K^+	Potassium ion
DF	Decapacitation factor	kHz	Kilohertz
DHA	Dehydro*epi*androsterone	L:D	Light:dark, e.g., 12L:12D = 12 hours light:12 hours dark
DHT	Dihydrotestosterone		
DHTP	Dihydrotestosterone propionate		

LDH	Lactate dehydrogenase	PG	Prostaglandin
LH	Luteinizing hormone	PGC	Primordial germ cell
LH-RH	Luteinizing hormone–releasing hormone	PGE	Prostaglandin E
		PGF	Prostaglandin F
LL	Continuous light	PGH_2	Prostaglandin H_2
LQ	Lordosis quotient	PGI_2	Prostaglandin I_2 = Prostacyclin
lx	Lux	PINX	Pinealectomized
MAN	Magnocellular nucleus of the anterior neostriatum	PMSG	Pregnant mare's serum gonadotrophin
		POA	Preoptic area of the brain
6-MBOA	6-methoxy-benzoxazolinone	PRL	Prolactin
MEE	Methoxyethynylestradiol	RA	Nucleus robustus archistriatalis
MPO	Medial preoptic area	Rh	Rhesus factor
MT	17α-methyltestosterone	RIA	Radioimmunoassay
MUA	Multiunit activity	RNA	Ribonucleic acid
Na^+	Sodium ion	SCA	Sperm-coating antigen
nXII	Nucleus of the hypoglossal nerve	SCGX	Superior cervical ganglionectomy
nXXIts	Tracheosyringeal portion of the hypoglossal motor nucleus	SCN	Suprachiasmatic nucleus
		SDH	Sorbitol dehydrogenase
oFSH	Ovine follicle-stimulating hormone	SGP	Second growth phase
$2\text{-OH-}E_2$	2 hydroxyestradiol	T	Testosterone
20α-OHP	20α-hydroxypregn-4-en-3-one (*also abbreviated* 20α-ol)	TC	Testosterone cypionate
		Tfm	Testicular feminization syndrome
OIH	Ovulation-inducing hormone	TP	Testosterone propionate
oLH	Ovine luteinizing hormone	TRH	Thyrotrophin-releasing hormone
oPRL	Ovine prolactin	TSH	Thyrotrophin (thyroid-stimulating hormone)
P_4	Progesterone		
pACTH	Porcine adrenocorticotropic hormone	VMH	Ventromedial hypothalamus
PFF	Porcine follicular fluid	ZP	Zona pellucida

Reproductive Physiology of Vertebrates

Sex and Its Determination

The overwhelming majority of vertebrates reproduce sexually. This statement is not as obvious as it appears at first inspection. We should ask what is meant by reproduction and what is meant by sexual.

METHODS OF REPRODUCTION

Cohen (1977) has made the important point that reproduction means that two parents produce new parents in the next generation. If a set of parents have offspring that reach sexual maturity but do not become parents, we can say that the parents have not reproduced. In a stable population each set of parents will produce, on the average, two offspring that reproduce (Cohen, 1977). In view of the large number of offspring born, nearly all sexually produced organisms in such a population must, therefore, fail to reproduce, although they may have bred, i.e., given birth to young. This distinction is an important one, although I and other authors cited in this book may frequently use the word ''reproduce'' in the traditional sense of ''producing offspring.''

In sexual reproduction, the offspring result from the fusion of the sperm and the oocyte nuclei. On close examination we will discover that not all cases in which sperm ''fertilize'' the oocyte are cases of true sexual reproduction. In some species of fishes, there are only females, as a result of the methods of reproduction known as gynogenesis and hybridogenesis (Schultz, 1971).

In *gynogenesis* (from the Greek word meaning birth of women), a sperm activates the oocyte, but the sperm nucleus does not fuse with the oocyte nucleus. This raises the question of where the sperm come from to activate the oocyte in these all-female populations. The answer is that males from different, but related, species provide the sperm. But why would males waste their sperm by mating with females of another species? This latter question is even more puzzling when we know that if males are given a choice between females of their own species and females of another species, they mate preferentially with females of their own species. The answer may lie in the fact that the males lowest in the reproductive hierarchy of a species mate with females of other species when the number of females of their own species is limited. According to Schultz, 1971, examples of gynogenesis are found in the following species of fish:

Poecilia formosa (fertilized by males of *P. latipinna* or *P. mexicana*).
Poeciliopsis 2 monacha-lucida (fertilized by *P. monacha*).
Poeciliopsis monacha-2 lucida (fertilized by *P. lucida*).

Poeciliopsis 2 viriosa-lucida (fertilized by *P. viriosa*).

Carassius auratus gibelio, a triploid goldfish (fertilized by normal goldfish or by carp).

In *hybridogenesis,* the oocyte is activated by the sperm of a male of another species, sperm and oocyte nuclei fuse, and a zygote is formed that has characteristics from both parents. At meiosis, however, the chromosomes do not segregate at random to polar body and oocyte; rather, the maternal chromosomes go to the oocyte and the paternal chromosomes go to the polar body. This phenomenon has been reported only in the genus *Poeciliopsis.* One of the hybrid matings of *P. monacha* and *P. lucida* gave rise to the all-female *P. monacha-lucida.* The oocytes are fertilized by *P. lucida* sperm, and each generation of hybrids produces *P. monacha* types only because of the distribution of the chromosomes to the oocyte (Schultz, 1971). Schultz (1971) listed the following all-female forms of *Poeciliopsis: P. monacha-lucida, P. monacha-2 lucida, P. 2 monacha-lucida, P. 2 viriosa-lucida, P. monacha (-lucida) occidentalis, P. monacha (-lucida)-latidens,* and *P. viriosa (-lucida) latidens.*

In addition to the gynogenetic and hybridogenetic species there are *parthenogenetic* species, e.g., among reptiles (as will be discussed in Chapter 9), in which the oocyte does not need to be activated by sperm for development to proceed.

Having established that all-female species exist, we ask why they are so successful. It may be that the advantage lies in the fact that, everything else being equal, they will produce two females for every one in a two-sex species and thus increase the population faster. In the case of hybridogenesis, the fish will have hybrid vigor, which gives advantages for nonreproductive functions.

It appears, however, that gynogenesis offers no colonizing advantages over parthenogenesis, since the gynogenetic species must remain in the same habitat as the "fertilizing" species.

So far the terms "male" and "female" have been used as if they were permanent and well-established entities, but they are not. For instance in the coral reef fish, *Labroides dimidiatus,* the "males" have a harem in which the larger "females" are dominant over the smaller "females." The largest and most dominant "female" in turn is dominated by the "male." If the "male" is removed, this dominant "female" *immediately* begins to change sex. The first element of the change toward becoming a "male," i.e., male aggressive behavior, occurs about two hours after the "male" has been removed; male sexual behavior appears within a few days, and sperm release can occur 14–18 days after the change has been initiated (Robertson, 1972). We will discuss possible reasons for the lability of fish gonads in general when we consider the embryology of the reproductive system in Chapter 2.

Another striking example of the difficulty of designating an animal as either male or female can be found in the fish *Serranellus subligarius.* In this species, an individual can contain mature oocytes and sperm simultaneously. The oocytes and sperm are released through separate openings, and external self-fertilization can occur. Clark (1959) observed that in the sea, an individual with a distended abdomen might initiate courtship behavior by forming an S curve with its body. It is then courted, usually by a dark-banded individual, and the S-curved fish blanches (loses its banding). During courtship, both fish may change color and may reverse their roles. This interaction is an example of synchronous functional hermaphroditism, which we shall discuss further in Chapter 3.

Important issues about the evolution of sexual reproduction have been discussed by Williams (1975) and Maynard Smith (1978). Briefly and oversimplified, the argument is as follows: In a sexually reproducing species, an oocyte at the time of sperm penetration (when meiosis is completed and the second polar body is extruded) contains one allele of each pair, having lost the other at meiosis; that is, 50 percent of the alleles are lost at meiosis. In a parthenogenetically reproducing species that produces unreduced oocytes, because there is no meiosis or the products of meiosis are reunited, all alleles of the mother are present in the daughters. After sexual reproduction there is on the average one male for each female produced, whereas after parthenogenetic reproduction there is the advantage of all female offspring. This part of the argument was discussed earlier in this chapter.

Given this loss of 50 percent of the alleles at meiosis in sexual reproduction, it appears unlikely that this disadvantage would be compensated for by the advantages of sexual reproduction, when asexual and sexual reproduction are compared.

The advantage of sexual reproduction results, of course, from the increased variability introduced into the population, which increases the fitness of some individuals in the population. Nevertheless, models can be developed (Williams, 1975) in which the advantages of sexual reproduction are large enough to balance or to outweigh the disadvantage of the 50 percent loss of alleles in meiosis. Such models assume a high intensity of selection and, as we discussed, such high intensities of selection prevail in stable populations in which individuals breed at a high rate. The fittest individuals in the population are more likely to be found among sexually produced progeny than among asexually produced ones (Williams, 1975). Hence, genes that control or induce sexual reproduction are themselves favored. We are discussing the short-term benefits of sexual reproduction here, and they can accrue only to species with a high rate of fertility. The long-term advantage of sexual reproduction, according to Stanley (1975), lies in the high rates of diversification that accompany sexual reproduction; these rates balance the normal extinction rates for most taxa.

SEX RATIO

Assuming that a species has two sexes, we may inquire what the ratio is of males to females. Let us consider three types of sex ratios: The primary sex ratio, the ratio of males to females at the time of conception; the secondary sex ratio, the ratio of males to females at birth; the tertiary sex ratio, the ratio of males to females in the entire population under consideration.

The primary sex ratio is difficult to determine except in some species of birds; there it can be determined in those clutches of hens in which there is 100 percent fertility and 100 percent hatchability. Hays (1945) found that for 39 such hens there were 432 male and 438 female offspring. By analysis of rabbit eggs at first cleavage division (Kaufman, 1973) or by determination of

the sex ratio of rat blastocysts (Fechheimer and Beatty, 1974), it has been shown that the primary sex ratio of these two mammalian species is close to unity.

In humans, the secondary sex ratio favors males 106:100 (Central Bureau of Statistics, Netherlands). McMillen (1979), on the basis of a study of vital statistics of the United States, concluded that a conservative estimate of the primary sex ratio in humans is 120 males:100 females.

The secondary and tertiary sex ratios for various vertebrate species are summarized in Table 1.1. It is clear that for most species the secondary sex ratio is close to equality. For humans, there are claims that the secondary sex ratio changes further in favor of males during times of war, but data collected in the Netherlands for a 45-year period, including World War II, show no such trend (Table 1.2).

Fisher (1930) tried to explain why the secondary sex ratio for most organisms is close to equality. In his view, if there were a genetic predisposition in males to produce male offspring, sooner or later the sons of such fathers could not find females to mate with, and thus they would not reproduce. Sires with a tendency to produce female offspring would then have an increased chance to have offspring, until the ratio was out of balance again, this time in favor of females, when the same argument would apply in reverse. Thus there would be a tendency towards a 100:100 male-to-female ratio. For a recent discussion of this problem, see Werren and Charnov (1978).

SEX DETERMINATION

How is sex determined? To answer this question, one must consider both phenotypic and genotypic sex. *Phenotypic* sex we will define for the purpose of this book as follows: A male is characterized by the presence of a testis or testes and the absence of ovaries; a female is characterized by the presence of an ovary or ovaries and the absence of testes; a hermaphrodite possesses ovarian and testicular tissue.

Genotypic sex is the sex as determined on the basis of the chromosome constitution of the individual. With some exceptions, to be noted later in this chapter, one pair of chromosomes, the sex

Table 1.1

Sex ratios of vertebrates (males per 100 individuals)*

Species	At birth	Adult	Species	At birth	Adult
Mammalia			*Mammalia (cont.)*		
Man, white, U.S.A.†	51.39		Mouse, albino	50(48–52)	
Man, non-white, U.S.A.†	50.28		Mouse, Japanese and	51(48–52)	
Ass (*Equus asinus*)	49(32–66)		waltzing		
Badger (*Meles meles*)		54(34–75)	Mule	44(42–47)	
Bat, big brown		68(32–78)	Muskrat (*Ondatra*	54(42–66)	56(55–57)
(*Eptesicus fuscus*)			zibethica)		
Bat (*Myotis sodalis*)		51(49–53)	Opossum (*Didelphis*	57	57(50–68)
Bat (*Pipistrellus*		74(68–80)	marsupialis)		
subflavus)			Rabbit (*Oryctolagus*	50(49–51)	
Beaver (*Castor*		52(45–59)	spp.)		
canadensis)			Rabbit, Flemish giant	57(46–68)	
Cattle (*Bos taurus*)			Rabbit, Polish	51(45–56)	
Single birth	52(43–58)		Raccoon (*Procyon lotor*)		52(46–58)
Twin birth	49		Rat (*Rattus norvegicus*)	50(49–51)	
Triplet birth	46		Rat, albino	50(49–51)	
Ayrshire	49(45–53)		Rat, brown	51(48–55)	41(40–42)
Brown alpine	50(47–52)		Rat, hybrid and piebald	51(50–52)	
Guernsey	44(41–48)		Seal, harbor (*Phoca*		51(48–54)
Hereford	51(49–53)		groenlandica)		
Holstein	49(47–50)		Sheep (*Ovis aries*)		
Jersey	52(48–56)		Single birth	50(44–55)	
Shorthorn	49(47–51)		Twin and triplet birth	49	
Welsh, black	50(48–52)		Quadruplet birth	43	
Zebu	51(50–52)		Cheviot	49(48–50)	
Dog (*Canis familiaris*)			Karakul	52(47–57)	
Collie, St. Bernard,	54(53–55)		Merino	47(44–50)	
spaniel			Navajo	49(48–50)	
German shepherd	55(51–58)		Swine (*Sus scrofa*)	52(51–53)	
Greyhound	52(51–54)		Berkshire	51(50–52)	
Schnauzer	51(49–53)		Duroc, Jersey	49(48–50)	
Terrier	56(53–59)		German improved	51(50–52)	
Elk, North American	51(46–56)		Inbred and linecross	52(51–53)	
Fox, red	50(41–59)	52(46–56)	non-inbred		
Fox, silver	53(51–55)		Weasel (*Mustela nivalis*)		73(63–83)
Goat, Angora	50(48–52)				
Goat, crossbreeds	51(48–54)		*Aves*		
Goat, Saanen (Britain)	55(50–61)		Canary (*Serinus* spp.)	44(36–51)	
Goat, Toggenburg	50(46–55)		Cuckoo (*Cuculus* sp.)		55(52–58)
Guinea pig (*Cavia*	52(51–53)		Crow (*Corvus*		56
porcellus)			brachyrhynchos)		
Hare (*Lepus*		54(52–56)	Dove (*Zenaldura*	52(50–54)	60(57–63)
americanus)			macroura)		
Horse (*Equus caballus*)	52(50–55)		Duck (*Anas* spp.)	50(49–52)	
Horse, Thoroughbred	50		Canvasback	51(47–55)	
Mink, ranch (*Mustela*	51(49–53)		Mallard	53(49–57)	52(48–56)
spp.)			Pintail	53(49–57)	66(64–69)
Mink, hybrid, pastel	50(48–52)		Redhead	53(48–58)	
Mole (*Scalopus*		58(55–60)	Fowl (*Gallus* spp.)	49(47–51)	
aquaticus)			Rhode Island Red	50(46–54)	
Mouse (*Mus musculus*)		52(50–54)	White Leghorn	49(48–51)	

Table 1.1—*Continued*

Species	At birth	Adult
Aves (cont.)		
Grackle (*Quiscalus* spp.)		30(26–35)
Grouse, ruffed (*Bonasa* sp.)	52(47–56)	52(42–61)
Grouse, sage		50(48–52)
Grouse, sharp-tailed		63(54–72)
Hawk (*Buteo* spp.)		50(46–54)
Parrot (*Psittacus* spp.)		57(50–63)
Partridge (*Perdix* spp.)	43(38–48)	60(58–62)
Pheasant, ring-necked (*Phasianus* sp.)	50(46–54)	52(50–55)
Pigeon (*Columba* spp.)	50(48–52)	
Quail, bobwhite (*Colinus* sp.)		(59–66)
Quail, California (*Lophortyx* sp.)	58(54–62)	
Redwing (*Agelaius* sp.)		77(73–81)
Sparrow (*Melospiza melodia*)		52
Starling (*Sturnus* spp.)		68
Turkey (*Meleagris gallopavo*)		50(48–55)
Reptilia		
Lizards		58(55–62)
Snakes		50(48–52)
Tortoises, turtles		43(38–48)
Amphibia		
Frogs, toads		45(43–47)
Salamanders, newts		56(51–61)
Pisces		
Elasmobranchs		47(45–48)
Sturgeons and spoonbills		17(8–26)
True bony fishes		
Minnow (*Gambusia holbrookii*)	50(48–53)	
Walleye (*Stezostedon vitreum*)		72

*Values in parentheses are estimates of the 95 percent range. Adult values should be considered with caution because of extreme variation caused by geographic area, species considered, time of year, and population conditions.

†Values based on 50 percent sample of 1952 registered live births.

Modified from Spector, 1956; reprinted with permission of W. B. Saunders Co.

Table 1.2

Sex ratio for births registered in the Netherlands, 1930–1974

Year	Boys/100 girls	Number of births
1930	106	182,310
1931	107	177,387
1932	106	178,525
1933	106	171,289
1934	106	172,214
1935	107	170,425
1936	106	171,675
1937	106	170,220
1938	106	178,422
1939	105	180,917
1940	105	184,846
1941	107	181,959
1942	105	189,975
1943	107	209,379
1944	107	219,946
1945	107	209,607
1946	107	284,456
1947	106	267,348
1948	107	247,923
1949	106	236,177
1950	107	229,718
1964	106	250,914
1974	105	185,982

Data from Centraal Bureau voor de Statistiek, The Netherlands, 1977.

chromosomes, can be distinguished by their unequal size in one of the sexes (the heterogametic sex). In mammals and in some species of other vertebrate classes, the male sex is heterogametic and the female is homogametic. This means that in the male, one chromosome of the sex chromosomes is different (usually smaller) from the homologous chromosome with which it pairs at meiosis, whereas in the female, the two chromosomes are identical in size and shape. For such species, we will generally use the symbols XY for the male and XX for the female. In birds and some other species of other vertebrate classes, the male is homogametic and the female is heterogametic; for these species we will generally use the symbol ZZ for males and ZW for females. One can often distinguish the heterogametic sex microscopically by the size difference between the two sex chromosomes. However, even if one were unable visually to determine the XY–XX or ZZ–ZW mechanism of sex determination, one can still determine which mechanism of sex determination prevails in a species, as we shall discuss below.

Although most mammalian species follow the XY–XX system, a number of species have unusual sex chromosome sets, some of which are listed in Table 1.3. The cytological explanation for some of these conditions is illustrated in Figures 1.1, 1.2, and 1.3.

The existence of both XX and XY females and of XY males in the wood lemming, *Myopus schistocolor* (Fredga et al., 1976), requires further explanation. The somatic cells of the XY females have an X and a Y chromosome, but the Y chromosome is absent from the germ cells. To explain this phenomenon, Wachtel et al. (1976) and Gropp et al. (1978) have postulated a double non-

Table 1.3

Mammalian species with unconventional inherited sex chromosomes

Species	Sex chromosomes	
Marsupials		
Macrotis lagotis	X Y$_1$Y$_2$	♂
Potorous tridactylus	X Y$_1$Y$_2$	♂
Wallabia (Protemnodon) bicolor	X Y$_1$Y$_2$	♂
Lagorchestes conspicillatus	X$_1$X$_1$X$_2$X$_2$	♀
	X$_1$X$_2$Y	♂
Edentates		
Choloepus hoffmanni	X$_1$X$_1$X$_2$X$_2$	♀
	X$_1$X$_2$Y	♂
Insectivores		
Echinops telfairi	X Y$_1$Y$_2$	♂
Sorex araneus	X Y$_1$Y$_2$	♂
Chiroptera		
Choeroniscus godmani	X Y$_1$Y$_2$	♂
Carollia perspicillata azteca	X Y$_1$Y$_2$	♂
C. subrufa	X Y$_1$Y$_2$	♂
Artibeus jamaicensis	X Y$_1$Y$_2$	♂
A. lituratus	X Y$_1$Y$_2$	♂
A. toltecus	X Y$_1$Y$_2$	♂
Rodents		
Gerbillus gerbillus	X Y$_1$Y$_2$	♂
Mus minutoides	X$_1$X$_1$X$_2$X$_2$	♀
	X$_1$X$_2$Y	♂
Microtus minutoides	X$_1$X$_2$Y	♂
M. oregoni[a]	XO	♀
	XY	♂
Myopus schisticolor[b]	XY	♀
	XX	♀
	XY	♂

Table 1.3—*Continued*

Species	Sex chromosomes	
Rodents (*cont.*)		
Ellobius lutescens[a]	XO	♂
	XO	♀
Carnivores		
Atilax paludinosus	X$_1$X$_1$X$_2$X$_2$	♀
	X$_1$X$_2$Y	♂
Herpestes auropunctatus[a]	XO	♂
Herpestes brachyurus	X$_1$X$_1$X$_2$X$_2$	♀
	X$_1$X$_2$Y	♂
Herpestes edwardsi	X$_1$X$_1$X$_2$X$_2$	♀
	X$_1$X$_2$Y	♂
Herpestes fuscus	X$_1$X$_1$X$_2$X$_2$	♀
	X$_1$X$_2$Y	♂
Herpestes ichneumon	X$_1$X$_1$X$_2$X$_2$	♀
	X$_1$X$_2$Y	♂
Herpestes javanicus	X$_1$X$_1$X$_2$X$_2$	♀
	X$_2$X$_2$Y	♂
Herpestes sanguineus	X$_1$X$_1$X$_2$X$_2$	♀
	X$_1$X$_2$Y	♂
Artiodactyla		
Tragelaphus angasi[a]	X$_1$X$_1$X$_2$X$_2$	♀
	X$_1$X$_2$Y	♂
Tragelaphus strepsiceros[a]	X$_1$X$_1$X$_2$X$_2$	♀
	X$_1$X$_2$Y	♂

Data from Fredga, 1970, except for [a] from Ohno, 1967, and [b] from Fredga et al., 1976.

disjunction that eliminates the Y chromosome and causes a duplication of the X chromosome. Early during the developmental stage of the XY female's gonad, during an early oogonial mitotic division, double nondisjunction of both daughter chromatids of X to one pole and of both chromatids of the Y to the other pole, with degeneration of the YY cells, is probably involved. The XX and XY female are not morphologically distinguishable, but most of the XY females have female offspring only, whereas XX females have male and female offspring. Gropp et al. (1978) assume that the X chromosome of these XY females carries a sex reversal gene that prevents the expression of the H-Y antigen (to be discussed below) gene located on the Y chromosome. As Table 1.4 illustrates, XY females do not have the H-Y antigen, and so it is probable that a gene on the X chromosome indeed has this effect (Gropp et al., 1978).

Figure 1.1 The type of Y autosome translocation which gives an apparent XO-constitution to the male. Such a translocation is maintained by the small Indian mongoose (*Herpestes auropunctatus*). The X is in solid black; the original Y is lined. A pair of autosomes involved in translocation are outlined. *Top:* A very terminal translocation attaches a great part of the original Y to the very tip of an autosome. A minute reciprocal product of the translocation carrying the Y centromere shall be eliminated. During male meiosis, the X and an intact autosome segregate against an autosome carrying the Y material (extreme right). *Bottom:* The chromosome constitution of the male (a circle at the left) and the female (a circle at the right). Since the original Y was so minute in size, an autosome carrying the Y material cannot be distinguished from its intact homologue. Thus, the male gives an appearance of having the XO-constitution. (From Ohno, 1967; reprinted with permission of Springer-Verlag.)

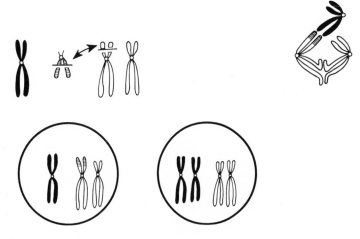

A few exceptional cases in humans and mice (*Mus musculus*) need discussion. Some human females have the XO constitution named Turner's syndrome. Such women have nonfunctional streaklike ovaries, frequently have difficulty with depth perception and are obviously sterile. In mice, however, XO females are fertile. In XX mice, the presence of an autosomal sex reversal gene *Sxr* causes the animals to be phenotypic males. The testes, however, are about 20 percent of those of normal XY mice and germ cells are absent. In such XX *Sxr* male embryos, the testes are normal till day 16, except for the absence of meiotic prophase, and spermatogonia present at

birth generally do not survive beyond day 5 after birth. In XO male mice with the *Sxr* gene, the testes are larger than those in the XX females with the *Sxr* gene, and spermatogenesis occurs, but is aberrant during meiosis. In XY males with the *Sxr* gene, spermatogenesis is abnormal, sperm concentrations in the ejaculate are low, and the incidence of abnormal sperm is high.

In humans, the presence of a sex reversal gene, presumably on the X chromosome, has been postulated to explain the occurrence of XY women with 46 chromosomes. Women with this phenotype, in contrast to the XY wood lemming females, have abnormal, streak ovaries as a result

Figure 1.2 The same type of Y-autosome translocation [shown in Figure 1.1] which gives the $X_1X_2Y/X_1X_1X_2X_2$-scheme of the sex-determining mechanism to a species [e.g., *Mus minutoides*] *Top:* If the species has the duplicate-type X and the proportionately large Y, a very terminal translocation between the Y and an autosome produces a new morphologically distinct chromosome, as an addition of the Y-chromosome material greatly increases the size of an autosome involved in the translocation. This new chromosome is now regarded as the Y, the original duplicate-type X as the X_1, and an intact autosome as the X_2. During male meiosis, the X_1 and the X_2 segregate against the new Y (extreme right). *Bottom:* The chromosome constitution of the male which is X_1X_2Y (a circle at the left), and the female which is $X_1X_1X_2X_2$ (a circle at the right). (From Ohno, 1967; reprinted with permission of Springer-Verlag.)

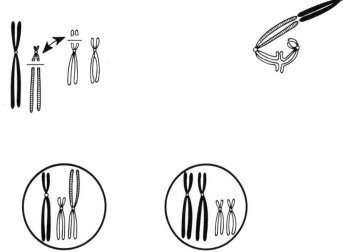

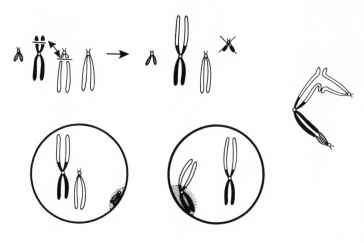

Figure 1.3 The type of X autosome translocation which gives the XY_1Y_2-scheme of the sex-determining mechanism to a species [e.g., *Gerbillus gerbillus*]. The original X is painted solid black, the members of an autosomal pair involved in translocation are outlined. The original Y is lined. *Top:* The nature of translocation. A very terminal translocation attaches a great part of an autosome to the tip of the X. A minute reciprocal product is eliminated. The original Y is now regarded as the Y_1 and an intact autosome as the Y_2, because both of them segregate against the new X during male meiosis (extreme right). *Bottom:* In the male somatic cell (left), only the Y becomes heterochromatic [shows maximal staining in the resting nucleus or in prophase and contains few genes], while the entire new X remains euchromatic [shows its maximum staining during metaphase and less dense staining in the resting nucleus; contains most of the genes]. In the female somatic nucleus (right) only a part of one of the two new X chromosomes which represent the original X manifests the heterochromatic condition. (From Ohno, 1967; reprinted with permission of Springer-Verlag.)

of early involution of their ovaries (German et al., 1978).

H-Y ANTIGEN AND THE Y CHROMOSOME

What is the function of the Y chromosome? The cases just discussed show that not all animals with a Y chromosome are phenotypic males. It has been found thus far that in mammals, the presence of the H-Y antigen is better correlated with the presence of testes than with the presence of the Y chromosome (see Table 1.4). In the wood lemming male, XY individuals possess the H-Y antigen, whereas female XY animals lack the antigen.

Animals with a Y chromosome do not always have testes and animals with testes do not always have a Y chromosome. For example, the XX mice with the autosomal *Sxr* gene have sterile testes, and they show male sexual behavior. However, XO mice with *Sxr* also have testes, germ cells are present, and the ejaculate contains sperm, albeit in low numbers and with a high incidence of abnormalities (Cattanach, 1975).

In goats, a double set of the autosomal gene for unhorned or polled (*P*) causes genetic females to become intersexes (animals with a mixture of male and female characteristics). Genetic crosses between horned, heterozygous polled and homo-

zygous polled billies and polled nannies with the theoretically expected and the actually observed offspring are summarized in Table 1.5. The observed data show an excess of males at the expense of intersexes, suggesting that some intersexes are considered at birth to be males. The intersexes have testicular gonads that secrete androgens, which cause enlargement of the clitoris, a greater anogenital distance than in normal females, and male sexual behavior. Some intersexes can be distinguished from normal males only by chromosomal analysis (Short, 1974). In pigs also, a genetic factor causes genetic females to have testicular tissue; however, neither the genetic basis nor the embryological development of the condition is well understood.

In cattle, the freemartin syndrome develops in an XX female that is a twin to a bull, if vascular anastomosis occurs between the twins' placentas. The syndrome is characterized mainly by the presence of testicular tissue, the absence of a Müllerian duct system (oviducts and uterus), and often an enlarged clitoris (see Marcum, 1974, for review). The tissues of freemartins show the presence of XY as well as XX cells (the animals are called "chimeras"). Ohno et al. (1976) determined that the H-Y antigen is present in the ovotestes of freemartins in concentrations similar to those found in the bull twin's testes. Appar-

Table 1.4

Distribution of H-Y antigen in various vetebrates

Species	Sex-determining mechanism	Phenotype	H-Y antigen present in bearer of
Rana pipiens (frog)	XX–XY	XX—ovaries; XY—testes	XY—testes
Xenopus laevis (toad)	ZW–ZZ	ZW—ovaries; ZZ—testes	ZW—ovaries
Gallus domesticus (chicken)	ZW–ZZ	ZW—ovary; ZZ—testes	ZW—ovary
Mus musculus (mouse)	XX–XY	XX—ovaries; XY—testes	XY—testes
	XX–XY and sex reversal gene (*Sxr*)	XX—testes; XY—testes	XX—testes; XY—testes
	XX–XY + testicular feminization gene (*Tfm*) on X chromosome	XX—ovaries; XY—testes*	XY—testes
	XO–XY	XO—ovaries; XY—testes	XY—testes
	XO–XY + *Sxr*	XO—testes; XY—testes	XO—testes; XY—testes
Myopus schisticolor (wood lemming)	XX–XY	XX—ovaries; XY—ovaries; XY—testes	XY—testes
Ellobius lutescens (mole–vole)	XO–XO	XO—ovaries; XO—testes	XO—testes
Homo sapiens (human)	XX–XY	XY—testes; XX—ovaries	XY—testes
	XX–XY and *Tfm*	XY—testes*; XX—ovaries	XY—testes
	XX–XY	XY—testes; XX—testes (exceptional)	XY—testes; XX—testes
	XX–XY	XY—testes; XX—streak gonads	XX—streak gonads; XY—testes
Bos taurus (cattle)	XX–XY	XX—ovaries; XY—testes	XY—testes
	XX/XY–XY (freemartin)	XX/XY—ovotestes; XY—testes	XX/XY ovotestes; XY—testes
Canis familiaris (dog)	XX–XY	XX—ovaries; XY—testes	XY—testes
	XX–XY	XX—testis, ovotestes (exceptional cases)	XX—testes

*The individuals with XX and the testicular feminization gene have internal (inguinal canal) testes, but have otherwise a completely female appearance. None of the tissues of these individuals respond to androgens.

Data from Ohno, 1979; Selden et al., 1978; Silvers and Wachtel, 1977; Wachtel, 1977; Wachtel et al., 1975.

ently, the H-Y antigen is secreted by the bull and the antigen can coat the XX cells of the female, which have receptors for the antigen.

The presence of H-Y antigen does not guarantee the differentiation of testicular tissue in mammals, a conclusion supported by the presence in humans of H-Y antigen and streak gonads that lack testicular tissue. Wachtel (1977) has postulated that not only is the H-Y antigen required but

also a gene that codes the cells for H-Y antigen receptors.

The African clawed toad (*Xenopus laevis*) and birds have the ZZ-ZW mechanism of sex determination. The H-Y antigen is expressed in chickens in both sexes, but to a greater extent in the female than in the male (Wachtel et al., 1975). After estradiol-induced sex reversal, the H-Y antigen is fully expressed in the gonads of genetic

Table 1.5

Effects of the polled (P) allele on sex of goats

Mating type 1	horned billy *p/p* × polled nanny *P/p*:				
		Expected ratio			Observed ratio
		Polled *P/p*	Horned *p/p*	Total	
	♂	25	25	50	No data given
	♀	25	25	50	
	⚥	0	0	0	
Mating type 2	polled billy *P/p* × polled nanny *P/p*:				
		Expected ratio			Observed ratio (594 kids examined)
		Polled *P/P*	Polled *P/p*	Horned *p/p*	Total	
	♂	12.5	25	12.5	50	♂54
	♀	0	25	12.5	37.5	♀39
	⚥	12.5	0	0	12.5	⚥ 7
Mating type 3	polled billy *P/P* × polled nanny *P/p*:				
		Expected ratio			Observed ratio (4,784 kids examined)
		Polled *P/P*	Polled *P/p*	Total	
	♂	25	25	50	♂64
	♀	0	25	25	♀24
	⚥	25	0	25	⚥12

Modified from Short, 1974; reprinted with permission of *Glaxo Volume.*

male (ZZ) embryos (Müller et al., 1979). This suggests that it is not the genes on the W chromosome that code for the H-Y antigen in chickens. The findings in *Xenopus laevis* and in birds that the H-Y antigen is more fully expressed in the females indicate that the H-Y antigen causes ovarian differentiation in species with the ZZ–ZW and testicular differentiation in species with the XY–XX mechanism of sex determination.

In chickens, occasionally two embryos develop in double-yolked fertilized eggs. The male ("freemartin") that develops in the same egg with a female embryo shows feminization of the left gonad, whereas the female has a normal left ovary, normal regression of the right gonad, and the absence of both Müllerian ducts, as in normal males (Lutz and Lutz-Ostertag, 1958). Presumably estrogens secreted by the female embryo developing in the double-yolked egg induce full expression of the H-Y antigen, which causes ovarian development of the left gonad in the male embryo. Correspondingly, secretions by the male's right gonad may be responsible for the regression of the two Müllerian ducts in the female twin. Moreover, it is, of course, possible that the female twin's ovary secretes the H-Y antigen and that the receptor cells

on the male twin's left gonad facilitate the further expression of the H-Y antigen, as proposed by Ohno et al. (1976) for freemartins in cattle. In the chick, therefore, there would be an endocrine and an immunological explanation for the effects observed for the gonad of the homogametic twin.

The marmoset (*Callithrix jacchus*), however, presents a problem. In this species the pregnant female always carries twins and placental anastomosis is common; yet in cases of male-female twins, there is no evidence of modification of either the male or female gonads. Ohno et al. (1976) speculated that either the male marmoset embryo does not secrete the H-Y antigen, or the antigen is not transferred to the XX cells, perhaps because they lack receptors.

On the basis of the findings in mice, goats, and pigs, we can state that the presence of a Y chromosome is not a prerequisite for testicular development. What then is the role of the Y chromosome with respect to sex determination? In most mammalian species there is a good correlation between the presence of the Y chromosome and the presence of the H-Y antigen, and in humans it has been shown that XYY men have more H-Y antigen than XY men (Silvers and Wachtel, 1977). If one assumes that the presence of the H-Y antigen is determined by the presence of the Y chromosome, how can one explain the absence of the H-Y antigen in female XY wood lemmings or the absence of the Y chromosome in mice and pigs?

In the case of the wood lemming, Silvers and Wachtel (1977) assume that the H-Y antigen gene on the Y chromosome has been inactivated because of an inhibitor gene on the X chromosome. The presence of testicular tissue in the absence of a Y chromosome can be explained for the XX mouse by the autosomal *Sxr* gene. Ohno (1979) has considered three different hypotheses for the origin of the autosomal *Sxr* gene: (1) The *Sxr* is on the Y chromosome, and a piece of Y chromosome containing the male determining gene has been translocated to an autosomal chromosome. (2) The H-Y antigen structural gene is located on an autosomal chromosome, and the Y chromosome carries its activator gene; in this case, the *Sxr* represents a constitutively mutated H-Y structural gene, and the Y-linked activating gene is no longer necessary. (3) The H-Y structural gene exists in multiple copies located on the Y chromosome, and the X chromosome carries a modulator gene; the *Sxr* gene then represents some of the multiple copies that have been translocated to an autosomal chromosome. According to Ohno (1979), the third explanation is the most attractive one because it can explain all the abnormal gonadal developments found in mammals.

In goats, the intersex condition can be explained similarly by the polled gene. The homozygous polled (*PP*) XX intersex goat has the H-Y antigen (Wachtel et al., 1978), while the heterozygous *Pp* XX goat is a normal female, so that in this case, the proposed Y translocation would act as a recessive for gonadal differentiation. In the mouse, however, the *Sxr* gene is dominant. There is, as Wachtel et al. (1978) have pointed out, an apparent disparity in the Y-to-autosome translocation being a recessive determinant in the mouse. Wachtel et al. (1978:280) explain this disparity as follows:

The disparity may be more apparent than real. Consider the arguments that the Y chromosomal testis-determining gene exists in multiple copies (Ohno, 1967), or that the H-Y locus may comprise a superfamily of genes

determining classes and subclasses of H-Y molecules that are structurally and functionally related. In either event, translocation of a critical portion of H-Y genes could give rise to a dominant mode of sex determination as in the mouse and translocation of subcritical portion could give rise to a recessive mode, as in the goat. The polled intersex condition could accordingly represent an accumulation of testis determinants inherited autosomally from both father and mother.

We have just seen that the gene for the H-Y antigen may be present in multiple copies on the Y chromosome and that in the case of the XY female wood lemming, a gene located on the X chromosome inhibits the expression of the H-Y antigen gene. There are, however, arguments against the assumption that the H-Y antigen gene is Y-linked. Steroid hormones can induce the expression of the H-Y antigen in XX (fishes) and ZZ (birds) individuals (Ohno, 1979). Thus consideration should be given to X linkage of the H-Y antigen gene, with an activator located on the Y chromosome. This hypothesis was first proposed by Hamerton (1968), although the H-Y antigen at that time was not identified and was hypothesized as a testis-determining gene product. According to Ohno (1979), the evidence is not available to decide whether the H-Y antigen is present in multiple copies on the Y chromosome or in single or multiple copies on the X chromosome.

YY CHROMOSOMAL CONSTITUTION

The use of sex-reversal techniques on the Japanese rice fish (*Oryzias latipes*) has made it possible to obtain YY males that have no X chromosome. Fineman et al. (1975) have investigated some aspects of the reproductive physiology of these males in comparison with XY males. In competition for females, YY males, which are larger than XY males, are more successful in mating: 88 percent of the spawning is done by YY males; behaviorally, the YY males are dominant over the XY males (Hamilton and Walter, 1968). (There is also evidence from mice that the Y chromosome contributes to aggressiveness [Sellmanoff et al., 1975]).

Fineman et al. (1974) found that XX females have a life span of 141 days, XY males of 328 days, and YY males of 246 days. The shorter life

Table 1.6

Reproductive performances of phenotypic males of the Japanese rice fish, *Oryzias latipes*

Treatment	Genotype	Number of fish studied	Reproductive intensity*	Reproductive capacity*
Untreated	XY	17	87	164.16
Untreated	YY	25	72	135.07
Androgenic	XX	23	32	52.50
Androgenic	XY	22	69	126.75
Androgenic	YY	18	55	106.32

*Reproductive intensity was the percentage of eggs fertilized per day; reproductive capacity was the reproductive intensity times the number of reproductive days.
Modified from Fineman et. al., 1975; reprinted with permission of Alan R. Liss, Inc.

span of the female may be correlated with egg production and with the effects of estrogens, for after estrogen treatment, the life spans were 185 days for females, 168 days for XX males, and 122 days for YY males. Reproductive performances of XX, XY, and YY genotypes were investigated after addition of either estrogens or androgens to the aquarium and under controlled conditions. Data presented in Table 1.6 show that, against expectations, the reproductive capacity of XY males was greater than that of YY males.

The XYY condition found occasionally in humans has given rise to sensational and erroneous news stories about an associated syndrome of antisocial behavior. We do know that the frequency of XYY males, and especially of XXYY males, in mental hospitals is higher than their incidence in the population at birth would lead one to expect. Generally XYY males are taller than most XY men, their blood testosterone concentrations are within the range normal for XY men, and their mental abilities vary over a wide range (Hook, 1973). According to Sharma et al. (1975), testosterone production and plasma concentrations are lower in XYY men than in XY men matched for height and age. With respect to antisocial behavior Hook (1973:147) states, "Discovery of an extra sex chromosome in a male hardly predicts antisocial behavior with the same confidence, for instance, that the observations of trisomy 21 predicts mental retardation." This cautious statement contrasts with the statement by Price and Whatmore (1967) that "it seems reasonable to suggest that their [nine XYY men] antisocial behavior is due to the extra Y chromosome."

DETERMINING THE HETEROGAMETIC SEX

We can determine whether a given species follows the XY–XX or the ZZ–ZW system of sex determination by considering various types of evidence. First, in karyotyping, we can study the morphology of the chromosomes to try to detect differences between the homologous chromosomes of one pair or to detect the absence of one chromosome of the pair. Second, we can investigate the pattern of inheritance of sex-linked characters, if these are present (i.e., criss-cross inheritance). Third, we can breed with sex-reversed animals; for example, by treating larvae of the Japanese rice fish (*Oryzias latipes*) with either estrogens or androgens, one can obtain all female or all male populations. These males and females are capable of breeding, and by the use of this method, it has been possible not only to establish the XX–XY pattern of sex determination but also to obtain YY males without an X chromosome. The reproductive capacities of such males have just been discussed.

Indirect evidence of the mode of sex determination, although not firm proof, can be obtained by several kinds of observations. First, if we can temporarily or permanently change the phenotypic sex by hormone treatment, for example, and if this treatment is successful in one of the sexes only, then the changed sex is usually the homogametic sex. Second, after gonadectomy of an embryo, the young usually comes to resemble the homogametic sex at birth in the secondary sex organs or the secondary sex characteristics or both. We shall discuss this effect in more detail when we discuss sexual differentiation in Chapter 2. Third, we can refer to Haldane's rule, which states: If, among the offspring of a cross of two races, one sex is missing, rare, or sterile, then that is the heterogametic sex (Beatty, 1964).

CONSEQUENCES OF THE SEX-DETERMINING MECHANISM

The consequences of the XY–XX and ZZ–ZW mechanisms vary. The time when the sex of the offspring is determined differs with the mechanism: in the ZZ–ZW mechanism, sex is determined at the time of ovulation; in the XY–XX mechanism it is determined at the time of fertilization. As discussed above, the sex-determining mechanism predicts reasonably well the morphology of animals exposed to sex hormones during development and the morphology in the absence of the gonads. And as we shall discuss in Chapter 2 in detail, the sex-determining mechanism plays an important role in determining how animals will behave as adults in response to administered doses of sex hormones. In general, for the few species that have been investigated, it appears that the homogametic sex will show a higher incidence of animals with sexual behavior appropriate to both sexes. For example, in cattle the females, which are homogametic, show male and female sexual

Table 1.7

Sex-determining mechanism in various vertebrates

Group	♂	♀	Reference
Mammals	XY	XX	Beatty, 1964
Birds	ZZ	ZW	Beatty, 1964
Reptiles			
Anolis bimaculatus	X_1X_2Y	$X_1X_1X_2X_2$	Gorman and Atkins, 1966
Anolis biporcatus	X_1X_2Y	$X_1X_1X_2X_2$	Gorman and Atkins, 1966
Anolis leachi	X_1X_2Y	$X_1X_1X_2X_2$	Gorman and Atkins, 1966
Anolis gingivinus	X_1X_2Y	$X_1X_1X_2X_2$	Gorman and Atkins, 1966
Anolis marmoratus ferus	X_1X_2Y	$X_1X_1X_2X_2$	Gorman and Atkins, 1966
Bothrops alternatus	ZZ	ZW	Benirschke and Hsu, 1973
Bothrops jararacussu	ZZ	ZW	Benirschke and Hsu, 1971
Bothrops moojeni	ZZ	ZW	Benirschke and Hsu, 1971
Chironius bicarinatus	ZZ	ZW	Benirschke and Hsu, 1971
Clelia occipitolutea	ZZ	ZW	Benirschke and Hsu, 1973
Cnemidophorus tigris	XY	XX	Cole et al., 1969
Crotalus durissus terrificus	ZZ	ZW	Benirschke and Hsu, 1971
Drymarchon corais corais	ZZ	ZW	Benirschke and Hsu, 1973
Eunectes murinus	ZZ	ZW	Benirschke and Hsu, 1971
Lialis burtonis	X_1X_2Y	$X_1X_1X_2X_2$	Gorman and Gress, 1970
Liophis miliaris	ZZ	ZW	Benirschke and Hsu, 1973
Mastigodryas bifossatus bifossatus	ZZ	ZW	Benirschke and Hsu, 1973
Philodryas olfersii olfersii	ZZ	ZW	Benirschke and Hsu, 1971
Philodryas serra	ZZ	ZW	Benirschke and Hsu, 1973
Polychrus marmoratus	X_1X_2Y	$X_1X_1X_2X_2$	Gorman et al., 1967
Sceloporus jarrovi	X_1X_2Y	$X_1X_1X_2X_2$	Cole et al., 1967
Sceloporus poinsetti	X_1X_2Y	$X_1X_1X_2X_2$	Cole et al., 1967
Spilotes pullatus anomalepsis	ZZ	ZW	Benirschke and Hsu, 1973
Thamnodynastes strigatus	ZZ	ZW	Benirschke and Hsu, 1971
Tomodon dorsatus	ZZ	ZW	Benirschke and Hsu, 1971
Uta (5 species)	XY	XX	Pennock et al., 1969
Xenodon merremii	ZZ	ZW	Benirschke and Hsu, 1971
Xenodon neuwiedii	ZZ	ZW	Benirschke and Hsu, 1973

(continued)

Table 1.7—*Continued*

Group	♂	♀	Reference
Amphibia			
Anura			
Bufo vulgaris	ZZ?	ZW?	Beatty, 1964
Rana arvalis	ZZ	ZW	Beatty, 1964
Rana japonica	XY?	XX?	Beatty, 1964
Rana pipiens	XY	XX	Wachtel et al., 1975
Rana temporaria	XY?	XX?	Beatty, 1964
Xenopus laevis	ZZ	ZW	Beatty, 1964
Urodeles			
Ambystoma sp.	ZZ	ZW	Beatty, 1964
Hynobius nebulosus	XY	XX	Beatty, 1964
Hynobius tokyoensis	XY	XX	Beatty, 1964
Oedipina (5 species)	XY	XX	Beatty, 1964
Siredon sp.	ZZ	ZW	Beatty, 1964
Triturus sp.[a]	XY	XX	Beatty, 1964
Teleosts			
Betta splendens	XY	XX	Benirschke and Hsu, 1973
	ZZ	XW[b]	
Gasterosteus aculeatus	ZZ	ZW	Beatty, 1964
Megupsilon aporus[c]	Y	XX	Benirschke and Hsu, 1973
Mogrunda obscura	XY	XX	Beatty, 1964
Oryzias latipes	XY	XX	Beatty, 1964
Platypoecilus (Xiphophorus) maculatus	XY	XX[d]	Beatty, 1964
	ZZ	ZW[e]	
Poecilia reticulata	ZZ	ZW	Beatty, 1964

[a]Exception may be *Triturus cristatus,* which may have ZZ–ZW (Gallien, 1965).
[b]Females were changed to males by androgens and crossed with untreated females. Eight such crosses yielded only one female offspring; three crosses yielded, respectively, 55 percent, 71 percent, and 96 percent females.
[c]Female 2n-48; male 2n-47 with one very large sex chromosome.
[d]Mexican race.
[e]British Honduran race.

behavior during estrus, whereas in birds the males, which are homogametic, may show male and female sexual behavior.

Table 1.7 lists the sex-determining mechanisms for various vertebrates and Table 1.8 for different groups of lizards. As can be seen, sometimes different races of the same species each have a different sex-determining mechanism. For example, in *Platypoecilus (Xiphophorus) maculatus,* the Mexican race has the XY constitution for the male and the XX for the female, whereas the British Honduran race has the ZZ constitution for the male and the ZW for the female. The cross of these two races using ZZ males yields all male offspring (ZZ ♂ × XX ♀ → ZX). Thus we know that ZX is male. Other crosses have been found to yield as follows:

$$XY\ \male\ \times\ XW\ \female\ \rightarrow\ 3\ \female : 1\ \male$$
$$XZ\ \male\ \times\ XW\ \female\ \rightarrow\ 3\ \female : 1\ \male$$
$$XZ\ \male\ \times\ WY\ \female\ \rightarrow\ 1\ \female : 1\ \male$$
$$XZ\ \male\ \times\ XX\ \female\ \rightarrow\ 1\ \female : 1\ \male$$

The case of the guppy (*Poecilia reticulata*) is an interesting one, because it has been possible by artificial selection to change one pair of autosomes into a pair of sex chromosomes, the original sex chromosomes into autosomes, and the sex-determining mechanism from XY–XX to ZZ–ZW (Beatty, 1964). The embryology of the gonads of teleost fishes, which differs from that of most other vertebrates, may explain the lability of sex determination in teleosts (to be discussed in Chapter 2).

Table 1.8

Sex-determining mechanism in families of lizards

Family	Number of species	Number of species karyotyped	Number of species with sex chromosomes	Nature of sex chromosome
Gekkonidae	650	54	2	ZZ–ZW
Pygopodidae	30	6	5	$X_1X_1X_2X_2–X_1X_2Y$
Xantusidae	15	10	—	—
Iguanidae	600	145	45	XX–XY; $X_1X_1X_2X_2–X_1X_2Y$
Agamidae	300	19	—	—
Chamaeleontidae	24	36	—	—
Scincidae	800	35	3	XX–XY; $X_1X_1X_2X_2–X_1X_2Y$
Teiidae	200	46	1	XX–XY*
Lacertidae	150	32	3	ZZ–ZW
Gerrhosauridae	50	1	—	—
Varanidae	32	18	3†	ZZ–ZW

*King attributes in his table a ZZ–ZW mechanism to the Teiidae, but in the text and cited source the mechanism is stated to be XX–XY.

†King lists in his table 4 varanids, but the cited sources list only 3 varanids that have sex chromosomes identified. I acknowledge Dan Blackburn's help in pointing out these contradictions.

Data from King, 1977.

DIAGNOSING GENETIC SEX IN INDIVIDUALS

. Once we know the sex-determining mechanism of a species, how can we determine the sex of a given individual? This problem clearly has important clinical implications. There are three methods for diagnosing genetic sex: (1) By examining the karyotype, we can diagnose the sex chromosome constitution of the cells. (2) By using special staining techniques (Caspersson et al., 1970), we can detect the presence of the Y chromosome in certain species, e.g., humans, and detect the W chromosome in some snakes (Beçak and Beçak, 1972). (3) Finally, after making preparations of epithelial cells (e.g., of the mouth cavity) and staining these cells with special stains (e.g., Feulgen stain), we can detect a dark body, if present, on the nucleus. This dark structure (Figure 1.4) is called the sex chromatin or the Barr body, after Murray L. Barr, its discoverer. In leukocytes, the sex chromatin is present as a drumstick-like appendage to the nucleus, as illustrated in Figure 1.4. The sex chromatin method of diagnosis of the genetic sex is not applicable to all species, although it is applicable to many mammals. Moore (1966) lists taxa in which no sex chromatin has been found.

SEX CHROMATIN

The discovery of the sex chromatin has led to some important other findings. For example, the number of sex chromatin bodies is usually one less than the number of sex (X) chromosomes present. Because in cases of triploidy this rule does not hold, a more general formula is $S = X - A^m$, in which S is the number of sex chromosomes and A^m is the number of sets of maternal autosomes (Brown and Chandra, 1973).

Another finding is that the sex chromatin is a condensed X chromosome. This finding has given rise to several hypotheses that together are called the Lyon hypothesis, named after M. F. Lyon, who proposed the hypothesis. Mittwoch (1973) has summarized this hypothesis as follows: (1) The genes on the X chromosome that form the sex chromatin are inactive. (2) In an XX cell one or the other X chromosome is inactivated at random, so that maternal and paternal X chromosomes have an equal chance of being inactivated. (3) The process of inactivation occurs early during embryonic development. (4) Inactivation of the X chromosome is not repaired; once inactivation has begun, descendants of the same chromosome (i.e., either maternal or paternal) will always be inactivated in descendants of a given cell.

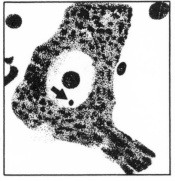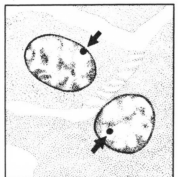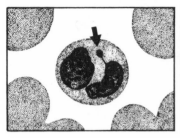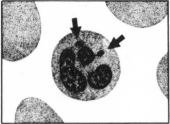

Figure 1.4 Intranuclear sex inclusion or Barr body (arrowed) in a female (XX) nerve cell (*top left*) and in two female fibroblasts (*top right*). One "drumstick" is seen in a normal female polymorphonuclear leukocyte (*bottom left*), and two in a polymorph from an XXX woman (*bottom right*). (From Short, 1972; reprinted with permission of Cambridge University Press.)

Let us consider the four aspects of the Lyon hypothesis.

1. *The genes on the sex chromatin are inactivated.* If one X chromosome in each XX cell is inactivated, one would expect that women with XX and XO constitutions would be similar except for the presence of ovaries in XX women and their absence in XO women (there is no sex chromatin in ovarian germ cells according to Lyon, 1974b, as neither X chromosome is inactivated in those cells). In fact, however, XO women have characteristic somatic deficiencies, such as short stature, a short webbed neck, and difficulties in depth perception, in much higher frequency than XX women. Thus it is possible that not all genes on the X chromosome are inactivated.

2. *The paternal or maternal X chromosome is inactivated at random.* For marsupials, there is considerable evidence that the paternal X chromosome is preferentially inactivated (Lyon, 1974a). In the gazelle (*Gazella gazella*), it is the maternal X chromosome that is inactivated. In women with Turner's syndrome, that is, with the XX constitution but with one structurally abnormal X chromosome, the abnormal X chromosome is always inactivated. In mice, it is the paternal X chromosome that is inactivated in the extra-em-

bryonal tissue (Takagi and Sasaki, 1975). Thus, although in most cases, the X chromosome is inactivated at random, there are exceptions to this tendency.

3. *Inactivation occurs early in embryonic development.* This hypothesis has been confirmed by investigations so far. Adler et al. (1977) have obtained persuasive evidence (by measuring the activity of the enzyme α galactosidase, which is determined by a gene on the X chromosome) that in blastocysts, one X chromosome is indeed inactivated but that there is no activation of one of two inactive X chromosomes. These observations thus confirm Lyon's hypothesis of early inactivation.

4. *Inactivation of the X chromosome is not repaired.* During spermatogenesis in heterogametic males, the X chromosome is inactivated during the first meiotic prophase, although it is subsequently reactivated (Lifschytz, 1972; Lifschytz and Lindsey, 1972). Furthermore it is active in the male's somatic cells, which explains the absence of sex chromatin in the somatic cells. However, in the female's germ cells both XX chromosomes are active, but one X chromosome is permanently inactivated in the somatic cells.

Migeon and Jelalian (1977) have pointed out that while primordial germs cells (PGCs) have a

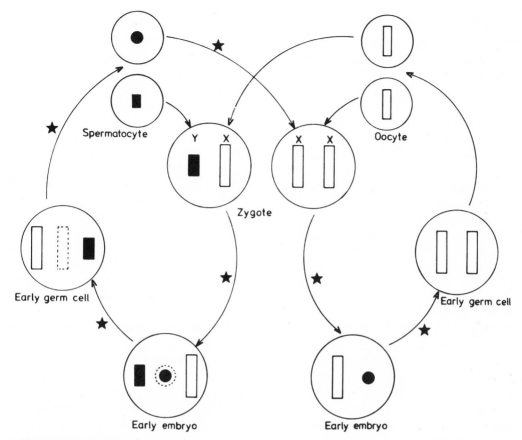

Figure 1.5 X-chromosome activity at different stages of the life cycle. (Modified from Lyon, 1974b; reprinted with permission of Plenum Publishing Corp.)

□ = active chromosomes
■ = inactive chromosomes
⬚ = supernumerary X chromosomes

• = condensed chromosome
⦂ = condensed supernumerary chromosome
★ = change of activity

condensed X chromosome, oocytes have two normal XX chromosomes. This must mean either that a condensed X chromosome is reactivated, as occurs in spermatogenesis, or that in some PGCs the X chromosome escapes inactivation. Figure 1.5 illustrates the activity of the X chromosome during various stages of the life cycle of mammals.

We see then, that the Lyon hypothesis must remain open to modification as new evidence about sex chromatin is uncovered. One finding that must be accounted for is that the sex chromatin in chickens, contrary to expectations, is present in the female, the heterogametic sex (Moore, 1966). This sex chromatin represents the condensed W chromosome (Bloom, 1974).

MODIFYING THE SEX RATIO

If it were possible to influence the primary sex ratio, the secondary sex ratio could presumably be modified without causing differential death of either male or female embryos. This would have significant implications for science, agriculture, and economics, and for ethics and society as well.

Several efforts have been made to influence the primary sex ratio. Most such efforts have been directed toward obtaining an enrichment of either X- or Y-bearing sperm. As it is known that sperm bearing Y chromosomes have a smaller mass than sperm bearing X chromosomes (Sumner and Robinson, 1976) methods to obtain enrichment can be based on physical principles.

Most of these attempts have failed to obtain either 100 percent male or 100 percent female offspring. There has been, at most, some enrichment of either male or female offspring. However, as H. W. Norton (Kiddy and Hafs, 1971:101, 102) has pointed out, these deviations from the normal sex ratio may not have been significant, because data were pooled, such data will overestimate the statistical significance of the results because sources of variance that are not binomial are not taken into account. The methods used to obtain enrichment of either X or Y chromosome–bearing sperm are: (1) separation of X- and Y-bearing spermatozoa by gravity (Lavon et al., 1971; Bahr, 1971; Schilling, 1971); (2) exposure of semen to reduced atmospheric pressure (Foote and Quevedo, 1971); (3) centrifugation (Lindahl, 1971), and (4) electrophoresis, based on the assumption (Bahr, 1971) that X- and Y-bearing sperm have different electric charges (Hafs and Boyd, 1971).

Two other methods have been used to control the sex ratio: genetic selection and selective destruction of Y chromosome–bearing spermatozoa. Genetic selection was originally used for two lines of blood pH, one line of high pH and one of low pH. In the line selected for high pH the sex ratio was 101 males : 100 females, in the line selected for low pH the ratio was 67 males : 100 females. Special diets were effective in changing blood pH, but were ineffective in changing the sex ratio (Weir, 1971).

Selective destruction of spermatozoa bearing Y chromosomes has been achieved in mice. The skin of a male was grafted to a female of the same inbred strain of mice. Most genes in the males and females are expected to be the same, except of course, for those on the Y chromosome. The female made antibodies against the gene products of the Y-linked genes. Semen was then exposed to the sera of such mice, and the semen was then used for artificial insemination (Bennett and Boyse, 1973). There were 45.4 percent males from 372 offspring after such inseminations compared with 53.3 percent males from 5,859 offspring after control insemination.

In nature, a curious instance of regulation of the sex ratio has been observed in the common grackle (*Quiscalus quiscula*). The incidence of female embryos decreases significantly as the breeding season proceeds. This decrease may be ecologically advantageous, because female offspring are smaller than males and, at the start of the breeding season, food may be in relatively short supply. The anticipatory change in sex ratio of the embryo means that a puzzling nonrandom segregation of Z and W chromosomes occurs in the ovary (Howe, 1977). An understanding of this mechanism could be of considerable significance to producers of poultry.

Table 1.9

Sex ratio of offspring and parental response to hepatitis B virus (HBV)

Parental response to HBV	Number of couples	Live births		Sex ratio (males/100 females)
		Male	Female	
Either parent HBsAg(+), anti-HBs(−)	33	60	24	250
Both parents HBsAg(−), anti-HBs(−)	29	51	35	146
Both parents HBsAg(−), either parent anti-HBs(+)	154	241	222	109
Neither parent HBsAg(+), anti-HBs(−) and either parent HBsAg(+), anti-HBs(+)	16	25	27	93
Both parents HBsAg(−), mother anti-HBs(−)		112	86	130
Both parents HBsAg(−), mother anti-HBs(+)		146	140	104
Father HBsAg(+)		13	7	186
Mother HBsAg(+) without fetal loss		17	5	340
Mother HBsAg(+) with fetal loss		10	9	110

Modified from Drew et al., 1978; reprinted with permission of J. S. Drew and *Science*. Copyright 1978 by the American Association for the Advancement of Science.

The sex ratio of human offspring is related to the response of parents to hepatitis B virus (Drew et al., 1978). The data in Table 1.9 illustrate this relationship. Drew et al. (1978) propose that the hepatitis B surface antigen (HBsAg) may cross-react with a male-associated antigen. If such cross-reactivity does exist, then sperm bearing Y chromosomes of HBsAg(+) men would presumably be protected, by the HBsAg in their semen, from any antibodies against HBsAg (anti-HBs) in the prospective mother. However, Y-bearing sperm from HBsAg(−) men might be impeded in fertilization by such anti-HBs in the prospective mother. Antibodies in pregnant women may also preferentially destroy the tissues of male fetuses. However, in HBsAg(+), anti-HBs(−) women tolerant of HBsAg, no sensitization against male tissue would occur, and the male fetuses would survive. The disproportionate loss of male fetuses in HBsAg(+) women could be explained by continued replication of the hepatitis B virus in male embryos or fetuses.

This chapter has shown that much that appears obvious in sexual reproduction is, on closer look, not fully understood.

Sexual Development

We will consider as a continuum an organism's development from a fertilized egg, in which genetic sex determination has occurred, to an individual with full reproductive ability. Along this continuum, for convenience, a number of successive events can be distinguished: genetic sex determination → differentiaion of the sexual phenotype (ovaries, testes, or ovotestes) → differentiation of secondary sex organs → differentiation of the brain in its morphology and its ability to respond to hormones in the future → prepubertal somatic growth → puberty (spermatogenesis or ovulation, full development of secondary sex organs and secondary sex characters, and sexual behavior).

DIFFERENTIATION OF THE GONAD

The primordial germ cells (PGCs), which can be distinguished by their size, structure, and staining qualities (e.g., they show strong alkaline phosphatase activity and are rich in glycogen), play an indispensable role in determining the fate of the gonad. In some species, the site at which PGCs finally settle determines whether the animal will be a phenotypic male or female; in other species, the constitution of the gonadal ridge determines whether the PGCs will become oocytes or sperm. In the absence of the PGCs, the gonad will be sterile.

The original location where PGCs are found in various vertebrates is listed in Table 2.1. In most mammals, the PGCs migrate from these sites to the gonadal ridge by ameboid movement through the tissues, but in birds, in the primitive reptile, *Sphenodon punctatus,* and possibly sometimes in the mammals, such as cattle, sheep, goats, and pigs, the PGCs migrate via the vascular system (Baker, 1972).

In the chick, the PGCs are freed into the space between endoderm and ectoderm, and most enter the vitelline blood vessels. They circulate within the embryo and eventually arrive at the site of the presumptive gonads. (Some of the chick's PGCs do not enter the blood vessels but migrate into the tissues, where they are probably capable of ameboid movements.) By day 3 of incubation, most of the PGCs have left the blood vessels, possibly using ameboid movements (Fujimoto et al., 1976), and have colonized the gonadal site.

After the gonads have been colonized by the PGCs, differentiation of the gonads can proceed. A somewhat arbitrary distinction can be made between gonadal differentiation in cyclostomes and teleosts, on the one hand, and in elasmobranchs, amphibia, and amniotes on the other.

Cyclostomes and Teleosts

In the classical concept, the gonad in cyclostomes and teleosts arises from a single primordium in the peritoneal epithelium. This

Table 2.1

Original location of primordial germ cells in various vertebrates

Vertebrate group	Original location of PGCs
Lampreys	Posterior endoderm
Teleosts	Posterior endoderm near tail bud anlage
Elasmobranchs	Laterally at junction of extra-embryonic and embryonic endoderm
Anurans	Endoderm below blastocoel cavity
Urodeles	Medial hypomere mesoderm
Turtles	Posterior crescent at junction of extraembryonic and embryonic endoderm
Other reptiles	Scattered in yolk sac endoderm
Birds	Anterior crescent at junction of extraembryonic and embryonic endoderm*
Mammals	Posterior crescent at junction of extraembryonic and embryonic endoderm

*According to Dubois (1967), the PGCs originate in the *posterior* part of the blastula, and the anterior crescent is a secondary formation.
Data from Ballard, 1964.

primordium is considered to be homologous with the gonadal cortex of elasmobranchs, amphibia, and amniotes (Hoar, 1969). The origin of the gonad from a single type of tissue presumably explains the lability of teleost gonads and the high incidence of hermaphroditism in this class of vertebrates.

Hardisty (1965b:371), on the basis of detailed investigations of lampreys, has stated that, "in spite of the absence of clearly defined cortical and medullary components of distinct embryological origin and with specific inductive effects on differentiation of gonia, there are some indications, especially from the observations made on (*Lampetra*) *planeri,* that somatic elements may play an important role in sex differentiation." He points out that the vertebrate mesodermal blastema* is pluripotential and that in lampreys the various derivatives of the coelomic wall give rise in succession to mesonephric, adrenocortical, and stromatic and follicular elements of the gonad. The

*Blastema: the mass of undifferentiated cells that give rise to an organ.

distinction that has traditionally been made between gonadal differentiation of lampreys and teleosts, on the one hand, and of elasmobranchs and tetrapods on the other, tends to be blurred by this concept of a pluripotential blastema with a succession of tissue that differentiates from it. Hardisty (1971) also points out that administering sex hormones to lampreys prior to differentiation does not influence differentiation in either the male or female direction; thus, the lamprey gonad is not as labile as that of many teleosts. Such teleost gonadal lability is demonstrated by goldfish (*Carassius auratus*) ovaries when cultured in vitro: cells that appear to be destined to become oocytes are transformed into cells destined to become spermatozoa. This transformation is promoted by testosterone, but not by testicular transplants (Remacle et al., 1976). Examples of the lability of gonads of teleosts in vivo are found in protogynous and protandrous fishes and will be discussed in Chapter 3.

According to Hardisty (1971), the development of the gonads of cyclostomes and probably of some teleosts consists of a progression of steps:

1. PGCs migrate to the gonadal ridge.
2. The gonadal ridge develops: it comes to contain a small number of germ cells and becomes covered by peritoneal epithelium. (This epithelium later gives rise to the follicular elements of both male and female gonads.)
3. The undifferentiated gonad slowly develops with germ cells dividing at a slow rate. In cyclostomes this period may last from 6 to 24 months.
4. A period of more rapid mitotic divisions follows, during which germ cells give rise to germ cell nests or cysts.
5. Meiotic prophase starts both in isolated germ cells and in the cysts. These cells may then start to grow or they may undergo atresia (degeneration and reabsorption) in the earlier stages of prophase.
6. Oogenesis occurs in all lamprey gonads, but in the gonads destined to become ovaries, oogenesis is more synchronous than in future testes, so that in the females the gonads come to contain only oocytes in the cytoplasmaic growth phase. In male lampreys those germ cells that have proceeded to the cytoplasmic growth phase

are eliminated by atresia. As a result of this elimination of the female germ cells, the testes become smaller during differentiation. Some undifferentiated germ cells then remain, which develop into nests of primary spermatogonia by the process of mitosis. In teleosts, mitosis of germ cells continues throughout life, but in lampreys oocytes do not divide mitotically after they reach sexual maturity (see Chapter 7).

In some fishes with ovotestes, i.e., in intersexes (see Chapter 3), each undifferentiated gonad has male and female somatic territories, in which testis and ovary, respectively, will develop (Harrington, 1974): the hilus of the gonad and one of its outgrowths, the lateral ramus, are the male territory; its other outgrowth, the medial ramus, is the female territory (Harrington, 1975).

Elasmobranchs, Amphibians, and Amniotes

In elasmobranchs, amphibia, and amniotes, the undifferentiated gonad consists, in most cases of (1) an inner medulla of mesenchyme, which arises from a medial cellular proliferation that also gives rise to the adrenocortical tissue, and (2) an outer mesodermal cortex, which arises from an elongated ridge of the peritoneal wall. In the undifferentiated state, no distinction can be made histologically between gonads of genetic males and females, unless one looks for sex chromatin in the somatic cells of the gonad. When the primordial germ cells colonize the gonad, where in the gonad they settle seems to determine the sexual phenotype of the animal. If the PGCs settle in the medulla, the gonad develops into a testis; if they settle in the cortex, the gonad becomes an ovary. Development into testis or ovary involves a proliferation of medulla or cortex, respectively.

In birds, the left gonad of males is generally larger than the right one, but both are functional; whereas in the females of most species, the left gonad is functional in the adult, the right gonad being rudimentary. Stanley and Witschi (1940) and van Limborgh (1966, 1968) have proposed that the difference in development between the right gonads of male and female is related to the distribution of PGCs between left and right gonad in each sex. Stanley and Witschi (1940) compared the ratio of PGCs that migrated to left and right gonads in chicken embryos, in which female adults have a rudimentary right gonad, with the ratio of PGCs in embryos of various species of hawks, in which the right gonad of the female is often functional. They found that in the chicken embryo the ratio of PGCs was 3 left : 1 right gonad; in hawks the corresponding ratio was 1–2 left : 1 right. This finding suggests that the number of PGCs in the gonad determines, to a large extent, the development and future size of the gonad. Using sex chromatin for diagnosis of genetic sex, van Limborgh (1966, 1968) studied the same ratio in male and female embryos of chickens and ducks and found that in both sexes of both species, the distribution of PGCs was asymmetric. In female duck embryos, 57.4 percent of the PGCs went to the left gonad and 42.6 percent went to the right. When the PGCs that went to the medulla and cortex were considered separately, then 58.9 percent went to the left cortex and 41.1 percent to the right, and 44.0 percent went to the left medulla and 56.0 percent to the right. In the male duck embryo, 44.3 percent of the total number of PGCs went to the right gonad; 43.6 percent of the PGCs that went to the cortex went to the right and 50.6 percent of the PGCs that went to the medulla went to the right (van Limborgh, 1966). In female chick embryos, 22.1 percent of the total number of PGCs went to the right gonad, 20.0 percent of the PGCs that went to the cortex went to the right, and 32.7 percent of the PGCs that went to the medulla went to the right. The corresponding percentages for male chick embryos are 33.8, 33.1, and 36.9 (van Limborgh, 1968). All these studies seem to indicate that the number of PGCs exert an important influence on the future size and function of the gonad.

Fargeix (1977) investigated the extent to which the ratio of PGCs in left and right gonads results from properties of somatic tissue, rather than from PGCs. For this study, two unincubated halves of blastoderm from Japanese quail were fused and then incubated. In the left embryo, the ratio of PGCs in left and right gonads was 2.8 : 1, and in the right embryo this ratio was 2.7 : 1. These results suggest that the somatic tissue largely determines the ratio of distribution of PGCs between left and right gonads.

The PGCs of chickens enter the vascular network by ameboid movement rather than by chemotaxic mechanisms (Dubois, 1969). They also enter the gonad by ameboid movements, but in this case under the influence of chemical attraction (Cuminge and Dubois, 1969). Once in the gonadal ridge, the PGCs undergo biochemical changes (loss of glycogen and lipids) and subsequent differentiation. Female germ cells (oogonia) cannot leave the gonad (Dubois, 1965; Tachinante, 1974), but male germ cells (spermatogonia) retain their ability to respond to the attractive stimulus of germinal epithelium; they will leave the gonad in which they are settled if an undifferentiated epithelium is juxtaposed to it.

Primordial germ cells will remain in the circulatory system if the gonadal ridge tissue is totally removed, and some of them remain in that system if part of the gonadal tissue is removed (Simon, 1960). These findings indicate a quantitative relationship between the attractant and the number of PGCs that settle in the gonad. The attractant does not appear to be species-specific, for PGCs from Japanese quail (*Coturnix coturnix japonica*) will colonize the presumptive gonads of chickens (*Gallus domesticus*), and even PGCs from mice (*Mus musculus*) will invade the chickens' gonadal ridges (Tachinante, 1974).

In mammals and in chelonian and lacertilian reptiles, the PGCs are formed in the blastomeres of the vegetative pole of the morula. They remain in this area, which is close to the future gonadal ridge, and thus their migration distance is short. In birds and in the viper, the site of PGC formation is further from the gonadal primordium (anlage), and in these animals the PGCs migrate in the blood vessels (Dubois and Cuminge, 1974).

Does the development of the presumptive gonad into an ovary result from information carried by the two XX chromosomes in female PGCs, or does it result from information contained in the somatic cells of the presumptive gonad? The answer to this question may vary with the species. Blackler (1965) transferred PGCs from genetic males to genetic females of the African clawed toad (*Xenopus laevis*) and from genetic females to genetic males, in each case destroying the host's own PGCs and controlling the experiment by using genetic markers. The results showed that the fate of the PGCs was determined by the gonad in which they settled. In male hosts,

the PGCs yielded sperm, and in female hosts, oocytes—regardless of the sex constitution of the donor. In subsequent experiments, Blackler and Gecking (1972) introduced PGCs from *Xenopus mülleri* into *X. laevis* larvae after removal of the *X. laevis* PGCs. The adult females that developed laid eggs (oocytes) closely resembling those of *X. mülleri,* and when such eggs were fertilized by *X. mülleri* sperm, they developed into individuals with *X. mülleri* characteristics. These experiments demonstrate that the genetic information that determines the fate of the PGCs is modified by the somatic gonadal tissue, so that PGCs that carry ZW chromosomes can yield sperm carrying Z and W chromosomes. However, the somatic genetic information contained in the PGCs is not jeopardized by passage through another animal, even if it belongs to another species.

The relationship between the germinal and somatic cells in mammals has been studied in chimeric mice, obtained by fusion of male and female embryos (O and Baker, 1978). In such chimeras it was found that XY PGCs can survive in XX somatic tissue, but that XX PGCs fail to survive in XY somatic tissue. In intersex goats, pigs, and cattle, the testicular part of the gonad is invaded by XX PGCs, which all degenerate (Short, 1974).

Ford et al. (1975) have reported on an XX/XY female mouse obtained by aggregation of morulae from an albino and an agouti strain. This animal was mated to an albino male of a different strain and yielded an albino son that was XXY and sterile. It was concluded on the basis of cytological and genetic data that the son must have originated from a Y-bearing oocyte. In a subsequent study, Evans et al. (1977) found a mouse oocyte with XY chromosomes at diakinesis. These findings for mice indicate a mechanism similar to that discussed above for *Xenopus laevis*.

If one assumes that the PGCs carry genes directing their development into either oocytes or spermatozoa, then one must ask how it happens that presumptive oogonia normally settle in the female's gonadal cortex and presumptive spermatogonia in the male's medulla. If, on the other hand, it is the properties of gonadal tissue that determine the fate of the PGCs, then the question is what causes the medulla to develop in the male and the cortex in the female. In each case the issue is: What causes differentiation of the indifferent gonad?

Witschi (1968) has proposed an explanation based largely on experiments with amphibia. In his hypothesis, illustrated in Figure 2.1, four hormones are proposed: medullarin and the anticortex hormone, which are secreted by the medulla, and corticin and the antimedulla hormone, which are secreted by the cortex. His hypothesis is based on transplantation experiments in which male and female frog gonads in different stages of development were juxtaposed either in vivo or in vitro. These proposed hormones have not been obtained even in crude form, but as we shall see, the H-Y antigen may have a similar function in mammals as the proposed medullarin of Witschi and a similar function in birds as his proposed corticin.

Experimental evidence indicates that the H-Y

Figure 2.1 Sequences in sex development (*Xenopus*). Interpretative diagrammatic presentation of pathways in sex development. Differentiation of gonad primordia into testes or ovaries occurs during stages 24–27. Androgens and estrogens are produced by the gonads beginning stage 33 (solid arrows). If these hormones are excreted earlier (broken lines) they are of uncertain, possibly of adrenal origin. (From Witschi, 1968; reprinted with permission of Academic Press, Inc.)

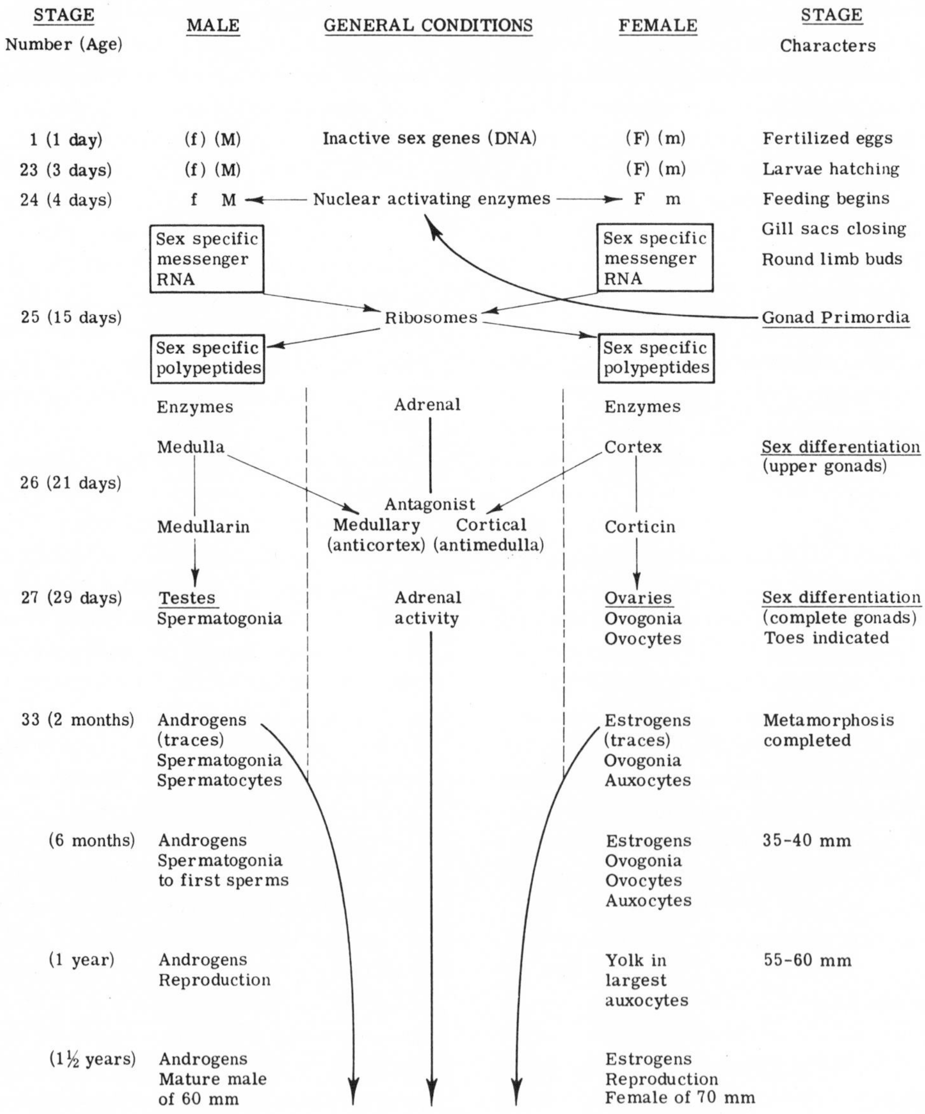

STAGE Number (Age)	MALE	GENERAL CONDITIONS	FEMALE	STAGE Characters
1 (1 day)	(f) (M)	Inactive sex genes (DNA)	(F) (m)	Fertilized eggs
23 (3 days)	(f) (M)		(F) (m)	Larvae hatching
24 (4 days)	f M	Nuclear activating enzymes	F m	Feeding begins
	Sex specific messenger RNA		Sex specific messenger RNA	Gill sacs closing / Round limb buds
25 (15 days)		Ribosomes		Gonad Primordia
	Sex specific polypeptides		Sex specific polypeptides	
	Enzymes	Adrenal	Enzymes	
	Medulla		Cortex	Sex differentiation (upper gonads)
26 (21 days)		Antagonist		
	Medullarin	Medullary Cortical (anticortex) (antimedulla)	Corticin	
27 (29 days)	Testes Spermatogonia	Adrenal activity	Ovaries Ovogonia Ovocytes	Sex differentiation (complete gonads) Toes indicated
33 (2 months)	Androgens (traces) Spermatogonia Spermatocytes		Estrogens (traces) Ovogonia Auxocytes	Metamorphosis completed
(6 months)	Androgens Spermatogonia to first sperms		Estrogens Ovogonia Ovocytes Auxocytes	35–40 mm
(1 year)	Androgens Reproduction		Yolk in largest auxocytes	55–60 mm
(1½ years)	Androgens Mature male of 60 mm		Estrogens Reproduction Female of 70 mm	

antigen is crucial for determining the differentiation of the gonad. After exposure to the H-Y antibody, suspensions of testicular tissue of newborn mice and rats (tested separately) organized into aggregates that strongly resembled ovarian tissue (Ohno et al., 1978; Zenzes et al., 1978).

Could the gonadal steroid hormones, estrogens and androgens, be involved in differentiation of the gonads? From the principle of parsimony, an argument can be made that the gonadal tissues produce steroids in the adult and may therefore also produce them in the embryo. The hypothesis that steroids produced by the undifferentiated gonad direct the morphological and biochemical differentiation of cortex and medulla has been extensively tested; the prevailing evidence is that estrogens and androgens are *not* the hormones that cause differentiation, although sex reversal can be obtained in various vertebrates (van Tienhoven,

1968) by administering relatively large doses of these hormones. However, there are discrepancies between transplantation experiments and such hormone administration experiments (Witschi, 1968). The discrepancies and the paradoxical effects obtained with steroid hormones—androgen-caused feminization and estrogen-caused masculinization—argue against the role of these steroids in gonadal differentiation.

In view of the crucial role played by PGCs in determining the fate of the gonad, the effect of gonadal steroids on PGC migration is of interest. After estrogen administration to chicken embryos, 54.2 percent of PGCs were in the left gonad, whereas in embryos treated with testosterone propionate 81.1 percent were in the left gonad (Swartz, 1975). Cyproterone acetate, an anti-androgen, had no effect on the distribution of PGCs to left or right gonads (Swartz, 1977). It is

Figure 2.2 Diagram to illustrate proposed scheme for sex determination [and gonadal differentiation] in mammals. M = gene for medullary stimulation, F = gene for cortical stimulation, C = controlling elements on the X and Y, P = autosomal gene or chromosome region controlling horn growth in the goat with a pleiotropic effect in sex determination, X = X chromosome, Y = Y chromosome, A = autosome, S = suppression. Hatched chromosomes are heterochromatic. Hatched areas in ovary indicate functional follicles; solid areas indicate atretic follicles. Hatched circles on the testis indicate tubules with germ cells; open circles, tubules without germ cells. (From Hamerton et al., 1969; reprinted with permission of J. L. Hamerton and *Nature*.)

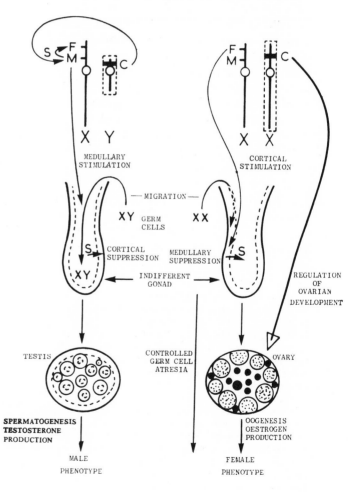

thus possible that the feminizing effect of estrogens on the left gonad of genetic male chicken embryos found by others (van Tienhoven, 1968) may be partly the result of their influence on the distribution of PGCs to the two gonads. The experiments by Müller et al. (1979) mentioned in Chapter 1 suggest that the principal effect of estrogen administration is probably the enhancement of the expression of the H-Y antigen. However, we do not know how the H-Y antigen causes differentiation, and so the regulation of the distribution of PGCs to left and right gonads may be one of its effects.

Figure 2.2 illustrates a scheme for sex determination and gonadal differentiation for mammals, as proposed by Hamerton et al. (1969). In this scheme, the genes of the sex chromosomes are activated by autosomal genes. In Figure 2.3, Hamerton et al. explain intersexuality in goats. Their scheme for male and female differentiation is similar to Witschi's (1968) scheme illustrated in Figure 2.1, but it is less detailed, probably because amphibia, which Witschi studied, are a more convenient animal than goats for this type of investigation, and more details are known about amphibian* than mammalian gonadal differentiation. In both schemes, the cortical suppression may be effected by the H-Y antigen.

The morphological differentiation of the mammalian testis consists of:

1. Development of the medullary tissue into medullary sex cords, which eventually give rise to the seminiferous tubules and the tunica albuginea, a connective tissue membrane surrounding the testis.

*For example, within one species of frog (*Rana temporaria*) there are a number of races that differ in the timing of male sex differentiation (Witschi, 1929): There are (1) undifferentiated races, in which the males differentiate after metamorphosis (at metamorphosis these races are 100 percent female); (2) differentiated races, in which the males differentiate before metamorphosis (at metamorphosis these races are 50 percent male 50 percent female); and (3) semidifferentiated races, in which the male differentiates during metamorphosis. In general, the differentiated races are found at higher altitudes (e.g., the Alps) and latitudes (the Baltic region), and the undifferentiated races at lower altitudes (e.g., the Netherlands) and latitudes (southern Germany) (Witschi, 1929). It is also easier to transplant gonads in amphibian than in mammalian embryos.

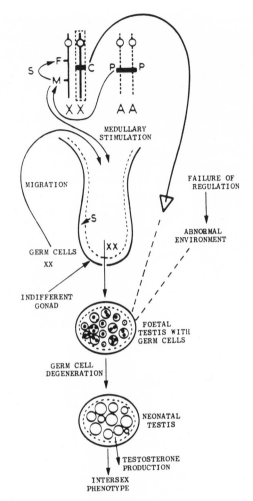

Figure 2.3 Diagram to show the proposed mechanism by which an autosomal gene or chromosome area (PP) acts by controlling the "medullary stimulating" gene M carried by the X chromosome resulting in the formation of a testis in genetic females in the goat (*Capra hirca*). For abbreviations see Figure 2.2. (From Hamerton et al., 1969; reprinted with permission of J. L. Hamerton and *Nature*.)

2. Development of mesenchyme, which contributes to interstitial and thecal tissues; the two outermost sheaths surrounding the ovarian follicle are the theca externa and theca interna.

3. Development of the mesonephros, the primitive kidney, which contributes cells to the developing gonad, and the juxtaposition of which to the gonads appears to be indispensable for the differentiation of the gonad (Baker, 1972).

The differentiation of the mammalian ovary consists of:

1. Development of the cortex into secondary sex cords, and the formation of follicles.

2. Development of the mesenchyme, which provides for the interstitial and thecal tissues (Baker, 1972).

Figure 2.4 illustrates gonad differentiation in amniotes, showing the origin and fate of cortical and medullary components. The differentiation of the mammalian testis precedes that of the ovary, and the number of mitoses in the testis during its early development is higher than in the ovary at the same stage. Mittwoch (1973, 1975) contends that this earlier mitosis results from the presence of the Y chromosome, and that this early growth protects the testes of the developing fetus against the high concentrations of female sex hormones of the mother.

The idea that the indifferent or undifferentiated gonad in mammals invariably has both a medullary and a cortical component has been challenged by Gropp and Ohno (1966), who conclude from histochemical evidence that there is one blastema only in each of the gonads of fetal cattle. Figure 2.5 illustrates how differentiation proceeds from one common blastema in the gonads. Thus it ap-

pears that the traditional concept of one common blastema for each gonad in cyclostomes and teleosts and of two separate origins (cortex and medulla) for each gonad in elasmobranchs, amphibia, and amniotes is undergoing some revisions; at least, there appear to be enough exceptions to preclude making any sharp distinction between the two types of gonadal differentiation.

In chickens, the female is the heterogametic sex and any maternal hormones to which the embryo may be exposed are present in the egg at the time the egg is laid. The amounts of hormones present in the egg are presumably small, therefore the development of the gonad is not jeopardized by maternal hormones, as it is in the mammalian embryo, in which the embryo is exposed to high concentrations of estrogen. Chicken testes are larger than chicken ovaries until about the seventh day of incubation. After this stage the ovary grows at a greater rate than the testes. Mittwoch (1973: 121, 131) has speculated that the ovary must "reach a certain size by a given stage of development in order to develop into a testis or ovotestes." It seems to me that cause and effect might be just the reverse of those implied by Mittwoch (1973), i.e., that the growth of either test-

Figure 2.4 Diagrams illustrating schematically the main features of gonad differentiation in amniote embryos with reference to the origin of the medullary and cortical components. *A*, Origin of the primary sex cords (medullary cords) from the germinal epithelium. *B*, Gonad at the indifferent stage of sexual differentiation; the well-developed primary sex cords represent the male or medullary component, whereas the germinal epithelium represents, potentially, the cortical component. *C*, Differentiation of a testis consists in the further development of the primary sex cords, and the reduction of the germinal epithelium to a thin, serous membrane, accompanied by development of the tunica albuginea. *D*, Differentiation of an ovary consists in reduction of the primary sex cords to medullary cords of the ovary, whereas the cortex is formed by continued development of cortical cords from the germinal epithelium. (From Burns, 1961; reprinted with permission. © 1961, The Williams & Wilkins Co., Baltimore.)

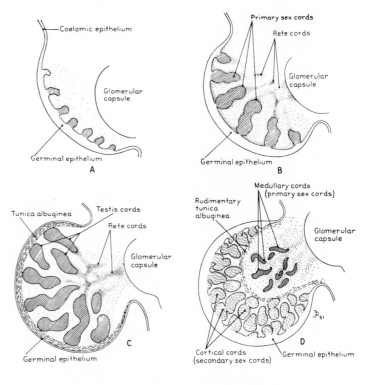

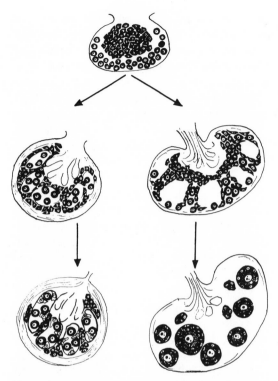

Figure 2.5 A schematic representation of the differentiation process of a common somatic blastema of an embryonic indifferent gonad into testicular interstitial cells on one hand, and ovarian follicular cells on the other. Germ cells of all stages are shown as relatively large round cells with a large white nucleus, whereas testicular interstitial cells, ovarian follicular cells, and their common progenitors are shown as smaller angular cells with a small white nucleus. *Top:* An indifferent gonad with primordial germ cells in the periphery and the somatic blastema in the center. *Middle:* The initial stage of sexual differentiation. In the male (left), primordial germ cells actively invade the mass of blastema cells, while blastema cells move slightly toward the periphery, leaving behind the rete area in the hilar region. The peripheral area becomes devoid of important elements and organizes the sheath of connective tissue (*Tunica albuginea*). In the female (right), the blastema sends out strands toward the periphery and meets with primordial germ cells, which are then incorporated into strands. *Bottom:* Now the testis (left) and the ovary (right) are fully organized. In the testis, the blastema cells are recognized as interstitial cells, while in the ovary they are recognized as follicular cells. (From Ohno, 1967; reprinted with permission of Springer-Verlag.)

es or ovary is the result of the presence or absence of the W chromosome and or H-Y antigen.

In most adult vertebrates the pituitary gland controls the activity of the gonads by its secretion of gonadotrophins; the question arises whether it controls the differentiation of the embryonic gonads also. Thus far, there is no evidence that removal of the pituitary gland (hypophysectomy) affects the differentiation of the gonads of some species, e.g., in sharks (*Scyliorhinus* sp.), frogs (*Rana temporaria, R. pipiens, R. sylvatica*), toads (*Bufo americanus*), lizards (*Lacerta vivipara*), chickens (*Gallus domesticus*), ducks (*Anas platyrhynchos*), rats (*Rattus norvegicus*), rabbits (*Oryctolagus cuniculus*), or mice (*Mus musculus*) (for references, see Dufaure, 1966; van Tienhoven, 1968). Injections with gonadotrophins, especially of mammalian luteinizing hormone (LH), have caused feminization of genetic males in several amphibia (*Rana temporaria, Rhacophorus schlegelii, Bufo arenarum*), in a lizard (*Lacerta vivipara*), a turtle (*Emys leprosa*), and chickens (*Gallus domesticus*) (Dufaure, 1966). However, the results of these studies are difficult to interpret, because the mammalian preparations used were not very pure. When undifferentiated gonads of Japanese quail (*Coturnix coturnix japonica*) were cultured in vitro, differentiation proceeded normally (Lutz-Ostertag, 1965), a result indicating the autonomy of the process.

The thyroid is indirectly implicated in the differentiation of the gonads in the amphibia, *Rhacophorus schlegelii* and *Rana temporaria*, because treatment with thiourea, which blocks synthesis of thyroid hormones, causes ovarian development in genetic males. However, as neither thyroidectomy nor hypophysectomy has this effect on males, it may well be that thiourea has a direct effect on the gonad that produces its effect via an extrathyroidal mechanism.

In some lower vertebrates, environmental factors can affect sexual differentiation. No primary males* are found in wild populations of the self-fertilizing fish, *Rivulus marmoratus*, which lives in the tropical waters off Florida and the Dutch Antilles. In the laboratory, primary males can be obtained by incubating the larvae at 19°C; at 25° and 31°C, hermaphrodites will be obtained (Harrington, 1975). Harrington (1975) speculates that temperature affects the mitotic rate in the gonads and that it has an effect similar to that hypoth-

*Primary males are males that have not passed through a hermaphroditic stage.

esized by Mittwoch (1973) for the Y chromosome in mammals. In some amphibia (*Rana sylvatica, Hynobius retardatus, Bufo vulgaris*) rearing larvae at 18°–21°C yields a normal sex ratio, but at 15°C yields hermaphrodites only, and at 10°C yields females only (van Tienhoven, 1968, for review).

The effect of incubation temperature on the sex ratio of various species of turtles is summarized in Table 2.2. It is difficult to predict from information on one species what the effect of a high temperature of incubation will be on the sex ratio of another species. It appears that exposure of eggs to cyclic temperature changes may generally yield sex ratios closer to unity than sex ratios obtained at constant temperatures. Japanese quail eggs incubated at temperatures 1.5°–2.5°C above

the normal incubation temperature of 37–38°C, beginning three to five days before the normal hatching time, yielded females with right gonads larger than normal (Lutz-Ostertag, 1965).

It is obviously difficult to change the temperature to which the true mammalian embryo is exposed. Torrey (1950), however, did this by transplanting fetal rat gonads to the anterior eye chambers of adult rats and leaving some of the eyelids open, so that the gonads would be at lower temperatures than normal, and sewing some shut, so that the gonads would be exposed to normal body temperature. The cortex of the fetal gonads of the former group was destroyed, whereas the gonads of the latter group developed normally.

The interruption of the sexual embrace of the

Table 2.2

Effect of incubation temperature on sex ratio of various species of turtles

Species	Temperature of incubation, °C	Percent male	Percent female	Percent unknown	Reference
		Sex			
Testudo graeca (Greek tortoise)	26–27 or 31	97.6	2.4	—	Pieau, 1971
Emys orbicularis (European pond turtle)	30	4.2	95.8	—	Pieau, 1971
	27–28	100	—	—	Pieau, 1973
	29–30	—	100	—	
	24–30*	100†	—		
	26–31*	50	50		
Chelydra serpentina (snapping turtle)	20	—	100‡	—	Yntema, 1976
	22	10	90		
	24	100	—		
	26	99	1		
	28	65	35		
	30	—	100‡		
Graptemys ouachitensis (map turtle)	25	90	—	10†	Bull and Vogt, 1979
	30.5	—	89	11†	
	20–30*	65	—	35	
	23–33*	—	59.6	40.4	
G. pseudogeographica (map turtle)	25	78	—	22	Bull and Vogt, 1979
	30.5	1.7	63.4	34.9	
	20–30*	68	—	32	
	23–33*	—	64	36	
G. geographica (map turtle)	25	80	—	20	Bull and Vogt, 1979
	30.5	—	74	26	
Chrysemys picta (painted turtle)	25	79	—	21	Bull and Vogt, 1979
	30.5	—	80	20	
Trionyx spiniferus (soft-shelled turtle)	25	40	41	19	Bull and Vogt, 1979
	30.5	31	28	41	

*Embryos were exposed to daily cyclic changes in temperature.
†Males plus intersexes.
‡Some gonads at 3 months of age were bisexual.

frogs *Rana esculenta* and *R. temporaria* causes failure of oviposition, so that when the eggs are finally laid, they have undergone an aging process. Such "overripe" eggs, when fertilized, yield a preponderance of males, without differential mortality of females (Witschi, 1952); however, the mechanism by which aging of eggs channels differentiation into the male phenotype is not known.

The ratio of potassium (K^+) to calcium (Ca^{2+}) ions in a Ringer's solution in which *Discoglossus pictus* tadpoles are raised affects the

Table 2.3

Urogenital homologies of mammals*

Male	Indifferent stage	Female
Testis	Gonad	Ovary
1. —		1. Cortex
2. Seminiferous tubules		2. Medulla (primary)
3. Rete testis		3. *Rete ovarii*
1. *Mesorchium*		1. Mesovarium
2. —		2. Suspensory ligament of ovary
3. *Ligamentum testis*	Genital ligaments	3. Proper ovarian ligament
4. *Gubernaculum testis* (in part)		4. Round ligament of uterus
5. *Gubernaculum testis* (as a whole)		5. —
6. —		6. Broad ligament of uterus
	Mesonephric collecting tubules	
Efferent ductules and *appendix epididymidis*	Cranial group	*Epoöphoron* and *vesicular appendices*
Paradidymis and *aberrant ductules*	Caudal group	*Paroöphoron*
1. Ductus epididymidis		
2. Ductus deferens	Mesonephric (Wolffian) duct	*Gartner's duct of the epoöphoron*
3. Seminal vesicle		
4. Ejaculatory duct		
1. *Appendix testis*		1. Uterine tube
2. —	Müllerian duct	2. Uterus
3. —		3. Vagina (upper part?)
Seminal colliculus	Müller's tubercle	Hymen (site of)
1. Bladder (except trigone?)	Vesico-urethral primordium	1. Bladder (except trigone?)
2. Upper prostatic urethra		2. Urethra
	Urogenital sinus	
1. Lower prostatic urethra	Pelvic portion	1. Vestibule (nearest vagina)
a. *Prostatic utricle* (or *vagina masculina*)		a. Vagina (lower part, at least)
b. Prostate gland		b. *Para-urethral ducts*
2. Membranous urethra		2. Vestibule (middle part)
3. Cavernous urethra	Phallic portion	3. Vestibule (between labia minora)
Bulbo-urethral glands		Vestibular glands (of Bartholin)
1. Penis	Phallus	1. Clitoris
a. Glans penis	Glans	a. Glans clitoridis
b. Urethral surface of penis	Lips of urethral groove	b. Labia minora
c. Corpora cavernosa penis	Shaft	c. Corpora cavernosa clitoridis
d. Corpus cavernosum urethrae		d. Vestibular bulbs
2. Scrotum	Labio-scrotal swellings	2. Labia majora
3. Scrotal raphé		3. Posterior commissure
4. —	Median swelling	4. Mons pubis

*Vestigal parts printed in italics.
From Arey, 1946; reprinted with permission of W. B. Saunders Co.

sex ratio. At a ratio of $K^+ : Ca^{2+} = 0.11$, 30 males and 16 females emerged; a ratio of 0.14 yielded 20 males and 47 females; a ratio of 0.48 yielded 39 males and 35 females; and a ratio of 0.03 yielded 33 males and 32 females. When tadpoles were raised in a high-calcium Ringer's solution ($K^+ : Ca^{2+} = 0.22$) for 20 days and then were transferred to a normal Ringer's solution (Bellec and Stolkowski, 1965), three out of 30 tadpoles had ovotestes.

Besides the gonads, the duct systems, which conduct the gametes or the products of conception to the exterior, must also differentiate. Table 2.3 and Figure 2.6 help to clarify which parts primordia (anlagen) of the embryo develop and which regress during differentiation of the duct system in each sex. Figure 2.7 indicates when female duct differentiation occurs in different mammalian species.

To learn what factors control these differentiations, one can gonadectomize the embryo before its secondary sex organs differentiate. Table 2.4 shows the results of such gonadectomy on amphibia with ZZ–ZW sex determination, birds (ZZ–ZW), and mammals (XY–XX). Table 2.5 presents tentative conclusions on the role of gonadal hormones in the differentiation of the gonoducts in mammals, birds, and reptiles. As Table 2.4 shows, the gonadectomized embryos take on some of the characteristics of the homogametic sex. This means that, in mammals, the ovaries, and thus perhaps the estrogens, are not essential for the differentiation of the female Müllerian duct system, but that the testes are required for the differentiation of the male (Wolffian) duct system and for the regression of the Müllerian duct system. In rats, it has been established by castration experiments, in which androgens and other possible confounding factors are eliminated, by the use of replacement therapy, and by injections of cyproterone acetate, an antiandrogen, that androgens cause differentiation of the male duct system, but that another testicular factor, x, is responsible for regression of the Müllerian ducts (Neumann et al. 1969). Figure 2.8 illustrates the results obtained after different experimental treatments.

In birds, the gonadectomized male and female embryos both show the presence of Wolffian and Müllerian ducts, a finding that indicates that testes are not required for the differentiation of the avian Wolffian duct system. The presence of both oviducts in the gonadectomized chicken embryos must mean that the testes produce an inhib-

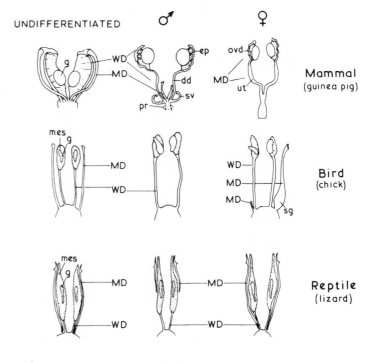

Figure 2.6 Schematic diagrams of undifferentiated and sex-differentiated stages of the reproductive ducts in fetal guinea pig (25.5 and 37 days), chick (6 and 18 days), and lizard (*Anguis*) (100–130 mg and 130–170 mg body weight). g = gonad, mes = mesonephros, WD = Wolffian duct, MD = Müllerian duct, ep = epididymis, dd = ductus deferens, sv = seminal vesicle, pr = prostate, ovd = oviduct, ut = uterus, sg = shell gland. (From Price et al., 1975; reprinted with permission of *American Zoologist*.)

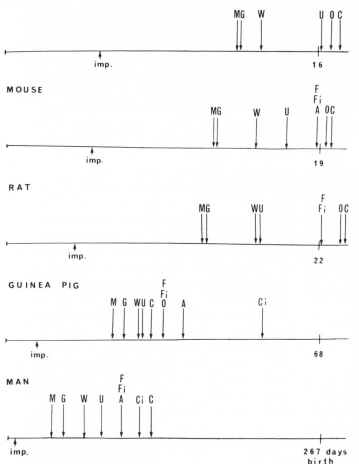

Figure 2.7 Comparative sequence of developmental stages in the female reproductive tract of different mammals. Arrows mark the beginning of certain important events, such as implantation (imp.), Müllerian duct development (M), gonadal sex differentiation (G), uterovaginal canal (U), and Wolffian duct degeneration (W), as well as the beginning of the following events in oviductal differentiation: O, demarcation from uterus; C, coiling; F, epithelial folds; Fi, fimbriae; A, ampulla differentiation; Ci, ciliated cells. (From Price et al., 1969, with permission of D. Price and E. S. E. Hafez. Reprinted from *The Mammalian Oviduct,* edited by E. S. E. Hafez and R. J. Blandau, by permission of The University of Chicago Press. Copyright © 1969 by The University of Chicago.

itor of oviduct differentiation (factor x'). Tran and Josso (1977), using chick Müllerian ducts cultured in vitro, found that neither testosterone nor mammalian fetal testicular tissue cocultured with the chick oviducts could inhibit the differentiation of the ducts. Chick testicular tissue, however, inhibited differentiation not only of chick oviduct, but also of rat oviduct. These results suggest that factor x of mammalian testes and factor x' of avian testes are not identical, but are probably structurally related. In cattle (Blanchard and Josso, 1974) and pigs (Tran et al., 1977), the anti-Müllerian gonadal hormone is secreted by the Sertoli cells, and Tran et al. (1977) stated that the presence of the hormone can be used as a functional marker of the Sertoli cell.

The reason for the regression of the right oviduct in female birds has not been entirely explained. There may be a difference in threshold between the left and right oviduct for the response to estrogens (see van Tienhoven, 1968). Thiebold (1975) has proposed that the ostial junction being initially less firmly attached to the pleuroperitoneal septum of the right oviduct, when compared to that of the left, causes regression of the right oviduct. According to this hypothesis, stretching during development causes the attachment of the right oviduct, but not of the left oviduct, to break. This stretching of the more firmly attached left oviduct presumably induces mitosis, whereas the failure of the right oviduct to stretch prevents growth.

We have seen that gonadal hormones are important for the differentiation of the male and female genital duct systems as they are for the

Table 2.4

Effects of early* gonadectomy on embryonic development
of secondary sex characteristics

Amphibia (ZZ-ZW sex determination)			
Müllerian ducts	Wolffian ducts		Cloacal glands
♂ Rudimentary	Persistent, sexually differentiated		Developed
♀ Developed	Persistent, functions as ureter		Absent
⚥ Present, undifferentiated	Present, undifferentiated posterior urogenital collecting duct M type		Absent
⚥ Present, undifferentiated	Present, undifferentiated posterior urogenital collecting duct M type		Absent

Birds			
Müllerian ducts	Wolffian ducts	Genital tubercle	Syrinx
♂ Absent	Present	M type	M type
♀ Left developed	Absent	F type	F type
⚥ Developed (both sides)	Present	M type	M type
⚥ Developed (both sides)	Present	M type	M type

Mammals					
Müllerian ducts	Wolffian ducts	Urogenital sinus	Prostate	Exterior genitalia	Mammary glands
♂ Absent	Developed	M type	Developed	M type	M type
♀ Developed	Absent	F type	Absent	F type	F type
⚥ Developed	Absent	F type	Absent	F type	F type
⚥ Developed	Absent†	F type	Absent	F type	F type

F = female; M = male; ♂ = normal male; ⚥ = gonadectomized male; ♀ = normal female;
⚥ = gonadectomized female.
*Before irreversible differentiation has occurred.
†In mice, partially persistent.
From van Tienhoven, 1968; reprinted with permission of W. B. Saunders Co.

differentiation of other secondary sex organs, as listed in Table 2.3. The pituitary controls the secretion of the gonadal hormones in the adult; does it influence the differentiation of the secondary sex organs of the embryo? Woods et al. (1977) hypophysectomized chicks by decapitating them and found that embryonic chick testes produce androgens independently of the pituitary until about day 13.5 of incubation. After that time, the pituitary becomes essential for androgen production, but itself operates independently of the hypothalamus. These findings explain why, in chicks hypophysectomized at 33–38 h of incubation, the male and female duct systems are normal (Fugo, 1940).

Decapitation of either rat or mouse fetuses had no great effect on the differentiation of the male and female genital duct systems in these species; however, it cannot be ruled out that maternal gonadotrophins maintained androgen secretion by the fetal testes (Jost, 1953). In rabbits, after decapitation of male fetuses on day 19, abnormalities appeared in some of the reproductive structures, increasing in severity in a posterior direction. Testes, epididymides, and deferent ducts were normal, and Müllerian ducts regressed normally; however, minor changes were seen in the seminal vesicles, the prostate was reduced, and the external genitalia were feminized. The later the decapitation was carried out, the less extensive its effects. These deficiencies were corrected by administering horse gonadotrophins to the fetus at the time of its decapitation (Jost, 1953).

In most cases, birds and mammals are born with gonads and gonoducts fully differentiated but not fully developed.* Marsupials are exceptional in that they are born quite immature, with gonads and gonoducts not fully differentiated. Between birth and puberty, here provisionally defined as the stage of first appearance of spermatozoa or

*Fully differentiated gonads and gonoducts can be recognized macroscopically and microscopically as testes, ovaries, oviducts, ductus deferentes, etc.

Table 2.5

Tentative conclusions on normal embryonic sex differentiation of Wolffian (WD) and Müllerian (MD) ducts in relation to gonadal hormones

Mammal

♂ XY
- WD maintenance dependent on testicular androgens
- MD regression caused by a testicular Müllerian duct-inhibiting factor (not characterized)

♀ XX
- WD regression independent of ovarian hormones: result of absence of testicular androgens
- MD maintenance independent of ovarian hormones

Bird

♂ ZZ
- WD maintenance independent of testicular hormones
- MD regression caused by testicular androgens or a Müllerian duct-inhibiting factor (not characterized)

♀ ZW
- WD retention independent of ovarian hormones
- L MD maintenance independent of ovarian hormones
- R MD regression caused by ovarian hormones

Reptile

♂ ZZ
- WD maintenance independent of testicular hormones
- MD regression caused by testicular hormones (androgens?)

♀ ZW
- WD retention independent of ovarian hormones
- MD maintenance independent of ovarian hormones

From Price et al., 1975; reprinted with permission of *American Zoologist*.

first ovulation with normal cycles, there is generally little development of gonads, secondary sex organs (e.g., oviducts and prostate) and secondary sex characteristics (e.g., beard and mammary glands). In some species, e.g., rabbits, oocytes enter prophase I of meiosis during the first few weeks after birth, but in many other mammalian species they enter this stage before the animals are born. At puberty the difference between males and females become accentuated.

Some important differences between the brains of males and females appear perinatally (i.e., around the time of birth), although their effects do not become apparent until puberty or later. Let us summarize these differences here and provide details later. The major brain-related differences between males and females in birds and mammals are:

1. In avian and mammalian species, females experience cyclic release of gonadotrophic hormones (GTH), whereas males generally do not show such cyclic releases, and the gonadotrophin release from day to day varies less. On close examination, this tonic secretion of GTH, at least in some species, has been shown to be episodic within each day.

2. The homogametic sex (males in birds, females in mammals) has a greater tendency to show bisexual behavior, even when exogenous hormone is not administered, than the heterogametic sex. This generalization is based on studies of a few species, such as Japanese quail, chickens, rats, and mice. In these species, it is also possible to induce bisexual behavior in gonadectomized animals by exogenous steroid sex hormones; e.g., male sexual behavior can be induced in ovariectomized rats treated with testosterone, and female sexual behavior can be induced in Japanese quail, in which the testes have been regressed by exposure to short photoperiods, by injections of estrogens. The generalization does not extend to amphibians, because in the African clawed toad (*Xenopus laevis*), in which the male is the homogametic sex, testosterone can induce male sexual behavior in females, but estrogens fail to induce female sexual behavior in males (Kelley and Pfaff, 1976).

3. Administration of either testosterone or estrogen to Japanese quail embryos results in demasculinization, i.e., lack of male sexual behavior in the males when they become adults. Female rats (*Rattus norvegicus*), mice (*Mus musculus*), and golden hamsters (*Mesocricetus auratus*), injected perinatally with either testosterone or estrogen, may show either defeminization, i.e., lack of female sexual behavior, or a higher incidence of masculinization, i.e., male sexual behavior, than their controls, or both defeminization and masculinization. The experimental evidence for these effects is discussed in the next sections.

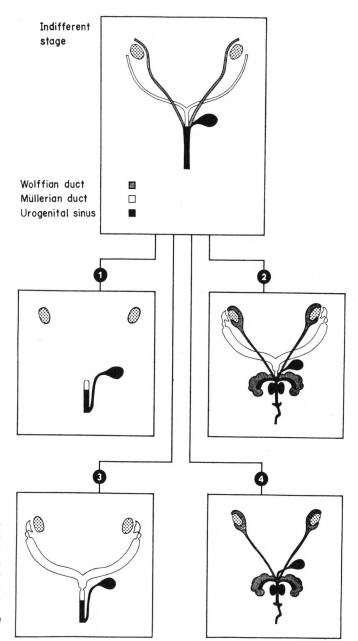

Indifferent stage

Wolffian duct
Müllerian duct
Urogenital sinus

Figure 2.8 The four possibilities of sexual differentiation [in mammals] depending on the hormonal conditions during fetal development. 1. After cyproterone treatment of fetus: factor x active, androgens not active. 2. After castration of fetus and subsequent treatment with androgens: androgens active, factor x not active. 3. After castration: neither androgens nor factor x active. 4. Control male: androgens and factor x both active. (From Neumann et al., 1969; reprinted with permission of F. Neumann and *Journal of Reproduction and Fertility*.)

DIFFERENCES IN GTH RELEASE BETWEEN MALES AND FEMALES

In female mammals, the periodic ovulations that occur during the breeding season are preceded by a release of luteinizing hormone (LH) from the pituitary. The details of the time sequence between LH release and ovulation and the factors that influence this LH release will be discussed in detail in Chapter 8.

In male mammals, the concentrations of LH and of the androgens secreted by the testes may fluctuate episodically during each day, but the day-to-day fluctuations are relatively small. Although when the fluctuations within a day are considered, the term "tonic secretion" is incorrect, the term has been used for LH and androgens to contrast it with the cyclic pattern of LH release found in females.

The tonic secretion of androgens may have adaptive significance, as suggested by the fact that the latency period between androgen injection and the appearance of male sexual behavior in castrated male mammals is longer than the latency period for female sexual behavior to be manifested after administration of ovarian hormones (estrogen and progesterone) in ovariectomized mammals. Moreover, males that are continuously capable of mating during the breeding season have an obvious advantage over males in which mating capability varies (Feder and Wade, 1974).

The difference between male and female hormone secretion patterns is determined perinatally. Pfeiffer (1936), in an elegant series of experiments, showed that follicles of rat ovaries transplanted to castrated males did not ovulate, unless the males had been castrated neonatally, in which case the ovaries underwent normal ovulatory cycles. It was not until the 1950's that many further investigations on this subject were carried out, but since that time, hundreds of studies have appeared on the subject. An attempt will be made to summarize them here.

The presence of testes in the normal male, or the administration of either estrogen or testosterone propionate to an intact female or a gonadectomized neonatal male or female rat results in the tonic pattern of LH release and in behavioral defeminization or masculinization or both when the animal is an adult. In the absence of testes neonatally, as in a gonadectomized male rat or in an intact or ovariectomized female, the cyclic release pattern of LH release occurs in the adult. As in the differentiation of the secondary sex organs and secondary sex characteristics, summarized in Table 2.4, we can generalize and state that the neutral pattern (found in the absence of the gonads) is similar to the pattern found in the female. Anatomically this means the presence of developed Müllerian ducts, the absence of Wolffian ducts, the absence of the prostate, a female type urogenital (UG) sinus, female type exterior genitalia, and female type mammary glands. Physiologically this means cyclic LH release in

the adult and the capacity for male and female sexual behavior as an adult when the appropriate gonadal hormones are either present or administered. In the presence of the testes or in the presence of testosterone, the Wolffian ducts will be developed, the UG sinus will be of the male type, the prostate will be developed, the external genitalia and mammary glands will be of the male type, the Müllerian duct will be regressed (but will be developed in the absence of the testes even when testosterone is present, as already discussed), the pattern of LH release will be the male type in the adult, and the adult will show either male behavior, a lack of female behavior, or both.

As a result of neonatal androgen exposure and the resulting lack of cyclic LH release, ovulations fail to occur, no corpora lutea are formed, and the animal is, of course, sterile as an adult. In such androgenized rats, the follicles are numerous and small, and they secrete estrogen, so that the blood estrogen concentration is similar to that found during diestrus in normal cyclic rats (Naftolin et al., 1972). The continuous estrogen secretion and lack of progesterone causes persistent vaginal estrus to be part of the anovulatory syndrome of such rats. In brief, this syndrome of the androgenized female rat is characterized by sterility, ovaries with many small follicles and no corpora lutea (CL), persistent vaginal estrus, and aberrant sexual behavior.

Usually testosterone propionate (TP) is used to obtain such androgenized rats. The severity of interference with the cyclic LH release depends on the dose of TP used; the higher the dose the earlier the syndrome appears. After a single 10 μg dose of TP on day 5, rats showed estrous cycles until about 120 days of age, the delayed anovulatory syndrome (DAS); whereas after a single 30 μg dose on day 5, the rats never showed cyclic LH release (Harlan and Gorski, 1977a).

Androgen exposure apparently does not interfere with the tonic secretion of LH, which is regulated by the ventromedial-arcuate complex of the hypothalamus, but it does interfere with the function of the preoptic-suprachiasmatic area of the hypothalamus, the area presumably involved in the cyclic release of LH (Barraclough, 1973). The model of tonic LH regulation by the ventromedial-arcuate nucleus and the regulation of cyclic LH release by the preoptic-suprachiasmatic area of the hypothalamus is based on numerous experiments, mainly with rats and mice.* These experiments include treatments such as lesions, electrochemical stimulation† of the hypothalamus under various hormonal conditions, and injections of steroids, especially estrogen, progesterone, and combinations of these. Such treatments have been followed by measurement of the number of ovulations and the LH concentrations in the plasma. This evidence from these studies has been extensively reviewed by Barraclough (1973).

Kubo et al. (1975), in studies in which the plasma LH concentration was measured after electrochemical stimulation of the preoptic area (POA) with varying levels of stimulation (20, 40, 80 μA) in normal and androgenized female rats under pentobarbital anesthesia (to block the spontaneous LH release from the pituitary), found no difference in LH release between normal and androgenized female rats. As the result of the greater responsiveness to LH of normal rat ovaries in comparison to androgenized rat ovaries, however, they found a higher incidence of ovulations in response to the increase in strength of the stimulus in the normal, but not in the androgenized rats. These results thus cast doubt on the hypothesis that neonatal androgen has its effect on LH release by its deleterious effect on the responsiveness of the POA. Kubo et al. (1975) point out that the POA may have an integrative role, required for the induction of the LH release necessary for

ovulation, and that this role is adversely affected by administration of neonatal androgen. The evidence is thus far convincing that androgenization interferes little, if at all, with the negative feedback of estrogen on LH (i.e., an increase in the dose of estrogen tends to lower LH secretion), but that it interferes with the stimulatory effect* on LH secretion (i.e., an increase in estrogen concentration results in increased LH secretion) of estrogen or progesterone, or both of these. Androgenization does not affect the capacity of the anterior pituitary (AP) to respond to endogenous luteinizing hormone–releasing hormone (LH–RH) (Mennin et al., 1974).

Mennin and Gorski (1975) showed that androgenized acyclic female rats do not release LH after estrogen priming and a subsequent injection of estrogen plus progesterone, whereas normal female rats do release LH after such treatment. Subsequently, Harlan and Gorski (1977a) compared rats showing the DAS syndrome with completely anovulatory rats for their ability to release LH in response to various doses of estradiol benzoate, given as a single injection 48 hours after ovariectomy at different ages. The completely anovulatory rats failed to release LH at ages of 32, 60, and 150 days, whereas the DAS rats released LH at ages 32 and 60 days but not at 150 days. These results thus show a correlation between the anovulatory syndrome and failure of the hypothalamo-hypophyseal system to respond to estrogen by LH release. In further investigations, Harlan and Gorski (1977b) found that in androgenized DAS female rats, plasma prolactin concentrations increased *prior* to the time the animal became acyclic, but that the change in response to estradiol benzoate coincided with the onset of acyclicity.

Mallampati and Johnson (1974) found a correlation between the extent of reproductive failure as a result of neonatal androgenization and prolactin concentrations in the pituitary and

---

*Rats and mice are particularly well-suited both for investigating the effect of gonadal hormones given perinatally on cyclic LH release in the adult, and for studying the effects of female sex hormones on the behavior of the adult, because the brains of rats and mice are not differentiated with respect to the release of LH and the behavioral response to steroids in the adult until the animals are about 5 days old. Hormones administered neonatally can, therefore, act directly in such species. In animals born with a differentiated brain, e.g., guinea pigs, the experimental treatment must be given to the pregnant mother, whose endocrine and metabolic systems may interfere with the effects of the exogenous hormones on the fetus.

†Stainless steel electrodes are used in electrochemical stimulation. As a result of the electric current, iron from the electrode is deposited, and these deposits have a stimulatory effect.

---

*The release of LH in response to estrogen has been called "positive feedback by estrogen." This term is erroneous, because LH release can be induced by estrogen administration in the absence of the ovary; thus it cannot be a positive feedback. The term "stimulatory effect of estrogen" should be used for the inrease in LH secretion after estrogen administration.

serum. It may be that androgenization increases the effect of the prolactin-inhibiting hormone or that it increases its secretion and the prolactin then interferes with the LH response to estrogen.

The inhibitory effect of perinatal androgens on the cyclic release seems to hold true for most other mammals investigated except the rhesus monkey and possibly primates in general. Women who have been exposed perinatally to androgens may show behavioral changes yet maintain normal ovulatory and menstrual cycles. Karsch et al. (1973) demonstrated that male rhesus monkeys (*Macaca mulatta*), when castrated as adults, implanted with estrogen (to reduce the LH concentration to that found in normal female rhesus monkeys during the menstrual cycle), and then injected with pulses of estrogen, released LH in a manner similar to normal females, in spite of the exposure of the male's brain to androgens before castration. Steiner et al. (1976) showed that in rhesus monkeys, the stimulatory effect of estrogen on LH release is similar in normal females, androgenized males, and androgenized females. These responses contrast with those of female rats, in which the hypothalamo-hypophyseal system of androgenized females is less sensitive to negative estrogen feedback than the normal female.

Species in which perinatal exposure of the brain to either endogenous or exogenous androgen causes acyclicity of LH release are: the mouse (*Mus musculus*) (Edwards, 1971), the golden hamster (*Mesocricetus auratus*) (Swanson, 1970), the guinea pig (*Cavia porcellus*) (Brown-Grant and Sherwood, 1971), the sheep (*Ovis aries*) (Karsch and Foster, 1975), and possibly the pig (*Sus scrofa*) (Hinz et al., 1974).

EFFECTS OF ANDROGENIZATION ON BEHAVIOR AS AN ADULT

The same type of treatment that causes acyclicity in rats also causes a change in the response of adult rats to gonadal hormones. If the brain is exposed to androgens during the perinatal period, the brain is defeminized or masculinized or both. Johnson (1975) has pointed out that the parts of the nervous system that underlie different sexual behavior patterns may develop independently of each other and that a hormone that af-

fects one part of the nervous system, and thus one behavior pattern, need not affect other parts of the neurobehavioral system.

After functional defeminization of the brain, a female animal will not show female sexual behavior as an adult (even when ovariectomized and given estrogen, or estrogen and progesterone, in doses and sequences that, in normal ovariectomized animals, facilitate normal female sexual behavior); however, after functional masculinization of the brain, the male animal will show male sexual behavior in response to male sex hormones.*

Adkins (1975, 1978) has proposed that members of the homogametic sex show a greater tendency for bisexual behavior than those of the heterogametic sex, and that, in the absence of perinatal sex steroids, the homogametic sex will be the "neutral" sex; that is, in the absence of these hormones, an individual will develop behaviorally and morphologically as the homogametic sex. This hypothesis is based on experimental evidence obtained with a few mammalian species and two avian species, Japanese quail and chickens. As we pointed out above, experimental evidence from the African clawed toad (*Xenopus laevis*) does not appear to support this hypothesis, and more evidence is needed. When one examines the evidence for mammalian species in detail, clear exceptions emerge. As will be discussed in detail in the discussion that follows, the males of the golden hamster and rhesus monkey can show male and female sexual behavior, but the male guinea pig, rat, and mouse show little female behavior, even after gonadectomy and appropriate injections of female sex hormones.

When one considers males and females, gonadectomized as adults and treated with either TP or estradiol benzoate plus progesterone (EB + P_4), it becomes clear that TP can induce mounting behavior and EB + P_4 can induce lordosis† behavior in the females of the species to be dis-

---

*Brown-Grant (1975) has cautioned that, in short-duration tests, one can determine the degree of receptivity but not the capacity for sexual behavior of rats treated neonatally with gonadal steroids, so that long-duration tests are required to determine this capacity. Most of the data discussed here have been obtained in short-duration tests.

†In lordosis, the animal makes its back concave, thereby exposing its genitalia.

cussed below. However, for genetic females, there are clear differences among species in the extent to which TP can induce mounting behavior; this behavior can be induced in female mice, rats, guinea pigs and dogs, but not in rhesus monkeys and hamsters. The degree to which estrogen plus progesterone can induce lordosis in genetic males is, in decreasing order: hamsters, monkeys, guinea pigs, rats, and mice (Goy and Goldfoot, 1973).

It appears that the male golden hamster is incompletely masculinized, because after neonatal TP treatment, the lordosis response in the adult castrated male treated with estrogen plus progesterone is inhibited (Swanson, 1971), and there is an increased aggressiveness, both in intact males and in castrated males treated with TP as replacement therapy (Payne and Swanson, 1973).

These observations may be explained by the fact that the golden hamster is born on the day that the sexual differentiation of the brain for future sexual behavior starts, whereas the rat is born 7 days after this brain differentiation has started. Thus the female rat may be exposed to testicular secretions by her brothers in the uterus during this brain differentiation. Clemens (1974) has tested the hypothesis that exposure in utero to the androgens secreted by male litter mates or "womb mates" might increase the male sexual behavior of female rats. His results (Table 2.6) indicate that the frequency of male sexual behavior in female rats depends on the juxtaposition of male fetuses and on the number of male fetuses in proximity to the female fetus.

Vom Saal and Bronson (1978) have found that masculinization of the genitalia in female fetuses of mice depends upon the proximity of male fetuses. Furthermore, female mice that were in proximity to male fetuses in utero were also more aggressive than females that were not close to male fetuses. In the case of the female hamster, however, the exposure of her nervous system to her fetal brothers' androgen may not last long enough to affect her behavior as an adult.

The idea that the female golden hamsters will not show male sexual behavior unless they are exposed neonatally to androgens is not entirely correct. If, instead of being injected, testosterone is implanted in Silastic, so that the hormone is

Table 2.6

Behavior of female rats as affected by male litter mates in utero

A. Relation of the number of males per litter to frequency of females showing mounting as adults

Males/ litter	Number of females tested	Percent of females mounting			
			after daily TP treatment for		
		Pretest	7 days	14 days	21 days
0	3	0	0	0	0
1	2	0	0	0	0
2	5	0	0	0	20
3	8	0	12	62	75
4	10	0	40	40	60
5	9	0	22	55	88

B. Relation of mean mount frequency in the female rat relative to the uterine loction of male sibs

Uterine position	Number of litters	Mean mount frequency*
F	3	1.9
FFFM	3	2.0
FFM	6	5.8
FM	12	5.1
MFM	3	10.4

F = female fetus, M = male fetus.
*The mean refers to the female italicized in the first column.
Modified from Clemens, 1974; reprinted with permission of Plenum Publishing Corp. and L. G. Clemens.

released slowly, male sexual behavior can be induced in female golden hamsters ovariectomized as adults (Noble, 1977). The latency period in these females is longer, and the frequency of mounting lower, than in adult-castrated males treated similarly; thus, there are quantitative differences between male and female golden hamsters that have been gonadectomized as adults and treated with sex steroids. There are also some qualitative differences (e.g., the ovariectomized hamsters do not show any ejaculatory response), but some of these may result from anatomical differences between the males and females: as Beach's informal dictum has it, "You cannot be a carpenter without a hammer." (Goy and Goldfoot, 1973:173).

DeBold and Whalen (1975) have shown that as little as 1 μg of testosterone propionate given at birth to female golden hamsters can masculinize the nervous system, so that when ovariectomized as adults and injected with 500 μg TP, they show mounting behavior. The dose of 1 μg TP at birth did not affect the duration of the lordosis behavior of the ovariectomized female treated with EB + P$_4$ indicating that no defeminization had occurred. After a neonatal dose of 5 μg TP, the lordosis response of ovariectomized females given EB + P$_4$ was partially inhibited, and a 50 μg TP dose neonatally reduced the duration of lordosis to less than 10 percent of that of the controls, i.e., female golden hamsters treated with oil neonatally. Thus, a small (1 μg) dose of TP neonatally is masculinizing but not defeminizing, whereas a large dose of TP can masculinize and defeminize female golden hamsters.

Male rhesus monkeys, like male hamsters, show a high degree of bisexual behavior, except that the expression of the behavior does not depend on the presence of either the male or female sex hormones. Both intact and castrated males show a high degree of presenting of the genitalia (Goy and Goldfoot, 1975). Defeminization of the male brain does not occur during normal development. In contrast to the male hamster, female rhesus monkeys do not show defeminization of the brain when the female fetus is exposed to high concentrations of testosterone, as occurs after testosterone administration to the pregnant mother. Even though the external genitalia are completely masculinized, female sexual behavior is not impeded in the adult (Goy and Goldfoot, 1975); however, the play behavior of female rhesus monkeys that have been exposed to high testosterone concentrations shows a greater frequency of male-type play than normal female monkeys of the same age (Goy and Resko, 1972).

Female dogs show little or no male sexual behavior after either ovariectomy or ovariectomy plus injections of TP (Beach et al., 1972). Male dogs show virtually no female sexual behavior either after castration or after castration and injections of estrogen plus progesterone (Beach and Kuehn, 1970).

In experiments in which bitches were ovariectomized as adults and then injected with estrogen plus progesterone, it was found that exposure of the animals to TP in infancy reduced their recep-

tiveness towards males below that of bitches that had not been treated with TP in infancy. Injection of pregnant bitches with TP in order to expose the fetuses to androgens, followed by injection of TP into the daughters proved to be nearly completely inhibiting to female sexual behavior. These daughters were behaviorally similar to males castrated as adults and then treated with estrogen plus progesterone (Beach and Kuehn, 1972).

The effect of perinatal androgen treatment on male behavior was tested by ovariectomizing bitches and then administering TP. Treatment of pregnant females with TP and subsequent injections of TP in infancy were most effective in inducing mounting behavior after hormone treatment in the adult: 80 percent of such females showed mount-and-thrust responses, compared with 100 percent in males castrated either at birth or as adults. Treatment of bitches with TP at birth had no effect (20 percent responded in this group as did untreated females ovariectomized as adults), and exposure of the female fetuses to androgen in utero resulted in 50 percent mount-and-thrust behavior in the adults (Beach et al., 1972). These experiments show that the stage of development at which TP is administered determines how sexual behavior in the adult will be affected by the treatment. They also show that bitches may be defeminized, but not masculinized, by TP treatment in infancy.

In female guinea pigs, prenatal exposure to testosterone (by injection of TP into pregnant females) results in a failure to show lordosis behavior as an adult in response to injections of estrogen plus progesterone. Injections of dihydrotestosterone propionate (DHTP) and androstanediol propionate into pregnant female guinea pigs did not affect the lordosis behavior of their adult daughters in response to estrogen plus progesterone. Prenatal treatments of TP and DHTP increased the mounting frequency of female guinea pigs above that of controls when they received TP as adults, and androstanediol propionate (ADP) neonatally had no effect on mounting behavior. Treatment with DHTP of adult female guinea pigs of four experimental groups (controls, TP, DHTP, ADP, prenatal treatment) failed to induce mounting behavior, although DHTP induced such behavior in male guinea pigs. These experiments thus show that, prenatally, TP causes defeminization and masculinization in guinea pigs, DHTP under the same conditions causes masculinization, but not defeminization, and ADP has no effect on either masculinization or defeminization. Furthermore, there is evidence that the sensitivity of the adult guinea pig to specific steroids may not be entirely mediated by steroid hormones during critical periods of embryological differentiation (Goldfoot and van der Werff ten Bosch, 1975).

Levine (1971) has pointed out that less TP is required to masculinize the brains of male rats neonatally castrated than of female rats neonatally ovariectomized. He ascribed this difference to possible prenatal hormonal differences between male and female fetuses; however, genetic differences between the sexes may partly account for this difference in response, a view that is given some support by the findings of Pollak and Sachs (1975). They injected pregnant rats with 2.0 mg TP/day on days 16 through 20 of gestation. They subsequently injected the male and female offspring of these rats with 0.5 mg TP on days 1, 3, 5, 7, and 9 of life. Neither the male nor the female offspring showed mounting behavior when tested at 85 days of age, whereas the controls, postnatally oil-injected male offspring of oil-injected pregnant females, showed normal mounting behavior. (There were no postnatally TP-injected male offspring of oil-injected pregnant females included in the experiment.) After gonadectomy at 100 d of age and implantation of a 25 mg TP pellet, the proportion of rats that mounted, and showed intromission and ejaculation was consistently highest for control males, somewhat lower for postnatally TP-injected females from TP-injected mothers, still lower for postnatally TP-injected male rats from TP-injected pregnant dams, and lowest for control females. These results thus show that perinatal TP administration tends to masculinize female rats and to demasculinize male rats. Morphologically the anogenital distance was 39.0 mm for control males, 34.0 mm for perinatally TP-treated females, 22.0 mm for perinatally TP-treated females, and 17.0 mm for control females.

The effects of perinatal androgen exposure on the sexual behavior of adult females after ovariectomy and treatment with female or male gonadal hormones have been studied in hamsters (Swanson, 1971), mice (Manning and McGill, 1974), guinea pigs (Phoenix et al., 1959), sheep (Clarke, 1977), and dogs (Beach and Kuehn, 1970), as well as rats. Hamsters and mice respond much as rats do, although the time when the androgen is most effective in affecting behavior in the adult animal differs among species. Guinea pigs must be treated in utero by injection of the pregnant mother; dogs must be treated in utero and postnatally for the most effective defeminization.

In some "natural experiments," female fetuses may be exposed to androgens in greater concentrations than normal. One such natural experiment, in which the blood vessels of male and female twin calf fetuses anastomose, produces the freemartin. As discussed in more detail in Chapter 3, the bull calf is phenotypically normal, although it may be genetically XY/XX; the female twin, the freemartin, is XX/XY, has ovotestes, an enlarged clitoris, and a blind vagina, and it lacks oviducts and a uterus. The brain of the freemartin has probably been exposed to androgen secreted by its bull twin. Greene et al. (1978) have investigated the behavior of freemartins after treating them with either estrone or testosterone from birth to 50 weeks. Untreated controls were given Silastic implants of estrone or estradiol from 50 until 55 weeks of age; at 68 weeks some other freemartins that had not been treated were injected with dihydrotestosterone (DHT). Testosterone and estradiol each induced agonistic behavior, interest in the vulva, and mounting behavior, but estrone and DHT did not. Flehmen, lip curling, which may facilitate pheromones reaching the vomeronasal organ, was induced by testosterone only. Greene et al. (1978) concluded from these findings, and from observations made on ovariectomized heifers, that no appreciable androgenization of the neural centers responsible for sexual behavior had occurred in utero in the freemartins.

Cunningham et al. (1977) investigated the LH-releasing effect of estrogen in freemartins and found that, in seven out of twelve animals, estrogen caused a smaller LH peak and a later response than in normal heifers. In the five remaining freemartins, neither estradiol nor LH–RH caused LH release, suggesting that the failure of estradiol to induce LH release may have resulted from failure at the site of the pituitary and not necessarily of the hypothalamus.

In humans, fetuses with a genetic defect that prevents the synthesis of cortisol are exposed to high concentrations of androgens, as is evident from the enlarged clitorises of such girls and the enlarged penes of such boys at birth. The high androgen concentration results from the lack of negative cortisol feedback on secretion of adrenocorticotrophic hormone (ACTH); the resulting high ACTH concentration stimulates adrenal androgen secretion. If this adreno-genital syndrome is correctly diagnosed, such children are given cortisone replacement therapy shortly after birth and for the rest of their lives, and their androgen concentrations then return to normal.

Money and Ehrhardt (1968) have made extensive and careful studies of women with the adreno-genital syndrome and compared them with carefully matched normal controls. Such women differed from their normal controls mainly in a higher incidence of a tomboy play pattern during their youth than normal women. One must take into account the difficulties inherent in studies of human behavior due to profound social and educational influences on behavior;* however, the results of the Money and Ehrhardt (1968) studies agree with studies in female monkeys after intrauterine androgen treatment (Goy and Resko, 1972), and the pattern may thus be general for primates.

The effects of perinatal androgen exposure in mammals are not limited to effects on gonadotrophin release and sexual or aggressive behavior; other effects are listed in Table 2.7. It is especially noteworthy that perinatal androgens can affect hepatic enzymes involved in steroid metabolism and so affect the endocrine environment to which the adult hypothalamus is exposed, even if the androgens had none of the effects described. Perinatal androgens can also affect enzymes involved in the inactivation of oxytocin, another effect that may influence hypothalamic function.

ACTION OF PERINATAL ANDROGENS ON THE NERVOUS SYSTEM

To investigate where androgens act in the central nervous system, small amounts of androgen can be placed in different parts of the central nervous system. The available evidence shows that in golden hamsters (Brayshaw and Swanson, 1973) and in rats (Nadler, 1972; Hayashi and Gorski, 1974; Lobl and Gorski, 1974), implants in the hypothalamus are effective in masculinization, defeminization, and changing the LH release pattern of the adult, whereas implants outside the hypothalamus are ineffective. The intrahypothalamic implants in the ventromedial-arcuate nucleus area and in the preoptic area were found to be effective if they remained in place at least 48 hours. This appears to be a long period, because after an injection of TP that effectively defeminizes and masculinizes, one may expect the steroid to be removed from the circulation in 24 hours or less. Why such a long exposure is necessary for an intrahypothalamic implant to be effective is not clear. It may be that not enough androgen diffuses to essential brain areas at some distance from the implant, or that the rate of androgen release is not rapid enough, or that androgens must be converted to estrogens in the circulation to be most effective (Hayashi and Gorski, 1974).

Perinatal androgens acting on the hypothalamus probably do not affect only sexual behavior. Hart (1968), studying rats castrated at four days of age found that 14 out of 17 did not ejaculate, whereas 11 rats out of 14 castrated at 14 days of age ejaculated. After spinal transection of these two groups of rats and subsequent androgen treatment, the 14-day castrates showed genital responses such as erection of the glans penis, flips of the glans penis, and ventral flexion of the pelvis, whereas the 4-day castrates failed to show genital respones. These results suggest that androgens have an important role in organizing the spinal neural substrate.

---

*Parents may have allowed girls with masculinized genitalia to wear boys' clothes; being a tomboy may then have been easier.

Table 2.7

Effects of perinatal androgens on adult physiological functions
other than GTH release and sexual behavior

Function	Effect of perinatal androgen	Animal	Reference
Behavior			
Territorial marking	Increase	Mongolian gerbil	Turner, 1975
Nearness to female	Increase	Rat	Myerson, 1975
Aggression	Increase	Mouse	Bronson and Desjardins, 1971
Avoidance	Decrease	Rat	Beatty and Beatty, 1970
Maternal behavior	Decrease	Rat	McCullough et al., 1974
	Decrease	Rabbit	Fuller et al., 1970
			Anderson et al., 1970
Saccharin preference	Decrease	Rat	Wade and Zucker, 1969
Open field behavior	Decrease	Rat	Blizard and Denef, 1973
Play by young	Masculinization of females' behavior*	Rhesus monkey	Goy and Resko, 1972
Urination posture	Increase in leg lifting	Dog	Beach, 1974
Susceptibility to attack	Increase	Mouse	Lee and Griffo, 1973
Secondary sex organs			
Seminal vesicles	Increased response to androgens	Mouse	Bronson et al., 1972
Seminal vesicles, coagulating gland, penis, penile spines, levator ani muscle	Increased response to androgens	Rat	Dixit and Niemi, 1973 Chung and Ferland-Raymond, 1975
Vagina	Ovary-independent cornification; earlier opening	Rat	Takasugi, 1976 Barraclough, 1966
Uterus	Decreased growth response to estrogen	Rat	Lobl and Maenza, 1975
	Higher incidence of ovary-independent hyperplastic or metaplastic epithelium	Rat	Kramen and Johnson, 1975
	Increased glycogen content at 66 days of age	Rat	Lobl and Maenza, 1977
Hepatic enzymes			
Steroid mixed function oxidases	Perinatal androgen required for presence in adult	Rat	Tabei and Heinrichs, 1974
2α Hydroxylase	Increased† activity	Rat	Gustafsson et al., 1975
7α Hydroxylase	Increased activity	Rat	Tabei and Heinrichs, 1975
7β Hydroxylase	Increased† activity	Rat	Tabei and Heinrichs, 1975
15β Hydroxylase	Decreased† activity	Rat	Gustafsson et al., 1975
16α Hydroxylase	Increased† activity	Rat	Gustafsson et al., 1975 Tabei and Heinrichs, 1975
5α Reductase	Decreased† activity	Rat	Gustafsson et al., 1975
5β Reductase	Increased† activity	Rat	Gustafsson et al., 1975
3β Hydroxy steroid dehydrogenase (HSD)	Increased† activity	Rat	Denef and De Moor, 1972 Gustafsson et al., 1975
17α HSD	Increased† activity	Rat	Gustafsson et al., 1975
20β HSD	Increased activity	Rat	Denef and De Moor, 1972
Hypothalamic enzymes			
Oxytocin-inactivating enzyme	Depressed activity	Rat	Griffiths and Hooper, 1972, 1973
Life span	Shortened	Rat	Dörner and Hinz, 1975

*Genetic females had masculinized genitalia and were classified as pseudohermaphrodites.
†The effects of perinatal androgens on these enzymes depends on the hypothalamo-hypophyseal system. The androgen presumably inhibits the pituitary's secretion of a feminizing factor on hepatic enzyme activity.

The possibility that neonatal androgens affect the sexual behavior of males via peripheral tissues should be considered, because a completely developed penis might be required for intromissions and ejaculations. Rats with incompletely developed penes (small spines, small penes) have fewer intromissions (Gorski and Goldman, 1971). Hart (1972) investigated this problem by castrating rats on day 4 and treating them either with TP or fluoxymesterone (FM), which does not maintain sexual behavior in adult castrates but does maintain the penis. The FM-treated rats intromitted but failed to ejaculate, whereas the TP-treated animals showed complete male sexual behavior. After spinal transection, both groups showed penile reflexes, an observation suggesting a difference between the central nervous tissue of the brain and that of the spinal cord. Androgens may need to be converted to estrogens in order to affect the organization of the neural substrate of the brain; FM is not converted to estrogens, and it may not need to be to affect the organization of the neural substrate of the spinal cord.

In female rats and golden hamsters, masculinization and defeminization after a single injection of an androgen occurs only if the androgen can be converted to an estrogen* (Clemens, 1974).

However, when implanted in Silastic capsules, to ensure slow absorption, testosterone, androsterone, and dihydrotestosterone (DHT) suppressed lordosis in female golden hamsters and in neonatally castrated male hamsters (Gerall et al., 1975). This result suggests that long exposure

---

*Testosterone, TP, and androstenedione can be converted to estrogens (dihydrotestosterone and dihydrotestosterone propionate cannot) as shown in the metabolic pathway below (Naftolin and Ryan, 1975):

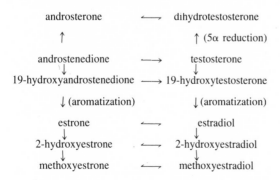

has effects different from those due to short exposure.

It has been found that in the rat and the hamster (1) the effects of neonatal estrogens can mimic those of neonatally administered androgens (McEwen et al., 1974), (2) the neonatal limbic system can convert androstenedione to estrone (Reddy et al., 1974); and (3) estrogen receptors in the cell nucleus are present in the differentiating brain in the areas where the aromatization enzymes are found (Westley and Salaman, 1976). All these findings provide strong arguments for the hypothesis that, in rats and hamsters, androgens, to be effective in defeminization and masculinization, must be converted to estrogens. In guinea pigs, masculinization does not require aromatization, but defeminization probably does. Defeminization probably does not occur in rhesus monkeys, and masculinization does not require conversion of androgens to estrogens (McEwen, 1981). Clemens (1974) reported that masculinization of female golden hamsters by neonatally administered testosterone is prevented by agents that block aromatization.

Callard et al. (1978) have emphasized that aromatase is present in the brain of representatives of all vertebrate groups except the Agnatha. In mammals, the aromatase is restricted to the limbic system, which includes the preoptic area, hypothalamus, septum, amygdala, and hippocampus. In nonmammalian vertebrates, aromatization occurs in preoptic-hypothalamic tissue and (except for sharks) in the telencephalon, which is represented in the limbic lobe of the mammalian brain. This widespread presence of aromatase suggests that conversion of androgens to estrogens in the brain may characterize vertebrates generally.

If rats do become demasculinized or defeminized by estrogens, then how is the female fetus protected against maternal estrogens? One protective mechanism may be provided by progesterone. Experimentally, the simultaneous neonatal injection of progesterone and estrogens protects against the defeminizing and masculinizing effects of the neonatally administered estrogens (see van Tienhoven [1968] for review).

An important known protective mechanism is provided by alpha feto-protein (AFP), which is present in the serum and brain of fetal and young rats of both sexes until about 15 days of age.

Since AFP has a high affinity for estrogens, but not for androgens, free estrogens probably do not reach the neural target tissue, but androgens do. When they reach such tissue, they are aromatized to estrogens and can affect the tissue (McEwen et al., 1975).

NEONATAL ANDROGENS

Effects on Pituitary Function

How neonatal androgens affect the hypothalamus has been investigated from different points of view. Androgens in most adult vertebrates inhibit gonadotrophin (GTH) secretion; therefore, it was logical to question whether the differentiation of the brain due to androgen resulted from the inhibition of GTH by androgen. By injection of 50 μg follicle-stimulating hormone (FSH) plus 50 μg LH together with 100 μg TP into neonatal rats, the onset of persistent vaginal estrus was delayed and its incidence reduced, and the percentage of animals showing lordosis was higher than in TP-injected rats.

In two experiments, the frequency of male sexual behavior after ovariectomy and TP administration in adult rats was affected differently by the prenatal treatments: in one experiment, the simultaneous injection of FSH plus LH and TP increased the number of mounts by the ovariectomized, TP-treated females above that obtained with females that received neonatal TP only; in the second experiment there was no difference due to these neonatal treatments (Sheridan et al., 1973). Sheridan et al. (1973) hypothesized that in normal neonatal female rats, high concentrations of gonadotrophins act on the "cyclic center," organizing it so that GTH release in adulthood will be cyclic; however, in males, the serum concentrations of GTH are too low, due to inhibition of GTH secretion by testicular androgen secretion, to induce such cyclicity. If this hypothesis is correct, it would explain the induction of acyclicity in adult female rats after neonatal injections of estrogens, testosterone, clomiphene (an anti-estrogen that depresses GTH secretion), progesterone desoxycorticosterone acetate, and human chorionic gonadotrophin (hCG), all of which suppress tonic FSH and LH secretion. However, DHT given in a single injection to neonatal female rats depresses serum FSH and LH

concentrations in the neonate, but does not cause behavioral defeminization (Korenbrot et al., 1975). This result seems to suggest rejection of the Sheridan et al. (1973) hypothesis.

Anti-gonadotrophic serum injected into neonatal male rats results in: (1) poor penile development, a condition which may explain the low observed frequency of intromission and the observed failure to impregnate females; (2) a high lordosis quotient, i.e., number of lordosis × 100/number of mounts, in males after castration and treatment with ovarian hormones, an effect indicating a lack of defeminization; (3) no cyclic GTH release in the castrated males (after implantation of the ovaries no corpora lutea were formed, a result indicating that normal masculinization of the GTH release pattern did occur); (4) low frequency of intromission in castrated rats treated with androgen, which may also have resulted from the poor penile development (Gorski and Goldman, 1971). These results can be explained by assuming that (1) the anti-GTH serum lowers androgen secretion by the neonatal testes, so that the animals become equivalent to castrates in their behavior and their response to androgens as adults; (2) the threshold amount of endogenous androgen necessary to inhibit the GTH release control system and the amount necessary to affect the behavioral system differ, so that small amounts of androgens secreted by the fetal testes in spite of the anti-GTH would still be sufficient to affect the GTH control system, but not sufficient to affect the behavioral system.

In view of the effects of LH–RH and gonadotrophins on androgen secretion by the adult rat testes, it is not surprising that neonatal injections of either LH–RH (Arai, 1974) or gonadotrophins (hCG, LH, FSH) (Arai and Serisawa, 1973) masculinize the brain of rats in the presence of the testes but not in their absence.

Effects on Nucleic Acid and Protein Synthesis

Neonatal androgens may act via the synthesis of DNA, RNA, or protein. A number of experiments support this view, among them the following:

1. Inhibition of DNA synthesis for 24 hours (by hydroxyurea or 5-bromodeoxyuridine) or inhibition of protein synthesis for 6 hours (by pur-

omycin or cycloheximide) in female neonatal rats prevents the androgen-induced persistent vaginal estrus (Barnea and Lindner, 1972). Intracranial implantation of cycloheximide in female neonatal rats blocked androgen-induced sterility, whereas systemic administration of cycloheximide failed to do so (Gorski and Shryne, 1972). Salaman and Birkett (1974) confirmed that hydroxyurea inhibited the effect of neonatal androgen on cyclic LH release in female rats, as did amanitin (a specific inhibitor of messenger RNA synthesis), actinomycin D (an inhibitior of ribosomal RNA synthesis), puromycin (an inhibitior of protein synthesis) and fluorouracil (an inducer of translation errors and the synthesis of nonfunctional proteins).

2. Three hours after injection, neonatal androgen increased the incorporation of tritiated uridine into the medial preoptic area and the amygdala of female neonatal rats, whereas it decreased this uptake in other areas of the brain in comparison with control rats (Clayton et al., 1970).

3. Namiki et al. (1972) reported small but consistent differences in brain RNA between male and female rats, and also between androgenized and control females; so that the males and androgenized females' brains were similar, and both were different from the control females' brains in the nucleotide composition of RNA. Experiments *2* and *3* thus strongly suggest an effect of neonatal testosterone on the quantitative and qualitative aspects of nucleic acid metabolism in the brain.

4. Neonatal estradiol benzoate inhibited the incorporation of[^{3}H] lysine at 10 weeks of age by neurons of arcuate, paraventricular, periventricular, and supraoptic nuclei (Litteria and O'Brien, 1975). This observation is relevant to the conversion of androgens to estrogens in the hypothalamus.

Effects on Neurotransmitters

Androgens may affect the synthesis of neurotransmitters. Hardin (1973) reported that 300 μg TP given at 5 days of age increased norepinephrine and decreased histamine concentrations (per unit of wet weight) in the brains of rats at 10 days of age, without significantly affecting the concentrations of dopamine, serotonin, and spermidine. The fact that pentobarbital (Nishizuka, 1976; Sutherland and Gorski, 1972), reserpine (Nishizuka, 1976), and the alpha blocker phenoxybenzamine (Nishizuka, 1976) can block the effects of neonatal androgen administration on behavior in the adult, while the beta blocker propanolol cannot (Nishizuka, 1976) tends to support the hypothesis that androgens given neonatally may affect biogenic amine concentrations or turnover at a later age. Further support for such actions of androgens and estrogens may be found in the evidence that links estrogen metabolism with that of catecholamines and links these in turn with brain differentiation. Estradiol is hydroxylated to 2-hydroxyestradiol (2-OHE$_2$); this intermediate competes with catecholamines for the enzyme catechol-O-methyl-transferase (COMT) (see Figure 2.9), with 2-OHE$_2$ having the higher affinity for the enzyme. This competition results in a slower breakdown of catecholamines (Breuer and Köster, 1975). Dörner et al. (1977) have shown that female rats treated nenonatally with pargylene (an inhibitor of monoamine oxidase) not only showed precocious puberty (an effect also observed after neonatal TP treatment), but also showed diminished male mounting behavior compared with controls at 4 months and similar male mounting behavior as controls at 8 months of age. These experiments by Dörner et al. (1977) indicate that neonatal changes in catecholamine metabolism can affect the adult behavior of the animal.

Effects on Morphology of the Nervous System

Neonatal androgens affect the structural organization of the nervous system. Toran-Allerand (1978) reviewed the sexual dimorphism in the central nervous system of vertebrates and stated that in the rat, the following structures in the brain are sexually dimorphic if androgens are present neonatally in one sex: (1) the neuronal nuclear and nucleolar size in the preoptic area, ventromedial area, and the amygdala; (2) the ultrastructural organelles and synaptic vesicles in the arcuate nucleus; (3) the synaptic organization

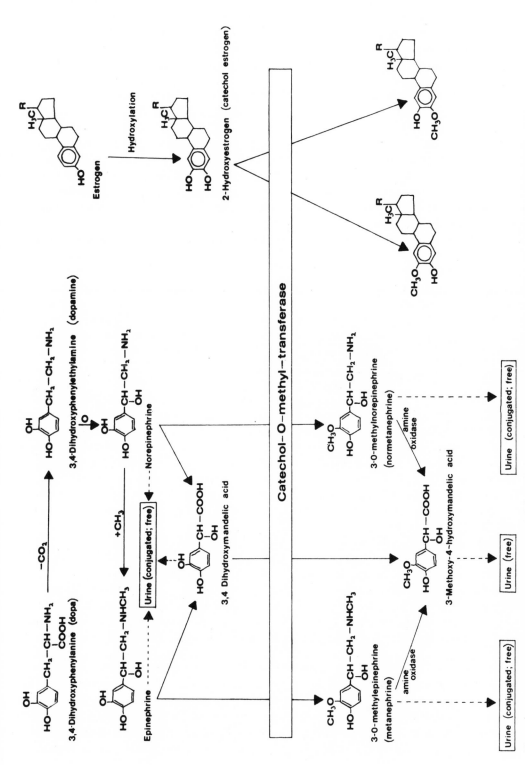

Figure 2.9 The central place of the enzyme catechol-O-methyl-transferase in catecholamine and estrogen metabolism. The affinity of the enzyme for 2-hydroxyestrogens is greater than the affinity for epinephrine and norepinephrine. The pathways depicted are according to Turner, 1966, and Breuer and Köster, 1975.

of the nonamygdaloid efferents in the preoptic area; (4) the dendritic branching patterns of the preoptic area and the suprachiasmatic area; and (5) the gross nuclear volume of the preoptic area. The dendritic branching pattern in the preoptic and suprachiasmatic areas of the hamster is similarly dimorphic. In canaries (*Serinus canarius*) and zebrafinches (*Poephila guttata*), the volumes of caudal nucleus of the hyperstriatum ventrale (HVc), of the nucleus robustus archistriatalis (RA) and the tracheosyringeal portion of the hypoglossal motor nucleus (nXIIts) are larger in males than in females. Furthermore, one area, area X, of the lobus parolfactorius is present in males only, and the size of the cell bodies in the HVc, RA, nXIIts and in the magnocellular nucleus of the antrior neostriatum (MAN) are larger in males than in females. These nuclei, except MAN, are associated with singing in these species, and normally only the males sing. However, when female zebrafinches are injected neonatally with estrogen, they are capable of singing as adults, provided they receive exogenous testosterone (control females treated with testosterone do not sing). The neonatal estrogen treatment also masculinizes the HVc, RA, MAN and area X (Arnold, 1980).

The sexual dimorphism of the nervous system is not limited to the brain; for instance, the motor neurons that innervate the striated muscles of the penis in the rat form a nucleus that is present in males but not in females (Arnold, 1980). In male cats, the number of sympathetic preganglionic neurons in the spinal cord is higher than it is in female cats (Henry and Calaresu, 1972).

In vitro addition of either estradiol or androgens to cultures of mouse hypothalamus promotes outward axonal growth of neuronal processes from the margins of specific areas of the preoptic-anterior hypothalamus. This response also includes enhanced maturation and myelinization (Toran-Allerand, 1976). When antibodies to estrogens are added to such cultures but no estrogens or androgens are added, in vitro development in these areas is retarded.

These anatomical findings are supplemented by neurophysiological findings: in response to electrical stimulation of the amygdala, the increase in the percentage of cells firing in the me-

dial basal hypothalamus was higher in males (56 percent) and in females treated with 1.25 mg TP on day 3 (41 percent) than in untreated females (24 percent) and in males castrated neonatally (31 percent) (Dyer et al., 1976). These findings show that a single neonatal injection of androgen can permanently affect neural transmission. Kawakami and Terasawa (1972) showed that, in androgenized female rats, after electrical stimulation of the medial amygdala, serum FSH concentration decreased but serum LH concentration did not change, whereas in control rats serum FSH and LH concentration increased. After electrical stimulation of the hippocampus, serum LH and serum FSH concentrations increased in androgenized female rats; however, in cyclic control rats the serum FSH concentration increased, but the serum LH concentration did not. Such hippocampal stimulation inhibits the ovulation induced by electric stimulation of the medial preoptic area in control female rats but not in androgenized female rats. Electrical stimulation of the medial preoptic area revealed no differences in serum FSH and LH concentrations between cyclic females and androgenized females. These results should be viewed with some caution because the controls and androgenized females were not used in a contemporaneous experiment. These experiments thus tend to support the hypothesis that connections between amygdala and hypothalamus are affected by neonatal androgens; however, the possibility cannot be excluded that such neonatal androgens affect the ability of LH–RH neurons to release their hormones.

Effects on Sex Steroid Concentrations in the Adult Brain

In view of the many effects of perinatal gonadal steroids, it does not seem wise to assume that there is only one mechanism by which their effects, on either GTH release or on male and female sexual and other behaviors, are mediated. Heffner and van Tienhoven (1979) found that [^{3}H] estradiol concentrations in the anterior and middle hypothalamus of adult female rats was lower in neonatally androgenized rats than in controls. Since this decrease, due to neonatal androgen treatment, was found in rats that had been neonatally ovariectomized and also in those that had been neonatally sham-operated, one may conclude that the lower estradiol uptake was not the result of changed ovarian function.

To what extent do events in untreated neonatal rats agree with the experimental findings that androgens and estrogens neonatally can masculinize the brain? Female neonatal rats do not normally secrete detectable amounts of estrogens that would masculinize the brain, but male neonatal rats secrete androgens during the first week; thereafter, androgen secretion decreases sharply until near puberty when it rises again (Resko et al., 1968). Thus the brain of the male rats would be appropriately masculinized and, as we have discussed earlier in this chapter, androgen secretion by male fetal rats can affect the sexual differentiation of their sisters in utero.

The differentiation of the avian brain with respect to sexual behavior has been investigated most thoroughly for the Japanese quail. Adkins (1975) showed that the fundamental sex difference between male and female sexual behavior lies in the male showing the masculine copulatory sequence consisting of headgrab, mount, and cloacal contact. Both sexes will show female sexual behavior (solicitation of mating, squatting) when tested with males and after having been treated with estrogens. The differentiation of the brain of the male for male sexual behavior is inhibited by administration of either androgens or estrogens to the embryo, thus these hormones demasculinize the male. Females were not masculinized by embryonic treatment with either androgens or estrogens (Adkins, 1975; 1978; Whitsett et al., 1977). Adkins postulated that the male quail is similar to the neutral sex; implying that estrogen secreted by the embryonic ovaries feminizes the brain of female quail. This hypothesis was tested by exposing quail embryos to an antiestrogen. Females so treated as embryos showed behavioral masculinization (Adkins, 1978). Such a behavioral role for estrogens would be compatible with the morphological data showing that, in birds, estrogens are required for feminization of the secondary sex organs. However, as discussed on p. 66, in canaries and zebrafinches neonatal estrogens and androgens masculinize the brain,

contrary to the hypothesis that estrogens defeminize the avian brain and that the male brain should be similar to the brain of a gonadectomized bird.

At ten days of incubation, the testes of chick embryos incubated in vitro produce little testosterone and no estrogens from the progesterone precursor, but the left ovaries of such embryos produce estradiol and estrone, and the right ovaries produce smaller amounts of estradiol than the left ovaries (Galli and Wassermann, 1973).

The gonadal steroids are not the only hormones that neonatally modify the adult function of the brain. The effects of gonadotrophin-releasing hormones and of gonadotrophins have been mentioned in the discussion of possible mechanisms of the action of gonadal hormones. Other instances are:

1. Thyroxine administration to neonatal rats results in inferior field maze performance, inferior growth during adolescence and adulthood, and a performance deficit in a passive avoidance test (Davenport and Gonzalez, 1973). Large doses of thyroxine given neonatally to rats reduce the ^{131}I uptake by the thyroid and lower the rate of thyroid-stimulating hormone (TSH) secretion after thyroidectomy below that of normal thyroidectomized control rats (Azizi et al. 1974). Bakke et al. (1975) found that neonatal thyroxine administration impaired body growth, lowered concentrations of TSH in the pituitary and serum of the adult rat, lowered serum thyroxine in the adult, and reduced the response to administration of propylithiouracil and of thyrotrophin-releasing hormone (TRH). These effects are probably the result of impairment of the hypothalamic secretion of TRH in the adult, due to neonatal thyroxine treatment.

2. Neonatally adrenal corticosteroids affect adult behavior and hormonal responses. Offspring of adrenalectomized pregnant rats have hypertrophied fetal adrenals and fetal plasma corticosterone concentrations higher than normal. As adults, the male offspring show fewer avoidances in avoidance tests and a smaller than normal increase in plasma corticosteroid concentration in response to stress (Levine, 1971). Open-field behavior of the offspring of adrenalectomized female rats does not decrease between days 1 and 2 of the test, whereas in male off-

spring of control females, a decrease does occur (Levine, 1971).

Neonatal corticosterone treatment increases the submissiveness of mice in adulthood (Leshner and Schwartz, 1977). Hydrocortisone administered neonatally inhibits the ether-induced increase in plasma corticosterone concentration at age 11 days and makes the response at age 25 days slower than in controls. Handling of neonatal animals may have partly the same effect as neonatal corticosteroid treatment, because handling increases plasma corticosterone concentrations in neonatals (Denenberg et al., 1967). Neonatal handling has a number of repercussions when the animal is adult. For instance, the elevation of plasma corticosterone in response to electric shock is more rapid and greater in rats handled neonatally than in control rats not handled (Levine, 1962). The neonatally handled rats also show a greater ascorbic acid depletion in the adrenals than controls (Levine et al., 1957). Denenberg and Rosenberg (1967) found that neonatal handling of female rats decreased the activity of their granddaughters in an open field test and decreased their body weight at weaning in comparison with granddaughters of nonhandled rats. These findings illustrate the importance of having complete information and control of the management of the animal breeding colony, especially when the animals are to be used in behavior or in growth experiments.

Ader (1975) has emphasized a number of important points about the effects of neonatal handling of rats: (1) the changes imposed on "emotional behavior" by neonatal handling can be obtained by neonatal handling only; in contrast, (2) the changes in adrenocortical activity due to neonatal handling can be obtained by post-weaning handling as well; and (3) the magnitude of stimulation of the adult does not affect the reduction in adrenocortical activity obtained by previous neonatal handling of the animals.

Reports vary concerning the effects of stress on sexual differentiation. Ward (1975) reported that prenatal stress reduced the frequency of copulation and ejaculation responses in intact males; but she found that after either (1) castration and TP treatment or (2) castration, TP treatment, and mild electric shock (to increase arousal), some of the stress-induced decrease in male copulatory behavior was restored, although not to the frequency found in either control or postnatally stressed rats. After receiving TP, 80 percent of the prenatally stressed castrated male rats showed lordosis, whereas none of the controls or postnatally stressed rat showed lordosis. The prenatally stressed castrated male rats also showed lordosis after $E + P_4$ treatment. Ward (1975) concluded that these males were behaviorally feminized and demasculinized. It seems more accurate to state that defeminization and masculinization had failed to occur.

According to Whitney and Herrenkohl (1977), prenatal stress does not affect copulatory and ejaculatory responses of otherwise untreated male rats, but it increases their lordotic response after castration and $E + P_4$ treatment. In other words, masculinization occurs in male rats in spite of prenatal stress, but defeminization fails to occur. This effect of prenatal stress is prevented by radiofrequency lesions of the anterior hypothalamus, a result indicating that the prenatal stress affected the differentiation of the male rat's brain (Whitney and Herrenkohl, 1977).

Neonatal handling of mice (*Mus musculus*) reduces their survival time as adults after injection of leukemic cells (Levine and Cohen, 1959). King and Eleftheriou (1959) compared the tractable deermouse, (*Peromyscus maniculatus gracilis*) with the wilder subspecies *P. maniculatus bairdii* for their adult responses after neonatal handling. Handling increased the activity of both subspecies when tested in an activity alley, but the subsp. *gracilis* performed better than the subsp. *bairdii* in an avoidance conditioning test; however what is more interesting, is that the subsp. *gracilis* mice performed better than their controls and the subsp. *bairdii* mice performed worse than theirs. It is thus very important in designing experiments in which brain differentiation is modified experimentally, to control for the genetic background of the animals and the behavior under control conditions.

Denenberg et al. (1973) have shown that handling of rabbits during the first 20 days of life (1) increases open-field activity and exploratory behavior; (2) causes maintenance of open-field activity and rearing during a four-day test period, whereas in controls both behaviors decline during this period; and (3) causes an increase in active waking and a decrease in quiet waking, with an increase in active sleep and rapid eye movement sleep as a percentage of total sleep (Denenberg et al., 1977).

Mice reared from days 4 to 21 in the presence of a nonlactating female rat, an "aunt,"* showed, as adults, less open-field behavior, greater body weight, and a smaller increase in plasma corticosterone in response to a new environment than did mice reared without a rat "aunt" (Denenberg et al., 1969).

The consequences of the fertilized egg having either XY or XX or alternatively ZZ or ZW are probably first reflected in the presence or absence (or in birds in the greater or lesser expression) of the H-Y antigen, which, in turn, governs the dif-

ferentiation of the gonad. Vertebrates may differ in whether the somatic tissue or the primordial germ cells determine whether the primordial germ cells become oocytes or sperm. The differentiation of the gonad determines the hormones to be secreted. These hormones, in turn, determine the differentiation or regression of the secondary sex organs and the differentiation of the future function of the brain in regulating the pattern of release of gonadotrophic hormones (except probably in primates) and in governing the behavioral response to gonadal hormones in the adult animal. It is hypothesized that, if an animal is without gonadal hormones when morphology and future behavioral capacity are differentiating, it will come to resemble the homogametic sex, which has a greater tendency for bisexual behavior than the heterogametic sex.

Between the neonatal period, on the one hand, and puberty on the other hand, relatively little sexual differentiation occurs. The onset of puberty will be discussed in Chapter 4.

---

*Such an "aunt" takes over much of the rearing of the young mice, except for the nursing.

3

Intersexes

We have mentioned earlier that intersexuality is found more frequently among fishes than among other vertebrates. Figure 3.1 illustrates the phylogenetic relationships among these groups in which functional hermaphroditism has been found to occur regularly in one form or another, e.g., synchronous, protogynous, or protandrous. Definitions of these and other terms are given in Table 3.1.

FISHES

Among the lower fishes (Figure 3.1A) functional hermaphroditism does not seem to be common. Here we briefly discuss a few cases that have been found.

Cyclostomes

The male and female Atlantic hagfish (*Myxine glutinosa*) each function as male and female respectively, although the gonad is often hermaphroditic. In other hagfishes, *Polistotrema stouti* (*Bdellostoma stouti*) and *Eptatretus burgeri* (*B. burgeri*), hermaphrodites are apparently rare (Atz, 1964). Lampreys are also not hermophroditic as adults.

Elasmobranchs

In this group hermaphroditism is rare. In *Squalus acanthias*, the spiny dogfish, it has apparently never been found, despite the thousands of animals used in comparative anatomy classes. In the lesser spotted dogfish (*Scyliorhinus canicula*), a few hermaphrodites have been described by Atz (1964), who also cites evidence that in the sharks *Notorynchus* and *Hexanchus*, the ovary is associated with a rudimentary testis.

Sturgeons

Hermaphroditism is rare among these.

Teleosts

A complete listing of the incidence of intersexuality in teleosts is found in Atz (1964) and Smith (1975). We select only a few interesting ones to discuss. The self-fertilizing cyprinodont, *Rivulus marmoratus*, which is characterized by internal fertilization, is of considerable interest, because it has genetically different groups of homozygous isogenic individuals. In the wild, males are rare, but in the laboratory, males can be obtained by incubating the eggs at low temperatures, as discussed in Chapter 2, or by short-day treatment of adult fish after previously exposing them to high temperature. This treatment brings about involution of the ovarian tissue and results in a secondary male. *Rivulus marmoratus* can thus be designated as a synchronous hermaphrodite with facultative diandry. Another species with potential self-fertil-

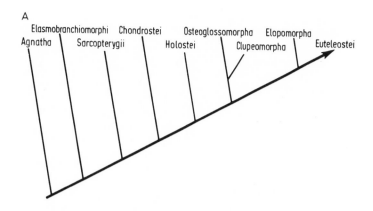

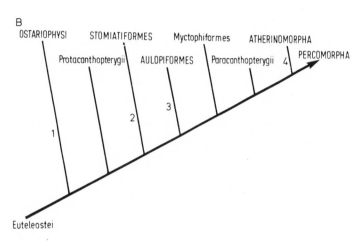

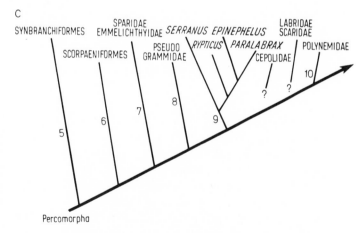

Figure 3.1 Phylogenetic relationships of hermaphroditic fishes. Compiled from various sources. A, Lower fishes, Agnatha to Euteleostei. B, Ostariophysi to Percomorpha. C, Percomorpha. Groups with hermaphroditic representatives are listed in capitals. Numbers indicate lineages in which hermaphroditism has arisen independently. (From Smith, 1975; reprinted with permission of Springer-Verlag.)

ization, *Serranellus subligarius*, has been discussed in Chapter 1.

Synchronous and protogynous species are found in the teleost family Serranidae quite commonly, but gonochoristic species also occur. The most complete hermaphroditism is found in the family Sparidae, which includes gonochorists, rudimentary hermaphrodites, and protandrous and protogynous hermaphrodites. A hypothetical scheme for the evolution of the gonad of a protogynous species is illustrated in Figure 3.2. According to this hypothesis, the ovary becomes an

Table 3.1

Distinguishing characteristics of intersexes or intersexual condition
(presence of male and female characteristics in the same individual)

Type of intersex or intersexual condition	Characteristics	Example
Hermaphrodite	Same individual has ovarian and testicular tissue.	
Synchronous	Ripe spermatozoa and oocytes are present simultaneously in the same individual. In some cases self-fertilization occurs.	*Rivulus marmoratus*
Protandrous	Individual functions first as a male (testis produces sperm) and subsequently as a female (ovary produces oocytes).	Teleost family of Sparidae
Protogynous	Individual functions first as a female (ovary produces oocytes) and subsequently as a male (testis produces sperm).	Teleost family of Labridae
Rudimentary	Individual functions as a male (testis produces sperm), but ovarian tissue is present; or the individual functions as a female (ovary produces oocytes), but testicular tissue is present.	*Amphiprion bicinctus*
Pseudohermaphrodite	Individual has gonads of one sex, but secondary sex organs and/or secondary characteristics of the other sex.	Testicular feminization syndrome in humans and mice
Sex reversal	Individual functions as a male, but acquires ovarian function and then functions as a female, or individual functions as a female and acquires testicular function and then functions as a male.	Protandry and protogyny
Sex inversion	Secondary sex characteristics of the individual change without a change in the gonadal tissue, therefore without sex reversal.	Adrenogenital syndrome in women
Diandry	Some individuals differentiate as males and remain males; others first pass through a functional female stage before transforming into males.	Undifferentiated races of frogs
Gonochorism		
Primary	Genetic sex agrees with phenotypic sex.	Most classes of vertebrates except teleosts
Secondary	Genetic sex and phenotype do not agree.	*Myopus schisticolor*

Data from Atz, 1964; Smith, 1975.

ovotestis after having become functional as a result of testicular tissue developing among layers of ovarian tissue. After this stage of ovarian intersexuality, the ovarian tissue regresses, and after an intermediate stage of testicular intersexuality, the gonad becomes completely testicular.

One of the best-studied cases of protogynous hermaphroditism is that of the swamp eel or rice-field eel, *Monopterus albus,* of the order Synbranchiformes. In this species, female function and male function are completely separated in time, both gametogenically and hormonally. The animal first spawns oocytes and secretes estrogens and later releases sperm and secretes androgens, but at no time are functional ovarian and testicular tissue present simultaneously in the same individual, so that the rice-field eel does not follow the hypothetical scheme of Figure 3.2. The change in secretion of sex steroids is illustrated in Figure 3.3. The increase in androgen secretion is highly correlated with the observation that 3β-hydroxysteroid dehydrogenase (3β-HSD), an enzyme that plays a key role in steroidogenesis, shows a weak reaction during the female phase and a strong reaction during the intersexual and male phase. During the female phase, enzyme activity is found in the granulosa cells of some large follicles and some scattered

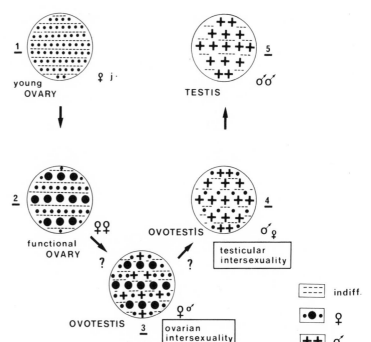

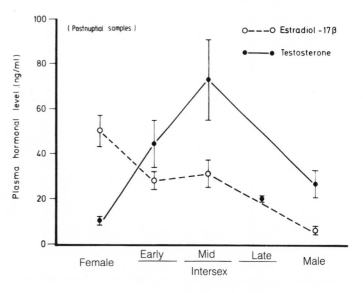

Figure 3.2 Hypothetical scheme of gonadal evolution of the fish *Epinephelus*. j = juvenile. (From Bruslé and Bruslé, 1975; reprinted with permission of Springer-Verlag.)

interstitial cells; during the intersexual and male phase, the enzyme activity is localized mainly in interstitial cells (Tang et al., 1975). The decrease in estradiol secretion and the increase in testosterone secretion associated with the sex reversal of the rice-field eel might be suspected to be causative factors of the structural changes in the germinal tissue. Experimentally, however, neither the administration of various androgens (testosterone, methyltestosterone, 11-ketosterone) nor the

administration of estrone causes changes in the sex differentiation, although androgens stimulate spermatogenesis, and estrone increases the incidence of small oocytes and causes the destruction of Leydig cells and spermatogenetic cells. Androgens do not affect either the cryptic spermatogonia or the ovarian follicles of females. It seems unlikely, in view of these experiments and others, in which estrogen synthesis was blocked by cyano-ketone without affecting sex reversal

Figure 3.3 Plasma levels of testosterone and estradiol-17β in *Monopterus* at various sexual phases. (From Chan et al., 1975; reprinted with permission of Springer-Verlag.)

(Chan et al., 1975), that sex steroids cause sex reversal.

Injections of mammalian LH cause precocious sex reversal in these eels by stimulating testicular lobule formation and by stimulating the 3β-HSD activity of the interstitial cells (Tang et al., 1975). Further investigations with fish gonadotrophins are clearly needed, but a tentative model for sex reversal of this fish is presented in Figure 3.4. Note, however, the many question marks.

In some protogynous genera, *Coris, Thalassoma,* and *Halichoeres,* precocious sex reversal can be induced by administration of androgens (see Chan et al., 1975, for references). Ovariectomy of the fighting fish, *Betta splendens,* and

Figure 3.4 A proposed scheme for the endocrine events of natural sex reversal in *Monopterus,* showing the various possible controls in the life cycle of the process, which involves the nature of the germ cells, the gonadal endocrine interaction, and the hypothalamo-adenohypophysiogonadal axis, with genetic and age factor as the ultimate basis of control. (From Chan et al., 1975; reprinted with permission of Springer-Verlag.)

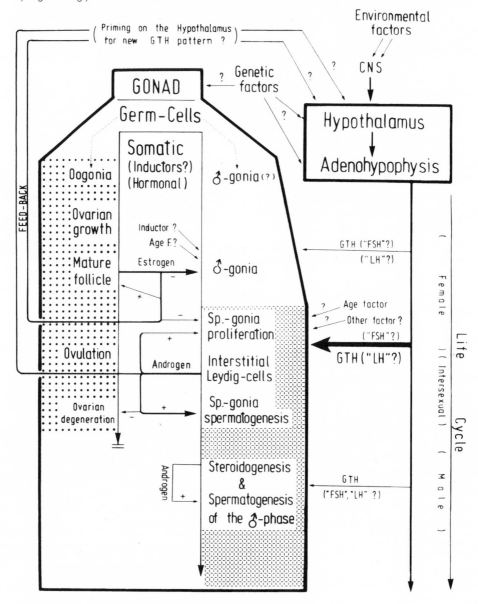

the paradise fish, *Macropodus opercularis,* causes a high incidence of sex reversals (Atz, 1965; Becker et al., 1975), but the reasons for this sex reversal are not understood.

Hermaphroditism has arisen several times independently in teleost fishes. Smith (1975) cites the following evidence:

1. *Separation of testicular and ovarian tissue.* This varies, from testes and ovaries each encapsulated in their own connective tissue (Aulopiformes) to male and female cells scattered throughout the germinal epithelium (*Epinephelus*).

2. *Location of testes with respect to the ovary.* In the Aulopiformes, testes lie dorsal to the ovaries, but in the Sparidae the testes are on the ventral wall of the gonad.

3. *Arrangement of gonoducts.* In the Aulopiformes, each testis has its own duct, which either leaves the body separately or after it has joined the contralateral duct. In *Epinephelus,* the sperm run through lacunae that develop in the outer wall as the gonad changes from ovary into testes. The lack of correlation among these features in different groups of fishes supports the idea of the independent origin of hermaphrodites in teleosts.

Warren (1975) has discussed the advantages of sequential hermaphroditism. The advantage is most obvious when there is an age-specific fecundity, which is not distributed in the same way for males and females. An animal can then function as that sex which has, at that age, the highest fecundity, and so gain an advantage over gonochorists. Protandry may have the greatest advantages when female fecundity increases with age and mating occurs at random. Protogyny would be most advantageous when female fecundity decreases with age or when male fecundity is low at early ages (as a result of either inexperience, territoriality, in which a smaller male has less chance to mate than a larger, older male, or female selection of mates).

Choat and Robertson (1975) have discussed protogynous hermaphroditism in detail for the family Scaridae. In 20 of 23 species, the larger individuals are more brightly colored than the smaller ones, with an overwhelming probability that a brightly colored individual is a male. Functionally, males that are drab colored are generally primary males.* Bright coloration is generally associated with pair spawning and may function for species and sex recognition. In group-spawning species, bright colors might be disadvantageous, because they might provoke attacks, whereas drab color would be advantageous because the animals might be considered females and not be attacked. It may thus be an advantage to an individual to start as a female, especially if few males are required for fertilization, and reproduce. When the individual becomes older and thus better able to defend itself, it may be advantageous for it to become a male and fertilize the eggs of many females.

Fricke and Fricke (1977) have described the sex reversal in coral-reef fishes with a restricted distribution. In two species, *Amphiprion bicinctus* and *A. akallopisus,* which live on anemones, males are born as males and females as females, but after birth both undergo first a male then a female phase. The testes of functional males always contain immature oocytes, but the ovaries of functional females contain no testicular tissue. The social units of these species consist of a large female, a small male, and a number of subadults and juveniles that are offspring of the adult pair. They are close, monogamous groups, with pair bonds and individual recognition. The female controls the sex and the growth of the subordinates. After removal of the dominant (α) female of *A. akallopisus,* the β male takes the α position and changes to a female in less than 63 days. In *A. bicinctus,* an original β male laid eggs 26 days after removal of the α female. The subadults are stunted in their growth, but develop rapidly after the adult dies. This system of socially controlled sex change and a large-sized female but a small male is advantageous because it allows for a constant supply of potential males and females from the subadult population; it reduces the necessity for migration and thus the risk of predation. The advantage of a large female is that it allows for greater oocyte production (body weight and ovarian weight are

*Primary males have large testes with centrally located sperm ducts; secondary males have lobulated testes and peripheral sperm ducts.

highly correlated), whereas the smaller male is able to fertilize and guard offspring, but requires less food than a larger individual.

AMPHIBIANS

In certain toads, e.g., *Bufo bufo,* the normally inactive Bidder's organ becomes a functional ovary after castration of normal males. This is, therefore, a form a protandry, although it does not normally occur in nature except possibly when the testes are destroyed by a disease that does not also kill the animal. We mentioned in Chapter 2 the undifferentiated races of frogs, which during their development go through a hermaphroditic stage, but they do not function as sequential hermaphrodites. Foote (1964) provides information about aberrant manifestations of hermaphroditism and experimentally induced hermaphrodites in amphibia.

REPTILES

Among reptiles, there are several unisexual parthenogenetic species (Cole, 1975), but no hermaphroditic species have been found. Intersexuality is an abnormal occurrence, although, as we mentioned in Chapter 2 in the discussion on embryology, intersexes will occur in some species as the result of incubation at noncyclic temperature variations (Pieau, 1975). Forbes (1964) has summarized much of the information on nonfunctional hermaphroditism among reptiles.

BIRDS

In birds, intersexes may be the result of genetic selection, embryological conditions, or unnatural treatment.

Genetic Selection

In pigeons, *Columba livia,* selection has resulted in a line of birds with a hermaphroditic left gonad in males (Riddle et al., 1942). In crosses of *Cairina moschata* ♂ × *Anas platyrhynchos* ♀, the left gonad of offspring consists of connective tissue, cords of undifferentiated cells, and large cells with 3β-HSD activity; the right gonad is rudimentary, consisting of mesonephric tubules.

In the reciprocal cross, the left ovary is normal and eggs are produced in the first year, but the ovarian tissue regresses, medullary tissue proliferates, and testicular tubules develop with or without gonocytes (Gomot, 1975). The right gonad initially contains oocytes, but later tubules that may be sterile or may contain spermatocytes develop.

Freemartins

We have discussed in Chapter 2 the "freemartins" that are found among embryos developing in double-yolked eggs with embryos of different sex. Some feminization of the testes occurs but little or no masculinization of the ovaries (Lutz and Lutz-Ostertag, 1975). Changes in development of the secondary sex organs of such "freemartins" are discussed in Chapter 2.

Removal of the Functional Left Ovary

This operation, in certain species, leads to the proliferation of the rudimentary right gonad into a testis or ovotestis. This change has been induced in chickens, *Gallus domesticus* (Gomot, 1975; Taber, 1964); muscovy ducks, *Cairina moschata* (Gomot, 1975); and ring doves, *Streptopelia risoria* (Taber, 1964); however, it does not occur in mallard ducks, *Anas platyrhynchos* (Gomot, 1975); turkeys, *Meleagris gallopavo;* pheasants, *Phasianus colchicus;* bobwhite quail, *Colinus virginianus;* or starlings, *Sturnus vulgaris* (see van Tienhoven, 1968, for documentation). In hybrids (*Anas platyrhynchos* ♂ × *Cairina moschata* ♀), ovariectomy results in the development of a small right gonad with germ cells present in 3/14 of the cases only (i.e., testis, 2 atretic follicles); whereas in the reciprocal cross, after ovariectomy one finds a small nodule, which consists mostly of mesonephric nodules and which is very similar to the right gonad found in nonovariectomized females of this hybrid.

In chickens, the available evidence indicates that the older the bird at ovariectomy, the greater the probability that the right gonad will become an ovotestis, which may contain follicles of ovulatory size and tubules with spermatozoa; the younger the bird at ovariectomy, the greater the probability that the right gonad will develop into a testis (van Tienhoven, 1968). The development of the right gonad in an ovariectomized hen can be inhibited by the injections of physiological doses of estrogens (van Tienhoven, 1968, for review) and also by a steroid-free extract of ovarian tissue (Gardner et al., 1964).

Ovariectomy of 13 *Cairina moschata* at 2–15 days of age resulted in the development of the right gonad with atretic follicles in 11 cases, testicular tissue in 1 case, and a structure with clear cells in 1 case. When the ovary was removed between 40 days and 6 months, in 1 of the 5 birds the right gonad was an ovary, 3 had spermatocytes, and 1 had clear cells. It thus seems that in the chicken and muscovy duck, age at ovariectomy has opposite effects on the development of the right gonad. In the Japanese quail (*Coturnix coturnix japonica*), the right gonad proliferates after ovariectomy into a structure consisting of medullary tubules with fat-laden cells (Kannankeril and Domm, 1968).

Mention should be made of the often misquoted paper by Crew (1923). Crew found a bird that laid eggs and later sired offspring. At autopsy, the animal was found to have *two* testes *and* a diseased ovary. This is thus a case of a true hermaphrodite, and it should not be confused with the *sex reversal* that occurs after loss of the ovary (experimentally or by disease). As the ductus deferens is lacking in such birds, even after development of the rudimentary gonad into a testis, such animals cannot be used as sires. Frankenhuis (1974) tried to obtain self-fertilization in chickens by grafting. He transplanted the right ovary of a female embryo 10 days old into a gonadectomized cockerel 10 days old and the left ovary into an ovariectomized pullet 10 days old. In theory, mating of the two animals having the transplants would amount to self-fertilization, if fertilization occurred. Unfortunately, the implantation of the ovary disturbed the oviduct, so that oocytes were not caught. Frankenhuis is pursuing this novel idea.

Pesticides and Herbicides

Feminization of the testes occurs in some species as the result of administration of the following compounds (Lutz and Lutz-Ostertag, 1975).

1. *Organophosphates:* Parathion (chicken, Japanese quail); azinphos (pheasant, gray partridge, red-legged partridge).

2. *Organo-chlorinated pesticides:* Aldrine (chicken, Japanese quail, pheasant); endosulfan or thodian (chicken, Japanese quail); DDT (chicken, Japanese quail); lindane (Japanese quail).

3. *Herbicides:* 2,4-D (pheasant, gray partridge, red-legged partridge, Japanese quail); 2, 4,5-T (chicken, Japanese quail); simazine (chicken, Japanese quail).

4. *Synergized pyrethrines* (Chickens, Japanese quail); no examples given by Lutz and Lutz-Ostertag (1975).

MAMMALS

We have discussed some cases of intersexuality in mammals in Chapter 1; in Table 3.2 we summarize some of the better known cases. It needs to be added that freemartins have occasionally been observed in pigs and sheep and that the XXY condition has been observed in species other than cats. The calico cat is, however, easily recognizable, and the fact that it is usually sterile, if it is a male, makes the coat color a diagnostic character.

We want to discuss the freemartin in some detail, because the explanation for its occurrence has evaded many investigators for a long time, although bit by bit, parts of the problem have been solved. In 1911, Tandler and Keller suggested that the freemartin was the result of the anastomosis of the central blood vessels of the bull and heifer fetus. Lillie (1917), who was apparently unaware of Tandler and Keller's work, came to the same conclusion and suggested that male sex hormone secreted by the bull fetus might be the cause of the modification of the ovary of the heifer calf. Subsequently, Jost et al. (1963) demonstrated that injections of androgens started as early as day 40–42 of gestation caused modification of the external genitalia of the heifer fetus (thus demonstrating that the androgens reached the embryo), but failed to modify the ovaries. It was later discovered that freemartins as well as their bull twins were chimeras with respect to the somatic cells, XX/XY for the heifer and XY/XX for the bull. That the chimerism is not the cause of the freemartin, however, is

Table 3.2

Genotype, gonadal phenotype, and other characteristics of mammalian intersex, and explanation for the condition

Species	Genotype	Phenotype	Explanation of condition
Goat (*Capra hircus*)	PP; XX (P = polled)	Testes; female appearance; enlarged clitoris	*P* sex-reverses gonads but not germ cells; primordial germ cells do not survive in absence of follicle cells
Pig (*Sus scrofa*)	XX	Testes	Unknown
Mouse (*Mus musculus*)	XX; *Sxr*	Testes; sterile, H-Y antigen present	*Sxr* causes progressive necrosis of germ cells
	XO; *Sxr*	Testes; germ cells present but few sperm, mostly abnormal	
	XO	Normal fertile female	
	XY; *Sxr*	Testes; low sperm count, high incidence abnormal sperm.	
Cattle (*Bos taurus*)	XX/XY (twin to bull with anastomosis)	Testicular gonads; no Müllerian duct derivatives	H-Y antigen coating of PGC in genital ridge causes testicular development
Cat (*Felis catus*)	XXY (calico cat)	Testes small and usually sterile, analogous with Klinefelter's syndrome (Table 3.3)	XX chromosomes cause gonia to be "female," they then degenerate in medullary environment
Dog (cocker spaniel) (*Canis familiaris*)	XX	Fertile female with ovotestes, with permanent testicular tubules, H-Y antigen positive	H-Y gene transmitted anomalously
Rat (*Rattus norvegicus*)	XY	Testes in inguinal canal; germical arrest; no Wolffian ducts; no Müllerian ducts	Insensitivity to androgens; testicular feminization syndrome

shown by the following evidence: (1) The germ cells of the female do not carry XY chromosomes and those of the male do not carry XX chromosomes; at least, the evidence that they do is either indirect or weak (Ford and Evans, 1977).* (2) There is no correlation between the incidence of somatic XY cells and the degree of freemartinism (Vigier et al., 1972). (3) By experimentally separating the blood vessels of cattle twins *after* the exchange of XX-XY cells, but *prior* to differentiation of the gonads, XX/XY chimeras were produced with normal ovarian tissue (Vigier et al., 1976).

*If this is indeed true, then the data on bulls that were twins to heifer calves and that are supposed to have sired a preponderance of females (Ford and Evans, 1977) are probably the result of some sampling error.

It now appears that the male twin may secrete H-Y antigen, which coats the germ cells (provided they have receptors for this antigen), and that this H-Y antigen then subverts the development of the gonad of the female into an ovary (Ohno et al., 1976). The "problem" of the marmoset, which does not produce freemartins, although the placental blood vessels anastomose and the twins are chimeras (Ford and Evans, 1977) may be explained by assuming that either the H-Y antigen is not secreted and distributed by the male fetus or that the germ cells lack the receptors for this antigen, or both. Intersexes in humans are listed in Table 3.3, together with a possible explanation of the observed phenomena.

The question may be asked why functional hermaphroditism occurs as part of the normal reproductive pattern of some species. Charnov et al. (1976) have, on theoretical grounds, concluded that low mobility favors hermaphroditism, because males will not need to develop structures or behaviors to hold females. This implies that one individual can serve as both a male and a female. Since low mobility also implies that only a few females can be fertilized, and thus there are few chances for spreading genes, it would be advantageous to function as a female simultaneously or sequentially. The exact details of the explanation are presented in the paper by Charnov et al. (1976).

Table 3.3

Intersexes in humans, genotype, characteristics and gonadal phenotype, and explanation for the condition

Genotype		Characteristics and gonadal phenotype	Explanation	Reference
Total number of chromosomes	Sex chromosomes			
45	XO	Turner's syndrome; streak gonads	Lack of the second X chromosome affects development of the ovary	German et al., 1978
46	XX	Turner's syndrome; streak gonads	One X chromosome is usually abnormal	Armstrong, 1964; Jacobs, 1966
46	XX	Resembles Klinefelter's syndrome (XXY); azoospermia; male sexual behavior	Mosaic XX/XY with loss of Y; translocation of either to X or to autosome; autosomal mutant with same effect as Y	Vague et al., 1977
46	XY	Testes in inguinal canal; no Müllerian ducts or Wolffian ducts; female external genitalia, breasts developed, female sexual behavior; testicular feminization syndrome	Somatic tissues are insensitive to androgens	Armstrong, 1964
46	XY	Streak gonads	Sex reversal gene on X chromosomes	German et al., 1978
46	XX or XX/XY	Ovary on one side; testes on other side	Unknown	Armstrong, 1964
47	XXX	Normal ovaries; sometimes menstrual problems	No effect on ovaries by the extra X chromosome	Armstrong, 1964; Jacobs, 1966
47	XXY	Klinefelter's syndrome; usually small testes, usually sterile; breast development	Regression of germ cells in medulla because of XX	Armstrong, 1964; Jacobs, 1966
47	XYY	Testes normal or small	No effect from extra Y	Armstrong, 1964; Jacobs, 1966
48	XXXX	Normal ovaries; normal menstrual history	No harmful effect of extra X chromosomes in ovary	Jacobs, 1966
48	XXXY	Testes small; sterile; breast development greater than in 47-XXY	Regression of germ cells in medulla because of the extra X chromosomes	Armstrong, 1964; Jacobs, 1966
48	XXYY	Testes small; secondary sex characters poorly developed	Unknown	Armstrong, 1964; Jacobs, 1966

4

Puberty

Puberty is here defined as the condition of being first capable of reproducing, or, as the period during which an organism is becoming capable of first reproducing. Several criteria can be used to mark the onset of puberty. In the male, for instance, one can use the presence, for the first time, of spermatozoa in the testes, or one may wish to use the ability to copulate and ejaculate mature sperm. In the female, one can use as criteria the opening of the vagina, or the opening of the vagina plus the onset of regular estrous cycles, or the occurrence of the first ovulation or laying of the first egg.

The river lamprey (*Lampetra fluviatilis*) shows a most curious phenomenon: it stops feeding in the fall when it migrates into fresh water and does not resume feeding. In the spring it spawns and dies. Larsen's (1978) summary of the findings indicate that (1) environmental cues, such as photoperiod and water temperature, do not seem to be important; (2) transplanting pituitaries into lampreys does not influence sexual maturity, for such lampreys reach sexual maturity at about the same time as intact controls; and (3) neither hypophyseal gonadotrophins nor administration of sex steroids induces precocious sexual maturity. Administration of gonadotrophins is effective in preventing the effects of hypophysectomy in delaying sexual maturity, and it may, therefore, be an in-

crease in the sensitivity of the gonads that allows sexual maturation to occur (Larsen, 1978). Larsen (1978) reports that although the intestine atrophies during the initial stages of starvation, extirpation of the intestine does not cause precocious sexual maturity. Similarly, although insulin secretion diminishes between the start of migration and spawning, insulin injections do not delay sexual maturity. In brief, the processes involved in the onset of sexual maturation in lampreys remain a mystery.

Most of the subsequent discussion will deal with birds and mammals; however, because of some unique features, the genetic control of sexual maturity in the platyfish, *Xiphophorus maculatus*, is discussed. This species shows considerable intraspecific variation in the age at which sexual maturity occurs. This variation is controlled by sex-linked multiple alleles, which determine the age at which the gonadotrophic zone in the adenohypophysis develops and gonadotrophin secretion presumably starts (Schreibman and Kallman, 1977). The sex chromosome is genetically marked by various codominant pigment genes, some of which are closely linked to the gene (P^e) for early differentiation of the gonadotrophs. It is, therefore, easy to determine which offspring carry the P^e gene. Breeding experiments have determined that there is a good correlation between the presence of the *Pe* gene and the early onset of sexual maturity in males and females.

We will limit the discussion that follows to a consideration of the changes in hormone concentrations found in association with puberty and the experimental factors that influence the age at which puberty is attained. For convenience, the male and female are discussed separately.

PUBERTY IN MALE BIRDS

The endocrine changes associated with the onset of sexual maturity have been studied in greatest detail in the Japanese quail, a bird that reaches sexual maturity, under appropriately long photoperiods, at about 6–7 weeks of age. Follett (1976) kept males on short days (8L : 16D) until they were 4 weeks old and then transferred them to long days. The FSH concentration in the plasma rose 12-fold, to 348 ng/ml, during the first 9 days after the transfer and then declined. At the time that testicular weight (3.693g) was at its maximum, the FSH concentration was about 100 ng/ml (a fourfold increase over the initial concentration on the day of transfer to long days). The LH concentration increased fivefold in 4 days, a concentration that was maintained in one experiment but declined in another experiment between day 11 and 28, when testicular weights were still increasing. Plasma testosterone concentrations reached a peak of 367 ng/ml at 7 days and thereafter fluctuated around a value of about 3 ng/ml. Testicular weights increased more than 10-fold at day 7, nearly 40-fold at day 11, more than 70-fold by day 15, and more than 200-fold by day 40. Somewhat surprisingly, the increases in plasma gonadotrophins were not accompanied by any considerable change in hypothalamic LH–RH concentrations (Bicknell and Follett, 1977), although LH–RH concentrations tended to increase with age and attainment of sexual maturity.

In the chicken, puberty is attained between 16–24 weeks of age. Sharp et al. (1977) found the following sequence of endocrine events before the animals were fully sexually mature: (1) a decrease in testicular androstenedione concentration, (2) an increase in hypophyseal LH concentration, accompanying the increase in rate of testicular and comb growth, (3) an increase in plasma LH and testosterone concentrations subsequent to the rise in testicular and comb size, and (4) following the initial increase in LH concentration, a rise in plasma testosterone concentration, associated with the completion of spermatogenesis and followed by a temporary decrease in LH secretion.

The pineal is involved in sexual development of cockerels for, after pinealectomy at the age of 8 or 9 days, sexual maturation is reached in about 7 weeks by some birds instead of in 11 weeks, as in the controls (Cogburn and Harrison, 1977). The pineal is thus dispensable for attaining sexual maturity, although its presence reduces the age at which sexual maturity is reached.

PUBERTY IN MALE MAMMALS

In mammals, the rat (*Rattus norvegicus*) has been most thoroughly investigated, and much of the present discussion is based on data obtained from this species. Ramaley (1979) has divided the development of the male rat into four periods: the neonatal (from birth to 7 days), the early juvenile (7 days to weaning at about 21 days), the late juvenile (weaning to 30 days), and the peripubertal (30 to 45 days). During the neonatal period, testosterone is the main androgen secreted by the testes. The early juvenile period is characterized by the degeneration or quiescence of Leydig cells that developed prenatally and by the secretion of androstenediol. The late juvenile period starts at about three weeks of age, when the first leptotene spermatocytes move into the inner germinal epithelium compartment. Sertoli cells that have proliferated until this time cease to proliferate. The peripubertal period is characterized by the resumption of androgen secretion and the secretory activity of the Leydig cells.

Figure 4.1, shows changes in LH, FSH, prolactin, and androgen concentrations in the blood plasma during development, as summarized by Ramaley (1979). The values reported by different investigators for the hormone concentrations around the time of puberty differ. The data attributed to Döhler and Wuttke (1975) in Figure 4.1, however, are not in accord with the data reported by these investigators; for example, dur-

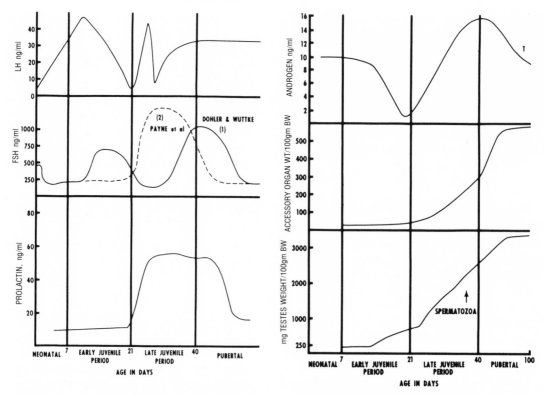

Figure 4.1. Gonadotropin and steroid profiles and testicular and accessory organ weights of rats during different stages of development. (From Ramaley, 1979; reprinted with permission of *Biology of Reproduction*.)

ing the early juvenile period, the original data show FSH concentrations of about 250 ng/ml, but show no values of around 700 ng/ml, and at 40 days concentrations of about 500 ng/ml are reported and not values of about 1000 ng/ml.

The data summarized in Table 4.1, illustrate the variability in plasma hormone concentrations reported by different investigators for the male rat around the time of puberty. Such differences may be explained for the pituitary hormones by possible differences between the purities of the preparations used as standards. Such an explanation does not apply, however, for the differences reported for testosterone concentrations, which do not vary as much as the gonadotrophin concentrations.

The pattern of change in hormone concentrations between the day of birth and puberty also differs among the different reports. The reader interested in the details of these changes should consult original publications, such as those listed

as references in Table 4.1. In young rats, guinea pigs and cattle, the testes secrete larger amounts of androstenedione than testosterone. As the animal approaches puberty, the production of testosterone increases in relation to androstenedione, and as a result, the secondary sex organs increase in weight (MacKinnon et al., 1978).

Bartke et al. (1974) have compared two early- and two late-maturing strains of mice with respect to their body weights, testicular weights, seminal vesicle weights, and concentrations of free and esterified cholesterol at 40, 50, and 60 days of age, and their plasma FSH concentration at 60 days. The only consistent difference they found was a higher testicular concentration of esterified cholesterol at 50 and 60 days in the early-maturing strain. This suggests that, at least in mice, some of the important mechanisms determining the onset of puberty may be in the ability of the interstitial cells to mobilize precursors for androgen production.

Table 4.1

Concentrations (ng/ml) of FSH, LH, prolactin, and testosterone
in the plasma or serum of male rats during the prepubertal and pubertal period
and in adulthood as reported by different investigators

Hormone	Age of rat (days)						Reference
	1–2	12–15	18–21	29–31	39–42	Adult	
FSH			450	1000	1200	750	de Jong and Sharpe, 1977
	370	420–600	300–400	80	50	130	Döhler and Wuttke, 1974
	270	200	225	270–300	350–400	275	Döhler and Wuttke, 1975
				300	200		Eldridge et al., 1974
			200	600	400	300	Gupta et al., 1975
	3.8	3.6	3.6	6.8	7.0	6.0	MacKinnon et al., 1978
LH		<1	<1	2.2	15	9	de Jong and Sharpe, 1977
	30–50	20–40	10–15	30–40	50–100	20	Döhler and Wuttke, 1974
	20	30–50	25	40	30–40	20	Döhler and Wuttke, 1975
				950	850		Eldridge et al., 1974
			40	75	15	80	Gupta et al., 1975
		<1–2	<1–7	<1–4	≅2–8	2–5	MacKinnon et al., 1978
		0.4	0.4	.65	.45	.8	McCann et al., 1974
Prolactin	15–20	10–15	20–25	110	100–130	25	Döhler and Wuttke, 1974
	35–40	15	25	80–100	100–125	40	Döhler and Wuttke, 1975
	20	≅10	≅15	≅25	≅20	40	MacKinnon et al., 1978
		4	4	15	15	32	McCann et al., 1974
Testosterone			<0.5	<0.5	1.5	1.8	de Jong and Sharpe, 1977
	0.5–1.0	0.7–1.0	0.7–1.0	0.6	0.4–0.6	1.2	Döhler and Wuttke, 1975
			0.5	2.2	2.0	2.5	Gupta et al., 1975
	.8	1.0	<0.5	<0.5	1.0	2.8–5.0*	Moger, 1977

Note: These values were read from graphs so they are not exact.
*Testosterone plus dihydrotestosterone.

It is always hazardous to draw conclusions about cause-and-effect relationships from survey data alone. In view of the differences just mentioned, not only with respect to concentrations of hormones, but also with differences in hormone profiles, the difficulties of interpretation are compounded. One should, therefore, view with caution the hypothesis of MacKinnon et al. (1978) that in the male rat, at least, puberty occurs as a gradual sequence of events. According to their hypothesis, the change in testicular metabolism, i.e., the shift from androstenedione to testosterone production, and the increased concentrations of FSH in the serum, which increase the number of LH receptors in the interstitial cells, combine to increase testosterone concentrations. The higher testosterone concentration stimulates certain aspects of spermatogenesis (to be discussed in Chapter 6), the development of secondary sex organs, secondary sex characteristics, and reproductive behavior.

Onset of Puberty

Smith (1978, unpublished) used a theoretical, mathematical approach to unraveling the various mechanisms involved in the onset of puberty. The model was based on the negative feedback system between the hypothalamo-hypophyseal system and the testes, and it led to the prediction that pulsatile releases of LH–RH and of LH would be necessary for testosterone production to increase to levels expected to occur at the onset of puberty. Indeed such pulsatile releases of LH–RH have been found in monkeys (Carmel et al., 1976) and pulsatile LH releases have been detected in Japanese quail (Gledhill and Follett, 1976), cockerels (Wilson and Sharp, 1975), and humans (Johanson, 1974; Nankin and Troen, 1971, 1972; Rowe et al., 1975). Pulsatile release of testosterone in adult animals has been found in bulls (Katongole et al., 1971). Johanson (1974) has expressed doubt that such pulsatile LH releases mark the onset of puberty. It is also not definitely established whether pulsatile LH releases occur mainly during sleep (Kulin and Santen, 1976).

In rams, in addition to the general rise in plasma FSH, LH, and testosterone concentrations, there are pulsatile releases of FSH and LH (especially of LH) when the animals are shifted from a long day (16L:8D) to a short day (8L:16D), a shift that stimulates sperm and testosterone production (Lincoln and Peet, 1977). This may mean that pulsatile releases of gonadotrophins occur both pre- and postpubertally and that the gonadotrophins may indeed be required for full gonadal development.

Smith's model also shows that the onset of puberty can be caused by any one or a combination of the following factors: (1) increased sensitivity of the anterior pituitary (AP) to LH–RH, (2) increased gonadal sensitivity to LH, and (3) increased sensitivity of the hypothalamic-AP system to negative feedback by testosterone. Smith (1978, unpublished) also mentioned that decreased clearance rates of LH–RH, LH, and testosterone, could also cause the onset of puberty. Each of these assumptions can and has been tested experimentally.

Increased Sensitivity of the AP to LH–RH. The hypothesis of a change in AP sensitivity to LH–RH around the time of puberty has been tested in rats by Debeljuk et al. (1972). They found a smaller release of LH at 15 and 240 days of age than at 25, 30, 35, 45, and 60 days, but the differences between 25 and 60 days were small. For FSH release, the highest response was obtained at 25, 30, and 35 days, with a gradual decrease from 35 days to 240 days. In boys, Grumbach et al. (1974) found that the LH release after LH–RH administration was 3–4 times higher around puberty than prior to puberty.

Increased Gonadal Sensitivity to LH. Evidence for an increased sensitivity of the testes to

gonadotrophin at puberty when compared with prepubescent animals is not clear-cut. Odell and Swerdloff (1974) reported a greater response, as measured by the percentage of increase in serum testosterone, after LH injections of 3.0 μg/100 g BW per day in 21-day-old rats than in either 41-day-old or 62-day-old rats. However, at a dose of 30.0 μg/100 g BW per day, the response was greater in 41-day-old than in either 21-day-old or 62-day-old rats. Moger and Armstrong (1974) found a smaller change in plasma testosterone concentrations after LH administration in immature rats rather than mature rats, suggesting a difference in threshold between mature and immature rats; however, this does not mean that puberty is caused or correlated with such a change in threshold.

Apparently, the LH receptors of the Leydig cells may be present from the time of formation of the cell (Greenstein, 1978). In fact, Pahnke et al. (1975) found that binding of hCG per Leydig cell was 5–6 times higher in the testes of 10-day-old rats than in 30- to 100-day-old rats, with no changes taking place between 30 and 110 days. Greenstein (1978) suggests, ''The immature mammalian testis may be fully equipped with the hormone receptor populations required for normal spermatogenesis and testosterone, but all the necessary enzyme systems may not be active when the cell is differentiated.''

Increased Sensitivity of Hypothalamic-AP System to Negative Feedback by Testosterone. A change in threshold of the hypothalamo-AP system to the feedback action of testosterone has been reported for rats. Lower amounts of exogenous testosterone and lower concentrations of testosterone are required to suppress gonadotrophin secretion in prepubertal than in postpubertal castrated rats (Davidson, 1974; McCann et al., 1974; Negri and Gay, 1976; and Smith et al., 1975). Bloch et al. (1974) not only found such a difference in threshold, but they also observed that in rats castrated at 40 to 70 days of age, testosterone (6 μg/100 g) had a stimulatory effect on LH secretion, whereas such an effect was absent at 10 days of age. This suggests that, in the intact rat, there may be at puberty, not only a change in threshold for the negative-feedback mechanism but also the appearance of a positive feedback between the anterior pituitary (i.e., LH) and the testis (i.e., androgen), thus raising the concentrations of both of these hormones until negative feedback becomes operative. This could result in the pulsatile releases of LH observed around the time of puberty. The sensitivity of the seminal vesicles of rats to androgens also increases at puberty. It thus appears that in at least parts of the reproductive system the threshold of sensitivity of target organs to their hormones declines at puberty. A summary of postulated changes in the endocrine system that result in the attainment of puberty in humans is illustrated in Figure 4.2 and Table 4.2.

Puberty and Prolactin Secretion

In the discussion so far, no account has been taken of the possible association between puberty and changes in prolactin secretion that may occur around the time of puberty. Figure 4.3 illustrates changes in serum prolactin before puberty and the concentrations attained in adult rats. Döhler and Wuttke (1974, 1975) report a concentration of about 25–50 ng/ml in the adult rat and a peak concentration of about 120–140 ng/ml at 41 days of age.

The importance of prolactin for testicular function is illustrated by the following findings:

1. Dwarf mice are genetically deficient in prolactin, have plasma FSH concentrations lower than normal, and are sterile. Exogenous prolactin or prolactin secreted by an anterior pituitary under the kidney capsule make such mice fertile (Bartke, 1965), elevate plasma FSH concentrations to those found in normal mice, and increase the in vitro production of testosterone in the presence of hCG in the medium (Bartke et al., 1977a).

2. In hypophysectomized mice, prolactin enhances spermatogenesis (Bartke, 1971) and testosterone secretion (Bartke and Dalterio, 1976), which are induced by LH treatments.

3. Prolactin restores the testosterone concentrations and the size of the testes to normal in hamsters in which testicular regression and a decrease in testosterone concentrations have been induced by exposure to short day lengths (Bartke et al., 1975).

4. In rats, hyperprolactinemia, induced by multiple AP grafts, (a) depresses the plasma LH concentration without affecting plasma testosterone concentrations, (b) decreases plasma FSH concentrations without affecting testicular weights, and (c) increases seminal vesicle weights. In mice, such grafts do not affect either testicular weight or testosterone concentration, although plasma prolactin concentrations are increased (Bartke et al., 1977b).

Figure 4.2 Schematic illustration of the change in set point of the hypothalamic gonadostat (denoted by the dashed and solid line) and the maturation of the negative and positive feedback mechanism from fetal life to adulthood [of humans] in relation to the hormonal changes of puberty. (From Grumbach et al., 1974; reprinted with permission of John Wiley & Sons, Inc.)

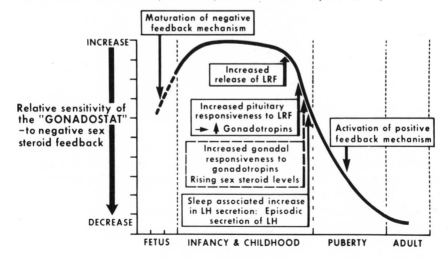

Table 4.2

Postulated ontogeny of hypothalamic-pituitary gonadotrophin-gonadal circuit (in humans)

Fetus
  Secretion of pituitary FSH and LH by 80 days gestation
  "Unrestrained" secretion of LRF (100 to 150 days)
  Maturation of negative sex steroid feedback mechanism after 150 days gestation—sex difference
  Low level of LRF secretion at term

Infancy
  Negative feedback control of FSH and LH secretion highly sensitive to sex steroids (low set point)
  Higher mean serum FSH and LH levels in females

Late Prepubertal Period
  Decreasing sensitivity to hypothalamic gonadostat to sex steroids (increased set point)
  Increased secretion of LRF
  Increased responsiveness of gonadotropes to LRF
  Increased secretion of FSH and LH
  Increased responsiveness of gonad to FSH and LH
  Increased secretion of gonadal hormones

Puberty
  Further decrease in sensitivity of negative feedback mechanism to sex steroids
  Sleep-associated increase in episodic secretion of LH
  Progressive development of secondary sex characteristics
  Mid- to late-puberty—maturation of *positive* feedback mechanism and capacity to exhibit an estrogen-induced LH surge
  Spermatogenesis in male; ovulation in female

From Grumbach et al., 1974; reprinted with permission of John Wiley & Sons, Inc.

As will be discussed in the next section, lesions in the limbic system or in the hypothalamus may induce precocious puberty in female rats; however, according to Ramirez (1973), similar lesions have not been effective in bringing about earlier puberty in male rats. Baum and Goldfoot (1974) report, however, that in immature ferrets (*Mustela furo*), lesions of the rostral mediobasal hypothalamus, which destroy the nucleus suprachiasmaticus, the median eminence, the anterior hypothalamic area, and the ventromedial-arcuate nuclear complex, result in: larger testes, a larger tubular diameter, a greater number of Leydig cells, and a higher testicular testosterone concentration than in sham-operated controls. Lesions of the amygdala, however, do not affect either testicular weight or Leydig cell index, or the onset of spermatogenesis in prepubertal male ferrets, but in prepubertal females, such lesions cause earlier estrus (Baum and Goldfoot, 1975). It has not been determined whether such lesions in the ferret cause an increase in prolactin secretion. In the rat, however, hypothalamic lesions do, indeed, cause an increase in prolactin secretion.

The hypothalamus can be selectively isolated from the influence of different parts of the central nervous system by deafferentation. Collu et al. (1974), using this technique, found that frontal deafferentation at 25 days of age resulted in larger testes, larger seminal vesicles and higher plasma testosterone concentrations than in the controls. Posterior deafferentation had no effect on these characteristics, but complete deafferentation resulted in smaller testes and seminal vesicles, and lower plasma testosterone concentrations than in controls. This suggests that there are inhibitory frontal inputs to the hypothalamus which prevent early puberty. The origin of these inputs and how they are modulated either by hormones or by other parts of the CNS have not been determined.

PUBERTY IN FEMALE BIRDS

Attainment of puberty in birds has been most extensively studied in the chicken and Japanese quail. These two species differ considerably in the age at which they reach sexual maturity. Japanese quail can lay eggs at about 5–7 weeks of age, whereas chickens usually are about 18–20 weeks old when they begin to lay eggs.

Information on the plasma hormone concentrations during sexual development of birds is rather limited. Sharp (1975) found that in chickens that started laying between 22 and 25 weeks of age there was a prepubertal LH peak 4–8 weeks prior to the laying of the first egg. Williams and Sharp (1977) obtained similar results and determined that progesterone concentrations did not rise until the day of laying of the first egg.

In so far as valid conclusions can be drawn from experiments in which mammalian gonadotrophic hormones have been used, it seems that the ovary of the chicken does not re-

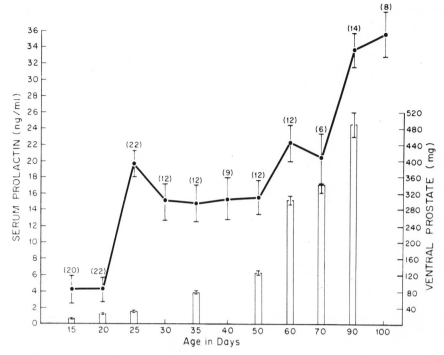

Figure 4.3 Plasma prolactin during development in male rats. (From McCann et al., 1974; reprinted with permission of John Wiley & Sons, Inc.)

spond to stimulation until about 20 weeks of age. Estrogen secretion can be stimulated and an increase in the weight of the ovary can be obtained in chickens less than 18–20 weeks old by using pregnant mare's serum gonadotrophin (PMSG) or mammalian FSH, but large follicles with yellow yolk cannot be obtained (see van Tienhoven, 1961, for review).

As will be discussed in Chapter 7, yolk precursors form in the liver as a result of estrogen stimulation. It has, however, not been possible to obtain large follicles in pullets pretreated with estrogen (i.e., so that hyperlipemia in the plasma is obtained) and subsequently treated with the extract of as many as 15 "broiler" anterior pituitaries per day (Phillips, 1959). It is, of course, possible that the use of crude AP extracts introduced not only gonadotrophins but also other hormones that might inhibit the effect of the gonadotrophins. Apparently, no experiments of this kind have been performed on chickens using purified chicken gonadotrophins. When mammalian gonadotrophins are administered, the ovary of the adult hen responds by the formation of many large follicles (see van Tienhoven, 1961, for review).

One of the sites of control for the onset of puberty in the chicken is apparently in the ovary. It remains refractory to gonadotrophin stimulation until about 18–20 weeks. It would probably be fruitful to investigate whether the concentration of gonadotrophin receptors in the ovary changes at 18 to 20 weeks and where such receptors are located in the ovary. In some mammalian species, puberty is correlated with the attainment of a threshold body weight, as in cattle and humans; in other species, puberty is correlated with the age of the animal, as in pigs. In chickens, puberty is correlated with age.

PUBERTY IN FEMALE MAMMALS

Much of the information and many of the hypotheses about the neuroendocrine causes for puberty in mammals are based on the sexual maturation of the rat. As discussed in Chapter 2, the ventromedial arcuate hypothalamic complex seems to regulate the tonic secretion of GTH, and the preoptic-anterior hypothalamus plays a key role in the cyclic release of GTH. In the rat and some other species, a rise in estrogen concentration precedes the release of GTH; therefore, this stimulatory action of estrogen may be required for normal GTH release.

Information about the external environment in many species—principally about the photoperiod—reaches the hypothalamus either directly via the retino-hypothalamic pathway (Gorski, 1974) or indirectly via other pathways. An example of the effects of photoperiod on the onset of puberty is the earlier vaginal opening of rats placed under continuous light when compared with rats under an alternating schedule of 14 hours of light and 10 hours of dark (Ramaley and Bartosik, 1975). The information about the external environment and that of hormone titers in the internal environment are integrated in the CNS, and the integrated information is transmitted to the hypothalamus. There, neurosecretory cells translate it into neurohormonal secretions that stimulate or inhibit anterior pituitary secretions, these, in turn, affect target organs that secrete steroid hormones. The steroid hormones not only influence the secondary sex organs but also induce appropriate sexual behavior. The activity of the hypothalamus is, in turn, modified by extrahypothalamic influences.

The amygdala plays an important role in the prepubertal female rat. Electrolytic lesions of the anterior mediocortical amygdala in 22-day-old rats result in precocious vaginal opening, suggesting an inhibitory role for this amygdaloid structure; however, central mediocortical amygdaloid lesions have no effect on the age of vaginal opening. Döcke (1974) found that lesions of the basal ventral hippocampus delayed vaginal opening. In an extension of those studies, Döcke et al. (1976) found that lesions of the anterior mediocortical amygdala in 21-day-old rats caused precocious puberty, and such lesions further advanced precocious puberty induced by lesions in the medial preoptic area. Similar lesions at 26 days had no effect, and such lesions at 32 days delayed puberty. Lesions of the posterior mediocortical amygdala in rats either 26 or 32 days old delayed puberty and blocked the effects of lesions of the ventromedial-arcuate complex

which caused early puberty in rats without lesions in the amygdala.

Gorski (1974) has proposed a tentative model (Figure 4.4) and documentation for the interactions among environment, extrahypothalamic and hypothalamic structures, the hypophysis, and gonads. The hypothalamus may be expected to play a crucial role in determining the onset of puberty, and much of the evidence for this has been reviewed (Donovan and van der Werff ten Bosch, 1965; Ramirez, 1973; and Davidson, 1974). Lesions made in the anterior hypothalamus result in advanced vaginal opening, but so do lesions in the mid- and posterior hypothalamus (Ramirez, 1973; Davidson, 1974), so that there is little anatomical specificity (Davidson, 1974). Anterior hypothalamic lesions are associated with hyperprolactinemia, which is, in turn, associated with precocious puberty. The effect of posterior hypothalamic lesions on prolactin secretion in prepubertal rats has not been determined.

Androgen treatment of neonatal rats results in early vaginal opening, but this is not usually followed by early regular estrous cycles (Davidson, 1974), so that vaginal opening may not be a good measure of precocious puberty. Dörner et al. (1977) reported that neonatal treatment of rats with pargyline (a monoamine oxidase inhibitor) resulted in precocious vaginal opening and first ovulation at 35–36 days versus 37–38 days for controls. Neonatal reserpine (a monoamine depletor) treatment resulted in delayed vaginal opening and first ovulation at 42 days versus 35–36 days for controls. Pyridostigmine (an acetylcholinesterase inhibitor) administered neonatally advanced puberty to about the same extent as pargyline. These two drugs had different effects on behavior, however, when the animals were adults. Pargyline-treated rats showed diminished male sexual behavior and decreased emotional responses (i.e., defecation rate), whereas pyridostigmine-treated rats showed increased male sexual behavior and decreased exploratory behavior compared with controls. The differentiation of the hypothalamus can thus affect the onset of puberty. Experimentally, precocious ovulation can be induced by the administration of LH–RH, PMSG, estradiol, and 5α-

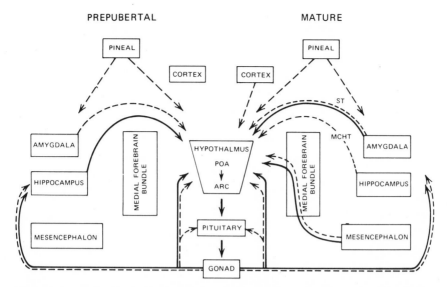

PREPUBERTAL   MATURE

Figure 4.4 Highly schematic diagram of the extrahypothalamic influences on the hypothalamic regulation of gonadotropin secretion in the prepubertal (*left*) and mature (*right*) rat. Solid lines designate facilitation; broken lines designate inhibitory influences; ARC = arcuate nucleus, MCHT = medial corticohypothalamic tract, POA = preoptic area, ST = stria terminalis. (From Gorski, 1974; reprinted with permission of John Wiley & Sons, Inc.)

androstane-3β, 17β-diol. These observations suggest that the anterior pituitary, the ovaries, and the hypothalamic mechanism involved in the stimulatory effect of estrogen and 5α-androstane-3β, 17β-diol are each capable of responding to the appropriate endocrine stimulus. We will consider the onset of puberty within this framework of environmental influences, extrahypothalamic influences, hypothalamic influences, and interactions with pituitary and gonadal hormones. However, a survey of changes in concentrations of gonadotrophic and gonadal hormones is necessary in order to evaluate what hormones may be associated with the age at which first estrus and ovulation occur.

Hormonal Changes

FSH. Figure 4.5 shows the changes in FSH concentrations during development of the rat. The changes shown are well in accord with the data reported by MacKinnon et al. (1978) in a recent review. As is the case with the male (Table 4.1), there is considerable variation among the concentrations reported for FSH in plasma or serum; however, there is good agreement on the pattern of change in FSH concentration with age.

LH. The values reported for plasma and serum concentrations of LH vary considerably. This variation may be explained by the pulsatile LH releases (Figure 4.5), which makes the exact time of sampling an important factor. The differences among reported concentrations makes it almost impossible to make a general statement (Döhler and Wuttke, 1974; Advis and Ramirez, 1977; MacKinnon et al., 1978; see Figure 4.5).

In an experimental study, Advis and Ramirez (1977) ascertained that lesions of the anterior hypothalamus, which accelerated ovarian weight increases and precocious puberty, did not affect FSH and LH concentrations, except for an increase in LH during the morning prior to vaginal opening. They suggest that possibly other hormones, specifically prolactin and growth hormone, may have been affected by the lesions.

Prolactin. Prolactin concentrations increase slowly from day 20 until a few days before puberty, when there is a sharp increase, according to MacKinnon et al. (1978), however, Döhler

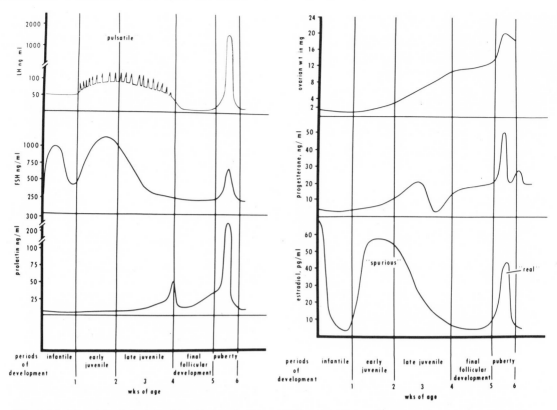

Figure 4.5 Gonadotropin and steroid profiles, and ovarian weights of rats during different stages of development. (From Ramaley, 1979; reprinted with permission of *Biology of Reproduction.*)

and Wuttke (1974) found a sharp increase at day 21 (about 40 ng/ml on day 20 and 80 ng/ml on day 21) and a further increase to about 170 ng/ml on day 37.

Various experiments have provided evidence of an important role of prolactin in determining the onset of puberty. Injection of prolactin (Wuttke et al., 1976), transplantation of the AP to the kidney (Wuttke and Gelato, 1976), lesions of the hypothalamus (Alvarez et al., 1977; Wuttke and Gelato, 1976), and administration of a dopamine receptor blocker, which prevents the inhibition of prolactin release by dopamine (Advis and Ojeda, 1978), all cause hyperprolactinemia and precocious puberty. Alvarez et al. (1977) investigated the specificity of the effect of anterior hypothalamic lesions on precocious puberty and associated prolactinemia by the use of the drug, 2-bromo-α-ergocryptine (CB-154), which inhibits prolactin secretion. They found that in the lesioned rats, ergocryptine prevented precocious

puberty. This suggests that the lesions had their effect on puberty principally by increasing prolactin secretion and not by increasing FSH and LH secretion. Advis and Ojeda (1978), using the drug sulpiride, which blocks the inhibition of dopamine on prolactin secretion, confirmed the finding that hyperprolactinemia, in the absence of elevated concentrations of either LH or FSH, was correlated with precocious puberty. Rats treated with sulpiride ovulated earlier, had higher serum prolactin and progesterone concentrations, and had heavier uteri than the control rats. Advis and Ojeda (1978) interpreted the heavier uteri as an indicator for greater estrogen secretion. Such an interpretation however, does not take into account the findings of Advis and Alvarez (1977) that prolactin increases uterine sensitivity to estradiol (see next paragraph). In vitro studies by Advis and Ojeda (1978) showed that hyperprolactinemia increased the responsiveness of the ovary, as measured by the release of estrogen and progesterone (but not of androgens) into the medium. This suggests that prolactin may cause precocious puberty by increasing the sensitivity of the ovaries to gonadotrophins. Prolactin in these experiments also stimulated the secretion of adrenal progesterone (Advis and Ojeda, 1978).

Advis and Alvarez (1977) proposed, on the basis of findings that the uterine response to exogenous estrogen in ovariectomized rats with lesions of the anterior hypothalamus was greater than the response in nonlesioned ovariectomized controls, that prolactin may not only play a crucial role in the onset of puberty, but may at the same time increase the sensitivity of the uterus to estrogens.

The effect of prolactin in causing precocious puberty does not need to be restricted to an increased sensitivity of the ovaries to gonadotrophins and of the uterus to estrogens. Beck and Wuttke (1977) and Wuttke and Gelato (1976) have proposed that chronic, high prolactin concentrations cause activation of the tubero-infundibular dopaminergic system. This activation would, in turn, cause an increase in dopaminergic traffic, which would result in a desensitization of postsynaptic dopamine receptors. Since dopamine inhibits LH release, probably by reducing LH– RH release, an increase in prolactin would eventually cause an increase in gonadotrophin secretion, resulting in ovulation. Advis et al. (1978) observed a decrease in dopamine turnover and an increase in norepinephrine turnover between prepubertal anestrus and the first proestrus; however, in their experiment, the prolactin concentration at anestrus was low.

Before assigning a crucial function to prolactin for the onset of puberty, one should take into account that inhibition of prolactin secretion, by stimulating dopamine receptors, does not result in a delay of puberty (Wuttke and Gelato, 1976). Ramaley and Campbell (1977) arrived at the conclusion that the increased prolactin concentration is not in itself required for the onset of puberty. They based their opinion on experiments in which adrenalectomy delayed puberty and prevented the increase in prolactin, but in which corticosterone replacement therapy resulted in vaginal opening at the same age as in intact controls, but without an increase in circulating prolactin concentrations.

Growth Hormone. In the discussion of puberty in humans later in this chapter, it will be mentioned that the percentage of fat per unit of dry weight may be correlated with the onset of puberty. Growth hormone plays an important role in growth, i.e., protein deposition. The role that growth hormone plays in the onset of puberty has been investigated in rats only recently. Ramaley (1979) infected prepubertal female rats with the tapeworm *Spirometra mansonoides* which produces somatomedin. This inhibits growth hormone secretion by the host but permits normal growth of the host. Rats so infected showed delayed puberty. Advis et al. (1981) implanted rat growth hormone pellets in the median eminence of prepubertal female rats, a procedure that inhibited the diurnal pulsatile release of growth hormone that occurred in control rats. They found that the growth rate of rats with implants was slower than that of the controls and that puberty was delayed about four days. The inhibition of growth hormone secretion seemed to affect mainly the ovary; FSH, LH and prolactin secretion were not affected. The ovarian weight was similar to that of the controls, but upon incubation of the ovaries, the secretion of estrogen and progesterone in response to the addition of hCG to the incubation medium was reduced. It appears, therefore, that growth hormone is necessary for the synthesis of LH receptors (Advis et al., 1981).

As will be discussed in Chapter 8, estrogen administration can stimulate LH release in the adult rat, and during the estrous cycle, and increase in estrogen concentration precedes the LH peak. The question arises whether estrogens can induce precocious puberty and whether a rise in the estrogen concentration precedes the LH release at the first proestrus. The answer is yes, because estradiol administered either systemically or implanted into the hypothalamus does, indeed, cause precocious sexual maturity (Döcke, 1975). A rise in estradiol prior to the LH release for the first ovulation has been reported by Meys-Roelofs et al. (1975b) and Parker and Mahesh (1976), although the concentrations reported differ considerably, e.g., 200 pg/ml and 37 pg/ml, respectively. The data show, however, that the preovulatory estradiol increase at first estrus is similar to the increase later in the adult.

The changes in gonadotrophin and estrogen concentrations during prepubertal development can be related to each other and to a proposed theory of change in the negative feedback threshold for estrogen in the inhibition of gonadotrophin secretion (McCann et al., 1974). The high concentrations of FSH and estradiol around 10 days of age (Figure 4.5) suggest a higher threshold before, rather than after, puberty for the negative feedback in the hypothalamo-AP-ovarian axis. This change in threshold is similar to that illustrated for the male in Figure 4.2. There is, however, an alternate explanation to that proposing a change in threshold. The blood of neonatal rats contains alpha feto-protein (AFP), which has a highly specific affinity for estradiol and, by binding it, makes the estradiol unavailable for action on the hypothalamus and pituitary. Andrews and Ojeda (1977) injected different doses of each of the following estrogens in 13-day-old female rats: estradiol, which is strongly bound by AFP; diethylstilbestrol (DES), which is weakly bound by AFP; and 11-β-methoxyethynylestradiol (MEE), which is not bound by AFP. Of the three estrogens, only estradiol failed to inhibit gonadotrophin secretion. None of the three estrogens stimulated secretion of FSH and LH in 13-day-old rats, but they all stimulated gonadotrophin secretion in 26-day-old rats. In 13-day-old rats, plasma prolactin concentrations did not increase in respone to estradiol, increased slightly in response to DES, and increased substantially in response to MEE. These findings suggest that the absence of AFP from the circulation may play a crucial role in controlling the onset of puberty in the rat. Changes in threshold in the hypothalamo-AP-ovarian axis occur as the animal matures.

Sensitivity of Pituitary, Ovary, and Uterus

The sensitivity of the anterior pituitary to LH–RH is greatest at about 15–25 days and is lowest at 35 days, according to MacKinnon et al.

(1978). Ovarian sensitivity is lacking during the neonatal period; rat ovaries do not respond to exogenous gonadotrophins. After this period, however, the ovary gains the capability of responding to gonadotrophin stimulation (Ramaley, 1979). The lack of response during the neonatal period can be correlated with a lack of gonadotrophin receptors. Experiments by Advis et al. (1977) have shown that anterior hypothalamic lesions that result in precocious puberty also result in an increased ovarian sensitivity to hCG, which may well be the result of the hyperprolactinemia in such lesioned rats.

The uterus is not part of the hypothalamo-AP-ovarian axis, but its response to estrogens is pertinent to this discussion. The response of the uterus to estradiol valerate is greater in prepubertal ovariectomized rats with anterior hypothalamic lesions than in ovariectomized controls (Advis and Alvarez, 1977). At normal puberty there is also an increased sensitivity of the uterus to estrogen (Ramirez, 1973).

Other Factors Affecting Onset of Puberty

Other factors besides the hypothalamo-AP-ovarian interactions have been found to affect the age at which puberty is reached. Among these factors the adrenal has been investigated most. The adrenal, in addition to secreting corticosterone, may also secrete gonadal steroid hormones. Meys-Roelofs and Moll (1978) conclude, on the basis of their own evidence and the evidence available in the literature, that the adrenals probably do not secrete estrogens, or at least not in amounts that are physiologically significant; they are, however, a source of androgens and progestins. Progesterone is secreted with a daily* rhythm that develops after the daily rhythm for corticosterone secretion has developed.

---

*We will use the words "daily rhythm" for a rhythm of 24 hours and "circadian rhythm" for a free-running rhythm with a period of *about* 24 hours. We are well aware that the daily rhythm is the result of a synchronization of the circadian rhythm by 24 hour signals of light and dark, but we use the term daily rhythm, as defined, for didactic purposes.

Progesterone may either inhibit or enhance the secretion of one or more gonadotrophic hormones, i.e., FSH, LH, or prolactin, in the rat. For example, after adrenalectomy, the plasma prolactin concentration of immature rats decreases (Gelato et al., 1976). Adrenalectomy at ages younger than 15 days may result in diminished FSH secretion, but after adrenalectomy at 15 days, FSH secretion may increase (Meys-Roelofs and Moll, 1978).

Ramaley (1974) reviewed much of the evidence for the role of the adrenal in the onset of puberty and concluded that the adrenal played an important role. Meys-Roelofs and Moll (1978), however, concluded that the adrenal corticosterone is required for the general health of the animal, because the delaying effect of adrenalectomy on puberty can be prevented by corticosterone administration and because the effects of adrenalectomy are not present if one takes into account the weight, instead of the age, of the animal when puberty is reached.

The thymus has also been implicated in the onset of puberty. Besedovsky and Sorkin (1974) noted that genetically athymic mice reached puberty between 35 and 50 days. The grafting of a thymus in such mice on day 2 of life resulted in earlier puberty, but injections of thymocytes were not effective. Lintern-Moore (1977) found that thymectomy of rats at birth resulted in a disruption in the distribution of follicles of different sizes and in the number of ovarian follicles during development, so that there were more primordial follicles in the thymectomized mice than in controls at day 10, but this difference changed gradually, so that by day 50, there were fewer primordial germ cells in the thymectomized than in control mice. The number of follicles with at least four layers of follicle cells was smaller in the thymectomized mice than in controls, although the difference was not significant for each age group that was sampled. At day 130 the ovaries were atrophied. The delay of puberty after thymectomy, according to Lintern-Moore (1977), is the result of an effect on the AP resulting in deficient gonadotrophin secretion. Experiments conducted by Dörner et al. (1977) cast doubt on the control of puberty by the thymus; neonatal pyridostigmine treatment, which caused precocious puberty, and neonatal reserpine treatment, which caused delayed puberty, both resulted in hypoplasia of the thymus at 15 days of age. Clearly more research is needed to determine the mechanism by which the thymus affects the ovary.

The evidence presented shows the great complexity of factors that determine the onset of puberty in the laboratory rat. It should be kept in mind, however, that experimental rats have been bred and maintained under rather standard conditions of photoperiod (12L : 12D or 14L : 10D usually), temperature, nutrition, etc., so that the factors that may be important for wild rats may have lost their significance as a result of selection for animals that breed well in captivity.

The neuroendocrine mechanisms involved in determining the onset of puberty in the rat are, of course, of great interest. It appears that the disappearance of the AFP, and thus the availability of estradiol for negative feedback, may be of crucial significance in lowering the FSH concentrations. Whether such high FSH concentrations early in life are necessary for normal development of the ovary needs to be determined.

Another factor determining the onset of puberty is the maturation of the hypothalamo-AP unit, which allows the stimulatory effect of estrogen on gonadotrophin release to be expressed. What this maturation involves is not known. An attractive hypothesis would be that enzymes that are available for the metabolism of estradiol change, so that more 2-OH-estradiol is formed. This estradiol would then compete with the catecholamines for the enzyme catechol-O-methyltransferase, as outlined in Chapter 2.

The changes in biogenic amine secretion could then increase prolactin secretion, which sensitizes the ovaries to FSH and LH, and the uterus to estrogens. Beck et al. (1977) have proposed that the steadily increasing concentrations of prolactin after day 20 inhibit LH release (a condition that they found in their experiments with rats having AP transplants), so that endogenous estradiol concentrations cannot evoke LH release. Prolactin has this function for a relatively short time, presumably because desensitization of the feedback (prolactin → LH → estradiol) occurs at puberty. This would then result in the release of

LH and of endogenous estradiol and further LH release to preovulatory surge concentrations.

Female Mammals Other than Rats

We will discuss experimental evidence obtained in species other than the rat, i.e., the sheep (*Ovis aries*), the rhesus monkey (*Macaca mulatta*) and the mouse (*Mus musculus*). In a recent study, Fitzgerald (1978) observed that lambs born in March started to show estrous cycles at 216–222 days at a body weight of 38 kg, whereas lambs born in July, first had estrous cycles at 170 days at a live weight of 38 kg. This suggests that body weight may be an important factor in determining when puberty is reached, provided that the photoperiod is appropriate. Fitzgerald (1978) also found that in both groups of lambs, prolactin concentrations decreased with the decrease in photoperiod, while Walton et al. (1977) found a decrease in prolactin concentrations during the transition from anestrus to first ovulation of the breeding cycle in adult sheep.

In lambs 4 weeks old (Fitzgerald, 1978), 7 weeks old (Foster and Karsch, 1975), or older, estrogen can induce LH release, and progesterone can inhibit such an estrogen-induced LH surge at 12 weeks of age (Foster and Karsch, 1975). The ovary can respond to PMSG in progesterone-treated 10-week-old lambs (Trounson et al., 1977). It thus appears that all parts of the hypothalamo-AP-ovarian system are functioning prior to puberty. It may be that, in prepubertal lambs, prolactin prevents LH release, as was suggested by Walton et al. (1977) for adult sheep during anestrus. Ryan and Foster (1980) have proposed that the daily pulsatile release of LH at about hourly intervals is necessary for the attainment of puberty in the lamb, and that such pulsatile LH release is prevented by an estrogen negative feedback, since such hourly LH surges occur after ovariectomy of lambs at any time after 1–2 months of age. With increasing age or, according to Fitzgerald (1978), with increasing body weight, the sensitivity of the hypothalamo-AP unit to estrogen decreases, allowing LH pulses to occur at hourly intervals.

The endocrinology of the prepubertal rhesus monkey differs in two important aspects from that of the rat and the sheep. Ovariectomy of immature monkeys does not result in an increase in LH until about 200 days after the ovariectomy, indicating that the negative feedback is not functional (Dierschke et al., 1974b).

Efforts to induce LH surges by administration either of estrogens (Dierschke et al., 1974a, b) or of 5α-androstane-3β, 17β-diol (Dierschke et al., 1974b) have failed in immature monkeys. In adult monkeys, the negative feedback and the stimulatory effect of estrogen on LH secretion are readily produced. Dierschke et al. (1974a) determined that administration of LH–RH resulted in FSH and LH release, indicating that the pituitary can respond. It thus seems that the hypothalamic mechanisms involved in inhibiting or facilitating LH–RH secretion are not functional in immature monkeys. Infusion of 1 μg of LH–RH per minute for 6 minutes every hour in monkeys 11–15 months old resulted in normal ovulatory menstrual cycles as long as the infusions continued (Wildt et al., 1980). These brilliantly conceived experiments demonstrate that the pulsatile release of LH–RH is the controlling factor for the onset of puberty in the rhesus monkey.

In mice (*Mus musculus*), the onset of puberty is dependent on the body weight (Bronson, 1981) and on the presence of a male. As will be discussed in Chapter 14, olfactory and tactile cues provided by the male can induce puberty to occur earlier when a male is present. Bronson (1981), after an extensive series of experiments with intact mice and ovariectomized mice implanted with different amounts of estradiolbenzoate in capsules, formulated the following tentative model for the attainment of puberty in the mouse. In a group of ovariectomized mice of different body weight, the secretion of LH is greater in mice of greater body weights and thus is independent of any gonadal steroid feedback. As the mouse gains in body weight, there is more flexibility in the requirements for estradiol necessary to induce LH release. For example, both the amount of estrogen implanted (to provide the estradiol for negative feedback) and the amount of estradiolbenzoate injected for stimulatory LH release are quite fixed in 13–14 g females, but at a body weight of 18–20 g, the requirements are more flexible. The requirements for adult females are the most flexible. The presence of a male is not required for the final maturation of the LH surge mechanism, both with respect to its sensitivity to estradiol and with respect to the amount of LH that can be released. Apparently, the male's cues cause two releases of estradiol by the ovary, one on each successive day; these surges then lead to the preovulatory LH surge.

For humans, our knowledge about the onset of puberty is dependent largely on evidence collected by surveys of changes in variables and on clinical evidence, not on experimental evidence. At puberty there are notable changes in growth rate, deposition of fat, development of secondary sex organs (such as uterus and vagina) and secondary sex characteristics (such as breast development, pattern of hair growth) (Marshall, 1978). The sex changes and the observation that during the last century, menarche (first occurrence of menstruation) has shown a tendency to occur at an earlier age, 3–4 months per decade in Europe (Frisch, 1974), have led to studies that support the idea that the percentage of fat per unit of dry weight is correlated with the onset of puberty (Frisch, 1974). This idea has been criti-

cized on statistical grounds (Marshall, 1978), but whatever the statistical relationship, it does not reveal the physiological causes of puberty, although it may indicate the direction that experimental research should take.

Grumbach et al. (1974) have summarized much of the endocrinological evidence pertaining to the onset of puberty in humans, and they have proposed the temporal changes that occur in the hypothalamo-AP-gonadal axis of humans from fetus to puberty (Figure 4.2). This proposal suggests that both the negative feedback mechanism of estrogen and the stimulatory effect of estrogen on LH release are functional earlier in humans than in rhesus monkeys.

Despite the relatively few species that have been investigated, there seems to be a common mechanism involved in the onset of puberty, i.e., the onset of pulsatile LH releases and the suppression of such releases by estrogen. The evidence we have reviewed in this chapter shows that there are differences among species in the manner in which the negative feedback operates and in the endogenous and environmental factors that influence the onset of puberty.

5

Anatomy of the Reproductive System

The gonads of vertebrates produce gametes and, in most cases, steroid hormones. These hormones serve various functions, such as inducing or facilitating sexual behavior and/or parental behavior, preparing the reproductive tract for receiving the gametes, caring for the zygote, and other functions.

Here we present a general outline of the anatomy of the male and female reproductive systems of representatives of the different classes of the vertebrates.

MALE REPRODUCTIVE SYSTEM

In most vertebrates, the testes are paired, but in some, e.g., the cyclostomes, the left and right testes are fused, and in some teleost species only one testis develops, as in *Notopterus notopterus*. In most classes of vertebrates the sperm produced in the testes are conducted to the outside by a duct system; in cyclostomes and the Salmonidae, however, such a duct system is lacking. In vertebrate species that do have a sperm duct system, this system may be separate from the urinary Wolffian duct system, as is the case in teleosts. In all other vertebrates, sperm are transported in tubules that originate as pronephric and mesonephric tubules. The basic scheme for the interrelation between testis and duct system is shown in Figure 5.1A.

Cyclostomes

In lampreys, the one testis consists of two gonads fused in the middle. It is located on the midline and is attached to the body wall by a mesentery. The testis consists of lobules, each containing several ampullae lined with germinal epithelium. When the lamprey becomes sexually mature and starts spawning, these ampullae break down, and the sperm are released into the coelom. There is no duct system, and the sperm leave the body through an opening secondarily connecting the coelom to the cloaca, as in *Lampetra* or through a median opening between the anus and the urinary opening, as in *Myxine* (Baer, 1964).

Elasmobranchs

The testes are paired and suspended from the body wall by mesorchia. Close to each testis lies the epigonal organ (Figure 5.2), which consists of lymphoid or hemapoietic tissue. The testes consist of ampullae, which are arranged in zones in the spiny dog fish, *Scyliorhinus canicula*. In the basking shark, *Cetorhinus maximus,* the testes consist of many lobules separated by connective tissue, each lobule corresponding to the entire testis of the spiny dog fish (Hoar, 1969). Only the cranial portion of the opisthonephros communicates with the testis (Figure 5.1E).

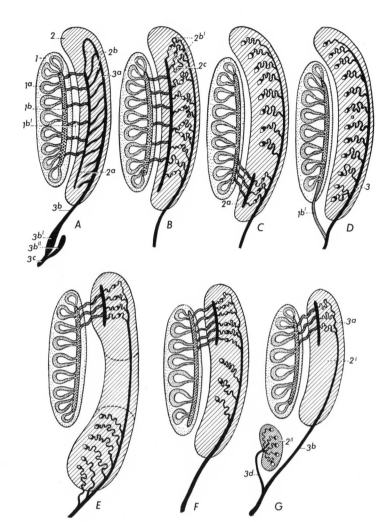

Figure 5.1 Interrelationships of the testis, epididymis and efferent ducts in vertebrates. *A*, Basic scheme; *B, Acipenser, Lepisosteus, Neoceratodus; C, Lepidosiren; D, Polypterus; E,* Selachians; *F,* Amphibia; *G,* Amniotes.

1 = testis, 1ᵃ = seminiferous ampullae or seminiferous tubules, 1ᵇ = rete testis, 1ᵇ′ = ductuli efferentes, 2 = opisthonephros, 2′ = mesonephros of aminotes, 2″ = metanephros of amniotes, 2ᵃ = kidney border canal, 2ᵇ = epididymal ductule, 2ᵇ′ = lobuli of kidney, 2ᶜ = glomerulus of kidney, 3 = primary urinary duct and its derivatives, 3ᵃ = epididymal duct, 3ᵇ = ductus deferens, 3ᵇ′ = ampulla of ductus deferens; 3ᵇ″ = vesicula seminalis, 3ᶜ = ejaculatory duct, 3ᵈ = ureter.

(From Giersberg and Rietschel, 1968; reprinted with permission of VEB Gustav Fischer Verlag.)

The number of efferent ducts* that lead to the ductus deferens* varies according to the species: it is 8 in *Centrina,* 5 in *Scyliorhinus canicula,* 2–3 in *Squalus,* 1 or 2 in *Mustelus vulgaris,* and 1 or 2 in *Galeus.* In the rays, there is one efferent duct (Gérard, 1957). Harder (1975) stated that the number may vary between 1 and 18, and Wourms (1977) stated that there were 2–6 of these *ductuli efferentes.* In the Greenland shark, *Laemargus borealis,* the Wolffian duct is absorbed in the adult (Hoar, 1969).

*The terms *vasa* efferentia and *vas* deferens are frequently used, but, according to anatomical nomenclature, they are incorrect, and the terms *ductuli* and *ductus,* respectively, should be used.

The sexual part of the opisthonephros of elasmobranchs functions as the epididymis. This leads into the ductus deferens, which may at its caudal end have diverticula that serve as sperm storage organs (sperm sacs), or the ductus deferens may have a wide ampulla (Figure 5.2). The cranial part of the kidney forms a Leydig gland (Figure 5.2), which secretes the seminal fluid. This Leydig gland is not homologous with the Leydig cells (which are sometimes called Leydig gland) of the higher vertebrates. The sperm may be packed into spermatophores (e.g., in the basking shark).

The copulatory organs of the elasmobranchs are very elaborate. They consist of claspers and a siphon sac in sharks or a clasper gland in skates and rays (Wourms, 1977). The claspers are modifica-

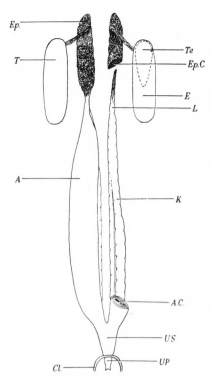

Ep.
T
A

Te
Ep.C.
E
L

K

A.C.

US

UP

Cl

Figure 5.2 Reproductive organs of the male basking shark (*Cetorhinus maximus*). Most of the ampulla on the right side of the drawing has been removed. A = ampulla ductus deferentis, A.C. = cut end of the ampulla, Cl. = cut wall of the cloaca, E = epigonal organ, Ep = epididymis, Ep.C. = cut end of the epididymis, K = kidney, L = Leydig's gland, T = testis and epigonal organ, Te = position of the testis (dotted line), UP = urogenital papilla, US = urogenital sinus. (From Matthews, 1950; reprinted with permission of The Royal Society.)

tions of the pelvic fin (Figure 5.3A). Wourms (1977) stated that the clasper may be considered part of the pelvic fin rolled up to form a tube and with edges of the fin overlapping. The siphon sacs are paired structures in the pelvic area between the skin and the abdominal wall (Wourms, 1977). They have muscular tissue and are lined by a secretory epithelium, which in skates, may be very glandular.

Holocephali

The urogenital tracts of male Holocephali and elasmobranchs are very similar. The ductuli efferentes are more numerous in the Holocephali than in the elasmobranchs and may extend into the mesorchium. Several ureters are present, some of

which open into the posterior part of the nephric duct or ampullae (Harder, 1975). The ductus deferens of the Holocephali has a complex system of chambered ampullae with the epithelium differentiated into regions (Wourms, 1977).

The chimeras or Holocephali have one pair of pelvic fin claspers homologous with the claspers of the elasmobranchs and a single clasper in front of the dorsal fin. In the Holocephali, both claspers are covered with dermal teeth, whereas in the elasmobranchs the claspers are smooth (Harder, 1975).

Dipnoi

In the Australian lung fish (*Neoceratodus forsteri*), some renal tubules throughout the length of the kidney form an efferent duct system (Figure 5.1B). In the South American and African lung fishes, the Leptosirenidae, however, only some posterior renal tubules are used (Figure 5.1C). No copulatory organs have been found in the Dipnoi.

Chondrostei

The renal-testicular tubule relationships in sturgeons, *Acipenser,* and gars, *Lepisosteus,* are similar to those in the Australian lung fishes. In the Polypteridae, the duct systems are separated (Figure 5.1D), and the sperm duct opens into the urinary duct.

Arnoult (1964) and Holden (1971) have presented evidence that the anal fin is used as a copulatory chamber in *Polypterus senegalus.* Holden (1971) speculates that this copulatory mechanism is required to conserve sperm, because the testes of these fish are quite small, even during the breeding season.

The testes of sturgeons are lobed (Harder, 1975) and are suspended by mesorchia. The male sturgeon has a well-developed Müllerian duct, but its function has not been determined (Harder, 1975).

Teleosts

There is no connection between testes and kidney in teleosts. In some species, the sperm duct has its own opening to the exterior, whereas in

Figure 5.3 A "Mixopterygia" (Clasper) of the cartilagenous fishes. A_1 = left pelvic fin of a shark (*Chlamydoselachus*), dorsal view; A_2 = left pelvic fin of *Dasyatis*, dorsal view; A_3 = right pelvic fin of Chimaera, (Holocephali), ventral view. 1 = pelvic plate, 1' = tenaculum, 2 = basal cartilage of metaptergii, 3 = connecting cartilage, 4 = appendix, 4a = marginal cartilage, 4b = terminal cartilage, 4c = ventral covering cartilage, 5 = radii, 5' = modified radius.

B, Priapum of Phallostethidae. B_1 = *Phallostethus,* lateral view; B_2 = *Solenophallus* in ventral view. 1a = Ctenactinium (fin ray of right pelvic fin), 1a' = desmactinium—fin ray of left pelvic fin, 1b = pelvic bone of the right side, 1c = toxactinium, 1d = 4–6 fin rays of the right pelvic fin, 1d' = fin rays of the left side, 2a = testis, 2b = epididymis, 2c = ductus deferens, 2d = vesicle of ductus deferens, 2e = genital papilla, 3 = anus.

C, Gonopodia in the Poeciliidae. C_1 = *Xiphophorus*, C_2 = *Gambusia*, C_3 = *Furcipenis*, C_4 = *Cnesterodon*, C_5 = *Girardinus*, C_6 = *Priapichthys*. III, IV, and V are anal fin rays that have been modified into gonopodia. (From Giersberg and Rietschel, 1968; reprinted with permission of VEB Gustav Fischer Verlag.)

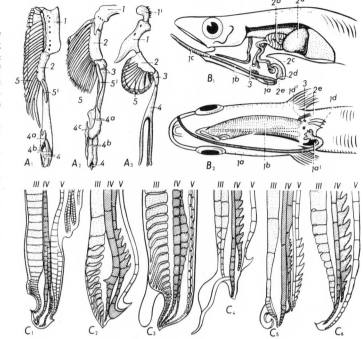

Figure 5.4 Genital organs of *Trachycorystes*. 1. *T. striatulus*, adult male. 2. *T. striatulus*. A = sperm-producing lobe of testis, B = gelatin-producing lobe of testis, A.f. = anal fin with pseudopenis. 3. Tail of juvenile male with incompletely developed pseudopenis (arrow). (From von Ihering, 1937; reprinted with permission of the American Society of Ichthyologists and Herpetologists.)

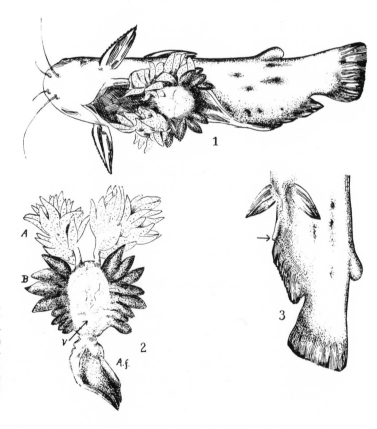

others there is no sperm duct, e.g., in the Salmonidae and Anguillidae (Hoar, 1969), and the sperm are released into peritoneal cavity and then leave via peritoneal canals.

The testes of most species of teleosts are paired, but sometimes only one testis is present; e.g., *Notopterus notopterus* has only a left testis (Harder, 1975). The testes are often oval or cylindrical, although in *Perca* and *Cyprinus* they are Y-shaped, whereas in the Ictaluridae, *Glyptosternum* and *Trachycorystes* the testes are branched, fingerlike organs (Figure 5.4). In this type of gonad, spermatogenesis takes place only in the anterior part. The posterior lobes (Figure 5.4) secrete a fluid, which, in *Trachycorystes,* seals off the vagina after mating (Harder, 1975); in the Ictaluridae and *Glyptosternum,* the function of this secretion is not known. The testes are suspended by mesorchia, but when the testes fuse, the mesorchium becomes attached to the mesentery of the intestine (Harder, 1975).

Acinar and Tubular Testes. Teleost testes have been classified on the basis of either the presence of acini (lobules) or tubules or on the basis of the distribution of spermatogonia in the tubules. The structure of an acinar type testis is illustrated in Figure 5.5 and that of a tubular type in Figure 5.6. The acinar type is reportedly found in the Clupeidae, Cyprinidae, Salmonidae, and Esocidae; the tubular type is found in Perciformes (Harder, 1975). Grier et al. (1980), after a critical review of the literature, concluded that the criteria for classifying teleost testes as either lobular or tubular were not well established and that the classification of Leydig cells, Sertoli cells, and lobule boundary cells had led to misconceptions when this dual classification was used. They proposed a classification based on the intratubular distribution of spermatogonia.

Intratubular Distribution of Spermatogonia. In the Salmoniformes and Perciformes, tubules form either anastomosing or branching networks, and spermatogonia are present along the entire length of each tubule. The spermatogonia and their associated Sertoli cells form nearly solid cords within the tubules when the testes are regressed. When spermatogenesis begins, the Sertoli cells form the

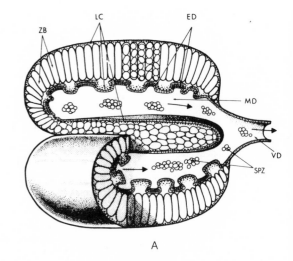

A

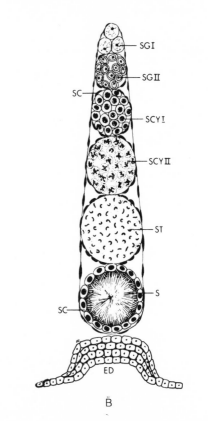

B

Figure 5.5 *A,* Schematic drawing of an acinar testis of teleost. *B,* Longitudinal section through a tubule. ZB = sperm tubule, LC = Leydig cell, ED = efferent duct, MD = main duct, VD = ductus deferens; SPZ = spermatozeugmata. SGI = primary spermatogonia, SGII = secondary spermatogonia, SCYI = primary spermatocytes, SCYII = secondary spermatocytes, ST = spermatids, S = sperm, SC = Sertoli cell. (From van den Hurk, 1975; reprinted with permission of R. van den Hurk.)

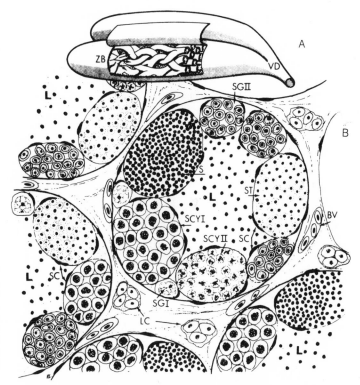

Figure 5.6 *A,* schematic drawing of a tubular testis of a teleost. *B,* cross section through tubules. SGI = primary spermatogonia, SGII = secondary spermatogonia, SCYI = primary spermatocytes, SCYII = secondary spermatocytes, ST = spermatids, S = sperm, L = lumen of tubule filled with sperm, SC = Sertoli cells, LC = Leydig cells, ZB = seminiferous tubule, VD = ductus deferens, BV = blood vessel. (From van den Hurk, 1975; reprinted with permission of R. van den Hurk.)

border of the cysts within which spermatogenesis occurs (Grier, 1981).

In the Atheriniformes, the spermatogonia are present only in the distal end of each tubule. The Sertoli cells form cysts by the time spermatogonia are transformed into primary spermatocytes (Grier, 1981). After spermiation, the Sertoli cyst cells become efferent duct cells.

The identity and homologies of Leydig, Sertoli, and lobule boundary cells has caused controversies. These have been reviewed by Grier et al. (1980) and Grier (1981). It appears from the available microscopic and ultrastructural evidence that the Leydig cells may be interstitial cells or they may be distributed around the efferent ducts at the periphery of the testes, as illustrated in Figure 5.5A. The Sertoli cells are tubular cells, which have several functions, i.e., nourishment of germ cells, production of steroids, phagocytosis of the residual bodies of spermatids, and in some cases, participation in the formation of efferent duct cells. The Sertoli cells have erroneously been called lobule boundary cells in earlier investigations (Grier, 1981). The true lobule boundary

cells of teleost testes lie outside the basement membrane and they may be myoid in nature, as they are in mammalian testes. The ultrastructural evidence argues against a steroidal capacity for these cells (Grier, 1981).

The posterior lobe of some fishes, e.g., *Clarias* spp. and *Heteropneustes* spp., both genera belonging to the Siluridae, secrete materials that participate in the formation of spermatophores (sperm packets) (Harder, 1974).

In Atheriniformes, which have internal fertilization and in which naked sperm bundles (spermatozeugmata) (Figure 5.5A) are produced, the Sertoli cell-spermatid or Sertoli cell-sperm evolves in slightly different ways in different species. In the Poeciliidae, sperm nuclei become embedded within the cytoplasmic recesses of the Sertoli cell, whereas in the Goodeidae, the flagella of the spermatid become associated with the Sertoli cyst cells (Grier, 1981). Spermatophore formation, at least in *Horaichthys setnai,* involves the Sertoli cells in the sense that their secretory product coalesce around the mature sperm.

Grier (1981) has pointed out the coevolution of

testes and anal fin in the various groups of teleosts, which has resulted in modifications of those into different structures for the transfer of sperm into the female reproductive tract. Generally, species that have a gonopodium (see below) have unmodified testes and do not produce either spermatozeugmata or spermatophores, although in the most primitive representative of the Anablepidae, *Anableps dowi,* partial spermatozeugmata are formed. Grier (1981) has speculated that the ancestral "type" from which the Anablepidae have evolved may have had a testicular structure in which spermatozeugmata were formed.

The sperm duct, the *ductus deferens,* does not produce secretions (Harder, 1975), but there are sometimes accessory sexual organs, such as the seminal vesicles (Figure 5.7A and B) of the Indian catfish, *Heteropneustes fossilis,* which do produce secretions. The term *seminal vesicles* has been used for different structures. For instance in the frillfin goby, *Bathygobius soporator,* the seminal vesicles do contain sperm and thus are truly seminal vesicles (Tavolga, 1955). So-called seminal vesicles, which are separate glandular structures, are found in the following teleost fish: the toad fish, *Opsanus tau;* the gobies, *Gobius niger, G. minutus, G. paganellus;* the loach, *Cobitis fossilis;* the goatfish, *Mullus barbatus;* the northern pike, *Esox lucius;* the mudsucker, *Gillichthys mirabilis;* the Indian catfish, *Heteropneustes fossilis* (Hoffman, 1963) (Figure 5.7A); *Ictalurus furcatus, I. catus, I. nebulosus* (Sneed and Clemens, 1963); and the South American catfish, *Trachycorystes striatulus.* In the Ictaluridae mentioned and the South American catfish, the seminal vesicles are part of, or are derived from, the testes (von Ihering, 1937; Sneed and Clemens, 1963). The secretions of the seminal vesicles reach the exterior via the sperm duct and genital papilla.

That the seminal vesicles of the mudsucker do not respond to androgens suggests that these seminal vesicles are not derivatives of the Wolffian duct system (Weisel, 1949). We cannot, however, use such a failure to respond to androgens as definitive evidence. For instance, in the musk shrew (*Suncus murinus*), the uterus does not respond to estrogen (Dryden and Anderson, 1977), although it is clearly a Müllerian derivative.

The sperm ducts of the Salmonidae and Anguillidae have secondarily disappeared (Hoar, 1969). The copulatory organ of teleosts varies from being absent in many groups (Breder and Rosen, 1966) to the bizarre priapus of the Phallostethidae (phallus = penis, stethus = chest) (Figure 5.3B) and Neosthethidae, in which ves-

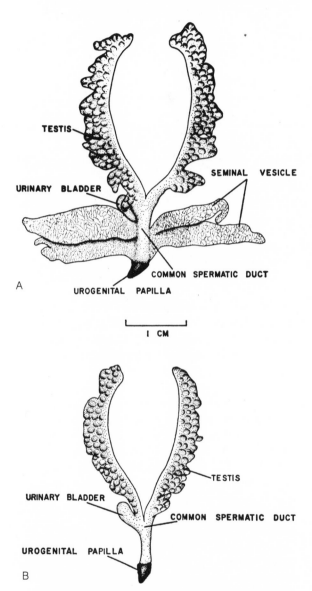

Figure 5.7 Male reproductive tract of the Indian catfish. *Heteropneustes fossilis* during the spawning season. *A,* intact. *B,* vesiculectomized. (From Sundararaj and Nayyar, 1969. Reprinted from "Effect of extirpation of 'seminal vesicles' on the reproductive performance of the male catfish, Heteropneustes fossilis Bloch" by B. I. Sundararaj and S. K. Nayyar, *Physiological Zoology,* by permission of The University of Chicago Press. Copyright © 1969 by The University of Chicago.)

tiges of the pelvic fin, the pelvic girdle, parts of the pectoral girdle, and the first pair of ribs form the copulatory organ (Bailey, 1936).

The copulatory organ may thus be variously formed. It may be formed by the genital papilla, which may be elongated, e.g., in *Zenarchopterus,* some Cottoidei, and in some species of the Blennoidei (Breder and Rosen, 1966). Some brotulid fish, e.g., of the genus *Typhlias, Stygicola, Dinematichthys, Brosmophycis,* and the species *Lucifuga subterranea* and *Dipulus caecus,* have a penislike structure (Hubbs, 1938). The morphology of this structure was studied by Turner (1946), who concluded that it was of an elaborately modified genital sinus.

The anal fin may be modified to form a copulatory organ called a gonopodium (Figure 5.3C). In the clinid fish, *Starksia,* the copulatory organ is formed by the anal fin, which is attached to the genital papilla (Böhlke and Springer, 1961). Figure 5.3C illustrates the formation of the gonopodium from rays 3, 4, and 5 of the anal fin in the Poeciliidae, and Figure 5.8 illustrates the changes in the shape of the gonopodium during erection. The relative sizes of this organ in different species of poeciliid fish is illustrated in Figure 5.9A and the mating behavior of such fish is shown in Figure 5.9B and C. In the poeciliid fishes, the anatomy of the gonopodium is used for taxonomic classification (Rosen and Bailey, 1963). The copulatory organ of the Jenynsiidae and Anablepidae is peculiar in that, in an individual, it can move in one direction only, left in some individuals, right in others. (The females correspondingly have either a left or a right genital opening.) Turner (1948) found that the copulatory organ of *Jenynsia lineata* is formed by elongation of anal fin rays 3, 4, 6, 7, and 8 (1, 2, and 5 are resorbed) and a

Figure 5.8 Changes in the external contours of the gonopodium of *Xiphophorus helleri* Heckel during erection, showing the manner in which a groove is formed along one side of the fin. In the resting phase, A, the much smaller rays 6, 7, 8 and 9 are depressed on the base of ray 5. [Rays are not shown in figure.] As the gonopodium begins its swing downward and out to one side (toward the reader). B and C rays 4 and 5 fall sideways and the tissue, which relates the smaller posterior rays, is distended. In C, the beginning of a groove along the side of the gonopodium facing the observer is revealed by the deeply shaded interior areas. In D, one edge of ray 5 is approaching the opposite edge of ray 3 preparatory to forming a closed channel, and the posterior rays are further separated in this region. Note position of spines on ray 3 in relation to the tip of the fin. Note also how both terminal holdfast elements are now facing in the same direction. In E, the gonopodium is fully erected and the opening at the base of the gonopodial groove is facing upward and slightly away from the reader. In their completely distended condition in E, the posterior rays together serve as a cup to receive discharged spermatophores. The position of the pelvic fin that is swung forward with the gonopodium is indicated in E. The base of this fin is indicated in outline in A to D. (From Rosen and Gordon, 1953; reprinted with permission of the New York Zoological Society.)

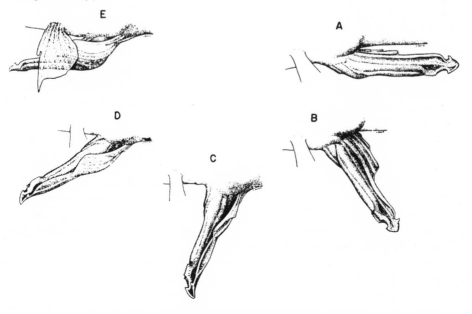

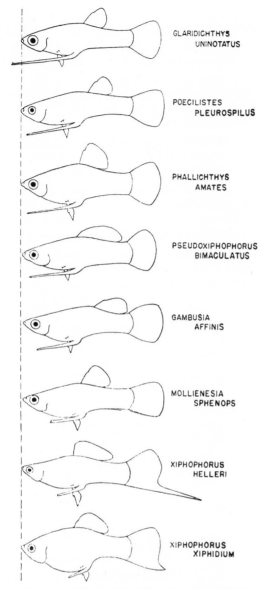

two rays of the anal fin are separate, and the first ray, which is much larger than the others, is prehensile and can be bent around the female, so that it can act as a clasper as well as an intromittent organ (Morris, 1956). In *Tomeurus gracilis,* the gonopodium, which extends beyond the tip of the head, is separated from the anal fin and is displaced under the pectoral fin. It is, however, *not* an intromittent organ, for the spermatophores are deposited near but not in the female's genital opening.

In the scorpaenid fish, *Sebastiscus marmoratus,* the mesonephric duct and urethra jut out behind the anus and form the copulatory organ (Mizue, 1959). In the Phallostethidae and Neosthetidae, the pectoral girdle, first pair of ribs, vestiges of the pelvic fins and of the pelvic girdle together help form the copulatory organ.

It is not clear what structures are involved in the formation of the "copulatory organ" of *Trachycorystes striatulus* (Figure 5.4). Such an intromittent organ occurs also in the genera *Asterophysis, Pseudoauchenipterus, Auchenipterichthys,* and *Tatia,* and probably in *Ceratocheilus.* Mohsen (1961) described a pear-shaped pseudopenis for *Skiffia lermae* (Goodeidae), into which the ducti deferentes and the urinary duct drain. It is not clear what the anatomical origin of this pseudopenis is and whether it is an intromittent organ.

In certain species of the genus *Corynopoma,* spoon-shaped appendages are found on the left side of the body. These organs are not intromittent organs, although they may function in establishing sexual contact. In *C. landoni,* these appendages develop from scales, whereas in *C. aliata* and *C. riisei,* they develop from the edge of the gill cover (Harder, 1975).

Amphibians

The testes are paired, and the sperm reach the exterior via the ductus deferens, which opens into the cloaca. Near the testes lie the fat bodies (Figure 5.10), which are essential for normal spermatogenesis, since their removal leads to a decrease in testicular weight in *Rana esculenta* (Chieffi et al., 1975) and *Triturus viridescens* (Adams and Rae, 1929).

As Figure 5.1F indicates, the connections be-

Figure 5.9A Comparison of the males of eight representative species of poeciliid fishes with their gonopodia swung forward into the erected position. The fish are drawn to the same size (standard length) to illustrate differences in the lengths and relative positions of their gonopodia. The left ventral fin is cut off at the base. (Figures 5.9A, B, and C from Rosen and Gordon, 1953; reprinted with permission of the New York Zoological Society.)

thickening with bilateral asymmetry of rays 6 and 7, so that the fin becomes converted into a tube into which the sperm duct opens. The anal fin of the Embiotocidae is enlarged and fleshy, so that it looks tubercular (Hubbs, 1918).

In the cottid fish, *Oligocottus snyderi,* the first

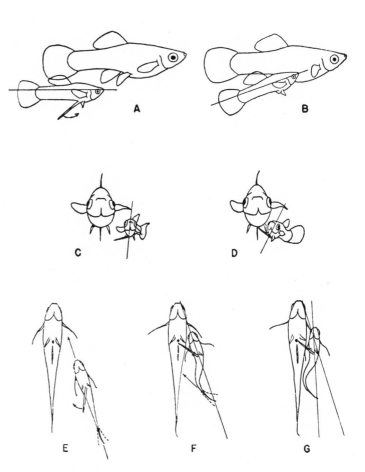

Figure 5.9B Spawning *Fundulus heteroclitus*; note interlocked anal and pelvic fins.

Figure 5.9C Orientation of male and female poeciliid fishes during copulation.

tween the kidney and testes are similar to those found in the elasmobranchs. A more critical examination, however, shows that in most Anura, the Wolffian duct serves as the urinary duct and the semen duct (Figure 5.11), except that in the genus *Alytes*, the two duct systems are completely separate. In the urodeles, the urinary and gonadal systems are not separated. In *Megalobatrachus* (Figure 5.11A$_4$), *Triturus* (*Diemictylus*) (Figure 5.11A$_3$) and *Hynobius*, the mesonephric tubules drain directly into the cloaca and not into the Wolffian duct.

The ductus deferens of some anurans have a bottle-shaped dilation (ampulla) at the caudal portion, which functions as a sperm storage organ. This has been found, e.g., in the toad (*Bufo bufo japonicus*), in the tree frog (*Hyla arborea japonica*), in *Rana nigramaculata*, in the bullfrog (*Rana catesbeiana*), in *Rhacophorus buergeri* (Figure 5.10B) and *Rh. japonicus* (Figure 5.10C). In other species, e.g., *Rana japonica, R. ornati-*

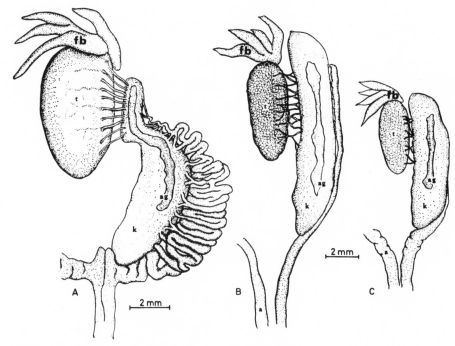

Figure 5.10 Ventral view of urogenital organs. *A, Rhacophorus aboreus; B, Rhacophorus buergeri; C, Rhacophorus japonicus.* a = ampulla, ag = adrenal gland, k = kidney, t = testis, fb = fat bodies. (From Iwasawa and Michibata, 1972; reprinted with permission of Zoological Society of Japan.)

ventris, R. tagoi, R. tsushimensis, and *R. okinavana,* there is a knoblike outgrowth of the outer wall of the Wolffian duct, which functions as a true seminal vesicle for storing sperm. In still other frogs, e.g., *Rhacophorus arboreus,* the part of the Wolffian duct lateral to the kidney is coiled (Figure 5.10A) and serves as a sperm storage organ (Iwasawa and Michibata, 1972).

The ductus deferens opens into the cloaca. Copulatory organs are rare in amphibia. In the primitive limbless Caecilia or Gymnophiona, fertilization is internal and the males have a modified cloaca (Figure 5.12) which they evert and introduce into the cloaca of the female (Giersberg and Rietschel, 1968). Among the Anura, fertilization is internal in the ''tailed'' frog, *Ascaphus truei,* which has a permanently everted cloaca that serves as a copulatory organ, and in the viviparous toad of East Africa (*Nectophrynoides vivipara*), in which no copulatory organ has been found (Noble, 1954).

In the urodeles, fertilization is internal*, but

---

*Fertilization is external in the two most primitive families, the Hynobiidae and Cryptobranchidae (Noble, 1954).

the sperm are not introduced into the cloaca by the male; they are deposited instead on the ground in spermatophores (Figure 5.13), which are subsequently picked up by the female by means of the cloacal lips. The formation of the spermatophore of the salamander requires two of the three sets of cloacal glands, i.e., the pelvic gland located in the roof of the cloaca and the cloacal glands that cover the cloacal wall. The third set of glands, the abdominal glands, which in some salamanders extend forward into the abdominal cavity, do not take part in spermatophore formation (Noble 1954).

The testes of the Anura are generally compact, and consist of convoluted seminiferous tubules. The interstitial tissue, which contains the Leydig cells, lies among the tubules. Spermatogenesis is uniform in all the tubules. The testes of the urodeles differ from those of the Anura; they consist of lobules connected by narrow bridges of tissue (Lofts, 1974), each of which drains into a short duct that joins the ductus deferens. These lobules of the urodeles are transient, and are replaced, in contrast to the seminiferous tubules of the Anura, which are permanent structures. The

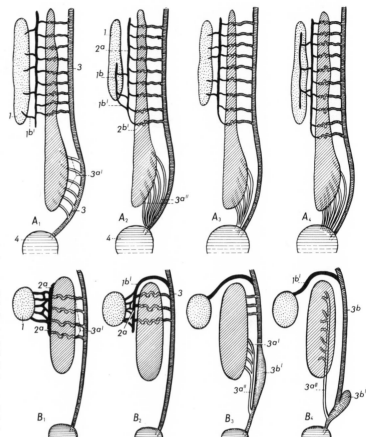

Figure 5.11 Urogenital connections in amphibia. A_1–A_4 = urodeles: A_1 = *Necturus*, A_2 = *Triturus*,* A_3 = Diemictylus,* A_4 = *Megalobatrachus*. B_1–B_4 = Anura: B_1 = *Rana*, B_2 = *Bombina*, B_3 = *Discoglossus*, B_4 = *Alytes*. 1 = testis, 1^b = rete testis, $1^{b'}$ = ductuli efferentes; 2^a = kidney border canal; $2^{b'}$ = kidney tubules; 3 = primary urinary duct, $3^{a'}$ = collecting tubes separated, $3^{a''}$ = collecting duct changed into secondary urinary duct, 3^b = ductus deferens, $3^{b'}$ = ampulla of ductus deferens, $3^{b''}$ = seminal vesicle; 4 = cloaca. (From Giersberg and Rietschel, 1968; reprinted with permission of VEB Gustav Fischer Verlag.)

*The difference between *Triturus* and *Diemictylus* is probably within a genus variability instead of a difference between genera, since *Triturus* and *Diemictylus* are now considered to be the same genus.

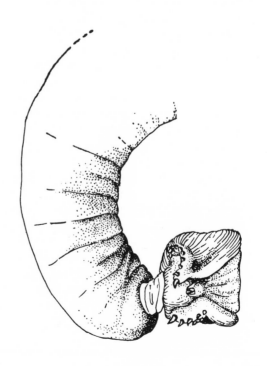

Figure 5.12 Intromittent organ of *Scolecomorphus ulugurensis*. (From Noble, 1954; reprinted with permission of Dover Publications, Inc.)

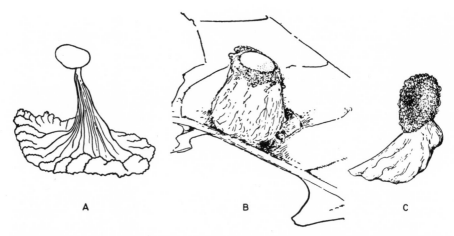

A   B   C

Figure 5.13 Spermatophores of common salamanders. *A, Triturus viridescens* (after Smith). *B, Desmognathus fuscus* (after Noble and Weber). *C, Eurycea bislineata.* (From Noble, 1954; reprinted with permission of Dover Publications, Inc.)

similarity of the testes of urodeles with the acinar testes of some teleosts is striking. The lobules are held together by connective tissue, and the entire testis is surrounded by a fibrous coat (Lofts, 1974). Further details of the testicular histology will be discussed in Chapter 6.

Reptiles

In the reptiles, the testes are paired organs suspended by mesorchia. In lizards, fat bodies are present (Adams and Rae, 1929), but these do not appear essential for spermatogenesis if the animal receives sufficient food (Cuellar, 1973). The testes contain seminiferous tubules and Leydig cells, which are usually located among the tubules. However, in all teiid lizards investigated (17 species of *Cnemidophorus* and one of *Ameiva*), the Leydig cells form a circumtesticular tunic (Lowe and Goldberg, 1966).

As in all amniotes, the Wolffian duct has lost its connection with the mesonephros and has become the ductus deferens. The genital region of the mesonephros functions as the epididymis (Figure 5.1G). The arrangement of the ductuli efferentes varies among reptiles. In snakes, several seminiferous tubules drain into one ductule, and the ductuli are arranged along the length of the testes; in lizards (Sauria), they are reduced to a single marginal canal. In the Anguidae, they are intermediate; several of the ductuli efferentes anastomose and drain into the ductuli epididymides. In the Cholomidae, there are many efferent ductuli that anastomose and form a network, the *rete testis,* which differs from the mammalian rete testis in being located outside the testis (Guibé, 1970).

Figure 5.14 Everted hemipenis of a rattlesnake (*Crotalus atrox*). (From Evans, 1978; reprinted with permission of W. B. Saunders.)

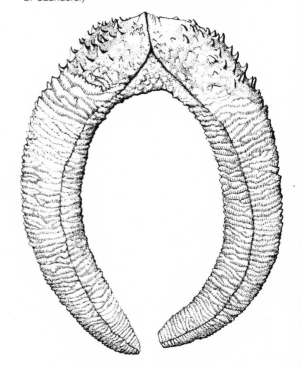

The epididymis is large in lizards (Sauria), worm lizards (Anguidae), and turtles (Chelonia); in snakes the size of the epididymis varies with the species. In reptiles, the kidney has a sexual segment, the size of which varies, there being a greater variation among lizards than among snakes (Fox, 1977). In snakes and *Varanus,* this sexual segment corresponds to the medial region of the distal convoluted tubules; in other Squamata, the collecting tubules form the main part of the sexual segment. In lizards and worm lizards, the sexual segment includes part or all of the ureter and, in lizards, increases under the influence of androgens.

The functions of this sexual segment and its secretions have not been determined, but the following functions (Fox, 1977) have been proposed, for the secretions: (1) They may help separate semen from urine by blocking renal tubules and the ureter during copulation. (2) They may assist in complete emptying of the ampullae and seminal grooves. (3) They may form the copulatory plug, which in garden snakes (*Thamnophis sirtalis*) prevents other conspecifics from copulating with the mated female (Devine, 1975). (4) They may contain nutrients for the sperm. (5) They may act as pheromones; evidence for this is the fact that male garter snakes do not court recently mated females (Devine, 1977; Ross and Crews, 1977).

In males, under the influence of androgens, a part of each kidney tubule thickens. The cells of

these tubules are very active during the breeding season, and their secretion may be important in either the transport or the nourishment of the sperm (Bellairs, 1970).

Male lizards and snakes have a paired hemipenis (Figure 5.14), the two arms of which are caudal extensions of the cloaca, each of which can be everted independently of the other into the cloaca of the female. During copulation, the sperm run along an external groove, and the spermatic sulcus of the hemipenis into the female's cloaca. Spines or ridges help to maintain intromission.

The penis of chelonians and crocodiles (Figure 5.15) is a modification of the floor of the cloaca. It consists of a median groove with ridges of erectile tissue along each side. The caudal end of the penis is a raised gland, which acts as an entering wedge prior to the filling of the erectile blood sinuses; consequently, intromission in turtles also involves an eversion of the cloaca. After testosterone propionate treatment of young crocodiles (*Crocodylus palustris*), there is no change in the sexual segment of the kidney tubules, a slight response of the Wolffian duct, but an enormous development of the penis (Ramaswami and Jacob, 1965).

Birds

The testes of birds (Figure 5.16A), which can be very large (e.g., 360 mg in a 30 g grass para-

Figure 5.15 Genital system of a male turtle (*Pseudemys* sp.). The cloaca has been opened dorsally to expose the entrance of the rectum and the dorsal groove of the penis. (From Evans, 1978; reprinted with permission of W. B. Saunders.)

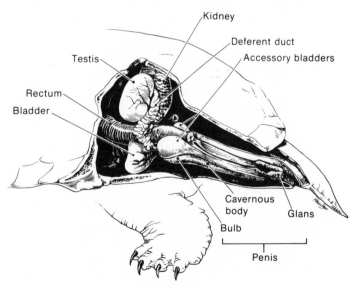

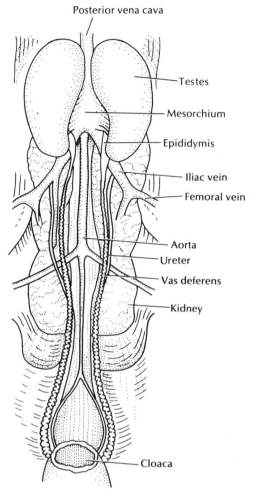

Figure 5.16A Reproductive organs of a rooster. (From Sturkie, 1976; reprinted with permission of Springer-Verlag.)

Figure 5.16B Copulatory organ of a rooster. II = second fold of cloaca, l = swollen lymph fold, g = longitudinal groove, p = erected phallus, III = third fold of cloaca, v = papillary process of vas deferens. (From Nishiyama, 1955; reprinted with permission of the Faculty of Agriculture of Kyushu University).

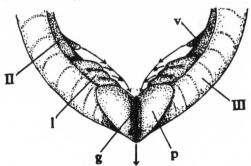

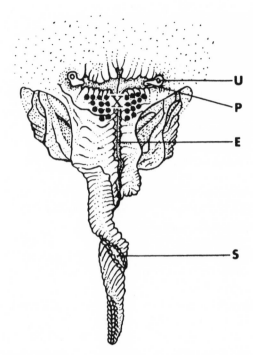

Figure 5.16C Copulatory organ of a drake, *Anas* sp. U = Opening of ureter, P = papilla of ductus deferens, E = ejaculatory groove, S = seminal groove, X = collecting point of fluid. Ejaculatory groove region is marked with large dots. (From Nishiyama et al., 1976; reprinted with permission of *Poultry Science*.)

Figure 5.16D Copulatory organ of *Rhea americana*. 1 = colon, 2a = urodeum, 2b = protodeum, 2c = bursa of Fabricius, 3 = penis, 4 = ductus deferens, 5 = ureter. (From Giersberg and Rietschel, 1968; reprinted with permission of VEB Gustav Fischer Verlag.)

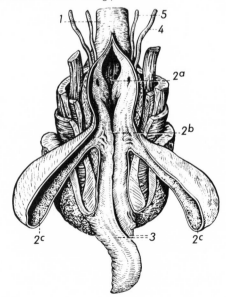

keet, *Melopsittacus undulatus*) are suspended by mesorchia. There are no tubuli recti, but the rete testis connects the seminiferous tubules to the ductuli efferentes (Tingari, 1971). The epididymis is small compared with the epididymis of mammals. It is homologous with the caput epididymidis of mammals, and different regions of the ductus deferens are homologous with the corpus and cauda epididymidis of mammals.

The ductus deferens, under the influence of either endogenous or exogenous androgens, becomes coiled. It has no glandular areas that might correspond to the accessory glands of the mammalian reproductive system. In some species there is an enlargement of the ductus deferens, called the "seminal vesicle" or "seminal glomus," which is also highly coiled and in which sperm are stored. Wolfson (1954) demonstrated that in a number of passerine species (*Junco hyemalis, Zonotrichia albicollis, Z. leucophrys,* and *Melospiza melodia*), the temperature of the seminal vesicle was 0.8–4.7°C below the deep body temperature. He proposed that such low temperatures might be beneficial for the sperm and might correspond to the effect of the scrotum on mammalian testes.

The copulatory organs of many birds are small, and they are not intromittent organs. During mating, the cloaca of the two sexes are brought into brief contact, the so-called cloacal kiss. An example of such a small copulatory organ is that of the rooster, illustrated in Figure 5.16B. Geese, swans, ducks, the Anatidae (Figure 5.16C) and ostriches, cassowaries, emus and rheas (Figure 5.16D) have an intromittent organ, which is called a penis but which is actually a pseudopenis.

The male Japanese quail has foam glands, which produce a foam that presumably aids in the deposition or transport of sperm. These glands, which respond to androgen treatment by an increase in secretory activity, are located between the stratified squamous epithelium of the proctodeum and the fibrous sheath of the cloacal sphincter muscle (Knight and Klemm, 1972).

Mammals

The location of the testes varies in mammals. Giersberg and Rietschel (1968) distinguish the following five different types of arrangement.

1. The testes remain in the body cavity, and there is no gubernaculum, as in the monotremes (duck-billed platypus, *Ornithorhynchus anatinus,* and the spiny anteaters, *Tachyglossus* sp., *Zaglossus* sp.), some primitive insectivora (the Macroscelididae or elephant shrews, and *Tenrec ecaudatus*), and the Hyraxes.

2. The testes remain in the body cavity, but a gubernaculum that later regresses is formed, as in the Sirenia (seacows), elephants, and the Edentata (sloths and anteaters).

3. The testes move caudad, a gubernaculum and inguinal canal are formed, but the testes do not enter the inguinal canal, as in the Cetacea (whales, dolphins, porpoises) and armadillos.

4. There is a gubernaculum, and the testes descend temporarily or permanently into a pouch of the cremaster muscle (Figure 5.17), but there is no true scrotum. The testes descend temporarily in *Solenodon,* in moles, hedgehogs, shrews, aardvarks (*Orycteropus afer*), and many rodents, and in the lagomorphs (rabbits). The testes descend permanently in the Phascolomidae (wombats), Tapiridae, Hippopotamidae, Rhinocerotidae, Manidae (pangolins), Pinnipedia (seals, seal lions, walruses), some land carnivora (e.g., hyaenas, *Crocuta*) and some Chiroptera,* e.g., the Vespertilionidae (McKeever, 1970).

5. A gubernaculum is present and the testes decend through the inguinal canal either temporarily (usually during the breeding season) or permanently into the scrotum. Figure 5.17 shows variations in the descent of the testes into the scrotum.

In most species in which the testes descend either into a cremaster muscle pouch or into a scrotum, the descent is permanent. In some species, the testes can return to the abdominal cavity or to the inguinal canal. Examples are: *Orycteropus* (aardvarks), some rodents (*Rattus, Sciurus, Tamias,* i.e., chipmunks) and Chiroptera,*

*Eckstein and Zuckerman (1956) list "most" bats as having periodically scrotal testes. Prasad (1974) lists all Chiroptera as having periodically scrotal testes, and Giersberg and Rietschel (1968) mention the Chiroptera only under their classification 5 (see above). These statements are not true for all Chiroptera (McKeever, 1970; also Wimsatt, personal communication, 1977). The scrotum of Chiroptera is not permanent (Eckstein and Zuckerman, 1956; Prasad, 1974).

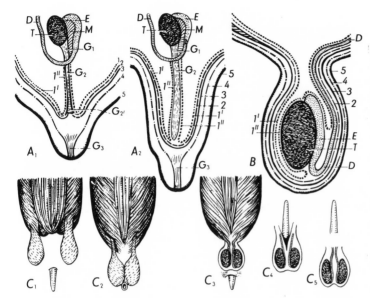

Figure 5.17 Descent of the testes. A_1 = with an inguinal cone; A_2 = without an inguinal cone. B = scheme of testes in scrotum after complete descent, sagital section. C_1–C_5 = extra abdominal position of the testes: C_1 = original position, cremaster pouch separate, penis cloacal; C_2 = *Tupaja*, scrotum with prepenial cremaster pouch; C_3 = Methatheria with prepenial scrotum; C_4 = Eutheria with prepenial scrotum; C_5 = Eutheria with postpenial scrotum and pendulous penis. 1 = peritoneum; 1' = parietal layer of 1; 1" = visceral layer of 1; 2 = transverse fascia and transverse muscle; 3 = musculus obliquus internis abdominis; 4 = M. obliquus externus abdominis; 5 = skin and hypodermis. G = gubernaculum of the testis; G_1 = ligament of the testis; G_2 = inguinal ligament; $G_{2'}$ = inguinal cone; G_3 = ligament of scrotum; T = testis; E = epididymis; D = ductus deferens; M = mesorchium. (From Giersberg and Rietschel, 1968; reprinted with permission of VEB Gustav Fischer Verlag.)

and some primates (*Loris* and *Perodicticus*), according to Prasad (1974).

The phenomenon of testes in a scrotum, in which they are exposed to temperatures lower than the body, appears for the first time in mammals. Bedford (1978a,b) has tried to explain this on evolutionary grounds. He makes a convincing argument that selection favored individuals in which the epididymis was exposed to temperatures lower than the body and that the testis, which descended with the epididymis, may secondarily have lost the capacity to function at body temperature. (As we shall discuss in Chapter 7, spermatogenesis is impaired when testes of normally scrotal mammals are exposed to body temperature.) There are, however, mammals without a scrotum, in which the testes clearly function normally at body temperature. Except for marine mammals, scrotal mammals are polygynous species, in which the dominant male sires many more offspring than males lower in the dominance hierarchy. Ascrotal species, however, in general seem to be monogamous or live in family units. The high number of ejaculations by dominant polygynous males may, therefore, have favored those males with epididymides in which sperm could be stored successfully. Such storage would probably be more successful at lower temperatures because metabolic rates would be lower; thus eventually, individuals with epididymides (and as a consequence testes) at temperatures lower than the body would have a selective advantage.

The apparent contradiction of polygynous ascrotal marine mammals may, in fact, be only an apparent contradiction. In the southern elephant seal (*Mirounga leonina*), the temperature of the testis is 6°C below body temperature (Bryden, 1967), apparently as a result of blood flow from the rear flippers (Bedford, 1978a). In some seals, e.g., *Callorhinus* and *Zalophus,* the testes are located in a scrotal pouch (Bedford, 1978a).

Many mammalian species have one or more well developed accessory reproductive glands. These glands (Figure 5.18) are:

1. *Ampullary glands*. These are derived from the Wolffian duct system and have a simple cuboidal, or columnar epithelium; their secretions are high in ergothioneine (Mann, 1964).

2. *Seminal vesicles*. More properly called vesicular glands, these are also derived from the Wolffian duct system. They have a simple or pseudostratified epithelium and secrete fructose, citric acid, proteins, and ascorbic acid (Mann, 1964).

3. *Paraurethral glands* (Prasad, 1974). Also called glands of Littré, these are probably synonymous with the glands named paraprostate glands by McKeever (1970) and Holtz (1972). These glands originate from the epithelium of the

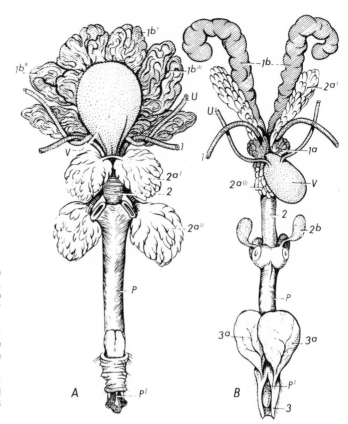

Figure 5.18 Accessory sexual glands of the male. *A*, Hedgehog (*Erinaceus europaeus*); *B*, mouse (*Mus musculus*). 1 = ductus deferens; 1a = ampullary gland; 1b = seminal vesicle; 1b'–1b''' = parts of the seminal vesicles; 2 = urethra; 2a = prostate gland; 2a'–2a'' = parts of the prostate gland; 2b = bulbourethral gland (note that the urethral glands in the wall of the urethra cannot be seen); 3 = preputium, 3a = preputial glands, P = penis, P' = glans penis; U = ureter (urinary duct); V = urinary bladder. (From Giersberg and Rietschel, 1968; reprinted with permission of VEB Gustav Fischer Verlag.)

urethra and are thus part of the urogenital sinus. They are usually small structures, the ducts of which drain into the urethra. According to Holtz (1972), there are two different types of paraprostate glands in the rabbit: (*a*) glands consisting of small clusters of fluid-filled vesicles, which, under microscopic examination, resemble the bulbourethral glands, and (*b*) small, compact, whitish glands, which are histologically identical with the prostate.

4. *Proprostate glands.* These are found in the rabbit, and appear to be part of the prostate; however, they are separated from the prostate by a connective tissue system. These glands originate, as does the prostate, from endoderm of the urogenital sinus (Holtz, 1972). Holtz (1972) emphasizes that some names given to these glands are not appropriate, e.g., coagulating gland (because the glands are not the source of the gel mass in the ejaculate), and prostate and glandula vesicularis (because they are not Wolffian duct derivatives).

The proprostate is an acinar gland, rich in smooth muscle and with a columnar epithelium. It secretes a white granular substance, which gives the gland a white color and has two large excretory ducts that lead to small urethral diverticula.

5. *Prostate gland.* This originates as part of the urogenital sinus and has simple or pseudostratified columnar epithelium. The secretions contain enzymes, such as diastase, glucoronidase, fibrinolysin, acid phosphatase as well as citric acid (Mann, 1964). The gland has ducts that drain into the urethra.

6. *Urethral gland.* This gland is found in some Chiroptera (Figure 5.19; Prasad, 1974) between the prostate and the openings of the ducts at the bulbourethral glands. It lies around the urethra, is covered by the muscles of the urethra, and its secretions drain via short ducts into the urethra.

7. *Bulbourethral glands* (Cowper's glands). These glands are also part of the urogenital sinus.

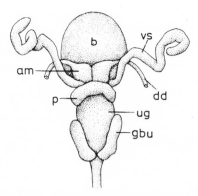

Figure 5.19 Reproductive organs of one of the Chiroptera, the giant fruit bat (*Pteropus giganteus*), dorsal view. am = ampullary glands, b = urinary bladder, dd = ductus deferens, gbu = bulbourethral glands, p = prostate, ug = urethral gland, vs = seminal vesicle. (From Prasad, 1974; reprinted with permission of Walter de Gruyter & Co.)

They have a simple columnar epithelium which secretes a viscous fluid containing sialoprotein (McKeever, 1970), and have ducts that drain into the urethra.

8. *Bulbar glands*. Also called the *inferior bulbourethral glands,* these glands seem to be present in some of the Sciuridae (Prasad, 1974). They are glandular developments of the bulbourethral glands inside the corpus spongiosum. The secretions drain via a glandular duct (ductus penis) at the ventral side of the urethra.

9. *Prostatic utricle*. Also called *uterus masculinus,* this is usually vestigial or even absent but can, according to Prasad (1974), be well developed. It is derived from the Müllerian duct, and the ducts drain into the urethra.

Since this is a derivative of the Müllerian duct system, it might be expected to be sensitive to estrogens. As estrogens have been suspected of inducing cancer of derivatives of the Müllerian duct system in women, it may be prudent to consider seriously the wisdom of the use of estrogens in treating cancer of the prostate.

10. *Preputial glands*. These are modified sebaceous glands with a stratified, squamous epithelium and with ducts draining at the surface of the glands. They are enormously developed in the beaver (Figure 5.20)

11. *Inguinal glands*. These are sebaceous glands and sweat glands found on the surface of the inguinal canal and in skin folds of the genital

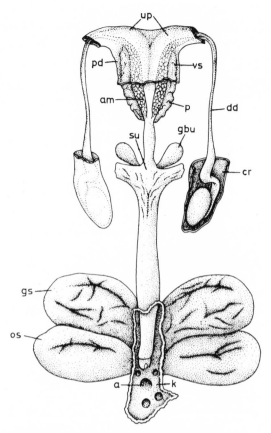

Figure 5.20 Reproductive organs of the beaver (*Castor fiber*). The rectum is opened in order to show the penis and ducts of the gland. a = anus, am = ampulla of ductus deferens, cr = cremaster pouch, dd = ductus deferens, gbu = bulbourethral gland, gs = preputial glands, k = cloaca, os = oil sac, p = prostate, pd = plica Douglasi, su = urogenital sinus, vs = seminal vesicle, up = prostatic utricle. (From Prasad, 1974; reprinted with permission of Walter de Gruyter & Co.)

region. Holtz (1972) suggests the name *perineal glands* for these in the rabbit.

Table 5.1 lists the presence of these various glands in some mammalian species. The effect of the presence of these accessory sex glands on fertility has been studied in the boar (*Sus scrofa*), which produces a large amount of fluid in the ejaculate (about 200 cc) 26 percent of which is contributed by the seminal vesicles, 56 percent by the prostate, and 19 percent by the bulbourethral glands. These large quantities of secretions make this species a very suitable animal for studying

Table 5.1

Comparative anatomy of the male reproductive structures*

Species	Testis	Ampullary glands	Seminal vesicles	Prostate gland	Paraprostate glands	Bulbo-urethral glands	Baculum	Urethral glands	Bulbar gland	Preputial glands	Inguinal glands
Bat	3 × 2	4 × 15	A	5 × 3	A	1 × 1	F	F	A	A	A
Cat	14 × 8	A	A	5 × 2	A	4 × 3	F	A	A	A	A
Dog	40 × 30	F	A	25 × 16	A	A	F	F	A	A	A
Gerbil	14 × 9	6 × 5	20 × 12	9 × 7	A	6 × 4	F	F	A	A	A
Guinea pig	25 × 15	A	115 × 7	15 × 8	A	8 × 5	F	F	A	PD	A
Hamster	14 × 11	3 × 3	11 × 6	8 × 6	A	4 × 3	F	F	A	3 × 1	A
Mink	11 × 8	F	A	10 × 5	A	A	F	A	A	A	A
Mouse	6 × 4	2 × 1	13 × 4	4 × 4	A	3 × 2	F	F	A	6 × 5	A
Opossum	15 × 11	A	A	40 × 9	A	19 × 15	A	A	A	A	A
Rabbit—European	35 × 15	15 × 4	19 × 7	19 × 6	6 × 2	6 × 3	A	A	A	A	15 × 7
Rabbit—cottontail	35 × 17	F	A	25 × 6	A	6 × 2	A	A	A	A	10 × 4
Rat	20 × 12	4 × 4	20 × 10	13 × 10	A	5 × 3	F	F	A	16 × 4	A
Shrew	3 × 2	3 × 1	A	3 × 1.5	A	1 × 1	A	A	A	Diffuse	A
Squirrel—tree	30 × 12	F	7 × 4	28 × 8	A	13 × 13	F	A	10 × 7	A	A
Squirrel—ground	14 × 8	F	8 × 7	18 × 18	A	8 × 6	F	A	10 × 8	F	A
Subhuman primate	10 × 7	A	14 × 7	6 × 5	A	A	F	A	A	Diffuse	A
Ferret	F	F	A	F	A	A	F	A	A	A	A

F, functional gland is present, but was not measured; A, the structure does not occur; PD, the gland is poorly developed.

*Measurements are in millimeters.

From McKeever, 1970; reprinted with permission of Lea & Febiger.

these glands. Removal of the seminal vesicles, or the bulbourethral glands, or both, had no effect on fertility (McKenzie et al., 1938). The effect of removal of the prostate was not reported for the boar, but apparently removal of the prostate in humans does not affect fertility.

In rats, removal of the seminal vesicles and the prostate and coagulating gland reduces fertility severely, but removal of the following glands or combination of glands has no effect on fertility: bulbourethral glands, seminal vesicles, prostate and coagulating gland, bulbourethral glands and seminal vesicles, bulbourethral glands and coagulating gland, seminal vesicles and prostate (Greenstein and Hart, 1964).

Prototheria. The penis of the Prototheria, like that of the reptiles and birds, is located on the floor of the cloaca, and lies in a preputial fold of the ventral wall of the cloaca. The penis conducts the semen but not the urine, to the outside, because the semen-urine duct gives off a branch to the cloaca before it enters the penis. The ductus deferens drains into the urogenital sinus, dorsal to the ureters, not ventral, as in the Metatheria and Eutheria (Giersberg and Rietschel, 1968). The penis contains fibrovascular tissue which is homologous with the corpus cavernosum penis of the Metatheria and Eutheria. The glans of the duck-billed platypus is bifurcated (Prasad, 1974).

Metatheria. The penis is located in a cloacal fold, which is surrounded, together with the anus, by a cloacal sphincter muscle, so that it is not visible generally except during erection. However, in *Phascogale* (marsupial pocket mice) and *Dasyurus* (Australian ''native cats'') the penis is free and not in a cloacal fold (Giersberg and Rietschel, 1968). The penis of the opossum *Didelphis virginiana* is bifurcated (Figure 5.21) as is the penis of *Dasyurus* and *Phascolomis,* although in these last two genera, the bifurcation is not as marked. The penis of the Macropodidae and Tarsipes is definitely not forked (Prasad, 1974).

In *Peramelidae* (bandicoots), the urogenital sinus conducts only the semen to the outside; the urine is voided via the cloaca. In *Didelphis* and *Caenolestes* (bat opossum), the distal opening of the urinary duct lies at the base of the glans and

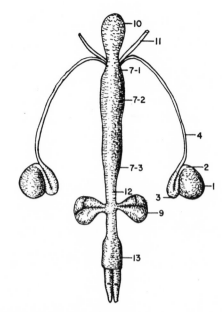

Figure 5.21 Reproductive organs of the opossum. Note the bifurcated penis. 1 = testis, 2 = caput epididymis, 3 = cauda epididymis, 4 = ductus deferens, 5 = ampullary gland, 6 = vesicular gland, 7 = prostate gland, 8 = anterior lobe of prostate, 9 = bulbourethral gland, 10 = bladder, 11 = ureter, 12 = urethra, 13 = penis. (From Hafez, 1970; reprinted with permission of Lea & Febiger.)

the ductus deferens ends in the tip of the penis. In *Myrmecobius* (numbats), *Thylacinus* (pouched wolf) and *Dasyurus,* the urethra divides into two branches, one branch going to one tip of the bifurcated glans, the other to the other tip.

Eutheria. The perineum divides the cloaca into the dorsocaudal rectum and the ventrocranial urogenital sinus, so that the penis lies on the ventral side of the anus. The distance between anus and penis can be small, as in beavers (*Castor* sp.) and hares (*Lepus europeaus*) (Giersberg and Rietschel, 1968). Prasad (1974) states that in beavers, the urogenital openings and the anus open into a cloaca (Figure 5.20), a situation also found in the Prototheria, Metatheria, Edentata, and in the Insectivora, and *Neomys* sp. In other species, the penis lies, with its preputial fold, near the umbilicus. The double skin fold, which surrounds the penis, partly or completely forms the prepuce (Figure 5.22), which in the horse consists of two circular preputial folds which are tele-

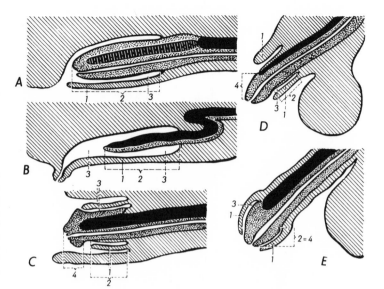

Figure 5.22 Medial view of the penis of some mammals.

A, seal (*Phoca*); B, bull (*Bos taurus*); C, horse (*Equus caballus*); D, long-tailed monkey (*Cercopithecus* sp.); E, human (*Homo sapiens*). Black = corpus cavernosum penis; stippled = corpus spongiosum penis and corpus cavernosum glandis; heavy black lines = os penis in A. 1 = preputium, 2 = pars intrapreputialis penis, 3 = preputial cavity, 4 = glans penis. (From Giersberg and Rietschel, 1968; reprinted with permission of VEB Gustav Fischer Verlag.)

scoped so that the penis can be pulled inside the folds. The pars intrapreputialis penis, shown in Figure 5.22C, corresponds to the glans penis in humans (Giersberg and Rietschel, 1968).

Several types of penes have been distinguished (Prasad, 1974).

1. *Indifferent type* (edentates and rodents). The penis is short, and the corpus spongiosum penis is not covered by a tunica. The corpus cavernosum can be either fibrous tissue or trabecular erectile tissue.

2. *Fibro-elastic type* (ruminants and whales). Usually the penis is long and fibrous and has a sigmoid flexure in the resting stage. There is little erectile tissue in the corpus cavernosum, the erection consisting of the straightening out of the sigmoid flexure. The corpus spongiosum is surrounded by a well-developed tunica.

3. *Vascular type* (Perissodactyla, Carnivora, Primates, and some Insectivora). The penis, if not in erection, is long and flexible. The corpus cavernosum penis and corpus spongiosum urethra are surrounded by a fibrous capsule, the tunica albuginea. Erection results from the corpora cavernosa and corpora spongiosa filling with blood.

4. *Intermediary type* (elephants, and the sirenia). This type combines features of types 2 and 3.

According to Slijper, as cited by Prasad (1974), there is a general correlation between the type of penis and the length of coitus. Animals with a fibro-elastic type penis have a short coitus, the ones with the indifferent type have a relatively short coitus, the ones with the intermediary type have a longer coitus than the other two types, but shorter than animals with the vascular type in which coitus may be fairly long. However, coitus in swine which have an intermediate type penis, may last as long as 30 minutes!

In most species of Chiroptera, Rodentia, Carnivora, Insectivora, and many primates (but not humans) there is a penile bone or baculum present. The variation in these is shown in Figure 5.23. The male reproductive systems of some mammalian species are illustrated, in situ, in Figure 5.24.

FEMALE REPRODUCTIVE SYSTEM

In most vertebrates, the female has two ovaries, but there are exceptions to this (Table 5.2), and among birds, most species have only one functional ovary, although some species have a high incidence of two functional ovaries (Table 5.3).

The female genital duct system of vertebrates parallels that of the males to the extent that in cyclostomes there is no genital duct system, in the

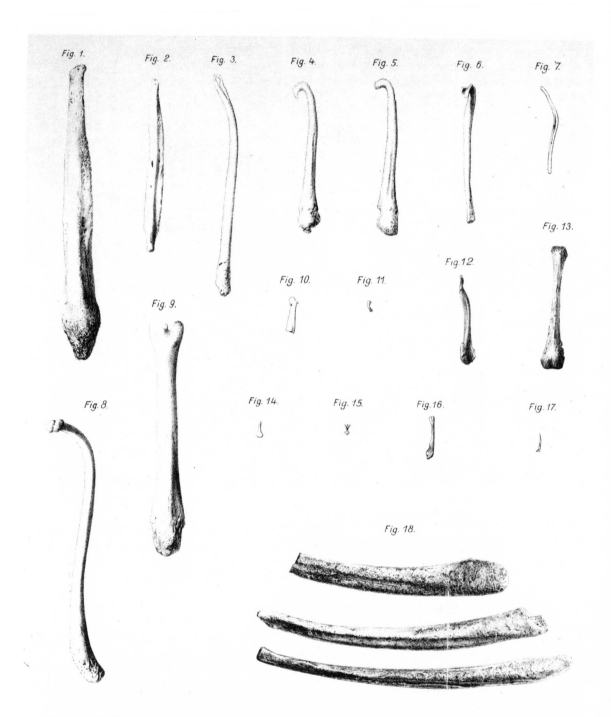

Figure 5.23 Os priapi, penile bones of different animals. 1, *Canis familiaris* (dog); 2, *Canis vulpes* (fox); 3, *Mustela foina* (European marten); 4, *Putorius foetidus* (mottled polecat); 5, *Putorius furo* (northern weasel); 6, *Putorius putorius* (common polecat); 7, *Putorius ermineas* (Mongolian ermine); 8, *Procyon lotor* (raccoon); 9, *Lutra vulgaris* (European otter); 10, *Sciurus vulgaris* (a tree squirrel); 11, *Spermophilus citillus* (a ground squirrel); 12, *Castor fiber* (young beaver); 13, adult beaver; 14, *Mus decumanus* (a small mouse); 15, *Arvicola arvalis* (water vole); 16, *Cavia cobaya* (guinea pig); 17, *Vespertilio murinus* (gray mouse-eared bat); 18, *Ursus spelaens* (fossil bear). (From Gilbert, 1892; reprinted with permission of the Akademische, Verlaggesellschaft Geest und Portig, K.G.)

Salmonidae the funnel is a secondary formation, and in the teleosts, the oviduct is not a Müllerian duct system but is formed from part of the ovary and mesentery (Webster and Webster, 1974).

Cyclostomes

The lampreys have one very large ovary, which originates from the fusion of the two gonad primordia. The ovary contains many follicles, (24,000–236,000), each of which consists of a fibrous theca externa, a glandular theca interna, and a granulosa layer (Dodd, 1972). The follicles are all in the same stage of development at any particular time, and as Dodd (1972) has pointed out, no oogonial nests remain at the time of metamorphosis. This unique feature may be related to the

fact that lampreys spawn and die shortly thereafter. The follicles form neither corpora atretica nor corpora lutea, although atresia does occur in *Petromyzon marinus*. Atresia consists of reabsorption of the yolk and follicle cells, so that only stroma cells are left.

The single large ovary of the Myxiniidae (hagfishes) results from the failure of development of one of the two gonad primordia (Hoar, 1969). There are few oocytes (1–21) (Dodd, 1972), and these are arranged in a row along the ovary. Each oocyte is enclosed in a tough shell which is secreted by the follicle. One can distinguish four layers in the follicular wall, an inner, simple, follicular epithelium; two layers of connective tissue; and an outer squamous layer (Dodd, 1972). In the hagfishes there are corpora atretica and "corpora

Table 5.2

Vertebrates with one functional ovary

Species	Explanation for one-ovary condition
Lampreys	Fusion of two gonads
Hagfishes	One gonad fails to develop
Perches, *Perca*	Fusion of two gonads
Pike perch, *Lucia-Stizostedion* sp.	Fusion of two gonads
Stone loach, *Noemacheilus* sp.	Fusion of two gonads
European bitterling, *Rhodeus amarus*	Fusion of two gonads
Japanese ricefish, *Oryzias latipes*	One gonad fails to develop
Guppy, *Poecilia reticulata*	One gonad fails to develop
Sharks	
Scyliorhinus	Left ovary becomes atrophic
Pristiophorus	Left ovary becomes atrophic
Carcharhinus	Left ovary becomes atrophic
Galeus	Left ovary becomes atrophic
Mustelus	Left ovary becomes atrophic
Sphyrna	Left ovary becomes atrophic
Viviparous ray, *Urolophus*	Left ovary functional
Ovoviviparous ray, *Dasyatis*	Right ovary absent
Blind worm snakes, *Typhlopidae*	Left ovary and oviduct absent; reason unknown
Birds	Left ovary functional in most species; right ovary regresses in embryos (for exceptions see Table 5.3)
Duckbill platypus, *Ornithorhynchus anatinus*	Left ovary functional
Bats	
Miniopterus natalensis	Left ovary functional
Miniopterus schreibersi	Right ovary functional
Rhinolophus	Right ovary functional
Tadarida cyanocephala	Right ovary functional
Molossus ater	Right ovary functional
Mountain viscacha, *Lagidium peruanum*	Right ovary functional, but after its removal left ovary becomes functional
Water buck, *Kobus defassa*	Left ovary functional

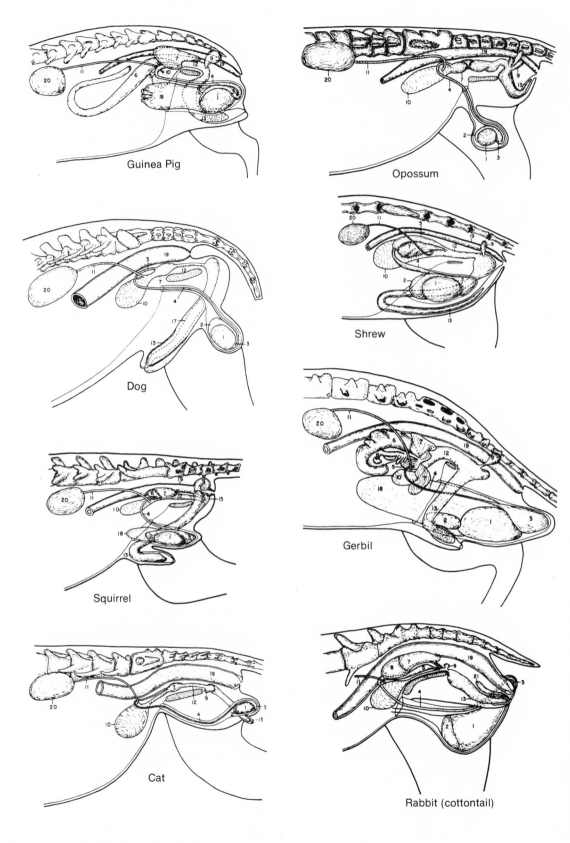

Guinea Pig

Opossum

Dog

Shrew

Squirrel

Gerbil

Cat

Rabbit (cottontail)

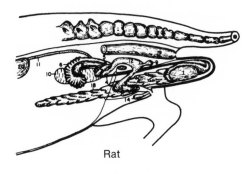

Rat

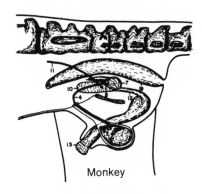

Monkey

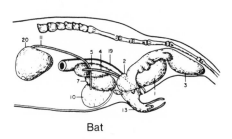

Bat

Figure 5.24 Reproductive organs of different mammalian species *in situ*. 1 = testis, 2 = caput epididymis, 3 = cauda epididymis, 4 = ductus deferens, 5 = ampullary gland, 6 = seminal vesicle, 7 = prostate gland, 8 = anterior lobe of prostate gland, 9 = bulbourethral gland, 10 = bladder, 11 = ureter, 12 = urethra, 13 = penis, 14 = preputial gland, 15 = bulbar gland, 16 = paraprostate gland, 17 = baculum, 18 = fat body, 19 = rectum, 20 = kidney. (From McKeever, 1970; reprinted with permission of Lea & Febiger.)

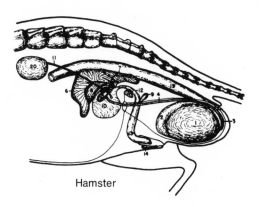

Hamster

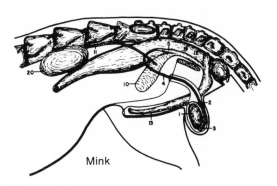

Mink

Table 5.3

Avian species in which 50 percent or more
of the individuals examined have two functional ovaries

Species	Number examined*	Incidence of two ovaries
Apteryx australis (kiwi)	2	2/2
Podiceps cristatus (great crested grebe)	4	2/4
Fulmarus glacialis (fulmar)	2	2/2
Cathartes aura (turkey vulture)	5	4/5
Circus aeruginosus (marsh harrier)	6	4/6
C. approximans (swamp harrier)	17	16/17
C. pygarus	5	4/5
C. cyaneus cyaneus	13	13/13
C. cyaneus hudsonius (marsh hawk)	31	22/31
C. macrourus	7	5/7
Accipiter gentilis atricapillus	14	13/14
A. cooperi (Cooper's hawk)	3	3/3
A. fasciatus	5	5/5
A. nisus	30	20/30
Ieracidea berigora	2	2/2
Buteo buteo (common buzzard)	11	6/11
B. jamaicensis (red-tailed hawk)	15	12/15
B. platypterus	5	4/5
Falco columbarius (pigeon hawk)	4	2/4
F. vespertinus	9	6/9
F. tinnunculus (European kestrel)	20	12/20
Eclectus roratus	4	2/4
Bubo bubo (eagle owl)	7	7/7
Strix aluco (tawney owl)	4	4/4

*Data are not included if findings on one bird only were listed
and if the number of birds examined was not given.
Modified from Kinsky, 1971.

lutea,''* which originate either from follicular
cells or from fluid-filled cysts (Dodd, 1972).
There are no gonoducts in either lampreys or hag-

---

*''Corpora lutea'' in quotation marks will be used for
structures originating from ruptured follicles and resembling mammalian corpora lutea, but not known to secrete
progestins.

---

fishes, and the gametes are shed in the body cavity
and leave via an orifice secondarily connecting the
coelom with the cloaca, in *Lampetra*, and through
a single median orifice between anus and urinary
opening in *Myxine* (Baer, 1964).

Elasmobranchs

The ovaries originally are paired in all species,
but during development they may become asymmetric. The left ovary atrophies, for instance, in
the sharks (*Scyliorhinus, Pristiophorus, Car-
charinus, Galeus, Mustelus,* and *Sphyrna*)
(Wourms, 1977), although both oviducts are present. Among the rays in the viviparous* genus
Urolophus, the left ovary is functional and the
right is nonfunctional, but each has an oviduct.
However, in the ovoviviparous ray, *Dasyatis*,
both the right ovary and the right oviduct are absent. In the skates, which are oviparous, both ovaries and both oviducts are functional.

The ovaries are suspended by mesovaria. They
are usually naked, i.e., *gymnovaria,* and the follicles develop from the germinal epithelium covering the ovary. However, the hollow ovary of the
basking shark, *Cetorhinus maximus,* which
weighs about 12 kg (the animal weighs about 3–4
tons), is invested by a fibrous coat. In this shark
the germinal epithelium forms a network of tubules which opens into a pocket into which the
oocytes are released. They then travel through the
peritoneum to the oviduct. This ovary is exceptional and should, according to Wourms (1977),
not be considered as typical for the viviparous
condition in sharks.

The ovary contains: (1) follicles, which in the
oviparous *Scyliorhinus canicula* are quite large,
but in the viviparous basking shark are not more
than 5 mm in diameter (Dodd, 1972) at most; (2)
corpora atretica, which are derived from egg-con-
taining follicles that have degenerated (Dodd,
1972); (3) ''corpora lutea'' which are ruptured
follicles that have reorganized into structures that
resemble mammalian corpora lutea; (4) corpora
lutea found, e.g., in the ovoviviparous *Squalus
acanthias* in which the granulosa shows evidence

---

*Wourms (1977) states that *Urolophus* is viviparous, but
Breder and Rosen (1966) state, ''The rays proper are all
ovoviviparous.''

of 3β-HSD activity (Lance and Callard, 1969); and (5) "functional corpora atretica", a name here used for lack of a better one, which are derived from atretic follicles but show luteinization and the presence of steroids. Functional corpora atretica are found in the ovoviviparous electric rays, *Torpedo marmorata* and *T. ocellata*. The ruptured follicles, "corpora lutea," do not show evidence of steroidogenesis (Chieffi, 1962) in these rays.

The (Müllerian) oviduct has four regions:

1. *Funnel* (anterior ostium tubae). This collects the ovulated oocytes. It is formed either by fusion of the anterior end of the left and right oviduct, as in the basking shark, or by asymmetric development of one primitive funnel, as in *Scyliorhinus* (Wourms, 1977).

2. *Shell gland or nidamental gland.* This structure is well developed in oviparous and ovoviviparous species, but may be vestigial in viviparous ones. It is a tubular gland which secretes albumin, mucus, and in those species producing an egg case, egg case proteins. In *Scyliorhinus,* sperm are stored in this part of the oviduct (Wourms, 1977).

3. *Isthmus.* This connects the shell gland to the uterus.

4. *Uterus.* This may play a role in the nutrition of the embryos. The left and right uterus may fuse posteriorly and open into a common vagina, or they may open separately into the cloaca. The common vagina, if present, also opens into the cloaca. In some cases a hymen is present at the posterior end of the oviduct (Dodd, 1972).

Holocephali

In the chimaera, *Hydrolagus colliei*, two ovaries and two oviducts are present. In general, the reproductive system resembles that of the elasmobranchs, except that in the Holocephali, the two oviducts always have one common funnel, and each oviduct has a separate opening into the urogenital sinus lateral to the excretory duct (Harder, 1975), whereas in the elasmobranchs the oviducts open via either a common vagina or separate vaginas into the cloaca. The uterus contributes to the formation of the egg case in *H. colliei,*

but does not do so in elasmobranchs (Wourms, 1977).

Chondrostei

The Acipenseridae (sturgeons) have a vestigial connection between ovary and kidney. The ovary starts out as a ribbonlike structure, as it does in teleosts, but it rolls up laterally and dorsally, so that it forms a hollow organ (Giersberg and Rietschel, 1968). The Müllerian oviduct is continuous with the ovary and opens into the cloaca next to the Wolffian duct (Baer, 1964; Harder, 1975). The oocytes, after release from the follicle, pass through the Müllerian oviduct to the cloaca.

Teleosts

The ovaria start as ribbonlike structures (Figure 5.25A), but they roll up in a lateral and dorsal direction, so that the side at which ovulations oc-

Figure 5.25 Ovaries and oviducts of teleosts. A_1–A_3 = cross section through ovaries: A_1 = ribbonlike ovary (original condition), A_2 = ovary rolled up on itself (entovarial), A_3 = ovary rolled against body wall (parovarial). B = ovary and oviducts of salmonidae. 1a = ovarium, 1a' = lamella of ovary, 1b = oviduct, 1c = genital funnel, 1c' = opening of genital funnel, 1c" = genital pore, 2 = coelom cavity, 2' = coelom pore, 3 = intestine, 3a = mesentery, 3a' = opening of 3a via 1c' (arrow). (From Giersberg and Rietschel, 1968; reprinted with permission of VEB Gustav Fischer Verlag.)

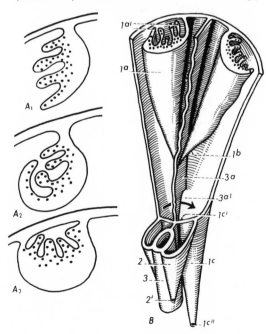

cur is turned inward and surrounds a tube, the oviduct. The oviduct is called *entovarial* (Figure 5.25A$_2$) when the free side of the ovary is against ovarian tissue, but *parovarial* (Figure 5.25A$_3$) when the free side of the ovary is against the wall of the peritoneal cavity. The oviducts are thus extensions of these peritoneal folds, which enclose the ovary during its development (Hoar, 1969).

The ovaries are usually paired, hollow organs suspended by mesovaria. The left ovary of the smelt (*Osmerus eperlanus*) is located anteriorly and the right ovary posteriorly (Harder, 1975). Only one gonad primordium develops in the Japanese rice fish (*Oryzias latipes*) and the guppy (*Poecilia reticulata*). In the perches (*Perca* spp.), the ovaries are fused completely, whereas in the pike perch (*Lucia,* formerly *Stizostedion, perca*), they are fused only caudally. The ovaries are fused medially in the stone loach (*Noemacheilus*

sp.) and in the European bitterling (*Rhodeus amarus*). Oocytes and embryos develop in the wall of the ovary (Lambert, 1970).

The teleost ovary may contain follicles, corpora atretica, pre- and postovulatory ''corpora lutea'' and pre- and postovulatory corpora lutea. The follicular wall may, in addition, form (1) a calyx from the theca closing the ovarian wall after rupture of the follicle, (2) a calyx nutricius from the follicular wall that secretes nutritious material (embryotrophe) for the embryo either into the ovary or into the follicle containing the embryo. Figure 5.26 illustrates the fate of an ovarian follicle in the European bitterling. The function of the follicle will be discussed in more detail in Chapter 7.

The oviducts, which, as noted, are *not* Müllerian ducts but peritoneal folds that have grown backward to form oviducts, are absent in some species. In the Salmonidae (salmons and trouts), the anterior part of the oviduct has disappeared,

Figure 5.26   Fate of the ovarian follicle of the European bitterling, *Rhodeus amarus*. (From Bretschneider and Duyvené de Wit, 1947; reprinted with permission of Elsevier Publishing Co.)

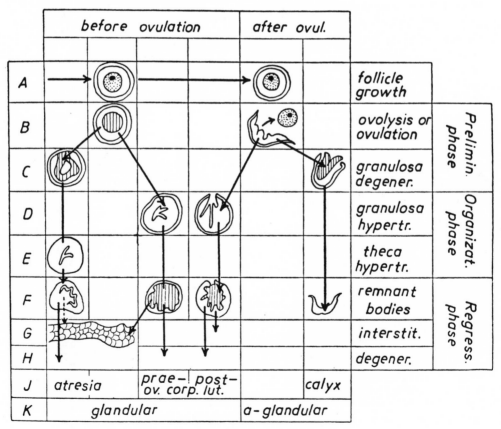

and the oocytes fall into the body cavity and then reach the funnel-like remnant of the posterior part of the oviduct (Figure 5.25B). In the Galaxiidae, Glyodontiidae, Notopteridae, Osteoglossidae, and Anguillidae, and in *Misgurnus,* the oviducts have completely disappeared, and the gametes leave the body cavity via a genital pore, which is not, however, homologous with the abdominal pore found in many sharks (Harder, 1975).

The genital opening may be in various locations. The midgullar position of the female's genital opening in the Phallostethidae is adapted to the specialized penis of the male. Teleosts have some specialized structures and associated behaviors, e.g., the ovipositor of the bitterling (*Rhodeus amarus*), a skinlike structure derived from the genital papilla. In this species, the growth of the ovipositor and the nuptial color of the male require the presence of a fresh water mussel, *Unio pictorum* or *Anodonta intermedia** (Banarescu, 1973). Males select the female with the largest ovipositor. At spawning, the females thrust their ovipositor into the *excurrent* syphon of the mussel and thus have to push their oocytes upstream (Wunder, 1931). Breder and Rosen (1966) state that, exceptionally, the female lays her oocytes in the incurrent syphon. The male subsequently spawns near the *incurrent* syphon, or the sperm would never enter the mussel (Wunder, 1931). The oocytes are fertilized inside the mussel and they develop in its gill chambers. After about a month the embryos hatch and the young bitterlings leave the mussel. The larvae of the mussel attach themselves to the gills and fins of the young fish and live there as parasites for a time (Wunder, 1931).

The Jenynsiidae and Anablepidae have their genital opening either on the left or right, corresponding to the males having a left or right gonopodium. Among the Jenynsiidae, three-fifths of the males are dextral and two-fifths sinistral,

---

*H. E. Evans has caught large numbers of the European bitterling at the Sawmill River near New York City. This species may have been released by some fancier or by a company discontinuing the use of this animal for bioassays. Apparently, therefore, the animals can use other mussel species than the ones mentioned above, for these species do not occur in the Sawmill River (personal communication, H. E. Evans).

whereas for the females the ratio is just the reverse (Breder and Rosen, 1966).

Amphibians

The ovaries are paired, lobed, hollow organs consisting mainly of a cortex. The outside of the ovary is covered with germinal epithelium; the cavity is lined with cells of medullary origin.

Anura. The ovaries of the Anura, which are largely oviparous, have follicles and scars from ruptured follicles, corpora atretica and "functional corpora atretica," that is, structures formed from nonovulated follicles, but resembling corpora lutea. In the viviparous toad *Nectophrynoides occidentalis,* Xavier (1974) identified corpora lutea formed from ruptured follicles. These corpora lutea contained cholesterol, showed a positive reaction for 3β-HSD, and were capable of synthesizing progesterone from pregnenolone. The quantity of progesterone synthesized paralleled the size of the corpora lutea, and both decreased as pseudopregnancy (ovulation without mating) progressed beyond four months.

Ruptured follicles appear to have a steroid secretory function in *Rana esculenta* and *R. cyanophlyctis,* which are oviparous, and in some ovoviviparous species. The females of some species carry the fertilized eggs either in a pouch, e.g., *Gastrotheca marsupiata* and *G. pygmaea,* or in separate chambers, e.g., *Pipa pipa* and *Cryptobatrachus evansi* (Figure 5.27).

The developing embryos probably derive nutrients from the mother, as we shall discuss in Chapter 10.

Urodeles. The ovaries resemble those of the Anura, with follicles, corpora lutea, and corpora atretica, which originate from follicles in which the yolk has been resorbed. According to Lofts (1974), there is little or no evidence that these structures secrete steroids. Corpora lutea, which result from a reorganization of the ruptured follicle into a progestin-secreting structure, are found in *Triturus cristatus* and *Salamandra salamandra* (Lofts, 1974).

The (Müllerian) oviducts of amphibia have three regions: the pars recta, the pars convoluta,

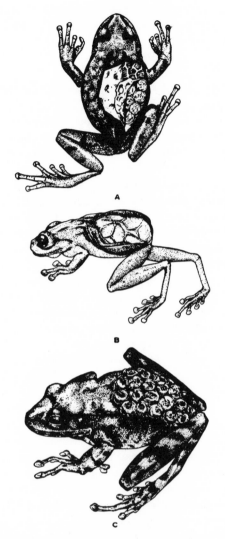

Figure 5.27 The brood pouch in females of different species of amphibians. *A, Gastrotheca marsupiata; B, Gastrotheca pygmaea,* eggs removed from the brood pouch; *C, Cryptobatrachus evansi.* (From Noble, 1954; reprinted with permission of Dover Publications, Inc.)

and the pars uterina (Lofts, 1974). The oviducts end in the cloaca, which in the urodeles, shows some specialization. In salamanders, three sets of glands are associated with the oviduct: (1) *pelvic glands,* which are located in the roof of the cloaca, and serve as reservoirs for the storage of spermatozoa; (2) *cloacal glands,* present in all ambystomids, salamandrids, and the primitive plethodontids (Noble, 1954), which may play a role in formation of the egg case; and (3) *abdomi-*

nal glands, present in *Ambystoma, Necturus,* and *Eurycea,* which are apparently rudimentary (Noble, 1954).

In the urodeles with internal fertilization, the female cloaca may have well developed cloacal lips, with which the spermatophores can be picked up from the ground (see Chapter 9).

Reptiles

The ovaries are paired, hollow organs, suspended by mesorchia and with squamous epithelium lining the ovarian cavity. The follicles project from short stalks, and the walls, which surround the large yolk, consist of a granulosa layer, a theca interna, and a theca externa. Corpora lutea are formed after ovulation as the result of the luteinization and proliferation and/or hypertrophy of the granulosa cells and supporting connective tissue from the thecal layers. In oviparous species, the corpora lutea regress soon after oviposition (Fox, 1977). In the lizard, *Anolis carolinensis,* follicles that have become atretic form corpora atretica. The corpora atretica seem to determine the response of the ovarian follicles to exogenous and probably endogenous gonadotrophins (Crews and Licht, 1975). After ovulation, the oocytes may according to Bellairs (1970), wander through the body cavity and enter either of the two oviducts.

Abdominal fat bodies have been found in all temperate zone reptiles that have been examined and in some anoline lizards of warmer regions (Fox, 1977). The lipids from these bodies pass to the liver and are incorporated into yolk precursors. In the lizard, *Uta stansburiana,* removal of the fat bodies inhibits follicular development or, when the follicles are developed, causes atresia of the developed follicles. There appears to be a reciprocal relationship between ovary and fat bodies, because ovariectomy prevents rapid fat mobilization from the fat bodies (Fox, 1977).

The oviducts, which are Müllerian ducts, are convoluted. They consist of (1) an infundibulum or funnel; (2) a convoluted segment with a layer of longitudinal and a layer of circular muscle, and a mucosa with many alveolar glands and ciliated epithelium (Fox, 1956; Guibé, 1970); (3) an isthmus; (4) a uterine segment, which has a strong musculature and is lined with ciliated and nonciliated columnar epithelium; and (5) a vagina which has mucous glands, and traverses the cloaca.

The cloacal openings for the paired oviducts are separate in *Lacerta viridis, Sceloporus* sp., *Hemidacytylus* sp., *Phrynosoma* sp., but united in *Gerrhonotus* sp. and *Cnemidophorus* sp. In the blind, limbless lizards (*Anniella*) both ovaries are functional and of the same size, but only the right oviduct is functional. The left oviduct is smaller than the right oviduct and has numerous glands, the secretions of which probably pass out via the cloaca (Coe and Kunkel, 1905). In blind worm snakes (Typhlopidae), the left oviduct is absent.

As we will see in Chapter 9, in some reptiles the sperm retain their fertilizing capacity for a considerable length of time after insemination. It is not surprising, therefore, that there are structures for storing sperm in the oviduct. In snakes, turtles and lizards, sperm are retained in ''seminal vesicles,''* which are specialized alveolar glands at the base of the infundibulum. From these glands, ducts lead into the oviductal lumen (Fox, 1977). In the snake, *Thamnophis elegans terrestris,* the sperm are retained in the posterior two-thirds of the oviduct, and the sperm do not reach the ''seminal vesicles'' until February–March, if the snake mates in the fall. If insemination occurs in the spring, however, the sperm reach the ''seminal vesicles'' in April. In the adder, *Vipera aspis,* the sperm remain in the vagina after fall mating and proceed to the ''seminal vesicles'' in the spring just before they fertilize the oocytes (Fox, 1977).

Iguanids have ''seminal vesicles'' in the anterior segment of the vagina, whereas gekkonids have these structures between uterus and infundibulum. The seminal receptacles, found in the caudal part of the albuminous region of the oviduct of the box turtle (*Terrapene carolina*), are morphologically unspecialized glands (Fox, 1977).

Birds

Most avian species have only a left ovary and a left oviduct, but some species have been found to

have two functional ovaries (Table 5.3), and in most of these species, it is the left oviduct that is functional. Oocytes ovulated from the right ovary will, in the absence of a right oviduct, drop into the body cavity and will be absorbed, since a mesentery separates the right ovary from the left oviduct, so that an oocyte from the right ovary cannot be picked up by the left oviduct.

The ovary is attached to the body wall by a mesovarium. It has many follicles embedded in a sparse stroma of connective tissue. During the laying season, there is a series of large follicles, which show a hierarchy of size (Fig. 5.28).

The follicular wall consists of a granulosa layer, a theca interna, and a theca externa. The large follicles are filled with yellow yolk; some of the smaller follicles are filled with either yellow or white yolk, depending on the stage of development. On the wall of the large follicles is a clear band, the stigma, which macroscopically is devoid of blood vessels and along which the follicle ruptures.

In addition to the large and developing follicles, one finds atretic and ruptured follicles. In atretic follicles, the yolk is reabsorbed, and the granulosa cells become dissociated from the follicular wall and move into the follicles (Erpino, 1973). Atretic follicles are not always present in laying chickens, and ruptured follicles rapidly disappear after ovulation. In laying chickens, they are difficult to find five days after their rupture. There are no corpora lutea in birds.

Usually one (the left) oviduct, which is a Müllerian duct, is present. By inbreeding, however, Morgan and Kohlmeyer (1957) were able to obtain chicken strains with two oviducts (and one ovary). The oviduct (Aitken, 1971; King and McLelland 1975; Romanoff and Romanoff, 1949) has five anatomical and functional regions (Figure 5.29).

1. *Infundibulum.* This region consists of a funnel and a tubular part. In the sexually mature chicken, the infundibulum is about 7–9 cm long. It has two muscle layers—a circular and a longitudinal layer—and a few muscle cells scattered in bundles. The wall of the funnel is thin and has low mucosal folds. The tubular part has a thicker wall and high mucosal folds. This part of the oviduct

*Quotation marks are used because the term *seminal vesicle* usually refers to an accessory sex organ of males.

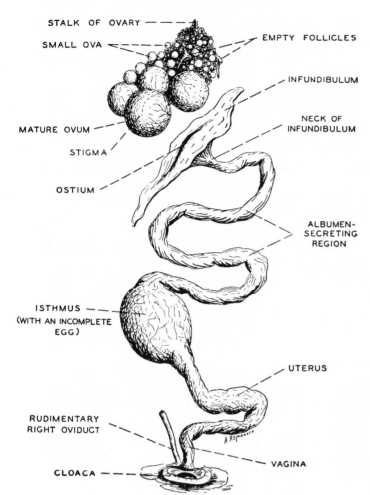

Figure 5.28 Reproductive organs of the hen. (From Romanoff and Romanoff, 1949; reprinted with permission of John Wiley & Sons, Inc.)

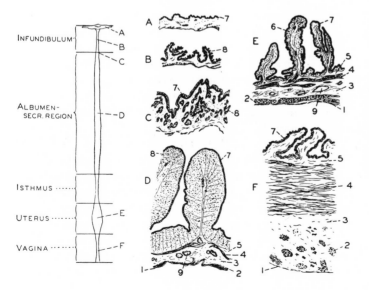

Figure 5.29 Transverse sections through different parts of the oviduct of the adult chicken. A = lips of infundibulum, B = neck of infundibulum, C = transition region from infundibulum to magnum, D = magnum or albumen secreting region, E = walls of the uterus, F = walls of the vagina. 1 = Peritoneal membrane, 2 = longitudinal muscle fibers, 3 = connective tissue, 4 = circular muscle layer, 5 = inner layer of connective tissue, 6 = thick layer of convoluted tubular glands, 7 = epithelium of the duct, 8 = ducts of the convoluted glands, 9 = blood vessels. (From Romanoff and Romanoff, 1949; reprinted with permission of John Wiley & Sons, Inc.)
Note: hen can be any female bird

secretes thick albumen around the yolk during its passage, a passage that takes 15–30 minutes (Aitken, 1971).

2. *Magnum or albumen-secreting region.* This, in the adult chicken, is about 32–34 cm long, white, and highly coiled. It has thick walls with longitudinal circular layers of muscle cells and a longitudinally folded mucous membrane lined with high, tubular, glandular, ciliated epithelium. The glandular epithelium gives this region of the wall its thickness. Thick albumen is secreted in this region around the yolk as it traverses the magnum (in about 2–3 hours).

3. *Isthmus.* This region is about 10 cm long in the adult chicken and is separated from the magnum by a clear translucent band of nonglandular tissue. The isthmus has a thicker circular muscle layer than the infundibulum and the magnum, and its glandular development is less marked than that of the magnum. The glands are tubular and secrete the proteinaceous egg membranes around the yolk and its surrounding thick albumen, as the egg is transported through the isthmus (in about 1 hour).

4. *Shell gland or uterus.* This, in the adult chicken, is about 11 cm long and has very thick muscular walls. It has a prominent longitudinal muscle layer and is lined with tubular and unicellular goblet-type cells. The cells of this lining secrete the thin albumen, the shell, and any pigments that may color the shell. In the chicken, the egg remains in this region 20–26 hours.

5. *Vagina.* This is about 10 cm long; it is separated from the shell gland by a sphincter muscle and ends in the cloaca. It has a longitudinal muscle layer that is easily distinguished and a circular muscle layer that is well developed. The mucous membrane has low narrow folds and is lined with ciliated and nonciliated cells. There are no glands and no secretion in the vagina proper (Aitken, 1971); the egg is "finished" in the shell gland.

In addition to the glands just mentioned in connection with egg formation, there are two areas where spermatozoa are stored and can retain their fertilizing capacity (about 7–14 days in chickens and 40–50 days in turkeys as we shall see in Chapter 9). These two regions are at the extreme ends of the oviduct, i.e., at the infundibulum and at the utero-vaginal junction (Aitken, 1971).

The ovary receives its blood from the gonadrenal artery, which is a branch of the dorsal aorta that sends a branch, the ovarian-oviductal artery, to the ovary. The blood is drained by a cranial ovarian-oviductal and a caudal ovarian vein, each of which drain into the vena cava. The venous system of the ovary is very well developed (Nalbandov and James, 1949).

The innervation of the ovary is complex; adrenergic and cholinergic fibers have been found. The adrenergic supply originates from the 5th, 6th, 7th, sometimes the 4th thoracic, and the 1st and 2nd lumbar ganglia of the sympathetic chain (Gilbert, 1971). The origin of the parasympathetic fibers is not known (Gilbert, 1971).

Mammals

Ovary. The ovaries are usually paired, but in some species (Table 5.2) one may be inactive; that is, either it does not ovulate or it ovulates only after the active ovary has been removed. For example, if the right ovary of the mountain viscacha (*Lagidium peruanum*) is removed, the left one becomes functional. The right ovary is also the functional one in the following bats: *Miniopterus schreibersi, Rhinolophus* sp., *Tadarida cyanophala* and *Molossus ater.* The left ovary is dominant in the bat, *Miniopterus natalensis,* and in the waterbuck, *Kobus defassa* (Wimsatt, 1975).

Each ovary is suspended from the dorsal body wall by the mesovarium or secondarily by the broad ligament. In many species (Table 5.4), the ovary lies in a membranous sac called the *ovarian bursa,* which is a fold of the mesosalpinx, the mesentery of the oviduct. The function of the bursa has not been determined (Mossman and Duke, 1973a).

The ovary consists of the stroma, the germinal epithelium, the germ cells and their associated structures, interstitial cells, and various vestigial structures. The germinal epithelium, which covers the ovary, may be either directly on the ovarian stroma (e.g., in shrews) or it may be on a fibrous ovarian tunica albuginea (Mossman and Duke, 1973a). It may consist of squamous or low columnar cells. The germ cells and their associated structures include oocytes, follicles in different stages of development, connective tissue, corpora

Table 5.4

Occurrence of ovarian bursa in mammals

Animals	Anatomy of bursa					Animals	Anatomy of bursa				
	1	2	3	4	5		1	2	3	4	5
Prototheria						Rodentia (*cont.*)					
Tachyglossus	+					Hystricidae		+			
Ornithorhynchus				+		Caviidae			+		
Methatheria						Dasyproctidae					+
Phalangeridae				+		Cetacea					+
Phascolomidae			+			Carnivora					
Macropodidae				+*		Canidae	+				
Eutheria						Ursidae	+				
Insectivora						Procyonidae	+				
Tenrecinae	+					Mustelidae	+				
Erinaceinae	+					Viverridae	+	+			
Soricinae	+					Hyaenidae		+			
Talpidae	+					Felidae		+	+		+
Chiroptera	+					Pinnipedia		+			
Primates						Hyracoidea					
Lemuridae			+			Procaviidae		+			
Hominidea			+	+	+	Perissodactyla					
Edentata						Equidae				+	+
Bradypodidae	+					Artiodactyla					
Dasypodidae		+				Suidae		+			
Lagomorpha						Camelidae	+				
Lepus europaeus					+	Cervidae			+	+	
Oryctolagus cuniculus		+				Bovidae		+			
Rodentia											
Sciuridae		+									
Castoridae		+									
Cricetidae	+										
Muridae	+										
Gliridae					+						
Dipodidae	+										

*Ovary end used by infundibulum.
1 = Complete bursa, with participation of infundibulum.
2 = Incomplete bursa, with participation of infundibulum wide opening to peritoneal cavity.
3 = Incomplete bursa without participation of infundibulum.
4 = Peritoneal pocket only; ovary lies outside this pocket.
5 = No peritoneal pocket.
From Strauss, 1964; reprinted with permission of Walter de Gruyter & Co.

lutea, old corpora lutea called *corpora albicantia,* and interstitial tissue.

1. *Follicles.* These can be characterized as small, medium and large follicles, according to Pedersen and Peters (1968). The main distinctions, as illustrated in Figure 5.30, are made on the basis of oocyte size, number of layers of follicular cells, and the presence of follicular fluid. The follicular wall has a granulosa layer, a theca externa, and a theca interna. The oocyte is surrounded by an acellular layer, the zona pellucida, which separates it from the surrounding follicular cells. In the large follicles, the oocyte is surrounded by granulosa cells, which form a struc-

ture, the cumulus oophorus, which connects it to the follicular wall. Immediately surrounding the zona pellucida is a layer of columnar or pseudo-stratified columnar cells, which is called the *corona radiata* (Mossman and Duke, 1973a).

2. *Atretic follicles.* These are degenerating follicles. The first sign of atresia is the pyknosis of the epithelial follicular cells, which are adjacent to the lumen of the large follicles and in layers just external to the corona radiata of secondary follicles (Mossman and Duke, 1973a). Subsequently, the cumulus oophorus shrinks, the granulosa cells disappear, and fibroblasts and leukocytes invade the atretic follicle.

3. *Corpora lutea.* These originate from rup-

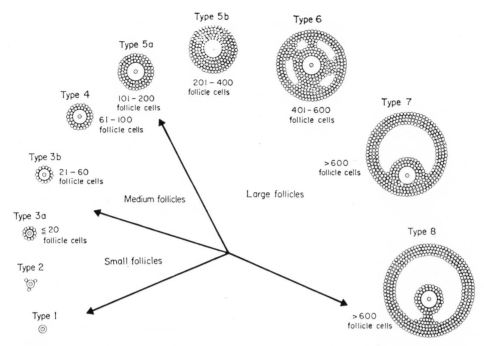

Figure 5.30 Classification of ovarian follicles of the mouse. (From Pedersen and Peters, 1968; reprinted with permission of T. Pedersen and *Journal of Reproduction and Fertility*.)

Inside the figure:

Type 5b
201 – 400
follicle cells

Type 6
401 – 600
follicle cells

Type 5a
101 – 200
follicle cells

Type 4
61 – 100
follicle cells

Type 7
>600
follicle cells

Type 3b
21 – 60
follicle cells

Medium follicles

Large follicles

Type 3a
≦ 20
follicle cells

Type 8

Type 2

Small follicles

Type 1

>600
follicle cells

tured follicles and are formed by enlargement and multiplication of the granulosa cells. The corpora lutea consist of a solid mass of polyhedral cells. Mossman and Duke (1973a) proposed the term ''luteal gland'' in preference to corpus luteum, but the use of the term *corpus luteum* is of such long standing that substitution of a new term seems to have little chance of adoption. The corpora lutea of the Madagascar hedgehog (*Setifer setosus*) develop from a follicle in which no antrum or cavity is formed. The oocyte is fertilized inside the follicle, and the zygote is extruded by contractions of the follicular sheath and a shift in position of the granulosa cells, so that the zygote becomes exposed on the surface of the follicle (Strauss, 1966; Mossman and Duke 1973a,b).

4. *Accessory corpora lutea.* These structures can have two different origins: (1) granulosa cells of atretic follicles, as in the North American porcupine (*Erithizon dorsatum*) and in the primates, *Tupaia* sp. and *Urogale* sp., or (2) the epithelium of medullary cords, as in juvenile eastern chipmunks (*Tamias striatus*), in which medullary follicles ovulate and form large corpora lutea.

5. *Corpora albicantia.* These are small masses of connective tissue which are the remains of degenerated corpora lutea.

6. *Thecal glands* (Mossman and Duke, 1973a). These consist of a layer or zone of glandular cells adjacent to the granulosa cells of mature follicles.

7. *Interstitial glands.* These may originate from the ovarian stroma, medullary cord cells, and the theca interna of atretic follicles. Mossman and Duke (1973b) stress the inaccuracy of calling these structures ''luteinized follicles.''

Some structures in the ovary are vestigial. The *rete ovarii* plays a role in oogenesis, as will be discussed in Chapter 7. It is a regular network of spaces lined with a simple low columnar epithelium. It is well developed in the tarsier, *Tarsius* sp. (a primate) and in pangolins, *Manis* sp. (Mossman and Duke 1973a). The *epoophoron* consists of vestiges of efferent ductules and is homologous with the epididymis of the male. *Gartner's duct* or the duct of the epoophoron is a vestige of the ductus deferens. The *paraoophoron* consists of the more rostral and the more caudal mesonephric tubules—or ductuli aberrantes. It is homologous with the (also vestigial) paradidymis of the male.

The ovary receives blood from the ovarian artery, a branch of the dorsal aorta, and blood leaves the ovary via the ovarian vein, which drains into the caudal vena cava. Some branches of the ovarian artery and vein anastomose with ovarian branches of the uterine blood vessels (Mossman and Duke, 1973a).

The ovary receives fibers from the vagus nerve and also adrenergic fibers. As we shall see in Chapter 7, adrenergic mechanisms are involved in ovulation.

Oviduct. The oviduct is suspended by the mesosalpinx on one side; the other side is attached to the mesotubarium superius, which connects the oviduct to the fimbria (Figure 5.31). The fimbria consist of a fringe of irregular processes (Hafez, 1970). The relationship between oviduct and mesenteries can vary considerably among species. Beck and Boots (1974) give a comprehensive summary of the various relationships in a large number of species.

Starting at the cranial end, the oviduct consists of: (1) infundibulum, with a fimbriated, funnel-like opening, (2) the ampulla, (3) the isthmus, (4) the utero-tubal junction. The wall of the oviduct consists of several layers: The *tunica serosa,* the outermost layer is a thin layer of connective tissue covered by a single layer of squamous epithelium (Beck and Boots, 1974). The *tunica muscularis* consists of smooth muscle, which may be ar-

ranged in four distinct patterns, i.e., (1) a predominant circular layer of smooth muscle without distinguishable longitudinal layers, (2) thin longitudinal layers on both sides of a thicker circular layer, (3) a thick circular layer with an outer longitudinal muscle layer, and (4) a circular layer with an inner longitudinal muscle layer (Beck and Boots, 1974). The *tunica mucosa,* the innermost layer, consists of a lamina propria of connective tissue and a lamina epithelialis, which has ciliated cells and three types of nonciliated cells: secretory cells, wedge-type cells called "peg cells," and basal cells along the base of the epithelium (Beck and Boots, 1974).

In species such as pigs, mice, rats, humans and rhesus monkeys (Beck and Boots, 1974; Brenner and Anderson, 1973; Hafez, 1973b) the presence of ciliated cells depends on the stage of the reproductive cycle; however, the occurrence of cyclic changes in cilia is controversial (see Beck and Boots, 1974 for review).

The anatomy of the utero-tubal junction varies among species. Beck and Boots (1974) described ten different morphological types of utero-tubal junctions. Type 1–3 lack mucosal projections of oviductal or uterine origin. Type 4–10 have specialized mucosal projections of oviductal or uterine origin. The reader is referred to this excellent review by Beck and Boots (1974) for more details. The oocytes or the zygotes are transported rapidly through the utero-tubal junction, which may play a role in sperm transport (Hafez, 1973a).

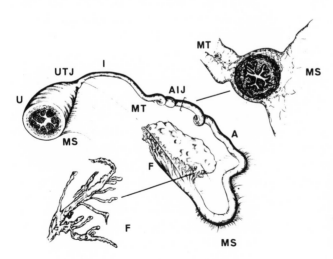

Figure 5.31 Diagram of the rabbit oviduct. A = ampulla, AIJ = ampullary-isthmic junction, F = fimbriae, I = isthmus, MS = mesosalpinx, MT = mesotubarium, U = uterus, UTJ = uterotubal junction. (From Hafez, 1973b; reprinted with permission of E. S. E. Hafez and the American Physiological Society.)

Uterus. The uterus is suspended by the mesometrium. There are different types of uteri which are illustrated in Figure 5.32. In the prototheria (Fig. 5.32A), the uteri are paired and open into the urogenital sinus; no vagina is present. In the metatheria, or marsupials, there are 2 vaginae, 2 cervices, 2 uteri (Figure 5.32). Sharman (1976) has illustrated some of the variations of this general plan and two such variations are illustrated in Figure 5.32 B_1 and B_2. The kangaroo (*Macropus* sp.) has an open midline vagina after the first parturition (Figure 5.32 B_2), so that it resembles the Eutherian midline vagina, but it has been formed in a different manner (Sharman, 1976).

Among the Eutheria one finds four kinds of uteri: (1) The duplex uterus (Figure 5.32 C_1) found in the rabbit has 1 vagina, 2 cervices, and 2 uterine horns. (2) The bicornuate uterus (Figure 5.32 C_2) found in the pig, has 1 vagina, 1 cervix, and 2 uterine horns. (3) The bipartite uterus found in cattle, sheep, and horses has 1 vagina, 1 cervix and either a small or a prominent uterine body, and 2 separate uterine horns. (4) The simplex uterus (Figure 5.32 C_3) found in most primates, including humans, has 1 vagina, 1 cervix, and 1 uterine body. In humans, various anomalies have been found; for example, all the three different types of uteri listed for the Eutheria (Arey, 1946; Giersberg and Rietschel, 1968; Hafez, 1970), a double uterus and vagina as a result of failure of the Müllerian ducts to fuse, a double uterus and vagina, and a subseptate uterus.

Small variations are found in the four different types of uteri. For example, the sloth (*Bradypus*

Figure 5.32 A = Prototheria: paired uteri, no vagina. B_1 = Metatheria: *Didelphis*, paired uteri, paired vaginae, single sinus vaginalis which ends blindly caudally; B_2 = Metatheria, *Macropus*, as B_1, but the sinus vaginalis is secondarily connection with the urogenital sinus. C_1 = Eutheria, duplex uterus; C_2 = bicornuate uterus; C_3 = simplex uterus; C_4 = uterus simplex with paired vagina as result of sagittal septum (*Bradypus*); C_5 = simplex uterus which ends into urogenital sinus because vagina is lacking (*Dasypus*).

1 = oviduct, 1' = opening of oviduct, 2 = uterus, 2a = body of uterus, 2b = uterine horn, 2c = cervix, 3 = vagina, 3a = paired vaginae, 3b = unpaired vaginae, 4 = urogenital sinus, 4a = vestibule of vagina, 4b = urethra, 5a = urinary bladder, 5b = ureter.

(From Giersberg and Rietschel, 1968; reprinted with permission of VEB Gustav Fischer Verlag.)

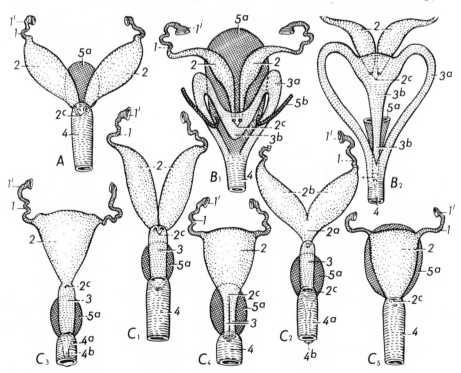

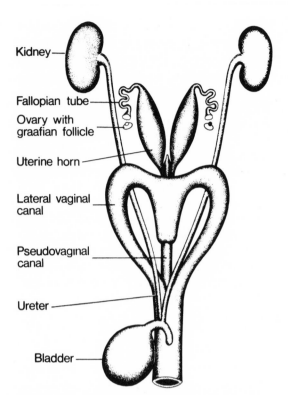

Figure 5.33 Peculiar anatomy of the female reproductive tract in marsupials. Spermatozoa ascend by way of the lateral vaginal canals, and the fetus is born through the central pseudovaginal canal, which is occluded at other times. (From Short, 1972; reprinted with permission of Cambridge University Press.)

Labels: Kidney, Fallopian tube, Ovary with graafian follicle, Uterine horn, Lateral vaginal canal, Pseudovaginal canal, Ureter, Bladder

sp.) has a simplex uterus, one cervix, but a paired vagina as the result of a sagittal septum (Figure 5.32 C_4); the armadillo (*Dasypus* sp.) has a simplex uterus with one cervix opening into the urogenital sinus (Figure 5.32 C_5) (Giersberg and Rietschel, 1968).

Sharman (1976) has pointed out that the peculiar vagina of marsupials (Figure 5.33) is the result of the position of the ureters of marsupials, which grow forward *between* the Wolffian and Müllerian ducts. This prevents the fusion of the vaginae in the midline (because of the intrusion of the ureters). The lateral vaginae serve for transport of the spermatozoa, but the birth of the young occurs via the midline birth canal, which connects the uterine openings *directly* with the urogenital sinus.

Cervix. The cervix is a sphincter at the posterior end of the uterus. Its wall is thick, and the inner wall has ridges. It consists of a serosa, an outer

longitudinal and an inner circular muscle, a submucosa, a mucosa consisting of columnar epithelium at the cranial end and squamous epithelium at the caudal end, and many goblet cells that secrete mucus. The consistency of this secretion varies during the menstrual cycle and is sometimes used to diagnose the time of ovulation in women, as is the crystallization pattern of the mucus.

Vagina. The relation of the Müllerian ducts, urogenital sinus, and the integument in the formation of the vagina is illustrated in Figure 5.34. The location of vaginal and urethral openings in a few species is illustrated in Figure 5.35. The vaginal wall consists of a serosa, an outer longitudinal and an inner circular muscle layer, a submucosa, and a mucosa consisting of stratified squamous epithelium.

In some laboratory species, especially the rat, mouse, and hamster, the stage of the estrous cycle is correlated with differences in the histological appearance of the vaginal epithelium, so that vaginal smears can be used to diagnose the stage of the

Figure 5.34 Contributions of Müllerian ducts, urogenital sinus, and integument in formation of the eutherian vagina. *A*, Hypothetical basic plan; *B*, Human; *C*, Mouse (*Mus* sp.); *D*, Mole (*Talpa* sp.). Dotted line = integument (ectoderm); thin line = intestine and urinary bladder; heavy line = urogenital sinus entoderm; dashed line = Müllerian duct, mesoderm. (From Giersberg and Rietschel, 1968; reprinted with permission of VEB Gustav Fischer Verlag.)

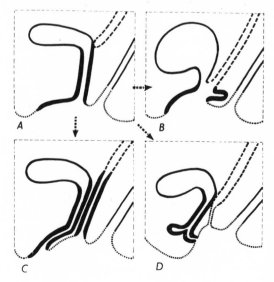

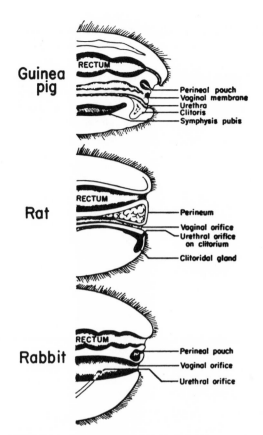

Guinea pig
RECTUM
— Perineal pouch
— Vaginal membrane
— Urethra
— Clitoris
— Symphysis pubis

Rat
RECTUM
— Perineum
— Vaginal orifice
— Urethral orifice on clitorium
— Clitoridal gland

Rabbit
RECTUM
— Perineal pouch
— Vaginal orifice
— Urethral orifice

Figure 5.35 Location of the urethral and vaginal orifices in the abdomen of the guinea pig (anestrous condition; vagina closed by an epithelial membrane), rat, and rabbit. (From Hafez, 1970; reprinted from E. S. E. Hafez, "Female reproductive organs," in *Reproduction and Breeding Techniques for Laboratory Animals,* ed. E. S. E. Hafez, pp. 74–106. Lea & Febiger, Publisher, Philadelphia, 1970.)

Table 5.5

Part of prostate found in female mammals

Species	Part of prostate
Insectivora	
Hemicentetes semispinosus	Dorsal-lateral
Erinaceus europaeus	Dorsal-lateral
Talpa europaea	Ventral
Chiroptera	
Taphozous	Dorsal
Nycteris luteola	Ventral
Coleura afra	Dorsal
Primates	
Macaca mulatta	Dorsal (?)
Cercopithecus aethiops	Dorsal (?)
Homo sapiens	Dorsal
Lagomorpha	
Oryctolagus cuniculus	Dorsal (?)
Sylvilagus floridanus	Dorsal
Rodentia	
Arvicola sapidus	Ventral (?)
Apodemus sylvaticus	Ventral
Arvicanthis cinerus	Ventral
Rattus norvegicus	Ventral
R. rattus	Ventral
Mastomys erythroleucus	Ventral
Mus musculus	Ventral

From Strauss, 1964; reprinted with permission of Walter de Gruyter & Co.

Figure 5.36 Exterior female genitalia of *Hyaena crocuta*. a = anus, b = perineum, c = false scrotum, d = clitoris not in erection and in erection (hatched), e = teat. (From Prasad, 1974; reprinted with permission of Walter de Gruyter & Co.)

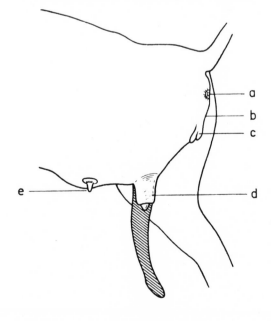

estrous cycle. During proestrus epithelial cells predominate; during estrus there are some remnants of epithelial cells and many large cornified cells; during metestrus there are many cornified and epithelial cells and leukocytes; during diestrus few epithelial cells remain and few to many leukocytes are found.

Prostate. Although the prostate is mostly a male secondary sexual gland, a female prostate is well developed in some species (Table 5.5). In laboratory rats, the "Witschi rat," developed by selective breeding, has a well-developed prostate.

External Genitalia. The external genitalia consist of the labia majora, labia minora, and the clitoris, but only the human has true labia (Hafez, 1970). Generally, the clitoris is poorly developed unless the fetus has been exposed to larger amounts of androgens than normal, e.g., in the adrenogenital syndrome, in the freemartin, or after experimental androgen treatment. However, in the hyaena (*Hyaena crocuta*), the clitoris is very well developed. This extreme development and the anatomy make it almost impossible to distinguish a male from a female, especially since the female also has a false scrotum (Figure 5.36).

Sexual Skin. Some primates have a so-called sexual skin, which is contiguous with the external genitalia. It consists of a highly vascularized edematous or pigmented dermis (Hafez, 1970). Its color and vascularization vary during the reproductive cycle.

6

The Testis

We have discussed the gross anatomy and some of the histological structure of the testis in Chapter 5. In this chapter we will concentrate on the physiological aspects of this organ. The two main functions of the testes are the production of gametes and the production of hormones, which perform important integrative functions, and in some species, are involved in spermatogenesis.

SPERMATOGENETIC FUNCTION

Spermatogenesis involves meiotic divisions as well as mitotic divisions, so that the resulting spermatozoa have n number of chromosomes. The simplest basic scheme of spermatogenesis is as follows:

Gametes	Number of chromosomes per somatic cell	DNA content of the cell
Spermatogonia mitotic multiplication ↓	2n	2C
Spermatogonia maturation ↓	2n	2C
Primary spermatocytes meiosis I ↓	2n*	4C
Secondary spermatocytes meiosis II ↓	n	2C
Spermatids spermiogenesis ↓	n	1C
Spermatozoa	n	1C

*At this stage there are 2n chromosomes, but each chromosome consists of two strands. This explains why the number of chromosomes is 2n, but the DNA content per cell is 4C.

There are two major types of spermatogenesis in vertebrates. In type 1, found in fishes and amphibians, spermatogenesis consists of the development of spermatogenic clones within cysts, which are located in tubular or lobular compartments. In type 2, found in amniotes (reptiles, birds, and mammals), spermatogenesis takes place in the seminiferous tubules, in several layers of epithelium, and there are no subcompartments (Roosen-Runge, 1977). In type 1 spermatogenesis, the cells within a cyst are all in the same stage of development (Fig. 5.4). The primary spermatogonia lie against the basal membrane and multiply several times in succession by mitotic divisions, as do the follicle cells, so that one follicle may contain several secondary spermatogonia. The secondary spermatogonia within a follicle divide synchronously, and after about 8 mitotic di-

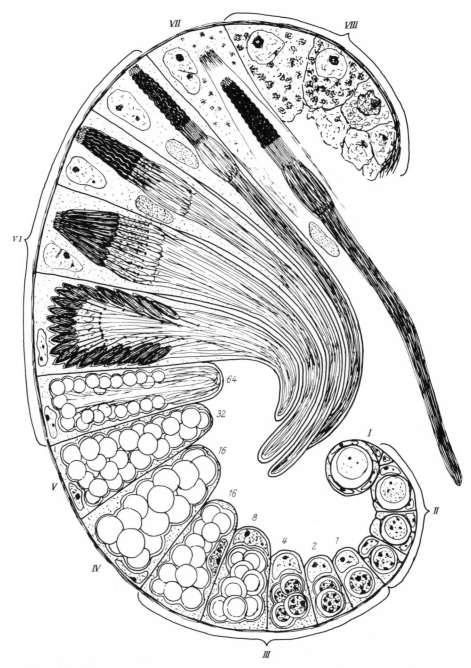

Figure 6.1 Diagram indicating developmental history of single spermatocyst [in a shark]. Development proceeds from lower right (I) around to the upper right (VIII). *I* indicates a newly formed seminiferous follicle. The segment marked *II* indicates the period of mitotic proliferation of spermatogonia and of follicle cells as separate individuals. *III* marks the period in which spermatogonial mitoses continue after engulfment by a follicle cell. The arabic numeral at the inner end of each spermatocyst indicates the number of germ cells contained within. Number 1 corresponds to the formation of a spermatocyst by engulfment of a single spermatogonium by a single follicle cell. The next developmental stage of a spermatocyst after proliferation of 16 spermatogonia involves their transformation into 16 primary spermatocytes (*IV*), 32 secondary spermatocytes (*V*), and 64 spermatids (*VI*). Stages in spermiogenesis (*VI*) are followed by spermiation (*VII*) and progressive degeneration of the retained somatic elements of the spent follicle (*VIII*). It should be remembered that an actual follicle contains many spermatocytes, all in the same stage of development. (From Stanley, 1966; reprinted with permission of Springer-Verlag.)

visions, the follicle contains more than 200 secondary spermatogonia, when maturation begins. The maturation process consists mainly of enlargement of the nucleus. The resulting cell is now called the primary spermatocyte. At this stage of development, the follicle cells become glandular in appearance.

The primary spermatocytes undergo meiotic division and form the secondary spermatocytes. At this time, spaces develop between these spermatocytes, and eventually each supporting follicle cell forms a separate pouch containing a clone of spermatids (Roosen-Runge, 1977). The unit of follicle cell and germ cells is called a spermatocyst. The development of such a spermatocyst in a shark is illustrated in Figure 6.1. The secondary spermatocytes divide once more mitotically and form the spermatids. The follicle becomes a hollow vesicle with secondary spermatocytes located along the inner wall, which is glandular in appearance, as shown by the well-developed endoplasmic reticulum and the many mitochondria. Spermiogenesis then begins. The vesicle bursts and the cells of the vesicle become attached to the boundary membrane of the tubules or lobules, so that the vesicle cells contain bundles of sperm with their heads in the cytoplasm of the old vesicle cells. The follicle (vesicle) cells now resemble Sertoli cells. Type 2 spermatogenesis will be discussed in the sections on birds and mammals. We shall discuss spermatogenesis and its control by the pituitary in the various classes of vertebrates that have type 1 spermatogenesis.

Cyclostomes

Spermatogenesis takes place in follicles. In the primitive hag fish (*Eptatretus stouti*) of the Pacific Ocean, hypophysectomy (AP) has little effect on spermatogenesis (Matty et al., 1976), although it results in an increased incidence of abnormal testicular follicles. In the lampreys (*Lampetra* spp.), the anterior pituitary (AP) is not required for spermatogenesis to proceed beyond the stage of the primary spermatocyte formation. The rate of spermatogenesis after AP is slower than in controls, but the process is otherwise normal (Dodd, 1975).

Elasmobranchs

Spermatogenesis takes place in follicles, as illustrated in Figure 6.1. The testis of the dogfish (*Squalus acanthias*) has a definite rostral caudal zonation (Hoar, 1969), which is the result of the migration of the ampullae, a migration closely related to the supporting cells of the ampullae. In the developing ampulla, the lumen is lined with prismatic epithelial cells; at this stage, the germ cells are secondary spermatogonia. After several cell divisions, the germ cells form eight layers, and the epithelial cells move between the spermatogonia to the periphery. The processes of the epithelial cells envelop some but not all spermatogonia. At this stage of development, the former prismatic epithelial cells are transformed into nurse cells, which may be analagous to the mammalian Sertoli cells. By the time the nurse cells are formed, the cytoplasmic processes contain 64 spermatids. The number of nurse cells equals the number of bundles of spermatozoa, about 424 per ampulla (Roosen-Runge, 1977). There are probably 10 spermatogonial divisions between the primordial cells and the formation of the spermatocyst, as shown in Figure 6.2 (Holstein, 1969).

Removal of the ventral lobe of the AP of the dogfish (*Scyliorhinus canicula*) results in resorption of those spermatogonia that are to undergo the last four series of mitotic divisions, although this resorption occurs only if the water temperature is above 13°C (Dobson and Dodd, 1977c; Dodd 1975). Injection of homogenates of ventral lobes prevents testicular degeneration, which normally follows ventral lobectomy, but hCG and PMSG are not effective (Dobson and Dodd, 1979). The main effect of the gonadotrophic hormone is exerted at the mitotic division that converts 8 spermatogonia into 16 primary spermatocytes. The breakdown zone that appears after ventral lobectomy consists of late spermatogonial ampullae, in which the degenerating germ cells are phagocytosed by the Sertoli cells (Dobson and Dodd, 1977b).

Administration of testosterone does not reverse the effects of ventral lobectomy (Dobson and Dodd, 1977a). This observation and the fact that

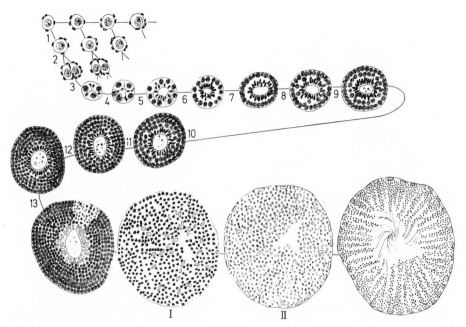

Figure 6.2 Schematic diagram of spermatogenesis in the dogfish (*Squalus acanthias*) from spermatogonia to spermatid formation. The last three ampullae contain primary spermatocytes (I), secondary spermatocytes (II), and spermatids. (From Holstein, 1969; reprinted with permission of Springer-Verlag.)

blood testosterone concentrations do not decrease after ventral lobectomy (Dobson and Dodd, 1977a) suggest that testosterone cannot maintain spermatogenesis. This, of course, does not imply that testosterone does not play some role in normal spermatogenesis. It is possible that testosterone is required for one of the stages following spermatocyte formation. This hypothesis could be tested by the use of anti-androgens or antisera against androgens. Complete AP leads to a statistically significant (P < 0.05) decrease in plasma testosterone concentration (12.10 ± 0.67 ng/ml in controls, 9.95 ± 0.98 ng/ml in hypophysectomized dogfish). These data indicate that the AP affects testosterone secretion to some extent. In higher vertebrates, AP usually results in nearly undetectable concentrations of plasma testosterone.

Teleosts

Spermatogenesis occurs in spermatocysts (also called follicles). The development of germ cells and the spermatocysts, with the germ cells developing synchronously within a spermatocyst, has been reviewed extensively by Roosen-Runge (1977). In general, type 1 spermatogenesis is found in teleosts, except in the sea perch (*Cymatogaster* sp.), in which the spermatogonia divide asynchronously, and in which spermatocysts are formed not during the spermatogonial stage but during the spermatocyte stage (Roosen-Runge, 1977). Grier (1975) found that in *Poecilia latipinna*, the spermatocysts develop when the primary spermatocytes become surrounded by a single layer of Sertoli cells at the periphery of the testis. As spermatogenesis proceeds, they move centrally in the testes toward the ductuli efferentes.

After AP, the testes regress and spermatogenesis does not continue beyond formation of spermatogonia in the species investigated, i.e., the goldfish (*Carassius auratus*) (Billard et al., 1970; Breton et al., 1973; Yamazaki and Donaldson, 1968, 1969); the guppy (*Poecilia reticulata*) (Breton et al., 1973); the lake chub (*Couesius plumbeus*) (Ahsan, 1966), the killifish (*Fundulus heteroclitus*) (Lofts et al., 1966; Pickford et al., 1972); the Indian catfish (*Heteropneustes fossilis*) (Sundararaj and Nayyar, 1967; Sundararaj et al., 1971; Nayyar et al., 1976); and the longjaw goby (*Gillichthys mirabilis*) (deVlaming, 1974).

The study of the control of spermatogenesis in teleosts has been difficult because of the specificity of the gonadotrophins; that is, spermatogenesis may require the gonadotrophin secreted by the species being investigated or the gonadotrophin of a related species. It is difficult, however, to obtain enough pituitaries and to purify a sufficient amount of gonadotrophin for injection purposes, except for some commercial fishes.

In the species, that have been investigated, gonadotrophins from the same or from related species—but usually not those from unrelated species (Breton et al., 1973)—are capable of restoring testicular weight and spermatogenesis (Ahsan, 1966; Billard et al., 1970; Breton et al., 1973; Sundararaj et al., 1971; Yamazaki and Donaldson, 1969). Mammalian gonadotrophin preparations have (except for FSH, which has no effect on either testicular weight or on spermatogenesis) varying effects, according to the species investigated. For example, human chorionic gonadotrophin (hCG) can restore testicular weight and spermatogenesis in AP Indian catfish (Sundararaj and Nayyar, 1967), but not in AP goldfish (Breton et al., 1973), guppies (Breton et al., 1973), or the longjaw goby (deVlaming, 1974). Similarly, varying results have been obtained with mammalian LH preparations. Androgen administration can restore spermatogenesis in AP Indian catfish (Sundararaj and Nayyar, 1967; Nayyar et al., 1976) and killifish (Lofts et al., 1966).

Amphibians

Spermatogenesis takes place in spermatocysts, which are located in ampullae in the urodeles and in tubules in the anurans. In the anurans (frogs and toads), three types of spermatogenesis can be distinguished as follows:

1. *Continuous spermatogenesis.* Spermatogonia mature into spermatocytes during the entire year.

2. *Discontinuous spermatogenesis.* Primary spermatogonia lose their capability to divide mitotically during part of the year, and administration of gonadoptrophins does not restore this capability.

3. *Potentially continuous spermatogenesis.* Primary spermatogonia continue to divide slowly, and the spermatocysts continue to develop; how-ever, development during part of the year does not continue beyond the primary spermatocyte stage, and these spermatocytes normally degenerate. Nevertheless, spermatogenesis can be stimulated during the season when spermatogenesis normally ceases, either by injection of gonadotrophins or by exposure of the animals to high temperatures, such as they experience during the normal period of spermatogenesis of the species. Table 6.1 lists the distribution of these three types of spermatogenesis among various species.

Spermatogenesis takes place wtihin spermatocysts and, when the flagella of the spermatids start to grow, the spermatocysts rupture, opening into the lumen of the tubule. At the peripheral pole of the spermatocyst, the spermatocyst cells are transformed into so-called supporting cells of Sertoli, to which bundles of spermatids (60–150) become attached. These supporting cells are the targets of gonadotrophic hormones, which cause the release

Table 6.1

Continuous, potentially continuous, and discontinuous spermatogenetic cycles in some Anura

Continuous cycles
Bufo paracnemis
B. arenarum
B. melanostictus
B. granulosus
Leptodactylus prognathus
L. ocellatus reticulatus
L. laticeps
Physalaemus fascomaculatus
Pseudis paradoxa
P. mantidactyla
Rana hexadactyla
R. esculenta (southern range)
Discoglossus pictus (southern range)
Discontinuous cycles
Rana temporaria
R. arvalis
R. dalmatina
Potentially continuous cycles
Rana esculenta (northern range)
R. tigrina
Discoglossus pictus (northern range)
Leptodactylus bufonius
L. ocellatus tyica (two breaks in spermatogenesis: winter and summer)
Leptobatrachus asper

Data from Lofts, 1974.

of the sperm cells (spermiation) by increasing the fluid intake of the supporting cells. Both ovine FSH (oFSH) and ovine LH (oLH) can induce spermiation. In the grass frog (*Rana pipiens*), oLH was more potent than oFSH, but in the Pacific tree frog (*Hyla regilla*) and in the coqui (*Eleutherodactylus coqui*), oFSH was about twice as potent as oLH. Neither the α nor the β subunits of LH nor the α unit of FSH induced spermiation, but the β unit of FSH retained about 25 percent of the activity of the native oFSH. Crude extract of AP of the bullfrog (*Rana catesbeiana*) had a higher potency than the most active bovine preparations (Licht, 1973). Both purified bullfrog FSH and bullfrog LH caused spermiation in AP bullfrogs (Muller, 1976).

Our understanding of the normal control of spermatogenesis is incomplete, because in many experiments, mammalian gonadotrophic hormones have been used. It is now known that mammalian and amphibian gonadotrophins differ biochemically and physiologically (Licht et al., 1977). The following information is relevant: First, AP results in degeneration of the testes and of the germinal epithelium. According to Rastogi et al. (1976), the primary spermatocytes are the most sensitive elements of the germinal epithelium of the green frog (*Rana esculenta*), whereas the spermatids are not affected. In the toad (*Bufo bufo bufo*), AP reduces the rate of multiplication of primary spermatogonia a few days after AP; however, the mitotic divisions of secondary spermatogonia are not affected for about a week and then diminish. The differentiation of the secondary spermatogonia fails to occur, and the spermatocysts with secondary spermatogonia degenerate. The primary spermatocytes and the subsequent stages of spermatogenetic development are not affected by AP. In general, one can state that AP affects premeiotic but not postmeiotic stages of spermatogenesis in this species (Guha and Jørgensen, 1978).

Second, after AP of green frogs, spermatogenesis can be restored either by injection of homogenates of amphibian pituitaries or by homiotropic, but not by ectopic pars distalis transplants (Rastogi et al., 1976). In AP bullfrogs, either bullfrog FSH or bullfrog LH stimulates spermatogenesis; however, androgen production is stimulated by bullfrog LH but not by bullfrog FSH (Muller, 1976). Third, PMSG or mammalian FSH can stimulate spermatogenesis up to the secondary spermatocyte state in AP larval salamanders (*Ambystoma tigrinum*), but hCG and mammalian LH are not effective (Licht et al., 1977).

Finally, the role of testosterone in the spermatogenesis of the Anura appears to be a complicated one. In the leopard frog (*Rana pipiens*), testosterone not only fails to restore spermatogenesis in AP frogs but inhibits the restoration induced by exogenous FSH plus LH (Basu et al., 1966). According to Rastogi et al. (1976), testosterone stimulates spermatid formation, but it does not affect other stages of spermatogenesis of green frogs (*Rana esculenta*) and thus cannot restore spermatogenesis in AP frogs. Experiments by Moore (1975) with AP larval *Ambystoma tigrinum* further illustrate the complex interaction between testosterone and FSH. After AP, treatment for 40 days with oFSH resulted in animals with large germinal cysts, many with secondary spermatocytes. Testosterone enhanced the effect of oFSH when the AP larvae were treated for 20 or 40 days with oFSH plus testosterone. Comparisons made with 40 days of oFSH treatment showed that testosterone *inhibited* the action of oFSH during the last 20 days (of the 40-day treatment) if the animals had *not* received testosterone during the first 20 days, but *enhanced* spermatogenesis if testosterone had been given during the first 20 days. Rastogi et al. (1976) have proposed that the interaction between the environment, the AP, and the gonadal hormones in *Rana esculenta* can be explained as shown in Figure 6.3. The sequence of events and hormone concentrations found in *Rana esculenta* around Naples, Italy, are illustrated in Figure 6.4.

Besides the control of spermatogenesis by the AP, there is an important influence on spermatogenesis by the fat bodies. In green frogs, removal of fat bodies in February leads to degeneration of the testes. This degeneration cannot be prevented by the administration of exogenous gonadotrophins but can be prevented by implanting fat bodies, obtained in March or October, or by injecting extracts of such fat bodies. Unilateral removal of fat bodies leads to degeneration of the ipsilateral testis, suggesting that there is a direct

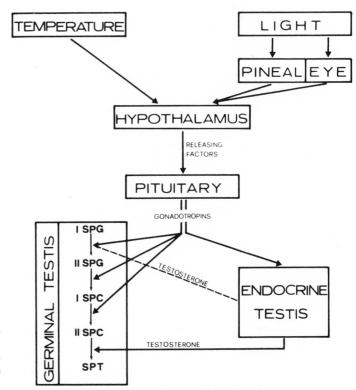

Figure 6.3 Spermatogenesis in *Rana esculenta.* Generalized scheme depicting the major relationships between the ambiental cues [light and temperature] hypothalamus, pituitary, pineal, eye, and endocrine tissue of the testis involved in the control of spermatogenesis. [Abbreviations: I SPG = primary spermatogonium, II SPG = secondary spermatogonium, I SPC = primary spermatocyte, II SPC = secondary spermatocyte, SPT = spermatid.] (From Rastogi et al., 1976; reprinted with permission of Alan R. Liss, Inc.)

transfer of lipid materials from fat body to testis. The mobilization of such materials appears to be under the control of the pars distalis. There is also a reciprocal effect of the testes on the fat bodies, because castration leads to a decrease in the weight of the fat bodies. The relationships illustrated in Figure 6.5 have been proposed by Chieffi et al. (1975) to explain experimental observations.

Urodeles. The urodeles have lobules filled with follicles, and all follicles within a lobule are in the same stage of spermatogenesis. The testes are divided into different zones, with spermatogenesis being most advanced in the posterior zone and least advanced in the anterior zone. In the newt, *Trituroides hongkongensis,* a yellow zone is formed by the hypertrophied walls of the lobules and the Sertoli cells. These two structures together form an interstitial gland that is eventually resorbed. The yellow zone is most prominent from February through April, when testicular weights are at their minimum. When spermatogenetic activity starts, the yellow zone begins to be resorbed, and the resorption is completed by May.

Spermatid formation starts in June, and spermiogenesis begins in July, followed by sperm evacuation in August and September (Lofts, 1974).

The ductuli efferentes conduct the sperm cells into the sexual segment of the kidney, after which the sperm are discharged via the ductus deferens into the cloaca. In all urodeles with internal fertilization (*Cryptobranchidae* and *Hynobiidae* have external fertilization), the sperm are packaged into spermatophores, which are described in Chapter 5.

Reptiles

In reptiles, birds, and mammals, type 2 spermatogenesis is found. Spermatogenesis takes place in tubules and several layers of germ epithelium are formed. The spermatogenetic process can be divided into three main periods: (1) the multiplication of spermatogonia, which after differentiation into spermatocytes move inward to form a new layer, the germinal epithelium (Roosen-Runge, 1977); (2) spermatocyte division,

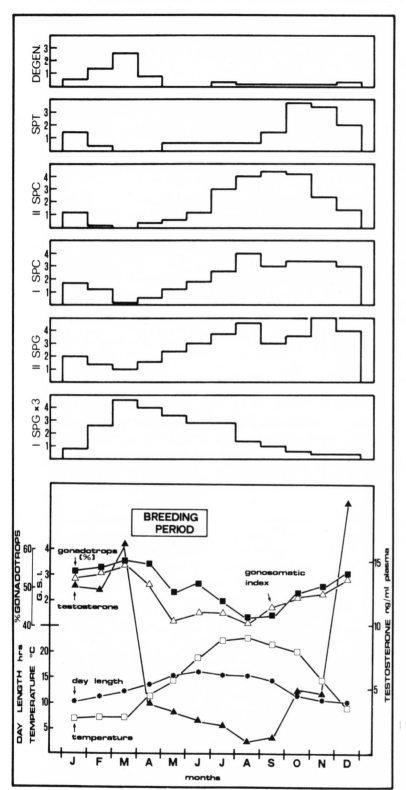

Figure 6.4 Seasonal variation in the spermatogenetic activity of *Rana esculenta* expressed in number of cell nests (ordinate in the upper part of the figure) per seminiferous tubule. Each value is the mean for 10–15 frogs. Lower part of the figure shows annual changes in the testicular weight (gonosomatic index), plasma testosterone level (based on data from d'Istria et al., 1974), pituitary gonadotropic activity (% gonadotropes, based on data from Rastogi and Chieffi, 1970) ambient temperature (°C) and day length (photoperiod in hours). I SPG = primary spermatogonia, II SPG = secondary spermatogonia, I SPC = primary spermatocytes, II SPC = secondary spermatocytes, SPT = spermatids, DEGEN = degenerating cell nests. (From Rastogi et al., 1976; reprinted with permission of Alan R. Liss Inc.)

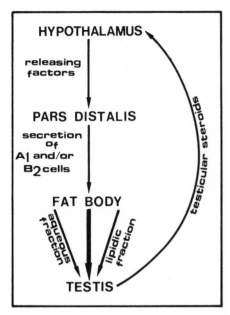

Figure 6.5 Schematic diagram showing possible position of frog fat bodies in the hypothalamo-hypophyseal-gonadal axis. [A₁ and B₂ cells are gonadotrophin cells] (From Chieffi et al., 1975; reprinted with permission of Springer-Verlag.)

which results in spermatid formation; and (3) spermiogenesis, the period in which spermatozoa are formed from the spermatids. In ducks and several mammalian species, different stages of spermiogenesis have been correlated with different cell associations of the germ cells, as will be discussed in the sections on birds and mammals.

The testis consists of seminiferous tubules, with Leydig cells among the tubules, as described in Chapter 5. Teiid lizards have a circumtesticular layer of Leydig cells. Before the breeding season, the tubules are enveloped by a peritubular membrane, which consists largely of myoid cells. At the end of the breeding season, these cells are replaced by fibroblasts (Roosen-Runge, 1977).

Little appears to be known about the quantitative aspects of spermatogenesis in reptiles, in contrast to the information available on some birds and especially mammals; e.g., Roosen-Runge's excellent review (1977) cites no data on any reptiles for the duration of the spermatogenetic cycle (the period between the appearance of the original spermatogonium and the release of the spermatozoa produced from this spermatogonium).

Control of spermatogenesis is largely exercised by the AP. Hypophysectomy results in the degeneration of the germinal epithelium in the snakes, *Thamnophis sirtalis*, the garter snake, *T. sauritus* (Licht, 1972a) and *T. radix*, (Schaefer, 1933); in *Anolis carolinensis*, the American chameleon, (Licht, 1968); and in the turtles, *Chrysemys picta*, the painted turtle, (Callard et al., 1976), *Kinosternon subrubrum*, the mud turtle, and *Pseudemys scripta*, the slider, (Licht, 1972b). The control of spermatogenesis by the AP has been studied in most detail for the American chameleon.

Although spermatogenesis is dependent upon the AP, it is also affected by the environmental temperature. At 31°C, AP resulted in complete gonadal regression in about 2 weeks, whereas at 20°C, even three weeks after AP, little effect could be detected (Licht, 1968; Licht and Pearson, 1969). The effect of temperature on testicular degeneration after AP is thus quite similar to that observed in *Scyliorhinus canicula*. After injection of oLH and oFSH, spermatogenesis was restored in AP *Anolis carolinensis* at 31°C, but the amount of LH required was several hundred times greater than the amount of FSH. When LH and FSH were given together, the effect on spermatogenesis was less than that obtained by FSH alone (Licht and Pearson, 1969). Testosterone retarded post-AP regression of the testes, but it did not restore spermatogenesis. Subsequent studies with *Anolis carolinensis* have shown that:

1. All the gonadotrophic activity for spermatogenesis is in the β subunit of oFSH (Licht and Papkoff, 1971).

2. After removal of residual LH by LH antiserum, oFSH retains its capacity to stimulate and maintain spermatogenesis of AP animals (Licht and Tsui, 1975).

3. Ovine LH retains its capability to stimulate spermatogenesis after removal of FSH activity (Licht and Papkoff, 1973).

4. Testicular growth and spermatogenesis can be maintained in AP animals with all species of amphibian, reptilian, and avian FSH and LH examined (Licht et al., 1977), although the LH of certain species may be less effective than the FSH (Licht et al., 1977). The greater effectiveness of

the FSH can be explained by the lower clearance rate of FSH in comparison with LH (Licht et al., 1977).

Investigation on the relationship between AP and testes of reptiles have so far shown that:

1. Snakes and lizards lack LH, as tested by induction of ovulation in ovaries of *Hyla regilla* and *Xenopus laevis* (Licht, 1974), but LH is present in the AP of turtles and several Crocodilia (Licht, 1974; Licht and Papkoff, 1974; Licht et al., 1976).

2. After AP, the testes of snakes and lizards respond to oFSH or to very high doses of oLH by restoration of spermatogenesis (Licht, 1972a). In neither *Anolis carolinensis* nor in *Thamnophis sirtalis* (Licht, 1972a) does testosterone restore spermatogenesis in the AP animal.

3. In snakes and lizards, the testis shows little specificity for FSH and LH, which may be related to a lack of specificity in gonadotrophin-binding sites for mammalian gonadotrophins (Licht and Midgley 1976a, b; 1977).

4. Ovine FSH is more potent than oLH in restoring spermatogenesis in AP *Pseudemys scripta, Kinosternon subrubrum* (Licht, 1972b) and *Chrysemys picta* (Callard et al., 1976).

As we shall discuss later in this chapter, exposure of the testes of scrotal mammals to higher than their normal temperature interferes with spermatogenesis. Exposure of the lizard *Urosaurus ornatus* for ten days to temperatures 1–2°C above its preferred environmental temperature leads to degeneration of the germinal epithelium (Licht, 1965). Thus, in reptiles and mammals, high temperature has a profound effect on reproduction.

In most species, the Leydig cells are located among the testicular tubules, but in teiid lizards, the Leydig cells form a circumtesticular capsule. In *Cnemidophorus tigris*, this capsule binds radio-iodinated human FSH intensely (Licht and Midgley, 1977). In the garter snake, the slider, and the American chameleon, the binding occurs in the interstitial cells and probably in the spermatogonia and Sertoli cells.

Birds

So far we have discussed spermatogenesis in poikilotherms, but birds and mammals are homeotherms. As we shall discuss in the section Mammals, exposure of the testes to body temperature, by interfering with spermatogenesis, causes sterility in many scrotal mammals. Not only are birds homeothermic, but they also generally have high body temperatures; e.g., the temperature of the domestic hen fluctuates daily between 39°C and 42°C. In the scrotum, the testes are exposed to temperatures lower than body temperature. In all birds, the testes are internal and the question has been asked whether exposure to temperatures either lower than or higher than body temperature would affect spermatogenesis. Exposure of testes of sexually immature roosters to temperatures lower than body temperature, by creating an artificial ''scrotum,'' caused precocious spermatogenesis (Williams, 1958a), and exposure of testes to temperature higher than body temperature (i.e., in vitro exposure to warm Ringer's solution followed by regrafting of the testes) interfered with spermatogenesis (Williams, 1958b). The air sacs, which might act as cooling organs for the testes, probably do not function in this way, because the testicular temperature is the same as that in the rest of the body cavity (Williams, 1958a), and destruction of the air sacs has been found not to affect spermatogenesis in roosters (Herin et al., 1960).

During spermiogenesis, 10 different stages (Figure 6.6) of spermatid development can be identified accurately in the mallard drake (*Anas platyrhynchos*). These 10 stages are correlated with 8 different types of germ cell associations in the drake, as illustrated in Figure 6.7 and Table 6.2. The study of such cell associations makes it possible to determine which spermatogonial cells are renewed.

Control of spermatogenesis is exercised by the hypothalamo-anterior pituitary system. Lesions made in the mammillary nuclei and the posterior part of the ventromedial nucleus resulted in testicular atrophy and regressed combs (20 percent of controls). Lesions in the posterior part of the

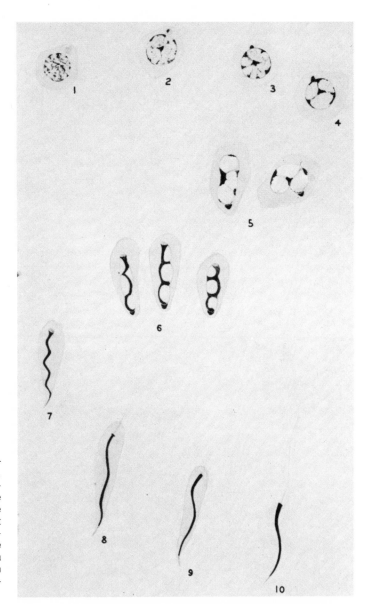

Figure 6.6 Spermiogenesis in the drake as observed in sections stained with PAS-hematoxylin, showing the series of changes which the spermatid undergoes. 1, young spermatid; 10, mature sperm. The proacrosomic granule (in the cytoplasm in light gray) is shown as a black dot (2), which is attached to the nucleus (3) and becomes a triangular acrosome (4–6). The acrosome becomes more lightly stained (7) but persists as a fine point at the tip of the nucleus (8–10). (From Clermont, 1958; reprinted with permission of Masson S.A., Paris.)

mammillary nuclei, impinging on the arcuate nucleus, caused degeneration of the germinal epithelium, but the combs regressed to 40 percent of the size of combs in controls (Ravona et al., 1973).

As expected, the testes of all birds investigated (Japanese quail, chickens, drakes, pigeons) regress after AP (van Tienhoven, 1961); although mammalian FSH can maintain spermatogenesis in AP roosters (Nalbandov et al., 1951).

Brown et al. (1975), using chicken FSH and

LH, showed that in AP Japanese quail, FSH stimulated testicular growth, the formation of spermatocytes to the pachytene stage, and the Sertoli cells; whereas LH administration resulted in differentiation of the interstitium, some spermatogonial differentiation, and partial differentiation of the Sertoli cells. These last two effects may have been the result of testosterone secretion stimulated by LH.

The effects of testosterone on the testes of AP

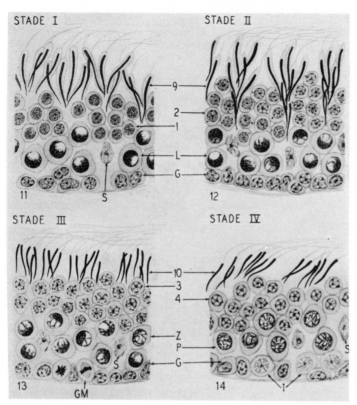

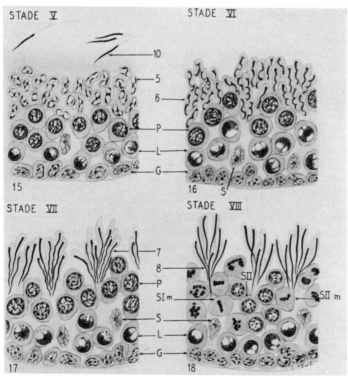

Figure 6.7 Schematic representation of the eight stages of the spermatogenetic cycle in the germinal epithelium of the drake. Numbers 1 through 10 indicate spermatids in different stages of spermiogenesis (Figure 6.8). S = Sertoli cell, G = spermatogonium, GM = spermatogonium in mitosis, I = primary spermatocyte in interphase, L = spermatocyte in leptotene stage, Z = spermatocyte in zygotene stage, P = spermatocyte in pachytene stage, SIm = primary spermatocyte in metaphase, SII = secondary spermatocyte, SIIm = secondary spermatocyte in metaphase. (From Clermont, 1958; reprinted with permission of Masson S.A., Paris.)

Table 6.2

Cell types found during different stages of the spermatogenetic cycle of the drake

Cell type	Stage of the cycle							
	I	II	III	IV	V	VI	VII	VIII
Spermatogonia (G)	G	G	G	G	G	G	G	G
Spermatogonia in mitosis (M)			M		M			M
Primary spermatocytes								
Interphase (I)				I				
Leptotene (L)	L	L			L	L	L	L
Zygotene (Z)	Z	Z	Z					
Pachytene (P)				P	P	P	P	
Diakinesis (D)							D	D
Metaphase (SIm)								SIm
Secondary spermatocytes (SII)								SII
Dividing spermatocytes (SIIm)								SIIm
Spermatids in stages 1–10 of spermiogenesis*	1	2	3	4	5	6	7	8
	9	9	10	10				

*See Figure 6.6.
Modified from Clermont, 1958; reprinted with permission of Masson S. A. Paris.

birds seem to vary with the species. Testosterone maintained and also restored spermatogenesis in AP pigeons (Chu, 1940; Chu and You, 1946); however, in Japanese quail, testosterone injections in doses that were effective in pigeons did not restore spermatogenesis in AP males (Baylé et al., 1970). Brown and Follett (1977), using Silastic implants of testosterone propionate (TP), found that the TP retarded the regression of the testes but did not prevent regression of the testes or restore regressed testes of AP Japanese quail. A combination of oFSH plus TP was more effective than TP alone in inducing testicular growth in AP immature quail (Brown and Follett, 1977).

Experiments in which TP is administered to intact birds, with or without enlarged testes, are difficult to interpret because of the possibility that even with high doses of TP, some FSH, LH, or FSH and LH are still secreted. If testosterone acts both to depress gonadotrophins and, at the site of the testes, to stimulate a step in spermatogenesis after FSH, LH, or FSH plus LH, then one is dealing with too many uncertainties to give a clear interpretation [see, however, Brown and Follett (1977) and Desjardins and Turek (1977)].

Testosterone injections restored spermatogenesis in the regressed testes of photorefractory African weaver finches (*Quelea quelea*) and in sexually immature house sparrows (*Passer domesticus*) (Brown and Follett, 1977). They maintained spermatogenesis in slate-colored juncos (*Junco hyemalis*) and white-throated sparrows (*Zonotrichia albicollis*) after transfer of the birds from long to short days, a transfer that normally causes involution of the testes (Wolfson and Harris, 1959). In roosters, androgen administration did not restore spermatogenesis in animals bearing hypothalamic lesions that cause testicular regression (Snapir et al., 1969). So far, the only bird in which testosterone has been shown to restore spermatogenesis in the AP animal is the pigeon.

The inherent disadvantage of using AP animals which are deficient not only in FSH and LH, but also in ACTH, TSH, GH, and prolactin, could be overcome by selective "GTH hypophysectomy," i.e., by antibodies to FSH or LH obtained either by passive or active immunization. However, this requires the availability of pure avian LH and FSH in quantities that so far have not been available.

Mammals

In mammals, as in birds, the occurrence of each stage of spermiogenesis (Figure 6.8) is correlated

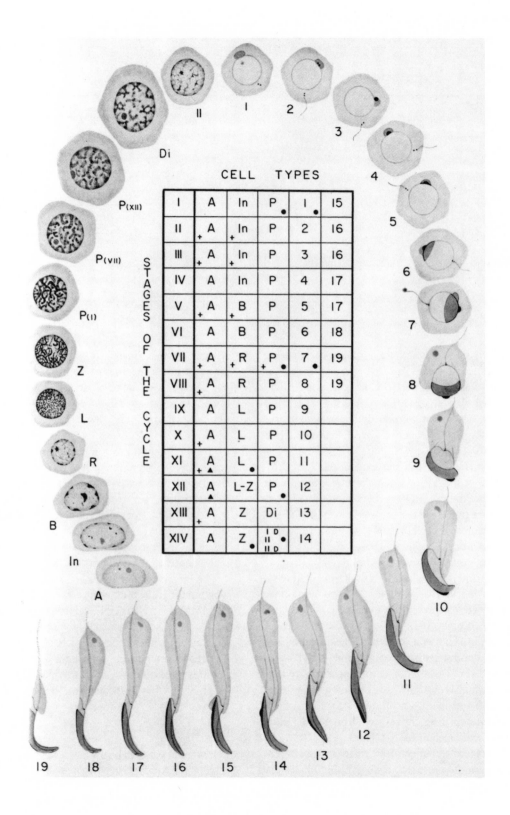

CELL TYPES

STAGES OF THE CYCLE

	A	In	P		
I	A	In	P •	1 •	15
II	A +	In +	P	2	16
III	A +	In +	P	3	16
IV	A	In	P	4	17
V	A +	B +	P	5	17
VI	A	B	P	6	18
VII	A +	R +	P + •	7 •	19
VIII	A +	R	P	8	19
IX	A	L	P	9	
X	A +	L	P	10	
XI	A + ▲	L •	P	11	
XII	A ▲	L-Z	P •	12	
XIII	A +	Z	Di	13	
XIV	A •	Z •	II D • / II D	14	

Figure 6.8 (*Opposite page*) Starting from the lower left and encircling the table, drawings illustrate the steps of spermatogenesis in the rat. A = type A spermatogonium; In = intermediate type spermatogonium; B = type B spermatogonium; R = resting primary spermatocyte; L = leptotene spermatocyte; Z = zygotene spermatocyte; $P_{(I)}$, $P_{(VII)}$, $P_{(XII)}$ = early, mid-, and late pachytene spermatocytes, respectively; roman numerals in brackets indicate the stages of the cycle at which they are found. Di = diplotene; II = secondary spermatocyte, 1–19 = steps in spermiogenesis. (From Leblond and Clermont, 1952.) The table in the center of the figure gives the cellular composition of the stages of the cycle of the seminiferous epithelium (roman numerals, I–XIV). I^D and II^D = first and second maturation divisions of spermatocytes (+), (●), (▲) = cells counted in the first, second, or third series of counts respectively. (From Clermont, 1962; reprinted with permission of Alan R. Liss, Inc.)

with certain cell associations in the germinal epithelium, as illustrated in Figure 6.9. The occurrence of these cell associations and the labeling of germ cells by [³H]-thymidine can be used to study several aspects of spermatogenesis quantitatively, such as: (1) the manner in which spermatogonial stem cells are renewed, as shown, for example, in Figure 6.10; (2) the duration of the seminiferous epithelial cycle, i.e., the time that elapses between two successive appearances of the same cellular associations; and (3) the duration of the spermatogenetic cycle, i.e., the interval between the appearance of the original spermatogonium and the appearance of the sperm produced from it. The number of spermatogonial mitoses for different species is given in Table 6.3; the duration of the spermatogenetic cycle is given in Table 6.4. In mammals, the spermatogenetic cycle lasts about four times as long as the epithelial cycle.

What we know about the control of spermatogenesis is based on the results obtained from in vitro and in vivo studies, largely of rat testes. Hypophysectomy of rats results in an increased incidence of degeneration of midpachytene spermatocytes and step* 7 and 19 spermatids (Figure 6.8) (Russell and Clermont, 1977). Administration of FSH does not affect the incidence of such degenerating cells, but LH injections in physiological doses reduce the incidence of degenerating germ cells in stages† VII and VIII of the cycle. Administration of FSH plus LH reduces the inci-

dence of degenerating cells to that of the intact rat (Russell and Clermont, 1977).

Raj and Dym (1976), using carefully characterized antisera, found that administration of LH antiserum to 20-day-old male rats for 14 days resulted in a reduction of the testicular weight to about one-third of that of the controls; a reduction in the weights of the epididymides, seminal vesicles, and ventral prostates; and a reduction in plasma testosterone concentration to about 20 percent of that of the controls. The histological appearance of the testes was similar to that seen in $\cancel{AP}$ rats, i.e., small-diameter tubules, few spermatids, pachytene secondary spermatocytes reduced in number to about 50 percent of that of controls, but normal spermatogonia. Antiserum to FSH re-

Table 6.3

Number of spermatogonial mitoses in different vertebrates

Species	Number of mitoses
Elasmobranchs	
Torpedo marmorata	4
Squalus caniculus	12
Teleosts	
Umbra limi	5–6
Phoxinus laevis	4–5
Perca flavescens	4–5
Brachydanio rerio	4–5
Sebastodes paucispinus	6?
Aequidens portalegrensis	5?
Gambusia holbrookii	9–11
Poecilia reticulata	13
Amphibia	
Ambystoma sp.	6
Proteus anguinus	6 and 7
Triton sp.	6
Mammals	
Bull	5
Ram	4
Gerbil	4
Mouse	5
Rat	5
Rabbit	4
Rhesus monkey (*Macaca mulatta*)	4
Long-tailed monkey (*Cercopithecus aethiops*)	4 or 5
Humans	3?

Modified from Roosen-Runge, 1977; reprinted with permission of Cambridge University Press.

*Step refers to the different types of germ cells.
†Stage refers to the different types of cell associations.

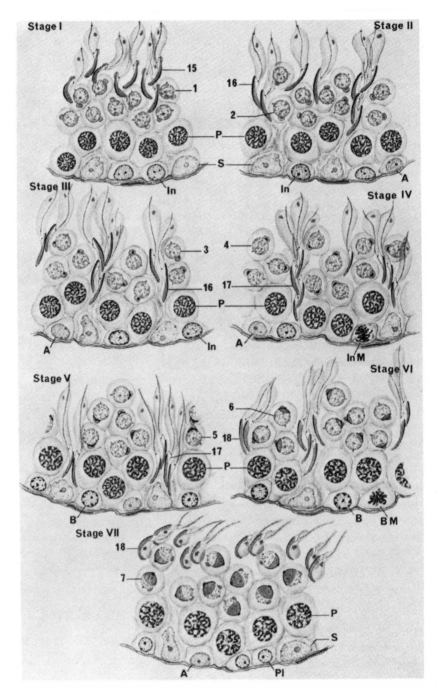

Figure 6.9 Series of drawings illustrating 14 stages (roman numerals I–XIV) of cycle of seminiferous epithelium of the rat as seen in PA-Schiff-hematoxylin-stained sections of testis. Each drawing represents a portion of seminiferous epithelium and gives cellular composition of the stage as well as arrangement of various generations of germ cells within the epithelium. Drawings represent complete series of successive cellular associations appearing in any one area of a seminiferous tubule (after stage XIV, stage I reappears, and so on). A, In, and B = type A, intermediate-type, and type B spermatogonia; InM, BM = intermediate type and type B in mitosis; Pl = preleptotene primary spermatocytes; L = leptotene spermatocytes; Z = zygotene spermatocytes; P = pachytene spermatocytes; S = sertoli cell; Di = diplotene spermatocytes; II = secondary spermatocytes; IM, IIM = primary and secondary meiotic divisions; 1–19 = spermatids at various steps of spermiogenesis; RB = residual bodies. PA-Schiff-positive acrosomic system of spermatids closely associated to nucleus is shown in black or deep gray. First 14 steps of spermiogenesis were utilized to identify the 14 stages of cycle. (From Clermont, 1972; reprinted with permission of Y. Clermont and The American Physiological Society.)

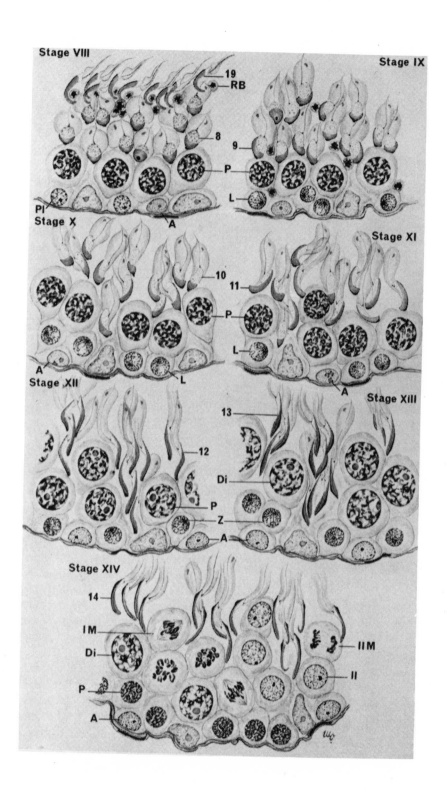

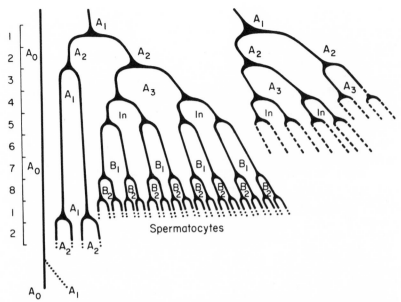

Figure 6.10 Model of spermatogonial renewal in bull recently proposed by Hochereau-de Reviers. According to this scheme there are three generations of type A spermatogonia (A_1, A_2, A_3), one of intermediate type spermatogonia (In) and two of type B spermatogonia (B_1, B_2). Nodal division, equivalent in nature, is taking place among type A_2 spermatogonia in stages 2–3 of cycle. Degeneration affecting mainly type A_3 and intermediate-type spermatogonia is indicated by the broken line. Some type A_0 spermatogonia are present (on left of diagram) and considered to divide slowly and sporadically (dotted line) to give rise to new type A_0 and type A_1 spermatogonia. [The numbers on the line at the left side of the diagram indicate the stages of the spermatogenetic cycle.] (From Clermont, 1972; reprinted with permission of Y. Clermont and the American Physiological Society.)

duced testicular weights to 55 percent of normal, but did not affect the weights of the epididymides, seminal vesicles, and ventral prostates or the concentrations of plasma testosterone. Histological examinations of the testes revealed a reduction in the number of spermatids to 83 percent of that of the controls, and of the number of pachytene spermatocytes to 65 percent of that of the controls.

Table 6.4

Duration of seminiferous epithelial cycle and spermatogenetic cycle in different species of mammals

Species	Epithelial cycle (days)	Spermatogenetic cycle (days)
Rat	12.3	49
Mouse	8.6	34
Rabbit	10.9	44
Bull	14.0	56
Ram	10.0	40–49
Boar	8.0	34
Humans	16	74

These investigations thus showed that lack of LH, probably because it reduced testosterone secretion, has a more severe effect on spermatogenesis in the prepubertal rat than a lack of FSH. The effect of FSH is probably mediated through its effect on the secretion of androgen-binding protein (ABP) by the Sertoli cells, to be discussed below.

In rats, testosterone, dihydrotestosterone, and 5α-androstane-3α, 17β-diol can maintain spermatogenesis in recently *AP* animals (Ahmad et al., 1975; Setchell, 1978), but these androgens cannot restore the germinal epithelium once it has undergone degeneration (Setchell, 1978). The ram is similar to the rat in this respect; i.e., testosterone does not prevent testicular degeneration in *AP* animals (Monet-Kuntz et al., 1977). In contrast, testosterone has been reported to restore spermatogenesis in *AP* monkeys and squirrels (Setchell, 1978).

Follicle-stimulating hormone acts primarily on the Sertoli cell, where it stimulates the secretion of

ABP. After $\cancel{AP}$, the ABP disappears from the testes, but its secretion is reinitiated by FSH administration in $\cancel{AP}$ rats (Means, 1977). The secretion of ABP is initiated by FSH, but not by testosterone, although testosterone stimulates the accumulation of ABP in the testes in $\cancel{AP}$ rats; in vitro, FSH and testosterone each stimulate ABP production. Cyproterone acetate, an antiandrogen, does not influence the FSH-induced initiation of ABP secretion in vitro, but it inhibits the testosterone-induced ABP production (Louis and Fritz, 1979).

In 15-day-old rats, FSH administration results in an increased ABP concentration in the testes and the epididymides. This effect of FSH on the testes is enhanced by concurrent administration of either testosterone or LH, but these two hormones each block the effect of FSH on the secretion of ABP into the epididymis (Kotite et al., 1978). This blocking effect of testosterone has been found in the immature rat but not in the adult rat. The significance of this finding is obscure.

Steinberger and Steinberger (1973) have proposed the tentative interpretation of the hormonal control of spermatogenesis in the rat, as illustrated in Figure 6.11. Testosterone, which according to this interpretation is required in the formation of spermatid step 1, is secreted by the Leydig cells under the influence of LH. According to the ex-

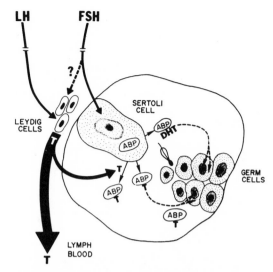

Figure 6.12 Schematic drawing illustrating the action of LH and FSH in the testis. FSH stimulation of the production of the androgen-binding protein (ABP) causes an accumulation of androgen (T = testosterone; DHT = 5α-dihydrotestosterone) in close proximity to the androgen-dependent cells within the germinal epithelium. (From French et al., 1974; reprinted with permission of International Planned Parenthood Federation.)

planation illustrated in Figure 6.12, testosterone enters the tubule, is bound by ABP, and is transported to the germinal epithelial cells, where, after it is released from the ABP, it acts on the appropriate germ cells. Figure 6.13 summarizes

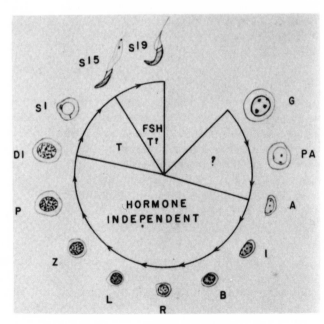

Figure 6.11 Proposed scheme of the hormonal control of spermatogenesis [in rats]. G = gonocytes, PA = primitive type A spermatogonia, A = type A spermatogonia, I = intermediate spermatogonia, B = type B spermatogonia, R = resting spermatocytes, L = leptotene spermatocytes, Z = zygotene spermatocytes, P = pachytene spermatocytes, DI = diakinesis, S1 = spermatid step 1, S15 = spermatid step 15, S19 = spermatid step 19, FSH = follicle-stimulating hormone, T = testosterone. (From Steinberger and Steinberger, 1974; reprinted with permission of E. Steinberger and the American Physiological Society.)

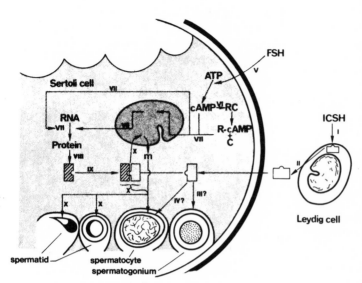

Figure 6.13 Possible molecular effects of the gonadotropins and testosterone in relation to control of spermatogenesis. I = stimulation of testosterone production in the Leydig cell by ICSH = LH; II = transport of testosterone into the seminiferous tubules and into Sertoli cells—the "active" androgen from here on is given as testosterone, but it may well be a metabolite of testosterone; III = possible direct interaction of testosterone with germinal cells to initiate a wave of spermatogenesis; IV = possible direct action of testosterone on spermatocytes to promote meiotic division; V = production of cyclic AMP (cAMP) in the seminiferous tubules as a result of FSH binding and stimulation of adenylate cyclase; VI = activation of cAMP-dependent protein kinase (RC) by cAMP; VII = increased RNA and protein synthesis effected by cAMP, a protein kinase component, or another FSH "messenger" by a variety of mechanisms; VIII = increased androgen receptor protein as a result of increased synthetic activity; IX = complex formation between testosterone and receptor protein; X = direct interaction of the hormone-receptor complex with germinal cells, resulting in the progression of spermatogenesis, or indirect interaction through a "message" (m), generated as a result of events in the Sertoli cell controlled by this complex. (From Steinberger et al., 1974; reprinted with permission of E. Steinberger and Plenum Publishing Corp.)

the possible molecular effects of gonadotrophins on spermatogenesis. Rommerts et al. (1976) have cast doubt on the biological significance of the ABP in spermatogenesis. Presumably, only free testosterone can act on the germ cells; therefore, binding of the free testosterone by ABP as it enters the tubules would tend to impede the action of testosterone on the germ cells.

There is evidence that in rats and mice, prolactin (PRL) potentiates the effects of LH in restoring spermatogenesis in AP animals, although it does not have this effect when it is administered with testosterone (Hansson et al., 1976). This finding suggests that PRL acts on the Leydig cells and not on the germinal epithelium. The idea that PRL acts on the Leydig cells is supported by the observation that Leydig cells have PRL receptors. Nevertheless, a PRL deficiency, such as occurs in dwarf mice, is associated with sterility, which can be overcome by exogenous PRL.

Several types of evidence suggest that there is negative feedback between the testicular seminiferous tubules and the AP, as will be discussed in the next section.

ENDOCRINE FUNCTION: ANDROGENS

The testes of most vertebrates have an endocrine function as well as a spermatogenetic one. It is useful to consider briefly what types of evidence can be used to demonstrate the secretion of hormones by gonads:

1. Arterio-venous (A-V) differences in the hormone concentrations, preferably in an unanesthetized animal, provide the best evidence.

2. Presence of the hormone in the peripheral circulation (which does not show which organ is secreting the hormone) and its disappearance after ablation of the gonad.

3. Presence of the hormone in extracts of the gonad.

4. Synthesis of the hormone by the gonad during in vitro incubation or tissue culture. There is evidence that some hormones are sometimes secreted in vitro, although they cannot be demonstrated by the A-V difference method.

5. Regression of secondary sex organs or secondary sex characters after gonadectomy.

6. Correlations between activity of either in-

terstitial cells or Sertoli cells, or both, and the condition of the secondary sex organs or sex characters.

7. Presence of smooth endoplasmic reticulum and/or enzymes involved in steroidogenesis, e.g., 3β-hydroxysteroid dehydrogenase (3β-HSD).

The pathways of androgen and estrogen synthesis make it possible to understand the endocrine function of the testes. Figure 6.14 shows some of the major pathways and the major enzymes. It shows that testosterone can be formed by two pathways. One pathway via steroids that have a double bond between C_4 and C_5 (for example, progesterone) is the Δ^5 pathway; the other pathway via dehydro*epi*androsterone (DHA) is the DHA pathway. Testosterone can either be converted to 5α-dihydrotestosterone (Figure 6.14) (and in some species to 5β-dihydrotestosterone), which cannot be aromatized to form estrogens, or testosterone can be aromatized to form estrogens. In Chapter 2, we discussed the relevance of aromatization of androgens in the developing brain for sexual differentiation of adult sexual and aggressive behavior and for the release of gonadotrophic hormones in adult males and females.

In discussing the evidence for the control of testicular androgen secretion by the AP for different classes of vertebrates, we shall see that during phylogenetic development, androgen secretion becomes more and more dependent upon the gonadotrophic secretions by the AP.

Cyclostomes

Although hagfishes lack secondary sex characters (Callard et al., 1978) and show no evidence of the presence of Leydig cells, testosterone and estradiol are present in the blood of *Eptatretus stouti*. Neither AP nor implants of hagfish pituitaries affect the concentration of these steroids, which range from 2.8 to 7.7 ng/ml, in the blood (Matty et al., 1976). It appears, therefore, that both the spermatogenetic and endocrine function of the gonads of this species are independent of the pituitary.

Lampreys have interstitial cells showing ultrastructural and histochemical evidence of steroid

secretion, and testosterone has also been identified in testicular extracts (Table 6.5). Indirect evidence (regression of secondary sex characters after AP), however, suggests that the pituitary may control the secretion of androgens (Callard et al., 1978).

Elasmobranchs

The presence of testosterone in testicular extracts and in the plasma of some elasmobranchs (Tables 6.5 and 6.6) is a good indication of testosterone secretion by the testes. During the year, the concentration of testosterone in *Scyliorhinus canicula* varies from 2 ng/ml in February to 6 ng/ml in August (Dobson, 1975). These values are considerably lower than those reported for skates (Table 6.6). The differences between the testosterone concentration in February and August may be the result of a higher rate of metabolism of testosterone at the higher temperature. It is almost certainly not mediated via the AP because, as discussed earlier in this chapter, removal of the ventral lobe of the AP does not affect plasma testosterone concentration.

Teleosts

As can be seen from Table 6.5 and 6.6, there is evidence that teleosts secrete androgens. The major androgen is 11-ketotestosterone, which seems to be limited to this class. The control of steroid production by the pituitary is suggested by the fact that the fish pituitary homogenates injected into AP fish stimulate steroidogenesis (Callard et al., 1978). As is the case for the control of spermatogenesis, the nature and mode of action of fish gonadotrophins have not been elucidated.

Amphibians

Evidence that the gonads secrete androgens is documented in Table 6.5 and 6.6. After AP, there is loss of interstitial cell cholesterol in *Rana temporaria,* a loss that is prevented by mammalian LH injections (Lofts, 1961). In intact frogs (*R. esculenta, R. pipiens*), mammalian LH increases 3β-HSD activity. In AP salamanders (*Pleurodeles waltlii*), mammalian LH is also more effective

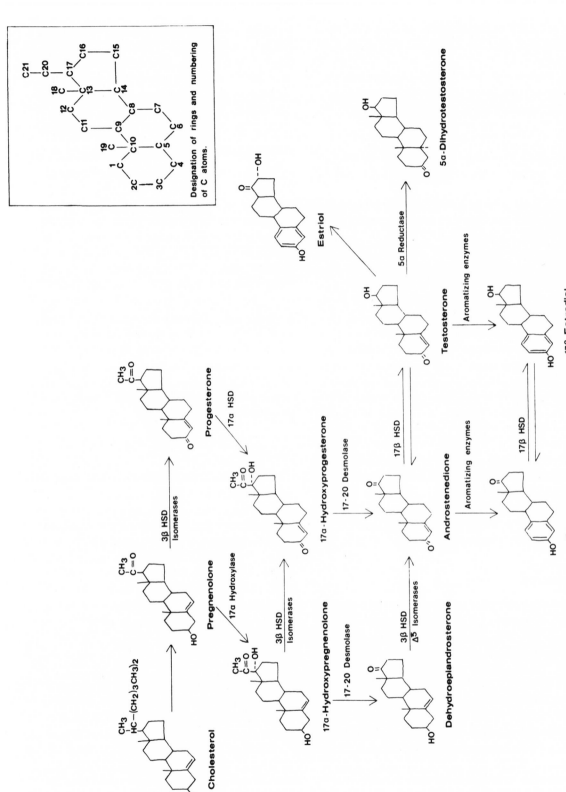

Figure 6.14 Major biosynthetic pathways for progestins, androgens, and estrogens. The progesterone pathway includes pregnenolone → progesterone → 17α hydroxyprogesterone → androstenedione → testosterone. The DHA pathway includes: pregnenolone → 17α hydroxypregnenolone → DHA → androstenedione ⇌ testosterone.

Table 6.5

Evidence for testicular steroid secretion in different cyclostomes, elasmobranchs, teleosts, amphibia, and reptiles

Species	Evidence
Cyclostomes	
Lampetra sp.	Testosterone in testicular extracts
	Presence of interstitial cells with steroidogenic morphology; weak 3β-HSD activity in interstitial cells
Eptatretus stouti	Testosterone in testicular extracts; no evidence for interstitial cells
Elasmobranchs	
Scyliorhinus canicula	Testosterone in plasma
S. stellaris	Testosterone, androstenedione, progesterone, estradiol-17β in extracts of testis; testosterone (?) in peripheral plasma
Squalus acanthias	Testosterone in incubates of testes with progesterone as precursor
Raja radiata	Testosterone in plasma
R. ocellata	Testosterone in plasma
Teleosts	
Serranus cabrilla	Testosterone in plasma
Pleuronectes platessa	Testosterone in plasma
Oncorhynchus nerka	11-ketotestosterone* in plasma
Salmo salar	11-ketotestosterone in plasma and in incubates of testes
	11-hydroxy-testosterone in incubates of testes
S. gairdneri	11-ketotestosterone in plasma
	11-ketotestosterone and 11-hydroxy-testosterone in incubates of testes
Pseudopleuronectes americanus	11-ketotestosterone in plasma
Microstomus kitt	11-ketotestosterone and 11-hydroxy-testosterone in incubates of testes
Poecilia latipinna	11-ketotestosterone and 11-hydroxy-testosterone in incubates of testes
Perca fluviatilis	11β hydroxy-testosterone and 11-oxo-testosterone in vitro
Cyprinus carpio	Testosterone in testes
Morone labrax	Testosterone (traces) E_1, E_2, and E_3 in testes; 3β-HSD in lobule boundary cells
Gobius paganellus	In vitro testosterone production from pregnenolone
Gasterosteus aculeatus	Testosterone, androstenedione, dehydro-*epi*-androsterone (DHA) in testicular extract
Esox lucius	Testosterone, DHA, androstenediol 11β-hydroxy-testosterone, and 11-oxo-testosterone in vitro
Cymatogaster aggregata	DHA, androstenedione in testes extract
Monopterus albus	Testosterone and androstenedione produced in vitro from pregnenolone
Amphibia†	
Pleurodeles waltlii	Testosterone and androstenedione in vitro from progesterone; testosterone in plasma
Triturus cristatus	Progesterone in testicular extract (no testosterone or androstenedione, trace of E_1); 11-ketotestosterone and 11-dihydro-testosterone from testosterone in vitro
Xenopus laevis	Androstenedione in vitro from 17α-hydroxy-progesterone
Rana esculenta	Testosterone in plasma
R. pipiens	Testosterone and DHT in plasma
R. catesbeiana	DHT in vitro from androstenedione; DHT and testosterone in plasma

Table 6.5—*Continued*

Species	Evidence
Amphibia (*cont.*)	
Bufo marinus	Testosterone in plasma
Necturus maculosus	DHT and testosterone in plasma
Reptiles	
Naja naja	In vitro production of androstenedione, DHA, and testosterone
Natrix sipedon	In vitro formation of testosterone from pregnenolone testosterone in testicular extract
Lacerta sicula	In vitro formation of testosterone from pregnenolone testosterone in testicular extract
Tiliqua scincoides	Testosterone in plasma
Chrysemys picta	Testosterone in plasma and produced in vitro
Anolis carolinensis	Testosterone in plasma

*11-ketotestosterone has about 10 times the potency of testosterone in promoting appearance of male secondary sex characters in female Japanese rice fish (Callard et al., 1978).

†Concentration of testosterone in urodeles 40–200 ng/ml and in anura 0.1–25 ng/ml (Callard et al., 1978).

Data from Callard et al., 1978; Chieffi, 1972; Garnier, 1972; Kime and Hews, 1978; Nandi, 1967; Weisbart et al., 1978.

Table 6.6

Plasma-serum testosterone concentrations in some nonmammalian and mammalian vertebrates

Class or order and species	Testosterone ng/ml mean or mean range ± SE (individual range)	Remarks	Method*
Chondrichthyes			
Raja radiata	36.0 ± 4.5 − 58.0 ± 11.0	Diurnal range	GLC
(skate)	74.0 (28.0 − 102.0)		GLC
Raja ocellata	101.0 (22.0 − 208.0)		GLC
(skate)			
Osteichthyes			
Oncorhynchus nerka	(10.0 − 108.0)		PIID/UV
(sockeye salmon)			
Salmo salar	5.0 − 10.0		DIDD
(Atlantic salmon)			
S. gairdneri	27.4 ± 5.4		CPBA
(rainbow trout)			
Amphibia			
Necturus maculosus	115.8 ± 37.8 (42.9 − 261.0)		DIDD
(mudpuppy)			
Rana esculenta	1.0 − 19.5		RIA
(common European frog)			
Reptilia			
Tiliqua rugosa	<5.0 − 32.5	Annual range	CPBA
(lizard)		Value may reflect other 17β-Oll steroids as well; uncorrected for 92.1 ± 2.4% recovery.	
Aves			
Anas sp.	0.5 ± 0.2 − 1.6 ± 0.3	Annual range ± SD	GLC
(domestic duck)	0.2 ± 0.1 − 2.1 ± 0.6	Annual range	GLC
	(0.7 − 2.7)		GLC
	0.30 ± 0.01 − 2.0 ± 0.4	Annual range	GLC

Table 6.6—*Continued*

Class or order and species	Testosterone ng/ml mean or mean range ± SE (individual range)	Remarks	Method*
Coturnix coturnix (quail)	(0.2 − 0.5)		GLC
Gallus sp. (domestic fowl)	(0.8 − 7.8)		GLC
	2.1 ± 0.4 − 2.4 ± 0.6	± SD	CPBA
	7.0 ± 0.6 − 11.3 ± 1.8	Diurnal range	RIA
Columba sp. (pigeon)	0.6 ± 0.2 (0.2 − 0.98)		DIDD
	(0.3 − 0.4)		GLC
Agelaius phoeniceus (red-winged blackbird)	(0.4 − 2.7)	Annual range	RIA
Sturnus vulgaris (starling)	0.1 − 0.4	Annual range	CPBA
Insectivora			
Suncus murinus (musk shrew)	1.5 ± 0.1		RIA
	1.6 ± 0.2		RIA
Chiroptera			
Nyctalus noctula (noctule bat)	12.9 − 102.0 (4 − 134)	Annual range	RIA
Pipistrellus pipistrellus (pipistrelle bat)	80.5; 119.8; 83.5	Aug.; Nov.; Feb	GLC
Myotis lucifugus lucifugus (little brown bat)	2.5 ± 0.3 − 11.7 ± 3.2	May−July	
	59.1 ± 9.2	mid-Aug.	RIA
Primates			
Macaca mulatta (rhesus monkey)	8.9 ± 2.0 (1.6 − 14.3)		DIDD
	3.2 ± 0.9 − 9.5 ± 0.9	Annual range	GLC
	4.4 ± 1.1; 4.8 ± 1.4	2 groups	GLC
	8.9 ± 2.0 − 15.5 ± 4.8 (6.0 − 24.0)	Diurnal range ± SD	CPBA
	10.3 ± 0.7 (2.2 − 24.1)		CPBA
	6.7 ± 3.0 (2.0 − 15.6)	± SD	CPBA
	5.0 ± 0.8 − 17.0 ± 1.5 (2.0 − 26.0)	Diurnal range	RIA
	5.9 ± 2.1 − 11.7 ± 5.9	Annual range ± SD	RIA
M. fuscata	(0.2 − 19.8)		RIA
M. irus	2.6 (0.7 − 5.6)		DIDD
Cebus apella (capuchin monkey)	21.8 (2.6 − 50.8)		DIDD
	2.3	n = 1	
Saimiri sciureus (squirrel monkey)	25.3 (11.4 − 39.3)		DIDD
	25.0 (11.0 − 44.5)		
Ateles geoffroyi (spider monkey)	21.7 (4.4 − 38.9)		
	42.4	n = 1	DIDD
Papio sp. (baboon)	10.4 (1.7 − 17.9)		DIDD
Homo sapiens (human)	7.4 ± 2.6 (4.4 − 13.0)	± SD	DIDD
	6.7 ± 0.3 (2.8 − 14.4)		DIDD
	8.0 ± 2.5	± SD	DIDD
	5.8 ± 1.5	± SD	DIDD
	4.2 ± 1.2 (2.6 − 5.8)	± SD	GLC
	6.4 (4.7 − 10.5)		GLC
	2.9 ± 0.4 − 7.8 ± 1.7 (1.3 − 17.7)	Diurnal range	GLC
	6.1 (2.4 − 12.9)		CPBA
	2.9 ± 0.1 − 4.4 ± 0.2 (1.5 − 7.2)	Diurnal range	CPBA
	6.0 ± 0.5 (5.1 − 7.9)		CPBA
	6.8 ± 1.8 (4.5 − 9.6)	± SD	CPBA
	5.3 ± 2.6 (2.4 − 10.0)	± SD	CPBA
	7.1 ± 1.4 (4.5 − 11.0)	± SD	RIA
	12.0 ± 0.5		RIA

Table 6.6—*Continued*

Class or order and species	Testosterone ng/ml mean or mean range ± SE (individual range)	Remarks	Method*
Homo sapiens (cont.)			
	5.9 ± 2.0	± SD	RIA
	4.6 − 6.8	Diurnal range	RIA
	0.8 − 5.5	Diurnal range	RIA
	(2.5 − 13.0)	8 h range for 3 subjects	RIA
	(2.0 − 7.0)	8 h range for 4 subjects	RIA
	3.2 − 7.4 (1.0 − 11.0)	Diurnal range	RIA
	3.0 ± 0.04 − 4.6 ± 0.1		RIA
Carnivora			
Dog	1.9 ± 0.4 (0.5 − 4.3)		CPBA
	0.8 (0.2 − 2.1)		RIA
Vulpes vulpes (English red fox)	0.2 ± 0.1 − 1.1 ± 0.4	Annual range	RIA
Mustela erminea (stoat)	4.5 ± 2.0 − 26.0 ± 4.0	Annual range	GLC
Mustela sp. (ferret)	(7.0 − 83.0)	Annual range	RIA
Martes foina (marten)	0.123 − 9.261	Annual range	RIA†
Perissodactyla			
Stallion	1.5 − 3.2 (? − 5.7)	Annual range	RIA
	(0.1 − 1.6)		RIA
Wild stallion	1.6 ± 0.3 − 3.0 ± 0.4	± SE?	CPBA
Artiodactyla			
Boar	2.9 ± 0.1		RIA
	4.0 − 7.0	At sexual maturity	CPBA
Bull	2.2 − 24.4		GLC
	0.5 ± 0.2	Puberty (1 yr)	GLC
	2.0 ± 0.2 − 23.9 ± 0.4	± SD	CPBA
	(2 − 20)	Diurnal range	CPBA
	9.8 (3.2 − 25.5)	After electroejaculation	CPBA
	1.0 ± 0.002		RIA
	1.8 − 6.2	Diurnal range	RIA
	6.5 ± 1.5 − 8.3 ± 1.3		RIA
	4.4 ± 3.0; 4.3 ± 2.3	12 mo; 15 mo; ± SD	RIA
Ram	11.5 ± 1.5 (6.2 − 20.8)		CPBA
	(0.5 − 10.0)	Jan.−Sept.	CPBA
	(3.0 − 28.0)	Oct.−Dec.	
	3.2 − 4.7	Feb.−Mar.	RIA
	2.0 − 2.6	June−Dec.	
	9.9 ± 1.2		RIA
	(1.6 − 26.0)	Diurnal range	RIA
	(1.0 − 14.0)	Annual range	RIA
	0.7 ± 0.2 − 9.3 ± 1.7	May diurnal range	RIA
	0.9 ± 0.1 − 8.8 ± 3.3	Aug diurnal range	
	4.3 ± 1.6 − 19.7 ± 0.4	Jan diurnal range	
	(0.2 − 15.9)	Diurnal range	RIA
Goat	6.2 ± 0.7 (3.2 − 8.1)		CPBA
Hyracoidea			
Procavia habessinica (rock hyrax)	5.3 (1.6 − 8.5)	Non-breeding	CPBA
	27.1 (15.1 − 58.0)	Breeding	

Table 6.6—*Continued*

Class or order and species	Testosterone ng/ml mean or mean range ± SE (individual range)	Remarks	Method*
Proboscidea			
Elephas maximus	0.7 (0.2 − 1.4)	Nonmusth	CPBA
(Asiatic elephant)	10.7 (4.3 − 13.7)	Premusth	
	44.7 (29.6 − 65.4)	Full musth	
	0.7 (0.2 − 1.2)	Postmusth	
Cetacea			
Tursiops truncatus	(1.5 − 24.0)	Annual range	CPBA
(Bottlenose dolphin)			
Rodentia			
Rat	3.3 ± 0.5 (2.0 − 4.8)		DIDD
	4.5		GLC
	3.9 ± 3.5; 1.5 ± 1.1	18 wks; 33 wks old	GLC
	3.1 ± 0.2		GLC
	7.7 ± 1.6; 5.1 ± 1.2;	60; 70; 80; 90 days	GLC
	5.3 ± 1.8; 4.2 ± 1.1		
	(1.1 − 2.0)		GLC
	7.2 (6.5 − 8.0)		CPBA
	1.8 ± 0.6 − 5.7 ± 0.6		CPBA
	1.8 ± 0.3 − 4.0 ± 0.6		CPBA
	5.8 ± 1.1 (1.8 − 15.3)		RIA
	1.7 ± 0.4 − 3.8 ± 0.9		RIA
	1.9 ± 0.1		RIA
	2.8 ± 0.8		RIA
	0.2 ± 0.03; 0.5 ± 0.1	Conscious; anesthetized	RIA
	1.7 ± 0.2		RIA
	0.8 − 6.0	Diurnal range	RIA
	2.5 − 7.5		RIA
	1.8 ± 0.3 − 2.2 ± 0.3	At sexual maturity	RIA
	6.2 ± 0.7		RIA
Mouse	2.1 ± 0.1 − 3.1 ± 0.3		CPBA
	18.7 ± 4.9 (0.9 − 38.3)	CD-1 strain	RIA
	1.0 ± 0.3 (0.3 − 2.2)	YS strain	
	9.4 ± 0.1		RIA
Guinea pig	3.1 ± 0.7 (1.3 − 5.2)		DIDD
	2.6 ± 0.3 (1.3 − 6.7)		GLC
Lagomorpha			
Rabbit	0.5 ± 0.1; 2.7	Baseline; postcopulation	DIDD
	3.8		GLC
	1.1 ± 0.1		RIA
	(0.5 − 10.0)		RIA
	0.5 − 9.0	Diurnal range	RIA
	3.1 ± 1.1 − 11.9 ± 1.1		RIA

*PHD/UV = phenylhydrazine derivative; DIDD = double isotope derivative dilution assay; RIA = radioimmunoassay; GLC = gas chromatography; CPBA = competitive protein binding assay.
†Audy 1978.
Slightly modified from Gustafson and Shemesh, 1976; reprinted with permission of *Biology of Reproduction*.

than FSH in stimulating testicular 3β-HSD, whereas FSH stimulates spermatogenesis (Andrieux et al., 1973). In AP bullfrogs, bullfrog LH injections but not bullfrog FSH injections increase plasma testosterone and dihydrotestosterone (DHT) concentrations. Neither bullfrog FSH nor ovine PRL affects this response to bullfrog LH (Muller, 1977b).

In vitro, the secretion of DHT and testosterone by bullfrog testes was stimulated by LH from bullfrogs, tiger salamanders, snapping turtles (*Chelydra serpentina*), alligators (*Alligator mississippiensis*), and sheep, but not by bullfrog or oFSH. Moreover DHT was present in the medium in greater amounts than testosterone (Muller, 1977a). This evidence all supports the concept of control of steroid secretion by the interstitial cells through LH and the control of spermatogenesis by FSH, while spermiation may be induced by either of these gonadotrophins.

Reptiles

The excellent review by Licht et al. (1977) is the main basis for the discussion on reptiles that follows. In AP *Anolis carolinensis* males, plasma testosterone can be substantially increased (from less than 0.6 ng/ml to more than 150 ng/ml) by oFSH (treated with LH antiserum). In vitro, oFSH and oLH stimulated androgen production by the testes of *Anolis* and snakes. After acute injection, oFSH and oLH were equally potent in increasing plasma testosterone, but after chronic injection oFSH was more potent.

When the FSH and LH of birds, alligators, sea turtles, snakes and bullfrogs were tested in AP lizards, both hormones increased the plasma testosterone concentration. Similarly, the plasma testosterone concentrations in AP turtles (*Kinosternon* spp.) were increased either by FSH or LH obtained from snapping turtles (*Chelydra*).

In vitro, the testes of all three orders of reptiles respond to FSH or LH from any of the three orders of reptiles tested by an increase in plasma testosterone concentration. The relative potency of a particular hormone, however, may vary among groups, with snake hormones, for instance, being more potent in snakes than in lizards and turtles. The ratio of FSH:LH potency varies with the

source of the hormones and the species used for incubation of the testes (Licht et al., 1977).

Clever use of the circumtesticular capsule of Leydig cells in the lizard *Cnemidophorus*, has shown that it is specifically the Leydig cells that produce the androgens, with oFSH and oLH having the same potency in stimulating androgen production in vitro (Licht et al., 1977).

Birds

The secretion of androgens by the testes of several species of birds is documented in Tables 6.5 and 6.6. Some interesting phenomena have been noted in different species. Nalbandov et al. (1951) noted that after AP, the combs of roosters decreased in size and vascularity. Mammalian LH injections maintained the comb only for a short period or could restore the comb for a short time, after it had atrophied, whereas crude AP extracts from roosters could maintain or restore the comb indefinitely. On the basis of these and other experiments, Nalbandov et al. (1951) proposed the existence of a specific avian LH, with properties different from mammalian LH, i.e., the existence of a third gonadotrophic hormone. This suggestion does not seem to have been tested since purified chicken LH has become available.

Lofts and Marshall (1959) found that AP induced the secretion of progesterone (as measured by the Hooker-Forbes bioassay) by pigeon testes. Progesterone induces incubation behavior in pigeons (Lehrman, 1961), so that the secretion of progesterone by males need not be surprising. However, Silver et al. (1974) found no fluctuations of the plasma progesterone concentration in male ring doves (*Streptopelia risoria*) during the reproductive cycle.

It would be of some importance to repeat Loft and Marshall's experiments, using RIA and competitive protein-binding assays for the progesterone determination in AP pigeon plasma and testes. The crucial experiments to date are those of Brown et al. (1975), in which purified avian FSH and LH were used in AP Japanese quail. In these experiments, LH stimulated the Leydig cells, but unfortunately no testosterone determinations were made.

Dispersed cells of rooster testes, when incubated in vitro produced androgens. The following hormones have been found to stimulate androgen production: toad (*Bufo* sp) AP extract, turtle (*Chrysemys* sp) AP extract, avian LH and FSH and human FSH and LH. Snake (*Natrix* sp.) AP extracts, however, failed to stimulate steroidogenesis under these conditions (Callard et al., 1978).

Mammals

In Chapter 5, we mentioned that many mammals have testes that are either permanently or seasonally in the scrotum and that in still other species the scrotal testes are pulled into the inguinal canal, apparently under the control of the animals' central nervous system. For several scrotal mammals, it has been demonstrated that exposure of the testes to body temperature—either by making the animal cryptorchid or by insulating the scrotum—leads to damage of the germinal epithelium, but leaves the Sertoli cells and the Leydig cells relatively undamaged (see Blackshaw, 1977, for references). More research is required, however, to assess possible damage to Sertoli and Leydig cells, since these cells clearly may make important contributions to normal spermatogenesis.

The concentration of plasma testosterone in different species varies over a wide range, as can be seen in Table 6.6. In the Chiroptera, the plasma testosterone concentration is higher than in any other mammalian order, whereas in Insectivora the concentration has the lowest range, although such a generalization may be premature, since only one species seems to have been investigated. The source of testosterone and androstenedione is the Leydig cells. The secretory activity of the Leydig cells is controlled by the AP and specifically by LH. The following evidence can be presented for this control: (1) specific LH receptors have been identified on Leydig cells (Hansson et al., 1976); (2) in isolated Leydig cells, LH stimulates androgen secretion in vitro; (3) after treatment of male rats with LH antiserum, plasma testosterone concentrations drop sharply from 616 pg/ml to 113 pg/ml, whereas FSH antiserum has no significant effect on plasma testosterone secretion (Raj and Dym, 1976); (4) in *AP* animals, LH

injections stimulate testosterone production (Bartke et al., 1978). This evidence, of course, does not exclude the possibility that other hormones may have an effect on testicular androgen production either alone or by interaction with LH.

There is evidence from various experiments that FSH may play a role in testicular steroid production; e.g., (1) plasma testosterone concentrations in humans are better correlated with plasma FSH than with plasma LH concentrations; (2) the testosterone response to hCG is better correlated with FSH than with LH concentrations; (3) in developing male rabbits, plasma testosterone concentration is correlated with plasma FSH, not with plasma LH concentration; and (4) in *AP* rats, FSH prevents the loss of testicular responsiveness to LH (Bartke et al., 1978). Moreover, FSH may induce LH receptor formation in the testes and thus may modulate the response of the testes to LH. In order to have these effects, however, FSH would need to bind to Leydig cells, and there is no evidence that it does (Bartke et al., 1978).

Another AP hormone that may play a role in the control of Leydig cell steroidogenesis is PRL. Bartke et al. (1978) cite the following kinds of evidence:

1. PRL potentiates the effect of LH in restoring spermatogenesis and in stimulating testosterone production in *AP* mice.

2. Testes of dwarf mice, which are deficient in PRL, show greater testosterone production in response to PRL plus hCG than in response to hCG alone.

3. PRL increases the binding of LH by Leydig cells of dwarf mice, of *AP* rats, and of the regressed testes of hamsters on short photoperiods.

4. Treatment of immature male rats with bromoergocryptine, which inhibits release of PRL, reduces the binding of LH by the testes.

5. Iodinated PRL becomes localized in the interstitial cells.

6. The concentration of esterified cholesterol in the testis increases after PRL administration. Esterified cholesterol is a precursor of cholesterol, and thus of testosterone. It is thus possible that the potentiating effect of PRL on testosterone production in LH-treated *AP* rats is the result of a greater availability of the precursor. Ewing and Brown

(1977) dismiss this as an unlikely mechanism without presenting data to support their position.

7. PRL administration increases plasma FSH concentrations in immature female rats and in dwarf mice, and thus, through its effect on FSH secretion, promotes steroidogenesis. Rats with the testicular feminization syndrome (Tfm) provide evidence that androgens may affect the Leydig cells. In such rats, the proportion of Leydig cells to other testicular cells is greater than in normal rats. Such rats also show the paradox of low plasma testosterone concentrations in spermatic vein blood, but elevated systemic plasma LH concentrations (Hansson et al., 1976). Such a discrepancy could be the result of a defective androgen biosynthesis, probably as the result of a sharp reduction of LH receptors on the Leydig cells of Tfm rats. Hansson et al. (1976) have presented evidence that androgens are required for differentiation of the Leydig cells at puberty and may be required for the induction of LH receptors on the Leydig cell membrane. In vitro incubation

of testes with pregnenolone results in production of testosterone by normal testes, but of androstenedione by Tfm testes, thus suggesting 17β HSD deficiency. Furthermore, adult Tfm rats have a higher 5α-reductase activity than normal adults. Thus the low testosterone concentrations in Tfm rats may result from a synthesis of testosterone lower than normal and a metabolism higher than normal.

Although the Sertoli cells have either a low capacity, or lack the capacity, to form androgens from either cholesterol or pregnenolone (Dorrington et al., 1978), they can convert testosterone to estradiol. This estradiol synthesis by the testes may be a two-cell compartment process (Figure 6.15), with the estrogen formed under the influence of FSH. It has been found that the in vitro formation of estradiol by Sertoli cells in 16–20 d old rats is considerably higher than in 30 and 40 d old rats. The physiological significance of estradiol in the male is still not clear.

ENDOCRINE FUNCTION: INHIBIN

After destruction of the germinal epithelium by either X-rays, cryptorchidism, or deficiency of vitamin A or vitamin E, so-called castration or signet ring cells appear in the AP, and FSH secretion increases (Setchell et al., 1977). Further investigation has revealed that the Sertoli cells (Steinberger, 1981) secrete what is probably a glycoprotein with a molecular weight of 15,000–20,000, with an isoelectric point between pH 5.0 and 6.0 (de Jong et al., 1981). In the rat, this protein, called *inhibin*, decreased the endogenous LH-RH concentration of isolated rat hypothalamic tissue incubated in vitro, decreased FSH secretion, decreased LH-RH-induced FSH secretion, and in higher doses decreased LH secretion (Franchimont et al., 1981). In addition to these endocrine feed-back effects, inhibin inhibits DNA synthesis of dividing spermatogonia of rats (Demoulin et al., 1981). In rams, injections of rete testis fluid, which has a high concentration of inhibin, results in a long-lasting depressing effect on FSH secretion, a short-lasting depressing effect on LH secretion, and disappearance of the pulsatile LH secretion pattern of cryptorchid

Figure 6.15 Model of the two-cell–two-gonadotrophin hypothesis for testicular estradiol synthesis. (From Dorrington et al., 1978; reprinted with permission of *Biology of Reproduction*.)

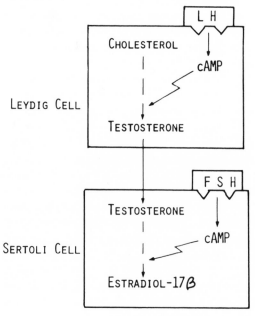

rams, in which FSH and LH secretion are higher than in intact controls, but in which pulsatile LH secretion patterns are similar to those in controls (Blanc et al., 1981). The role of inhibin in normal male reproduction remains to be determined. Tentatively one can speculate that it may have a function in seasonally breeding species, in which the testes regress at the end of the breeding season. An increase in inhibin secretion at the end of the breeding season would inhibit GTH secretion and spermatogonial mitosis.

The Sertoli cells, as we have seen, have many functions, such as phagocytosis (Chapter 5), nourishment of germ cells, secretion of ABP, secretion of inhibin and, in the embryo, secretion of the Müllerian duct-inhibiting factor (Chapter 2). Inhibin secretion is reduced after AP, but can be restored by either FSH, LH, or testosterone treatment, the effects of FSH and testosterone being direct and the effect of LH probably being indirect by stimulating testosterone secretion. Exposure of the testes to high temperatures by experimental cryptorchidy reduces the ability of Sertoli cells to secrete inhibin in in vitro cultures (Steinberger, 1981). The decreased inhibin secretion during cryptorchidy explains the changes in the AP and the increased secretion of FSH, but one would expect an increase in dividing spermatogonia; instead, the germinal epithelium degenerates. Steinberger (1981) has speculated that impairment of other Sertoli functions may be the cause of the degeneration of the germinal epithelium. One of these impaired functions, is the secretion of ABP (Hagenas and Ritzen, 1976), which may partly explain the damage to the germinal epithelium.

FUNCTION OF PAMPINIFORM PLEXUS AND SCROTUM

In scrotal mammals, the testes are at a temperature that is lower than that of the body cavity. Two principal mechanisms are involved in the process of maintaining this lower-than-body temperature. First, in most species, the tunica dartos, a smooth muscle under the scrotal skin contracts at low temperatures, thus bringing the testes closer to the body, and relaxes at high temperatures, thus allowing greater heat loss through the scrotal skin. Interestingly, however, in Afrikaander cattle—living in very high environmental temperatures (above body temperature)—the tunica dartos muscle contracts and brings the testes closer to the (cooler) body (Bonsma, 1940, as cited by Hodson, 1970; Waites, 1970). The cremaster muscle (Figure 6.16) is a striated muscle that is not capable of sustained contraction, and its role in temperature regulation is not certain (Waites, 1970). Second, there is an efficient heat exchange mechanism between the cooler testicular vein blood and the warmer internal spermatic arterial blood. This heat exchange is possible because of the extensive coiling of the artery, as illustrated in Figure 6.16. This coiled artery and associated vein is called the *pampiniform plexus*.

The significance of maintaining the testes at lower-than-body temperature is shown by the following effects of body temperature: Exposure of the testes of scrotal mammals either for a short time (e.g., 46–47°C for 15–30 minutes in the guinea pig) or for a longer duration (1 to 2 weeks, depending on the species) to body temperature, or for 2½ months (in sheep) by insulation of the scrotum, results in damage to the germinal epithelium, as does high environmental temperature or fever. Indeed, exposure to body temperature may lead to irreversible damage and permanent sterility (VanDemark and Free, 1970).

In a species that normally has scrotal testes, some animals have testes that remain in the body cavity or in the inguinal canal; this is called *cryptorchidism*. In some species, e.g., the rat, the inguinal canal remains open after testicular descent, and one can make rats cryptorchid experimentally by pushing the testes into the inguinal canal and ligating the canal. Studies with cryptorchid bulls have revealed that: (1) plasma LH concentrations are higher than in controls; (2) episodic releases of LH have greater amplitude and occur with greater frequency than in controls; (3) plasma FSH concentrations are higher than in controls; (4) after administration of LH-RH, the plasms FSH and LH concentrations increase more rapidly and to higher values than in similarly treated controls; (5) plasma testosterone concentrations are not significantly different from those of controls; and (6) the increase in plasma testosterone after LH administration is similar to cryptorchid and normal bulls (Schanbacher, 1979).

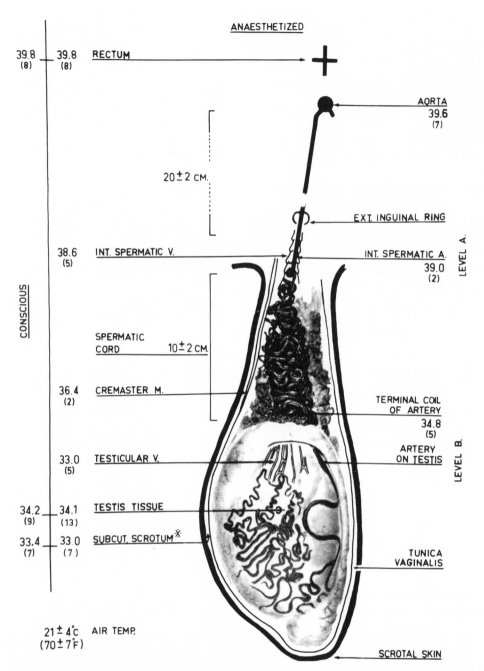

Figure 6.16 Anatomical relationships and experimental temperature readings at various sites in the scrotum of conscious and anesthetized rams. The internal spermatic artery had been filled with neoprene, and the cast had been exposed by removal of the pampiniform plexus and tunica albuginea over the artery on the testis. Figures in parentheses give the number of temperature readings from which the average was obtained. (From Waites and Moule, 1961; reprinted with permission of G. M. H. Waites and *Journal of Reproduction and Fertility*.)

*Subcutaneous scrotum temperature readings were taken beneath the posterior skin and not the anterior (shown here for purposes of illustration only). The temperature in the artery of the testis was 34.4°C., based on seven observations.

EFFECTS OF ANDROGENS

Androgens have many important coordinating functions, some of which are listed here:

1. They inhibit gonadotrophin secretion and also affect the type of gonadotrophin secreted. Bogdanove et al. (1975) found that the FSH secreted by intact male rats differed from the FSH secreted by castrated males with respect to (*a*) a change in the ratio of potency as determined by the bioassay to potency as determined by radioimmunoassay (RIA), such that FSH of castrated rats had a higher biological activity of FSH with respect to the RIA than the FSH of intact rats; (*b*) the molecular weight of the FSH of castrated rats being higher than that of the FSH of intact rats; and (*c*) the clearance rate, and thus the survival, in the circulation of FSH from castrated rats being greater than that of FSH from intact rats.

2. In killifish, the Indian catfish, pigeons, and the mammals investigated, androgens can either maintain or restore spermatogenesis in *AP* animals.

3. Androgens play an important part in regulating the physiology of the epididymis. In the lizard *Lacerta vivipara,* the secretion of granules that are mixed with the sperm is androgen dependent (Gigon-Depeiges and Dufaure, 1977). The function of these granules is not known. In bulls, Amann and Ganjam (1976) showed that androgens are transferred across the pampiniform plexus. This transfer raises the concentration of androgens in the epididymis to twice that found in the peripheral blood. In the epididymis, the spermatozoa gain in fertilizing capacity (sperm maturation). This sperm maturation is androgen-de-pendent (Orgebin-Crist et al., 1975) and is probably exerted through secretions by the epididymal epithelium, but what these substances are and how they mediate their effect is not known.

4. In mammals, androgens are required for the maintenance of the fertilizing capacity of sperm in the ductus deferens; in chickens this does not appear to be necessary (van Tienhoven, 1968).

5. Androgens are required for maintenance of the seminal vesicles of Indian catfish (Chapter 5) and of the secondary sex glands of mammals.

6. The thermoregulatory function of the scrotum is androgen dependent (Waites, 1977).

7. The development of the anal fin, which serves as a copulatory organ in some fishes, is androgen dependent.

8. Many secondary sex characteristics, such as the nuptial colors of fish, the comb of the rooster, the development of the thumb pads, the vocal apparatus, and the muscular development of the forelimbs of frogs, and the development of the antlers of deer are all androgen dependent. One needs little imagination to understand the adaptive value of these characteristics.

9. Androgens, at least in birds and mammals, increase nitrogen retention and thus promote muscular development. The use of metabolic androgens (which have a high metabolic potency but a low masculinizing effect) by weight lifters and other athletes and by older people who may suffer from muscular weakness is based on this nitrogen-sparing effect.

10. Androgens have profound effects in a number of classes of animals (possibly with the exception of amphibia) on aggressive and sexual behavior (Chapter 14).

The Ovary

7

Like the testis, the ovary performs two main functions, gametogenesis and hormone production. There are other similarities, as well as some differences between testes and ovaries, some of which are summarized in Table 7.1.

OOGENESIS

We must first discuss the process of meiosis in the ovary (Figures 7.1 and 7.2). The oogonia undergo mitotic divisions and yield primary oocytes, which undergo a meiotic division that results in a secondary oocyte (2n chromosomes) and a polar body. The secondary oocyte by a meiotic division yields the ovum and a second polar body each with n chromosomes. This series of processes, which is concerned with the nuclear divisions, is called oogenesis, and should not be confused with growth of the follicle and with vitellogenesis (yolk formation), which may be quite spectacular, as in avian and reptilian ovaries, for example.

Over the years there has been some controversy about whether in some species (cyclostomes, elasmobranchs, a few teleosts, possibly a few reptiles, all birds, and all or most mammals), mitoses of oogonia occur after sexual maturity has been achieved (Zuckerman and Baker, 1977). For rabbits, Kennelly and Foote (1966) clearly demonstrated, in a classical paper, that neogenesis of

oocytes did not occur. They labeled the oogonia by injecting [^{3}H]-thymidine systemically on the day of birth; then at different ages they collected the ovaries or, after sexual maturity, secondary oocytes after superovulation. Since 90 percent of the oocytes in these rabbits between age 12 and 20 weeks and 80 percent of the ovulated oocytes carried the [^{3}H]label, it was clear that the labeled thymidine had not been diluted. If mitosis had occurred, the labeled thymidine should have been diluted. This convincing evidence has not been countered by other experiments (Zuckerman and Baker, 1977).

The primary oocyte is enclosed in a follicle and, depending on the species one is considering, either follicular growth or the deposition of yolk are the major changes that occur between birth and the preovulatory changes. We shall discuss the control of follicular growth and oocyte maturation for different classes of vertebrates, although our knowledge, especially for nonmammalian classes, is fragmentary, and it is speculative to state that a pattern exists. If one is willing to speculate, however, it appears that the ovary is more autonomous in the lower vertebrates than in the higher vertebrates. Much more evidence needs to be obtained, however, especially in teleost fishes, before we can move from the realm of speculation to that of evolutionary trends.

Table 7.1

Comparison of the ovary and the testis

Characteristic	Ovary	Testis
Location	Abdominal in all species	Abdominal in inframammalian animals, scrotal in many mammals
Symmetry	In many reptiles, most birds, only left ovary developed; in duckbill platypus and some bats only 1 ovary developed	In most species, two testes function
Origin in elasmobranchs and tetrapods	Cortex of indifferent gonad	Medulla of indifferent gonad
Seasonal cycles	In many species	In many species
Cyclic phenomena during breeding season	In many mammals, pronounced cycles	No pronounced cycles
Hormones secreted	Estrogens Progestins Androgens	Androgens Progestins Estrogens
Control by gonadotrophins	Follicular growth by FSH + LH Interstitial cells by LH Corpus luteum by LH (?)	Spermatogenesis by FSH or LH (?) Interstitial cells by LH
Mitosis	Mitosis continues in teleosts, amphibians, most reptiles after sexual maturity No mitosis after sexual maturity in cyclostomes, elasmobranchs, birds, and most or all mammals	Mitosis after sexual maturity

From van Tienhoven, 1968; reprinted with permission of W. B. Saunders.

Cyclostomes

In the lamprey (*Petromyzon*), hypophysectomy (AP) prevents the increase in dry weight of the oocytes, and the oocytes remain morphologically immature, but there is *no atresia* of the follicles. Growth of the oocytes and vitellogenesis are possible in AP *Lampetra fluviatilis* (Larsen, 1965). In the female hagfish, *Eptatretus stouti*, AP has, as in the case of the male, little effect on the gonads. Ovulations and estrogen secretions are not different for AP fish, controls, and fish treated with hagfish pituitaries (Matty et al., 1976).

Elasmobranchs

Total hypophysectomy of the lesser spotted dogfish (*Scyliorhinus canicula*) does not affect the rate of vitellogenin (a yolk precursor) formation or the rate of synthesis of yolk granules; but it does delay removal of vitellogenin from the plasma (Craik, 1978). Craik (1978) obtained similar results after removal of the neurointermediate-median-rostral lobe complex as after total hypophysectomy, but ventral lobectomy did not have this effect.

Removal of the ventral lobe of the hypophysis resulted in atresia that could be detected about three weeks after the operation. Resorption of the yolk from the atretic follicles may take as long as 14 months (Dodd, 1975), and Craik (1978) proposed that the ventral lobe may control the transport of yolk into the follicles. Ovulation is, according to Dodd (1975), probably under control of the hypophysis. It is not known which AP hormones are involved in the control of either yolk deposition or ovulation. So far only one gonadotrophic hormone has been isolated from dogfish AP (Sumpter et al. 1978).

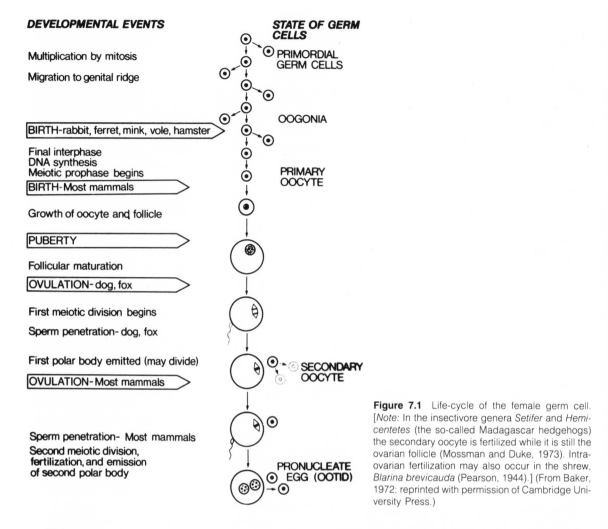

DEVELOPMENTAL EVENTS

Multiplication by mitosis

Migration to genital ridge

BIRTH-rabbit, ferret, mink, vole, hamster

Final interphase
DNA synthesis
Meiotic prophase begins
BIRTH-Most mammals

Growth of oocyte and follicle

PUBERTY

Follicular maturation
OVULATION-dog, fox

First meiotic division begins

Sperm penetration- dog, fox

First polar body emitted (may divide)
OVULATION-Most mammals

Sperm penetration- Most mammals
Second meiotic division,
fertilization, and emission
of second polar body

STATE OF GERM CELLS

PRIMORDIAL GERM CELLS

OOGONIA

PRIMARY OOCYTE

SECONDARY OOCYTE

PRONUCLEATE EGG (OOTID)

Figure 7.1 Life-cycle of the female germ cell. [*Note:* In the insectivore genera *Setifer* and *Hemicentetes* (the so-called Madagascar hedgehogs) the secondary oocyte is fertilized while it is still the ovarian follicle (Mossman and Duke, 1973). Intra-ovarian fertilization may also occur in the shrew, *Blarina brevicauda* (Pearson, 1944).] (From Baker, 1972; reprinted with permission of Cambridge University Press.)

Teleosts

The AP probably controls oogonial proliferation since AP reduces oogonial proliferation in the plaice (*Pleuronectes platessa*) and eliminates it in the goldfish (*Carassius auratus*) (deVlaming, 1974). The proliferation of oogonia in AP goldfish can be restored by carp or salmon AP extracts (deVlaming, 1974). The growth of yolkless oocytes is apparently independent of the AP (deVlaming, 1974). In vitro culture of goldfish ovaries suggests that oogonial mitoses, maturation to the meiotic prophase, and previtellogenic growth can occur, in succession, independent of pituitary gonadotrophins; however, survival of vitellogenic oocytes was improved by parabiosis of AP and ovary of goldfish, and by carp AP extract, and to a

lesser extent by hCG and testosterone (Remacle et al., 1976).

The most extensive investigations of the hypophyseal control of the ovary have been carried out with the Indian catfish (*Heteropneustes fossilis*). Hypophysectomy, during the early spawning period, when ovaries are large and the follicles are filled with yolk, results in atresia. This atresia can be prevented by the injection of as little as 5 μg of partially purified salmon GTH (Sundararaj et al., 1972a), or 1 μg of purified GTH from the puntius carp (*Puntius gonionotus*) (Sundararaj et al., 1976).

Ng and Idler (1978) were able to extract two gonadotrophic hormones from the pituitaries of *Hippoglossoides platessoides,* both of which ex-

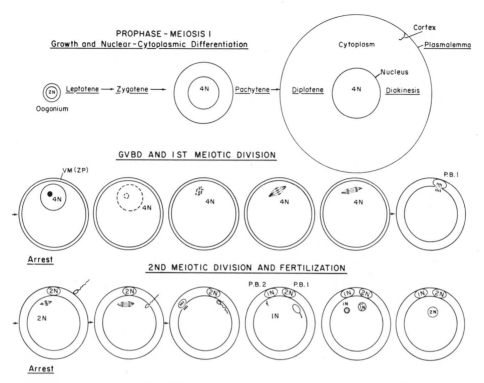

Figure 7.2 Diagrammatic representation of the scope of the oocyte maturation process. Figure illustrates the major structural (nuclear and cytoplasmic) changes that occur in oocytes from the time of oogonium formation through reconstitution of the two N or diploid state following fertilization. Based on the oocyte maturation process in amphibians and most mammals. GVBD = germinal vesicle breakdown, P.B.1 = first polar body, P.B.2 = second polar body, VM = vitelline membrane, ZP = zona pellucida. (From Schuetz, 1974; reprinted with permission of *Biology of Reproduction*.)

isted in two forms, with molecular weight of 28,000 and 62,000. One gonadotrophin either had a low carbohydrate content or was a nonglycoprotein, and both forms stimulated vitellogenesis and ovarian growth in AP winter flounders (*Pseudopleuronectes americanus*). The other gonadotrophin was a glycoprotein and both forms were capable of stimulating oocyte maturation and ovulation in AP winter flounders. This is the first physiological evidence in teleosts for the existence of two gonadotrophins (Ng and Idler, 1978). The evidence for two gonadotrophins in teleosts is further substantiated by ultrastructural evidence of the presence of two types of gonadotrophic cells in the pituitary of the three-spined stickleback (*Gasterosteus aculeatus*) (Slijkhuis, 1978).

In vivo, estradiol (E_2), estrone (E_1), and estriol (E_3) can each partly maintain the follicles of AP gravid (with large yolky follicles) catfish, but testosterone, progesterone, and corticosteroids are marginally effective (i.e., maintenance is better than in controls, but not as good as with the estrogens), whereas mammalian LH (20 µg), hCG (25 IU) and to a somewhat lesser extent PMSG (40 IU) can maintain the follicles so that they are only slightly smaller than in intact controls (Anand and Sundararaj, 1974).

It is extremely interesting that the gonadal hormones (pregnenolone, 17α-OH pregnenolone, 17α-OH progesterone, testosterone, E_1, E_2, and E_3) were either ineffective or only marginally effective in inducing oocyte maturation of Indian catfish in vitro, but that hydrocortisone and deoxycorticosterone were quite effective at doses of 1–10 µg/ml (Goswami and Sundararaj, 1971). Experiments by Truscott et al. (1978) have shown that injections of oLH or of porcine ACTH in Indian catfish with either regressed follicles or large yolky follicles, were followed by an increase in plasma cortisol concentrations. Injections of purified salmon gonadotrophin into gravid Indian catfish also increased plasma cortisol concentrations as did oLH and porcine ACTH injections in ovariectomized animals. These results suggest that the adrenal of this fish is stimulated to secrete cortisol not only by a mammalian ACTH, but also by oLH and a fish gonadotrophin. As we shall discuss later in this chapter in the section on ovulation, corticosteroids play an important role in ovulation in some teleosts.

In the Japanese rice fish (*Oryzias latipes*), cortisol is probably secreted by the ovary and is more potent than progesterone in inducing oocyte maturation. In the rainbow trout (*Salmo gairdneri*), the northern pike (*Esox lucius*), and the goldfish, 17α-hydroxy-20β-dihydroprogesterone is the strongest inducer of oocyte maturation (Jalabert, 1976). It is thus of considerable interest that 17α-OH progesterone (which also induces oocyte maturation in rainbow trout, northern pike, and goldfish) and 17α-hydroxy-20β-dihydroprogesterone (17α-20β Pg) have been found in the plasma of prematuring and maturing winter flounder. Either of these two steroids added to plasma will induce maturation of the oocytes of this species in vitro. The changes in concentration of these steroids between premature fish and mature fish suggest that 17α-OH-progesterone is more important than 17α-OH dihydroprogesterone (Campbell et al., 1976). Goetz and Bergman (1978) tested the effect of progestins, androgens, and corticosteroids on oocyte maturation in vitro on brook trout (*Salvelinus fontinalis*) and yellow perch (*Perca flavescens*). They concluded that androgens and 11-oxygenated corticosteroids were less effective than progestins and the other corticosteroids. The events occurring during oocyte maturation in the rainbow trout are presented in Figure 7.3.

Figure 7.3 Tentative scheme for the control of oocyte maturation in trout. GVBD = germinal vesicle breakdown. (From Jalabert, 1976; reprinted with permission of *Journal of the Fisheries Research Board of Canada*.)

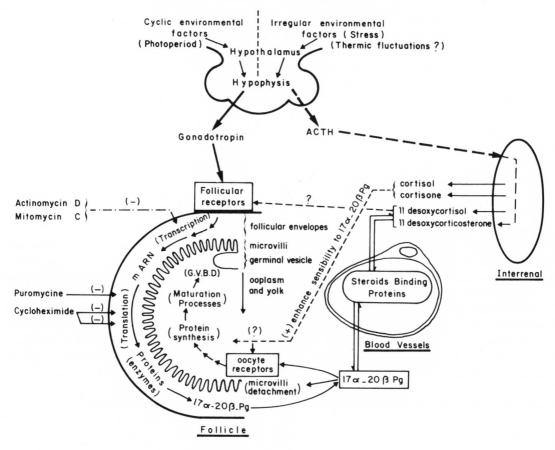

Amphibians

The follicular growth in the ovary of the toad (*Bufo bufo bufo*) has been particularly well studied, and the following account is largely taken from the publications of Barker-Jørgensen (1974, 1975). The smallest oocytes are surrounded by a single layer of flat follicle cells. This is the previtellogenic growth phase, or first growth phase (FGP), during which the cell diameter may increase ten times and the cortical cytoplasm develops microvilli toward the follicle cells. This first growth phase ends with the appearance of these microvilli.

The second growth phase (SGP) consists of yolk accumulation, during which the oocyte becomes surrounded by an epithelial, a thecal, and a granulosa layer. The oocyte may increase another 10 times in size as a result of this yolk accumulation (Barker-Jørgensen, 1974). The hormonal control of vitellogenesis, i.e., the synthesis of yolk precursors, which is essential for yolk accumulation to occur, will be discussed under ''vitellogenesis'' in this chapter.

The effects of hCG on the ovaries of *AP Bufo bufo bufo* depend on the state of the ovary. When no SGP follicles are present, hCG promotes the recruitment of FGP follicles to SGP follicles (Barker-Jørgensen, 1974, 1975). If, however, SGP follicles are present, hCG maintains the SGP follicles but does not induce the recruitment of FGP follicles to SGP follicles (Barker-Jørgensen, 1974).

In a recent publication, Browne et al. (1979) provided evidence that in the African clawed toad (*Xenopus laevis*), hCG regulates either the opening or the assembling of gap junctions between follicle cells and oocytes, or both. The function of these gap junctions has not been determined, but Browne et al. (1979) speculate that a factor that affects endocytotic uptake of vitellogenin by the oocyte may be transmitted from follicle cells to oocyte via these gap junctions.

The exact role of amphibian gonadotrophins in the regulation of follicular growth and oocyte maturation awaits experiments either with purified amphibian gonadotrophins in *AP* amphibia or experiments with specific antisera to amphibian gonadotrophins.

Maturation of amphibian oocytes will be dis-

cussed in some detail in the section on ovulation in this chapter. It is sufficient here to state that mammalian LH-induced resumption of meiosis seems to be mediated by progesterone and/or its metabolites 20β-progesterone and 17α, 20β-hydroxy-progesterone (Thibault, 1977).

Reptiles

In reptiles, five patterns of ovulation can be distinguished (Smith et al., 1972): (1) In *mono-allochronic* ovulation, one oocyte is released from an ovary and later another oocyte is released from the other ovary, but the release is so regulated that there is never more than one egg in the oviduct. This pattern is regularly found in the genera, *Anolis, Chamapleosis, Chamaelinorops, Phenacosaurus,* and *Tropidodactylus.* (2) In *poly-allochronic* ovulation, more than one oocyte is released from one ovary. According to Smith et al. (1973), this pattern may be hypothetical because it has not been found in this class. (3) In *polyautochronic* ovulation, more than one oocyte is released simultaneously or nearly simultaneously from the two ovaries. Despite the presence of only one functional oviduct, this pattern is found in some species of *Tantilla* and all investigated species of *Ramphotyphlops, Typhlops, Leptotyphlops, Helminthophis,* and *Anomalepis.* (4) In *monoautochronic* ovulation, one oocyte is released synchronously from each of the two ovaries. This pattern has been found in most geckos and in *Xantusia vigilis* and *Carlia (Leiolopisma) rhomboidalis.* (5) In *monochronic* ovulation, only one ovary is functional; I do not know whether this pattern occurs in reptiles.

These patterns of ovulation may be determined to some extent by the patterns of ovarian growth. Jones et al. (1978) have pointed out that each ovary contains a small region on its dorsal surface, the so-called *germinal bed,* from which growing follicles originate during embryonic development. Some species of lizards, for example, *Xantusia vigilis* and *Carlia rhomboidalis,* have two such germinal beds per ovary (Jones et al., 1978). In these two species, many follicles grow, but the number that might ovulate is reduced by atresia, so that the ovulation pattern is reduced to the monoautochronic pattern.

Jones et al. (1978) have analyzed the follicular growth of *Lepidodactylus lugubris,* a member of the Gekkonidae, which has one germinal bed per ovary. In this species, the hierarchy of follicular sizes is maintained by the recruitment from the smaller follicles, if a large follicle is ovulated or has become atretic. Atresia in this species occurs in about 17 percent of the large follicles, whereas in *Xantusia vigilis* it occurs in 54 percent of the large follicles (Jones et al., 1978).

In reptiles, the control of the ovary by the AP has probably been studied most extensively in the American chameleon (*Anolis carolinensis*). Hypophysectomy leads to atresia of the yolky follicles, which is more effectively prevented by administration of mammalian FSH than by mammalian LH. Tokarz (1978), using AP American chameleons during the reproductively quiescent season, found that oFSH, but not oLH, caused incorporation of [³H]-thymidine into surface epithelial cells, prefollicular cells, and oogonia (in this species the oogonia undergo mitotic divisions after the animal is sexually mature). Injections of oFSH also increased the number of oogonia and the ratio of labeled to unlabeled oogonia. These effects of oFSH are apparently not mediated by estradiol, because administration of estradiol-17β did not have any of the effects of oFSH.

After AP, the small follicles of the American chameleon do not become atretic (Jones et al., 1976), suggesting that the maintenance of small follicles does not require gonadotrophins, but that their growth and the subsequent maintenance of the yolky condition do. In intact juvenile American chameleons, oFSH (1) stimulates differentiation and enlargement of oocytes, (2) stimulates formation of follicles and differentiation of the granulosa cells, and (3) maintains the hierarchy of the follicles. This effect of FSH on the ovary is mediated by estrogens and is the result of a differential increase in the vascularity of the largest follicles (Licht et al., 1977).

Birds

In the domestic chicken, the ovary undergoes a spectacular increase in weight at the time of sexual maturity; it increases from about 1 g at 120 days old to 20 g at 180 days (Nalbandov and James, 1949). In the earlier-maturing modern strains of chickens, this change probably occurs earlier, and the rate at which the ovary increases in weight may be even greater. The increase in weight of the ovary is the result of yolk deposition, which is discussed in the section on vitellogenesis in this chapter. As pointed out in Chapter 4, such yolk deposition in the chicken does not occur before about 100–120 days, even when vitellogenesis has been induced by estrogens and the estrogen-treated pullet is injected with chicken pituitary extracts.

In the sexually mature chicken, AP results in atresia, but injections of either mammalian gonadotrophins or of chicken pituitary extract prevent the atresia, and ovulation by such follicles can be induced with either of these two types of preparations (Nalbandov, 1976). In intact birds, the effects of mammalian gonadotrophin treatment seem to depend on the species. In chickens with developed ovaries, mammalian gonadotrophins (FSH, PMSG) injections stimulate the growth of a large number of follicles, so that the hierarchy of these structures is lost and spontaneous ovulations stop. In song birds, the administration of PMSG stimulates the follicles, so that the hierarchy among the follicles normally found during the breeding season results (see van Tienhoven 1961 for review). The reason for this difference in response between the chicken and song birds are not clear.

Although purified chicken FSH and LH have been prepared (Stockel-Hartree and Cunningham, 1971), no experiments seem to have been carried out with such preparations on AP hens. The development of the right rudimentary gonad after sinistral ovariectomy has been discussed in Chapter 5.

Mammals

In the mammalian ovary (at least in the cat, mink, ferret, and mouse), the rete ovarii plays an important role in oogenesis and in follicle formation (Byskov, 1975, 1978). The rete cells, which are probably forerunners of the follicular cells, specifically of the granulosa cells, are secretory and secrete a meiosis-inducing substance by which they presumably affect the germ cells, since the germ cells in contact with the rete cells are the first to begin meiosis.

In addition to this intraovarian control of oogenesis, the control by the anterior pituitary is of particular importance. We have already pointed out in Chapter 6 the difficulties of interpretation of experiments in which AP animals are used, and we have emphasized the advantages of using FSH and LH antisera in arriving at an understanding of the role of these hormones in regulation of ovarian function.

Schwartz (1974), in an excellent critical review, has considered (1) the control by the AP of follicular growth, maturation, and ovulation, and (2) whether FSH and LH are secreted as one gonadotrophic complex or as two separate hormones. The salient points of her findings are:

1. At the end of oogonial mitosis, the number of primordial follicles is at its maximum and the subsequent decrease in these follicles is the result of (a) atresia, which occurs after the follicle has reached the 21- to 69-cell stage (Figure 5.30) or has progressed further, and (b) the maturation of the follicle to a type 8 follicle (Figure 5.30) and the subsequent ovulation, after which the follicle is converted to a CL. These two pathways are illustrated in Figure 7.4.

2. The early postnatal growth of the follicle of the mouse may be independent of the AP, but after AP, the initial development of follicles and atresia continue, although the rate of loss of follicles decreases (Schwartz, 1974). Nakano et al. (1975), using [3H]-thymidine labeling found that in immature AP rats, the growth of follicles continued and that the regressions of the number of labeled granulosa cells on the number of granulosa cells in the follicle were not different between intact and AP rats. Schwartz (1974) has suggested that GTH increases the rate of entry of follicles from the nonproliferating into the proliferating pool of follicles (Figure 7.4).

3. In the absence of gonadotrophins, no follicles grow beyond type 5 (Figure 5.30), according to Schwartz (1974). Lunenfeld et al. (1975), using antiserum to rat gonadotrophin, found that in the mouse, few follicles grew beyond the 40-cell stage, although the total number of growing follicles was not changed; however, injections of FSH or human menopausal gonadotrophin (which has FSH and LH properties) increased the number of follicles growing beyond the 40-cell stage. These observations are in accord with the findings of Edwards et al. (1977) in AP rats treated with either FSH or hCG; FSH was effective in restoring follicular growth in the absence of stimulation of steroidogenesis.

4. The use of LH antiserum and FSH antiserum after day 5 did not permit a definite conclusion about the role of either LH or FSH in early postnatal follicular growth.

5. In the AP juvenile prepubertal rat, oFSH can cause antrum formation, and LH can restore the interstitial cells of the ovary without inducing uterine growth or follicular growth. (FSH and LH

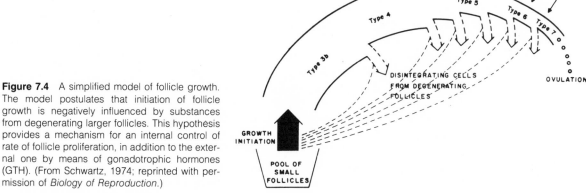

Figure 7.4 A simplified model of follicle growth. The model postulates that initiation of follicle growth is negatively influenced by substances from degenerating larger follicles. This hypothesis provides a mechanism for an internal control of rate of follicle proliferation, in addition to the external one by means of gonadotrophic hormones (GTH). (From Schwartz, 1974; reprinted with permission of *Biology of Reproduction*.)

are required for estrogen secretion, as will be discussed later in this chapter.)

6. Follicular growth in adult mammals requires the presence of the AP, or, in the AP̸ animal, it requires administration of gonadotrophins. However, the exact role of FSH or LH, or the necessity of a synergism between them, has not been fully elucidated. Peters et al. (1975) have proposed that FSH prevents resting large follicles from becoming atretic without inducing growth in such follicles.

7. Hirschfield and Midgley (1978) have presented persuasive evidence that the release of FSH during one estrous cycle of the rat may stimulate the growth of follicles of about 200–400 μm which are destined to be ovulated during the next cycle.

In addition to a possible direct effect of GTH on the follicles, there is probably an indirect effect, through stimulation of steroid secretion. There is evidence that gonadal hormones can affect the follicles. First, diethylstilbestrol (DES) prevents atresia and the decrease in ovarian weight after AP̸ of rats, and it allows the formation of medium-sized follicles (Louvet et al., 1975). In AP̸ rats, estradiol 2 mg/day) stimulates preantral growth of follicles, and increases the estrogen receptors in the ovary and the response of granulosa cells to FSH (Richards, 1975). Injection of LH into such estrogen-treated AP̸ rats causes follicular atresia, whereas administration of FSH + LH + estradiol to AP̸ rats induces luteinization. Treatment of AP̸ immature female rats with DES also stimulates the production of a protein, with the characteristics of an androgen receptor, in the cytosol of the ovaries (Schreiber et al., 1976; Schreiber and Ross, 1976). As mentioned below, androgens may be important in the preantral follicular development, so that DES would thus also provide for the transport mechanism of androgens. Second, this effect of DES is inhibited by injections of small doses of hCG (0.3-3 IU), LH (3 IU), but not by FSH (Louvet et al., 1975). Third, this countereffect of LH and of hCG to DES is, in turn, prevented by the administration of antiserum to testosterone (Louvet et al., 1974). Finally, LH or hCG injections into intact immature female rats result in a decrease in ovarian weight, presumably because

the secretion of androgen in response to these hormones inhibits follicular growth (Louvet et al., 1975).

VITELLOGENESIS

In species of oviparous groups (e.g., elasmobranchs, teleosts, amphibia, reptiles, birds, and monotremes), the follicles contain substantial amounts of yolk, and the spectacular increase in size of the follicles in such animals consists largely of the deposition of yolk. Yolk precursors are formed in the liver in response to estrogens, or when fat bodies (Chapter 6) are present, estrogens may mobilize fats to restructure these precursors in the liver. A generalized scheme for the control of amphibian vitellogenesis by gonadotrophins is presented in Figure 7.5.

Gonadotrophic hormones stimulate estrogen secretion by the ovary; even in immature chickens, in which follicular growth cannot be stimulated, there is evidence of ovarian estrogen secretion in response to exogenous gonadotrophins. The estrogens stimulate vitellogenesis by their effect on the liver, and in amphibians, they promote the mobilization of lipids from the fat bodies. Estrogen administration to cockerels increases the concentration of a large number of yolk precursors in the blood. Table 7.2 illustrates some of the most prominent changes in blood composition after estrogen administration. An increase in blood phospholipids has also been observed in the bass (*Paralabrax clatharus*), the bullfrog (*Rana catesbeiana*), the turtle (*Pseudemys scripta troostii*), and the mouse (*Mus musculus*). The concentration of calcium in the blood after estrogen injection increases substantially in the bass, bullfrog, turtle, but not in the mouse. This increase in the calcium concentrations of the blood in the three nonmammalian species is associated with the large amounts of yolk precursors, which are not formed in the mouse.

In AP̸ winter flounders and in intact South African clawed toads (*Xenopus laevis*), yolk precursors accumulate in the blood after estrogen administration (Campbell and Idler, 1976; Ng and Idler, 1978). Incorporation of yolk precursors in AP̸ winter flounders is stimulated by the nonglycoprotein fraction of teleost APs; in intact *Xenopus*

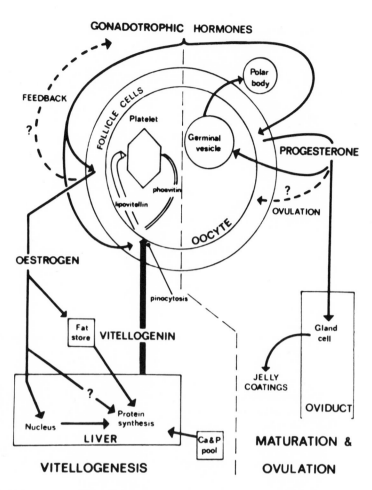

Figure 7.5 In this generalized scheme [of female amphibian reproduction], the major relationships between the pituitary gonadotropins and some of the processes involved in the production of an egg suitable for fertilization are depicted. On the left side of the figure, the primary responses involved in the vitellogenesis are shown, while on the right the involvement of progestins in maturation, ovulation, and the release of oviducal jelly coatings is summarized. Although the diagram has been constructed primarily from research on *Xenopus laevis* and *Bufo bufo*, it is likely that in other species, part of all of the diagram is applicable. The pituitary-hypothalamic relationships are not shown. (From Redshaw, 1972; reprinted with permission of *American Zoologist*.)

laevis, incorporation of yolk into the follicles is stimulated by mammalian FSH (Redshaw, 1972; Follett and Redshaw, 1974). In the AP lizard, *Dipsosaurus dorsalis*, estrogen administration does not induce vitellogenesis, but a combination of estrogen and ovine growth hormone does (Callard et al., 1973). This observation is difficult to interpret because growth hormones show a great deal of species specificity.

In captive mallard ducks (*Anas platyrhynchos*), the ovarian follicles frequently fail to develop beyond 2–3 mm because yolk is not deposited. Injections of estrogens result in the expected increase in yolk precursors, such as phospholipids, but neither estrogen alone nor estrogen pretreatment followed by injections of chicken pituitary powder bring about yolk deposition in the follicles (Phillips and van Tienhoven, 1960).

It is probably safe to assume that the following sequence takes place: gonadotrophin → stimulation of steroid production (especially estrogens) → estrogens or estrogens plus AP hormones → stimulation of synthesis of yolk precursors by the liver and mobilization of yolk precursors from stores → deposition of yolk in follicles under influence of AP hormones. Follett and Redshaw (1974) discuss vitellogenesis in detail.

OVULATION

After the follicle has reached the appropriate size, ovulation can occur. In most species, nuclear changes either precede or accompany ovulation. In most mammals, the secondary oocyte is released at ovulation; however, in the mare, bitch, and vixen, the primary oocyte is released, and in the tenrec (*Setifer setosus*) and the shrew (*Blarina brevicauda*), the sperm penetrates the primary

Table 7.2

Changes in blood composition of chickens after estrogen administration

Blood component	Sex	Control	Estrogen-treated
Lipids			
Total lipids (mg/100 ml plasma)	M	1,100	14,210
Phospholipids (mg/100 ml plasma)	M	162	934
Sphingomyelin (mg/100 ml plasma)	M	22	54
Cephalin (mg/100 ml plasma)	M	34	214
Cholesterol (mg/100 ml plasma)	M	235	1,136
Proteins			
Total protein (gm/100 ml serum)	M	3.90	7.40
Albumen (gm/100 ml serum)	M	1.00	0.60
Globulin (gm/100 ml serum)	M	2.90	6.80
Vitellin (dilution at which detected)	F	0	40
Hemoglobin (gm/100 ml blood)	F	9.0	5.6
Fat soluble vitamins			
Total Vitamin A (μg/100 ml)	F	5.1	46.8
Vitamin A ester (μg/100 ml)	F	0.9	42.8
Vitamin A alcohol (μg/100 ml)	F	4.2	4.0
Water soluble vitamins			
Riboflavin (p.p.m.)	F	trace	1.22
Biotin (ng/ml)	F	1.3	8.3
Minerals			
Calcium (mg/100 ml)	M	10	97
Ultrafilterable Ca. (mg/100 ml)	M	6.50	8.00
Magnesium (μg/100 ml)	F	Not detected	13.8
Inorganic phosphate (mg/100 ml)	M	6.20	20.00
Total sulfate (mg/100 ml)	M	5.80	1.70
Iron (μg/100 ml)	F	100	700

From van Tienhoven, 1968; used with permission of W. B. Saunders.

oocyte in the follicle before ovulation takes place. As we shall see later in this chapter, in teleosts, a number of permutations of the sequence, ovulation to fertilization to birth, occur.

Cyclostomes

As we mentioned earlier in this chapter, according to Matty et al. (1976), ovulations occur in the *AP* hagfish (*Eptatretus stouti*). In *AP* river lampreys (*Lampetra fluviatilis*), ovulations occur if large follicles are present at the time of *AP* (Larsen, 1973).

Elasmobranchs

Not enough is known about the control of ovulation in elasmobranchs, but Dodd (1975) has speculated that LH release occurs before ovulation.

Teleosts

Ovulation in teleost fishes is under the control of the AP through its regulation of steroid secretion by either the ovary or the interrenal gland (Figure 7.3). The hypophyseal-gonadotrophic control of ovulation in the goldfish is indicated by the release of GTH, as measured by a peak in plasma GTH concentration, during the late part of the light phase and the start of the dark phase in fish kept on a 16L : 18D photoperiod (Stacey et al., 1979). The effects of GTH administration on ovulation in the Indian catfish is probably mediated by corticosteroids secreted by the interrenal gland; however, the effect of GTH on the interrenal gland may not be a general phenomenon in teleosts, for in the Japanese rice fish (*Oryzias latipes*) the GTH appears to stimulate corticosteroid production by ovarian (follicular?) tissue (Hirose, 1976). The secretion of corticosteroids by ovarian tissue is in accord with the finding of Colombo et al. (1973) that ovaries of *Leptocottus armatus, Gillichthys mirabilis,* and *Microgadus proximus* produced 11-deoxycortisol and 11-deoxycorticosterone in vitro. Hirose (1976) observed that ovulation of Japanese rice fish oocytes could be induced in vitro by cortisol, but that removal of the follicle prevented the cortisol-induced germinal vesicle breakdown (GVBD), the indicator of meiosis. He proposed that cortisol has its effect on the oocyte nucleus through an agent other than cortisol, produced by the follicular wall. In the Indian catfish and the Japanese rice fish, the maturation of the oocyte (i.e., GVBD) and ovulation are intimately linked and have not been separated experimentally. In the rainbow trout, in vitro, ovulation does not occur at the end of oocyte maturation, and the two processes are controlled by different mediators. As we saw earlier in this chapter, 17α-20β Pg induces oocyte maturation in the rainbow trout, and also in the pike and goldfish, and corticosteroids enhance the sensitivity of the follicle to the effects of GTH and 17α-20β Pg (Jalabert, 1976).

In female *Tilapia aurea,* spawning is accompanied by large increases in the plasma concentrations of testosterone, 11-ketotestosterone, 11β-hydroxytestosterone, and deoxycorticosterone. The increase in deoxycorticosterone may be either the result of the secretion of corticosteroids required for ovulation to occur (as discussed above for the Indian catfish, rainbow trout, pike, and Japanese rice fish) or it may be the result of a release of corticosteroids associated with release of the oocytes into the water. As we shall discuss later in this chapter, egg laying in chickens is accompanied by an increase in plasma corticosterone concentration, and a similar process may prevail in *Tilapia aurea,* although this has not been investigated specifically. Hirose (1976) noted that GTH increases the water uptake by adult female ayu (*Plecoglossus altivelis*) and proposed that this water transfer is accompanied by movement of Na^+ into the animal. Corticosteroids are regulators of water and mineral metabolism, and the secretion of corticosteroids at the time of ovulation or spawning may thus be one of regulating the transfer of ions that are, in turn, regulating the process of ovulation. Different so-

lutions have evolved for secretion of corticosteroids; in the Indian catfish, they are secreted by the interrenal gland in response to GTH; in the Japanese rice fish, they are produced by the ovary in response to GTH, whereas in the goldfish, pike, and rainbow trout they are produced by the interrenal gland but probably under the influence of ACTH (Figure 7.3).

A tentative sequence of events that follows the release of GTH and the subsequent secretion of steroid hormones, such as corticosteroids and/or progestins, is illustrated in Figure 7.6 for the rainbow trout. The evidence that the sympathetic nervous system plays a role in this sequence is reasonably strong, but indirect. Epinephrine, one of the neurotransmitters of the sympathetic nervous system, can induce ovulation in vitro, an effect that can be inhibited by the α site blockers dibenamine and dibenzyline, and phentolamine, whereas β site blockers have no effect. The evidence that prostaglandins play a role in ovulation in the rainbow trout is mainly derived from experiments in which indomethacin, which inhibits prostaglandin synthesis, inhibited in vitro ovulation, and from experiments in which prostaglandin $F_{2\alpha}$ ($PGF_{2\alpha}$) induced ovulation in vitro (Jalabert, 1976). The $PGF_{2\alpha}$ may act on musclelike structures in the follicle and thus cause expulsion of the oocyte.

Amphibians

In amphibians, the processes of oocyte maturation and resumption of meiosis are linked, so that under many experimental conditions they have not been separated. The control of ovulation has been most thoroughly investigated in *Rana pipiens* and *Xenopus laevis.* As in the teleosts, gonadotrophins induce oocyte maturation and ovulation by steroidogenesis. In this case, progesterone is secreted and it or its metabolites 17α-hydroxyprogesterone, 20α-hydroxyprogesterone or 17α, 20β-hydroxyprogesterone (Thibault, 1977) induce meiosis, resulting in GVBD and ovulation (Schuetz, 1977). Germinal vesicle breakdown requires that protein synthesis occur (Guerrier et al., 1977), but this protein synthesis does not require the presence of the cell nucleus. Several factors

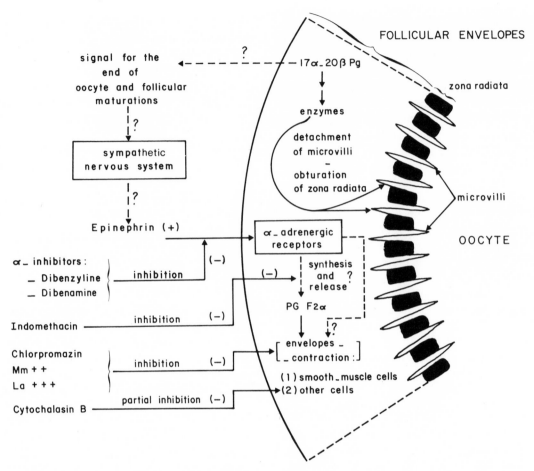

Figure 7.6 Tentative scheme for the [endocrine] control of ovulation in trout. (From Jalabert, 1976; reprinted with permission of *Journal of the Fisheries Research Board of Canada.*)

have been isolated in the induction of meiotic maturation (Schuetz, 1977): a maturation factor; a cytostatic factor arresting mitosis and cell cleavage; a pseudomaturation factor, inhibiting protein synthesis and causing abnormal synthesis; a chromosomal condensation factor, inducing chromosomal condensation; and a DNA synthesis factor, initiating DNA synthesis.

Biochemically, progesterone-induced maturation is accompanied by a decrease in cyclic 3′ 5′ adenosinemonophosphate (cAMP) in the follicles (Morrill et al., 1977; Speaker and Butcher, 1977), a decrease which may be necessary for normal maturation to proceed (Speaker and Butcher, 1977). The addition of dibutyryl-cAMP (db-cAMP) to the medium inhibits progesterone-induced GVBD (Schatz and Morrill, 1972). Pro-

gesterone-induced meiosis is also inhibited by the catalytic subunit of cAMP-dependent protein kinase, but meiosis can be induced by injection of the regulatory subunit of this protein kinase (Maller and Krebs, 1977). The inhibition of progesterone-induced GVBD by db-cAMP requires the presence of Ca^{2+} (Morrill et al., 1977).

The following tentative scheme can be proposed: gonadotrophin → progesterone synthesis → (progesterone metabolism?) → uptake of progesterone or its metabolites → increase in free Ca^{2+} → activation of phosphodiesterase → decrease in cAMP → decrease in catalytic unit of proteinkinase → decrease in phosphorylated protein that blocks meiosis.

It remains to be investigated to what extent the tentative sequence of events that has been proposed for the meiotic and ovulatory processes in the rainbow trout (Figure 7.6) is also applicable to these processes in amphibians.

Reptiles

The mechanisms involved in the meiotic maturation and in the ovulation of reptilian oocytes seem not to have been investigated to any great extent. The large amount of yolk in the follicle makes it more difficult to study the biochemical events that accompany the maturation of the oocytes in reptiles than in fishes and amphibians. It is known, however, that the AP is required for ovulation and that turtle FSH is more potent than turtle LH in inducing ovulation in American chameleons (Licht et al., 1977).

Birds

Ovulation can be induced in chickens by injection of mammalian LH preparations. As in fishes, catecholamines are involved in the ovulatory response. Ovulation can be blocked by injection of the catecholamine synthesis inhibitor α methyl metatyrosine (MMT), and the α adrenergic blocking agents dibenzyline and phentolamine, with the β adrenergic blocking agents not being effective. The fact that the anti-adrenergic blockage could be overcome by injection of 125 μg FSH plus 125 μg LH into the follicle should be interpreted cautiously in view of the very high doses of hormones used. (For comparison, in our laboratory, 16 μg LH given intravenously is 90–100 percent effective in causing ovulations in control hens.) The phentolamine blockage could be overcome by epinephrine (E), norepinephrine (NE) and phenylephrine, an α-site stimulant. Epinephrine was the only catecholamine that overcame the effect of MMT, and none of the above adrenergic drugs mentioned overcame the effect of dibenzyline. However, cAMP injection into the follicle caused ovulation of dibenzyline-blocked follicles (Kao and Nalbandov, 1972; Bahr et al., 1974). In our laboratory, cAMP was effective in vitro in inducing ovulation when injected directly into the follicle (unpublished results).

Bahr et al. (1974) proposed the scheme presented in Figure 7.7 and implied that it might be applicable to birds and mammals; however, evidence that prostaglandins are involved in ovulation in birds is scanty. Samsonovitch and Laguë (1977) found that neither injection of indomethacin, which blocks PG synthesis, nor of PGE_1 or PGE_2 into the follicle affected ovulation. Day and Nalbandov (1977) found an increase in PGF from near zero to about 8 ng/100 mg in the largest follicle of chickens at about 10 and 6 h prior to ovulation, and an increase to about 26 ng/100 mg at 1 h after ovulation. Indomethacin, injected intramuscularly, decreased the PGF peaks but did not prevent ovulation, although it caused a delay in oviposition. Indomethacin also did not affect steroidogenesis by the follicle. In our laboratory, we have found that follicles from hens pretreated with PMSG, ruptured incompletely after the injection of $PGF_{2\alpha}$ into the follicle.

The pathway in Figure 7.7 via prostaglandins may not be present in hens. The necessity of steroidogenesis for ovulation to occur has also not been demonstrated, nor has it been demonstrated that steroids injected into the follicle cause ovulation.

The physical processes involved in ovulation have been better studied in mammals than birds. Ischemia has been proposed as one of the causes for the weakening of the follicular wall. Support for this hypothesis, albeit indirect, is found in the experiments of Opel and Nalbandov (1962), in which they found that after AP, follicles become

Figure 7.7 Sequence of events postulated to occur after simulation of the ovarian follicle by LH and resulting in ovulation in birds and mammals and in luteinization in mammals. OIH = ovulation-inducing hormone. (From Bahr et al., 1974; reprinted with permission of *Biology of Reproduction*.)

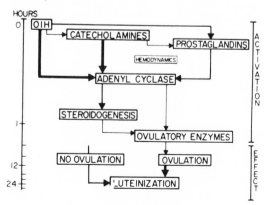

more sensitive to LH, and multiple ovulations result, although in untreated hens multiple ovulations cannot be obtained by LH injections.

Mammals

In mammals, there appears to be a reciprocal relationship between the oocyte and the follicular granulosa cells. Oocytes removed from the follicle have resumed meiosis even when no cumulus cells were attached to the oocyte (Thibault, 1977), thus suggesting that follicular cells inhibit meiosis. Furthermore, El Fouly et al. (1970) reported that in rabbits, removal of the oocyte (ooectomy) resulted in luteinization of the follicle.

In vitro experiments have confirmed the observation that in the presence of oocytes, granulosa cells do not differentiate into luteal cells, whereas in the presence of degenerating oocytes or in the absence of oocytes, the granulosa cells do differentiate into luteal cells (Nekola and Nalbandov, 1971; Stoklosowa and Nalbandov, 1972). Rothchild (1979) has pointed out, however, that the evidence for this interrelationship is still controversial; further investigations are needed before definite conclusions can be drawn.

Hunter et al. (1976) have shown that in pigs, ovulation can be induced before meiosis in the oocytes has been resumed. Injection of hCG on day 17 of the estrous cycle resulted in a high number of ovulations (18.6/pig) with 18.2 percent of the oocytes being primary oocytes. In contrast, injection of hCG on day 19 of the estrous cycle resulted in 12.9 ovulations/pig but only 2.1 percent of the oocytes were primary oocytes. The prematurely ovulated oocytes of the pigs treated on day 17 showed a frequency of 81.2 percent polyspermy, whereas the oocytes ovulated after hCG injection on day 19 showed a frequency of only 2.4 percent. These results show that the hCG treatment on day 17 of the estrous cycle interfered with the polyspermy-blocking mechanisms of the oocyte.

MATURATION OF OOCYTES

Oocyte maturation involves the changes that occur in the nucleus, cytoplasm, and membranes of the oocyte, and in the capacity of the oocyte to be penetrated by the sperm cell and respond by appropriate changes in the nucleus, membranes, and cytoplasm after the sperm has entered the oocyte, which is then an ovum.

When the oocytes are collected from antral follicles of juvenile animals, resumption of meiosis generally fails to occur during in vitro culture; such oocytes have been called incompetent oocytes (Wassarman et al., 1979). In contrast, oocytes from preovulatory follicles of sexually mature animals resume meiosis during in vitro culture; such oocytes have been called competent oocytes. Morphologically, competent oocytes differ from incompetent oocytes by their greater size (60–80 μm versus 20 μm); the thicker more dense zona pellucida (ZP), the large number of microvilli per unit of surface area, the more vesicular smooth endoplasmic reticulum, round mitochondria instead of elongated and dumbbell-shaped mitochondria, a Golgi apparatus consisting of parallel lamellae instead of flattened stacks of lamellae in a parallel arrangement, and a larger, fibrillar, dense nucleolus instead of a rapidly enlarging fibrillo-granular nucleolus (Wassarman et al., 1979). In addition to these ultrastructural differences, competent oocytes show a greater protein synthesis during meiotic maturation; however, the concommitant protein synthesis at the time of resumption of meiosis is not required for GVBD (Wassarman et al., 1979).

Investigations about the endocrine control of meiosis have revealed that AP in rats results in resumption of meiosis in oocytes removed from the follicles, but in 98 percent of the cases, meiosis stops at anaphase I, whereas oocytes removed from intact proestrous rats proceed to metaphase I. Erickson and Ryan (1976) have stated that FSH may be the factor required for meiosis to proceed from anaphase I to metaphase I.

Tsafriri (1979), comparing isolated rat oocytes that resume meiosis spontaneously with follicle-enclosed oocytes treated with LH to induce resumption of meiosis, found that in the follicle-enclosed oocyte, the LH-induced resumption of meiosis is probably mediated by a protein synthesized by the granulosa cells in response to the LH. As might be expected, puromycin inhibited the LH-induced resumption of meiosis, but did not inhibit the resumption of meiosis of isolated oocytes in vitro. Tsafriri (1979) contends that in the maturation of isolated oocytes "some of the regulatory events which lead to the resumption of meiosis following the LH-triggered maturation are bypassed.''

Some investigators have reported that pig ovarian follicular fluid (PFF) contains a factor or factors, probably peptides, which inhibit the resumption of meiosis by isolated oocytes (Channing, 1979a,b; Tsafriri et al., 1976). Hillensjö et al. (1979) have provided suggestive evidence that PFF inhibition of oocyte maturation in pigs requires the presence of cumulus oophorus cells capable of progesterone secretion.

The maturity of the oocyte determines to a large extent whether sperm can penetrate the oocyte. Hamster oocytes matured in vitro are not penetrated by sperm in vitro under conditions under which oocytes recovered from the Fallopian tube are penetrated. Presumably the ZP requires maturation in the follicle, which is GTH dependent. In some species (rabbit, pig, cow), sperm do penetrate the oocyte matured in vitro, but decondensation of the male pronucleus fails to occur, presumably because an oocyte cytoplasmic factor that regulates this decondensation is either absent or present. Treatment of oocytes matured in vitro with steroids, especially E_2-17β, 17α-OH-progesterone, and testosterone, induces the conditions that allow decondensation of the male pronucleus. Rabbit and human oocytes matured in vitro without follicle cells are capable of parthenogenetic development after suitable activation.

Experimental evidence and correlations with normal events in the reproductive cycle of the mammals investigated strongly suggest that the maturation required to obtain a normal oocyte, i.e., fertilizable and capable of normal development, is dependent upon gonadal steroid actions. Schuetz (1979) has compared the maturation of oocytes in mammals with that of animals other than mammals. He concludes that in nonmammalian vertebrates there is an increasing concentration of meiosis-inducing substance(s) which eventually reaches sufficient concentration(s) to result in resumption of meiosis. In mammals, however, there is a gradual decrease of meiosis-inhibiting factor(s) required before meiosis can continue.

In both amphibia and mammals, the oocytes remain encapsulated and in contact with somatic cells (Schuetz, 1979). In the frog, there is a single layer of follicle cells from which macrovilli project into the vitelline membrane to come into contact with the plasmalemma of the oocyte. Under the influence of the appropriate stimulus (LH, progesterone), the macrovilli retract, and loose clumps of cells form on the vitelline membrane. Such changes can also be induced by injection of a cytoplasmic maturation-promotion factor into the oocyte. Inhibition of protein synthesis prevents the maturation changes, suggesting that proteins, which may originate in the oocyte, are required for oocyte maturation (Schuetz, 1979). In mammals, the cumulus oophorus cells interdigitate with the ZP. During oocyte maturation, the germ cells and somatic cells disengage, an event accompanied by an increased sensitivity of the cumulus cells to the enzyme hyaluronidase and changes in the gap junctions between cumulus cells and oocyte (Schuetz, 1979).

Ovulation has probably been studied in more detail in mammals than in any other group. For the purpose of this discussion we will arbitrarily divide ovulation into two parts: (1) the endocrine events bringing about ovulation and (2) the rupture of the follicular wall and expulsion of the oocyte.

By use of antisera it has been shown that LH is the principal ovulation-inducing hormone in rats (Schwartz, 1974); in hamsters, FSH may be the ovulation-inducing hormone (Greenwald, 1971). However, LH, unlike FSH, induces steroidogenesis by the follicle of mammals. In most species investigated, ovulation is preceded by a surge of LH *and* FSH, although in the rabbit, and sometimes in women, there is an LH surge without an increase in FSH concentrations. In AP animals, either FSH or LH can induce ovulation, provided mature follicles are present. The question whether FSH and LH act in an additive or in a synergistic fashion has not been answered definitively. Lindner et al. (1977) state that FSH may prepare the granulosa membrane for estrogen production and that FSH plus estrogen sensitize the follicle for LH action by increasing the binding of LH.

The sequence of events and interactions between different ovarian biochemical mediators is presented in Figure 7.7. The evidence that ovulation-inducing hormones (OIH) act via catecholamines is based on the use of blocking agents and the subsequent injection of various catecholamines to overcome the blocking effect. Most of the evidence indicates that α receptors take part in this process.

Prostaglandins (PG) are produced in the ovary, and their production is stimulated by LH (Lindner et al., 1977); however, oocyte maturation, steroidogenesis, and luteinization proceed when PG synthesis is inhibited, although follicular rupture requires PG. The PGs presumably act on the contractile elements in the cells of the theca interna cells (Lindner et al., 1977); thus by causing contraction of these elements, $PGF_{2\alpha}$ may aid in the expulsion of the egg. The contractile elements of the follicle wall also respond to sympathomimetic drugs. The effect of these drugs may explain part of the effects of catecholamines on ovulation.

Adenylcyclase may be activated directly by OIH and indirectly by PG and the catecholamines. The increase in cAMP (in amphibians, oocyte maturation and ovulation require a *decrease* in cAMP) stimulates steroidogenesis, which may or may not be required for ovulation, through protein synthesis. The evidence that steroids are required for ovulation is still controversial. Bullock and Kappauf (1973) were able to dissociate ovulation from steroidogenesis in the rat, thus providing evidence that steroids were not required for ovulation in this animal. However, Takahashi et al. (1977) successfully induced ovulation in *AP* rats by repeated injections of progesterone, thus indicating that *in the absence of gonadotrophic hormones* exogenous progesterone can induce ovulation, suggesting that this hormone is part of the final common pathway for inducing ovulation. It is possible that steroids may facilitate or cause ovulation, but that they are dispensable. We should not be surprised that in such an important process as ovulation, which is required for survival of the species, there is redundancy in the processes that bring about the rupture of the follicle. Figure 7.7 illustrates that some event in the sequence can be induced in more than one manner; for example, adenylcyclase activation can occur as the result of a direct action of OIH or by an indirect action of OIH through either the catecholamines or the PGs. Similarly the synthesis of the PGs can be stimulated either directly by OIH

or indirectly through the catecholamines. Moreover, the different agents can have several effects; for example, OIH stimulates adenylcyclase activation, catecholamine release, and prostaglandin synthesis.

Morphologically many changes precede and accompany the rupture of the follicular wall. Bjersing and Cajander (1974 a–f) have presented beautiful photomicrographs, scanning electron micrographs, and transmission electronmicrographs of changes in the rabbit's ovarian follicle about the time of ovulation. It is not possible to do justice to these photographs, and so the interested reader should consult these papers. In brief, they observed that after GTH release, surface epithelial cells became larger and electron-dense bodies, which might represent lysosomal enzymes, appeared. The granulosa cells became less dense, and the granulosa cell projections increased in number and size, and penetrated into a partly fragmented basement membrane. Spherical inclusions became numerous in the granulosa cells about 8 h after the GTH release and the number of gap junctions decreased from 6 hours after GTH release until the time of ovulation, 10 h after GTH release. The decrease in the number of gap junctions correlated with a decrease in the cohesive force between granulosa cells, so that expansion took place. The influx of fluid and rapid follicular growth were probably facilitated by the changes in the basement membrane and the open channels to the antrum. The theca interna contained collagen fibers and smooth muscles in the apical region of the follicle. About 8 h after GTH release, edema became recognizable below the germinal epithelium covering the follicles. The edematous area contained dead or degenerating cells, while the tunica albuginea contained degenerating fibroblasts and fragmented collagen.

The release of GTH also increased the incidence of small fenestrations of the endothelium of the blood capillaries. Close to the time of ovulation large perforations developed; consequently the fragmentation of the basement membrane allowed free passage of the fluid from the capillary to the interstitium, resulting in edema. Pressure and fluid eventually were transmitted from the capillaries to the interstitium and to the antrum. This did not occur, however, until the tensile strength of the follicular wall had decreased and the subsequent influx of fluid had caused follicular expansion and eventually ovulation.

The rupture of the follicular wall has recently been compared with an inflammatory process (Espey, 1980). On the basis of several similarities between the inflammatory processes and ovulation, a tentative model was developed. In this model, the PGs are assigned a key role: $PGF_{2\alpha}$ stimulates the activation of quiescent fibroblasts to proliferating fibroblasts, which become the source of the proenzyme procollagenase and of the serine proteases that stimulate the conversion of procollagen to collagenase. Moreover, PGE_2 acts on the preovulatory follicle, in collaboration with histamine, serotonin, and possibly bradykinin, to make the preovulatory follicle a hyperemic inflamed follicle. This inflamed follicle is also a source of serine proteases. The collagenase, formed under the action of the serine proteases, acts on the collagen components of the follicular wall, and the subsequent weakening of this wall results in its rupture. The role of collagenase in ovulation has been substantiated by experiments in which collagenase injected into the follicle facilitates its rupture (Espey, 1974, 1980). In chickens, ovulation can also be induced by injection of either collagenase or nonspecific proteases (Nakajo et al., 1973).

OVARIAN HORMONE SECRETIONS

Steroid Hormones

In the lamprey (*Petromyzon marinus*), the ovary contains E_1, E_2, and P_4; in the ovary of the river lamprey (*Lampetra fluviatilis*), 3β-HSD has been found in the granulosa cells. Additional, indirect evidence for estrogen secretion under the influence of the AP is found in the fact that secondary sexual characters do not manifest themselves in either ovariectomized or AP animals (Dodd, 1972).

For the elasmobranchs, teleosts, amphibia, and amniotes, there is evidence from different sources that estrogens and other steroids are secreted by the ovary (Table 7.3). The cellular sources for the production of different steroid hormones have been most elegantly ascertained by in vitro stud-

ies, in which the different cells have been separated and incubated either alone or in different combinations with each other. This has not been done in all cases to be reported here, but other types of evidence can be used to determine which cells produce the various steroids.

Fortune et al. (1975) showed that pieces of *Xenopus laevis* ovary produced P_4 in vitro, and that P_4 production was increased by the addition of either hCG or frog pituitary homogenate to the medium and also by in vivo injections of hCG. Mulner et al. (1978) subsequently showed that in this species vitellogenic follicles (0.8–1.0 mm in diam.) and full-grown follicles (1.2 mm or larger) transformed pregnenolone to progesterone and then further to 17α-hydroxyprogesterone, 17α, 20α-dihydropregn-4-en-3-one, androstenedione, and testosterone. Aromatization of androstenedione and testosterone was 10–50 times higher in vitellogenic than in mature follicles. The presence of 3β-HSD and aromatase activity in the follicular walls only provide evidence that it was indeed the follicular wall that had the steroid-producing capacity. Colombo et al. (1977) have reported that upon incubation of ovarian tissue of the newt, *Triturus alpestris alpestris,* no steroid was synthesized from acetate; however, when progesterone was used as a precursor, 11-deoxycorticosterone, 11-deoxycortisol, 17α-hydroxyprogesterone, androstenedione, and testosterone were formed, with the yields of 11-deoxycorticosterone and 11-deoxycortisol being higher than those of the androgens. Upon superfusion of ovarian tissue, 11-deoxycorticosterone was produced from endogeneous precursors, with the production increasing after the addition of hCG to the medium.

Dispersed follicular cells from the ovary of the turtle, *Chrysemys picta,* produced progesterone and testosterone in vitro, a production that was increased by the addition of either oFSH or oLH. Estradiol was also produced, but an increase in production was only obtained after addition of FSH, not of LH. Corpora lutea cells secreted P_4 in vitro, a steroid production that was increased by the addition of db-cAMP (Lance and Callard, 1978).

The walls of the preovulatory follicle of the chicken contain progesterone, testosterone, and

Table 7.3

Evidence of gonadal steroid production in female inframammalian vertebrates

Species	Hormones*	Site	Evidence
Cyclostomes			
Petromyzon marinus	E_1, E_2, P	Ovary	Ovarian extract
Lampetra fluviatilis	Steroid	Granulosa	3 β-HSD
Elasmobranchs			
Torpedo marmorata	Steroid	Atretic follicle	3 β-HSD
	E_2, E_3, P	Ovary	Ovarian extract
Scyliorhinus stellaris	E_1, E_3, P, T	Ovary	Present in blood
	Steroid	Postovulatory follicle	3 β-HSD
S. canicula	E_1, E_2	Ovary	Ovarian extract
Squalus suckleyi	E_1, E_2, P	Ovary	Ovarian extract
S. acanthias	E_1, E_2	Ovary	Ovarian extract
	Steroid	Granulosa cells of follicle	3 β-HSD
	Steroid	Granulosa cells of CL	3 β-HSD
	Steroid	Atretic follicles	No 3 β-HSD
Scoliodon sorrokowah	Steroid	Interstitial cells	Indirect evidence (histochemical)
Mustela canis	Steroid	Atretic follicles	Indirect evidence (histochemical)
	Estrogen	Ovary	Ovarian extract
Raja erinacea	P	Ovary	In vitro synthesis
Lung fishes			
Protopterus annectens	E_2, P	Ovary	Ovarian extract
Teleosts			
Salmo irideus	E_2	Ovary	Extract
Cyprinus carpio	E_2, E_3, 16-epi-estriol	Ovary	Extract
Xiphias gladius	Estrogen†	Ovary	Extract of ovary
Pseudopleuronectes americanus	Estrogen†	Ovary	Extract of ovary
Lophius piscatarus	Estrogen†	Ovary	Extract of ovary
Gadus callarias	E_1, E_2	Ovary	Extract of ovary
Ictalurus punctatus	E_1, E_2 (α and β), E_3, 16-keto-E_2, epi-estriol	Ovary?	Plasma extract
Poecilia reticulata	Steroid	Granulosa cells of follicles during vitellogenesis	3 β-HSD, 17 β-HSD, 3 α-HSD
Scomber scomber	Steroid	Thecal cells of mature follicles	3 β-HSD
		Thecal cells of ruptured follicles	
Oncorhynchus nerka	E_2, P	Ovary	Extract of ovary
Pagellus acarne	11-keto-testosterone, 11-β-hydroxytestosterone	Ovary	In vitro biosynthesis

	Steroid	Tissue	Remarks
Mugil capito	Steroid	Follicular cells of secondary oocytes	3 β-HSD, 16 β-HSD, 17 β-HSD
	DHEA	Ovary	In extract
	E_2, 11-keto-testosterone	Ovary	Biosynthesis in vitro and in ovarian extract
M. cephalus	DHEA, E_2, 11-keto-testosterone	Ovary	In vitro biosynthesis
Coris julis	11-keto-testosterone, 11-β-hydroxytestosterone	Ovary	In vitro biosynthesis
Conger conger	P	Ovary	Ovarian extract
Tilapia aurea	E_1, E_2, 11-keto-testosterone, 11-β-hydroxytestosterone	Ovary	Ovarian extract
Heteropneustes fossilis	3α-hydroxy-5β-pregnan-20-one, 5β-pregnane-3α-20 α diol (?)	Ovary	Ovary mince or homogenate incubated with pregnenolone
Amphibia			
Rana temporaria	E_1, E_2, E_3	Ovary	Blood extract
R. pipiens	17 α-OH-P, A	Ovary	In vitro biosynthesis
R. esculenta	Steroid	Thecal and granulosa cells of follicles and postovulatory follicles	3 β-HSD
			3 β-HSD
Bufo bufo	E_2, E_3	Ovary	Blood extract; absent in ♀
	P	Ovary	Ovarian extract
B. vulgaris	E_1, E_2, E_3, P	Ovary	Ovarian extract
	P	Bidder's organ	Organ extract
Xenopus laevis	E_1, E_2	Ovary?	Blood extract
	E_1, E_2	Ovary	In vitro biosynthesis, stimulated by GTH
	Steroid	Follicular layer of mature oocytes	17 α, 17 β, 3 α and 3 β-HSD
		Postovulatory follicles	17 α, 17 β, 3 α, and 3 β-HSD
Reptiles			
Triturus cristatus	E_1	Fat bodies	In vitro biosynthesis
	E_1	Ovary	In vitro biosynthesis
Necturus maculosus	17 α-OH-P, A	Ovary	In vitro biosynthesis
Nectophrynoides occidentalis	P, 17 α-OH-P, T, A, E_1, E_2	Ovary	In vitro biosynthesis
Salamandra salamandra	Steroid	Follicular layer of mature oocytes	3 β- and 17 β-HSD
Thamnophis radix	P (bioassay)	Ovary?	Blood extract
Natrix sipedon sipedon	P (bioassay)	Ovary?	Blood extract
N. s. pictiventris	P	Ovary?	Blood extract
	P	Ovary	In vitro biosynthesis
Bothrops jararaca	Steroid	Ovary	3 β-HSD
	P (bioassay)	Ovary	Ovarian extract
Crotalus terrificus terrificus	P (bioassay)	Ovary	Ovarian extract

Table 7.3—*Continued*

Species	Hormones*	Site	Evidence
Vipera aspis	Steroid	Ovary	3 β-HSD
Coluber constrictor	P	Ovary	In vitro biosynthesis
Sceloporus cyanogenys	P	Ovary?	Blood extract
	Steroid	Granulosa cells of growing follicle	3 β-HSD
Lacerta sicula	E_2	Ovary	Ovarian extract
	Steroid	Ovary	3 β-HSD
L. vivipara	Steroid	Ovary	3 β-HSD
Dipsosaurus dorsalis	P, E_2, T, A	Ovary	In vitro biosynthesis
Birds			
Lobipes lobatus	T, A, P, E_1	Ovary	Ovarian extract
Meleagris gallopavo	E_1, E_2, E_3	Ovary	Blood extract
Gallus domesticus	P, E_1, E_2, T	Ovary	In blood and follicles; concentrations vary with ovarian cycle
	Estrogen	Ovary	Ovarian extract
	P, P, E_2, A	Ovary	In vitro biosynthesis
	T	Interstitial cells of ovary	Immunofluorescent antibodies, 3 β-HSD
	Steroid	Thecal cells	EM
		Granulosa cells not steroidogenic	EM
		Thecal cells	17 β-HSD
Rattus norvegicus	T, P, E_1, E_2-17β	Ovary	Peripheral blood and ovarian extracts
	T, DHT	Thecal cells	In vitro production under influence of LH
	E_2-16	Follicular granulosa cells	In vitro production from androgens after FSH induction
	P	Corpus luteum	In vitro production, CL extracts
Homo sapiens	P, E_1, E_2, T, A, 17 α-OHP	Ovary	Ovarian extracts

*A = androstenedione, DHT = dihydrotestosterone, E_1 = estrone, E_2 = estradiol 17 β or 17 α, E_3 = estriol, T = testosterone, P = progesterone.

†Type of estrogen not specified.

Data from Callard et al., 1972; Dodd, 1972; Eckstein, 1977; Fortune and Armstrong, 1977, 1978; Redshaw, 1972; Reinboth, 1972; van Tienhoven, 1968; Ungar et al., 1977.

estrogen (Shahabi et al., 1975), whereas the postovulatory follicle contains progesterone, androstenedione, testosterone, estradiol, and estrone (Dick et al., 1978). The distribution of these steroids between the thecal and granulosa layers is shown in Table 7.4.

Huang and coworkers (Huang and Nalbandov, 1979a,b; and Huang et al., 1979) have studied in vitro the contributions that chicken thecal and granulosa cells make to the production of different steroid hormones. Thecal cells from either pre- or postovulatory follicles, when incubated alone, failed to produce any of the steroid hormones (P_4, T, E) assayed. The addition of either oFSH or oLH stimulated a small amount of T production, amounts similar to those produced by the granulosa cells of the same (second and third largest) follicles. Ovine FSH and oLH also increased the estrogen production by the thecal cells of these follicles, but progesterone production, which was less than 1 percent of that produced by granulosa cells, was not stimulated by either of these two gonadotrophic hormones. (Huang and Nalbandov, 1979a; Huang et al., 1979).

The granulosa cells of the three largest follicles all were capable of T and P_4 synthesis, with the granulosa cells of the second largest follicle showing the highest production of T and P_4 at 15–16 h before ovulation of the largest follicle. Granulosa cells of the largest follicle showed their highest P_4 production capacity at the time of the preovulatory LH surge. The T-producing capacity of the granulosa cells of the largest follicle were not significantly affected by endogenous LH (Huang and Nalbandov, 1979b). Granulosa cells of postovulatory follicles were capable of T and P_4 synthesis in vitro, suggesting that in the chicken, production of these hormones does not require luteinization. The progesterone-producing capacity decreased significantly at 6–7 h after ovulation, but the T-producing capacity showed a bimodal distribution, with a decline at 4–5 h and a subsequent increase at 11–12 h, followed by a sharp decline at 19–20 h after ovulation (Huang and Nalbandov, 1979b).

When testosterone was provided as a substrate, thecal cells of the second and third largest follicles (F_2 and F_3) were capable of secreting estrogens, but thecal cells of the follicle destined to ovulate (F_1) and thecal cells of the postovulatory follicle lacked the aromatizing capacity. The thecal cells of F_2 and F_3 follicles produced testosterone when incubated with progesterone as a substrate.

Thecal and granulosa cells incubated together produced less P_4 than granulosa cells alone. The addition of oLH to the medium increased the P_4 production by thecal and granulosa cells when compared with thecal plus granulosa cells without oLH added; however, this P_4 production remained less than that of granulosa cells alone. Thecal and granulosa cells incubated together produced more T and E than either granulosa or thecal cells alone, and the T production was stimulated by oLH (Huang and Nalbandov, 1979a; Huang et al.,

Table 7.4

Steroid concentration or content in the granulosa and thecal layers of pre- and post-ovulatory follicles of the chicken

Follicle tissue	Steroid				
	P (ng/g)	E_1 (ng/g)	E_2 (ng/g)	T (ng/g)	A (ng/g)
Preovulatory theca	3.13 ± 6.7	0.71 ± 0.08	4.76 ± 0.08	7.7 ± 1.3	<1
Postovulatory theca	64.9 ± 11.2 ng/foll.	4.3 ± 0.85 ng/foll.	11.2 ± 3.8 ng/foll.	1.6 ± .03 ng/foll.	5.0 ± 2.9 ng/foll.
Preovulatory granulosa	2670 ± 1364	0.42 ± 0.05	4.5 ± 1.9	<1	3.0 ± 1.3
Postovulatory granulosa	59.7 ± 6.7	0.76 ± 0.29	1.5 ± 0	1.48 ± 0.33	5.6 ± 2.2

P = progesterone, E_1 = estrone, E_2 = estradiol, T = testosterone, A = androstenedione.
Data from Dick et al., 1978.

1979). On the basis of these findings, Huang et al. (1979) formulated the model illustrated in Figure 7.8.

The interstitial cells of the ovary produce androgens. Woods and Domm (1966), using immunofluorescent techniques, showed that androgens were present in the theca interna and granulosa layer of the developing and postovulatory follicle and in interstitial cells of the ovary. They also demonstrated the presence of androgens in the cells of the medullary cords and tubules, the rete testes, and the interstitial cells of the hypertrophied right gonad of sinistrally ovariectomized hens.

In mammals there is evidence in several species (pig, horse, cow, sheep, rabbit, and rat) that granulosa cells can produce estrogens, provided that androgen is supplied as a substrate (see Anderson et al., 1979 for references). Luteinizing hormone stimulates androgen secretion by thecal cells in the rat, and FSH stimulates the granulosa cells to convert androgen to estrogen (Fortune and Armstrong, 1977, 1978). The thecal cells in monkeys and humans seem to be capable of estrogen secretion in the absence of granulosa cells, although the presence of granulosa cells increases the production of estrogens (Channing, 1979a).

Thecal cells of cattle ovaries produced progesterone. Estradiol-17β (E_2) added to the medium in concentrations of 0.001 μg/ml more than doubled the P_4 production at 24–48 h of culture, but this effect was not as pronounced at 48–72 h of culture. At concentrations of 0.1, 1.0 and 10 μg/ml, E_2 inhibited P_4 secretion (Fortune and Hansel, 1979).

In vitro, granulosa cells from pig follicles (Ledwitz-Rigby and Rigby, 1979; Schomberg, 1979) and rats (Armstrong and Dorrington, 1976; Lucky et al., 1979) can also secrete progesterone. This secretion of P_4 is enhanced by androgens (Schomberg, 1979; Armstrong and Dorrington, 1976; Lucky et al., 1977), FSH, LH, FSH plus LH (Ledwitz-Rigby and Rigby, 1979), and by follicular fluid from preovulatory porcine follicles (Ledwitz-Rigby and Rigby, 1979). Progesterone secretion by follicular granulosa cells of the pig is inhibited by a fraction of the fluid from 1–2 mm pig follicles (Ledwitz-Rigby and Rigby, 1979). Rothchild (1979) has pointed out that the findings from in vitro experiments that follicular granulosa cells can synthesize progesterone and estrogens are not in accord with ultrastructural and cytochemical evidence that these cells lack the systems presumably required for steroid biosynthesis.

The interrelationships between the thecal and granulosa cells in steroid production, and the inhibitory and stimulatory effect of these steroids and of follicular fluid (to be discussed below) suggest that in addition to regulation by gonadotrophic hormones, intrafollicular and intraovarian regulation is important in modifying follicular function. The quantitative aspects of these regulatory interrelationships in vivo need to be established before it is feasible to formulate a quantitative model of the physiology of the follicle at different stages of development.

The mammalian CL secretes principally progesterone, but also other hormones. For instance, bovine CL secretes progesterone plus 20β-hydroxy-4-pregn-3-one and pregnenolone, but no estrogen (Savard, 1973), whereas the human CL secretes, in addition to progesterone, 17-hy-

Figure 7.8 Model of the two-cell-type hypothesis for [chicken] follicular testosterone and estrogen synthesis. Thick arrow represents major route and thin arrow represents minor route. (From Huang et al., 1979; reprinted with permission of *Biology of Reproduction*.)

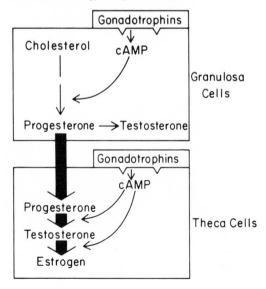

droxyprogesterone, 4-androstenedione, pregnenolone, 20α-hydroxypregn-4-en-3-one, estrone, and estradiol-17β. The CL of the rabbit, pig, sheep, horse secrete progesterone, 17α-hydroxyprogesterone, and 4-androstenedione, but not the estrogens, estrone and estradiol (Savard, 1973). The CL of the estrous cycle of the hamster during day 2 of the cycle elaborates progesterone and androstenedione in about equal amounts (Terranova et al., 1978).

The CL of pseudopregnant and pregnant rats secrete progesterone, 20α-hydroxypregn-4-en-3-one, and estrogens (Hilliard, 1973). The evidence for estrogen secretion by the CL is based on: (1) in vivo experiments, in which indirect evidence of estrogen secretion was obtained, e.g., cornified smears and increased uterine weight in AP rats with an AP autotransplant after LH administration (MacDonald et al., 1966), and (2) blastocyst implantation in AP rats provided with an exogenous or endogenous source of progesterone (MacDonald et al., 1967). However, in experiments, in which the follicles were destroyed by X-irradiation, morphologically normal CL did not secrete estrogen, as determined by vaginal smears (MacDonald et al., 1969). The presence of estradiol in the CL of pregnant rats at the end of pregnancy (Waynforth and Robertson, 1972) is further evidence of estrogen secretion by CL, as is the in vitro production of estrogen (at 1/20–1/200 of the rate of nonluteal ovarian tissue of the same rat ovaries that yielded the CL) by the CL of pregnant rats (Elbaum and Keyes, 1976). The fact that the CL can aromatize testosterone may indicate that testosterone serves as the substrate for estrogen formation by the CL at the end of pregnancy (Elbaum and Keyes, 1976).

Steroid hormones in the ovary are produced not only by the follicles and CL but also by the cells of the interstitial gland. In the rabbit, the interstitial gland, which is well developed, produces 20α-hydroxypregn-4-en-3-one (Dorrington and Kilpatrick, 1966; Hilliard et al., 1963). In the human and rat ovary, the interstitial cells secrete androgens or contain them (Savard, 1973; Woods and Domm, 1966).

The possible role that steroids produced by the various ovarian compartments play in the re-production of the different species will be discussed in Chapter 8.

Nonsteroid Hormones

Besides the steroid hormones, nonsteroid hormones, such as relaxin, follicular hormones, and prostaglandins, may affect reproduction in mammals.

Relaxin. This is a protein hormone with a molecular weight of about 6300 consisting of two subunits linked by disulfide bonds (Sherwood and O'Byrne, 1976). The concentrations of this hormone, which can be determined by RIA (Sherwood et al., 1975) varies during the estrous cycle of sheep. It is high during day 15 and 16 and at 24 h after the onset of the LH peak and at the time of ovulation (Chamley et al., 1975). In some species, the relaxin concentration in the blood increases during gestation, e.g., sheep (Bryant and Chamley, 1976), pigs (Sherwood et al., 1975), humans (Weiss et al., 1976), hamsters (O'Byrne et al., 1976), and rats, mice, and guinea pigs; and relaxin is present in high concentrations in pregnant dogs and monkeys (O'Byrne and Steinetz, 1976). In sheep, there is a sharp peak in the relaxin concentration at the time of parturition, when the lamb's head is out of the vagina of the ewe (Bryant and Chamley, 1976).

This hormone is secreted by the CL (Anderson et al., 1973; Weiss et al., 1976), specifically in the lutein cells (Kendall et al., 1978; Anderson et al., 1975). However, the ovary is not the only source of relaxin, because substantial concentrations of relaxin are present in the blood of ovariectomized pregnant sheep (Bryant and Chamley, 1976). The rat's uterine metrial gland is, however, not a source of relaxin (Anderson and Long, 1978).

The significance of the fluctuations in relaxin concentrations during the estrous cycle is not at all clear. The high concentrations of relaxin during pregnancy are probably related to the effects of relaxin on the reproductive tract, e.g., softening and relaxation of the cervix in hamsters (O'Byrne et al., 1976) and relaxation of the pelvic ligament in guinea pigs, which permits the large young to pass through the pelvis.

Relaxin also inhibits spontaneous uterine motility, increases the uterine water and glycogen content, and, together with estrogen and progesterone, stimulates uterine growth (see Anderson and Long, 1978, for documentation).

Follicular Fluid Factors. The follicular fluid of pigs (PFF) contains several factors that affect the physiology of the follicle and oocyte. These factors, their characteristics, and their effects (see Channing, 1979b and Channing et al., 1978, for documentation) are listed here:

1. *Oocyte maturation inhibitor*. This inhibitor has a MW of about 2,000 daltons. The concentration of this factor is higher in the PFF of small (1–2 mm) follicles than in the PFF of large (6–12 mm) follicles. The factor inhibits the spontaneous maturation of isolated pig oocytes surrounded by their cumulus, as well as progesterone secretion by the cumulus cells, and both effects are dose dependent. The effect of the inhibitor on oocyte maturation and progesterone secretion by the cumulus cells is reversible. This inhibitor also causes the outgrowth of the cumulus cells in vitro.

2. *Luteinization inhibitor*. This factor, present mainly in PFF of small follicles has a molecular weight greater than 100,000 (Ledwitz-Rigby and Rigby, 1979). It inhibits the spontaneous luteinization of granulosa cells cultured in vitro and the progesterone secretion by such cells. It also inhibits the stimulation of cAMP production by LH of these cells, possibly by decreasing the receptors for LH and hCG.

3. *Folliculastatin or ovarian inhibin*. This factor, which is present in higher concentrations in small follicles than in large follicles, has a MW greater than 10,000 daltons. In rats and pigs, it prevents the secondary rise in FSH concentration that follows ovulation; in rats, it prevents the rise in FSH that follows castration (a rise that cannot be inhibited by estrogen) and the LH-RH-induced increase in FSH concentration, although it does not inhibit the LH-RH-induced rise in LH concentration. Administration of folliculastatin results in a decrease in FSH concentration in the intact rat and in the intact and the ovariectomized rhesus monkey.

Prostaglandins. Earlier in this chapter we mentioned that gonadotrophins induce ovulation, probably by stimulating PG synthesis. Recently, Dodson and Watson (1979) have reported that in superfused preovulatory pig follicles, FSH addition to the medium resulted in a threefold increase in production of PGE and $PGF_{2\alpha}$, whereas LH addition resulted in about a twofold increase in the production of these PGs. Estradiol-17β addition increased PGE_2 production fourfold and $PGF_{2\alpha}$ production nearly tenfold. Since FSH and LH act synergistically in stimulating estrogen production, the results reported above suggest that FSH and LH each stimulate estrogen production to a suboptimal level and that gonadotrophin-induced PG production requires the intermediate estradiol production.

Functions of Ovarian Steroids. We wish to mention here some of the important functions of estrogens, progesterone, and androgens in female reproduction. The physiological functions of these hormones in the chicken may serve as an example. Just prior to sexual maturity, defined as ovulation or laying of the first egg, the plasma estradiol concentration rises from about 70–80 pg/ml 10 days before the first egg is laid to about 350 pg/ml at two days before the first egg is laid and then drops to about 125 pg/ml (Senior, 1974). Estrogens have several effects important for the initiation and maintenance of sustained egg production.

1. Estrogens increase food intake from 180 calories/day to about 300 calories/day, which is clearly adaptive, because it ensures a sufficient caloric intake to sustain egg production. (A hen laying 270 eggs/year produces 540 g of calcium, 1900 g of protein, 5400 g of solids—2 times its body weight—1600 g of lipids, and 21,600 calories.) Of the 300 calories consumed, 80 are deposited in the egg. The mechanism by which estrogens stimulate appetite is not established. It can either be a direct effect on the hypothalamic centers that regulate food intake, or it can induce changes in the blood constituents, which might then change the hypothalamic regulation, and, of course, a combination of these mechanisms might operate.

2. Estrogen in concert with progesterone is necessary for the priming of the hypothalamo-hypophyseal system, so that progesterone can induce LH release in ♀ hens (Wilson and Sharp, 1976).

3. Estrogens are probably involved in the regulation of female sexual behavior, e.g., crouching (van Tienhoven, 1968).

4. As we discussed under vitellogenesis, estrogens change the blood composition by acting on the liver, thus making precursors for yolk formation available.

5. Estrogens stimulate the deposition of calcium as medullary bone in the long bones. This calcium serves as a reserve during the periods of high egg production, when calcium cannot be taken in with food in enough quantity for egg production.

6. Estrogens stimulate growth and cytodifferentiation of the oviduct, priming it for the action of progesterone. Administration of estrogen to immature chicks results in the secretion of ovalbumin, conalbumin, and lysozyme (Schimke et al., 1975). The excellent reviews by O'Malley et al. (1975) and Schimke et al. (1975) should be consulted for extensive discussions of the molecular aspects of estrogen (and progesterone) action on the cellular metabolism of the oviduct.

7. Estrogens affect the plumage, so that both color and feather structure can be used as indications of estrogen secretion; ovariectomized hens and castrated and intact males all have the same type of feathers, i.e., the neutral type (Vevers, 1977).

Progesterone administered alone causes differentiation of the surface epithelium of the oviduct, with formation of goblet cells and ciliated epithelial cells (Schimke et al., 1975), but protein is not secreted. When estrogen administration stops, ovalbumin synthesis stops, but it can be reinitiated by the injection of either estrogens or progesterone (Schimke et al., 1975). Progesterone is essential for the induction of avidin synthesis by the estrogen-primed oviductal goblet cells (O'Malley et al., 1975; Schimke et al., 1975). Administration of progesterone to laying hens between 22 h and 12 h before ovulation induces LH release and premature ovulation (Wilson and Sharp, 1975), as is discussed further in Chapter 8. Progesterone administered in high doses can precipitate a molt in laying hens (Shaffner, 1955), probably by stimulation of the feather papillae (Shaffner, 1954), and cause atresia of the large follicles.

The chicken ovary secretes androgens, as is externally evident from the growth and reddening of the comb and wattles at the time of sexual maturity. Plasma androgen concentrations show a peak in concentration two hours prior to ovulation and so might be thought to be involved in the induction of ovulation. However, injections of testosterone at doses comparable to effective doses of progesterone (0.1–1.0 mg) are not as effective, i.e., about 50 percent of the hens ovulate after injection of 1 mg of TP and 16 mg need to be injected to obtain nearly 100 percent ovulations (Croze and Etches, 1977). Androgens induce protein synthesis in estrogen-primed oviducts, but the mechanism is not as well studied as for progesterone.

Several reviews are helpful in understanding the integrative action of ovarian hormones in bird species other than chickens or in other classes (Bentley, 1977; Eayers et al., 1977; Rowlands and Weir, 1977; Steinetz, 1973; Vevers, 1977). The concept of the integrative actions of hormones, together with a knowledge of the natural history of the species of interest, usually makes it possible to deduce the integrative action of these hormones for that particular species.

8

Reproductive Cycles

In reproductive cycles, we need to distinguish between (1) seasonal cycles of reproduction alternating with sexual quiescence and (2) cyclicity within the period of reproductive activity.

In a classic paper, ''The Evolution of Breeding Seasons,'' Baker (1938) laid the theoretical basis for the evolution of seasonal reproduction. He pointed out that the *ultimate* causes for a species adopting a certain breeding season are the favorable climatic conditions under which the young will grow up. The stimuli that cause or stimulate the parents to reproduce are the *proximate* causes. Thus the availability of grass and favorable temperatures in the spring may be the ultimate factors for the birth of lambs and fawns in the spring, but decreasing day length in the fall may have been the proximate cause that stimulated the hypothalamo-hypophyseal-gonadal axis of the parents.

In order to put seasonal rhythmicity in perspective, however, it is necessary to consider two rhythms that are endogenous—circadian and circannual rhythms. *Circadian* rhythms are rhythms that persist in a cueless environment and that have a period close to, but never exactly of, 24 hours. Such circadian rhythms are usually synchronized to 24-h (daily) rhythms or, in some marine animals, to tidal rhythms by external signals such as light and dark or the rising and falling of the water levels. The daily and tidal rhythms, as we shall see, play an important role in determining the response of an organism to an external stimulus. *Circannual* rhythms are rhythms with a period of nearly 365 days. They are, of course, more difficult to demonstrate, but they do exist and they may be important for transequatorial migrant birds (Gwinner, 1975). Such circannual rhythms are synchronized to yearly rhythms by the yearly changes in photoperiod and/or temperature.

The proximate causes that regulate the reproductive seasons of vertebrates may be expected to be different, e.g., for marine animals, for which temperature of the water is probably the most important cue, than for terrestial animals, for which the length of the photoperiod or the change in the photoperiod is probably the most consistent predictor and signal. As nothing is ever as simple as it appears, we can anticipate that photoperiod and temperature will interact, at least in some species, and that other factors will also modulate the effects of temperature and photoperiod.

The reproductive cycles during the breeding season are to a large extent regulated internally, through interactions, e.g., between the hypothalamo-hypophyseal system and the ovary, among various components of the ovary, and between the dam's neuroendocrine system and her offspring. With these various interactions in mind, we shall attempt to describe the various seasonal cycles and the reproductive cycles within seasons for different classes of vertebrates and the proximate factors that act as signals for seasonal reproduction.

CYCLOSTOMES

The lampreys, *Lampetra planeri* and *L. fluviatilis,* metamorphose during the fall of their fourth year and reproduce the next spring. The growth of follicles and therefore the time of spawning are probably largely regulated by the water temperature. After spawning about 100,000 oocytes, the animal dies, so no cyclic phenomena are found within the reproductive season. As Rowlands and Weir (1977) have pointed out, during the vitellogenetic growth of the follicles, the gut shrinks, and after spawning it atrophies and death follows.

In contrast to lampreys, the hagfish (*Myxine glutinosa*) releases few oocytes, since the ovary contains only 1-21 follicles. Ovulation occurs over several seasons and corpora atretica and "corpora lutea" are found, but so far their function remains to be established (Dodd, 1972).

ELASMOBRANCHS

Three types of cycles can be distinguished according to Wourms (1977): (1) a cycle in which species, e.g., *Scyliorhinus, Chlamydoselachus,* and *Heterodontis,* reproduce throughout the year; (2) a partially defined annual cycle; and (3) a well-defined annual or biennial cycle. Species with a partially defined cycle may reproduce at any time of the year, but they tend to show either one or two peaks of reproductive activity; for example, *Raja erinacea* shows two peaks, but *R. eglanteria* has only one peak. Among species with a well-defined cycle are, e.g., *Squalus acanthias* and *Mustelus canis. S. acanthias* has a gestation period of 22 months, i.e., a biennial cycle, although the males display an annual cycle. Parturition occurs in the fall and is followed shortly thereafter by copulation. The females of *M. canis* have a gestation period of 11 months, with parturition occurring in May and ovulation during the first three weeks in June with mating taking place before ovulation.

All these species are migratory, and it is probable that water temperature is the proximate factor.

I have been unable to find evidence about the pattern of ovulation during the breeding period for any elasmobranch, although Dodd (1975) states that ovulation appears to be inhibited by the presence of eggs in the oviduct.

The function of the corpora lutea, "corpora lutea," (see footnote on p. 122) and corpora atretica in the timing of events of reproduction in elasmobranchs remains to be investigated.

TELEOSTS

Oviparous Fishes

Fishes of the temperate and arctic zone waters show definite breeding seasons. Billard and Breton (1978) have divided the reproductive cycle into two parts: gametogenesis and spawning. In the temperate and arctic zones, several patterns of gametogenesis can be distinguished:

Gametogenesis. In salmonids, gametogenesis occurs during the summer and fall, at the time of decreasing photoperiod. In the brook trout (*Salvelinus fontinalis*) and rainbow trout (*Salmo gairdneri*), decreasing photoperiod and temperatures of about 16°C accelerate spermatogenesis (Peter and Crim, 1979). The beginning of multiplication of type A spermatogonia (see Chapter 6) and the appearance of some type B spermatogonia in the rainbow trout are associated with a small increase in plasma GTH, and at the time of the sharp rise in testicular weight and activity, there is also a sharp increase in plasma GTH concentration (Billard and Breton, 1978). In female rainbow trout, there is a slight rise in plasma GTH concentration and a high estrogen concentration at the start of oogenesis (Billard and Breton, 1978). Estrogens, as discussed in Chapter 6, may play an important role in vitellogenesis.

Crim et al. (1975) found that in the brook trout, which spawn in the fall, plasma GTH concentrations were low during the slow phase of spermatogenesis of males and early development stages (August) of follicular development. In Chilko sockeye salmon (*Oncorhynchus nerka*), plasma GTH was low in fish caught in sea water and in fish that had recently entered fresh water, yet in both groups spermatogenesis was well advanced. In the females of this species, plasma

GTH was low (3.3–4.1 ng/ml) during preliminary vitellogenesis and nearly doubled after the fish arrived in fresh water when vitellogenesis was well under way. The concentrations of GTH during vitellogenesis are low, compared with those found during spawning (280–364 ng/ml) (Crim et al., 1975). In the land-locked Atlantic salmon (*Salmo salar*), the pattern was similar to that found in the sockeye salmon (Crim et al., 1975).

The three-spined stickleback (*Gasterosteus aculeatus*) shows a pattern of gametogenesis similar to that of the spring-spawning salmonids; i.e., gametogenesis is complete towards the end of November. The proximate causes are photoperiod and temperature (Baggerman, 1978).

In some species of cyprinids gametogenesis begins in the fall, continues at a low rate during the winter, and ends in the spring, e.g., in the carp (*Cyprinus carpio*), the roach (*Rutilus rutilus*), in *Couesius plumbeus,* the bream (*Abramis brama*), *Hypseleotris galii, Agonus cataphractus,* and *Dicentrarchus labrax* (Billard and Breton, 1978). In other cyprinids, such as the tench (*Tinca tinca*), the black sea perch (*Cymatogaster aggregata*), *Pylodictus oliveris, Paralabrax clatharus, Notropis stramineus,* the mackerel (*Scomber scomber*), *Chromis chromis, Cynoscion regalis,* and *Epinephelus* sp., spermatogenesis starts in the late winter or early spring. In the tench, there is a slight rise in plasma GTH as spermatogenesis starts; and it is high as spermatogenesis progresses and it decreases slightly after spawning (Billard and Breton, 1978).

In the male goldfish, the pattern is similar to that found in the male tench (Billard and Breton, 1978). In the female goldfish, androgens and estrogens are present during the spawning season and, as mentioned in Chapter 7, during ovulation there is a peak in plasma GTH (Stacey et al., 1979).

Among the pleuronectids, e.g., in the winter flounder (*Pseudopleuronectes americanus*), the gonadosomatic index (weight of gonad as percentage of body weight) of the ovary reaches a plateau in November and then undergoes a slight rise in June, whereas in the males there is a continuous rise in this index from September to June. In both males and females, cortisol-cortisone and cortisone concentrations were high (about 12.5 μg/100 ml) during the period of gonadal quiescence and low (5 μg/100 ml) during the period of the plateau phase in the female and of the testicular growth in the male. However, the cortisol-cortisone concentrations rose between April and June, when spawning took place. In males and females, testosterone concentrations were about the same (except in June) and in both sexes the concentration was low (< 5 μg/100 ml) between July and November and then gradually rose to a concentration of 5 μg/100 ml in males and about 13 μg/100 ml in females. In males, the increase in 11-ketotestosterone concentration more or less paralleled the rise in the gonadosomatic index, reaching a concentration of about 15 μg/100 ml in June; in females, the 11-ketotestosterone concentration was quite low, 0.02 μg/100 ml or less (Campbell et al., 1976).

Spawning. Different patterns of reproductive cycles are reflected in the spawning patterns of teleosts.

Fall Spawning. For salmonids, the principal cue for spawning appears to be the photoperiod (Poston, 1978). In some species, e.g., the Pacific salmons of the genus *Oncorhynchus* migrate to the spawning waters, build a nest or nests from small pebbles, spawn and die. The dead bodies of the adults provide nutrients for the microorganisms that may be food for the young after they hatch. Many salmonids spawn in the fall, e.g., brown trout (*Salmo trutta*), Atlantic salmon, brook trout, rainbow trout, and several species of *Oncorhynchus*. It should be noted that in rainbow trout (Dr. H. A. Poston, personal communication) and brook trout (Breder and Rosen, 1966), there are fall-spawning and spring-spawning populations.

In the brook trout, brown trout, land-locked Atlantic salmon (Crim et al., 1975), and the rainbow trout (Billard and Breton, 1978), the GTH concentrations in the plasma are high (16–26 ng/ml) at the time of ovulation. In the Chilko sockeye salmon, concentrations of 280–364 ng/ml have been found in spawning females (Crim et al., 1975). The concentration of GTH in plasma at the time of spermiation was 6.2 ng/ml in male land-locked salmon, 9.0 ng/ml in brook trout, 17.3 ng/ml in sockeye salmon (Crim et al., 1975), and about 6 ng/ml in rainbow trout (Billard and Breton, 1978). In rainbow trout, the estradiol-17β concentration at spermiation was about 1.5 ng/ml, compared with about 2.5 ng/ml during the first stages of spermatogenesis (Billard and Breton, 1978).

It should be kept in mind that radioimmunoassays have revealed only one GTH in fish plasma; however, in several species, e.g., the three-spined stickleback (*Gasterosteus aculeatus*), the eel (*Anguilla anguilla*), the Atlantic salmon, and *Lepomis cyanellus,* there is cytological evidence for two types of gonadotrophin cells in the hypophysis (van Oordt, 1978). This morphological evidence is complemented by the isolation of two gonadotrophic hormones from the pituitaries of *Hippoglossoides platessoides,* as was mentioned in Chapter 7. If the assays for GTH reveal only one of these gonadotrophins, or if the assay shows

cross reactions for two gonadotrophins (if these exist in salmonids), then the data cited above may require reinterpretation.

Spring and Summer Spawning. This is found in the three-spined stickleback in the Netherlands (Baggerman, 1978), in the tench and the roach in France (Billard and Breton, 1978), the tarpon (*Megalops atlanticus*) in Florida, and a large number of species in the Chesapeake Bay region (Table 8.1).

Spring Spawning. This is found in France in the pike (*Esox lucius*) and the perch (*Perca* sp.), and in several species in the Chesapeake Bay region (Table 8.1). The American shad spawns during the spring in Maryland but during November

Table 8.1

Teleost species in the Chesapeake Bay region that spawn in the spring-summer, spring, or summer

Common name	Scientific name	Spawning season
Gizzard shad	*Dorosoma cepedianum*	Spring-summer
Bay anchovy	*Anchoa mitchilli*	Spring-summer
Carp	*Cyprinus carpio*	Spring-summer
Golden shiner	*Notemigonus crysoleucas*	Spring-summer
Bridle shiner	*Notropis bifrenatus*	Spring-summer
Iron color shiner	*N. hudsonius*	Spring-summer
White sucker	*Catostomus commersoni*	Spring-summer
Creek chubsucker	*Erimyzon oblongus*	Spring-summer
Northern redhorse	*Moxostoma macrolepidotum*	Spring-summer
Gafftopsail catfish	*Bagre marinus*	Spring-summer
White catfish	*Ictalurus catus*	Spring-summer
Yellow bullhead	*I. natalis*	Spring-summer
Brown bullhead	*I. nebulosus*	Spring-summer
Channel catfish	*I. punctatus*	Spring-summer
Tadpole madtom	*Noturus gyrinus*	Spring-summer
Blueback herring	*Alosa aestivalis*	Spring
Hickory shad	*A. mediocris*	Spring
Alewife	*A. pseudoharengus*	Spring
American shad	*A. sapidissima*	Spring
Striped anchovy	*Anchoa hepsetus*	Spring
Eastern mud minnow	*Umbra pygmaea*	Spring
Redfin pickerel	*Esox americanus americanus*	Spring
Chain pickerel	*E. niger*	Spring
Silvery minnow	*Hybognathus nuchalis*	Spring
Fall fish	*Semotilus corporalis*	Spring
Quilback	*Carpiodes cyprinus*	Spring
Silver anchovy	*Anchoviella eurystole*	Summer
Comely shiner	*Notropis amoenus*	Summer
Margined madtom	*Noturus insignis*	Summer

Data from Mansueti and Hardy, 1967.

in Florida. Some populations of brook trout and rainbow trout spawn in the spring, as mentioned above.

Summer Spawning. In the Chesapeake Bay region, three species (Table 8.1) have been reported to spawn during the summer, e.g., the silver anchovy (*Anchoviella eurystole*).

Winter Spawning. The round herring (*Etrumeus teres*) may be one of the few fishes that spawns during the winter. It spawns in the Gulf of Mexico and in the waters off the northern Florida coast (Mansueti and Hardy, 1967).

Year-Round Spawning. This occurs in populations of Atlantic menhaden (*Brevoortia tyrannus*) and the Atlantic herring (*Clupea harengus*) in the bay of Maine, and possibly in the inshore lizard fish (*Synodus foetens*) and the bone fish (*Albula vulpes*) (Mansueti and Hardy, 1967).

The South American cichlid fish *Aequidens portalegrensis,* which lives in the shallow waters off some parts of Brazil, Bolivia, Paraguay, and Uruguay, showed continuous breeding in the laboratory when kept at 25°C at 12L : 12D of artificial light in a room in which daylight could penetrate. Individual pairs spawned sometimes every 2–3 weeks for at least 250 days, the intervals between spawnings varying from 5 to 70 days (Polder, 1971).

Intermittent Breeding. This has been reported for the goldfish in the Chesapeake Bay region (Mansueti and Hardy, 1967) and in the carp in the USSR (Shikhshabekov, 1974).

Variations. Shikhshabekov (1974) has found that the duration of the spawning season varies with the latitude. In the USSR in Arakum Lake (which is about 43–44°N) the spawning season of the bream (*Abramis brama*) is 58–60 days, but in the Volga delta (46–47° N) it is 16 days. In the wild carp (*Cyprinus carpio*), the spawning period is 3 months in the Arakum Lake region and 1 month in the Volga delta. The time of spawning for some species is largely determined by the temperature of the water; the minimal temperature at which spawning still will occur is 14°C for the bream, 17°C for the carp and 10°C for the Caspian roach (*Rutilus rutilus caspicus*). The carp, in addition, requires newly flooded areas with meadow vegetation. Shikhshabekov (1974) ascribes the longer duration of the spawning season in the north compared to the south to different degrees of maturity of individual fish in the population of mature fish, not to an extended ovulation period of individuals.

The spawning period of females of different species can vary. In some species, all the oocytes are released from the ovary and spawned, whereas in others, one group of follicles matures and the oocytes are released while a next group of yolk-free follicles undergoes vitellogenesis, matures, and is released later. This pattern is found, e.g., in the mackerel, carp, and goldfish. An extreme example is the mekada or Japanese rice fish (*Oryzias latipes*), in which oocytes are released at intervals throughout the summer. In the mackerel, maturation of the second set of follicles stops during the spawning of the group of ripe follicles (Rowlands and Weir, 1977).

Patterns in Oviparous Tropical Fishes. In tropical and equatorial fishes, four patterns (see Billard and Breton, 1978) have been found. In one, e.g., *Tilapia* sp., there is a definite breeding season, with water temperature, food availability, and possibly rainfall acting as proximate causes. For equatorial populations, change in photoperiod is clearly not a proximate cause. In another pattern, there is a breeding season, with rainfall acting as the apparent proximate factor, as in the cyprinid *Barbus liberiensis,* which breeds during May to July, and the Indian catfish (*Heteropneustes fossilis*) which spawns during the monsoon season from July to August (Sundararaj and Vasal, 1976). However, in another pattern, fishes, e.g., Indian catfish, will respond to photoperiod manipulation. Ovarian recrudescence can be advanced by lengthening the photoperiod. There is, however, an important interaction of water temperature and photoperiod: with 12 hours light and 12 hours dark (12L : 12D), a greater response was obtained at 34° C than at either 25° or 30° C; however, with 14L : 10D, the response was significantly greater at 30° C than at either 25° C or 24° C, and greater at 34° C than at 25° C.

During the postspawning period when the hypothalamo-hypophyseal-gonadal axis is apparently photorefractory; that is, it does not respond to long photoperiods unless the organism is first subjected to short photoperiods (this topic is discussed in Chapter 14), female Indian catfish show a greater ovarian recrudescence after 60 days exposure to 9L : 15D at 30°C than at either 14L : 10D at 30°C or to 9L : 15D or 14L : 10D at 25° C. Under laboratory conditions, Indian catfish do not ovulate; instead follicular atresia occurs. This atresia cannot be prevented but can be delayed by about 40 d by keeping fish in water at 30° C, whereas at 25° C atresia occurs faster than at ambient temperature, which varied between 25° C and 30° C (Sundararaj and Vasal, 1976). Caribbean reef fishes seem to spawn mainly between February and April, when water temperatures are at their lowest (about 28° C). In the third pattern, there are two spawning periods, as in the sergeant major *Abudefduf trochschelii* (= *A. saxatilis*?) of the tropical waters of the east Pacific (Graham, 1972). This fish spawns during March to April and July to September (Billard and Breton, 1978). In the fourth pattern, spawning occurs throughout the year in the population, as in two groups of tropical Carangidae and Lutjanidae, and in the mosquito fish *Gambusia* sp. (Billard and Breton, 1978). In the guppy (*Poecilia reticulata*), continuous spermatogenesis can be obtained under laboratory conditions, in 27° C water and with 12L : 12D (Billard and Breton, 1978).

Not much seems to be known about the endocrine rhythms associated with gametogenesis and spawning except in the few species mentioned above.

Viviparous Fishes

Among viviparous species, the sequence and time intervals between ovulation (O), fertilization (F), hatching (H), and parturition (P) differ. Three principal patterns have been found: (1) In the *Zoarces* type pattern, found in *Zoarces viviparus*, ovulation occurs first and fertilization follows almost immediately. About midway during the gestation period of about 25–30 days, the embryos hatch but are retained until parturition occurs. (2) In the *Jenynsia* type pattern, found, for example,

in the genus *Jenynsia*, fertilization occurs in the follicle, and the fertilized oocyte is released. The young hatch about 5–8 days later, and parturition occurs about a month after fertilization occurred. (3) In the *Gambusia* type pattern, found for example, in the genus *Gambusia*, the oocyte is fertilized in the follicle and at the end of the gestation period of approximately a month, the oocyte is released from the follicle. Hatching and parturition then follow in relatively quick succession. Within these three patterns there can be, of course, be different cycles. We will describe briefly one such cycle, that of the black seaperch or shiner seaperch (*Cymatogaster aggregata*). The following description is based on the excellent paper of Wiebe (1968). Temperature and photoperiod are the principal proximate causes for this fish. This fish starts to form distinct schools in June and in July, when old and young, males and females school together, and courtship and insemination occur. During the postspawning period, schools consist of either males or females, and either young or old fish. Sperm can be found in the female's reproductive system between July and December, but fertilization occurs in November. This late fertilization is due to mitotic divisions in the ovary starting in June, but vitellogenesis not being completed till November–December. The follicles that contain oocytes that are not fertilized become atretic. The follicles around the fertilized oocytes, however, undergo important changes (see Wiebe, 1968, for details), which allow for nutrition and oxygen supply to the developing embryo. In June–July, the young are born, with the males being sexually mature at birth. The ovaries regress after parturition, but in August, vitellogenesis can recommence and by mid-November–December, the follicles are mature enough for fertilization.

In other species of teleosts there can be several ovarian cycles during one reproductive season. For example, in *Neoophorus diazi* of the family Goodeidae, follicular growth occurs during the spring, summer, and early fall. Sperm are present in the male throughout the year, but the female shows three cycles of ovulation, gestation, and parturition during the breeding season. Superfetation does not occur in this species or other Goodeidae, because gravid fish store neither sperm nor mature follicles (Mendoza, 1962). In species that produce more than one brood, (1) time may elapse between parturition and maturation of the follicles and fertilization, e.g., in the guppy; (2) fertilization may occur immediately after birth of the preceding brood, e.g., *Quintana atrizona;* or (3) superfetation may occur, so that as little as two groups of embryos of different ages are present, as in *Poeciliopsis pleurospilus,* or as many as six, as in *Heterandria formosa.*

AMPHIBIANS

Anura

Oviparous Species. In the frog *Rana temporaria,* the interstitial cells of the testes undergo cyclic changes. The size of the nucleus of the Leydig cells, which reflects the secretory activity of the cell, shows a peak from February to March when the secondary sex characters, such as the thumb pads, become well developed, and decrease from March to April, when the thumb pads regress (Lofts, 1974).

In this frog, spermatogenesis is seasonal, and it starts in the spring *after* the breeding season is over, so that by the end of summer the seminiferous tubules contain many spermatocysts in various stages of development (Lofts, 1974). In the fall, the formation of new cysts slows down, but the cysts present produce bundles of mature sperm, so that the male undergoes hibernation with sperm "ready to go" for the spring. This pattern of male reproduction is adaptive, since spermatogenesis, if it were to occur in the spring, would probably proceed at a slow rate because of low temperatures that prevail. Frogs that have mature sperm upon emerging from hibernation would have a clear advantage. With this pattern there is desynchronization of spermatogenesis and male sex hormone production. The species of anurans that have continuous spermatogenesis (see Chapter 6) still can show seasonal reproduction, because the females show seasonal breeding, and male sex hormone production may be seasonal.

In the female of *R. temporaria* and in other species of amphibia living in the temperate zone,

maturation of the eggs may take as long as three years (Lofts, 1974). During the winter of the first year, oogonia divide and produce the cell nests from which the young follicles develop during the following summer (Lofts, 1974). During the second year, previtellogenic growth of the follicle occurs, and during the last 9 months before the spring spawning, vitellogenic growth takes place (Lofts, 1974). At the time of breeding, therefore, there are four distinct groups of germ cells present (Lofts, 1974): (1) cell nests, (2) previtellogenic follicles, (3) vitellogenic follicles, and (4) fully grown ovulable postvitellogenic follicles.

The main proximate cause for timing of reproduction of Anura in the temperature zone seems to be temperature, whereas in subtropical, tropical, and arid regions, rainfall is the important factor, which may act directly or indirectly by making food available (Salthe and Mecham, 1974). Since, therefore, the breeding seasons of a particular species depend on temperature, they may thus be early at a more southern latitude and later at a more northern latitude. Most species breed in the spring, though not necessarily every year, and some species breed in spring and fall, with the spring being the main breeding season, i.e., as determined by the greater reproductive activity, e.g., *R. pipiens berlandieri* and *Pseudacris clarkii* (Salthe and Mecham, 1974).

Viviparous Species. The only viviparous frog known is the East African frog *Nectophrynoïdes occidentalis.* These females ovulate and mate in October, at the end of the rainy season. In November, they estivate during the dry season, which lasts about 6 months, and the frogs emerge in April and give birth to the young in June. During the estivation period the development of the embryos is slow, and at the end of this period the embryos are in stage Ib, characterized by moderate development of the forelimbs. The second stage of development, which takes place after the rains start, is more rapid, and the embryos develop to the final stage IV (characterized by regression of the tail) between April and June. During this period the corpora lutea, which are capable of producing progesterone during in vitro incubation (Xavier and Ozon, 1971), first regress and then degenerate. These corpora lutea are indispensable during the first gestation, since ovariectomy is followed by abortion of the young. If the frog has once been pregnant and has given birth, ovariectomy early during either a second, third, or fourth pregnancy is not followed by abortion, but by accelerative development of the young and parturition during late April or May (Zuber-Vogeli and Xavier, 1973). Progesterone implantation in such ovariectomized frogs retards the rate of embryonic development in comparison with that of embryos of ovariectomized frogs (Bentley, 1976). The maintenance of the pregnancy in frogs ovariectomized after their first pregnancy and the abortion of the young if ovariectomized during their first pregnancy may be the result of a different response of the AP to ovariectomy in the younger frogs in comparison with the older ones. For instance, in young ovariectomized females erythrosinophilic cells are absent and gonadotrophic cells are shrunk as is the case in young virgin intact females, whereas in older ovariectomized pregnant females, gonadotrophin cells are normal in size, but they have fewer granules than their corresponding controls (Zuber-Vogeli and Xavier, 1973).

The oocytes of the pregnant female start to develop when the corpora lutea start to regress in May. This growth of the oocytes is correlated with an increase in the secretory activity of the gonadotrophs in the AP. After parturition there is a sudden release of gonadotrophic hormones from the AP, and the secretory activity of the AP decreases. In September, the AP again starts to secrete gonadotrophins, and the final growth of the oocytes and the uptake of yolk follow. After maturation of the follicles, ovulation occurs. If the oocytes are not fertilized, they are retained in the uterus of the female for as long a period as embryos. The corpora lutea of the nonpregnant female regress early, in April, and the oocytes retained in the uterus are lysed. The ovarian oocytes, which in the pregnant female do not start to grow rapidly until September, develop rapidly and are mature in July, so that the ovarian cycle lasts 9 months in nonpregnant and 12 months in pregnant females (Xavier, 1974). During the ovarian cycle, the steroidogenic capability varies. During the preovulatory period, it synthesizes mostly estrogens and during gestation, it synthesizes mostly progesterone (Xavier, 1974). Figure

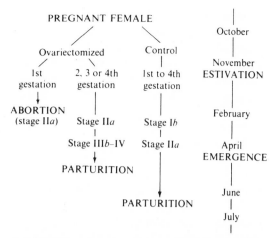

Figure 8.1 Gestation in the viviparous frog, *Nectophrynoides occidentalis,* in relation to seasons. Normally these frogs ovulate and become pregnant in October. The dry season commences in November, when they estivate, and from which they emerge, with the onset of rain, in April. The young are born in June. If these frogs are ovariectomized early in pregnancy, either they may abort, if the animals are young (and this is their first pregnancy), or, if they are large and it is the second, third, or fourth time of gestation, the development of the young is accelerated and they are born much earlier than usual. Ovarian progesterone is thought to delay the development of the young during the period of estivation. [For explanation of stages of developments see text.] (From Bentley, 1976; reprinted with permission of Cambridge University Press.)

8.1 summarizes the sequence of events in intact pregnant and ovariectomized viviparous frogs.

Urodeles

The seasonal variation in the testes of a seasonal urodele has been described in Chapter 6. This pattern is found in most of the urodeles. The anuran pattern, i.e., a brief spawning period, usually in the spring, and spermatogenesis completed before copulation, is found in *Desmognathus fuscus, Eurycea bislineata, Triturus viridescens, T. alpestris, T. palmatus, T. vulgaris, T. cristatus, Taricha torosa* and *Salamandra maculosa* (Lofts, 1974).

In *T. honkongensis,* the lobule boundary cells show evidence of steroidbiosynthesis, such as a smooth endoplasmic reticulum and 3β-HSD activity, before the sperm are discharged, suggesting that the secondary sex organs and sex characteristics are probably stimulated by androgens before the breeding season starts. Toward the end of the breeding season, the lobule boundary cells degenerate, and the secondary sex organs and secondary characters regress. As in *R. temporaria,* spermatogenesis and hormone production are desynchronized. The cycle of most female oviparous urodeles apparently resembles that of oviparous Anura.

The seasonal breeding patterns found in most temperate zone urodeles are as follows: Mating takes place in the fall, with the mating season ending when cold weather sets in. The sperm are stored in the spermatheca, and fertilization occurs in the spring. Vitellogenesis takes place during hibernation, and in the spring mating occurs again, followed by fertilization and oviposition in spring to summer. This pattern is found in: *Salamandra salamandra, Necturus maculosus, Plethodon cinereus, Euproctus asper,* and possibly *Notophthalmus viridescens* (Salthe and Mecham, 1974). In the genus *Proteus,* eggs may either be laid or retained; in the latter case the young salamanders have completely metamorphosed and resemble their parents when they are born, whereas in the former case, the metamorphosis occurs outside the body of the mother (Noble, 1954).

In very cold regions, there is no fall mating, and spring mating commences as soon as the weather becomes favorable. Ovulation and oviposition take place early in the spring after mating, although neither ovulation nor oviposition are dependent on mating. This pattern is found in *Ambystoma maculatum, A. gracile,* in the genera *Ranodon* and *Salamandrella,* and in *Hynobius nebulosus.*

In the cyclostomes, elasmobranchs, and teleosts that we have discussed so far, ovulation and oviposition could occur either in isolated females or in females that had been courted, but insemination was not a prerequisite for ovulation. In the anurans, there are no examples of such induced ovulators; however, among urodeles, the rough-skinned newt (*Taricha granulosa*) is an induced ovulator. Moore et al. (1979) showed that females ovulated only if they were exposed to courtship and were inseminated. Females of the rough-skinned newt have high plasma progesterone concentrations (about 10 ng/ml) at the start of the breeding season. In an experiment by Moore et al.

(1979), this high concentration decreased to 60 pg/ml in 6 hours, independent of courtship by a male. Moore et al. (1979) speculated that the change may have been related to a change in water temperature from 4° C to 13° C during this period. The high progesterone concentration at the start of the breeding season may make the females sexually more attractive to males. Moore (1978) has shown experimentally that progesterone injections of females make them more sexually attractive. After courting and insemination, a large number of oocytes may be released; as many as 38 oocytes have been found in the oviduct (Moore et al. 1979).

The courtship by the male increased plasma progesterone to about four times that found in noncourted control females, but courtship did not induce ovulation. The females that received spermatophores after courtship, ovulated, but their plasma progesterone concentrations were not different from those that had been courted but not inseminated. According to Moore et al. (1979), induced ovulations have been reported also in *Cynops (Diemyctylus, Triturus) pyrrhogaster,* and *Ambystoma mexicanum,* but the evidence is not very strong. Salthe and Mecham (1974) also list *Triturus vulgaris* and *Eurycea tynerensis* as induced ovulators.

The male rough-skinned newt shows two peaks in plasma androgen concentration, one 40 ng/ml in March to April during the mating season when spermatophores are released, and another one in October to November a little over 50 ng/ml, when the males clasp the females but probably do not deposit spermatophores (Specker and Moore, 1980). The testicular weights are low between mid-October through mid-July, when they rapidly recrudesce, but reach their maximal weight at the end of August. Spermatogenesis starts in late spring, and is completed prior to the increase in circulating androgens in August. Spermiogenesis occurs during August, and spermiation (release of sperm into the lumen of the lobules) starts in late September to early October and continues into April (Specker and Moore, 1980).

Two species of urodeles lay their eggs in the fall instead of in the spring, i.e., *Andrias davidianus* and *Ambystoma opacum,* which lay their eggs in the high mountain streams of China (Salthe and

Mecham, 1974). This is an unexpected phenomenon, because the zygotes have to withstand rather severe weather conditions before the embryos emerge in the spring.

For the urodeles of the temperature zone, the proximate causes that time the breeding cycle are temperature, humidity, and photoperiod (Salthe and Mecham, 1974). Tropical and subtropical amphibia may either breed the year round or show breeding seasons that are probably timed by rainfall (Salthe and Mecham, 1974, for details).

REPTILES

As we shall see in Chapter 9, spermatozoa have an unusually long functional survival time (several years in some species) in the reptilian female reproductive system; therefore, synchronization of ovulation and copulation are not essential. In some reptiles, the spermatogenetic activity and endocrine activity may be desynchronized. In the turtle, *Chrysemys picta,* mating occurs in April to May, spermatogenesis in June to October, and the highest plasma testosterone concentrations are found in December and April (no data are available for Jan. to March) (Callard et al., 1976). Testosterone is probably produced by the Sertoli cells, because, by histological criteria, the Leydig cells appear to be inactive at the time of high testosterone titers.

In temperate and subtropical zones, temperature and photoperiod interact in the timing of the reproductive cycles of reptiles; however, endogenous timing mechanisms can also be dominant. In the American chameleon (*Anolis carolinensis*), photoperiod has a permissive effect in facilitating testicular recrudescence in response to high (i.e., 32° C) temperatures. There is, however, evidence that this recrudescence may be initiated by an endogenous trigger (Licht, 1967a). In the lizard, *Lacerta sicula,* the endogenous rhythmicity of the sensitivity to thermal stimulation is an important factor, in that such stimulation during one part of the day is less effective than during another part of the day (Licht, 1972). We shall return to the aspects of the relative importance of different exogenous factors in Chapter 14.

Breeding in the temperature zone is, as expected, seasonal, but some species breed every other year, e.g., the rattlesnakes *Crotalus atrox* and *C. viridis* (Rowlands and Weir, 1977). Most species copulate in the spring after the end of hibernation; eggs are laid in May to June, and the young are usually born towards the end of the summer (Lofts, 1978). Some species of *Sceloporus,* however, mate and ovulate in the fall before hibernation, and the young are born in the spring or summer (Lofts, 1978).

The relative abundance of insects used as food by female reptiles may be a proximate cause for certain species (Lofts, 1978). Rainfall is apparently not an important cue in the timing of the reproductive cycles of reptiles in the temperate zone. In the desert, however, rain, by increasing the food supply for insects, increases their abundance and thus provides food for reptiles. Rain, therefore, may be an important proximate factor for desert reptiles, e.g., *Sceloporus orcutti, Uma inornata,* and *U. scoparia* (Rowlands and Weir, 1977). Correlations between egg production and rainfall have been found for *Anolis grahami, A. lineatopus,* and *A. sagrei,* and for the Chinese cobra (*Naja naja*) (Lofts, 1978). For *Anolis sagrei* there is experimental evidence that high (75 percent) relative humidity may be an important cue for the timing of egg laying (Brown and Sexton, 1973).

Tropical reptiles generally also show seasonal breeding patterns, although some species show continuous breeding, e.g., the three Javanese geckos, *Hemidactylus frenatus, Cosymbotus platyurus,* and *Peropus mutilatus* (Church, 1962), and the Malaysian snake, *Hopolopsis bucatta* (Lofts, 1978). However, the female equatorial lizard *Agama agama lionotus* produces eggs only in July and August, although the male has mature sperm in the testes throughout the year.

The growth of follicles and the release of oocytes show considerable variation among reptiles. For example, in the horned toad (*Phrynosoma solare*), follicular growth to ovulatory size takes about two weeks, but in the yuka night lizard (*Xantusia vigilis*), it may take as long as three years. According to Rowlands and Weir (1977), most reptiles have one ovulatory cycle in each breeding season, but in areas in which the climate is favorable and the breeding season is thus extended, there may be more than one ovulatory cycle, as in the teiid lizard *Cnemidophorus tigris,* and the skink, *Lygosoma laterale.*

Most is known about the reproductive cycle of the oviparous American chameleon (*Anolis carolinensis*). The males and females are reproductively inactive from September to January in their breeding range in the southeastern United States. In late January, the males emerge from their winter hiding places and establish breeding territories. The females emerge in late February, and vitellogenesis starts in March. Yolk is deposited in one follicle for about 10 to 14 days, and when the follicle is about 8 mm in diameter, the oocyte is released. Yolk deposition in a follicle in the other ovary starts, and ovulation occurs about two weeks after the previous ovulation. The alternation of ovulation between ovaries continues until late August when follicles become atretic. "Corpora lutea" are formed from the ruptured follicles from which yolks have been released, and functional corpora atretica are formed from the atretic follicles. The accumulation of such corpora atretica at the end of the breeding season makes the female refractory to environmental stimuli, which in the spring would cause recrudescence of the ovary. This refractoriness is the result of the failure of the ovary to respond to the exogenous gonadotrophins, which in the spring cause follicular growth and ovulation (see Chapter 5).

The female American chameleon is receptive to the courtship by the male only when a large preovulatory follicle is present. Such a follicle probably secretes estrogen or estrogen and progesterone hormones, which, in turn, increase the receptivity of nonreceptive females. After mating, the female becomes unreceptive to male courtship until the next follicle matures. As will be discussed in Chapter 14, photoperiod and ambient temperature are the principal proximate causes, but aggression among males can have an inhibitory effect on the development of the yolky follicles (Crews, 1975).

In oviparous species, "corpora lutea" are formed that affect the retention of the eggs in the oviducts. The longer the "CL" are maintained, the longer the retention of the eggs (Rowlands and Weir, 1977). What the control mechanisms are for maintaining the "CL" needs to be investigated. In viviparous reptiles, CL are formed; such CL can be removed from pregnant snakes *without* re-

percussions on pregnancy itself, but the young are retained instead of being expelled at the end of the gestation period. In the viviparous lizard *Chalcides ocellatus,* the CL can be removed at any stage of pregnancy without any effect on pregnancy (Rowlands and Weir, 1977).

BIRDS

Breeding Cycles

In birds, one can distinguish several types of breeding periodicity.

Continuous Breeding. Some birds in the *population* are reproductively active (laying eggs, incubating eggs, caring for the young) during every month of the year, although individual birds do not lay every month of the year, e.g., the tropical cormorants, *Phalocrocorax africanus* and *P. carbo.* A Senegalese bird, *Lagnosticta senegala,* apparently has the record of laying five successive clutches of eggs. Domesticated birds may lay continuously for a year or more; for example, Khaki Campbell ducks may lay 365 (or more) eggs in 365 days, and chickens of commercial egg-laying strains lay, on the average, more than 270 eggs in 365 days, with individual hens occasionally laying 365 eggs in 365 days.

Breeding Cycles of Less Than a Year. On some tropical islands, such as Ascension (7.5° S, 15° W), some species have intervals between breeding periods of 9.6 months, as in the sooty tern (*Sterna fuscata*), whereas in the brown booby (*Sula leucogaster*) and the lesser noddy (*Anous tenuirostris*), the intervals are 8 months. However, on Christmas Island, the sooty tern hens that have raised at least one young have an interval of a year between clutches; if no young are raised because of mortality, the interval is six months. The proximate factors for the intervals of 8 and 9.6 months in the bird populations of Ascension Island are not known. There is synchronization within the populations, so that the entire population of a species breeds simultaneously. The irregular cycles may give the advantage of making the breeding and molting, when the birds are particularly vulnerable to predation, unpredictable for potential predators (T. Eisner, personal communication).

Six-Month Cycle. This type of cycle has been found in the equatorial *Zonotrichia capensis,* which lives in the Andes. The males are capable of reproduction for 4 months, then have a two-month rest period and become reproductive again. Egg-laying females can be found throughout the year, but there are peaks every 5 and 7 months (Miller, 1965). The proximate causes are photoperiod and rainfall.

Opportunistic Breeders. These are found in deserts, where favorable conditions for care of the young may not occur regularly, and the parental generation must reproduce when conditions become favorable. Rainfall may make seeds available for food to birds, e.g., the budgerigar or grass parakeet (*Melopsittacus undulatus*), or it may make grass available for the building of nests for such species as the red-billed dioch (*Quelea quelea*). Under laboratory conditions, neither the testes nor the ovaries of these two species regress completely. If this incomplete regression also prevails in nature, then it would be adaptive because, as conditions become favorable, the testes and ovaries can produce sperm and oocytes in a shorter period than they could if the gonads were completely regressed. For some species of waterfowl in Australia, the water level is the proximate cause. For instance, the gray teal, the black swan (*Cygnus atratus*) and the pink-eared duck

(*Malacorhynchus membranaceus*) all breed only when the water rises to certain levels. Other species have a regular breeding season, but they breed again when water levels rise to a certain height, e.g., the black duck, the Australian shoveler (*Anas rhynchotis*), the hard head (*Aythya australis*), and the freckled duck (*Stictonetta naevosa*) (Firth, 1973).

Yearly Cycles. Most of the species that breed in the temperate, subarctic and arctic zones and that lay eggs in the spring, have a yearly cycle. The American goldfinch (*Carduelis* sp.), which lays eggs in August, is an exception, the proximate cause being apparently the availability of thistle down (P. Mundinger, personal communication). Among the yearly breeders, there are sometimes additional variations, e.g., in the King penguin (*Aptenodytes patagonica*), which lays early (December) one year but late (mid-February) the next year.

The proximate cause for many species with a yearly cycle is the photoperiod, which may either synchronize an endogenous circannual rhythm, e.g., in transequatorial migrants (Gwinner, 1975), or stimulate the hypothalamo-hypophyseal-gonadal axis. These effects of light will be discussed in more detail in Chapter 14.

An example of the behavioral, plumage, endocrine, and testicular changes in a seasonally re-

Figure 8.2 Phenology of major events in the annual cycle of starlings. The width of each horizontal bar is roughly proportional to the frequency or intensity of each event. Data on photorefractoriness are from Burger, 1947. (From Temple, 1974; reprinted with permission of S. A. Temple and Academic Press, Inc.)

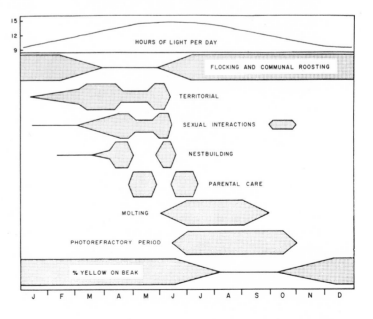

producing male in the temperate zone, i.e., the starling (*Sturnus vulgaris*), is shown in Figures 8.2 through 8.4. Figure 8.2 suggests that with increasing photoperiod there is an increase in territorial behavior and in sexual interactions, the sexual interactions being interrupted by nest building and parental care and being terminated by a molt. The photorefractory period will be discussed in Chapter 14; suffice it to state that its existence can be determined by exposing the birds to long days. The yellow color of the beak is androgen dependent, and there is a good correlation between the brief resumption of sexual activity and yellowing of the beak (Figure 8.2) and the increase in plasma testosterone concentration in the fall (Figure 8.3). Spermatozoa are found with greatest frequency during the period of sexual interactions (Figure 8.4). In Chapter 14, it will be shown that by artificial manipulation of the photoperiod, it has been possible to have spermatogenesis continue throughout the year in male starlings.

Wingfield and Farner (1978a,b) have studied the annual cycle of endocrine changes in the short-distance migratory subspecies of the white-crowned sparrow, *Zonotrichia leucophrys pugetensis*, and the long-distance migratory *Z. leucophrys gambelii*. Their findings are summarized here:

(1) *Zonotrichia leucophrys pugetensis* can have two broods between late April and mid-July, whereas *Z. leucophrys gambelii*, which breeds further north, has only one brood. (2) In *Z. leucophrys pugetensis*, testicular weights are maintained from the end of April till about mid-June; however, the plasma testosterone concentrations

decline rather rapidly after copulations are completed, and the plasma LH concentrations remain high until shortly before the courtship for the second brood starts. In *Z. leucophrys gambelii*, testicular weight and plasma LH concentrations are high from the time of arrival at the breeding grounds in mid-May until the early postnuptial molt during the first half of July; plasma testosterone concentrations are high on arrival at the breeding grounds, but decline sharply after courtship and copulations are over during the middle of May to early June. (3) In the female *Z. leucophrys pugetensis*, the ovaries grow rapidly immediately after arrival at the breeding grounds and, of course, decline after ovulations have been completed and incubation of the eggs has started. This decline is followed by a second increase in ovarian weight before the resumption of egg laying for the second brood, which is followed by a decrease in ovarian weight similar to to that for the first brood. Plasma estrone and LH concentration change in advance of but parallel with the ovarian weight changes. However, estradiol-17β concentrations fluctuate relatively little. Plasma testosterone concentrations parallel the ovarian weight change for the first brood, but after their decline remain low and do not rise during the resumption of ovarian growth. Plasma 5α-dihydrotestosterone concentrations, although not as pronounced as those of LH, parallel the LH fluctuations. Plasma corticosterone concentrations in males and females show a greater increase during courtship for the second brood than for the first brood. (4) In *Z. leucophrys gambelii*, ovarian weight increases rapidly after arrival at the breeding grounds and declines after completion of the laying of the

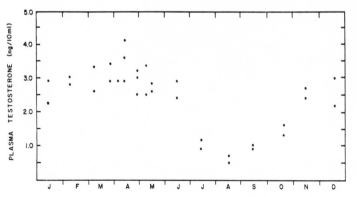

Figure 8.3 Seasonal variations in plasma testosterone titers of male starlings. Each point represents the mean of duplicate determinations on a 1:1 pool of plasms from two birds. From March to September each point is located at the midpoint of a 5-day collection period; all other points represent collections on a single day. (From Temple, 1974; reprinted with permission of S. A. Temple and Academic Press, Inc.)

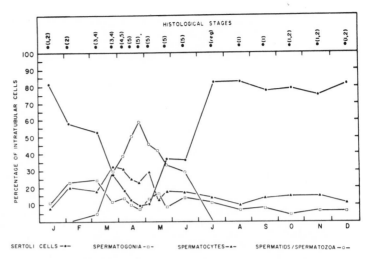

Figure 8.4 Seasonal variations in the condition of the starling's germinal epithelium. Each point represents the mean percentage of 1000 randomly selected microscopic fields (5 birds and 200 fields per bird). Histological stages are assigned after Selander and Hauser (1965): Stage 1: single row of spermatogonia and a few spermatocytes; Stage 2: two or three rows of spermatogonia and spermatocytes; Stage 3: spermatocytes common; Stage 4: spermatids present; Stage 5: many spermatozoa present; Regression (reg.): cellular debris and bundles of spermatozoa in tubules. (From Temple, 1974; reprinted with permission of S. A. Temple and Academic Press, Inc.)

clutch. Plasma LH, estradiol-17β, and estrone levels change slightly ahead of and parallel with ovarian weight changes. There is a sharp increase in plasma 5α-dihydrotestosterone concentrations in February, while the birds are still on the wintering grounds in California. Plasma testosterone concentrations show an increase before and a peak during spring migration. There is thus no correlation between the plasma LH concentrations and androgen concentrations in females, but there is a good correlation between LH and estrogen concentrations. The role of the androgens with respect to migration will be discussed in Chapter 13.

Determinacy of the Ovulatory Cycle

Within a breeding season, one can distinguish determinant and indeterminant layers.

Determinate Layers. In this type, the female lays a certain number of eggs (clutch) which does not increase when eggs are taken away, although the number may be reduced if bad weather prevails and atresia of the follicles occurs. Barry (1962) found that in the brant (*Branta bernicla hrota*) and snow (blue) geese (*Chen caerulescens*), the elapsed days between arrival at the breeding grounds and laying of the first egg was correlated with clutch size, and that the "missing" eggs could be accounted for by the atresia of the largest follicles.

Two ovarian mechanisms appear to determine the clutch size. In some species, e.g., the brant,

and snow goose, only a certain number of follicles grow to ovulatory size, and this number corresponds to the specific clutch size. In other species, e.g., *Agapornis* sp., the number of follicles that mature is larger than the size of the clutch. When the last egg of the clutch is laid, atresia occurs and the "extra" follicles are resorbed (W. C. Dilger, personal communication).

Indeterminate Layers. In this type, the clutch size may show little variation if the eggs are left in the nest, but if eggs are removed, the number of eggs exceeds the normal clutch size. The American flicker (*Colaptes auratus*) has been induced to lay as many as 50 eggs in 73 days (see Lehrman, 1961). Even more interesting is the case, also cited by Lehrman (1961), of the cuckoo (*Cuculus canorus*), which normally lays a total of 5–7 eggs per clutch, but which can be induced to lay a total of 20–25 eggs by removal of the eggs of the *host* species, so that the hosts build new nests, relay, and thus make more nests available for the cuckoo to lay her normal number of eggs per nest.

Extreme cases of indeterminate layers are the Khaki Campbell ducks, Japanese quail, chickens, and turkeys, all domesticated species in which egg production has been increased by selection and breeding. This is true to a somewhat lesser extent in turkeys, which are bred for meat production, but in which some selection for egg production is an economic necessity. Among a population of Japanese quail studied by Planck and Johnson

Table 8.2

Examples of day and time of laying of eggs by domestic chickens during different length of sequences*

Number of eggs in sequence	Day of week														
	S	M	T	W	T	F	S	S	M	T	W	T	F	S	S
2	10	14	—	10	15	—	10	14	—	10	16	—	10	15	—
3	9	11	15	—	9	12	16	—	9	11	16	—	9	12	15
4	9	10	13	16	—	8	10	11	15	—	9	10	12	—	9
5	8	10	12	13	17	—	9	10	11	12	16	—	8	9	11
6	8	9	10	11	13	17	—	8	9	10	12	13	15	—	8
7	8	9	9	10	11	12	16	—	8	9	10	10	11	12	17
8	8	9	9	10	10	11	13	17	—	8	9	9	9	11	12
9	8	8	9	9	10	10	11	12	16	—	8	8	9	10	10

*Lights on 0600–2000.
Data from Fraps, 1955, and Cornell University laying records.

(1975), some hens were found that laid (and ovulated?) at 24-hour intervals either from the start of laying or after a time interval, and some hens that had an egg-laying pattern that deviated from the 24-hour period.

The laying cycle of the domestic chicken has been studied in great detail. After the hen starts laying, she lays eggs in long sequences, but after about 6–10 weeks, the sequences decrease in length. A sequence is defined as a series of eggs laid on successive days. The timing of the laying of eggs in sequences of different length is illustrated in Table 8.2.

Endocrine Events and the Ovulatory Cycle of the Chicken

The endocrine events associated with the ovulatory cycle have been studied in great detail in the chicken. The following discussion of these endocrine events is based largely on the chicken. For reasons of convenience, the endocrine changes that occur during the laying of a 3- to 5-egg sequence have been studied, not those of the longer and shorter sequences. In Table 8.3, the time of the occurrence of peaks in the concentrations of LH, estrone (E_1), estradiol (E_2), progesterone, testosterone, dihydrotestosterone (DHT), corticosterone, the time of different stages of egg formation in the oviduct, and the time of laying are summarized. The timing of the peaks for the

different hormones reported in Table 8.3 agrees generally with the timing reported in the literature. We will briefly discuss these agreements and disagreements. There is some disagreement about the timing of the E_2 peaks. Peterson and Common (1972) found peaks at 22–18 h and at 6–2 h before ovulation. Senior (1974) and Senior and Cunningham (1974) reported a peak at 6 h before ovulation, with the peak of E_2 preceding the LH peak by 2 h. Laguë et al. (1975) reported that the E_2 and progesterone peaks coincided and that the LH peak preceded the progesterone and E_2 peak. Shodono et al. (1975) found LH, progesterone, and E_2 peaks at 4–5 h before ovulation. It appears that there is more variability in the timing of the E_2 peak with respect to ovulation than there is for the progesterone, testosterone, and LH peaks, and that taking average values may shift peaks or flatten them (Johnson and van Tienhoven, 1980a). Shodono et al. (1975) and Laguë et al. (1975) agree that there is an additional E_2 peak at about 24 h before ovulation. For E_1, a similar situation seems to prevail as for E_2; in general, E_1 and E_2 peaks seem to coincide (Peterson and Common, 1972; Senior, 1974; Laguë et al., 1975), although the early peak of E_1 preceded the early E_2 peak by about 2 h in the experiments of Peterson and Common.

The crepuscular, small, LH peak occurs at the time lights are turned off, according to Johnson and van Tienhoven (1980a) and 1 hour after lights

Table 8.3

Time of endocrine and egg formation events in a four-egg sequence of a laying chicken (lights on 0600–2000)

Event	Day						
	1	2	3	4	5	6	7
Estrone (E_1) peak	1400	1700	1900	2215	—	1400	1700
Estradiol (E_2) peak	1400	1700	1900	2215	—	1400	1700
Crepuscular LH peak (small)	2000–2100	2000–2100	2000–2100	2000–2100	2000–2100	2000–2100	2000–2100
Testosterone (T) peak	2200	—	0100	0300	0615	2200	—
Dihydrotestosterone (DHT) peak	—	Midnight	0300	0500	0815	—	Midnight
Preovulatory LH peak	—	Midnight	0300	0500	0815	—	Midnight
Progesterone peak	—	0200	0500	0700	1015	—	0200
E_2 peak	—	0200	0500	0700	1015	—	0200
E_1 peak	—	0200	0500	0700	1015	—	0200
Follicle ovulated		C_1	C_2	C_3	C_4		C_1
Ovulation with small corticosterone (C) peak	—	0600	0900	1100	1415	—	0600
Egg in magnum of oviduct; secretion of thick albumen	—	0610–0910	0910–1210	1110–1410	1425–1725	—	0610–0910
Egg in isthmus; secretion of membrane	—	0910–1010	1210–1310	1410–1510	1725–1825	—	0910–1010
Egg in shell gland; high PGE in plasma; secretion of thin albumen	—	1010–1400	1310–1700	1410–1900	1825–2215	—	1010–1400
Calcification	—	1400–0745	1700–0945	1900–1245	2215–1600	—	1400–0745
Oviposition with a peak of corticosterone, and a peak of vasotocin	—	—	0745	0945	1245	1600	—

Sources: For crepuscular LH, preovulatory LH, E_1, E_2, DHT, T, and oviposition C peak, Johnson and van Tienhoven, 1980a; for ovulatory C peak, Beuving and Vonder, 1981; for PGE in plasma, Hertelendy and Biellier, 1978; for egg formation, Warren and Scott, 1935; for vasotocin peak, Sturkie and Lin, 1967.

are turned off, according to Wilson and Sharp (1973), Williams and Sharp (1978) and Scanes et al. (1978).

The values reported for testosterone in Table 8.3 agree both with respect to the timing of the peak and the values obtained with the data reported by Peterson et al. (1973) and Etches and Cunningham (1977); however, Shahabi et al. (1975a) have reported a testosterone peak at the same time as the progesterone peak, and the values reported are about 30–50 percent of those found by others. There seem to be no other reports than that of Johnson and van Tienhoven (1980a) on peaks of DHT in relation to ovulation in the domestic hen.

The timing of the preovulatory LH peak agrees with the findings of some researchers (Senior and Cunningham, 1974; Laguë et al., 1975; Shodono et al., 1975; and Etches and Cunningham, 1976, 1977). Whether there are additional LH peaks at times other than 4–6 h before ovulation has not been entirely resolved. Huang and Nalbandov (1979) indicate peaks of similar magnitude at about 21, 13, and 7 h prior to ovulation. These peaks are apparently based on bioassays using the rat ovarian ascorbic acid depletion test, but the specificity of this assay is questionable (van Tienhoven and Planck, 1973). Recently, however, Ax (1978), using minced rooster testes as a bioassay for steroidogenesis in response to LH, found a small, but not significant (P > 0.05) peak at 22–20 h before the C_3 ovulation (Table 8.3), a "peak" at 12 h before the C_3 ovulation, and a large peak at 5 h before the C_3 ovulation. The 12-h peak occurred during a period of declining basal LH concentration. The ratio of bioassay to RIA values was 1.1, indicating a very good agreement between the two assay methods. It appears that the LH peak 12 h before ovulation corresponds to the crepuscular LH peak. Since the higher LH concentration at 22 h before ovulation was not significant, the controversy about the existence of one or two LH peaks seems to be resolved. It is clear from Table 8.3 that the interval between the crepuscular and ovulatory LH peak increases for each successive ovulation within a sequence.

The timing of the progesterone peak agrees with the data reported by some researchers (Cunningham and Furr 1972; Kappauf and van Tienhoven, 1972; Furr et al., 1973; Peterson and Common, 1971; Haynes et al., 1973; Laguë et al., 1975; Shahabi et al., 1975a; Shodono et al., 1975; Etches, 1979), and the values in Table 8.3 are in the same range as those reported in the papers cited.

The corticosterone peak at the time of ovulation has been investigated and reported by Beuving and Vonder (1981) and has probably not been detected by others because blood has not been sampled at short enough intervals; the width of the peak is about 20 minutes. The corticosterone peak at the time of laying is in agreement with the findings of Beuving and Vonder (1977).

Two gonodotrophins are not listed in Table 8.3, i.e., FSH and prolactin (PRL). Two reports have appeared on FSH concentrations. Imai and Nalbandov (1971) used pooled plasma samples from hens killed at different intervals before the expected C_3 ovulation and assayed these samples by the Steelman-Pohley ovarian weight augmentation test in rats; Scanes et al. (1977a) used RIA. Imai and Nalbandov (1971) report a variation of 600 percent between the highest and lowest plasma concentrations, whereas Scanes et al. (1977a) found a variation of only 22 percent. The two studies show agreement on the presence of an FSH peak at 14–11 h before ovulation. Scanes et al. (1977a) pointed out that their own two studies showed a lack of consistency in the timing of the

FSH peaks. Clearly more information is needed on FSH variations and their physiological significance.

The concentration of PRL varies inversely with that of LH, so that it is low at the time of preovulatory LH peak and high at 4 h after ovulation and thus about 22 h prior to the next ovulation (Scanes et al., 1977b). The significance of these variations in PRL concentration is not clear.

Hypothalamo-Hypophyseal-Ovarian Axis

After the survey of the timing of endocrine events during the ovulatory cycle, it is important to discuss the experimental evidence that has shed light on the significance of these hormonal changes in the regulation of hen's ovulatory cycle. One of the principal hormones, progesterone, acts through the hypothalamus, and therefore, consideration needs to be given to the role of the hypothalamus in this regulation.

Hypothalamus. The hypothalamus is crucial in the regulation of the ovulatory cycle, as various experiments show. Lesions in the hypothalamus prevented both spontaneous ovulation (Ralph, 1959) and progesterone-induced ovulation (Ralph and Fraps, 1959). Injections of progesterone in the hypothalamus induced premature ovulations (Ralph and Fraps, 1960) and injections of progesterone into the third ventricle of the brain result in LH release, progesterone and testosterone secretion, and premature ovulation (Johnson and van Tienhoven, 1980b). A progesterone receptor was found in the hypothalamus (Kawashima et al. 1978), and changes in the concentration of this receptor occurred during the ovulatory cycle, with peaks in the receptor concentration at 18 h and 8 h before ovulation (Kawashima et al., 1979). The hypothalamus contains material, presumably avian luteinizing hormone-releasing hormone (LH-RH), which, upon infusion into the anterior pituitary, results in premature ovulation (Opel and Lepore, 1978). Injections of antibodies against mammalian LH–RH into laying hens interrupted egg laying and prevented progesterone-induced LH release (Fraser and Sharp, 1978). Injections of mammalian LH-RH resulted in increases in plasma LH concentrations (Bonney et al., 1974, and Fraser and Sharp, 1978) and premature ovulation (Bonney et al., 1974, and van Tienhoven and Schally, 1973). The chicken hypothalamus contains a factor that causes prolactin release in vitro from chicken pituitaries (Hall and Chadwick, 1979).

Gonadotrophins. The role of FSH, LH, and prolactin are discussed separately.

FSH. As mentioned in the discussion of Table 8.3, there is a lack of information about avian FSH in female reproduction, and it should be noted that in the experiments discussed here, mammalian FSH was used. Injections of FSH into laying chickens destroyed the follicular hierarchy and made many of the large follicles ovulable, so that after LH injection in FSH-primed hens, multiple ovulations occurred (Fraps, 1955; Ogawa et al., 1977). Multiple ovulations also occurred when LH was administered shortly after AP of laying hens (as was discussed in Chapter 7). Ogawa et al. (1976, 1977), in a series of experiments, showed that injection of either PMSG or FSH prior to LH injection into hens, which had been hypophysectomized 8 h earlier, prevented multiple ovulations when FSH was injected not later than 4 h after the AP. These results suggest that the supply of FSH to a particular follicle is crucial; on the one hand, withdrawal of FSH makes the follicle too sensitive to LH, but on the other hand, FSH seems to increase sensitivity of the follicle. The destruction of the hierarchy of the follicles does not occur in all avian species, for Witschi (1955) obtained a normal hierarchy of follicles after injection of mammalian gonadotrophins in song birds. The reason for this difference among species awaits further investigation.

LH. The excellent correlation between the time of the ovulatory LH peak and the subsequent ovulation both during the ovulatory cycle and after either LH-RH, testosterone, or progesterone administration provides persuasive evidence that LH is the ovulation-inducing hormone in the chicken. Injections of mammalian LH induced

ovulation (Fraps, 1955, Ogawa et al., 1976, 1977) and resulted in an increase in plasma testosterone and progesterone concentration but not in estrogen concentration (Shahabi et al., 1975b).

PRL. There are few experiments in which PRL has been injected at different intervals before ovulation in laying hens. Tanaka et al. (1971) found that oPRL injected with 250 μg oLH in phenobarbital-injected hens (to block the endocrine feedback to the hypothalamus) decreased the incidence of LH-induced premature C_2 ovulations, but when the dose of oLH was 500 μg, prolactin did not have an effect on the incidence of such premature ovulations (Tanaka et al. 1971). In view of the changes in prolactin concentrations during the ovulatory cycle, more experiments need to be carried out, in which the plasma prolactin concentration is experimentally changed at different time intervals before the next expected ovulation.

Gonadal Hormones. We will now discuss the effect of gonadal hormones on ovulation.

Testosterone. Table 8.3 shows that the peak concentration of testosterone precedes the LH peak by 2 h. It is, therefore, attractive to speculate that such a peak of testosterone triggers the release of LH, which then, in turn, may stimulate the secretion of progesterone; however, the experimental evidence argues against this speculation. Testosterone injections will induce premature ovulations and LH release, but the dose necessary to obtain nearly 100 percent premature ovulation is unphysiologically high. Several researchers (Etches and Cunningham, 1976a; Croze and Etches, 1980; and Johnson and van Tienhoven, 1981) have found that the amount of testosterone injected to obtain premature ovulations resulted in plasma testosterone levels that were about 15 to 30 times higher than the normal testosterone peak concentration.

Johnson and van Tienhoven (1981) found that injection of as much as 80 μg of testosterone into the third ventricle did not result in premature ovulations, and injections of 20 μg failed to induce LH release, progesterone and testosterone secretion, or premature ovulations, whereas 20 µg of progesterone induced LH secretion, progesterone and testosterone secretion, and premature ovulations (Johnson and van Tienhoven 1980b). Testosterone fails to induce LH release in ovariectomized estrogen-progesterone-primed hens, whereas progesterone injections result in LH release in such hens (Wilson and Sharp, 1976b). These data argue against a triggering of LH release by testosterone during the ovulatory cycle. However, a number of other observations suggest that testosterone may be involved in some regulatory function of the hypothalamo-hypophyseal-ovarian axis. For example, LH injections result in an increase in testosterone secretion by the large follicles (Shahabi et al. 1975b). Injection of antiserum to testosterone at 12 h before ovulation has been reported to block ovulation (Furr and Smith, 1975), although, in our laboratory (van Tienhoven, unpublished observations), we have been unable to obtain such a blockage with several different antisera to testosterone.

Progesterone. It has been known for some time that progesterone can induce premature ovulations in laying hens (Fraps, 1955, for early literature) and that the dose necessary to induce such ovulations is within the physiological range (Etches and Cunningham, 1976a,b). We have stated above that infusion of progesterone into the hypothalamus or third ventricle results in LH release, progesterone secretion, and premature ovulation. Antiserum to progesterone injected 12 h before ovulation blocks or delays ovulation (Furr and Smith, 1975), suggesting that progesterone is essential for ovulation. In our laboratory, we have been unable to obtain either a blockage or a delay of ovulation, although we have used at least three different antisera to progesterone (Johnson and van Tienhoven, unpublished). Progesterone injections in ovariectomized estrogen-progesterone-primed hens result in a LH release at least as great as that in intact hens (Wilson and Sharp, 1976b). This suggests that progesterone has a stimulatory effect on LH release, but that it is not a positive feedback action; since the ovary is absent in the ovariectomized estrogen-progesterone-primed hens, no feedback is possible. The data cited above for progesterone-induced

LH release and for LH-induced progesterone secretion show a reciprocal relationship between LH and progesterone, which we will discuss later in the chapter in the section on rhythmicity.

Estrogens. Estrogens are necessary for the priming of the hypothalamo-hypophyseal system for the release of LH, as has been demonstrated by the experiments of Wilson and Sharp (1976b) with ovariectomized hens that were primed with different hormones before the injection of progesterone to induce LH release. During the ovulatory cycle, injections of estradiol do not induce premature ovulations at times when progesterone does. Estrogens do not act synergistically with progesterone in inducing premature ovulations (Laguë et al., 1975), and injections of antiserum to estrogens do not block or delay ovulation (Furr and Smith, 1975). These experiments thus suggest that estrogens may not be involved in the timing of the ovulatory cycle. Recent experiments by Wilson and Cunningham (1981), using the estrogen receptor blocker, tamoxifen, showed that injection of 2 mg of tamoxifen for two successive days during midcycle, inhibited LH release and ovulation of the third expected ovulation after the beginning of the treatment. Injection of LH-RH restored the increase in LH, and ovulation proceeded normally. Tamoxifen treatment resulted in a rise in the basal LH concentration and reduced the effectiveness of progesterone administration in inducing ovulation; as a consequence of the tamoxifen treatment, the sensitivity of the pituitary to exogenous LH-RH stimulation was increased. It thus seems that estrogens decrease the basal secretion of LH and facilitate the effect of progesterone on LH release. This effect is similar to the priming effect reported by Wilson and Sharp (1976b).

Dihydrotestosterone. The function of dihydrotestosterone in the regulation of ovulation has not been determined. Injections of 5α-dihydrotestosterone did not result in a dose-related response curve with respect to the induction of premature ovulations (Croze and Etches, 1980).

Corticosterone. The small corticosterone peak which occurs at the time of ovulation may be associated with the bursting of the follicle and possibly with pain associated with it. This pain may function as a stressor, resulting in an increase in corticosterone. In some humans, ovulation is associated with Mittelschmerz, a pain severe enough to be confused, sometimes, with the pain of an appendicitis. It would be interesting to determine whether ovulation under anesthesia with an anesthetic that does not cause an increase in corticosterone and does not block ovulation, is accompanied by the corticosterone peak found by Beuving and Vonder (1981). There is no evidence that a corticosterone peak occurs before or together with the LH, progesterone, and estrogen peaks.

Experimentally, premature ovulation can be induced with nonphysiological doses of ACTH (100 IU) (van Tienhoven, 1961) and of corticosterone (van Tienhoven, 1961; Sharp and Beuving, 1978). Etches and Cunningham (1976a) determined the ED_{50} for corticosterone to be 659 μg/hen. The fact that dexamethasone, which lowers plasma corticosterone concentrations (Etches, 1976), blocks ovulation (Soliman and Huston, 1974) also suggests that corticosterone may affect ovulation in the hen.

Meier (1972) has emphasized the role that adrenal steroids may play in phase-setting of the response to various hormones. Sharp and Beuving (1978) have investigated this aspect in laying hens. They found no evidence for a change in corticosterone concentration during the first 6 h of darkness, either during the night when there was no preovulatory LH release, or during the next night when there was a preovulatory LH release. Wilson and Cunningham (1980) investigated the effect on ovulation of metyrapone, a drug that lowers plasma corticosterone concentrations by blocking the response of the adrenals to injections of ACTH. Injection of metyrapone about 12 h before the expected ovulation of the C_1 follicle resulted in an increase in plasma LH and progesterone concentrations and in premature ovulations. Wilson and Cunningham (1980) attribute this sequence of events to the secretion of deoxycorticosterone by the adrenals, in which corticosterone synthesis is blocked by metyrapone. Wilson and Sharp (1976) have shown that deoxycorticosterone injections will induce LH release

and ovulation thus lending support to Wilson and Cunningham's interpretation.

When hens were injected with metyrapone 8 to 9 h after lights on for five successive days, ovipositions (and presumably the ovulations that preceded them by about 26 h) changed from occurring only during the light period to occurring also during the dark period.

As Table 8.3 shows, there is a daily crepuscular LH release, which occurs when lights are turned off. This small release occurs independent of any subsequent ovulation. Table 8.3 also shows that the preovulatory LH peak occurs during a restricted part of the 24-h day, that is, between midnight and 8:15. Fraps (1955) has called this the "open period" because LH release could occur. The open period starts about 2 to 3 h after the onset of dark and lasts for about 10 h, under a light regime of 16L : 8D. After the injections of metyrapone for 5 days the preovulatory LH release starts to occur outside the open period. Thus it appears that secretions by the adrenal gland may set the phase for LH release.

Model of Rhythmicity of the Chicken's Ovulatory Cycle

A tentative model to explain the ovulatory cycle of the chicken will be presented here. This model incorporates hypotheses proposed by Fraps (1966) and Williams and Sharp (1978), and although it is speculative, certain aspects can be tested. The adrenal corticosteroids, in response to the light-dark signals of the photoperiod, may regulate the sensitivity of either the hypothalamus to progesterone or the sensitivity of the AP to LH-RH or both. Under normal photoperiods, this regulation may set the phase for the crepuscular LH release. This release of LH stimulates the follicle to secrete progesterone, which, in turn, stimulates further LH secretions, so that it reaches a sufficient concentration for ovulation to occur. The model, as stated so far, does not explain the shift of about 2 h for LH release, progesterone and estrogen concentration, and ovulation. Johnson (1981, personal communication) has found that at 14 h prior to expected ovulation of the C_1 and C_2 follicle, the injection of LH-RH is followed by different responses. In the case of the C_1 follicle, there is a small initial increase in LH concentration, about 20 minutes after LH-RH injection similar to the crepuscular LH peak. This increase is followed by an increase in LH and progesterone concentrations, similar to the preovulatory increases during the ovulatory cycle, and a premature ovulation follows. In the case of the C_2 follicle, the initial increase in LH is similar to the one seen for the C_1 follicle; however, there is no increase in progesterone secretion or LH secretion and no premature ovulation; the ovulation occurs at the normally expected time. It thus appears that the follicle has to be capable of secreting progesterone in order to stimulate further LH release, which is necessary for ovulation. We speculate that the LH released at the onset of darkness can bind to the follicle to be ovulated and can induce progesterone secretion when the follicle has the capacity to secrete progesterone. The follicle may lose its ability to bind the LH long enough when the interval between crepuscular LH release and the time the follicle becomes capable of secreting progesterone becomes too long, i.e., more than 10 h. We do not have enough information to speculate about what regulates the capacity of the follicle to secrete progesterone. It is possible that the corticosteroids affect this capacity, in a way somewhat similar to the way corticosteroids regulate ovulation in fishes, as discussed in Chapter 7. However the capacity of the follicle to secrete progesterone may be regulated, the model developed above makes the ovary an integral part of the "clock" that regulates the ovulatory rhythm. Several aspects of the proposed model require verification. For example, it should be determined whether in the absence of the crepuscular LH peak, ovulation still occurs. It should also be determined whether the small LH peak occurs in hens kept under continuous light, when there are thus no light-dark signals and when hens ovulate and lay eggs at 26–28 h intervals. It should also be investigated whether LH binds to the largest follicle and retains its capacity to induce progesterone secretion until the follicle is ready to respond.

It has not been determined whether progesterone production is required for the ovulatory process itself, as it is, for instance, in some amphibians. Experiments by Huang et al. (1979) (Figure 7.8) have shown that the granulosa cells produce progesterone, which presumably diffuses to the thecal cells. There it becomes the precursor for testosterone formation, and testosterone can then be aromatized to estrogens by the thecal cells. As the follicle matures, the steroidogenic capacity of the thecal cell diminishes under the influence of gonadotrophins, and that of the granulosa cells increases, thus leading to a shift to increased progesterone production and decreased testosterone and estrogen production. These observations thus are in good agreement with the observations that progesterone synthesis is required either for stimulation of LH secretion or for ovulation itself. It is possible that the diminishing prolactin concentrations influence the shifting from progesterone, testosterone, and estrogen synthesis to progesterone synthesis by the maturing follicle. This possibility, however, has not been investigated.

Tanabe and Nakamura (1980), in a comparison among chickens, Japanese quail (*Coturnix coturnix japonica*), and domestic mallard ducks (*Anas platyrhynchos domestica*) found that under 14L : 10D (lights on at 0500 and off at 1900), chickens laid their eggs between 0600 and 1400, Japanese quail between 1400 and 1900 and ducks between 0200 and 0600. The endocrine changes described for plasma LH, progesterone, and estradiol in Table 8.3 were also found by Tanabe and Nakamura (1980). The plasma LH and estradiol concentration peaks occurred at 6 h and the progesterone concentration peak at 4 h before ovulation in the Japanese quail and peaks of LH, estradiol, and progesterone occurred at 3 h before ovulation in the duck. Tanabe and Nakamura (1980) suggest that for chickens and ducks, the onset of darkness sets the phase of these endocrine changes but that for Japanese quail, it is the onset of light that sets the phase. It would be valuable to determine whether there is a small LH peak at the onset of light in Japanese quail.

MAMMALS

Prototheria

The duck-billed platypus (*Ornithorhynchus anatinus*) breeds during the fall to winter in Australia. In the male, the testicular weight starts to increase in May, reaches a maximum in August, then starts to regress in September to October and reaches a minimum during December to April. Spermatogenesis starts in the winter and proceeds to completion; in October, regression of the tubules starts. The size of nucleus of the Leydig cells indicates that the activity of these cells is associated with the spermatogenetic cycle in the tubules. When the testes are at their maximal size, however, the Leydig cells are restricted to the small spaces between the large tubules; when the testes are small, the Leydig cells are inactive and enclose the tubules. During the inactive stage, the Leydig cells contain lipid droplets (Griffiths, 1978), which may be precursors of the male sex hormones.

Starting in June, the left ovary of the female platypus starts to increase in weight, as do the left and right uteri, but no changes take place in the right ovary. The left ovary reaches its maximal weight in August, when ovulation occurs, and then starts to regress, reaching its minimal weight in December.

After ovulation, fertilization takes place in the infundibulum of the oviduct, and during passage through the oviduct, a two-layered coat, which is either albumen or a mucopolysaccharide, is deposited around the yolk (Griffiths, 1978). Farther down, in the posterior end of the Fallopian tube, in the tubal gland region, the basal layer of the shell is deposited. (The gestation period and the function of the corpus luteum will be discussed in Chapter 10).

Estradiol concentrations in the peripheral blood of the nonpregnant platypus has been reported as 10 pg/ml in one case and 90 pg/ml in another. No estrone or estradiol was found, and for one female, a progesterone concentration of 2.1 ng/ml has been reported (Griffiths, 1978).

The spiny anteater, *Tachyglossus* sp., has a breeding season similar to that of the duck-billed platypus, breeding in July to August and sometimes until October. The testes show a reduction in weight by about mid-September. The breeding season of an anteater of the genus *Zaglossus* is ''a matter of conjecture'' (Griffiths, 1978).

Metatheria

Most of the Australian marsupials breed between January and June or sometimes into early September. Rainfall and availability of food (Rowlands and Weir, 1977) and photoperiod (Hearn et al., 1977) are probably the most important proximate causes (Rowlands and Weir, 1977).

Young are born very immature after a gestation period that may be shorter than the duration of the estrous cycle. Many species of marsupials are polyestrous but have monovular ovulations; however, polyovular ovulations occur. The American opossum (*Didelphis virginiana*), for example, gives birth to about a dozen young (which fit together in a tablespoon!).

Figures 8.5 and 8.6 illustrate the relationship between duration of the estrous cycle and of the gestation period in different species. In the brush possum (*Trichosurus vulpecula*) and other species not in the Macropodidae (wallabies and kangaroos), the gestation period is shorter than the estrous period, and when the young attach themselves to the teats in the pouch, they inhibit estrus and ovulation. In all Macropodidae, the young are born at the end of the luteal phase, and the gestation period and estrous cycle are of the same length, as in the red-necked wallaby. There is no postpartum estrus and the suckling young inhibits ovulation. In some species, the gestation period is longer than the estrous period, as in the swamp wallaby (*Wallabia bicolor*). The young wallaby is carried during estrus and is born after estrus. Pregnancy does not inhibit cyclicity, but suckling

Figure 8.5 Relationship between gestation length and estrous cycle length in three marsupials. In the brush possum, postpartum estrus is inhibited by the suckling stimulus of the pouch young. In the wallabies, the young do not reach the pouch in time to prevent estrus; they have evolved a lactation-induced embryonic diapause. (From Short, 1972; reprinted with permission of Cambridge University Press.)

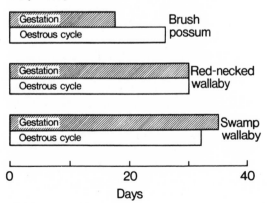

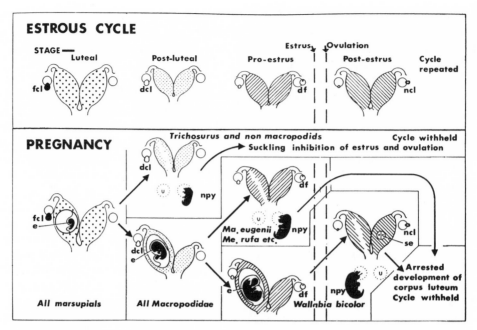

Figure 8.6 Estrous cycle and pregnancy in marsupials. Diagrammatic summary of size and functional relationships of ovary and uterus. Three different patterns in the reproductive cycle of the marsupials are shown: nonmacropods, like *Trichosurus* (Australian possum), and macropods (kangaroos), which have two types of cycle shown by: *Macropus eugenii* (the tammar wallaby) and *Megaleia rufus* (the red kangaroo), and *Wallabia bicolor* (the swamp wallaby). Estrous cycle: fcl = functional corpus luteum, dcl = degenerating corpus luteum, df = developing Graafian follicle, ncl = new corpus luteum. Pregnancy: e = intrauterine embryo, npy = newborn pouch young attached to teat, se = segmenting egg. (From Sharman, 1970; reprinted with permission of G. B. Sharman and *Science*. Copyright 1970 by the American Association for the Advancement of Science.)

does. Ovulation alternates rigidly between left and right ovaries and may result in a dividing zygote in one uterus while the young wallaby from the other uterus is born.

Figures 8.7 and 8.8 illustrate the relationship between embryo, suckling young in the pouch and the young that still nurses but moves in and out of the pouch in the red kangaroo. Figure 8.9 shows a flow diagram of the various alternate types of cyclicity that are found in the female quokka (*Setonix* sp.), depending on the occurrence of fertilization and loss of young.

In none of the marsupials is the function of the corpus luteum affected by pregnancy. The corpora lutea of the American opossum and of the Australian brush possum are not influenced by either gestation or lactation. In the bandicoots (*Peramelidae*), the corpora lutea are secretory during lactation and regress if lactation is terminated, a change that suggests that prolactin may be the luteotrophic hormone in this group of marsupials.

In most Macropodidae, the function of the corpus luteum is arrested by lactation, except in the western gray kangaroo (*Macropus fuliginosus*) and the brush possum (Tyndale-Biscoe and Hawkins, 1977). In the tammar wallaby (*Macropus eugenii*), the AP secretes a substance that inhibits the corpus luteum, which then permits the diapause of the embryo (Hearn, 1974). Tyndale-Biscoe and Hawkins (1977) found that, in this species, prolactin is probably the AP hormone that inhibits the corpus luteum.

The ovarian cycle can be inhibited by lactation by two mechanisms (Figure 8.7). In the first, the suckling young either inhibits or stimulates the AP, so that the ovarian cycle is interrupted, as in the brush possum and in early stages of lactation of the gray kangaroo (*Macropus giganteus*). Then after the suckling young is removed, estrus follows shortly. In the second mechanism, suckling stimulates or inhibits the AP, so that the corpus luteum is maintained and follicular development

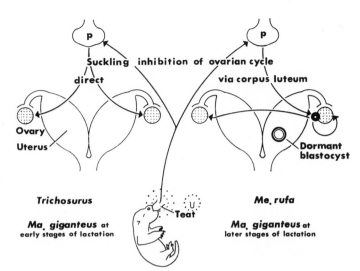

Figure 8.7 Mechanism by which suckling may inhibit the ovarian cycle in marsupials. *Trichosurus vulpecula* and *Macropus giganteus* in early stages of lactation, return to estrus soon after the young is removed from the pouch. In these animals, inhibition seems to result from a direct inhibitory effect on the ovary, mediated by the suckling stimulus. *Megaleia rufa* and *Macropus giganteus* at later stages of lactation do not return to estrus until about a month after the young is removed from the pouch. In these marsupials, a functional corpus luteum is necessary for the inhibition, which, as a result of suckling-pituitary stimulation, releases an "inhibitory factor." P = pituitary gland. (From Sharman, 1970; reprinted with permission of G. B. Sharman and *Science*. Copyright 1970 by the American Association for the Advancement of Science.)

in either ovary is inhibited, as in the gray kangaroo during late stages of lactation and in the red kangaroo (*Megaleia rufa*). Then if the suckling young is removed, estrus does not follow immediately, but after about a month; if, however, the corpus luteum is removed, the female will return to estrus immediately (Sharman, 1970).

Hearn et al. (1977) have proposed that the control of embryonic diapause in the Macropodidae is principally controlled by the suckling stimulus by the young. This stimulus decreases the secretion of the prolactin inhibiting factor (PIF) and simultaneously increases the sensitivity of the hypothalamus to the negative feedback effects of steroid hormones. The AP, in response to the decreases of LH-RH and PIF, secretes more prolactin and less gonadotrophic hormones. The corpus luteum, as a consequence, secretes little progesterone, and so

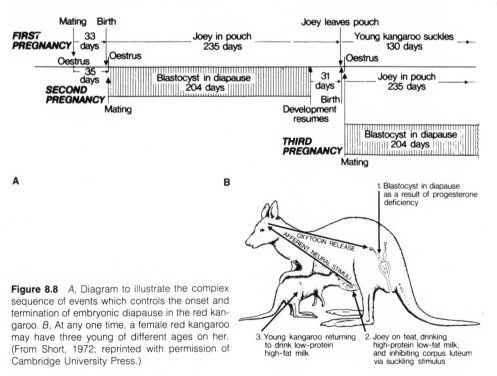

Figure 8.8 *A*, Diagram to illustrate the complex sequence of events which controls the onset and termination of embryonic diapause in the red kangaroo. *B*, At any one time, a female red kangaroo may have three young of different ages on her. (From Short, 1972; reprinted with permission of Cambridge University Press.)

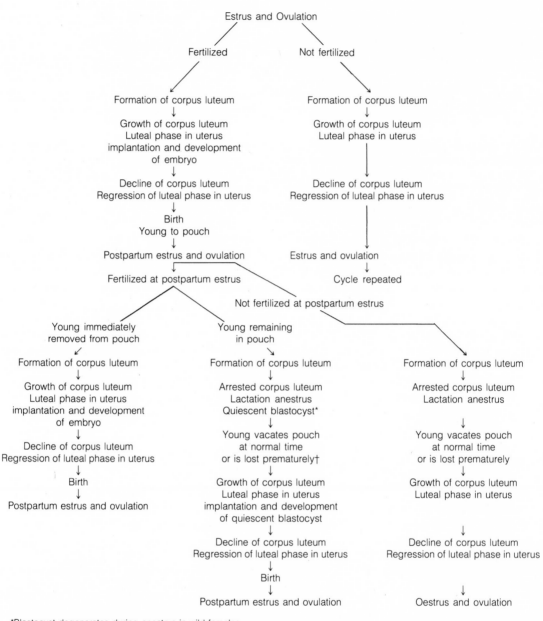

Estrus and Ovulation

Fertilized   Not fertilized

Formation of corpus luteum   Formation of corpus luteum
↓   ↓
Growth of corpus luteum   Growth of corpus luteum
Luteal phase in uterus   Luteal phase in uterus
implantation and development
of embryo
↓
Decline of corpus luteum   Decline of corpus luteum
Regression of luteal phase in uterus Regression of luteal phase in uterus
↓
Birth
Young to pouch
↓
Postpartum estrus and ovulation   Estrus and ovulation
↓   ↓
Fertilized at postpartum estrus   Cycle repeated

Not fertilized at postpartum estrus

Young immediately Young remaining
removed from pouch   in pouch

Formation of corpus luteum Formation of corpus luteum Formation of corpus luteum
↓   ↓   ↓
Growth of corpus luteum Arrested corpus luteum   Arrested corpus luteum
Luteal phase in uterus Lactation anestrus   Lactation anestrus
implantation and development Quiescent blastocyst*
of embryo   ↓
↓   Young vacates pouch   Young vacates pouch
Decline of corpus luteum at normal time   at normal time
Regression of luteal phase in uterus or is lost prematurely†   or is lost prematurely
↓   ↓   ↓
Birth   Growth of corpus luteum Growth of corpus luteum
↓   Luteal phase in uterus   Luteal phase in uterus
Postpartum estrus and ovulation implantation and development
of quiescent blastocyst
↓
Decline of corpus luteum   Decline of corpus luteum
Regression of luteal phase in uterus Regression of luteal phase in uterus
↓
Birth
↓   ↓
Postpartum estrus and ovulation   Oestrus and ovulation

*Blastocyst degenerates during anestrus in wild females.
†Growth of the corpus luteum does not occur after the young vacate the pouches of wild females, as these have entered anestrus.

Figure 8.9 Diagrammatic representation of the changes during the estrous cycle, pregnancy and lactation in *Setonix*. (From Sharman, 1955b; reprinted with permission of *Australian Journal of Zoology*, CSIRO Editorial and Publications Service.)

prevents the reactivation of the blastocyst in the uterus. Removal of the young results in increased PIF and LH-RH secretion, and through decreased prolactin and increased gonadotrophin secretion by the AP, in increased progesterone secretion and blastocyst activation. Short photoperiods can, however, have similar effects as a suckling young; i.e., they can decrease PIF and LH-RH secretion, so that if a young is removed in the fall, the blastocyst may not be reactivated. The secretion of gonadotrophins by the AP is low throughout the year, but it is sufficient to stimulate follicular growth, if the corpus luteum is removed; it appears that the corpus luteum is under a dual control, i.e., by a luteotrophic LH and a luteolytic prolactin.

Eutheria

With few exceptions, e.g., some bats and the Echidna, the sperm of mammals do not survive long in the female reproductive tract. It is known that neither the oocyte nor the sperm retain their fertilizing ability, and it is also known that an oocyte fertilized with a delay after insemination becomes a less viable zygote than one fertilized without a delay. This necessitates that ovulation and insemination be synchronized quite rigidly to yield the most viable offspring. This synchrony can be accomplished in two ways: Either copulation causes ovulation to occur, as in many species (Table 8.4), or copulation is permitted around the time of ovulation only, as in many other species. Notable exceptions are some monkeys and humans. However, in some monkeys, e.g., the rhesus monkeys (*Macaca mulatta*)—in which copulation may occur at any time during the menstrual cycle—there is a peak of frequency of copulations around the time of ovulation (Michael and Welegalla, 1968). In humans there is no evidence to indicate such a peak of intercourse around the time of ovulation (Udry and Morris, 1977).

Another factor in the reproduction cycles during the breeding season is whether or not the CL

Table 8.4

Species which display reflex ovulation by coitus

Order	Species	A	B	C	D	E
Monotrema						
Marsupialia	*Potorous tridactylus*	+				
	Didelphis azarae	+				
Insectivora	*Erinaceus europaeus* (hedgehog)	+				
	Neomys sodicus bicolor	+				
	Blarina brevicaudata (short-tailed shrew)	+				
	Scalopus	+				
	Sorex palustris navigator	+				
	Suncus murinus (Asian musk shrew)	+?				
Dermoptera						
Chiroptera	*Pteropus*	+				
Edentata						
Rodentia I	*Citellus tridecemlineatus*	+				
	All Sciuridae	+				
	Microtus californicus	+				
	M. guentheri	+				
	M. agrestis	+				
	M. ochrogaster	+				
	M. pennsylvanicus	+				
	M. pinetorum (pine vole)	+				
	Clethrionomys glareolus (bank vole)	+				
	Dicrostonyx groenlandicus (collared lemming)	+				

Table 8.4—*Continued*

Order	Species	A	B	C	D	E
Rodentia II	*Myocaster* (beaver)	+				
Rodentia III	*Mus musculus* (mouse)		+[a]			
	Rattus norvegicus (rat)		+[a,b]	+	+	
	Phenacomys	+				
Lagomorpha	*Oryctolagus cuniculus* (rabbit)	+				
	Lepus europaeus (hare)	+				
	Ochotona rufescens rufescens (Afghan pica)	+				
Cetacea	*Tursiops truncatus* (bottleneck dolphin)	+				
	Most Odontocetes	+				
Proboscidea						
Carnivora and Pinnipedia	*Lutra lutra* (lynx)	+				
	Felis domesticus (cat)	+				
	Mustela nivalis (weasel)	+				
	Herpestes auropunctato (mungo)	+				
	Ursus arctos horribilis (grizzly bear)	+?				
	U. americanus	+?				
	Procyon lotor R. (racoon)	+				
	Mustela furo (ferret)	+				
	M. vison	+				
	Mirounga leonina (sea elephant)	+				
	Canis familiaris (dog)			+[c]		
Perissodactyla						
Artiodactyla	*Bos taurus* (cattle)			+		
	Ovis aries (sheep)			+		+
	Sus scrofa (domestic pig)			+		
	Camelus bactrianus (camel)	+				
	C. dromedarius (dromedary)	+				
	Lama glama L. (llama)	+				
	Lama vicugna L. (vicuna)	+				
	Lama pacos (alpaca)	+				
Primates	*Macaca mulatta* (rhesus monkey)		+	+?		
	Homo sapiens (man)		+	+?		

A, reflex ovulators: coitus alone induces ovulation. B, facultative reflex ovulators: coitus alone, under certain conditions, can induce ovulation. C, species in which coitus during the standing heat period hastens ovulation. D, species in which coitus during the standing heat period increases ovulation rate. E, species in which coitus during the standing heat period increases conception rate. (Use of teaser animals in connection with artificial insemination.)

[a] In strains with spontaneous permanent estrus (Jöchle, 1973).

[b] In animals with constant light or postnatum androgen treatment induced permanent estrus, or in animals conditioned during proestrus with estrogens (Jöchle, 1973).

[c] In the bitch, not coitus but daily exposure to males hastens preovulatory LH release significantly; as a result, ovulations coincide with the first and second day of standing heat.

From Jöchle, 1975; reprinted with permission of W. Jöchle and *Journal of Reproduction and Fertility*.

formed after spontaneous ovulation will be functional spontaneously or whether they will become functional only as the result of the stimulus of coitus (or other stimulation of the genitalia).

As will become evident from the remainder of this chapter, the CL has a key function in the regulation of the estrous cycle. If the CL is maintained, the next expected estrous period fails to occur; if the CL regresses or stops the secretion of progestins, the next estrous period will occur, unless other factors interfere. The ways in which the CL can be maintained are illustrated in Figure 8.10. In the following discussion of the various types of estrous cycles, we will mention how the

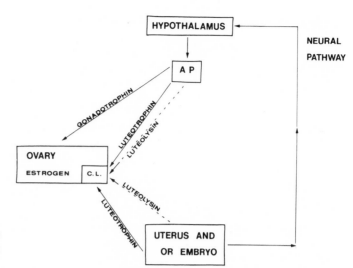

Figure 8.10 Pathways for the control of the CL of metatheria. Solid lines indicate stimulation; broken lines, inhibition. (Adapted from Donovan, 1967.)

CL of a particular animal species is maintained or how its demise is regulated.

Conaway (1971) has proposed a tentative classification of different types of reproductive cycles of Eutheria (Table 8.5). It should be pointed out that different species may be classified under the same type, but that the physiological events that result in similar cycles may be quite diverse.

Relatively Short Cycles with Spontaneous Ovulation and Spontaneous Pseudopregnancy. This is Type IA of Conaway (1971). Ovulation is spontaneous, the CL is maintained spontaneously (which is equivalent to a spontaneous pseudopreg-

nancy), and the cycle lasts about 2–5 weeks. This type is found in ungulates, the guinea pig, and higher primates.

Sheep (Ovis aries). Sheep are seasonal breeders, at least in the temperate zone, and the proximate cause is the photoperiod. Legan et al. (1977) have found that during the breeding season the decrease in progesterone, as the result of the regression of the CL, causes an increase in LH. This, in turn, results in an increase in estradiol (E_2), which stimulates further LH release, so that a preovulatory LH surge results. At the transition to the anestrous season (long days), the initial LH

Table 8.5

Classification of nonpregnancy reproductive cycles of Eutheria

Type*	Ovulation	Pseudopregnancy (functional CL)	Duration of cycle	Examples
I A	Spontaneous	Spontaneous	2–5 weeks	Ungulates, guinea pig, higher primates
I B	Spontaneous	Spontaneous	>5 weeks	Dog
II A	Induced; estrus also induced by social stimulation	Induced	<1 month	Soricidae, Microtini, Lagomorpha
II B	Induced; estrus spontaneous	Induced	4–8 weeks	Cat, ferret, tenrecs
III	Spontaneous	Induced	4–6 days if no coitus occurs; 2 weeks if coitus occurs	Small species of Muridae and Cricitidae

*Data on classification of types from Conaway, 1971.

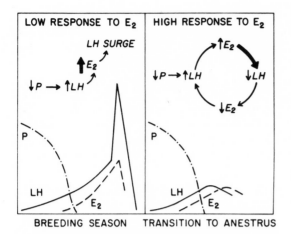

BREEDING SEASON   TRANSITION TO ANESTRUS

Figure 8.11 Working hypothesis for the endocrine basis of seasonal breeding in intact ewes. The *left* panel shows the sequence of events that lead to the preovulatory LH surge during the breeding season. The *right* panel shows that the LH surge does not occur during the transition to anestrus, because the increased response to the negative feedback action of estradiol (E_2) prevents the E_2 trigger for the LH surge. P = progesterone. (From Legan et al., 1977. Copyright © 1977 by The Endocrine Society. Reprinted with permission of S. J. Legan and Williams & Wilkins.)

increase occurs, but the ensuing E_2 increase now causes a suppression of LH, because the threshold for negative feedback to E_2 is lower. Figure 8.11 summarizes these concepts. At the start of the breeding season (short days), the threshold for the negative feedback to LH by E_2 increases and the LH surge can occur. The first ovulation of the breeding season is often not accompanied by estrus (because there is no estrus). This is erroneously called silent heat (Rowlands and Weir, 1977); it might better be designated a quiet ovulation. It is probably caused by a lack of exposure of the brain to progesterone, so that the estrogens secreted by the follicle do not cause behavioral estrus.

During the estrous cycle of the ewe, the surge of FSH and LH occurs on the same day, as illustrated in Figures 8.12 A and B. Ovulation occurs the next day, and progesterone concentrations increase on day 2, reaching a maximum on day 7. A high concentration is maintained until day 14, when it drops precipitously. Estradiol concentrations increase from day 12, reach a peak at day 16, and drop to a low concentration at day 2. Two small estrogen concentration peaks are reported to occur at day 4 and 10 (Robertson, 1977).

Pregnancy prevents the drop in progesterone concentration on day 14. After unilateral hysterectomy, the CL ipsilateral to the absent horn is maintained for as long as 5 months (the duration of the

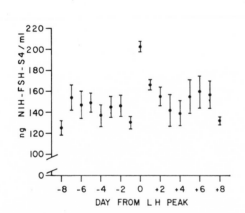

Figure 8.12A Mean daily serum FSH levels of three ewes throughout the estrous cycle. FSH has been measured with the OH radioimmunoassay using the antiserum no. 620. (From L'Hermite et al., 1972; reprinted with permission of *Biology of Reproduction*.)

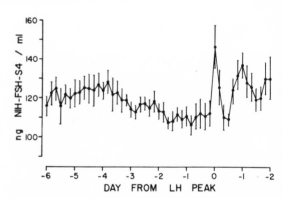

Figure 8.12B Mean serum FSH levels of eight cyclic ewes bled at 4-hour intervals. FSH was measured with the OH radioimmunoassay using antiserum no. 620. (From L'Hermite et al., 1972; reprinted with permission of *Biology of Reproduction*.)

gestation period). These observations have led to the concept of the secretion of a uterine luteolytic factor that causes the demise of the CL of the cycle. This luteolytic factor is presumed to be $PGF_{2\alpha}$ (Goding, 1974).

The transport of the luteolytic factor from uterus to ovary occurs via a local uterovarian pathway, which has the uterine vein as the uterine component and the ovarian artery as the ovarian component. The transfer appears to occur in areas where the ovarian artery and uterine vein are in close contact with each other (Mapletoft et al., 1976).

By the use of hysterectomized, X-irradiated (to remove follicles and thus diminish estrogen secretion) ewes, Gengenbach et al. (1977) have shown that a combination of estrogen and $PGF_{2\alpha}$ was more effective than either hormone alone in inhibiting the CL, but that $PGF_{2\alpha}$ does not absolutely require estrogen for its effect.

When the ewe is pregnant, the embryo may secrete a chemical that is antiluteolytic. Martal et al. (1979) found that extracts of 14–16-d sheep embryos injected into cyclic sheep prevented luteolysis and that the active component of the extract was probably a protein, which they named *trophoblastin*. Injections of such an extract maintained the CL for about the duration of a normal pregnancy. They surmise that another luteo-

trophic agent is necessary for maintenance of the CL of pregnancy. Apparently, the transport of the antiluteolytic factor occurs via a pathway similar to the pathway for the transport of the luteolytic factor (Mapletoft et al., 1976).

After the regression of the CL and the decrease in progesterone, there is a sharp increase in estrogen, which induces LH release (Figure 8.13). Whether LH release is the result of the rate of increase in estrogen, the reaching of threshold value by estrogen, or the decrease or rate of decrease of estrogen, needs investigation. Concannon et al. (1979a) have obtained evidence in dogs that the decrease in estrogen may be the signal for LH release; it remains to be seen whether this is true for species such as sheep.

Besides the preovulatory peak of estrogen, there is a smaller peak of estrogen during the first few days of the cycle and a peak at about midcycle (Hansel et al., 1973). The first peak is probably correlated with follicular growth, but the reason for the midcycle peak and its physiological significance are not well established. According to Hauger et al. (1977), the temporal relationships among progesterone, LH, and E_2 concentrations, as illustrated in Figure 8.13, can explain the estrous cycle of the ewe.

According to Nalbandov (1973), the CL requires the presence of the AP, but Denamur

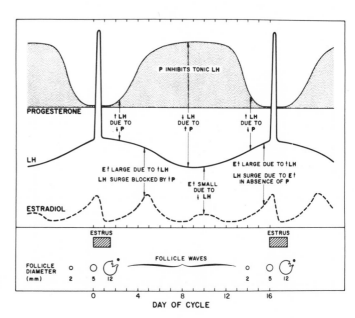

Figure 8.13 Model for hormonal control of the estrous cycle in the ewe. Release of LH by the combined effects of estrogen (E) and progesterone (P) during the ovine estrous cycle. (From Hauger et al., 1977. Copyright © 1977 by The Endocrine Society. Reprinted with permission of F. J. Karsch and Williams & Wilkins.)

(1974) has stated that after day 2, the CL persists in AP ewes. Another controversy is whether LH (provided it is perfused immediately after AP) is the only luteotrophic hormone (Nalbandov, 1973) or whether prolactin and LH (Denamur, 1974) are required for maintenance of the CL.

Cattle (Bos taurus). Domestic cattle, which have provided most of the data on cyclicity and endocrine changes, do not show overt seasonal breeding. The estrous cycle is 21 days, and the endocrine changes during the cycle are summarized in Figure 8.14. Although Robinson (1977) reported an LH peak at midcycle, no such peak has been reported by either Schams and Schallen-

berger (1976) or Hansel et al. (1973). The midcycle estrogen peak occurs in ewes and cattle, and is probably associated with the growth and regression of a follicle. The cycle of cattle and sheep appear to be similar except for some details, such as the longer duration of estrus in sheep than in cattle.

In sheep, the uterus has a luteolytic effect on the ipsilateral CL, because after unilateral hysterectomy the CL persists, thus showing a local effect of the uterus on the CL (Hansel et al., 1973). However, in a preliminary report Ward et al. (1976) state that removal of the uterine horn ipsilateral to the CL plus contralateral hemiovariectomy resulted in normal regression of the CL. As

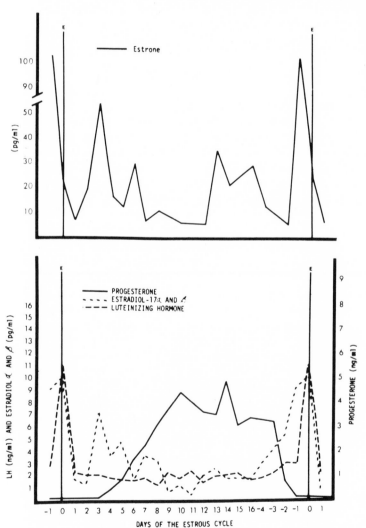

Figure 8.14 Peripheral plasma levels of progesterone, estradiol (17α and β), estrone, and luteinizing hormone in Holstein cows. LH values are expressed in terms of NIH-LH-B7 (1.16 × NIH-LH-S1). [The vertical line marked E indicates estrus.] (From Hansel et al., 1973; reprinted with permission of *Biology of Reproduction*.)

Fernandez-Baca et al. (1979) point out, this suggests that local transfer of the uterine luteolytic factor from uterus to ovary is required to induce regression of the CL. Three types of evidence indicate that the regression of the CL in the cyclic cow may be the result of a lack of a luteotrophic effect and not the result of a luteolytic effect of PGs produced by the ipsilateral uterus. (Conversely, the CL of pregnancy appears to be maintained by the production of a luteotrophic factor and not by a factor that inhibits the production or action of PGs.) First, in the pregnant cow, plasma progesterone concentrations increase above the concentrations found in cyclic cows between day 6 and 8 after insemination, that is, about 10 days before the progesterone secretion in the cyclic cow starts to decrease (Lukaszewska and Hansel, 1980). Second, extracts and homogenates of blastocysts recovered 18 days after the insemination of cows, stimulate progesterone secretion by dispersed bovine CL cells in vitro (Beal et al., 1981). Third, the plasma PGF concentration of ovarian arterial blood is similar in pregnant and cyclic cows, but the plasma PGF concentration in uterine vein blood is lower in pregnant cows. This lower concentration is attributed to a dilution effect as the result of the greater blood flow to the uterine horn in pregnant cows. This observation and interpretation do not negate the finding that twice-daily injections of $PGF_{2\alpha}$ on days 2 and 3 or on day 4 inhibit progesterone secretion without affecting the length of the estrous cycle and that twice daily injections on days 3 and 4 inhibit progesterone secretion and induce early estrus (Beal et al., 1980). Finally, the precursor of $PGF_{2\alpha}$ is prostaglandin H-2 (PGH-2) which is also the precursor of prostaglandin I-2 (PGI-2) or prostacyclin (Kelly, 1981). Injections of prostacyclin into the CL on day 12 or 13 of the estrous cycle resulted in an increase in progesterone secretion (Milvae and Hansel, 1980). This suggests that some prostaglandins may be luteotrophic, whereas others may be luteolytic. It will be necessary to evaluate which prostaglandins reach the ovary and CL, and at what time during the cycle, before the regulation of the function of the CL can be fully understood.

The role of the AP in maintenance and regression of the CL has not been assessed adequately, because no experiments have been reported on AP cows treated with purified bovine gonadotrophins. Hansel and Siefart (1967) have concluded that LH is the luteotrophic hormone of the cow because (1) LH can overcome the oxytocin-induced CL regression and decrease in progesterone secretion, (2) antiserum to LH causes partial regression of the CL of hysterectomized heifers, and (3) LH increases progesterone synthesis of the CL in vitro. In both sheep and cattle, estrogens can have a luteolytic effect (Hansel et al., 1973).

There are some marked differences in beef cattle, between cows that nurse their calves and cows that have no calf to nurse: First, in the nonnursing cows, the interval between parturition and the appearance of estrous cycles is 10-33 d (about the same as in dairy cows), but in nursing cows, this interval is at least 14 weeks (Radford et al., 1978). Second, plasma LH concentrations during the first 30 d postpartum are lower in nursing than in non-nursing cows. Finally, after injection of a $PGF_{2\alpha}$ analogue (Cloprostenol), followed by injection of estradiol benzoate 27 h later, at about 6 weeks portpartum, nonnursing cows showed regression of the CL, estrous behavior, and the formation of a new CL, whereas nursing cows showed no evidence of CL regression, no estrous behavior, and no ovulation (Radford et al., 1978). These differences between the two groups of cows could not be ascribed to differences in prolactin concentrations, however, because the variability was too large and the data presented by Radford et al. (1978) show that at 6 weeks postpartum the differences for prolactin concentration are small. It appears that nursing, which one would, of course, also find in nondomesticated cattle, inhibits LH release and extends the anestrous period postpartum of beef cattle.

Swine (Sus scrofa). Swine have no overt breeding season. Puberty is reached between 4 and 9 months of age, and the estrous cycle lasts 21 days. The changes in concentrations of progesterone, total estrogens, and LH are illustrated in Figure 8.15, and there is little variation in the FSH concentration during the cycle (Dziuk, 1977).

There are some important differences between sheep and cattle on the one hand and pigs on the

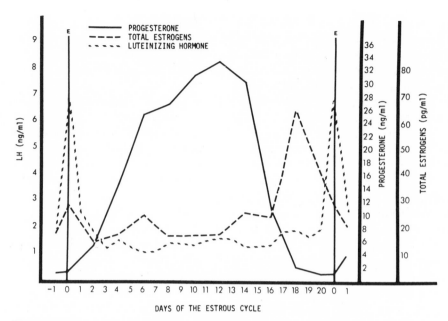

Figure 8.15 Peripheral plasma levels of luteinizing hormone, progesterone, and estrogens during the porcine estrous cycle. [The vertical line marked E indicates estrus.] (From Hansel et al., 1973; reprinted with permission of *Biology of Reproduction*.)

other hand. First, the number of ovulations in pigs (8 to 20) is generally larger than it is in cattle (1 to 2) and sheep (1 to 8). Second, the maintenance of the CL of the cycle does not require the support of the AP, for after AP, the duration of the functional life of the CL is similar to that of intact sows (Anderson et al., 1965). In hysterectomized AP sows, the CL regresses in about 10 days, but this regression can be prevented by administration of either hCG, bovine or equine LH, or dessicated porcine pituitaries. However, in nonhysterectomized AP pigs, the normal regression cannot be halted by these hormones (Anderson and Melampy, 1967). Finally, estrogens, which are luteolytic in sheep and cattle, are luteotrophic in pigs (Hansel et al., 1973). In otherwise intact pigs, hysterectomy results in maintenance of the CL and, as is the case in sheep and cattle, there is a local effect of the uterus on the CL (Anderson, 1973). The luteolytic hormone, according to Horton and Poyser (1976), is $PGF_{2\alpha}$. The luteolysin secretion is inhibited by the presence of embryos. If the number of embryos in the uterus drops below 4–6, the CL regresses and the remaining embryos are aborted. For maintenance of the CL of pregnancy, the AP is required (Anderson, 1973).

Horse (Equus caballus). Horses are seasonal polyestrous breeders. Unfortunately, breeding, racing, and show associations have made regulations which make it advantageous for foals to be born in January rather than later in the year. Thus breeders attempt to breed their horses before March, whereas light may be the proximate cause, but this hypothesis is based more on correlative evidence than on sound experimental evidence. The average length of the cycle is 21–22 days, but there is considerable variability (Stabenfeldt and Hughes, 1977).

The reproductive physiology of the mare has some unusual features and the combination of them is unique. We shall discuss these briefly:

1. The surface germinal epithelium is confined to an ovulation fossa, consisting of an ovulation pit and its rim, and a thin medulla surrounds the cortex. A similar ovarian morphology is found in the nine-banded armadillo (*Dasypus novemcinctus*) (Mossman and Duke, 1973). Ovulation occurs only from this ovulation fossa.

2. According to Stabenfeldt and Hughes (1977), several follicles may, however, mature and continue to grow after ovulation of one of the

follicles. Such follicles may (*a*) be ovulated within 24 h of ovulation, (*b*) be ovulated during the luteal phase of the cycle, or (*c*) regress.

The follicles ovulated after the first follicle might be the origin of accessory corpora lutea of pregnancy, but this has not been verified by the careful histological methods recommended by Mossman and Duke (1973). The primary CL do not regress, but remain functional until about 160–180 days of gestation, and accessory CL appear between days 40–60 (Squires et al., 1974).

3. Mares have a long period of estrus, 2–11 d with an average of 6 d (Stabenfeldt and Hughes, 1977). Furthermore, a mare may also be in estrus for several days, then not be in estrus one or more days, and then be in estrus again. This is the so-called split estrus. Ovulation occurs usually within 1–2 d after the end of estrus.

4. Unfertilized oocytes are retained in the oviduct (Betteridge and Mitchell, 1974). As we shall see in Chapter 10, this occurs in some other species, too.

During the estrous cycle of the mare, GTH concentrations are high during days 6–18, and LH reaches its peak 1–2 d *after* ovulation. At the end of estrus, FSH concentrations rise before the progesterone concentration increases. Estradiol concentrations reach a peak and decrease prior to the time of the LH peak (Stabenfeldt and Hughes, 1977; Miller et al., 1980). Progesterone concentrations drop prior to LH release, so this may relieve an inhibition of LH secretion; however, once LH secretion has started, progesterone is not effective in inhibiting it.

Testosterone has also been detected in the plasma of cyclic mares. In 4 out of 6 mares, there was a testosterone peak (15–70 ng/ml) at about the time of behavioral estrus and another peak at 11–13 d before estrus. The significance of these peaks of testosterone concentration for the reproductive physiology of the mare has not been determined.

Control of the CL has not been studied in AP horses; therefore, much of the discussion about such control is speculative. The uterus has a luteolytic effect, probably via $PGF_{2\alpha}$ (Stabenfeldt and Hughes, 1977); however, the pathway of transfer from uterus to ovary remains to be deter-

mined. If the uterus is indeed important in the control of life span of the CL, then the horse may have two mechanisms for maintaining the CL of pregnancy: (1) the presence of a zygote may inhibit the luteolytic effect of the uterus; or (2) the pregnant mare's serum gonadotrophin secreted by the fetus may maintain or stimulate the primary CL, a topic to be discussed in Chapter 10.

Rhesus Monkey (*Macaca mulatta*). The rhesus monkey has a menstrual cycle of about 28 d. A menstrual cycle differs from an estrous cycle in that the endometrium is sloughing off and results in bleeding. This bleeding is more easily detected than either ovulation or, in some species, estrous behavior (e.g., in humans). Consequently, the onset of menstruation has been considered the beginning of the reproductive cycle in monkeys and humans, whereas estrous behavior has been considered the beginning in domestic and laboratory species.

Knobil (1974) has presented data that show that in female rhesus monkeys, unlike many other species, the peak concentration of estradiol (about 300 pg/ml) does not precede the peaks of LH concentration (about 35 ng/ml) and FSH concentration (about 100 ng/ml), but coincides with them (day 0). The progesterone concentration starts to increase on day 1, reaches a peak (of about 4 ng/ml) on day 8, and starts to decline on day 10.

Much of the discussion that follows about the neuroendocrine control of the cycle is based on the reports of Knobil and his coworkers, as reviewed by Knobil and Plant (1978). In brief, the hypothalamus secretes LH-RH partly under the control of catecholaminergic neurons. The secretion of LH-RH seems to be controlled within the medial basal hypothalamus, because surgical isolation of this area affects neither the spontaneous LH release nor estrogen-induced LH release. Lesions of the arcuate nucleus result in the cessation of LH-RH secretion. In monkeys with lesions of the nucleus arcuatus, normal secretion of FSH and LH can be restored by administration of LH-RH, provided the LH-RH is given in a pulsated manner (6 minutes every hour), and not continuously.

The pituitary is the site of LH and FSH secretion and plays a crucial role in the hypothalamo-AP-ovarian axis. The stimulatory effect of estrogen on the release of LH occurs at the site of the AP, as is clearly demonstrated by the brilliant experiments of Knobil and his coworkers. In the absence of the ovary, LH and FSH are discharged in pulses with about 1-h intervals (circhoral rhythm). In hypothalamic-lesioned ovariectomized monkeys given circhoral LH-RH infusions, estradiol injection was followed by a discharge of LH and FSH. The concentrations of LH and FSH and the biphasic pattern of rise and fall in these monkeys are quite similar to those found during the spontaneous preovulatory LH and FSH release. Thus estradiol also provides the negative feedback at the level of the AP. Knobil and Plant (1978) did not exclude the possibility that under normal conditions estradiol might also act at the neuronal level.

The ovary secretes estrogens and progesterone. As indicated, estrogen inhibits the circhoral pattern of LH release, but estrogens also reduce the concentrations of LH. As the experiment with lesioned, ovariectomized, LH-RH perfused monkeys showed, estradiol by itself could trigger the release of LH; however, progesterone might act synergistically with estrogens, although by itself progesterone seemed to have little or no effect when given in physiological doses.

Knobil (1974) provided evidence that the 28-day clock required for the 28-day menstrual cycle is located in the ovary (the "pelvic clock") and that the hypothalamus-AP component responds to the peak in estrogens by a discharge of LH and FSH every 28 days. This release of FSH and LH then results in ovulation and the formation of a CL. Apparently, the CL has no influence, or very little influence, on the timing of the infertile reproductive cycle; its important effect starts when a fertilized ovum is present and develops. The CL of pregnancy, as will be discussed in Chapter 10, in contrast to the CL of pregnancy of the cow, is rescued by a luteotrophic influence from the embryo, after the CL has started to diminish its progesterone secretion towards the expected end of the cycle (Lukaszewska and Hansel (1980).

Humans (*Homo sapiens*). Knobil and Plant (1978) have pointed out the menstrual cycles of rhesus monkeys and humans resemble each other on many aspects. Figure 8.16 illustrates one important difference: i.e., that the estrogen peak precedes the preovulatory peaks of FSH and LH in humans, whereas in rhesus monkeys the estradiol and gonadotrophins peak on the same day. The discharges of FSH and LH differ in the rhesus monkey and the human, in that the discharge shows a circhoral rhythm in intact monkeys, but not in humans (Yen et al., 1975). In spite of these apparent differences, Yen et al. (1975) concluded that the controlling mechanisms that regulate the menstrual cycles of these two species are similar.

There is no good evidence that sexual receptivity during the menstrual cycle shows a peak around the time of ovulation. It is clearly desirable, both for optimal development of the zygote (see Chapter 10) as well as for purposes of contraception (for those who do not wish to use chemical or physical means of contraception) that the time of ovulation be predictable. A large number of variables have been studied, e.g., increase in basal body temperature, changes in the crystalization pattern of the cervical mucus, and the consistency of cervical mucus, but none appears to be satisfactory in consistently predicting the time of ovulation accurately. Figure 8.16 illustrates the changes in a number of hormones during the menstrual cycle. If it were possible to detect the estradiol peak by a simple "do it yourself method," then this should be a promising predictor of ovulation.

Hysterectomy does not affect the duration of the menstrual cycle in rhesus monkeys or humans. Prostaglandins seem to *stimulate* progesterone secretion when given in low doses and to inhibit it at high doses in rhesus monkeys and in the red hussar monkey (*Erythrocebus patas*). The inhibitory effect of such high doses of $PGF_{2\alpha}$ can be overcome by hCG (Auletta et al., 1973). The stimulatory effect of $PGF_{2\alpha}$ may be a physiological effect, because Shutt et al. (1976) found that increases in $PGF_{2\alpha}$ were associated with both the growth of the CL and its regression. It may well be that the effect of $PGF_{2\alpha}$ depends on the physiological status of the CL and on the amount of $PGF_{2\alpha}$ that reaches the CL. It appears that PGF and LH both have a permissive function with respect to the CL of the cycle and that both hormones thus can affect the expression of the full potential of this gland, but that these hormones do not control its fate.

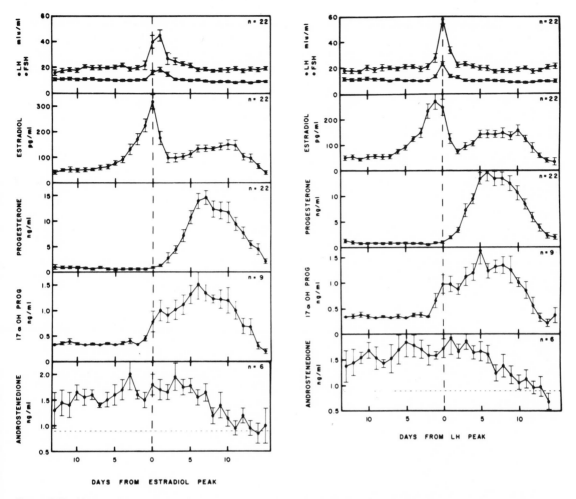

Figure 8.16 Mean and standard errors of LH, FSH, 17-hydroxyprogesterone, E₂, progesterone, and androstenedione oriented around the day of both the LH and E₂ peaks [during the menstrual cycle of humans]. The dashed line on both graphs for androstenedione indicates the upper limit of androstenedione levels in samples obtained from ovariectomized women. (From Ribeiro et al., 1974; reprinted with permission of W. O. Ribeiro, D. R. Mishell, Jr., and C. V. Mosby Company.)

Guinea Pig (Cavia porcellus). The estrous cycle of the guinea pig lasts about 15–18 d with an estrous period that lasts 6–11 h. Estrus occurs usually during the night (Asdell, 1964). The follicular phase of the cycle is short and the luteal phase lasts about 12–13 days.

In investigations of the hypothalamo-AP interaction with the ovary, injection of 10 μg of estradiol resulted consistently in a decrease in LH concentration, from about 2 ng/ml to about 1 ng/ml immediately after injection (demonstrating a negative feedback) and an increase in LH to a peak of about 5 ng/ml at about 36 h after injection. The timing of the surge of LH depended on the dose

(generally the larger the dose, the shorter the latency), and the interaction with the light-dark interphase, with LH release more frequently and consistently taking place in the dark (Terasawa et al., 1979a). This estrogen-induced LH surge is clearly not a positive feedback action of estrogen, but a stimulatory action.

In the absence of estrogen priming (1.5 μg, a dose that induces 12.5 percent LH surges), injection of 1 mg of progesterone can also induce an LH surge. The difference between the estrogen- and the progesterone-induced surge is that the former is not blocked consistently by barbiturate anesthesia, whereas the latter is (Terasawa et al.,

1979b). Estrogen and progesterone are required for the induction of estrous behavior in ovariectomized guinea pigs, (Young, 1961), and at day 18 of the cycle just prior to estrus, the concentrations of estrogen and progesterone are high (Joshi et al., 1973), thus providing experimental evidence and correlative evidence for the function of these hormones in inducing estrus.

The control of the function of the CL is influenced by the AP and uterus, but they seem to act permissively, in the same way that the AP acts in rhesus monkeys, instead of exerting a rigid control, as in sheep. Heap et al. (1967) made a systematic study of the effect of AP, hysterectomy, and AP plus hysterectomy on the CL of guinea pigs and on the concentration of progesterone in the plasma. Hypophysectomy early in the cycle resulted in the persistence of about half of the CL for at least 5 weeks. These CL were histologically normal and reached a volume similar to the CL of the cycle. Hypophysectomy late in the cycle did not delay regression of the CL, but the progesterone concentration in the CL was as high as that in the CL of pregnant guinea pigs, although plasma concentrations were low. Some of the data are summarized in Table 8.6. Hysterectomy also resulted in persistence of the CL, with a progesterone content similar to that of the CL of pregnancy and plasma progesterone levels that were not higher than those found during the cycle in intact females. Hysterectomy performed at different intervals after AP resulted, in some animals, in CL of the size found in pregnant guinea pigs, but in other animals, in CL of the size of those in the cycle. Plasma progesterone concentrations were low.

These results suggest that early in the cycle, the AP may have a luteolytic effect (after AP, progesterone content increases), but that the AP may also be luteotrophic, because growth of the CL is slower than in intact females. Later in the cycle, the AP may be considered luteolytic, because the progesterone content of the CL is higher in AP than in intact females; however, with respect to the regression of the follicle, the CL is more or less independent of the AP. The uterus is luteolytic and appears to exert its effect at about day 14–15 of the cycle, when the CL starts to regress. Hysterectomy after day 15 results in normal regression of the CL (Hilliard, 1973). The luteolytic effect of the uterus is probably mediated via prostaglandins (Blatchley et al., 1972; Bagwell et al., 1975).

It has not been satisfactorily determined which hormone(s) makes up the luteotrophic complex. The FSH may be part of this complex because,

Table 8.6

Effects of hysterectomy, hypophysectomy (AP), and pregnancy on CL size and progesterone content and on the arterial progesterone concentrations in guinea pigs

Condition	Day	Day of cycle	CL weight (ng)	Luteal progesterone content (ng)	Plasma progesterone conc. (ng/ml)
Normal cycle		5–6	2.8	8.6	9.8
		9	2.4	7.6	9.5
		15	0.9	0.7	3.9
Pregnant		15	4.6	95	6.3
		20–22	5.2	152	167.6
Hysterectomy	5	30	4.9	260	6.6
	5	60	4.8	144	2.5
	2	5 (after AP)	2.6	82	7.0
	2	19 (after AP)	1.6	52	1.5
	2	65–83 (after AP)	—	—	8.5
	2	7	2.6	82	7.0
	2	21	1.6	52	1.5

Data from Heap et al., 1967.

experimentally, FSH partly counteracts the luteolytic effect of estrogen injected early in the cycle (Hilliard, 1973). However, prolactin may also be part of this complex, since in females hypophysectomized on day 10 of the cycle, prolactin administration increases the size of the CL (Illingworth and Perry, 1971).

Estrogen has a luteolytic effect, exerted via the uterus, when given within the first 5 days after ovulation. This lytic effect is reduced by AP and prevented by hysterectomy. Estrogens stimulate the secretion of $PGF_{2\alpha}$ (Blatchley et al., 1972), and probably have their lytic effect through the action of $PFG_{2\alpha}$ on the CL (Illingworth and Perry, 1973). When estrogen is administered after day 9, it has an antiluteolytic effect; i.e., it prevents regression of the CL, and a luteotrophic effect; i.e., it stimulates CL growth and/or progesterone secretion. After estrogen administration, the CL are maintained for 3–4 weeks, and reach the size of and acquire the progesterone content found in CL of hysterectomized animals. Hypophysectomy does not inhibit either the antiluteolytic effect or the luteotrophic effect. After hysterectomy, estrogens can stimulate growth and progesterone secretion, i.e., increase the progesterone content of the CL and the plasma progesterone concentration. Illingworth and Perry (1973) injected large doses of estrogens (e.g., 20 mg estradiol, 25–75 mg stilbestrol) into intact guinea pigs at different times during the estrous cycle, and into hysterectomized and AP animals. They concluded that the luteolytic effect of estrogens was independent of the age of the CL but was dependent on factors from the AP and/or the uterus.

Elephant (*Loxodonta africana*). The African elephant has a definite breeding season; most of the conceptions occur during the wet season, at the time that the rainfall reaches its peak. This suggests that food and water may be the most important proximate causes (Hanks, 1977). The estrous cycle lasts about one month. Despite abundant amounts of CL, the plasma progesterone concentration and progesterone content of the CL are low (Rowlands and Weir, 1977).

It is clear that the cycles of the species discussed above, which fall in Conaway's Type IA, are quite similar, but that there are considerable differences in the manner in which the regulation is achieved. It is difficult to find selective advantages of one mechanism over another, so that the evolution of these mechanisms is not easily understood.

Relatively Long Cycles with Spontaneous Ovulation and Spontaneous Pseudopregnancy. This is Type 1B of Conaway (1971). This type of cycle differs from the one previously discussed (Type IA) by the duration of the interestrous interval.

The endocrine events in the dog (*Canis familiaris*) have been studied in detail during the last decade. According to Stabenfeldt and Shille (1977), there is little or no evidence that domesticated dogs show seasonal breeding. The length of the estrous cycle of bitches is about 60–70 days; this duration of the cycle is similar for unmated, sterile-bred, pseudopregnant, and pregnant animals. The interestrous interval varies between 18 and 55 weeks, with the greatest frequency at 24–36 weeks. They report further that pregnant bitches have a longer mean interrestrous interval (32 weeks) than nonpregnant ones (29 weeks). The estrous period lasts about 12 days; however, its length also depends on the criteria used to define estrus (Figure 8.17).

The plasma concentrations of LH, progesterone, and total estrogens during the estrous cycle and their relationship with reproductive behavior are shown in Figure 8.17. Wildt et al. (1979) determined the concentrations of estradiol-17β (E_2) and estrone (E_1) in the plasma of bitches and found that both estrogens had higher concentrations before the LH peak and both declined coincidentally with the peak, although the variability for E_1 was greater than for E_2. They suggest that E_1 does not have a controlling function, whereas E_2 does, and that E_1 plays a supportive role. There is good agreement about the timing and duration of the LH peak (Concannon et al., 1975; Mellin et al., 1976; Wildt et al., 1979). The duration of the elevated LH levels is longer than is found in reproductive cycles of other species discussed so far. Prolactin concentrations do not show a distinct pattern during the dog's estrous cycle (see Figure 8.18). The rise in progesterone prior to the preovulatory peak (Figure 8.19) is the result of

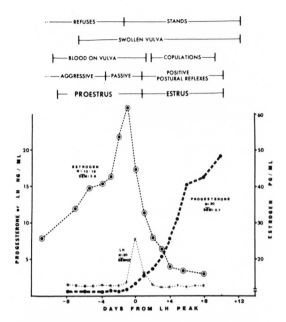

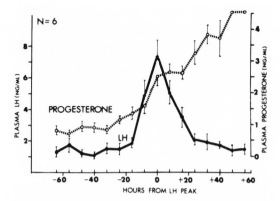

Figure 8.19 Mean (± SE) plasma LH and progesterone in beagle bitches with peak preovulatory LH levels aligned to a common hour 0. (From Concannon et al., 1977; reprinted with permission of *Biology of Reproduction*.)

Figure 8.17 Mean plasma levels of estrogen, LH, and progesterone during proestrus and estrus in beagle bitches. The horizontal bars represent the mean time of onset and termination of the parameter indicated. (From Concannon et al., 1975; reprinted with permission of *Biology of Reproduction*.)

preovulatory luteinization of the mature follicle (Concannon et al., 1975, 1977; Wildt et al., 1979). The role of rising concentrations of progesterone and diminishing concentrations of E in influencing LH release (Concannon et al., 1979a) and in inducing estrous behavior (Concannon et al., 1979b) has been investigated in ovariectomized bitches. The evidence strongly suggests that the preovulatory LH release and estrous behavior seen in intact bitches can be best obtained in ovariectomized females by extended exposure to estrogen, followed by a decreasing estrogen concentration by withdrawing estrogen treatment,

and a simultaneous increase in progesterone concentration by injection of progesterone.

In dogs, fertilization occurs 2–3 days after ovulation of the *primary* oocyte. This suggests that dog sperm retain their fertilizing capacity for a longer period than most mammalian sperm (Stabenfeldt and Shille, 1979). It would be of some interest to investigate whether deleterious effects on the embryo due to aging of gametes are less frequent than in other domestic and laboratory mammals.

False pregnancy, as manifested by deposition of abdominal fat, mammary gland development, nest building, and lactation in any combination occurs in nonpregnant bitches. The frequency reported varies between 3 and 40 percent (Stabenfeldt and Shille, 1977).

Relatively Short Cycles, Sometimes with Induced Estrus, Induced Ovulation, and Pseudopregnancy. This is Type IIA of Conaway (1971). In animals with this type of cycle, the CL is func-

Figure 8.18 Individual profiles of prolactin concentrations in the serum of three beagle bitches throughout the estrous cycle. Day 0 = first day of proestrus. (From Gräf, 1978; reprinted with permission of K. J. Gräf and *Journal of Reproduction and Fertility*.)

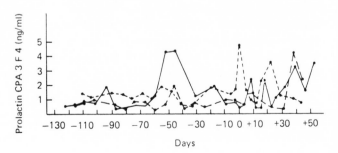

tional for about a month or less after a nonfertile mating. The best-studied representative of this group is the rabbit.

Rabbit (Oryctolagus cuniculus). Wild rabbits are seasonal breeders; in Wales, for example, the main breeding season lasts from January to June, and virtually no breeding occurs in November (Asdell, 1964). Domesticated laboratory rabbits tend to breed throughout the year, but during one or two months, they may not have ovulatory follicles and although they may copulate, no ovulations take place.

During the breeding season, follicles mature and become atretic in waves. The pattern of the changes in hormone concentrations is illustrated in Figures 8.20 and 8.21. It can be seen that immediately after coitus there is a rise in LH and in the progestin 20α-hydroxy-Δ^4 pregnen-3-one (i.e. 20α-dihydroprogesterone - 20α-ol). This hormone is probably secreted by the ovarian interstitial gland (Koering and Sholl, 1978). Hilliard (1973) has reviewed the evidence that suggests that there might be a true positive feedback between AP and ovary, with LH stimulating 20α-ol production which, in turn, stimulates LH production, etc. However, recently Goodman and Neill (1976) found that (1) one month after ovariectomy, coitus failed to induce preovulatory LH release patterns either in estrogen-primed or in estrogen-primed 20α-ol-treated does; (2) a normal preovulatory LH was obtained in does ovariectomized 15 minutes postcoitus; (3) the correlation between LH and 20α-ol concentrations in individual rabbits was variable; and (4) in long-term ovariectomized, estrogen-primed rabbits, which were mated and injected within six minutes of ejaculation with 20α-ol, no preovulatory LH release was found in 7 of 9 does. These results, which are not in accord with those reported by Hilliard et al. (1967), cast doubt on the concept of a positive feedback control mechanism between the AP (LH) and ovary (20α-ol).

Figure 8.22 shows the pattern of progesterone and 20α-ol (20α-OH) during the ensuing pseudopregnancy after an infertile mating or an artificially induced LH release. In intact does, pseudopregnancy lasts about 17 d (compared to 30–32 d of pregnancy). After hysterectomy,

pseudopregnancy lasts about 27 d. There seems to be no local effect of the uterus on the CL (Ginther, 1974), as there is in pigs and sheep.

The CL is dependent upon the AP, and specifi-

Figure 8.20 Postcoital patterns of LH, FSH, and 20α-dihydroprogesterone (20αP) in peripheral plasma of normal estrous rabbit. The number of animals at each time point is shown in parentheses. P = before mating. Vertical bars indicate SE. (From Goodman and Neill, 1976. Copyright © 1976 by The Endocrine Society. Reprinted with permission of A. L. Goodman and Williams & Wilkins.)

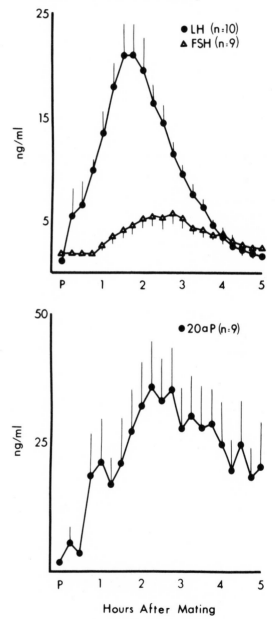

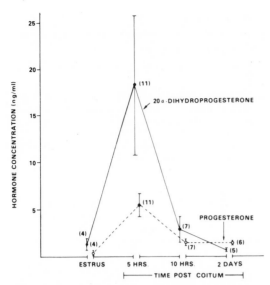

cally LH, because in AP rabbits the CL regresses, but the regression is prevented by LH administration (van Tienhoven, 1968). The effect of the AP is, however, not a direct luteotrophic effect on the CL but an effect exerted through estrogen secretion by the follicles. Evidence for this is provided by Hilliard (1973), who found that: (1) in rabbits the CL could be maintained by estrogen; (2) in the absence of follicles (e.g., destruction by X-irra-

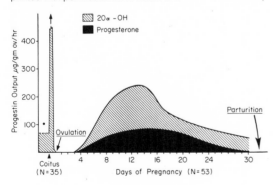

diation, luteinization by gonadotrophins, or cauterization) the CL regressed rapidly; (3) LH caused estradiol-17β release by tertiary follicles; and (4) antiserum to LH caused regression of the endometrium and the CL. Furthermore, Lee et al. (1971) found that the CL has estrogen receptors with characteristics similar to those of the uterus. The concentration of these receptors is high at the midluteal phase (day 8) and decreases before morphological regression of the CL is evident. The CL is independent of estrogen until day 5; then it becomes dependent on estrogen (Miller and Keyes, 1978), with the day of transition marked by the day when estrogen binding by the CL is highest.

According to Hilliard (1973), the effect of hysterectomy can be explained by two different mechanisms, when account is taken of the fact that the prolongation of luteal function is roughly proportional to the amount of uterine tissue removed. One mechanism is that the uterus removes estrogen from the circulation, thus removing the luteotrophic factor; the other mechanism is that the endometrium secretes a luteolytic factor, which, by analogy with other species, might be $PGF_{2\alpha}$.

Experiments by Keyes and Bullock (1974) showed that $PGF_{2\alpha}$ injections had a luteolytic effect, and that this effect could be reversed by estrogen treatment, while there was no evidence that $PGF_{2\alpha}$ affected estradiol concentrations in ovarian venous blood. This suggested that $PGF_{2\alpha}$ had a direct luteolytic effect on the CL and that it interfered with the luteotrophic effect of estrogens. The results that they obtained were similar for ovarian CL and for CL explanted to the kidney (to remove the CL from the luteolytic effects of the uterus). As no effects were noted on estrogen secretion, it appears that the second proposed mechanism may be operative. Of course, it must be demonstrated that $PGF_{2\alpha}$ concentrations in ovarian arterial blood change at the time of regression of the CL or that the sensitivity of the CL to PG changes. It is also possible that both mechanisms can operate simultaneously.

Progesterone concentrations during pseudopregnancy, which lasts 24 d in the European hare (*Lepus europaeus syriacus*), show a gradual increase after coitus, reaching a maximum of about

35 ng/ml on day 18 and declining to very low concentrations by day 24 (Stavy et al., 1978). The cottontail rabbit (*Sylvilagus floridanus*) has the same type of reproductive cycle as the rabbit, but social stimulation plays an important role in synchronizing the breeding time of the population (Conaway, 1971).

The prairie vole (*Microtus ochrogaster*) also shows the strong influence of social stimulation on the induction of estrus. The presence of a male is the most potent stimulus, probably through olfactory cues (Richmond and Stehn, 1976). Apparently, the male has to be "strange" to the female, e.g., a male left in the same cage with his daughters will not induce estrus in them, but after he has been removed for 8 days or more and then returned to the cage, he will induce estrus in them, if they are 36–45 days or older. Estrus can be maintained for at least 30 days, and copulation induces ovulation, which occurs 5–10 h later (Conaway, 1971).

In the related field vole (*Microtus agrestis*), it has been found that LH release occurs only after successful copulation and not in response to copulation attempts by the male. A maximum LH concentration of 61.0 ± 5.7 ng/ml was found 5 minutes after copulation, and the concentration dropped to precoital levels in about 2 h. Females that had successfully copulated and were allowed to copulate again 90 minutes later, showed a change in plasma LH concentration pattern very similar to that after the first copulation (Charlton et al., 1975).

Camelidae. At least five members of this family—the camel, dromedary, vicuna, llama, and alpaca (Table 8.4)—and probably the sixth member, the guanaco (*Lama guanicoe*), all show coitus-induced ovulation. Of these five species, the alpaca (*Lama pacos*) has been studied in most detail. The alpaca ovulates in response to mounting and intromission and also in response to hCG injections (Fernandez-Baca et al., 1970a). Among unmated females, about 5 percent ovulate spontaneously. The CL reaches its maximal size at day 8–9 and then starts to regress rapidly by day 13, regressing completely by day 18. The progesterone secretion shows a similar pattern: at day 8 the plasma concentration of progesterone is

422.5 $\pm$ 79.9 ng/100 ml and of 20α-ol is 336.1 $\pm$ 114.6 ng/100 ml, whereas by day 13 these concentrations are 10.5 ± 1.3 and 12.9 ng/100 ml respectively (Fernandez-Baca et al., 1970b).

The control exerted by the uterus over the CL differs between left and right horn, with the right horn having a local effect only and the left horn having a local and a systemic effect (Fernandez-Baca et al., 1979). The left and right uterus in this species seem to have different physiological capacities. Fernandez-Baca et al. (1970a) have reported that most pregnancies are in the left horn and that transuterine migration from right to left horn occurs. In 928 pregnancies, CL were found in equal percentages in left and right ovaries, but 98.4 percent of the pregnancies were in the left horn; in 50 percent the pregnancies were supported by a CL in the right ovary, confirming the high incidence of transuterine migration (Fernandez-Baca et al., 1979).

Cycles of Relatively Long Duration, with Spontaneous Estrus, Induced Ovulation, and Induced Pseudopregnancy. This is Type IIIB of Conaway (1971). Estrus is generally spontaneous. Its duration is more fixed than in Type IB and the CL is functional for about 4–6 weeks. This type is found in the domestic cat (*Felis domesticus*) and ferret (*Mustela furo*).

Cat (Felis domesticus). The breeding season has two peaks, one in the fall and one in the spring. The proximate causes that regulate the timing of these breeding periods have apparently not been established. Estrus lasts about 6 d in cats that ovulate and 8 d if ovulation fails to occur. This suggests that the CL represses the expression of estrus. Ovulation occurs between 25 h and 50 h after coitus, with an average of 45 h. The CL is functional for about 5 weeks, with a range of 30–75 days (Stabenfeldt and Shille, 1977).

The hormone concentrations found during estrus and during pseudopregnancy and pregnancy are summarized in Figures 8.23 and 8.24. The concentrations of progesterone are lower during pseudopregnancy than during pregnancy, and the life span of the CL, as judged by progesterone concentrations in the blood, is about 21 days. It is clear that after ovulation, estradiol concentrations

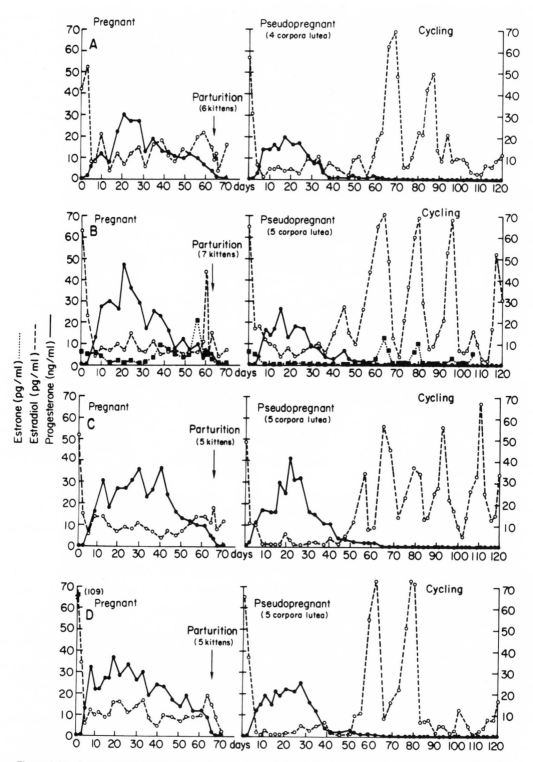

Figure 8.23 Profiles of plasma estradiol and progesterone in four individual cats and estrone in one cat (B) during pregnancy, pseudopregnancy, and polyestrus. (From Verhage et al., 1976; reprinted with permission of *Biology of Reproduction*.)

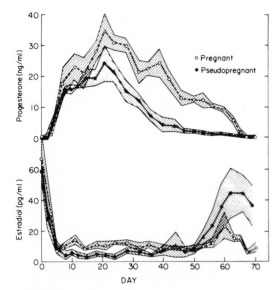

Figure 8.24 Mean (± SE) levels of estradiol and progesterone in pregnant and pseudopregnant cats (n = 4 except for Δ where n = 3) (From Verhage et al., 1976; reprinted with permission of *Biology of Reproduction*.)

are low compared to the concentrations found in the estrous cat. As far as I am aware, no results have been published on the experimental induction of LH release in ovariectomized queens treated with steroid hormones.

Ferret (Mustela furo). The ferret is a seasonal breeder (March–August) that breeds in the spring. Photoperiod is the proximate cause that times the breeding. Ovulation occurs about 30 h after coitus, which lasts about 2 h, and oocytes remain fertilizable for about 30 h (Asdell, 1964). The Cl of pregnancy and of pseudopregnancy do not differ either with respect to life span or to progesterone secretion. Figure 8.25 illustrates the progesterone concentrations found in the plasma of pregnant and pseudopregnant ferrets. The CL is AP dependent; AP causes regression of the CL, but sectioning of the pituitary stalk does not affect the CL during the first half of pseudopregnancy (Donovan, 1963). Neither the uterus nor the pres-

Figure 8.25 Plasma progesterone concentrations in pregnant and pseudopregnant ferrets. (*a*) Values in pregnant (solid symbols) and pseudopregnant (open symbols) females measured by fluorimetry (●, ○), competitive protein binding (■) and radioimmunoassay (▲, △). (*b*) Mean values ± S.E.M. of figures obtained by different assays. Figures in parentheses are the number of observations. (From Heap and Hammond, 1974; reprinted with permission of R. B. Heap and *Journal of Reproduction and Fertility*.)

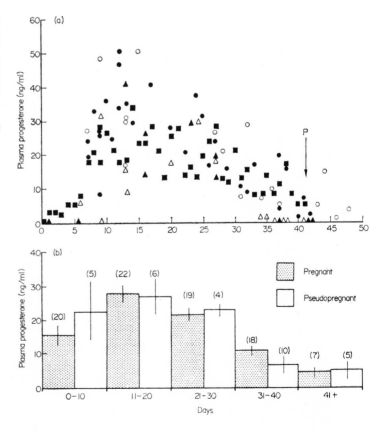

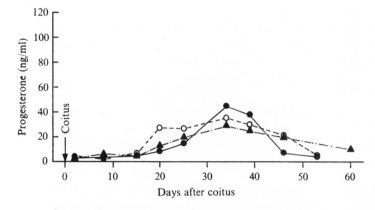

Figure 8.26 Plasma progesterone concentrations after coitus in [three] female mink mated once and failing to give birth. (From Møller, 1973; reprinted with permission of *Journal of Endocrinology*.)

ence of a conceptus affects the life span of the CL (Deanesly, 1967). In the mink (*Mustela vison*), the progesterone concentrations in the peripheral plasma are similar to those of the ferret (Figure 8.26).

According to Conaway (1971), the reproductive cycle of the primitive insectivores, the Tenrecidae, may meet the criteria of the Type IIB. The occurrence of this cycle in these insectivores can be used as an argument in favor of the concept that a long pseudopregnancy and induced ovulation are the primitive pattern. Conaway (1971), on the basis of the different derivation of the Soricidae (which have a medium-long pseudopregnancy) and the Tenrecidae, states that the differences in the duration of pseudopregnancy between these two groups of insectivores may not be related to their present ecological adaptations.

Short Cycles, with Spontaneous Estrus, Spontaneous Ovulation, and Induced Pseudopregnancy. This is Type III of Conaway (1971). Ovulation is spontaneous, but pseudopregnancy is induced; consequently, cycles are short when no mating occurs and about 2 weeks if pseudopregnancy is induced. This type of cycle, found in the laboratory rat (*Rattus norvegicus*), house mouse (*Mus musculus*), and golden hamster (*Mesocricetus auratus*), is probably the most intensively studied mammalian reproductive cycle, but it is not representative of the family Muridae. The family consists of 101 genera of which 5 species have clearly been shown to have this type of cycle (Conaway, 1971). In the family Cricetidae, which contains 99 genera, 8 genera have one or more species with this type of reproduction, for example, the golden hamster.

Rat (*Rattus norvegicus*). In rats, 4- and 5-day cycles are found, with the same rat sometimes showing both types of cycles. The main events of the 4-day cycle are summarized in Figures 8.27 and 8.28. The rat has been used in experimental reproductive research, partly at least, because during vaginal cycle, cells can be found in vaginal smears that can be correlated with the stages of the cycle.

The concentrations of LH, FSH, prolactin (PRL), estradiol (E_2), progesterone (P), 20α-hydroxypregn-4-en-3-one (20α-ol or 20α-OHP), and the change in uterine intraluminal water during the 4- and 5-day cycle of the rat are illustrated in Figures 8.29 through 8.31. The estradiol peak on day 3 of the 4-day cycle that precedes the LH and PRL peaks apparently is required for the release of these pituitary hormones, for injection of an estrogen antibody on day 2 at 1000 prevents the occurrence of the LH and PRL peak and also prevents behavioral estrus and the appearance of vaginal cornified squamous epithelial cells in the vaginal smear (Neill et al., 1971).

The following series of events may occur:

The central nervous system–AP system receives or generates a daily signal for the release of LH (Legan and Karsch, 1975), but in the presence of the ovaries this signal is expressed only once every 4–5 d. Then at proestrus, the rising titers of estradiol cause a preovulatory LH surge by acting on the preoptic hypothalamic area and limbic structures above the medial preoptic area, especially the bed nucleus of the stria terminalis and the lateral septum (Kawakami et al., 1978). In ovariectomized rats, a single injection of estrogen results in a daily LH surge for at least three days (Legan and Karsch, 1975; Legan et al., 1975).

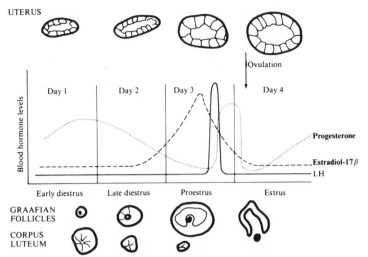

UTERUS

Day 1   Day 2   Day 3   Ovulation   Day 4

Blood hormone levels

Progesterone

Estradiol-17β

LH

Early diestrus   Late diestrus   Proestrus   Estrus

GRAAFIAN FOLLICLES

CORPUS LUTEUM

Figure 8.27 Estrous cycle of the laboratory rat. The levels of progesterone, estradiol-17β, and LH in the blood are shown in relation to estrus and ovulation. Also included are a representation of changes in the stages of development of the uterus, Graafian follicles, and corpus luteum. It can be seen that the corpus luteum does not persist throughout the entire estrous cycle of the rat and that the preovulatory increase in progesterone is due to the secretion of this hormone by the ovarian interstitial tissue. (From Bentley, 1976; reprinted with permission of Cambridge University Press.)

The daily release of LH in intact rats is blocked by the progesterone present in the system (Banks and Freeman, 1978), and the low concentration of progesterone on the day of proestrus permits the LH surge to occur in response to the E_2 secreted by the follicles. As Figure 8.29 indicates, there is a small LH release on the day of diestrus, but the stimulatory action of E_2 is lacking to translate this release into an LH surge. Banks and Freeman (1978) have found that progesterone can block the response of the AP to LH–RH. One of the mechanisms that may prevent a preovulatory LH surge during metestrus and diestrus is the concentration of progesterone on these days.

The principal difference between the 4- and the 5-day cycle appears to be a prolonged secretion of progesterone during the 5-day cycle (Figure 8.30). Whether such prolonged progesterone secretion is the cause or the result of 5-day cycle needs to be investigated. The data available for the control of the rat's estrous cycle strongly suggest that a ''clock'' in the nervous system times the cycle, whereas the clock for the timing of the cycle of the rhesus monkey is the ovary.

Pseudopregnancy is induced by stimulation of the cervix, which is normally provided by copulation. In the rat, copulation consists of a well-defined sequence of events which is summarized in Figure 8.32. The male and female investigate each other, and if the female is in heat, the male will mount and dismount about 10 times. Intromission of the penis occurs during some mounts and ejaculation follows a protracted pelvic thrust, which may aid in lodging the vaginal plug (formed by the seminal fluids) in the vaginal–cervical junction (Matthews and Adler, 1978). After sperm deposition, a vaginal or seminal plug is formed by the components of the seminal fluid. The more fully the perimeter of the rostral end of the seminal plug is lodged in the vaginal cervical junction, the greater the transport of the sperm from vagina to the uterus. Sperm transport was poorer after copulation with males that maintained contact for less than one second than after copulation with males that maintained contact for at least one second. After ejaculation, a refractory period of about five minutes sets in, during which the male will not copulate. This refractory period is

Figure 8.28 Principal events in the rat estrous cycle in relation to the time of the day. This cycle is precisely timed on the basis of a diurnal rhythm. Ovulation can be seen to occur shortly after midnight. Other events include mating behavior, the LH "surge," and the development of the uterus. (From Bentley, 1976; reprinted with permission of Cambridge University Press.)

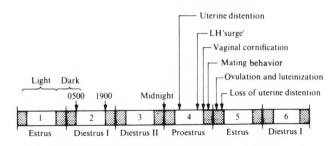

Uterine distention

LH 'surge'

Vaginal cornification

Mating behavior

Ovulation and luteinization

Loss of uterine distention

Light   Dark

0500   1900   Midnight

1   2   3   4   5   6

Estrus   Diestrus I   Diestrus II   Proestrus   Estrus   Diestrus I

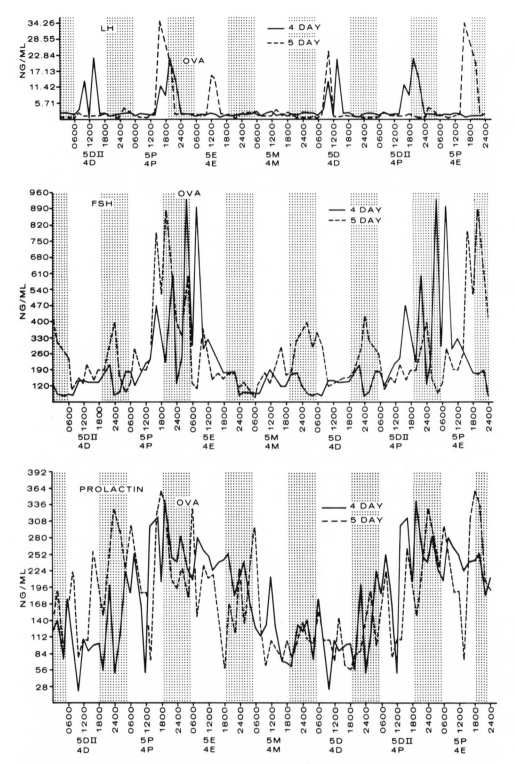

Figure 8.29 Concentrations of LH, FSH, and PRL in the serum of 4- and 5-day cyclic rats. Each hormone was measured at 2-hour intervals. M = metestrus, D = diestrus, DII = second day of diestrus, P = proestrus, E = estrus. Stippled area indicates dark period. Time is indicated on a 24-hour clock; 2400 = midnight. (From Nequin et al., 1979; reprinted with permission of *Biology of Reproduction*.)

Figure 8.30 (*Opposite page*) Concentrations of estradiol, progesterone, and 20α-hydroxypreg-4-en 3-one (20α-OHP) in serum of 4- and 5-day cyclic rats. For details, see Figure 8.29. (From Nequin et al., 1979; reprinted with permission of *Biology of Reproduction*.)

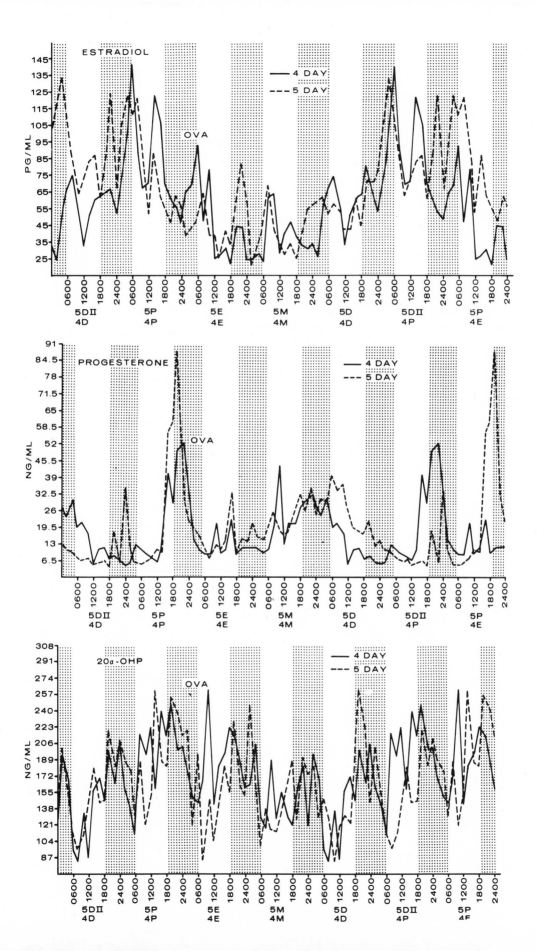

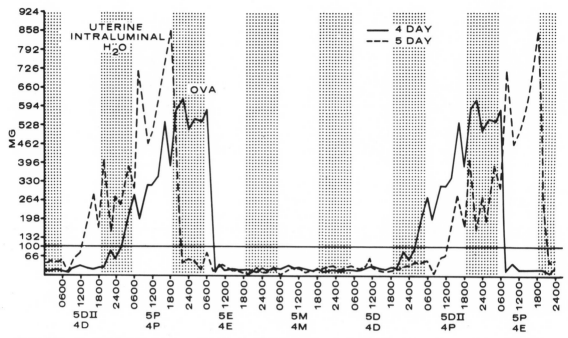

Figure 8.31 Uterine intraluminal water content in 4- and 5-day cyclic rats. For details of time axis, see Figure 8.29. (From Nequin et al., 1979; reprinted with permission of *Biology of Reproduction*.)

important, for sperm transport is adversely affected if a female copulates within five minutes of a previous copulation, because of the dislodging of the seminal plug by the second copulation (Adler, 1974). After five minutes, most sperm transport has occurred and so is not affected by subsequent copulations.

In the female, the sensory field area of the pudental nerve is larger and more sensitive in estrous

Figure 8.32 Schematic representation of the interaction between male and female rats during copulation. (From Adler, 1974; reprinted with permission of N. T. Adler and Plenum Publishing Corp.)

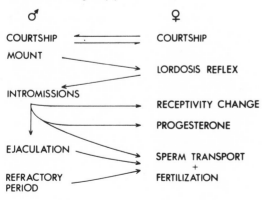

than in diestrous animals (Adler et al., 1977). This has been found to be the result of estrogen secretion (Komisaruk, 1974). The sensory field of this nerve extends from the base of the clitoral sheath to the base of the tail and laterally to the thigh. During mounting and attempts of intromission, the penis contacts this field and thus may facilitate the intensity of the female's lordosis and/or her orienting for intromission (Adler et al., 1977), in response to palpation of the flanks and the perineum and to cervical probing (estrogen does not facilitate lordosis in response to either of these stimulations alone, but only to the combination) (Komisaruk, 1974). The pelvic nerve receives sensory inputs from the vagina, cervix, and rectum, and the genitofemoral nerve receives sensory inputs from areas rostral to the pudental field with some overlap in the area of the clitoris (Adler, 1974). It seems that the pudental and genitofemoral field are involved in eliciting and orienting lordosis, that all nerves are involved in the lordosis reflex itself, and that the pelvic nerve is required for the initiation of the neuroendocrine reflex required for maintenance of CL (Adler, 1974). Adler (1974) has shown that five intromissions or more are required for the cessation of behavioral estrus and presumably for induction of progesterone secretion, and that multiple intromissions are necessary for sperm transport to the site of fertilization.

Gilman et al. (1979) have found that the female rat, if allowed to escape from the male, is capable of pacing the intromissions. When such pacing was allowed, 9/14 of the rats became pseudopregnant after five intromissions, whereas among females not allowed to escape, 3/14 became pseudopregnant after five intromissions. There were no differences between the two groups of rats with respect to the incidence of pseudopregnancy when either 10 or 15 intromissions occurred.

In the male, hormone concentrations change in conjunction with copulation. Males introduced in an arena with an estrous female show a doubling of plasma testosterone concentration, an almost threefold increase in plasma LH concentration,

and a larger than threefold increase in plasma PRL concentration. Following mating, experienced males showed higher concentrations of testosterone and PRL than after exposure to an estrous female, whereas the LH concentrations were about the same. Naive males showed lower concentrations of testosterone, LH, and PRL after mating than did experienced males under these conditions (Kamel et al., 1975).

In the female, stimulation of the cervix is followed by a release of PRL consisting of *two* daily surges, one diurnal and one nocturnal (de Greef et al., 1977), as illustrated in Figure 8.33. Such peaks have been confirmed by other investigators (Freeman et al., 1974; McLean and Nikitovitch-Winer, 1975; Hsueh and Voogt, 1975; de Greef et al., 1977; de Greef and Zeilmaker, 1978; Freeman and Sterman, 1978). In intact rats, these twice-daily surges continue until day 11, after the rat has been stimulated at the cervix and has become pseudopregnant (Figure 8.33). In the ovariec-

Figure 8.33 Serum prolactin concentrations at various times of the day in pseudopregnant rats following sham operation (open bars) or traumatization of both uterine horns (hatched bars) on day 4 of pseudopregnancy. The number of rats is indicated on top of the columns. The last prolactin surge was observed at 0300 h on day 11 of pseudopregnancy in sham-operated rats, whereas both prolactin surges were still present on day 16 of pseudopregnancy in the animals with decidual tissue. (From de Greef et al., 1977. Copyright © 1977 by The Endocrine Society. Reprinted with permission of W. J. de Greef and Williams & Wilkins.)

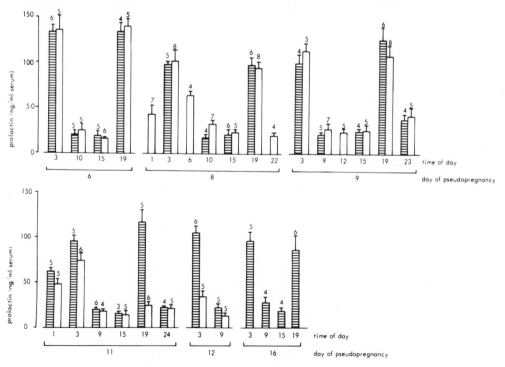

tomized rat, the PRL surges last 6 days only (Freeman and Sterman, 1978). By implanting either E_2 or progesterone in ovariectomized rats, Freeman and Sterman (1978) were able to establish that progesterone accentuated the nocturnal surge of PRL and that E_2 accentuated the diurnal and inhibited the nocturnal surge.

In the intact pseudopregnant rat, the two daily PRL peaks can be sustained after day 11 by implanting progesterone, so that plasma levels are in the range of 70–90 ng/ml (a concentration well within the physiological range; Figure 8.33). On the basis of these and other findings, de Greef and Zeilmaker (1978) suggest that PRL releases at the end of pseudopregnancy cease because of a decline in luteal function. Why this decline occurs, however, is not clear.

The PRL releases induced by cervical stimulation may have either a luteolytic or a luteotrophic effect. They may have a luteolytic effect on CL from the previous cycle. This interpretation is supported by the almost doubling of the number of CL after injection of antisera to PRL receptors from diestrus II (Figure 8.28) following diestrus I (Bohnet et al., 1978). They have, however, a luteotrophic effect on the CL of the cycle during which the copulation occurs. These two effects have been proposed in the light of the dependence of the structural luteolytic effect and the luteotrophic effect of prolactin on the time that elapses between AP and prolactin injections (Malven, 1969).

The concentrations of LH, FSH, and progesterone during pseudopregnancy in the rat are presented in Figure 8.34. It appears from these data that progesterone starts to decrease at day 9, prior to a decrease in LH or FSH concentrations or in the amplitude of the prolactin surges. From investigations of Takahashi et al. (1978), it appears that the life span of the CL of pseudopregnancy is "programmed" between days 2 and 4 after ovulation by LH secretion. If the in situ AP is removed in animals carrying an additional AP, isografted under the kidney capsule (to provide PRL secretion continuously), on day 0 or day 1, progesterone secretion is sustained at high concentrations beyond day 9 to day 18; this effect is overcome by LH injections. Progesterone secretion is also sustained in AP-isografted rats by in-

jection of LH antiserum from day 0 to day 5 and removal of the in situ AP on day 5, but after removal of the in situ AP on day 5, progesterone secretion diminishes. Thus, in this experimental model, LH has a luteolytic effect.

The endocrine factors which bring about the maintenance of the CL until day 9 of pseudopregnancy and its gradual decrease in progesterone secretion at day 9 and which result in structural regression by day 12 and the resumption of a new estrous cycle at day 13 have been studied for many years, yet a totally integrated model is still not established. Greep (1971) has called the problem of luteal function in the rat a "bugbear." There is no disagreement that AP causes regression of the CL and a decrease in progesterone secretion (Nalbandov, 1973). The difficulty has been to determine the luteotrophic hormone(s). On the one hand, evidence can be marshalled that PRL is the luteotrophic hormone, e.g.: (1) An AP transplanted to the kidney capsule secretes only PRL and causes maintenance of the CL in the AP rat, if the transplant is made shortly after ovulation and provided the AP is complete (Maslar, 1977). (2) If injected shortly after ovulation, PRL can maintain the CL structurally, but it does not stimulate progesterone secretion directly, and either estrogen or LH are required to support a maximal decidual response in AP rats (Hilliard, 1973). (3) High doses of estrogen that inhibit LH secretion maintain the CL of pseudopregnancy, presumably by stimulating prolactin secretion. A direct effect of estrogens on the CL cannot be ruled out, however.

On the other hand, evidence can also be found that LH is the luteotrophic hormone, e.g.: (1) Luteinizing hormone stimulates progesterone secretion in vitro and in vivo, whereas prolactin does not do this. (2) Luteinizing hormone can maintain pregnancy in the AP rat. (3) Antiserum to LH, administered on day 8, reduces the secretion of progesterone but stimulates secretion of 20α-ol. The fate of the CL in the pseudopregnant rat depends probably on the action of both prolactin and LH acting in the proper sequence and sometimes in combination.

In the rat, removal of the uterus results in extension in the life of the CL to 18–25 d from the normal life span of 14 days (Anderson, 1973). This effect of hysterectomy is dependent upon hy-

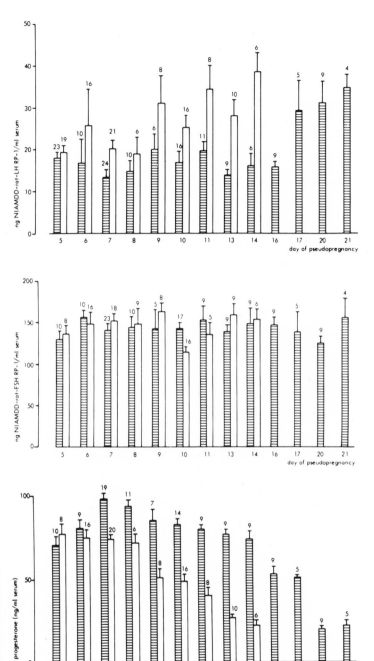

Figure 8.34 Concentrations of LH, FSH and progesterone in pseudopregnant rats, sham-operated (open bars) and traumatized to induce decidual response (hatched bars). The number of rats is indicated on top of the columns. For LH the difference between sham-operated and traumatized animals was significant on day 7 and from day 10 on; for FSH the difference between the two groups of rats was significant from day 10 on; for progesterone this was the case from day 7 on. (From de Greef et al., 1977. Copyright © 1977 by The Endocrine Society. Reprinted with permission of W. J. de Greef and Williams & Wilkins.)

pophyseal LH secretion, because injection of antiserum to LH on day 9, 12, 15, or 18 of pseudopregnancy in hysterectomized rats causes a precipitous drop in progesterone secretion (Akaka et al., 1977), but antiserum to LH injected on day 7 or 8 has no effect (Rothchild et al., 1974). Roth-

child et al. (1974) have suggested that the CL *becomes* dependent on LH and that the level of LH to which the CL is exposed may be a cause for this dependency.

Gibori et al. (1977) have shown that 100 μg of estradiol-17β can prevent the decrease in plasma progesterone secretion induced by antiserum to LH in the pregnant rat, suggesting that LH may play an important role in estrogen synthesis and that the estrogens can then maintain progesterone secretion. The dependency of the CL on LH may be explained by the requirement of LH for ovarian estrogen secretion and by the possibility that this requirement may increase as the CL develops.

The dependency of the CL of pseudopregnancy in the hysterectomized rat on LH, supports Nalbandov's (1973) general contention that the concept of a luteolytic factor need not be invoked to explain the extension of the life of the CL by hysterectomy. The extension of the life of CL by hysterectomy can be explained by the greater availability of estrogens in the absence of the uterus, which, as a target tissue, might be expected to bind estrogens that can now rescue the CL, but this explanation does not explain the local effect of the uterus on the CL. However, the lack of a demonstrated pathway for the transport of PG from uterus to ovary also fails to explain the local action of a luteolytic uterine factor. Clearly more experimentation is needed to solve these problems.

It is thus the contention of some investigators that the control of the CL is exercised by the AP, specifically by the secretion of LH, which makes estrogens that have their luteotrophic effect on the CL available. Others contend that the control of the CL is exercised by the uterus, specifically by $PGF_{2\alpha}$ (Lau et al., 1979), which interferes with the luteotrophic process. One should, of course, remember that the two mechanisms are not mutually exclusive and that the endometrium may secrete a luteolytic factor and at the same time may bind estrogen so that it is not available for the CL.

Golden Hamster (Mesocricetus auratus). The golden hamster has a remarkably constant 4-day cycle. Progesterone production reaches a peak at day 2, and the CL degenerates rapidly on day 3. This degeneration is independent of uterus and AP (Hilliard, 1973).

The pattern of LH, FSH, and E_2 concentrations during the cycle resembles that found in the 4-day-cycle rat with progesterone concentrations highest on day 4 (Greenwald, 1975). The sources of this preovulatory progesterone are the interstitial

gland (80 percent) and the antral follicles (20 percent) (Saidapur and Greenwald, 1979). Details of the hormonal fluctuations on day 4 of the cycle are illustrated in Figure 8.35. It can be seen that the rise in LH results in a sharp increase in plasma testosterone concentration and that this concentration reaches a peak prior to the LH peak. The increase in progesterone is more sustained than the increase in testosterone, while E_2 shows only a relatively minor transient increase. Testosterone is produced by both the antral follicles and the nonantral follicle part of the ovary, but the relative contributions by each of these compartments is not clear (Saidapur and Greenwald, 1979). The estrogens are largely produced by the antral follicles.

In pseudopregnant, hysterectomized hamsters, the demise of the CL can be induced by preventing PRL release (see below). After the progesterone concentration drops below 4 ng/ml, ovulations occur (Terranova and Greenwald, 1978), suggesting that, as in the rat, the concentrations of progesterone may determine the fate of the maturing follicles. Coitus induces pseudopregnancy, after infertile matings. The CL is maintained for 8–10 days; in the hysterectomized hamsters, it is maintained for 16–23 d (Anderson, 1973). The maintenance of the CL of pseudopregnancy is dependent on the AP, specifically on the luteotrophic complex secreted by the pituitary and the uterus. This luteotrophic complex consists of FSH, PRL, and trace amounts of LH (Hilliard, 1973). Large doses of LH are luteolytic in the hamster (as they are in

Figure 8.35 Interrelationship between the steroid hormones (P, T, E_1, and E_2) and the gonadotrophins (LH and FSH) on day 4 (proestrus) of the hamster cycle. (From Saidapur and Greenwald, 1979; reprinted with permission of *Biology of Reproduction.*)

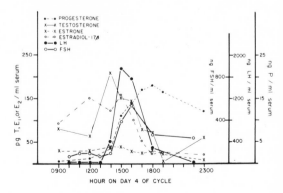

the rabbit and sheep), because they cause ovulation of antral follicles and deprive the CL of the estrogens required.

The importance of PRL for maintenance of the CL of pseudopregnancy is illustrated by experiments in which the CL rapidly declined after the inhibition of a single nocturnal surge of PRL by administration of ergocryptine, on day 8 of pseudopregnancy(Terranova and Greenwald, 1978). To maintain a number of follicles necessary as a source of estrogen, which helps to maintain the CL, FSH may be required. In this respect the hamster, rabbit, and sheep show similarities in the CL requiring estrogen.

House Mouse (*Mus musculus*). The estrous cycle of the mouse resembles that of the rat, except that the CL of the cycle lose their lipid more rapidly in the mouse than in the rat. Prolactin is luteotrophic and is probably the hormone secreted in response to stimulation of the cervix during copulation. Hysterectomy extends the life of the CL of pseudopregnancy from a 10–11 d average to 15–17 d.

COMPARISON OF TYPES OF REPRODUCTIVE CYCLES OF MAMMALS

One of the reasons for going into details about the different kinds of cycles and for discussing different species, illustrating each type, has been to point out the variation not only among these types of cycles but also within types. It appears that none of the animals discussed has the same mechanism for maintaining the CL. Even when there appear to be superficial similarities, there are usually differences as more data become available.

It is difficult to understand the kind of selective advantage of each of the mechanisms discussed. Conaway (1971) has attempted to explain some of the advantages for each of the types of reproductive cycles. Type IIA and Type III seem to be characteristic of prey species, which have a high reproduction rate and high mortality. Although the special adaptive value of Type III as compared to Type IIA is not clear, it has been observed that the species having Type III do not show the great amplitudes in population density found in species of Type IIA. Type IA species are generally long-lived prey species; therefore, there is no great pressure to have a cycle shortly after the current cycle is sterile. Type IB and IIB are restricted to large predator species, in which low but consistent reproduction rates are probably more advantageous than high reproduction rates, since large numbers would exhaust the resource of prey species, leading to mortality from starvation in the predators.

9

Insemination and Fertilization

SPERM CHANGES IN THE MALE REPRODUCTIVE TRACT

Between the shedding of the sperm into the lumen of the seminiferous tubule and the fertilization of the oocyte, depending on the species, the sperm undergo various changes.

In cyclostomes, teleosts, and anurans, sperm obtained from the testes are capable of fertilizing oocytes. Motility in some of these species may require passage through the excurrent duct system, e.g., in the skate (*Raja eglanteria*), the water snake (*Natrix sipedon*), the turtle (*Pseudemys scripta*), and the American chameleon (*Anolis carolinensis*) (Bedford, 1979).

Rooster sperm obtained from the testes have a poor fertilizing capacity when used for artificial insemination. Munro (1938) found that 5 out of 39 hens laid fertile eggs after insemination with testicular sperm; 5 out of 39 after insemination with epididymal sperm, and 57 out of 77 after insemination with sperm from the ductus deferens. Bedford (1979) found that 2–5 percent of the sperm obtained from the testes of roosters, drakes, or pigeons were motile when placed in Tyrode solution and that motility was acquired after passage through part of the excurrent duct. In the song sparrow (*Melospiza melodia*), however, Bedford found that the sperm needed to be exposed to the environment of the seminal vesicle (which has a lower temperature than the body cavity) to be motile when placed in Tyrode solution. In therian animals, i.e., metatherian and eutherian animals, sperm become motile only after they have undergone maturation changes in the epididymis. These maturation changes and the survival of the sperm do not require androgens, i.e., they are not affected by castration in reptiles and birds, with the possible exception of passerine birds. In therian animals, sperm maturation and survival are androgen-dependent except in the musk shrew (*Suncus murinus*) (Bedford, 1979). It should be recalled that the size of the uterus of the female musk shrew is not estrogen-dependent (Chapter 5).

Some of the changes in the sperm that occur in the epididymis and that may be associated with the acquisition of fertilizing capacity are: loss of the middle-piece cytoplasmic droplet; the development of forward motility; changes in resistance to heat, cold shock, harmful alkaloids, acids, and alkalis; and an increase in specific gravity (Bedford, 1975). In bulls, a forward mobility protein (FMP), which is a glycoprotein, is found in seminal plasma. This FMP induces forward motion in immotile epididymal sperm in the presence of high concentrations of cyclic AMP and seems to originate in the epididymis (Brandt et al., 1978). Bedford (1979) has described the changes that occur in the acrosome during the passage of the sperm through the epididymis and ductus deferens.

TIMING OF SPERM AND OOCYTE RELEASE

There are two ways for bringing male and female gametes together: (1) sperm and oocytes may be released at about the same time, as in many mammals and in some fishes and amphibians; (2) either the sperm or the oocytes retain the capacity to fertilize or be fertilized for an extended period of time. Among vertebrates, there do not seem to be species in which oocytes retain this latter capacity, although there are species in which the sperm retain their fertilizing capacity for an extended period, sometimes as long as several years (Table 9.1). The only species in which the fertilizing capacity of sperm is retained for an extended time are those that have "internal" fertilization. According to Cohen (1977), the terms

Table 9.1

Retention of fertilizing ability of sperm for extended periods in genital system of female in different species*

Genus or species	Time fertilizing capacity retained	Reference
Teleosts		
Cymatogaster sp.	Several months	Turner, 1947
Heterandria formosa	10 months	Turner, 1947
Poecilia reticulata	4 months	Parkes, 1960
Reptiles		
Malaclemmys centrata	4 years	Parkes, 1960
Terrapene carolina	4 years	Parkes, 1960
Microsaura pumila pumila	6 months	Parkes, 1960
Uta stansburiana	81 days	Cuellar, 1966
Vipera aspis	Over winter	Parkes, 1960
Causus rhombeatus	5 months	Parkes, 1960
Ancistrodon contortrix	11 days	Parkes, 1960
Crotalus viridis viridis	Over winter	Parkes, 1960
Tropidoclonion lineatum	Over winter	Parkes, 1960
Coronella austriaca	Over winter	Parkes, 1960
Thamnophis sirtalis	3 months	Parkes, 1960
Natrix natrix	Over winter	Parkes, 1960
N. vittata	1½ years	Parkes, 1960
N. subminiata	5 months	Parkes, 1960
Storeria dekayi	4 months	Parkes, 1960
Drymarchon corais couperi	4½ years	Parkes, 1960
Xenodon merremii	1 year	Parkes, 1960
Leptodeira annulata polysticta	6 years	Parkes, 1960
L. albofusca	1 year	Parkes, 1960
Boiga multimaculata	1 year	Parkes, 1960
Birds		
Goose	9.7 days	Johnson, 1954
Chicken	10–14 days	Nalbandov and Card, 1943
Turkey	45.5 days	van Tienhoven and Steel, 1957
Ringdove	8 days	Riddle and Behre, 1921
Mallard duck	7–10 days	Elder and Weller, 1954
Ring-necked pheasant	22 days	Schick, 1947
Bobwhite quail	8.3 days	Kulenkamp et al., 1965
Mammals		
Myotis l. lucifugus	138 days	Wimsatt, 1942
Eptesicus f. fuscus	156 days	Wimsatt, 1942
Horse	4–6 days	Parkes, 1960

*In teleosts the sperm are stored in the ovary, whereas in reptiles, birds, and mammals the sperm are stored in parts of the oviduct or the vagina.
From van Tienhoven, 1968; reprinted with permission of W. B. Saunders.

internal and *external* fertilization are incorrect. For instance, fertilization in mouth breeders is really internal, but it has generally been considered external. In my opinion, however, it is a vain hope to expect the existing terminology to be changed. I will continue to use the existing terminology, but I will restrict the term *internal fertilization* to fertilization that occurs inside the reproductive tract and *external fertilization* to fertilization outside the reproductive tract.

EXTERNAL FERTILIZATION

There are several strategies that increase the chances of male and female gamete encounters, and examples of these follow.

Gregarious sexual congress. Male and female gametes are released into the surrounding medium in large numbers, not only by individuals but also by the synchronized spawning by an entire group of males and females. Among the Clupeiformes, with the herring *Clupea harengus* as an illustration, spawning occurs in schools with pairing or polyandry occurring within the school, but with synchrony of the spawning among individuals in the school. In consequence, such large numbers of gametes are released that the water may turn milky white (Breder and Rosen, 1966). This is a strategy of sheer numbers and the waste of gametes is quite high, as is the waste of zygotes because there is no parental care.

Proximity of shedding of male and female gametes. Male and female gametes are brought in proximity, as is the case in spawning by lampreys and some teleost fishes. Lampreys, for instance, wrap the ends of their tails around each other at the time of spawning, so that the genital openings are close to each other. In the teleost family Syngnathidae (pipe fishes and seahorses), the female deposits the oocytes via an ovipositor into the pouch of the male. The genital opening of the male lies at the same height as the upper border of the pouch, so that when the sperm are released they are in proximity of the deposited oocytes.

Except for the East African frog (Chapter 8), Anura do not have internal fertilization but do have a sexual embrace (amplexus). In frogs, such as the leopard frog (*Rana pipiens*), the vocalizations of the male attract the females (Chapter 13).

These vocalizations of the male and the development of the thumbs and thumb pads, which facilitate holding the female, are under the control of androgens secreted by the testes. The males attempt to grasp receptive and unreceptive females and also other males. Unreceptive females and males will utter a release call, and the grasping male will release them. Receptive females, as well as unreceptive females that have had the abdomen distended artificially, do not give the release call (Diakow, 1977). Vasotocin, which causes distension of the abdomen, inhibits the release call or croak (Diakow, 1978) and may thus have an important coordinating function in the reproductive behavior of these frogs. If the female is receptive, the male and female will remain in sexual embrace, sometimes for days (Cohen, 1977); the female eventually lays her eggs (oocytes covered with egg jelly), after which the male releases his sperm.

Frogs in amplexus can move around and feed, so that the sexual embrace does not prevent the pair from escaping predators, although amplexus still endangers the pair to some extent. The advantage of bringing the gametes close together apparently has outweighed the disadvantage of the slightly higher mortality due to predation.

Spatial restriction of the release of gametes. The strategy used by seahorses incorporates a strategy other than proximity of the germ cells; it includes confinement; that is, dispersal of gametes by water currents is reduced by the spawning in a confined area. Other examples of spatial restriction for release of gametes are found in the three-spined stickleback, the European bitterling, cichlid fishes, and the Surinam frog. The male three-spined stickleback (*Gasterosteus aculeatus*), after building a nest (Chapter 13), chases the gravid female into the nest, after which the female and male spawn in succession. The female European bitterling (*Rhodeus amarus*) deposits her oocytes through an ovipositor into the excurrent opening of a fresh water mussel, and the male spawns close to the incurrent opening of the mussel. Fertilization occurs within the gill chambers of the mussel and the embryo develops there. Among cichlid fishes, mouth breeding occurs in many species. In *Haplochromis wingatii,* the female deposits her oocytes (eggs) in a pit dug by the male in the bottom of the aquarium. The male has spots that resemble eggs on his anal fin. These "egg-dummies" are more conspicuous than the eggs in the bottom of the pit. When the female grasps the egg-dummies on the anal fin, she is inseminated, so that any eggs that she already has scooped up from the pit and kept in her mouth can be fertilized (Wickler, 1962). In another cichlid fish, *Tilapia macrochir,* the female enters the male's spawning pit on the bottom of the aquarium. The male has white tassels near his genital opening and during courtship he discharges white threads of spermatophores, which the female takes into her mouth; she may also take the tassels into her mouth, and so act as a releaser for spawning by the male. During the interplay between male and female, the female may spawn and then pick up the eggs and keep them in her mouth. The taking in of the spermatophore or of the male's genital tassels and subsequent spawning is followed by fertilization of the oocytes (Wickler, 1965). In the Surinam frog (*Pipa pipa*), the female spawns in a horizontal upside-down position while in amplexus. The male catches the oocytes in the folds of his belly, transfers the eggs to the incubation chambers on the female's back (Chapter 5), and then spawns.

INTERNAL FERTILIZATION

Internal fertilization does not necessarily involve copulation. We have discussed in Chapter 5 that lizards may deposit their sperm in packages as spermatophores, which are picked up by the female with the cloacal labia and deposited in the genital tract. Copulation by means of a true intromittent organ is found among elasmobranchs, teleosts, amphibians, reptiles, birds, and mammals (Chapter 5). The site of fertilization seems to be the ovary (as in many teleosts that have internal fertilization and in the mammalian tenrec) and the anterior part of the oviduct.

Internal fertilization, which has developed independently several times, has generally been accompanied by a reduction in the number of oocytes that are released simultaneously. The number of sperm released, however, has remained high. The provocative reviews by Cohen (1971, 1977) should be consulted by those interested in the problem of the apparent difference in waste of oocytes and sperm. Cohen (1971, 1977) argues that in the female a large number of oocytes are lost in the ovary as a result of atresia (Table 9.2) and that postovulatory losses are relatively small, whereas in the male most of the sperm loss occurs after ejaculation. He contends that both the oocyte losses and the failure of sperm to reach the site of fertilization are correlated with chiasmatic failure. Wallace (1974) has rejected both Cohen's data and the theoretical basis of his arguments. It would, however, still be of some interest to compare the incidence of atresia in mammals that ovulate large numbers of oocytes, e.g., 800 in the plains viscacha (*Lagostomus maximus*) and 120 in the elephant shrew (*Elephantulus myurus jamesoni*), with the incidence of atresia in closely related species that ovulate fewer oocytes.

Table 9.2

Incidence of atresia of oocytes in some species of mammals

Species	Oocytes		
	Maximum number	Number at birth	Number at puberty
Guinea pig	105,000	45,000	13,000
Rat	71,000–160,500	52,000	23,000–27,000
Cow	2,700,000	60,000	21,000
Lemur	170,000	10,000	51,000*
Rhesus monkey	1,100,000	260,000	20,000
Human	6,100,000	2,000,000	250,000

*Cohen states that in the ewe and lemur ovogenesis occurs into adulthood. In view of the evidence cited in Chapter 7, this seems an incorrect inference.
Modified from Cohen, 1971.

Insemination by Copulation

In elasmobranchs, the claspers form the intromittent organs. In the spiny and in the smooth dog fish, one clasper is flexed medially, inserted and anchored in the oviduct by a complex of cartilages at the tip of the clasper. The sperm pass from the urogenital papilla into the groove of the clasper and are washed into the oviduct by seawater and secretions from a siphon sac, an epithelium-lined bladder filled with seawater. The filling occurs by flexing of the clasper prior to copulation (Gilbert and Heath, 1972).

In teleost fishes with internal fertilization and in which the males possess an intromittent organ (Chapter 5), such as in some of the Doradidae, in the Poeciliidae, Goodeidae, Jenynsidae, Anablepidae, Phallostethidae, and Neostethidae, the semen is (presumably) deposited in the oviduct. Among the Anura, the only species with known internal fertilization is the East African frog, but the method of insemination is not known. Reptiles show some special adaptations. The American chameleon (*Anolis carolinensis*) uses the left and right hemipenis (Chapter 5) alternately, depending on the sensory feedback from the hemipenis and to some extent from the ipsilateral testis (Crews, 1978). In some snakes, e.g., garter snakes, a copulatory plug is formed from kidney secretions (the *sexual* segment of the reptilian kidney) (Devine, 1975). This plug not only prevents probable sperm leakage, but it also prevents other males from inseminating the same female. Apparently, other males sense that a female has been inseminated because such females are not courted (Devine, 1977; Ross and Crews, 1977). In birds, the semen is usually deposited by the male's cloaca on the everted oviduct of the female, except in the Anatidae and ostriches, cassowaries, emus, and rheas, which have a copulatory organ (Chapter 5) that deposits the semen (presumably) in either the vagina or the shell gland. In mammals, ejaculation is preceded by mounting, erection of the penis, and intromission (Chapter 8). Gerall (1971) has reviewed the structures of the nervous system involved in mounting and erection, and some of the data are summarized in Table 9.3. This information shows that electrical stimulation of either the preoptic area, the me-

dial hypothalamus, areas of the thalamus, or the medial septopreoptic region can induce erection. Table 9.3 also shows that lesions of either the temporal lobe or of the amygdala are associated with greater frequency of mounting and erections. A flow sheet of the events that lead to ejaculation in humans is presented in Figure 9.1. Sexual behavior and its control will be discussed in Chapter 13. In the mammals investigated, insemination occurs either in the vagina (as in the rat, rabbit, and human) or in the cervix (as in horses, cattle, and swine) (Overstreet and Katz, 1977). Walton (1960), in his classification of patterns of insemination, included intrauterine insemination. According to Overstreet and Katz (1977), intrauterine insemination does not occur, at least not in the animals used by Walton as examples. They suggested that Walton's procedures for assessing the occurrence of intrauterine insemination may have introduced an experimental artifact.

SPERM TRANSPORT AND STORAGE

Transport

After insemination the sperm have to reach the site of fertilization, i.e., the ovary or upper oviduct. During this passage many sperm are lost; that is, they do not reach the site of fertilization. Table 9.4 illustrates this for mammalian species together with the time interval that it takes the sperm to reach this site. Much of the following account on sperm transport is based on the excellent review by Overstreet and Katz (1977).

Sperm transport seems to consist of two phases: a phase of rapid transport and a phase of slow transport. The phase of rapid transport is probably mainly the result of uterine contractions, and the vast majority of the sperm transported during this phase do not survive. The phase of slow transport, at least in the rabbit, occurs in the interval between the end of the coitus-induced internal contractions (which last about 2–5 min) and the contractions that occur 1–3 h thereafter. The sperm migrate into the uterus and oviduct from the cervix. It is beyond the scope of this book to discuss the evidence and theory of the sperm reservoir theory and the possible selection of sperm that may occur. The interested reader is referred to the review by

Table 9.3

Relationship between limbic system and reproductive behavior in male and female *Macaca mulatta* and *Saimiri sciureus*

Structure	Treatment	Male	Female
Putamen*	S	Erection and complete copulatory sequence in social situation	
Putamen*	R		No specific unit response to vaginal stimulation
Caudate nucleus†	S	No consistent effect on erection	
Cingular gyrus*	L	No permanent effect on sexual behavior	No effect on receptivity
Posterior cingular gyrus*	S	No effect on penile erection	
Anterior cingular gyrus*	S	Penile erection	
Cingular gyrus†	S	Low-rated erection	
Hippocampus*	R	No specific effect on electroencephalogram (EEG) of genital manipulation in immature subjects before or after androgen injection	
Hippocampus*	S	No effect on penile erection	
Temporal lobe resection (large lesions)*	L	Hypersexuality: increased frequency of copulatory activity, indiscriminate choice of objects in lab situation, increased orality, hypermetamorphosis, lessened fear, psychic blindness, alteration of social behavior (Kluver-Bucy phenomenon). Hypersexuality eliminated by castration	Increased tameness; effects on sexual behavior either lacking or temporary; pregnancy and parturition normal; offspring neglected or abused
Medial temporal lobe, anterior half of right and ventral half of left amygdala, rostral hippocampus*	L	Normal heterosexual activity	
Bilateral two-thirds of amygdaloid complex*	L	Apathetic, but frequent normal heterosexual contact	
Medial surface temporal lobe, left side rostral and middle third amygdaloid complex*	L	Temporary tameness; normal heterosexual behavior	
Right uncus and amygdala, left temporal cortex, basal portion of amygdala*	L	Apathetic; normal sexual behavior	

Table 9.3—*Continued*

Structure	Treatment	Male	Female
Basal lateral medial cortical nuclei*	L	Temporary tameness; hypersexuality in approximately two months	
Amygdala and uncus*	L	Juveniles; pretestosterone injections: greater frequency of mounting; exhibit erection when mounted; post-testosterone injections: increased mounting and grooming	
Amygdala*	L	Neonates; no hypersexuality at 1–2 years of age	
Amygdala*	R		Excitation of unit activity and both excitation and inhibition in animals with estrogen in response to vaginal stimulation
Amygdala and temporal lobe*	R	No specific effect on EEG of genital manipulation in immature subjects before or after androgen injection	
Amygdala corticomedial nucleus*	S	Penile erection	
Stria terminalis and its bed nucleus*	S	Penile erection	
Stria terminalis†	S	Negative effect on erection	
Basal septal region*	L	No effect on copulatory behavior	No effect on copulatory behavior; interference with maternal behavior
Septum, not including area of bed nucleus of stria terminalis*	S	No effect on erection	
Septum*	R		Unit response to vaginal stimulation; same as amygdala
Medial septopreoptic region†	S	Nodal point for erection	
Midline nuclei, thalamus*	S	Penile erection	
Anterior and midline thalamic nuclei†	S	Penile erection	Clitoral enlargement
Mediodorsal nucleus†	S	Penile erection	Clitoral enlargement
Centralis lateralis nucleus†	S	Seminal discharge independent of scratching	

Region		Effect
Preoptic*	S	Erection, pelvic thrusting, and ejaculation exhibited when restrained; erection in free social situation and suppression of sexual mounting and intromission
Preoptic*	R	Spindle EEGs evoked in immature subjects by genital stimulation after estrogen priming
Preoptic and suprachiasmatic*	R	Inhibition of single units in estrogen-primed immature and adult animals in response to vaginal stimulation
Anterior hypothalamus and ventromedial nucleus*	L	No effect
Rostral supraoptic region*	R	High amplitude slow waves in immature and androgenized males in response to genital stimulation
Anterior hypothalamic area*	R	Frequent inhibition of single unit activity in estrogenized females in response to vaginal stimulation
Ventromedial region*	L,S	Disrupted menstrual cycles
Ventromedial region*	R	Inhibition of single unit activity in estrogen-primed animals
Ventromedial region*	S	No penile erection
Medial hypothalamus*	S	Penile erection
Dorsomedial region*	L	Temporary and minimal effects on menstrual cycle
Dorsomedial region*	S	Disrupted menstrual cycles
Anterior tuberal region*	S	No effect on menstrual cycles
Anteromedial region of median eminence*	S	Accelerated onset of either vaginal cornification or flow depending on time of cycle
Ependymal cells in tuber cinereum*	M	Selective cytological reaction to estrogen
Medial hypothalamus, dorsal to VMN, lateral hypothalamus*	S	Penile erection
Posterior hypothalamus*	L	Temporary disruption of menstrual cycles
Posterior hypothalamus*	S	Prolonged cycles for duration of stimulation

Table 9.3—*Continued*

Structure	Treatment	Male	Female
Posterior hypothalamus*	S	Penile erection	
Posterolateral hypothalamus*	R		Increased tendency toward unit activity following vaginal stimulation regardless of hormone condition
Mammillary bodies*	S	No effect on penile erection	
Mammillary bodies*	R	High amplitude slow waves following genital stimulation in immature males injected with estrogen for at least 5 d	No distinctive pattern of unit response evoked by genital stimulation
Medial forebrain bundle*	S	Nodal tract for erection	
Mammillothalamic tract*	S	No effect on penile erection	
Medial preoptic, anteromedial nucleus, and supraoptic†	S	Penile erection; no emission	Clitoral enlargement
Ventromedial area†	S	Aversive effects	
Lateral hypothalamus†	S	Positive loci for erection	
Mammillary bodies†	S	Penile erection	
Mammillothalamic tract†	S	Nodal tract for erection	Clitoral enlargement
Medial forebrain bundle†	S	Nodal tract for erection	Clitoral enlargement
Nucleus tegmenti ventralis, nucleus opticus tegmenti, medial segment of substantia nigra†	S	Penile erection	
Regions near spinothalamic tract†	S	Genital scratching and manipulation often accompanied by emission	

S = electrical stimulation, R = recording, L = lesion, M = microscopic analysis.
*Macaca mulatta.
†Saimiri sciureus.
From Gerall, 1971; reprinted with permission of Charles C Thomas, Publisher, Springfield, Illinois.

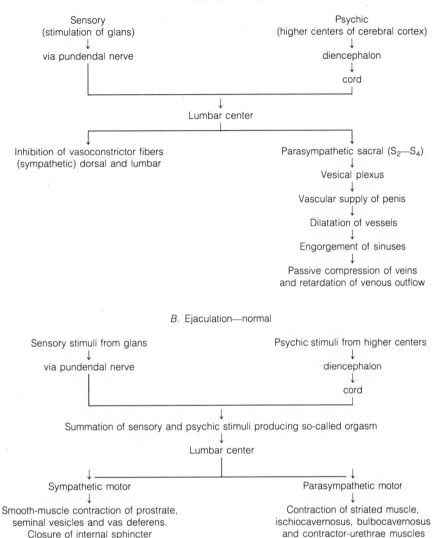

A. Erection—normal

Sensory
(stimulation of glans)
↓
via pundendal nerve

Psychic
(higher centers of cerebral cortex)
↓
diencephalon
↓
cord

↓
Lumbar center

Inhibition of vasoconstrictor fibers
(sympathetic) dorsal and lumbar

Parasympathetic sacral (S_2—S_4)
↓
Vesical plexus
↓
Vascular supply of penis
↓
Dilatation of vessels
↓
Engorgement of sinuses
↓
Passive compression of veins
and retardation of venous outflow

B. Ejaculation—normal

Sensory stimuli from glans
↓
via pundendal nerve

Psychic stimuli from higher centers
↓
diencephalon
↓
cord

↓
Summation of sensory and psychic stimuli producing so-called orgasm
↓
Lumbar center

Sympathetic motor
↓
Smooth-muscle contraction of prostrate,
seminal vesicles and vas deferens.
Closure of internal sphincter
↓
Emission

Parasympathetic motor
↓
Contraction of striated muscle,
ischiocavernosus, bulbocavernosus
and contractor-urethrae muscles
↓
Ejaculation

Figure 9.1 Suggested neural pathways involved in erection and ejaculation in [humans]. (From Whitelaw and Smithwick, 1951; reprinted with permission of *New England Journal of Medicine*.)

Overstreet and Katz (1977) and a recent publication by Fischer and Adams (1980).

In chickens and turkeys, the site of fertilization is reached by some sperm in about 15 min after insemination (van Tienhoven, 1968); however, sperm are released continuously from uterovaginal storage glands (see below) and, unless they are subsequently stored in the infundibular storage glands, may be lost for fertilization purposes.

The mechanism of sperm transport is probably a mixture of passive transport by the contractions of the female reproductive tract, by movement of the cilia in this tract, and by the swimming movements of the sperm. However, the swimming movements of the sperm are not required, because dead spermatozoa also reach the site of fertilization in a short time.

Different strategies can be imagined which

Table 9.4

Some characteristics of ejaculate and sperm transport
in different mammalian species

Species	Site of insemination	Number of sperm ejaculated	Number of sperm in ampulla	Time between insemination and appearance of sperm in oviduct (min)
Mouse	Uterus	5×10^7	17+	15
Rat	Uterus	6×10^7	5–100	2–30
Hamster	Uterus	8×10^7	Few	2–60
Rabbit	Vagina	$25–50 \times 10^7$	250–500	180–360
Guinea pig	Uterus	8×10^7	25–50	15
Dog	Uterus	12.5×10^7	5–100	2 min to several hours
Cat	Vagina and cervix	5.6×10^7	40–120	No data
Pig	Cervix and uterus	500×10^7	Few	30
Cattle	Vagina	$400–500 \times 10^7$	4,200–27,500	2–13
Sheep	Vagina	80×10^7	600–5,000	2–30
Human	Vagina	125×10^7	Few	30

Modified from Blandau, 1973; used with permission of R. J. Blandau and the American Physiological Society.

might induce contractions of the reproductive system. The semen may contain substances that cause contractions. This may be the case in the spiny dogfish (*Squalus acanthias*), in which serotonin, which induces uterine contractions in the female dogfish, is present in high concentrations. In mammals, prostaglandins present in the semen may have a similar function, although it should be kept in mind that different prostaglandins have different effects on uterine contractions (van Tienhoven, 1968). The act of insemination might cause the release of hormones that cause uterine contractions. In cattle, either natural or artificial insemination causes release of oxytocin, which in turn stimulates uterine contractions (Gwatkin, 1977; van Tienhoven, 1968). Fuchs (1972) found that in rabbits, the uterine contractions after mating resembled those induced by epinephrine injections and not those induced by either oxytocin administration or by $PGF_{2\alpha}$ injections. She also found that α-adrenergic blocking agents prevented the uterine response to coitus, but that alcohol, which blocks the oxytocin response, did not block the response to mating.

The cervix is a potential barrier to sperm transport in those species in which semen is deposited in the vagina. At the time of estrus, the cervical mucus is more easily penetrated by the sperm than at other times. In ruminants, mucoproteins secreted by the cervix facilitate sperm transport through the cervix (Austin, 1974b). The uterotubal junction is the most efficient barrier, preventing many sperm from reaching the site of fertilization. The anatomy of this junction, as described in Chapter 5, explains how in some species the flexures in the oviduct and in other folds inside the oviduct function as barriers.

Once inside the oviduct, sperm can be transported by cilia. This transport by cilia poses the question of how oocytes are transported towards the uterus and sperm towards the ovary. There may be two different mechanisms that can explain these opposite movements. In the rabbit oviduct there is a small strip of ciliated epithelium, in which the cilia beat in the opposite direction from cilia in the rest of the oviduct, and this may partly explain the opposite directions in which sperm and oocytes are transported (Austin, 1974b). Recently, Jansen (1978) discovered that during estrus the lumen of the isthmus of the oviduct in the rabbit contains a sticky mucus that might permit sperm to reach the anterior part of the oviduct on their own motile activity. Cilia probably play little or no part in this transport, because there are few present and some cilia are caught in strands of mucus. After ovulation, the mucus disappears and the cilia become so prominent that they dominate the mucosa of the isthmus. These cilia probably function in the transport of the oocytes or zygotes.

In the tammar wallaby (*Macropus eugenii*), sperm transport in the parturient side is about 5–50 percent of that on the nonparturient side, when the female mates at postpartum estrus. This selective transport of sperm to the oviduct, ipsilateral to the side at which ovulation occurs, takes place at the cervix, which serves as a reservoir for the semen for about 24 h after ovulation (Tyndale-Biscoe and Rodger, 1978).

Sperm Storage in the Genital System of the Female

The genital systems of females of some species, in which the sperm retain their capacity to fertilize oocytes for an extended period after insemination, have special structures in which sperm are stored. Some of these structures have been mentioned in Chapter 5. In the urodeles, sperm are stored in the spermatheca, which are differentiated from the cloaca. In reptiles, there are (1) vaginal tubules, as in the genera of *Anolis* and in the Iguanidae, which project like fingers of a glove, and (2) tubular folds, which develop from alveolar glands in the upper part of the oviduct, as in the adder (*Vipera aspis*) and garter snakes (*Thamnophis* spp.). In birds, there are so-called sperm nests or sperm storage glands, both at the uterovaginal junction and in the infundibulum, also called the fimbria or funnel, of the oviduct. There may be benefits from such a redundancy, but there is some controversy about which of these storage sites is the most important (Bobr et al., 1964). The liberation of the sperm from the infundibular sperm nests may occur as the result of the passage of the yolk (Grigg, 1957). Release of stored sperm from the uterovaginal glands is apparently independent of ovulation, oviposition, and follicular secretions (Compton and Van Krey, 1979), and it is probably a slow, continuous process (Burke and Ogasawara 1969; Compton and Van Krey, 1979). In mammals, prolonged storage of sperm in the female reproductive system is found most frequently among bats, but it also occurs normally in the hare (*Lepus europaeus*) and sometimes in house mice

(*Mus musculus*) (Racey, 1979). Prolonged sperm storage is found among bats in the family Vespertilionidae subfamily Vespertilioninae, and in the family Rhinolophus. Vespertilionid species living in the temperate zone as well as in the tropical zone show the prolonged sperm storage (Racey, 1979).

The duration of sperm storage compatible with retention of fertilizing capacity varies from 16 d in the tropical bat *Pipistrellus ceylonicus* to 198 d in *Nyctalus noctula* of the temperate zone. According to Racey (1979), sperm are stored for a much shorter time in tropical species than in hibernating temperate zone species.

The sperm may be stored either in (1) the oviduct (e.g., *Tylonycteris pachypus*, *T. robustula*, *Pipistrellus ceylonicus*, *Chalinolobus gouldi*, *Rhinolophus hipposideros*, and *R. ferrumequinum*), (2) the uterotubal junction (e.g., *Myotis lucifugus*, *M. daubentoni*, *Miniopterus schreibersi fuliginosus* and *Scotophilus heathi*), or (3) the uterus (e.g., *Pipistrellus abramus*, and *Nyctalus noctula*).

Racey (1979) has emphasized that in mammals generally the uterus is invaded by leucocytes after insemination, and that this occurs also in bats that do not show prolonged semen storage. However, the reproductive tracts of female bats that show semen storage are virtually free of leucocytes.

CAPACITATION

In mammals, spermatozoa that have just been ejaculated or that have been obtained from the epididymis are not capable of fertilizing an oocyte and therefore need to be capacitated. Capacitation normally occurs in the female's reproductive tract, usually either in the uterus or in the uterus and oviduct (mouse, hamster), and takes about 1–11 h to be completed (Bedford, 1970), depending on the species.

Capacitation normally occurs in the uterus and/or oviduct, but it can also occur in vitro in such media as follicular fluid and blood serum, provided that these body fluids have been heated to destroy factor(s) detrimental to sperm. Purification experiments by Austin (1974b) indicate that the capacitation factor may consist of two components: (1) a dialyzable heat-stable factor that helps to maintain sperm motility, and (2) a nondialyzable heat-labile factor that causes the acrosome reaction (Figure 9.2), although Austin (1974b) stressed that this terminology would tend to cause confusion between capacitation and the acrosome reaction. The enzymes β-amylase and β-glucuronidase can cause partial capacitation of rabbit sperm in vitro in Locke's solution (Johnson and Hunter, 1972).

Capacitated sperm can be decapacitated by exposing them to seminal plasma. A bioassay for the decapacitation factor (DF) is illustrated in Figure 9.3. The DF is nondialyzable, fairly heat-stable, and may be part of a glycoprotein complex, from which it can be separated by exposing the complex to pronase. The active component has a molecular weight of 300–500 (Austin, 1974b). Among mammalian species, the DF is not species specific (Bedford, 1970).

Since capacitation is reversible and since, e.g., in the guinea pig, epididymal sperm upon incubation in a defined medium release strongly antigenic components capable of decapacitation (Aonuma et al., 1973), capacitation may involve the removal of antigens from the sperm surface. Additional evidence for this is found in the experiments of Oliphant and Brackett(1973), who showed that antibodies against seminal plasma of rabbits reacted with sperm surface antigens that were not removed by repeated washing but were removed when sperm were incubated in the uterus. This removal of antigens paralleled an increase in the fertilizing capacity of sperm incubated in the uterus.

In the rabbit, the success of capacitation in the female's genital system is influenced by the hormonal status of the female. When progesterone secretion is high, capacitation is impeded in the uterus, but it can still occur in the oviduct (Bedford, 1970). Aonuma et al. (1973) found also that the amount of antigen released by sperm incubated in the uterus was influenced by the endocrine status of the female, with more antigen being released in an estrogen-dominated uterus and less being released in a progesterone-dominated uterus.

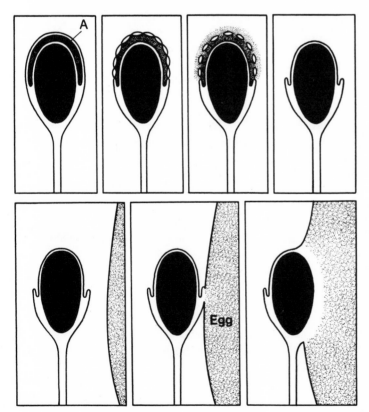

Figure 9.2 Diagram showing the pattern of the acrosome reaction (*above*), and the first steps in sperm-egg fusion (*below*). The outline of the spermatozoon represents its plasma membrane; the nucleus is solid black. A = acrosome. (From Austin, 1972; reprinted with permission of Cambridge University Press.)

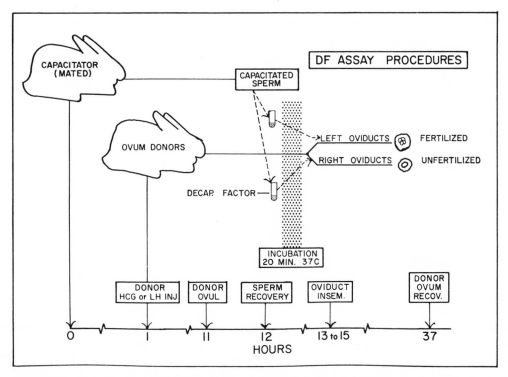

Figure 9.3 Assay for sperm decapitation factor (DF). (From Srivastava and Williams, 1970; reprinted with permission of Almqvist & Wiksell International, Stockholm, Sweden.)

Capacitation is required, in all mammalian species that have been investigated but, at least in turkeys (Howarth and Palmer, 1972) and chickens (Bakst and Howarth, 1977), the evidence indicates that capacitation may not be necessary in birds. In amphibians, a kind of capacitation occurs when the sperm traverse the jelly coat of the egg; without such passage, the sperm are incapable of fertilization.

ACROSOME REACTION

Capacitation of mammalian sperm is associated with modifications of the plasma membranes (Koehler, 1976) that facilitate the acrosome reaction. This reaction, which occurs in mammalian sperm prior to fertilization, consists of the fusion of sperm membranes covering the anterior part of the head of the sperm (Austin, 1974b), as illustrated diagrammatically in Figure 9.2.

The number of membranes or layers that surround the mammalian oocyte varies according to the species being considered. Figure 9.4 illustrates this for a number of animals. Figure 9.5 illustrates the sequence of events that occurs in fertilization of a rat oocyte.

Before sperm penetrate the membranes surrounding the oocyte, two other processes may occur: (1) attachment of the sperm to the oocyte, an attachment that is not species specific (Gwatkin, 1977); and (2) binding of the sperm to the zona pellucida (ZP) of the oocyte. This process is species specific and can be prevented by a trypsin-acrosin inhibitor (Gwatkin, 1977) and is Ca^{2+} dependent in the mouse (Heffner, 1979).

The acrosome reaction is accompanied by the release of hyaluronidase, which facilitates the penetration of the sperm through the cumulus oophorus, although in some species, e.g., cattle and sheep, the cumulus oophorus and the corona radiata are removed in the oviduct without the presence of sperm. The sperm head also contains acrosin, a trypsinlike enzyme, which is not released into the surrounding medium but remains

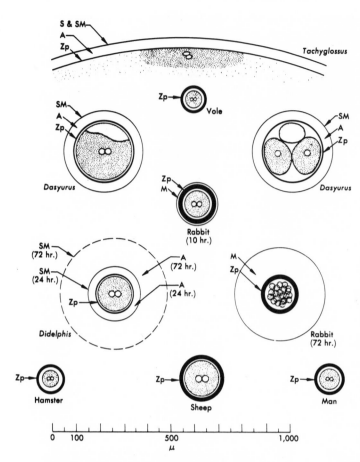

Figure 9.4 Various mammalian eggs, showing relative sizes and identity of membranes and investments. Developmental stages shown are pronuclear syngamy for each animal and, in addition, a stage of cleavage for the Australian native cat *Dasyurus* and the rabbit. Accretion of the albumin coat (A) by the egg of the opossum *Didelphis* and of the mucin coat (M) by the rabbit is indicated by the inclusion of outlines for two periods after ovulation (24 and 74 h) for *Didelphis* and (10 and 72 h) for the rabbit. Notable is the much greater thickness of the zona pellucida (Zp) in the placental mammals. S = shell, SM = shell membrane. (From Austin, 1965, as redrawn from Austin and Amoroso, 1959; reprinted with permission of C. R. Austin and Pergamon Press.)

attached to the inner acrosome membrane and locally liquifies the ZP so that the sperm can then enter the vitellus of the oocyte (Austin, 1972). Bedford and Cross (1978), however, have questioned the absolute requirement of acrosin for sperm penetration of the oocyte.

FERTILIZATION

After penetration of the ZP, the microvilli of the oocyte and sperm-plasma membrane fuse. This fusion is Ca^{2+} dependent in the hamster, guinea pig, and human (Yanagimachi, 1978b). Figure 9.2 illustrates the process of fusion diagrammatically. Reviews by Austin (1974a,b, 1978) and Epel (1978) provide details on sperm-oocyte fusion. Yanagimachi (1978a) has used fertilization of ZP-free oocytes in vitro as a model for the in vivo fertilization of oocytes with ZP. He states that there seem to be no differences with respect to the oocyte-sperm fusion between these two types

of oocytes, although the former becomes polyspermic and the latter monospermic. After the sperm make contact with the oocyte surface, they continue to make swimming motions, but after 5–15 seconds, the flagella start to beat more slowly and 15–25 seconds after making contact they become motionless (Yanagimachi, 1978a).

The microvilli of the oocyte and the sperm plasma membrane of the postacrosomal region fuse. The inner acrosomal membrane, which covers the anterior half of the sperm head, does not fuse with the oocyte plasma membrane but is incorporated into the oocyte phagocytotically (Yanagimachi, 1978a). Sperm that have not undergone the acrosome reaction cannot fuse with oocytes (Yanagimachi, 1978a).

In fishes, the egg may be surrounded by an impenetrable coating (Austin, 1978), which has special entryways for the sperm. For instance, lamprey eggs have a tuft (Figure 9.6) through which the sperm enter. Teleosts have one or more open-

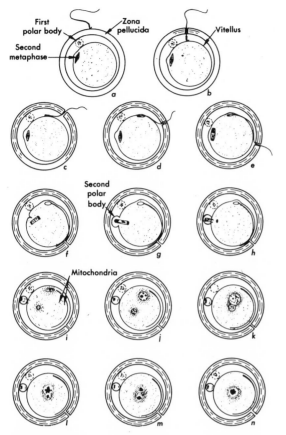

First polar body
Zona pellucida
Vitellus
Second metaphase
Second polar body
Mitochondria

Figure 9.5 Fertilization of the rat egg [oocyte]. (*a–d*) Entry of the spermatazoon. The shading in the zona pellucida denotes occurrence of the zona reaction. The outline of the first polar body is broken because, in the rat, it commonly disintegrates before ovulation. (*d–h*) Completion of second meiotic division. (*i–l*) Pronuclear development. (*m*) Reappearance of chromosome groups. (*n*) First cleavage metaphase. (From Austin, 1965, as redrawn from Austin and Bishop, 1957; reprinted with permission of Cambridge University Press.)

ings in the chorion called *micropyles*, through which the sperm enter (Figure 9.7). The micropyle may increase the motility of the sperm (as in the herring) as they come in contact with the chorion close to the micropyle (Figure 9.8), so that it appears that a chemotaxic mechanism operates in these species. The chorion itself also has a chemotaxic effect in some teleosts, as in Japanese rice fish, bitterling, fat-minnow, salmon, trout, and sturgeon (Austin, 1965). In fishes that have oocytes with a micropyle or that have oocytes not surrounded by impenetrable coats, the sperm do not show the acrosomes; however, in insects with

such oocytes, the sperm may have acrosomes (Austin, 1978).

In many species of invertebrates, as well as vertebrates, only one sperm enters the vitellus, and there is evidence that polyspermy (entry of more than one sperm into the vitellus) is detrimental for normal development. Several mechanisms contribute to the reduction or prevention of polyspermy: (1) the reduction of sperm numbers between the site of insemination and the site of fertilization; (2) the *zona reaction,* consisting of a decrease in the sperm-binding properties of the ZP (Epel and Carroll, 1975; Gwatkin, 1977; Yanagimachi, 1978a); (3) the *cortical reaction,* which consists of a fusion of granules embedded in the oocyte cortex with the oocyte plasma membrane followed by a discharge of the cortical granule contents into the space between plasma membrane and overlying surface coats (the perivitelline space) and which has cortical repercussions on the outer coats; and (4) the *vitelline reaction,* which is caused by the cortical reaction and which changes the vitelline membrane and makes it impenetrable to other sperm (Gwatkin, 1977). These reactions have been found in vertebrates and invertebrates, although the structural details may be somewhat different (Figure 9.9).

After the sperm nucleus has entered the oocyte,

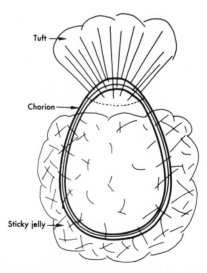

Tuft
Chorion
Sticky jelly

Figure 9.6 Egg of the lamprey *Lampetra* sp. Sperm entry occurs only through the tuft at one pole. (From Austin, 1965, as redrawn from R. A. Kille, 1960; reprinted with permission of Academic Press.)

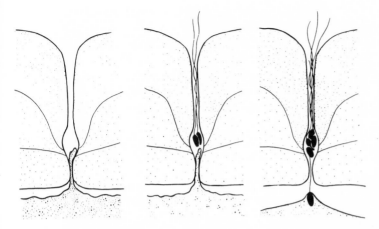

Figure 9.7 Entry of spermatozoa into the egg of the sturgeon. The micropyle traverses the three layers of chorion, and into its lower reaches projects a tongue of vitelline cytoplasm. Contact with and attachment to the cytoplasm appears to be necessary for the spermatozoon to complete its journey through the micropyle. (Redrawn from Ginsburg, 1959.)

it undergoes a series of transformations and forms the male pronucleus. The nucleus of the ovum also is changed into a female pronucleus, which eventually fuses with the male pronucleus. For ultrastructural aspects of fertilization, the reader is referred to the recent reviews by Yanagimachi (1978a) and Longo and Kunkle (1978). For biochemical and biophysical aspects of fertilization in animals, especially of invertebrates, the reader should consult the review by Epel (1978).

As Figure 9.5 shows, the meiotic division of the female germ cell, the secondary oocyte, is not

complete until the sperm penetrates the vitelline membrane, so that the ovum and second polar body separate after sperm penetration. Usually the nucleus of the sperm and of the ovum fuse so that a zygote is formed. An ovum in the strict sense of the word, therefore, usually exists only for a short time, unless the two nuclei do not fuse. The words *ovum* and *egg* as commonly used are thus, in terms of nuclear divisions, incorrect. Of course, the word *egg* also denotes an oocyte or zygote after it

Figure 9.8 Behavior of spermatozoa near the chorion of the herring (*Clupea*) egg, after removal of the vitellus. The spermatozoa are motionless until they drift onto the surface of the chorion near the micropyle. They then show vigorous motility and soon swim into and through the micropyle. (From Austin, 1965; as redrawn from Yanagimachi, 1957; reprinted with permission of the Zoological Society of Japan.)

Figure 9.9 Surface-associated events of fertilization compared between mammalian and echinoderm and amphibian embryos. *A*, Initial attachment of sperm to outer egg coats shortly after sperm contact. *B*, Beginning of cortical granule release following the sperm-egg membrane fusion. *C*, Completed alterations of egg surface and surface coats resulting in loss of sperm attachment sites on outer egg coats. (From Epel and Carroll, 1975; reprinted with permission of International Planned Parenthood Federation.)

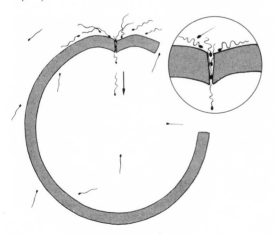

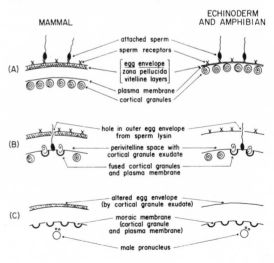

has been surrounded by egg jelly (as in amphibians) or albumen, shell membrane, and shell (as in birds).

AGING OF GAMETES

We mentioned in Chapter 2, that retention of frog oocytes in the oviduct results in a preponderance of male offspring. In *Rana pipiens* and *Xenopus laevis,* overripeness of oocytes prior to fertilization resulted in chromosomal abnormalities (i.e., lack of one or more chromosomes at cleavage) and abnormal embryos. Some abnormal embryos had no detectable chromosomal abnormalities (Witschi and Laguens, 1963). Salisbury and Hart (1970) reported that aging of the sperm of *Rana pipiens* led to the arrest of embryonic development at gastrulation. They ascribed this arrest to an alteration in the genome translation mechanism.

Aging of oocytes in mammals can be studied conveniently in the rabbit by artificial insemination at various intervals after ovulation has been induced by hCG. Austin (1967) established that a delay of 8 h increased the incidence of hypodiploidy from 26 percent (found in rabbits inseminated at the time of hCG injections) to 63 percent in 6-day-old embryos. Austin (1967) found that as little as 7 h after ovulation, about 60 percent of unfertilized eggs showed evidence of pyknosis. In the rat, delay of ovulation for 48 h by barbiturate treatment on the day of proestrus increased the incidence of chromosomal abnormalities a little more than threefold over the frequency found in controls (18 out of 390 versus 6 out of 410), decreased the number of implantation sites and the number of embryos below that of controls, and increased the number of degenerating implantation sites above that of controls (Butcher and Fugo, 1967). Delay of ovulation for 48 h dramatically increased the incidence of small embryos, and degeneration sites found at day 11 of pregnancy (Butcher et al., 1969). Kaufman (1978) has found that postovulation aging of oocytes increases the frequency of activation of oocyte development (after incubation in a medium containing hyaluronidase) and changes the frequency among different types of parthenogenetically developing oocytes. Such aging decreases the incidence of oocytes with one pronucleus with a second polar body, but increases the incidence of immediate cleavage of secondary oocytes and of oocytes in which one or two pronuclei form without extrusion of second polar body.

The effect of aging of spermatozoa in vivo has been studied in chickens (Nalbandov and Card, 1943) and turkeys (Hale, 1955). After a single insemination, the incidence of fertile eggs laid decreases with time, as expected, but the hatchability of fertile eggs also decreases. This indicates that although aged sperm may fertilize the oocytes, more embryos die as the result of fertilization by such sperm. According to Dharmarajan (1950), after fertilization by aged sperm, abnormalities occur in chicken embryos, mainly in the vascular and nervous system.

The effect of aging of spermatozoa in vitro has been studied mainly with sperm used for artificial insemination of cattle. Depending on the temperature at which the sperm are stored, there is a more or less rapid increase in the mortality of embryos as the sperm are stored longer than two days, although there is a higher mortality of embryos after use of semen used on the day of collection than of semen stored for one day (Salisbury and Hart, 1970). The initial increase in fertility and survival of embryos between the day of collection and the first day of storage may be the result of selective death of abnormal sperm (Salibury and Hart, 1970).

The findings that aging of sperm and oocytes of vertebrate species affects embryonic development unfavorably have clear implications for human reproduction, because of the lack of synchronization of insemination and ovulation. It is obvious that any technique that can be found that predicts the time of ovulation accurately can have great beneficial effects both for those who wish to become parents and for those who do not wish to do so and who cannot use or do not wish to use contraceptives or undergo sterilization.

THE *T* LOCUS

In our discussion so far, we have assumed that the genome of the sperm has no effect on their fertilizing capacity. In fact, the entire concept of Mendelian ratios is based on this assumption, but an important exception has been found in the house mouse (*Mus musculus*). On chromosome number 17 at the *T* locus there is a series of alleles, which are *T* (brachyury or short tail)-dominant to the normal (+) allele and lethal in the homozygous (*TT*) condition. The heterozygote (*T+*) has a short tail and some other skeletal abnormalities. Another *t* allele is classified in a large number of subgroups. These *t* alleles in combination with *T* result in tailless mice, but in combination with + result in normal mice. Mice homozygous for one *t* allele or heterozygous for any of the subgroup *t* alleles die early after fertilization.

In reciprocal matings of mice carrying different alleles of *T* and *t* at the *T* locus, the Mendelian ratio was distorted in a way that indicated that the sperm carrying one of the alleles had a better chance of fertilization than sperm carrying the other allele (Table 9.5). Table 9.5 also shows that for some genotypes, late mating, with respect to ovulation, tends to restore the ratio closer to the

one expected, e.g., Tt^0, Tt^2, but in others this effect is absent, e.g., Tt^3, or it makes the deviation from the expected ratio even larger, e.g., Tt^9.

Braden (1972) also showed that the *t* carried by the male had an effect, but that, in addition, there was intermale heterogeneity (the result depending on which male one selected from a population with the same *T* locus genotype), intramale heterogeneity (the result depending on the female to which the male is mated), and an effect of the genotype of the oocyte. The effect of the *T* locus alleles can be related to the fact that for each *T* locus allele tested a recognizable antigen has been found on the sperm surface (Yanagisawa et al., 1974; Bennett, 1975).

Some of the sperm of $t^{w32}/+$ mice are abnormal, but the frequency of about 8 percent is not high. Olds-Clarke and Becker (1978) observed that sperm from $t^{w32}/+$ males appeared to penetrate oocytes sooner than sperm from $+/+$ males. The t^{w32} is transmitted to about 75 percent of the

Table 9.5

Effect of time of mating in relation to ovulation on transmission ratio in $Tt\male \times + + \female$ matings

Male genotype and number	Number of progeny and transmission ratio (s)				Significance of difference (P)
	Normal mating		Late mating		
	+t:+T	s	+t:+T	s	
Tt^0 87	34: 8	0.81	27: 37	0.42	<0.001
88	90: 32	0.76	31: 43	0.42	<0.001
122	192: 60	0.76	52: 49	0.51	<0.001
124	212: 40	0.84	169:114	0.60	<0.001
Tt^3 77	129:297	0.30	71:205	0.26	<0.3
81	79: 99	0.44	60:114	0.34	<0.1
83	215:331	0.39	164:269	0.38	<0.7
Tt^9 82	61: 96	0.39	37:112	0.24	<0.01
86	53:127	0.29	23:125	0.16	<0.005
73	55: 83	0.40	39:144	0.21	<0.001
Tt^{p1} 168	163: 17	0.91	321: 43	0.88	<0.5
102	218: 11	0.95	129: 18	0.88	<0.01
177	104: 14	0.88	29: 31	0.48	<0.001
175	111: 3	0.97	78: 25	0.76	<0.001
Tt^{p2} 178	130: 77	0.63	48: 63	0.43	<0.001
103	118: 30	0.80	143: 48	0.75	<0.5
123	251:103	0.71	81: 47	0.63	<0.2
138	76: 45	0.63	33: 40	0.45	<0.025
136	66: 50	0.57	46: 74	0.38	<0.005
125	88: 73	0.55	76: 64	0.54	<0.95

From Braden, 1972; reprinted with permission of A. W. H. Braden.

offspring; thus an earlier penetration of the oocytes by t^{w32}-carrying sperm is in accordance with the transmission ratio data. The exact mechanism by which the t^{w32} gene has its effect needs to be investigated further.

GYNOGENESIS

The occurrence of gynogenesis in different species of Teleosts has been discussed in Chapter 1. Gynogenesis also has been found to occur in two salamanders, *Ambystoma jeffersonianum* and *A. laterale* in the northeastern United States. Both diploid bisexual and triploid all-female populations of these two species exist. The triploid form of *A. jeffersonianum* is called *A. platineum*; the triploid of *A. laterale* is called *A. tremblayi*. The oocytes of the triploid forms are activated by sperm of the diploid forms. Apparently the triploid *A. tremblayi* originates from hybridization of *A. laterale* and *A. jeffersonianum*, with *A. tremblayi* having two sets of *A. laterale* and one set of *A. jeffersonianum* chromosomes (Lofts, 1974).

PARTHENOGENESIS

The development from an unfertilized oocyte, or parthenogenesis, occurs in several genera of reptiles, such as *Lacerta armeniaca, L. rostambekovi, L. dahli,* and *L. unisexualis;* 11 species of Teiid lizards; 10 species in the genus *Cnemidophorus;* in *Gymnophthalmus underwoodi*; in some geckos (Cuellar, 1977b); and in the xantusid lizard, *Lepidophyma flavimaculatum.* Some of these species are diploid, e.g., *Cnemidophorus neomexicanus,* and some are triploid, e.g., *C. uniparens* (Cuellar, 1974). Skin grafting experiments with *C. velox* and *C. neomexicanus* revealed that *C. neomexicanus* was generally identical over a range of 250 km, suggesting that this species originated from a single individual. In *C. velox,* there was some rejection of skin grafts between different clones (Cuellar, 1977a). The diploid lizard *Lacerta armeniaca* is apparently the result of nondisjunction at meiosis II (see Darcey et al., 1971, for documentation).

In domestic turkeys, parthenogenesis was first discovered by Olsen in 1952 (see Olsen, 1960). It occurs occasionally in chickens (Poole and Olsen,

1958), although not as frequently as in turkeys. The incidence was increased in turkeys by a selection program, and viable poults hatched. All parthenotes that hatched and all embryos that could be sexed have always been diploid males. Of four such males that reached sexual maturity, three proved to be homozygous for the inheritance of plumage color. However, one male, from a dam heterozygous for bronze and white plumage, proved to be heterozygous because half of his offspring were white and half were bronze (Olsen and Buss, 1972). In skin transplant studies the dams accepted grafts from their parthenogenetic sons, but the sons rejected grafts from their dams (Healey et al., 1962). Cytological investigations strongly suggest that development starts with a haploid oocyte and that the diploid number of chromosomes is restored later (Darcey et al., 1971). Such restoration can occur by nuclear division at mitosis I in the absence of a corresponding cytoplasmic division or development, or the embryo can start by the fusion of two haploid cells that developed parthenogenetically. In either case, one would expect complete homozygosity. To explain the exceptional male, one needs to invoke two exceptional events: (1) crossing-over between centromere and the locus for the color in meiosis and (2) failure of the second meiotic division. If the heterozygosity was not exceptional, then the explanation for the turkey parthenotes would be either suppression of the expulsion of the second polar body or its reentry; either of these two processes would allow for crossing-over to occur. Reentry or suppression of the first polar body is not a possibility because such a mechanism would yield females.

Fowl pox vaccination has been observed to increase the incidence of parthenogenetic development in chicken and turkey eggs from hens that lay such eggs (Olsen, 1956). An outbreak of lymphomatosis also increased the incidence of parthenogenesis in chickens (Olsen, 1966).

Occasionally, parthenogenetic development occurs in mammals; the incidence can be increased by temperature shock, osmotic shock, electrical stimulation, removal of the cumulus oophorus, and by the divalent inophore A23187 (Gwatkin, 1977). All or nearly all parthenotes die early in development for reasons that are not known.

10

Care of the Embryo and Fetus

After fertilization, the zygote may be either retained by the mother (or in the case of seahorses and pipefishes in the pouch of the father) or released into the environment to take care of itself without receiving additional food, so that it lives on the yolk of the egg. However, parents may protect the embryo against predation and furnish it with heat or oxygen, or remove waste products from its environment. The main categories we will use for patterns of care of the embryo (fetus) are: (1) *oviparity,* in which the young emerge from the egg outside the dam's reproductive system and (2) *viviparity,* which is aplacental or placental. In aplacental viviparity the young are retained in their mother's reproductive system, but the embryos depend (*a*) solely on yolk reserves, (*b*) on oophagy—a form of intrauterine cannibalism, or (*c*) on placental analogues, such as uterine milk and trophonemata (to be discussed later in this chapter under Teleosts). (The term *aplacental viviparity* replaces an older term, *ovoviviparity.*) In *placental viviparity,* various forms of placentas provide for the embryo. Wourms (1977) discusses in some detail the evolutionary and ecological factors that may account for the apparent independent evolution of viviparity. We will discuss the different forms of care for the different classes.

CYCLOSTOMES

Prior to mating and spawning, lampreys build nests over sandy bottoms. They construct small walls from rocks downstream from the nest so that water runs relatively calmly over the nest. During spawning, they stir up sand to which the adhesive eggs may become attached, thus increasing their chance for survival by decreasing predation (see Breder and Rosen, 1966, for documentation on various species of lampreys). Beyond this anticipatory care of the young, there is no further care of the embryos since the parents die after spawning.

ELASMOBRANCHS

Table 10.1 shows the occurrence of viviparity and oviparity in different families, while Table 10.2 shows the gradation between oviparity and viviparity. The details of embryonic development will not be discussed here; instead we refer the reader to the excellent recent account by Wourms (1977). Table 10.3 gives a summary of the time it takes for embryonic development to be completed, i.e., until the young are hatched or born.

Aplacental species of elasmobranches have

Table 10.1

Modes of reproduction in chondrichthyan fish

Class Chondrichthyes	Suborder Squaloide
Subclass Elasmobranchii	14. Family Squalidae (dogfish sharks)—viviparous
Order Squaliformes	15. Family Pristiophoridae (saw sharks)—viviparous
Suborder Hexanchoidei	16. Family Squatinidae (angel sharks)—viviparous
1. Family Hexanchidae (cow sharks)—viviparous	Order Rajiformes
2. Family Chlamydoselachidae (frill sharks)— viviparous	Suborder Pristoidei
Suborder Heterodontoidei	17. Family Pristidae (sawfishes)—viviparous
3. Family Heterodontidae (bullhead or horn sharks)—oviparous	Suborder Rhinobatoidei
Suborder Lamnoidei	18. Family Rhinobatidae (guitarfishes)—viviparous
4. Family Odontaspidae (sand sharks)—viviparous	19. Family Rhynchobatidae (guitarfishes)— viviparous
5. Family Scapanorhynchidae (goblin sharks)— viviparous	Suborder Torpedinoidei
6. Family Lamnidae (mackerel sharks)—viviparous	20. Family Torpedinidae (electric rays)—viviparous
7. Family Cetorhinidae (basking sharks)— presumed viviparous	21. Family Narkidae (electric rays)—viviparous
8. Family Alopiidae (thresher sharks)—viviparous	22. Family Temeridae (electric rays)—viviparous
9. Family Orectolobidae (carpet or nurse sharks)— oviparous and viviparous	Suborder Rajoidei
10. Family Rhincodontidae (whale sharks)— oviparous	23. Family Rajidae (skates)—viviparous
11. Family Scyliorhinidae (cat sharks)—oviparous, one viviparous species	24. Family Arhynchobatidae (skates)—oviparous
12. Family Carcharhinidae (requiem sharks)— viviparous	25. Family Anacanthobatidae (skates)—oviparous
13. Family Sphyrnidae (hammerhead sharks)— viviparous	Suborder Myliobatoidei
	26. Family Dasyatidae (sting rays)—viviparous
	27. Family Myliobatidae (eagle rays)—viviparous
	28. Family Mobulidae (manta rays)—viviparous
	Subclass Holocephali
	Order Chimaeriformes
	29. Family Callorhynchidae (chimaera)—oviparous
	30. Family Chimaeridae (chimaera)—oviparous
	31. Family Rhinochimeridae (chimaera)—oviparous

From Wourms, 1977, as modified from Breder and Rosen, 1966, and Budker, 1971; reprinted with permission of *American Zoologist.*

different ways of caring for the embryo. When the yolk with which the embryo is provided serves as the entire source of nourishment, then the care is essentially not much different from that found in oviparous species except, of course, for the protection provided by the mother. A second method for aplacental species to obtain food during gestation is the eating of yolks or of embryos that follow the oldest or strongest embryo in the uterus, e.g., in porbeagle shark (*Lamna* sp.), the sand shark (*Odontapsis taurus*), and possibly some mackerel sharks (*Lamnidae*) and thresher sharks (*Alopiidae*).

The embryo of *Lamna* sp. absorbs its own yolk sac early and then proceeds to eat yolks of oocytes that are ovulated later. The cardiac stomach becomes distended by such yolks, and this distension gives the stomach the appearance of an enormous yolk sac. The sand shark embryos swim open-mouthed in the oviduct and appear to eat everything that is not oviduct wall. This pattern of nutrition has a double advantage for the young shark; it feeds itself, and it also receives training in finding food, so that it is born an accomplished predator (Wourms, 1977).

The aplacental viviparous sharks with placental analogues have two different strategies for providing the young with nourishment: (1) the secretion of uterine "milk," which may have an organic matter content of 13 percent and a total fat content of 8 percent (Wourms, 1977) and (2) tufts of uterine epithelium or trophonemata (long glandular appendages), which enter the embryo through the spiracles and pass into the esophagus. Nutrients secreted by the trophonemata enter the gut. This pattern of feeding, which is

Table 10.2

Changes in weight and composition of selachian embryos during development in oviparous, aplacental viviparous, and placental viviparous species

Characteristics in relation to embryo	Scyliorhinus canicula (dogfish)	Torpedo ocellata (electric ray)	Mustelus vulgaris (common smooth nursehound)	Dasyatis (Trygon) violacea (sting ray)	Mustelus laevis (nursehound)
Egg status	Oviparous	Aplacental viviparous	Aplacental viviparous	Aplacental viviparous	Viviparous
Gestation (months)	6	4	10	2	5
Weight of ovulated egg (g)	1.31	6.78	3.93	1.9	5.54
Weight of "completed" embryo (g)	2.69	13.37	60.16	118.0	189.6
Percent weight change in embryo	+105%	+97%	+1432%	+6105%	+3326%
Percent change in water in embryo	+213%	+251%	+2490%	+10412%	+5608%
Percent change ash in embryo	+292%	+157%	+2800%	+10250%	+7609%
Percent change organic substances in embryo	−21%	−23%	+356%	+1628%	+1064%
Placental type or type of nutrition	None	Milk from villi on uterine wall, into gut of embryo	Uterine secretion copious	Villi thick, milk-secreting and long into gut of embryo	Intimate yolk-sac

Modified from Cohen, 1977, based on data of Amaroso, 1960; reprinted with permission of Jack Cohen.

found in *Dasyatis violacea, Myliobatus bovina* and *Gymnura micrura* (Wourms, 1977), allows for a 17–50-fold gain in organic matter by the embryo.

Aplacental nutrient transfer is not always as efficient as just illustrated. In the electric rays, *Torpedo ocellata* and *T. marmorata,* the uterine milk is watery (1.2 percent organic matter, 0.1 percent fat), and the embryos show a net loss of 22 percent organic matter and 34 percent fat during development (Wourms, 1977).

Placental development is found within two families, the requiem sharks (Carcharhinidae) and the hammerhead sharks (Sphyrinidae), but not all genera, or species within genera, have a placental way of nutrient transfer; e.g., *Mustelus canis* has a placenta, but *Mustelus antarcticus* does not. In *Mustelus canis,* the embryo is dependent on its yolk during the first three months of

its development and then establishes a placental connection. The placenta that develops incorporates the egg capsule and consists of yolk sac and uterine epithelium, which eventually interdigitate with each other. Such interdigitation occurs also in most other species, such as, *Prionace glauca, Sphyrna tiburo*; however, in some, e.g., *Carcharhinus dussumerieri* and *C. falciformis,* there is no interdigitation, and the placenta rests on a vascularized region of the uterine wall (Wourms, 1977). In *C. falciformis,* there are five layers involved in the exchange between mother and fetus: (1) maternal endothelium, (2) maternal epithelium, (3) egg capsule, (4) fetal epithelium, and (5) fetal endothelium, although some of these layers may be reduced or lost. After hypophysectomy (AP) of *Mustelus canis* at the beginning of gestation, development of the young is not disturbed (Hoar, 1969). Placental transfer can be

Table 10.3

Duration of development in elasmobranchs from fertilization to independence of embryo from mother

Pattern and species	Duration of development (months)
Oviparous	
Skates	
Raja clavata	4.5–5.5
R. naevus	8
R. marginata	15
R. eglanteria	2¼–3
Sharks	
Heterodontis sp.	9–12
Scyliorhinus canicula	6–8
Viviparous	
Sharks	
Mustelus canis	8–12
Squalus acanthias	22–24
Rays	
Myliobatus bovina	3
Urolophus halleri	3
Dasyatis (Trygon) violacea	2

Data from Wourms, 1977.

quite efficient, as attested by the 8- to 10-fold increase in organic matter in the embryo during its development (Wourms, 1977).

TELEOSTS

After external fertilization, the zygotes may simply be abandoned (e.g., herrings), or they may be protected, or protected and nourished.

Protective Strategies

Protective strategies include nest building, mouth breeding, skin incubation, and incubation in other animals.

Nest building. Nests may be made in a slit in a rock or an empty shell, as found in many littoral fishes, e.g., the Blennidae and Cottidae. The fish *Pholis gunnellus* may lay her eggs in such a hiding place, but if she cannot find one, she will coil around the big ball of fertilized eggs, which, of course, contain developing embryos, and guard them in this manner. Nests may also be hollowed out in gravel or stone, as is done by trout and salmon, or they may be made of plant material

that is glued together by special secretions of the kidney (see Chapter 13), as is done by the three-spined stickleback (*Gasterosteus aculeatus*). Floating nests may be made from air bubbles, as in *Macropodus* spp. Bertin (1958) gives a rather complete account of the various nests used by fishes. In some species, the parents not only build a nest, but they may also fan the developing embryos to facilitate gaseous exchange and removal of wastes, e.g., *Crenilabrus ocellatus* (van Tienhoven, 1968).

Mouth breeding. In a number of Cichlidae, the females incubate the eggs in their mouth (in *Tilapia macrocephala*, the male incubates the eggs in his mouth). Females also incubate eggs in some species of Cyprinodontidae, Osteoglossidae, Osphrominidae and Siluridae. In representatives of the Ariidae, Cyclopteridae, Apogonidae, Cichlidae, Opisthognatidae, and Belontiidae, the male incubates the eggs. Mouth breeder parasitism, similar to the nest parasitism of cuckoos, is found among cichlid fishes. The female *Haplochromis polystigma* cares for her own offspring and those of a sympatric species, *H. chrynosotus*. Two other species, *H. macrostoma* and *Serranochromis robustus*, also guard broods consisting of a mixture of their own and another species (Ribbink, 1977).

Skin incubation. In the banjo catfishes (Bunocephalidae), the fertilized eggs are attached to the skin in various ways, although it is not clear whether this is only for protection or also for nutrition in some cases.

Incubation in another animal. This is found in the bitterling (*Rhodeus amarus*). In this species, fertilization occurs within a mussel, with the young growing inside the mussel's gill chamber. It is also found in *Aulichthys japonicus*, in which the eggs are incubated in the peribranchial cavity of ascidia and in *Careproctus* sp., which uses the gill chamber of the crab, *Paralithodes camtschatica*, as a place of incubation (Breder and Rosen, 1966).

Protective and Nutritional Strategies

Strategies that include protection and nourishment are feeding on skin secretions, pouch breeding, and nutrition from pseudoplacentas.

Feeding on skin secretions. Under the influence of prolactin, the male and female secrete a mucus that the young use as nourishment, e.g., in *Symphysodon discus* and *Aequidens latifrons*. In both of these species, prolactin administration results in the parents fanning the eggs in the nest (see van Tienhoven, 1968, for documentation).

Pouch breeding. In pipe fishes and seahorses, the female deposits her eggs in the pouch of the male; indeed, several females may lay eggs in the brood pouch of one male. The brood pouch maintains an environment of different composition than sea water, which is harmful for the embryos of *Hippocampus erectus,* according to Linton and Soloff (1964). Hypophysectomy of the seahorse (*Hippocampus hippocampus*) leads to an increased frequency of malformations and premature expulsions from the pouch; however, some embryos are unaffected, and these are "born" at the normal time (Boisseau, 1964). The proliferation of the epithelial lining of the brood pouch of seahorses, which occurs when he is incubating the young, is under the apparent control of prolactin; mammalian prolactin administration results in stimulation of this epithelium in intact and in ₳P seahorses (Clarke and Bern, 1980).

After internal fertilization, the zygotes may be discharged and abandoned, or they may be taken care of in the body of the female, depending upon whether they are oviparous or viviparous. In oviparous species, the young emerge from the egg outside the reproductive system and live on the yolk provided in the yolk sac and on nutrients from the environment, as in *Trachycorystes* sp., in which the zygotes are discharged shortly after fertilization, and in *Sebastodes* sp., in which some development takes place inside the ovarian cavity. Since emergence occurs outside the body of the mother, we still classify this pattern as oviparous. Since *Sebastodes* sp. carries about 2 million (Moser, 1967) embryos, there is a problem of providing sufficient oxygen and nutrients. These are provided, however, by a specialized double vascularization, in which anterior and posterior ovarian arteries are continuous with each other and thus form an arterial loop that provides an adequate oxygen supply (Moser, 1967).

In viviparity, which in teleosts is aplacental,

the embryo may be dependent mainly or solely on yolk reserves, e.g., in Poeciliidae, such as *Xiphophorus helleri,* or the embryo may use dead embryos, dead sperm, and/or desquamated cells which are absorbed through a hypertrophied yolk sac, as in the sea perch (*Cymatogaster aggregata*). Gaseous exchange occurs through spoon-shaped enlargements of the dorsal fin.

Nutrition from pseudoplacentas. The embryo may be served by several types of pseudoplacentas, four of which are described briefly. (1) Portal circulation of the yolk sac may serve principally for gas exchange, since the yolk seems to provide sufficient nourishment, e.g., in the guppy (*Poecilia reticulata*) and in *Heterandria formosa*; (2) The surface of the follicular wall contains villi, which are in contact with the embryo's ectoderm. The ectoderm is separated from the wall of the gut by a yolk sac space, and the gut is enormously expanded to form the belly sac, which, presumably, serves for absorption of nutrients. (3) Trophotaeniae, which are found in, e.g., the Goodeidae, are fetal processes that increase the absorbing surface and so enhance the absorption of nutrients from the follicular fluid. (4) Trophonemata similar to those described for elasmobranchs are found in the Jenynsiidae. In teleosts these are extensions from the ovary, but in elasmobranchs, they are extensions from the uterus, which enter the gills and pharynx of the young.

Not much is known about the endocrine influences on gestation in fishes. It is impossible to study the effect of ovariectomy, because gestation is in the ovary. Hypophysectomy does not affect embryonic mortality in *Poecilia latipinna* and increases it in *Gambusia* sp. (Hoar, 1969). For an excellent review and summary of the different ways in which various species care for the embryo, see Breder and Rosen (1966).

AMPHIBIANS

Among the Anura, as mentioned in Chapter 9, there are two species with internal fertilization. In *Nectophrynoides vivipara,* more than 100 embryos are present in the viviparous female. The embryos have long slim tails, which "function as so many pipe lines which bring oxygen to the larvae kept away from the uterine wall by the bodies of their brothers and sisters" (Noble, 1954).

In one of the Urodeles (a salamander, *Salamandra atra*) with internal fertilization, the embryos use their own yolk, but also practice oophagy and hemophagy (ingestion of blood present in the uterus as a result of hemorrhages in the uterus). Only one embryo out of a total of 40–60 ovulated fertilized oocytes survives, and it eats its other sibs (Amoroso, 1952).

In Anura with external fertilization, the young are protected and nourished in some interesting ways. In the Surinam frog (*Pipa pipa*) embryos are incubated in chambers or a pouch on the back of the mother and the marsupial toad (*Gastrotheca marsupiata*), for example, the larvae of *Pipa pipa* have very vascular, large tails, which act as a placenta by making contact with the vascular lining of the chambers. In the marsupial toad, the larval gills have vascular bell-shaped extensions, which spread out under the blood vessels of the pouch (Amoroso, 1952). The male Chilean toad (*Rhinoderma darwinii*) incubates the eggs in two vocal pouches lateral to the esophagus, and the young frogs leap out of the father's mouth when they have sufficiently developed. The female Australian frog, *Rheobatrachus silus*, swallows the recently fertilized eggs and incubates them in her stomach where they are protected, although they live on the yolk nutrients (Corben et al., 1974), until they are mature enough, when they leave the mother through the mouth. It is not understood how the female's digestive juice is prevented from digesting the eggs and embryos. The male midwife toad (*Alytes obstetricans*) catches the eggs which are laid in strings and then twists them around his legs. He carries these eggs for several weeks until the larvae are ready to hatch and he then delivers them to a pond, where there are no other tadpoles present (Goin and Goin, 1962).

REPTILES

Although most reptiles are oviparous, nearly half of the extant families contain some live–bearing species, and five of these families are completely viviparous (Blackburn, 1981). Viviparity is not necessarily confined to living forms, however; some fossil specimens of *Ichthyosaurus,* an extinct marine reptile of the Mesozoic, contain skeletons of unborn young in the body cavity (Romer, 1966). Reptilian viviparity is often the aplacental type, although representatives of at least nine families of lizards and snakes are reported to form yolk sac or chorio–allantoic placentas. According to Blackburn (1981), the minimum number of times viviparity could have evolved in reptiles is 75 (compared to 10 times in the Chondrichthyes, 12 times in the Osteichthyes, 4 times in Amphibia, and once in Mammalia). In lizards, viviparity has evolved at least 30 separate times; in snakes, at least 17 times, and in the amphisbaenids, probably only once. All crocodiles, turtles, and the tuatara are oviparous.

Care of the embryos or young is not common in reptiles, with some exceptions, however. The skinks of the genus *Eumeces* lay their eggs and protect them, and also warm them by basking in the sun and then coiling around the eggs to transfer heat to them (Goin and Goin, 1962; H. E. Evans, personal communication). Some pythons, e.g., *Python reticulatus,* coil themselves around the eggs and brood them for about 100 days, leaving them only to eat, drink, and molt (Honegger, 1975). Cobras and the American mudsnakes (*Farancia abacura*) also guard their eggs (Goin and Goin, 1962). The American alligator guards the nest and, when the young hatch and make a grunting noise, the mother may help them out of the nest.

Among reptiles with internal fertilization, there are nine families that have placentas. Amoroso (1952) has classified these placentas as yolk-sac placentas and chorio-allantoic placentas.

Yolk-sac placenta. This type of placenta is found in some Australian lizards, in the Italian lizard (*Chalcides tridactylus*), and in the garter snake (*Thamnophis sirtalis*). Fetal tissue does not invade the maternal tissue, although the omphalochorionic membrane and endometrium may show interlocking folds. The yolk-sac placenta becomes isolated as the extra-embryonic mesoderm develops in a manner peculiar to placental reptiles. The chorio-allantoic region of the placenta is simple in structure, and the fetal and maternal capillaries remain separated by layers of uterine and chorionic epithelium and a thin shell membrane. In lizards, the omphalo placenta, once formed, regresses gradually and disappears as the allantois expands around the yolk. In snakes, the yolk-sac placenta does not increase in size during development. The allantois spreads beneath the isolated omphalopleure and adheres to it (Hoffman, 1970), as illustrated in Figure 10.1.

Chorio-Allantoic Placentas. These are of three types. In one type, maternal circulation and fetal circulation may be close to each other as the result of reductions in uterine and chorio-allantoic tissue. This type is found in the skink, *Chalcides ocellatus,* and the Australian snakes, *Denisonia superba* and *D. sata.* In *Denisonia,* the allantoic membrane fuses with the isolated yolk endoderm of the yolk-sac placenta to form a chorio-omphalo allantoic membrane, as in *Enhydrina* and *Hydrophis.* In garter snakes, the allantois fuses with the yolk-sac placenta at its margins only. This vascularization of the yolk-sac placenta is (indirectly) allantoic. Na$^+$ and glycine pass to the fetus from the mother. The yolk-sac placenta apparently allows glycine to pass through it, whereas its permeability to P and Fe is low. The shell membrane seems to function as a dialyzing membrane.

In species of skinks of the genus *Lygosoma,* maternal capillaries bulge at the surface of folds of the uterus. Between these folds there are grooves lined with a glandular epithelium. Enlarged epithelial cells of the chorion interlock with the uterine folds, and the cytoplasmic processes of the chorionic epithelium penetrate between uterine epithelial cells, thus providing a relatively intimate contact between fetal and maternal tissues.

In the skink, *Chalcides tridactylus,* the folded maternal tissues overlie the main longitudinal uterine blood vessels. These folds indent the fetal placenta. The cuboidal uterine epithelial cells and

Figure 10.1 Diagrammatic view of cross sections through the chorionic vesicle [of the garter snake] at various developmental stages to illustrate the formation of the omphaloplacenta. *A,* Zehr stage 15–20. *B,* Zehr stage 24. *C,* Zehr stage 27 and after. (From Hoffman, 1970; reprinted with permission of Alan R. Liss, Inc.)

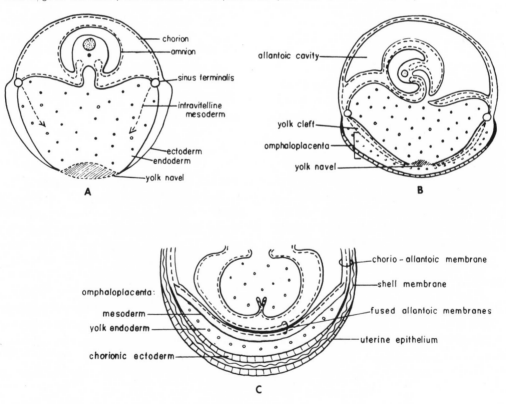

ectodermal chorion cells are enlarged, suggesting increased activity.

BIRDS

All birds are oviparous, and the eggs need to be warmed and guarded. Parents or substitute parents incubate the eggs by sitting on them and transferring heat from their body to the eggs. In many species, such an exchange is enhanced by the formation of a so-called incubation patch of the ventral skin. This usually consists of a bare skin. The feathers are either permanently absent, as in doves, or are pulled out by the parent, as in water fowl, or fall out through hormonal action, as in most birds. In addition, there is usually a thickening of the epidermis, and an edema and increased vascularization of the skin.

Incubation Patterns

Parental Incubation. This may be carried out by one parent or by both. When both sexes incubate the eggs, each develops an incubation patch, as in the Podicipediformes, Procellariiformes, Columbiformes, Piciformes, some of the Charadriiformes, Gruiformes, Falconiformes, and Passeriformes (Drent, 1975). When the female incubates, she develops an incubation patch, as in the Galliformes, Strigiformes, most Falconiformes, Apiclioformes, and some Passeriformes (Drent, 1975). When the male incubates, he develops an incubation patch, as in Phalaropes, Jacanas, some sandpipers, the Turmcidae, and the Tinamiformes (Drent, 1975). Not all species develop an incubation patch, however—for example, Cassin's auklet (*Ptychoramphus aleutica*) does not—and not all males develop an incubation patch when they are incubating, as, for example, bank swallows (*Riparia riparia*) (Drent, 1975).

Communal nesting. In this form of incubation, several females all lay their eggs in one nest (as in rheas, tinamous, anis, and ostriches). In the groove-billed anis (*Crotophaga sulcirostus*), both males and females help incubate the eggs (Vehrencamp, 1977).

Nest parasitism. In this incubation pattern, a species other than the parents incubate the eggs. Cuckoos and cowbirds are such nest parasites.

The eggs that are laid are strikingly similar in color and shape to the egg of the host, but the eggs of the parasite are larger. Wickler (1968) has excellent color photographs showing these similarities. Different females within a species of nest parasites may lay different colored eggs. Females within a species that lay similar-colored eggs belong to the same gen (plural gentes). Intriguing questions are: How does the parasite female "know" the color of the egg she is going to lay? One may propose that she may "know" which host species' nest to lay her eggs in because she is imprinted at the time of hatching on the nest and egg shell colors of the host's eggs and the egg from which she hatched. Then the question arises whether the male with which the mother mated influences the egg shell color of his daughter. It is possible that males and females from the same gen mate with each other only and not with those of other genste (Baker, 1942).

Incubation without nest sitting. The following account is mainly taken from Welty (1962). The Megapodiidae of Indonesia and Australia use the heat of the sun, the heat from rotting leaves, and a combination of both to provide heat for incubation. The maleo (*Megacephalon maleo*) of Celebes lays her eggs (after a migration of as long as 32 km) in the black sand of the ocean beach or in warm soil next to hot springs or volcanic steam fissures. Females may lay eggs in communal nests.

The brush turkey (*Alectura lathami*) piles up rotting vegetation to provide heat. The mallee fowl (*Leipoa ocellata*) of the semi-desert region of Australia uses a combination of rotting vegetation and heat from the sun (which is maximized by the rearranging of the sand by the male) to incubate the eggs. The Egyptian plover (*Aegyptius pluvialis*) uses a combination of sitting on the eggs during the night and the heat of the sand in which the eggs are buried to provide the necessary heat.

Physiology and Endocrinology of Embryo Care

The physiology and especially the endocrinology of nest building or incubation activity have been studied in most detail in canaries (*Serinus canarius*), ring doves (*Streptopelia risoria*), chickens (*Gallus domesticus*), and turkeys (*Meleagris gallopavo*). In canaries, estrogen is the principal hormone that results in nest building. For defeathering and increased vascularity of the incubation patch, a synergistic interaction between prolactin and estrogen is necessary for optimal results. Estrogen, probably in conjunction with progesterone, increases the sensitivity of the skin receptors of the incubation patch to tactile stimuli (Hinde, 1973); however, the building of a nest of normal size and composition or the lack of interest in nest building that the females normally show as egg laying approaches (Kern and Bushra, 1980) do not depend on the sensitivity of these receptors. In some other species there is evidence that estrogen and prolactin induce the development of the incubation patch, e.g., starlings (*Sturnus vulgaris*), red-winged blackbirds (*Agelaius phoeniceus*), English sparrows (*Passer domesticus*), and California quail (*Lophortyx californicus*), although the details of the action of these hormones may vary from species to species (Drent, 1975). Interestingly, in two species of phalaropes, *Phalaropus tricolor* and *P. lobatus,* in which the males are the only incubators, the incubation patch can be induced by a combination of testosterone and prolactin, but not by estrogen and prolactin. Induction of incubation behavior has been accomplished in chickens by the injection of prolactin (Wood-Gush, 1971). The studies by Lehrman and his colleagues carried out with ring doves have shown that progesterone induces incubation behavior (Lehrman, 1965; Komisaruk, 1967; Cheng, 1975) and that incubation behavior itself may then enhance prolactin secretion, which can maintain, but not initiate, incubation behavior (Lehrman, 1965).

The incubation-inducing effect of progesterone is facilitated by previous experience with nest-building phases of behavior (Michel, 1977). Early studies in the endocrine control of incubation behavior of chickens indicated that prolactin was the principal hormone that controlled the onset of this behavior. The evidence for this was the higher concentration of prolactin in the pituitaries of incubating or broody hens than in nonbroody hens. Furthermore, injections of mammalian prolactin preparations, which, in retrospect were probably not very pure (Proudman and Opel, 1981), induced incubation behavior (Lehrman, 1961). With the use of a sensitive radioimmunoassay for prolactin, Sharp et al. (1979) found that plasma prolactin concentrations were about 23 percent higher in incubating than in nonincubating bantam* hens, whereas plasma progesterone and LH concentrations were lower in the incubating than in nonincubating hens. The data available for the relationship between plasma prolactin concentrations and incubation behavior in turkeys vary considerably. Etches et al. (1979) found no significant difference in plasma prolactin concentration between broody and nonbroody hens in the same period after the start of egg laying. They observed that broodiness was always preceded by an increase in plasma prolactin concentration, but that such an increase was not always followed by broodiness. Burke and Dennison (1980) reported that in 6 of 7 hens that became broody (out of a total of 10 hens) the prolactin concentration increased and, on the average, reached a concentration of 500 ng/ml on the first day of broodiness and a peak concentration of about 1200 ng/ml. In the three hens that did not become broody, the peak concentrations of plasma prolactin, in the same period after the first ovulation as for the broody hens, were between about 500 ng/ml and 800 ng/ml for two hens and < 100 ng/ml for the third hen. For two of the three hens, as in the finding of Etches et al. (1979), an increase in plasma prolactin did not predict the onset of incubation behavior. Burke and Papkoff (1980), El Halawani et al. (1980), and Proudman and Opel (1981) have reported four- to ninefold higher plasma prolactin concentrations in broody than in nonbroody hens. Such a finding, of course, does not indicate a causal relationship. El Halawani et al. (1980) showed that not making nests available to incubating turkey hens resulted in a precipitous decrease in plasma prolactin concentration. The concentration of this hormone increased again after the hens were returned to the nest, suggesting that nesting behavior or exposure to a nest results in prolactin secretion. It remains to be in-

---

*By selection for high egg production and against broodiness, the broody behavior in commercial strains of chickens has been essentially eliminated. This necessitates the use of varieties of chickens, such as bantams, in which broodiness still occurs.

vestigated whether administration of turkey prolactin induced incubation behavior and, if so, under what circumstances.

MAMMALS

Prototheria

The duckbill platypus (*Ornithorhynchus anatinus*) incubates her eggs in a nest, and the echidna (*Tachyglossus aculeata*) incubates her eggs (after a sojourn in the uterus estimated to be about 12–28 d) in a pouch. In both cases the young lives on the yolk provided at ovulation and the albumen laid down in the oviduct; the mother provides heat and protection.

Metatheria

Examples of the reproductive cycles of different marsupials are presented in Figure 10.2. The marsupial oocyte is surrounded by a cell membrane and has more yolk than the oocyte of placental mammals. The zygote gives rise to a laminar blastocyst and most of the dividing cells of the embryo are parts of the developing germ layers, whereas in placental animals an invasive trophoblast develops (Parker, 1977).

The marsupial embryo takes up nutrients secreted by the endometrium during the preimplantation phase, which lasts more than half of the gestation period (Parker, 1977). There is no intimate contact between maternal and fetal tissues except in the bandicoot (*Isoodon macrourus*). This intimate contact between maternal and fetal tissues results in a more efficient transport of nutrients and waste products, and it is correlated with a shorter gestation than in related species that lack the intimate contact. Tyndale-Biscoe (1973) has speculated that the shorter gestation is the result either of an allograft rejection or of a response to avoid such a rejection. As we shall see in Chapter 12, in the Eutheria, a mechanism has developed that prevents the immunological

Figure 10.2 The reproductive cycles (▲–▲), gestation lengths (▲–■), and neonatal weights of 11 marsupials compared by reference to the day of birth. Subsequent lactation suppresses the reproductive cycle at approximately the stage reached at parturition, and removal of the pouch young (RPY) releases it from inhibition, so that the suppressed events are completed. The successive stages of the cycle are represented in the horizontal bars as follows: preluteal, blank; luteal, horizontal hatch; postluteal/proestrous, stipple. (From Tyndale-Biscoe, 1973; reprinted with permission of Edward Arnold Publishers.)

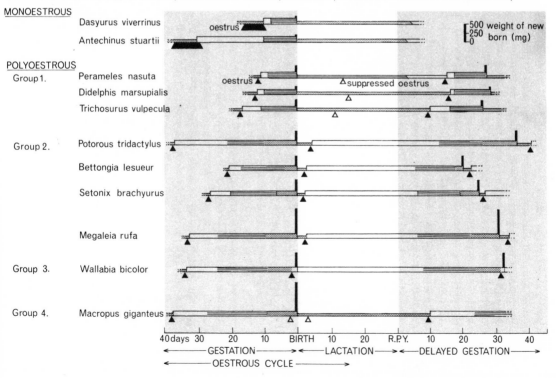

allograft rejection and so allows for a longer sustained pregnancy and more development prior to birth. Marsupials do not possess such a mechanism that prevents allograft rejection, and the more intimate contact of maternal and fetal tissues in the bandicoot than in other marsupials leads to earlier rejection and thus birth of the young. Figure 10.3 illustrates the earlier stages of development of marsupial embryos, and Figure 10.4 illustrates the two types of placentas found in marsupials, i.e., the yolk-sac and the yolk-sac allantoic placenta.

In Chapter 8, we discussed the delay in implantation of the blastocyst in several marsupials while there is a young in the pouch that is nursing frequently. Table 10.4 lists species in which implantation is delayed and the blastocyst undergoes a diapause during which either all development stops or the blastocyst grows at a very slow

rate. In the quokka (*Setonix brachyurus*), the tammar wallaby (*Macropus eugenii*), and the red kangaroo (*M. rufus*), removal of the pouch young results in histological changes in the embryo followed by changes in the endometrium during the luteal phase, followed by divisions and enlargement of CL cells (Tyndale-Biscoe, 1973). Seven days after removal of the pouch young of the tammar wallaby, a reduction in prolactin secretion, a reactivation of the CL, an increase in endometrial secretion, and reactivation of the blastocyst have been observed to occur in the sequence listed (Tyndale-Biscoe, 1973). In the red-necked wallaby (*M. rufogriseus banksianus*), the same sequence, reactivation of the CL, increase in endometrial secretion, and reactivation of the blastocyst has been found.

Removal of the CL before day 2 of gestation results in a prolongation of the diapause, whereas

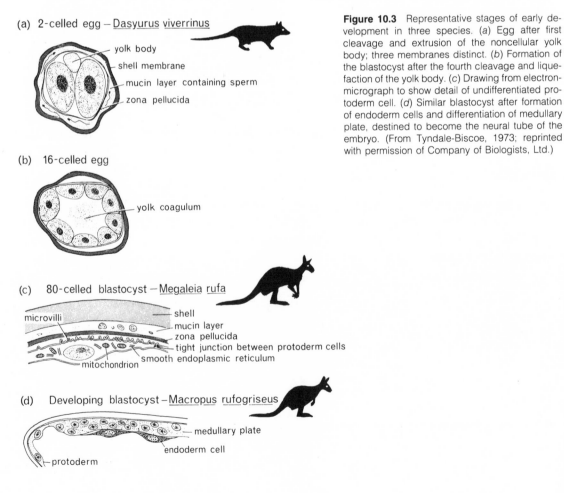

(a) 2-celled egg – Dasyurus viverrinus

yolk body
shell membrane
mucin layer containing sperm
zona pellucida

(b) 16-celled egg

yolk coagulum

(c) 80-celled blastocyst – Megaleia rufa

microvilli
shell
mucin layer
zona pellucida
tight junction between protoderm cells
smooth endoplasmic reticulum
mitochondrion

(d) Developing blastocyst – Macropus rufogriseus

medullary plate
endoderm cell
protoderm

Figure 10.3 Representative stages of early development in three species. (*a*) Egg after first cleavage and extrusion of the noncellular yolk body; three membranes distinct. (*b*) Formation of the blastocyst after the fourth cleavage and liquefaction of the yolk body. (*c*) Drawing from electronmicrograph to show detail of undifferentiated protoderm cell. (*d*) Similar blastocyst after formation of endoderm cells and differentiation of medullary plate, destined to become the neural tube of the embryo. (From Tyndale-Biscoe, 1973; reprinted with permission of Company of Biologists, Ltd.)

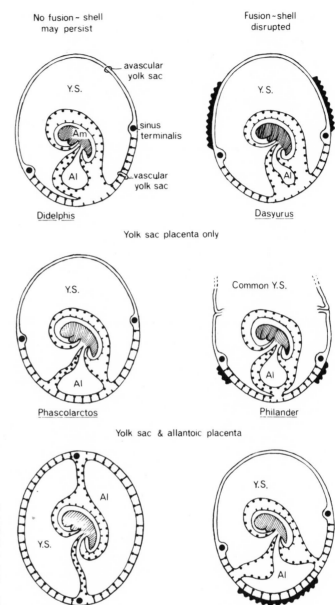

No fusion ~ shell may persist

Fusion ~ shell disrupted

avascular yolk sac

Y.S.

Am

sinus terminalis

Al

vascular yolk sac

Didelphis

Y. S.

Al

Dasyurus

Yolk sac placenta only

Y.S.

Al

Phascolarctos

Common Y.S.

Al

Philander

Yolk sac & allantoic placenta

Al

Y.S.

Tachyglossus

Y.S.

Al

Perameles

Figure 10.4 Disposition of the fetal membranes in five marsupials and in the monotreme egg, after laying. The several patterns differ according to the proportion of the chorion that is vascularized, and hence respiratory; according to whether the allantois participates in this; and according to the region, if any, of the chorion that invades the uterine epithelium. Al = Allantois, Am = Amnion, Y.S. = yolk-sac. (From Tyndale-Biscoe, 1973; reprinted with permission of Edward Arnold Publishers.)

removal of the CL between day 3 and 6 does not prevent the initiation of embryonic development but does result in the collapse of the embryo at the vesicle stage; removal of the CL after day 6 has no apparent effect on development, and the young are born at term.

Progesterone injections can initiate the development of the blastocyst in lactating kangaroos and tammar wallabies and in anestrous wallabies;

continued progesterone injections can maintain such an embryo for as long as ten days. By blastocyst transfer, it has been shown that a blastocyst in diapause can resume development when transferred to the uterus of a recipient up to 6 days more advanced in development than the recipient, but a blastocyst 2 days more advanced than the recipient will not develop. Transfer of tammar wallaby blastocysts into a recipient

Table 10.4

Metatheria in which embryonic diapause has been found

Common name	Scientific name
Potoroo	*Potorous tridactylus*
Boodie	*Bettongia lesueur*
Brush-tailed bettong	*B. penicillata*
Rat kangaroo	*B. gaimardi*
Rufous rat kangaroo	*Aepyprymnus rufescens*
Tammar wallaby	*Macropus eugenii*
Red-necked wallaby	*M. rufogriseus*
Eastern grey kangaroo	*M. giganteus*
Parma wallaby	*M. parma*
Whiptail wallaby	*M. parryi*
Euro	*M. robustus*
Western black-gloved wallaby	*M. irma*
Agile wallaby	*M. agilis*
Red kangaroo	*M. rufus*
Swamp wallaby	*Wallabia bicolor*
Western hare wallaby	*Lagorchestes hirsutus*
Banded hare wallaby	*Lagostrophus fasciatus*
Red-necked pademelon	*Thylogale thetis*
Tasmanian pademelon	*T. billardieri*
Plain rock wallaby	*Petrogale penicillata inornata*
Yellow-footed rock wallaby	*P. xanthopus*
Quokka	*Setonix brachyurus*
Noolbenger (honey possum)	*Tarsipes spencerae*
Mundarda (pigmy possum)	*Cercartetus concinnus*
Pigmy glider	*Acrobates pygmaeus*

Modified from tables of Renfree and Calaby, 1981.

ovariectomized on day 8 showed that one-day-old as well as 8-day-old blastocysts developed in intact and ovariectomized females. This suggests that the blastocyst does not require a direct stimulation by the CL. Tyndale-Biscoe (1973) concluded, on the basis of various experiments, that "embryonic survival in these species beyond the blastocyst stage is completely dependent upon the onset of the luteal phase in the uterus, and that the corpus luteum provides a secretion which synchronizes the two phenomena." From these results it appears that the CL is necessary to induce uterine changes that are required for the development of the blastocyst, but that once these changes are induced, the CL is dispensable.

Eutheria

Ovum Transport.* In most eutherian species, unfertilized eggs (oocytes) and fertilized eggs (zygotes) are transported, mainly by ciliary action (Dukelow and Riegle, 1974), through the oviduct into the uterus after delays at the isthmo-ampullar and the tubo-uterine junctions. The duration of tubal transport for different species is summarized in Table 10.5. Interestingly, for the three genera of bats in which the zygote remains in the oviduct for an extended time, there is a marked accentuation of secretory activity, after ovulation, in the oviduct containing the zygote or oocyte. This secretory activity may be a special adaptation for providing the zygote with a supportive environment (Wimsatt, 1975). In the mare (*Equus caballus*), the long-tongued bat (*Glossophaga soricina*), and possibly in the little bulldog bat (*Noctilio albiventris*), zygotes are transported to the uterus, but unfertilized oocytes are retained in the oviduct (Rasweiler, 1977). Why these three species differ from most other mammals and what mechanism is involved in the retention of the oocytes in the oviduct is unclear.

The incidence of oocytes that remain unfertilized is high in the American opossum (*Didelphis marsupialis*), which ovulates up to 56 oocytes of which 30 percent are not fertilized, and in the plains viscacha (*Lagostomus maximus*), which may ovulate as many as 800 oocytes of which 96 to 99 percent remain unfertilized (Wimsatt, 1975). The mechanisms that account for these low rates of fertilization are unknown. Is there a deficit of capacitated sperm, or are the oocytes defective in some manner so that fertilization cannot occur?

Ovum transport through the oviduct can be affected by hormonal treatments, the effect depending on the species and the dose of hormone used, as indicated in Table 10.6. In the rabbit, progestins seem to decrease the rate of ovum transport during the first 2.5 days of pseudopregnancy (Dukelow and Riegle, 1974). In spite of experiments reported in Table 10.6, it is not clear whether the ovary is required for normal ovum transport. For instance, ovariectomy of rats 7–20

*Ovum here means either oocyte or zygote.

Table 10.5

Chronology of preimplantation egg transport and development

Species	Tubal transport (days)	Development stage entering uterus	Implantation (days post fertilization)	Development stage at implantation
Marsupials				
Opossum (*Didelphis*)	1	1 cell, pronuclear	None, *strictu sensu*	—
Native cat (*Dasyurus*)	1 (?)	1 cell, pronuclear	?	Probably late primitive streak stage
Insectivores				
Elephant shrew (*Elephantulus myurus*)	1–2 (?)	1-, 2- and 4-cell stages	?	Early blastocyst
Tree shrew (*Urogale everetti*)	?	Unilaminar blastocyst	?	Bilaminar blastocyst
Tree shrew (*Tupaia longipes*)	5–6 (est.)	Bilaminar blastocyst	?	Bilaminar blastocyst
Shrew (*Sorex* sp., *Blarina brevicauda*)	?	Small blastocyst	?	Bilaminar blastocyst
Chiroptera				
Vampire bat (*Desmodus*)	16+	Blastocyst, some entoderm differentiated	not less than 16 days	Blastocyst, bilaminar partially
Long-tongued bat (*Glossophaga*)	12–14	Blastocyst	15	Blastocyst, bilaminar partially
Short-tailed bat (*Carollia*)	13–16	Blastocyst	?	?
Little brown bat (*Myotis lucifugus*)	?	Morula	not less than 10 days	Blastocyst, bilaminar
Free-tailed bat (*Tadarida brasiliensis*)	?	Advanced morula	?	Blastocyst, bilaminar
Primates				
Rhesus monkey (*Macaca*)	4	16 cells	8–9	Blastocyst
Baboon (*Papio* sp.)	4–5	32 ± cells	8–9	Blastocyst
Man (*Homo*)	3–4	?	6–7	Blastocyst

Table 10.5—*Continued*

Species	Tubal transport (days)	Development stage entering uterus	Implantation (days post fertilization)	Development stage at implantation
Rodents				
Rabbit (*Oryctolagus cuniculus*)	3–3½	20+ cell morula	7	Bilaminar blastocyst
Mouse (*Mus musculus*)	3–3½	16+ cell morula	6	Blastocyst, some entoderm
Rat (*Rattus norvegicus*)	4	Morula	6	Blastocyst, some entoderm
Hamster (*Cricetus auratus*)	3½	4–8 cells	4⅓	Blastocyst, some entoderm
Guinea pig (*Cavia porcellus*)	3½	8 cells	6	Unilaminar blastocyst
Mt. viscacha (*Lagidium*)	?	?	less than 11	Blastocyst
Plains viscacha (*Lagostomus*)	3–4	1–2 cells	?	Blastocyst
Carnivores				
Dog (*Canis familiaris*)	4–5	16 cell to young blastocyst	11–12*	Bilaminar blastocyst, 2500 μm diam.
Cat (*Felis catus*)	6–7	28–30 cell morula	13–14	Bilaminar blastocyst
Ferret (*Mustela putorius*)	5–6	32+ cells	12±	Bilaminar blastocyst
Ungulates				
Horse (*Equus caballus*)	?	?	49–63	Elongated chorionic vesicle with allantois
Pig (*Sus scrofa*)	2½–3¾	3–8+ cells	11–20 (progressive)	Elongated chorionic vesicle with developed allantois
Sheep (*Ovis aries*)	2½–2¾	?	15–17 (progressive)	15–19 cm chorionic vesicle with allantois
Cow (*Bos taurus*)	4–4½	8–16 cells	40+ (progressive)	Elongated chorionic vesicle with allantois

*Elapsed time from end of vaginal cornification (within 1–2 days of ovulation).
From Wimsatt, 1975; reprinted with permission of *Biology of Reproduction*.

Table 10.6

Effect of estradiolcyclopentylpropionate on ovum transport
in different laboratory animals

| Species | Dose required for: | | | |
	Accelerated transport (µg)	Retention in oviduct (µg)	Interruption of pregnancy in 80% of animals (µg)	Time eggs enter uteri of controls (day post coitum)
Guinea pig	50–100	250	10	4
Hamster	100	250	25	3 (afternoon)
Rat	10	No dose	10	5
Mouse	1	1	1	4
Rabbit	25	100	50	3*

*From literature.
Data from Greenwald, 1967. Table from van Tienhoven, 1968; reprinted with permission of
W. B. Saunders.

h after the presence of the vaginal plug or ovariectomy of guinea pigs 2 d after mating does not seem to affect ovum transport (van Tienhoven, 1968).

Development of the Zygote. During the sojourn in the oviduct, the oocyte may undergo changes; for example, the rabbit ovum loses its cumulus oophorus and corona radiata, but acquires an albumen layer. Furthermore, in most mammals, of course, fertilization takes place in the oviduct, with the exception of the Madagascar hedgehog (*Setifer setosus*), in which fertilization takes place in the ovarian follicle. The development of the zygote in most mammals thus starts in the oviduct. The extent to which the zygote develops in the oviduct is roughly proportional to the duration of its stay in the oviduct.

After reaching the uterus, oocytes and zygotes are transported by contractions of the myometrium. In polyovular mammals, the zygotes are transported in such a way that they are equidistant from each other at the time of implantation; this distance is thus inversely proportional to the number of zygotes that arrives in each uterine horn. The exact mechanism that makes such equidistant distribution possible is not known. In the rabbit, the stimulus for spacing is blastocyst expansion, which distends the uterus. Such blastocyst expansion does not occur in rats, mice, and other mammals, however; therefore, another mechanism must be involved, as Wimsatt (1975)

has pointed out. The rabbit blastocyst eventually expands so much that it is not propelled further and the blastocyst is arrested. The expansion eventually causes muscular relaxation at the antimesometrial side and a ballooning of the wall to form an implantation dome (Wimsatt, 1975). The role of sterioid hormones in the spacing of embryos is not clear, although it may be expected to differ among species, as is the case with the effect of these hormones on transport of zygotes in the oviduct.

Zygotes in the uterus may not only be transported within the uterine horn in which they arrived, but they may also be transported to the other uterine horn. In some species, implantation is always in the right horn, although ovulation is either bilateral or sinistral, thus providing evidence that transuterine migration is common in these species. Transuterine migration has been found (1) in the bat (*Miniopterus natalensis*), which has a dominant left ovary, and in the little brown bat (*Myotis lucifugus*), with equal ovaries, and (2) in the following Artiodactyla with equal ovaries: the Uganda kob (*Adenota kob*), the common duiker (*Sylvicapra grimmia*), the dik dik (*Rhynotrachus kirki*), the impala (*Aepyceros melampus*), the muntjak (deer) (*Muntiacus muntjak*); and (3) in the water buck (*Kobus defassa*), which has a dominant left ovary (Wimsatt, 1975). Transuterine migration has also been observed in species in which ovulation can occur from both ovaries, and implantation can occur in either horn or both horns of the uterus. Such migrations have been found among ungulates, carnivores, chiropteras, and insectivores.

During the preimplantation period, the embryo takes up nutrients that are provided by oviductal secretions, by interstitial fluid and peritoneal fluid which enter the uterus, and by secretions from the endometrial glands and luminal epithelium of the uterus (Hafez, 1972). In the rabbit, a protein, with a molecular weight of about 14,000–16,000 (Beato, 1977), is secreted by uterine epithelial cells into the uterine lumen under the stimulation of progesterone. This protein, uteroglobulin, is secreted during the preimplantation phase and penetrates into the blastocyst; however, the function of this protein is still an enigma (Beier, 1978).

Site of Implantation. In most species, the entire endometrium of both horns of the uterus is receptive for implantation of the embryo. However, in some species the areas of receptiveness are restricted and the receptiveness may vary from only one horn to a small part of endometrium within a horn (Figure 10.5 and 10.6). The restricted areas in the uterus of ruminants, the so-called caruncles, are illustrative of such restricted sites.

Both blastocyst and uterus have to be ready for implantation, and the synchrony of development of each for implantation needs to be very close. In rabbits, the transfer of zygotes that are more than 2 or 3 days out of phase with the uterus of the recipient results in a high percentage of failures of implantation and embryonic mortality. Similar types of observations have been made for rats, mice, and sheep (van Tienhoven, 1968). It is curious that the rather close synchrony required for implantation in the uterus does not apply to ectopic implantation. Mouse blastocysts transferred to such unlikely places as the anterior eye chamber, surface of the kidney, spleen or testis (!) can become implanted regardless of the endocrine status of the host (van Tienhoven, 1968).

Before discussing the sequence of the events involved in implantation, we should return to the embryo and consider the endocrinology of its survival. In the rat, the embryo can develop from the morula to the blastocyst stage after ovariectomy. In vitro two-celled rabbit embryos can develop

into blastocysts in a defined medium without hormones (Kane and Foote, 1970). Progesterone can, however, increase the survival time of the blastocyst prior to implantation in ovariectomized rats and stimulate mitotic activity in the blastocysts. However, morulae transferred to ovariectomized rats treated with progesterone show a poor rate of survival (Dickmann, 1967). These data indicate that, although early embryos can survive and grow in an abormonal environment, their destiny is influenced by the hormones present.

The questions whether the blastocyst contains steroids and whether it synthesizes steroid hormones have not been answered unequivocally, as has been pointed out by Bullock (1977) and Sauer (1979). In rabbits, rats, and mice, steroid hormones may be present in the blastocyst, but even on this point there is disagreement. Even if it is accepted that the blastocysts contain steroids, then they may have their origin in uterine fluids and may not have been formed within the blastocyst (Bullock, 1977; Sauer, 1979).

As we shall see later in this chapter, in the rat, as little as 2–4 ng estradiol-17β injected locally into the uterus results in implantation of blastocysts in ovariectomized progesterone-treated rats. The question whether estrogen produced by or present on the blastocyst can induce implantation is thus an important one. After studying the concentration of receptors at the site of implantation and at interimplantation sites, in ovariectomized progesterone-treated rats injected systemically with 0.25 μg estradiol, Martel and Psychoyos (1981) concluded that the blastocyst did not secrete estrogens. Steroidogenesis has been established in the pig blastocyst (Bullock, 1977), however, the estrone secreted by the pig blastocyst, as early as day 12 of gestation, is probably sulfated in the endometrium, then transported to the ovary, where it is hydrolyzed, and the resulting estrone then stimulates the maintenance of the CL (Sauer, 1979).

Implantation. Implantation can be defined as the establishment of contact between the trophoblast of the differentiating blastocyst and the uterine wall, with the subsequent erosion of the uterine epithelium, and invasion and decidualization of the stroma by the blastocyst, so that it becomes

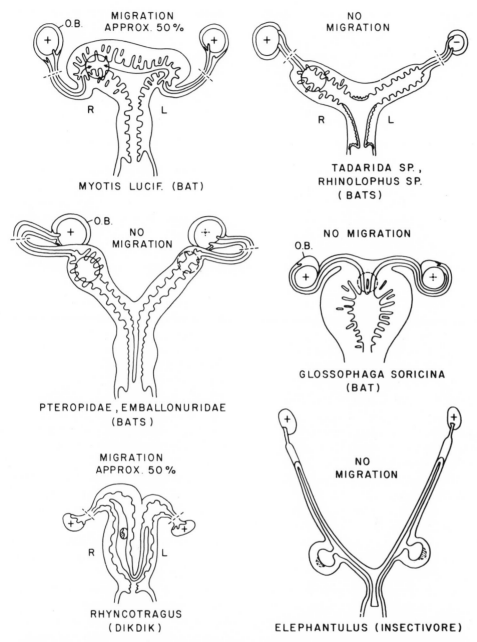

Figure 10.5 Some examples of specialized implantation sites in mammals. Solid circles within the diagrams indicate the site of implantation. Dashed circles indicate the site of implantation in a succeeding pregnancy in those cases where right and left ovaries ovulate alternatively. The plus and minus signs (+, −) on the ovaries indicate their competence for ovulation; a dashed plus sign indicates alternating ovulation between ovaries. The right side lies to the left of the figure in all cases. O.B. = ovarian bursa. (From Wimsatt, 1975; reprinted with permission of *Biology of Reproduction*.)

embedded in the uterine wall. In swine, sheep, cattle, and horses, no direct contact between trophoblast and endometrial stroma becomes established, and stromal decidualization does not occur, and in these cases the relation between

embryo and uterus will, therefore, be called attachment and not implantation (Sauer, 1979).

In the preparation of the uterus prior to implantation in the rat and mouse, the lumen takes on a slitlike appearance after ovulation, as the result

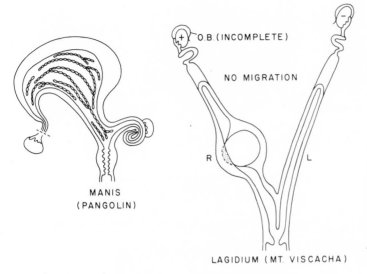

O.B. (INCOMPLETE)

NO MIGRATION

R   L

MANIS
(PANGOLIN)

LAGIDIUM (MT. VISCACHA)

Figure 10.6 Specialized implantation sites in the pangolin and the mountain viscacha. The beaded strips on the uterine wall in the pangolin indicate the location of the placental stripes of this species. The right horn lies to the left side in these figures. Note that the left ovary in the mountain viscacha is nonfunctional. O.B. = ovarian bursa. (From Wimsatt, 1975; reprinted with permission of *Biology of Reproduction*.)

of the loss of uterine fluid. Eventually, on day 5, the microvilli from opposing luminal surfaces are interlocked, so that the lumen is closed (Finn, 1977). On day 6 the luminal surface loses the regular microvilli and comes in close contact with the surface of the cells of the opposite wall. The lumen at this time is only about 1500 nm wide. The surface of the epithelial cells resembles that found during the attachment reaction of the trophoblast to the surface of the epithelium. Proliferation of the endometrial glands, which secrete a mucopolysaccharide may serve to nourish the embryo (Finn, 1977). There is little or no stromal differentiation prior to implantation, but after implantation the so-called decidual reaction, or decidualization, is principally the result of differentiation of stromal cells.

The preparation of the endometrium for implantation is under hormonal control and is dependent upon a complicated interaction between progesterone and estrogen. Progesterone is required for preparation of the uterus for implantation. In rats, estrogen made available endogenously or exogenously subsequent to this progesterone-induced preparation causes a short period of receptivity for the blastocyst, so that implantation can occur. If implantation fails to take place, the uterus becomes either nonreceptive or toxic to zygotes. This explains the precise synchrony between zygote and preparation of the uterus. The nonreceptive state is maintained as

long as progesterone is available to the uterus. When secretion or administration of progesterone is terminated, the uterus recovers, and a neutral state prevails until the secretion of progesterone reestablishes the receptivity of the uterus (Psychoyos, 1973).

During the reproductive cycle of the rat, coitus causes maintenance of the CL and progesterone secretion, and at day 2 estradiol concentrations in the plasma increase and remain elevated until about day 8. The details of the hormonal regulation have not been worked out in as great detail in the mouse, but the situation may be similar (Finn, 1977). It seems from an evolutionary point of view, however, most likely that the timing and the control of implantation in different species is correlated with the hormonal concentrations found in early pregnancy.

At the site of implantation there is a considerable increase in vascular permeability in the endometrium in response to the presence of the blastocyst (Sauer, 1979). In the rat, there is an estrogen surge at the time of implantation. This surge and the ability of estrogen to induce implantation in the case of lactation-induced and experimentally induced delayed implantation suggests that estrogen induces changes in the blastocyst-endometrial relationship, although the nature of the blastocyst stimulus is still unknown (Sauer, 1979). There is evidence that histamines and/or prostaglandins (PGs) may be involved in changes in vascular permeability in the uterus. The observation by Shemesh et al. (1979) that the bovine blastocyst can synthesize PGs, is suggestive for a possible role of these hormones in changing endometrial permeability. The estrone secreted by the pig embryo may serve as a luteotrophic signal, but it can also affect the uterine endometrium and thus affect the attachment of the blastocyst.

Preimplantation losses of about 30 percent of the embryos are found in the American opossum and the Australian native cat (*Dasyurus viverrinus*), which ovulate on the average of 22 and 20 to 25 oocytes, respectively. In the pronghorn (*Antilocapra americana*), the preimplantation loss of embryos is about 50 percent. The chorionic vesicles develop a necrotic tip at the oviductal end. The 4–5 embryos are initially about evenly distributed between a distal portion of the lumen nearer to the oviduct and a proximal portion near the uterine end. As the proximal ves-

icle expands, its necrotic tip eventually destroys the distal vesicles, so that only two embryos survive (Wimsatt, 1975).

In most mammals, implantation of the embryo occurs when it has reached the blastocyst stage (Table 10.5 and Figures 10.7 and 10.8), except in ungulates, in which an elongated chorionic vesicle with allantois becomes attached to the endometrium. Implantation can be classified according to the degree of penetration of the uterine mucosa by the blastocyst, or according to the orientation of the blastocyst in the uterus, or according to the point of first attachment of the trophoblast. The degree of penetration of the uterine mucosa by the blastocyst can be: (1) centric or superficial, (2) eccentric, (3) partly interstitial, and (4) interstitial, as illustrated in Figure 10.9. Orientation of the blastocyst in the uterus can be such that the direction of the germ disc is: (1) mesometrial, with the disc facing toward the mesometrium, (2) antimesometrial with

Figure 10.7 Patterns of blastocyst development in mammals. E.M.C. = entodermal mother cell, ENT. = entoderm, G.C. = germinal cells, M.PL. = medullary plate cells, P.TR. = primitive trophoblast, T.C. = trophoblast cells. (From Wimsatt, 1975; reprinted with permission of *Biology of Reproduction*.)

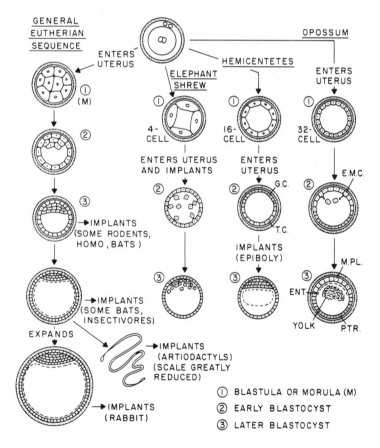

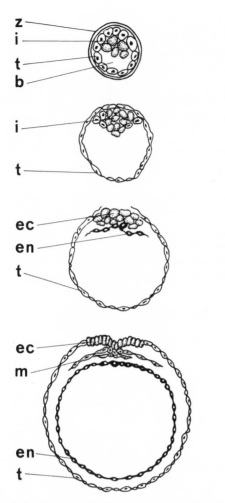

Figure 10.8 Diagrammatic illustration of the differentiation of a mammalian blastocyst. b = blastocoelic cavity, ec = ectoderm, en = endoderm, i = inner cell mass (embryonic disc), m = mesoderm, t = trophoblast, z = zona pellucida. (From Hafez, 1972; reprinted with permission of Charles C Thomas, Publisher, Springfield, Illinois.)

the disc facing away from the mesometrium, and (3) lateral, with the disc facing a position between 1 and 2, as illustrated in Figure 10.9. According to the point of first trophoblastic attachment, there are diffuse, antimesometrial, mesometrial, lateral, bilateral, circumferential (or equatorial), and cotyledonary attachments, as illustrated in Figure 10.10, with the cotyledonary type being a special case of diffuse attachment (Wimsatt, 1975). The advantages of these various patterns of implantation are unknown. The reader is referred to Wimsatt's (1975) excellent critical review and interpretation of the occurrence of the different patterns of orientation of

first trophoblastic attachment and disc attachment. In the Insectivora and Chiroptera (Rasweiler, 1979), these show a great deal of variation among orders and suborders, in contrast to the lack of variation in rodents and primates. A summary of the patterns found in different groups is given in Table 10.7.

We have used the terms *diffuse* and *superficial* implantation and pointed out that *cotyledonary* implantation is a special case of diffuse attachment (Steven and Morriss, 1975).

The initial phase of implantation consists of adhesion of the blastocyst to the uterine epithelium. In some species, e.g., the rat and the armadillo, the earliest stages of implantation consist of interdigitation of the microvilli of the blastocyst and the uterine epithelium. In these species the blastocyst is free from the zona pellucida (ZP). This shedding of the ZP can take place by lysis of the ZP or by escape of the blastocyst through an opening in the ZP, which may be either a narrow escape route or a wide gap.

As we mentioned, blastocysts can become implanted in almost any well-vascularized site outside the uterus, regardless of the hormonal status of the host. It is, therefore, not likely that the uterine hormones are required for the shedding of the ZP. The differentiation of a blastocyst is illustrated in Figure 10.8. Bergström (1971?) provides beautiful micrographs illustrating the differentiation of the mouse blastocyst.

In some species, such as the ferret and the rabbit (Steven and Morriss, 1975), the ZP is not shed, and trophoblastic processes penetrate the ZP and adhere to the apical cell membranes of the uterine luminal epithelium (Schlafke and Enders, 1975). Implantation, after shedding of the ZP, consists of three steps. First, the stickiness of the blastocyst changes, as the result of a change in the electrical charge of the blastocyst. This is correlated with a decrease in neuroaminic acid, which may increase the deformability of the blastocyst (Nilsson et al., 1974). Nilsson et al. (1975) have since shown that estrogen treatment of mice, in which implantation was prevented by progesterone, induced a decrease in the surface charge of the blastocyst and decreased its antigenicity. The second step consists of interdigitation of the microvilli of the trophoblast and uter-

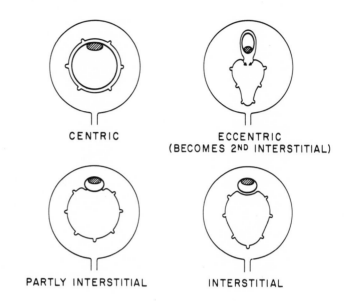

CENTRIC

ECCENTRIC
(BECOMES 2ND INTERSTITIAL)

PARTLY INTERSTITIAL

INTERSTITIAL

Figure 10.9 Implantation patterns and orientations. The upper four drawings illustrate the common implantation patterns observed in mammals. The lower three drawings illustrate the orientation of the embryonic cell mass in relation to the mesometrium. The mesometrium is directed toward the bottom of all figures. (From Wimsatt, 1975, with legend slightly changed; reprinted with permission of *Biology of Reproduction*.)

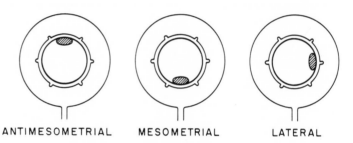

ANTIMESOMETRIAL   MESOMETRIAL   LATERAL

Figure 10.10 Diagrams showing variations in the points of first trophoblastic attachment observed in mammals. The points of first attachment are indicated by a thickening of the line designating trophoblast. The mesometrium is toward the bottom of all figures. (From Wimsatt, 1975; reprinted with permission of *Biology of Reproduction*.)

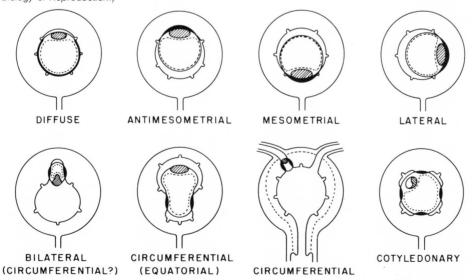

DIFFUSE   ANTIMESOMETRIAL   MESOMETRIAL   LATERAL

BILATERAL
(CIRCUMFERENTIAL?)

CIRCUMFERENTIAL
(EQUATORIAL)

CIRCUMFERENTIAL

COTYLEDONARY

Table 10.7

Implantation characteristics of eutheria

Classification	Implantation type			Orientation of disc		
	Centric	Eccentric	Interstitial	Mesometrial	Antimesometrial	Lateral
Insectivora						
Tenrecidae	+					+
Chrysochloridae	+			+		
Erinaceidae			+		+	
Macroscelididae		+		+		
Soricidae	+				+	
Talpidae (*Talpa*)	+				+	
Talpidae (*Scalopus*)	?				?	
Tupaiidae	+				+	
Chiroptera						
Pteropidae		?	+	+		
Megadermatidae	+			+(?)		
Noctilionidae	+					+
Phyllostomatidae						
(*Glossophaga*)			+		+	
(*Desmodus*)			+		+	
Vespertilionidae	+				+	
Molossidae	+				+	→
Primates						
Lemuridae	+					+
Lorisoidea	+					+
Tarsiidae	+					+
Cercopithecoidea	+				+	?*
Hominoidea			+		+	?*
Edentata						
Dasypus novemcinctus	"fundic"				+	
Lagomorpha						
Leporidae	+			+		
Rodentia						
Sciuromorpha		+		+		
Geomyoidea		+		+		
Myomorpha		+		+		
Hystricomorpha		Some	+	+		
Carnivora						
Fissipedia	+				+	
Pinnipedia	+				+	
Perissodactyla						
Equidae	+				+	
Artiodactyla						
Suiformes	+				+(?)	
Cervoidea	+				(?)	
Bovoidea	+				(?)	

*Luckett (1974) has proposed that the anterior and posterior walls of the (simplex) uterus are homologous to the lateral walls of a bicornuate uterus. If this is true then disc orientation in higher primates would be comparable to the lateral orientation of prosimians.

From Wimsatt, 1975; reprinted with permission of *Biology of Reproduction*.

Classification	Orientation of first attachment			Depth of implantation		
	Mesometrial	Antimesometrial	Lateral	Superficial	Partly interstitial	Interstitial
Insectivora						
Tenrecidae			+	+		
Chrysochloridae			+	+		
Erinaceidae		+				+
Macroscelididae	+					+
Soricidae			Circumferential	+		
Talpidae (*Talpa*)		+		+		
Talpidae (*Scalopus*)		?		+		
Tupaiidae			Bilateral	+		
Chiroptera						
Pteropidae	+				+	
Megadermatidae	+			+		
Noctilionidae			+	+		
Phyllostomatidae						
(*Glossophaga*)			+			Secondary
(*Desmodus*)		+				+
Vespertilionidae		+		+		
Molossidae		← Diffuse →		+		
Primates						
Lemuridae			Circumferential	+		
Lorisoidea			Circumferential			
Tarsiidae	+			+		
Cercopithecoidea		+	?*	+		
Hominoidea		+	?*			+
Edentata						
Dasypus novemcinctus		+		+		
Lagomorpha						
Leporidae	+			+		
Rodentia						
Sciuromorpha	+				+	
Geomyoidea	+				+	
Myomorpha	+					Secondary
Hystricomorpha	+					Some
Carnivora						
Fissipedia		+		+		
Pinnipedia		+		+		
Perissodactyla						
Equidae		← Diffuse →		+		
Artiodactyla						
Suiformes		← Diffuse →		+		
Cervoidea		← Cotyledonary →		+		
Bovoidea		← Cotyledonary →		+		

ine epithelium. The third step is the distortion of the surface of the trophoblast and of the uterus, when, except in the case of diffuse implantation, the trophoblasts invade the uterine epithelium. This process has been classified, according to the nature of the involvement of the uterine epithelium, by Schlafke and Enders (1975) as intrusion, displacement, or fusion implantation (Figure 10.11). In intrusion implantation, as found, e.g., in the ferret (*Mustela putorius*), the trophoblast penetrates between the uterine luminal epithelium to the basal lamina of the uterine epithelium and extends beneath it. In displacement implantation, as found, e.g., in the rat and mouse, the uterine luminal epithelium is readily dislodged from the basal lamina, and the trophoblast comes to lie along areas previously occupied by these displaced uterine cells. In fusion implantation, as found, e.g., in the rabbit, the trophoblast penetrates by fusion of an area of the syncytial trophoblast to individual uterine luminal cells.

The target of the trophoblast of carnivores is the glands of the uterus (Figure 10.12), with a secondary massive formation of an invasive trophoblastic syncytium that destroys the intraglandular stromal tissue (Wimsatt, 1975). In the rabbit, the point of origin of the trophoblastic invasion corresponds to the distribution of subepithelial capillaries, which are the target of the invading embryo (Figure 10.12). In shrews of the genera *Blarina* and *Sorex,* the surface of the epithelium prior to the attachment of the trophoblast shows the formation of cylindrical crypts (Figure 10.12), which forms the routes of invasion (Wimsatt, 1975).

Delayed Implantation. In most eutherian species, implantation or attachment to the endometrium starts shortly after the arrival of the embryo in the uterus. There are, however, a fairly large number of species in which implantation of the embryo is delayed. The species showing such delayed implantation can be divided into two groups, species with obligatory delayed implantation and species with facultative, usually lactation-induced, delayed implantation.

Obligatory delayed implantation is found in the species listed in Table 10.8. During the period of delay, the embryo is in embryonic diapause, and little growth or expansion occurs during this period in most species; in others, such as the roe deer (*Capreolus capreolus*), the rate of growth of the embryo is slow (Renfree and Calaby, 1981).

The endocrinology of implantation in most species with obligatory delayed implantation, except the mink (*Mustela vison*), is still largely a mystery. In the mink, which has a short and variable period of delay, if mating is early, the delay is long, if mating is late, there may be no delay (see Table 10.8). Implantation can be induced by

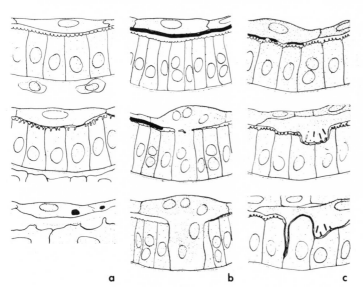

Figure 10.11 Three types of interaction of trophoblast with uterine epithelium during penetration of this epithelium. In all three types, apposition and adhesion precede penetration. However, in (a), displacement penetration (rat, mouse), the uterine luminal epithelium is readily dislodged from the basal lamina, and the trophoblast comes to lie along areas previously occupied by the displaced uterine cells. In (b), fusion penetration (rabbit), the syncytial trophoblast fuses with a uterine luminal epithelial cell. In (c), intrusion penetration (ferret, others?), projections of the syncytial trophoblast penetrate between uterine epithelial cells. (From Schlafke and Enders, 1975; reprinted with permission of *Biology of Reproduction*.)

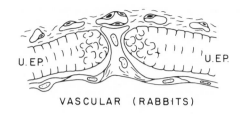

VASCULAR (RABBITS)

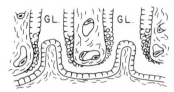

GLANDULAR (CARNIVORES, TENRECS)

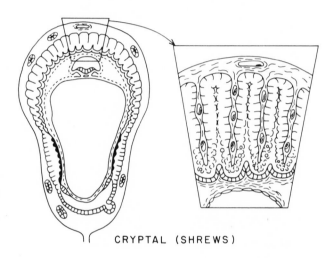

CRYPTAL (SHREWS)

Figure 10.12 "Targets" of trophoblastic invasion of the endometrium. [In the rabbit the targets are the blood vessels (U.EP. = uterine epithelium). In carnivores and tenrecs the targets are the endometrial glands (GL.). In shrews the targets are the crypts in the endometrium.] (From Wimsatt, 1975; reprinted with permission of *Biology of Reproduction*.)

prolactin administration in intact (Papke et al., 1980; Martinet et al., 1981; Murphy et al., 1981) and in hypophysectomized (Murphy et al., 1981) mink. Administration of bromocryptine, an inhibitor of prolactin secretion, delayed implantation in intact mink (Papke et al., 1980; Martinet et al., 1981). This experimental induction of implantation by prolactin agrees well with the observations that the initiation of implantation is preceded by activation of the CL and an increase in progesterone secretion, and that prolactin administration results in increased progesterone secretion (Martinet et al., 1981). The time of implantation in mink can be advanced by increasing the photoperiod (Martinet et al., 1981).

In the European badger (Canivenc and Bonnin, 1981), the western spotted skunk (Mead, 1981),

and the short-tailed weasel (Gulamhusein and Thawley, 1974), implantation is preceded by an increase in progesterone secretion; however, administration of progesterone has failed to induce implantation in the European badger (Canivenc and Bonnin, 1981) and the western spotted skunk (Mead, 1981, Mead et al., 1981). In the European badger, neither prolactin nor LH administration has induced premature blastocyst implantation, although LH concentrations increase prior to spontaneous implantation (Canivenc and Bonnin, 1981).

In the nine-banded armadillo, plasma progesterone increased just prior to implantation, but implantation could be induced experimentally by ovariectomy (Enders and Given, 1977; Wimsatt, 1975); however, unilateral ovariectomy, inde-

Table 10.8

Eutheria with delayed implantation of the embryo
and estimate of the dormant period

Species	Dormant period	Species	Dormant period
Chiroptera		Malayan sun bear, *Helarctos malayanus*	?
Fruit bat, *Eidolon helvum*	4–6 months	**Pinnepedia**	
Fruit bat, *Artibeus jamaicensis**	0–2½ months	Cape fur seal, *Arctocephalus pusillus*	3–5 months
Long-winged bat, *Miniopterus schreibersi*	6 months	Northern fur seal, *Callorhinus ursinus*	3–5 months
Big-eared bat, *Macrotus waterhousei*	4† months	South American sea lion, *Otaria flavescens*	≥3 months
Funnel-eared bat, *Natalus stramineus*	1½† months	Steller sea lion, *Eumetopias jubata*	3½ months
Edentates		Bearded seal, *Erignathus barbatus*	2–3 months
Nine-banded armadillo, *Dasypus novemcinctus*	4 months	Southern elephant seal, *Mirounga leonina*	4 months
Muleta armadillo, *Dasypus hybridus*	4 months	Harbor seal, *Phoca vitulina*	2–3 months
Three-banded armadillo, *Tolypeutes tricinctus*	4 months	Ringed seal, *P. hispida*	3–4 months
Carnivora		Harp seal, *P. groenlandica*	3 months
American badger, *Taxidea taxus*	6–6½ months	Hooded seal, *Cystophora cristata*	5 months
European badger, *Meles meles*‡	10 months	Grey seal, *Halichoerus grypus*	101–109 days
Chinese hog badger, *Arctonyx collaris*	?	Crabeater seal, *Lobodon carcinophagus*	4 months
American marten, *Martes americana*	7–8 months	Weddell seal, *Leptonychotes weddelli*	50–60 days
Sable, *M. zibellina*	7–8½ months	Mediterranean monk seal, *Monachus monachus*	?
Fisher, *M. pennanti*	9½–10½ months	**Insectivora**	
Beech marten, *M. foina*	7–8 months	Siberian mole, *Talpa altaica*	≥7 months
Mink, *Mustela vison*	0–37 days§	**Rodentia**	
Short-tailed weasel, *M. erminea*	9–10 months	Plains viscacha, *Lagostomus maximus*	11 days
Long-tailed weasel, *M. frenata*	6–10½ months	**Artiodactyla**	
Wolverine, *Gulo gulo*	6–8 months	Roe deer, *Capreolus capreolus*‡	4–5 months
River otter, *Lutra canadensis*	6 months		
Sea otter, *Enhydra lutris*	8–9 months		
Western spotted skunk, *Spilogale putorius latifrons*	6–7½ months		
Striped skunk, *Mephitis mephitis*	0–14 days		
Black bear, *Ursus americanus*	5 months		
European brown bear, *U. arctos arctos*	?		
Grizzly bear, *U. arctos horribilis*	?		
Himalayan black bear, *Selenarctos thibetanus*	?		
Polar bear, *Thalarctos maritimus*	7½–8 months		

*Blastocysts conceived in July to September show retarded development, but those conceived in the spring do not.

†In these species the embryo becomes implanted shortly after arrival, but formation of the placenta is slow and early embryonic development of the embryo is retarded.

‡Sometimes ovulation occurs at the time implantation normally takes place; if mating occurs also, implantation may occur without delay.

§The duration of the delay depends on the time of mating; the earlier in the season the mating occurs the longer the delay.

Data from van Tienhoven, 1968; Wimsatt, 1975; Enders and Given, 1977; Daniel, 1981; Renfree and Calaby, 1981; Mead, 1981.

pendent of the presence of a CL on the removed ovary, did not have this effect. The findings that administration of progesterone (24 mg/d) or estrone (0.2 mg/d) inhibited ovariectomy-induced implantation makes it difficult to reconcile these findings with the increased progesterone concentration at the time of implantation.

In the roe deer, an increase in plasma estrogen and progesterone concentration is associated with implantation, but according to Aitken (1981), these changes are a consequence of implantation and not the cause of it. Throughout the period of embryonic diapause, the ovaries look active histologically (Wimsatt, 1975).

It has been possible to induce premature implantation of the blastocyst in the western spotted skunk, the American pine marten, the long-tailed weasel, and the sable by exposing the pregnant females to *increasing* day length (Mead, 1981) and in the European badger (Canivenc and Bonnin, 1981) and possibly the river otter (*Lutra canadensis*) by exposing the pregnant females to *decreasing* day length.

From the available data, it is not possible to formulate a generalization about the endocrine causes of implantation in species with obligatory delayed implantation. It is not as difficult to understand why in the western subspecies of the spotted skunk (*Spilogale putorius latifrons*), which mates in the fall, delayed implantation is part of the reproductive cycle, but in the eastern subspecies (*Spilogale putorius ambarvalis*), which breeds in the spring, implantation is not delayed. Both subspecies give birth in the spring and, apparently for the western subspecies, mating in the fall was advantageous, whereas for the eastern subspecies, mating in the spring was advantageous. The ecological basis for the difference between these two subspecies with respect to the time of mating is probably quite complicated.

Facultative delayed implantation is induced by lactation and is found in the species listed in Table 10.9. The endocrinology of this type of delayed implantation has been studied in most detail in the rat and mouse. If the mouse or rat mates and conceives at postpartum estrus, the oocytes do not become implanted if the mother is nursing a critical number of pups. This critical

Table 10.9

Eutherian species with facultative delayed implantation

Rodentia
  Lesser short-tailed gerbil, *Dipodillus simoni*
  Mongolian gerbil, *Meriones unguiculatus*
  Shaw's jird, *M. shawi*
  Fat-tailed gerbil, *M. crassus*
  Bank vole, *Clethrionomys glareolus*
  Red-tree mouse, *Phenacomys longicaudus*
  Oldfield mouse, *Peromyscus polionotus*
  Deer mouse, *P. maniculatus*
  White footed mouse, *P. leucopus*
  Cotton mouse, *P. gossypinus*
  Pinon mouse, *P. truei*
  Northern grasshoppermouse, *Onychomys leucogaster*
  House mouse, *Mus musculus*
  Laboratory rat, *Rattus norvegicus*
  Australian brush rat, *R. fuscipes*
  New Holland mouse, *Pseudomys novaehollandiae*
  Fawn hopping-mouse, *Notomys cervinus*
Insectivores
  Common shrew, *Sorex araneus*
  Pygmy shrew, *S. minutus*
  European water shrew, *Neomys fodiens*

Data from Renfree and Calby, 1981; Vogel, 1981.

number varies probably among strains within the two species. The lactation induced and maintained by the suckling young inhibits FSH release and thus prevents the surge of estrogen secretion, which normally occurs on day 4 post coitum and which is necessary for implantation. The length of the delay depends on (1) the number of young being nursed, (2) the parity of the mother, the delay becoming shorter with increased parity, and (3) the continued presence of the young, temporary removal reducing duration of the delay (Gidley-Baird, 1981).

Experimentally, delay can be induced by either hypophysectomy or ovariectomy, prior to the afternoon of day 4 post coitum, thus preventing the estrogen release mentioned above, or by administration of a progesterone, such as medroxyprogesterone acetate. After ovariectomy, implantation can be induced by systemic administration of progesterone and estrogen in sequence, or by systemic administration of progesterone followed by local injection of as little as 2–4 ng estradiol-17β into the uterus. In the ovariectomized golden hamster (*Mesocricetus auratus*), progesterone administration alone is sufficient for the induction of implantation, but administration of progesterone and estrogen is followed by a higher incidence of implantations than progesterone administration (van Tienhoven, 1968).

In the *AP* mouse, administration of FSH is followed by implantation, whereas in the rat apparently FSH and LH are required for the induction of implantation (Gidley-Baird, 1981), although MacDonald et al. (1967) obtained implantation with LH alone. Gidley-Baird (1981) has concluded that in the rat and the mouse, prolactin and LH are required for the progesterone secretion necessary for implantation, and that in the mouse FSH, and in the rat FSH and LH are required for the estrogen secretion.

In the common shrew (*Sorex araneus*) and the European water shrew (*Neomys fodiens*), the delayed implantation that occurs after a postpartum conception is in a sense an obligatory delayed implantation, because the gestation period is shorter than the (for the suckling young) indispensable lactation period, by four days for the common shrew and by eight days for the European water shrew. Implantation has to be deferred to prevent the necessity of having to nurse, simultaneously, two litters of different ages (Vogel, 1981).

The endogenous or exogenous estrogen, which results in implantation in the rat and mouse, also increases the vascular permeability, as mentioned earlier in this chapter.

Immunological Aspects of Implantation and Gestation. An important fundamental question is why the embryo is not rejected by the uterus, because it is—except possibly for highly inbred lines—a foreign graft. We discuss briefly some of the possible mechanisms and the evidence for and against each proposed mechanism:

1. Since the fetus has 50 percent of the genetic endowment of the mother, the rejection may not occur because of genetic similarity. This argument is invalid, because: (*a*) mothers do reject skin grafts of their offspring, and (*b*) eggs transferred from a totally unrelated donor become implanted and are carried to term normally.

2. One can consider that the mother lacks an immunological response, because the uterus is an immunologically privileged site, similar to the anterior eye chamber or the cheek pouch of the hamster. Arguments against this explanation are: (*a*) that most of the immunologically privileged sites lack lymph vessels, whereas the uterus has lymph vessels, and (*b*) that ectopic pregnancies, such as occur on the outside of the intestine or elsewhere in the abdominal cavity, are not rejected. Moreover, blastocysts placed under the kidney capsule experimentally are not rejected.

3. Pregnancy can be seen as reducing the maternal immunological response. The main argument against this explanation is that skin grafts are equally rejected by pregnant and nonpregnant cows and mice, although there is some evidence that skin grafts survive longer in pregnant than in nonpregnant rabbits (Borland, 1975).

4. The trophoblast, and later the fetus, cannot induce an immune response. This explanation seems invalid because transplantation antigens can be detected early during embryonic development (Borland, 1975).

5. There is a noncellular barrier with a high sialic acid concentration, which protects the embryo against the immune response of the mother; however, trophoblast and other fetal antigens do reach the immunological system of the mother.

The Placenta. The placenta takes over the nutrition of the embryo, the exchange of gases between mother and embryo, and the removal of waste products from the embryo. In addition, the placenta serves as a mechanical barrier against shocks and, depending on the species, may secrete gonadotrophic hormones (as in the horse and human), a lactogenic hormone (as in humans), and steroid hormones. The placenta develops either as a chorio-vitelline (yolk-sac) placenta or a chorio-allantoic placenta. The chorio-vitelline placenta is usually a transient structure in the Eutheria, but, as we have seen earlier in this chapter, is the definitive placenta of marsupials.

Placentas can be classified in different ways, for instance, according to shape or according to the number of layers of tissue that separate fetal and maternal blood streams.

Classification by Shape. Classification by shape gives the following groups (Steven, 1974):

1. *Diffuse Placenta.* Most of the outer surface of the chorion is covered by small villi or folds, which are in intimate contact with corresponding depressions of the endometrium. This type is found in swine, horses, and donkeys.

2. *Cotyledonary or Multiplex Placenta.* The chorionic villi (cotyledons) are restricted to well-defined circular or oval areas on the chorion. The areas between these villi are relatively smooth. The fetal cotyledons, which overlie the caruncles of the uterus, form a hand-in-glove arrangement with the caruncles. This type is found in cattle, sheep, and goats.

3. *Zonary Placenta.* The chorionic villi are present as a band of placental tissue, which encircles the equatorial region of the chorionic sac. Complete girdles of villi are found in the dog, cat, and spotted hyaena (*Crocuta crocuta*); incomplete girdles of villi are found in the mink, the ferret, the polar bear (*Thalarctos maritimus*),

and the brown bear (*Ursus arctos*). The placenta of carnivores shows the presence of so-called hemophagous organs. These consist of either marginal or central effusions of maternal blood, which is in close contact with cells of the chorion that apparently ingest the red blood cells. Marginal effusions are found in the cat, the dog, the coyote (*Canis latrans*) and the red fox (*Vulpes vulpes*); central effusions are found in the river otter (*Lutra canadensis*), the Eurasian otter (*L. lutra*), the badger (*Meles meles*), the raccoon (*Procyon lotor*), the wolverine (*Gulo gulo*), the ferret, and the mink.

4. *Discoid Placenta.* The placenta is disc-shaped and may be single, as in the human, or double, as in the primate families Ceropithecidae and Cebidae.

Classification by Number of Layers. For classification of the placenta based on the number of layers that separate maternal and fetal circulations, we use the one given by Steven (1975), because it agrees well with the data obtained in

Table 10.10

Classification of Eutherian placentas, based on number of layers of tissue separating fetal and maternal circulation

Type of placenta	Layers of tissue	Animals in which found
Epitheliochorial	Endothelium of fetal capillaries, basement membrane of fetal capillaries, basement membrane of chorion, cells of chorion indented by fetal capillaries, zone of interdigitation of fetal and maternal microvilli, maternal epithelium, basement membrane of maternal epithelium, layer of scattered collagen fibers, basement membrane of maternal capillaries, endothelium of maternal capillaries	Artidactyla, Perissodactyla, Cetacea, Lemuridae
Endotheliochorial	Endothelium of fetal capillaries, basement membrane of capillary (in late stages of gestation), basement membrane of trophoblast, layer of syncytial trophoblast, interstitial membrane, endothelium of maternal capillaries	Carnivora, Pinnepedia
Haemomonochorial	Endothelium of fetal capillaries, basement membrane of fetal capillaries, basement membrane of trophoblast, cytotrophoblast, syncytiotrophoblast	Humans, armadillos, guinea pigs
Haemodichorial	Endothelium of fetal capillaries, basement membrane of fetal capillaries, basement membrane of trophoblast, inner layer of trophoblast, outer layer of trophoblast	Rabbits, chipmunks
Haemotrichorial	Endothelium of fetal capillaries, basement membrane of fetal capillaries, inner layer of trophoblast, middle layer of trophoblast, outer layer of trophoblast	Rats, mice, hamsters

Data from Steven, 1975.

investigations of the fine structure of the placenta of different groups. The types of placenta and the components of the barrier between the maternal and fetal circulation are listed in Table 10.10. This table illustrates the considerable variation in the number of layers that form the barrier between maternal and fetal circulations. Steven (1975) emphasized that it should not be concluded that the efficiency of placental transfer increases as the number of layers decreases. In fact, he emphasized that the greater the number of placental layers, the greater the development of the animal at birth, suggesting a greater efficiency of nutrient transfer as the number of placental layers becomes larger.

Pregnancy Recognition. The secretion of gonadotrophins by the fetus may function as a signal to the ovary that the CL needs to be maintained for maintenance of the pregnancy; alternatively, steroid secretion by the placenta can serve to maintain pregnancy independent of the ovary. For some species, such as the dog, cat, and ferret, (see Chapter 8), the secretion of progesterone for maintenance of pregnancy is largely independent of the presence of a fetus.

We have discussed the possible mechanisms that may be involved in maintenance of the CL in general (Figure 8.10). In order to understand how gestation is maintained in various species, it is necessary to know whether the hypophysis and/or ovary are required for maintenance of pregnancy. Table 10.11 shows the species in which pregnancy is continued after ovariectomy or AP. To this should be added species in which the CL disappears during gestation, as in the African bats *Nycteris luteola* and *Traeniops afer*, the black rhinoceros (*Diceros bicornis*), and the horse (Amoroso and Perry, 1977).

In the cases in which gestation is maintained after oophorectomy, it may be assumed that the placenta produces the hormones required for ges-

Table 10.11

Effect of removal of the ovaries or pituitary, in first and second halves (approximate) of pregnancy, on the maintenance of gestation in various species

Animal	Length of gestation (days)	Ovariectomy		Hypophysectomy	
		First half	Second half	First half	Second half
Sheep	148	−	+	−	+
Human	280	+	+	+	+
Monkey	165	+	+	+	+
Guinea pig	68	±	+	+	+
Rat	22	−	±	−	+
Mouse	23	−	−	±	+
Hamster	16	−	−	−	+
Cat	63	−	±	n.d.	±
Cow	282	−	±	n.d.	n.d.
Ferret	42	−	±	−	±
Horse	350	−	+	n.d.	n.d.
Dog	61	−	n.d.	−	±
Goat	148	−	−	−	−
Sow	113	−	−	−	−
Rabbit	28	−	−	−	−
Opossum	16	−	−	n.d.	n.d.
Blue fox*	52	−	−	n.d.	n.d.
Armadillo	150	Implantation may occur	−	n.d.	n.d.

+, Fetuses survive; ±, some fetuses survive; −, fetus aborts; n.d., not determined.
*From Møller, 1974.
Slightly modified from Heap, 1972, and added data from Møller, 1974; reprinted with permission of Cambridge University Press.

tation. There are alternate possibilities for those cases in which pregnancy is maintained after hypophysectomy; i.e., the CL is independent of the pituitary (as we discussed in Chapter 8), or the placenta produces luteotrophic hormone, maintaining the CL, or the placenta produces the steroid hormones required for maintenance of gestation. On the basis of the available evidence, it appears that luteotrophins are produced by the placentas in monkeys, humans, horses, rats, mice, sheep and goats (Thorburn et al., 1977).

Figure 10.13 illustrates the endocrine relationships in the maintenance of pregnancy in (1) the rabbit (Figure 10.13 A), which is AP- and ovary-dependent throughout pregnancy; (2) the rat (Figure 10.13 B), which during the first half of pregnancy is AP- and ovary-dependent and during the second half is AP-independent but ovary-dependent; (3) the sheep (Figure 10.13 C), which is ovary- and AP-dependent during the first half, but ovary- and AP-independent during the second half of pregnancy and, and (4) the human (Figure 10.13 D), who is early in gestation independent of AP and ovary for maintenance of pregnancy.

In the rabbit, the principal luteotrophin is probably LH. This maintains the estrogen secretion by the follicles, which, in turn, stimulates progestin secretion by the CL necessary for maintenance of of gestation. In the rat, the stimulus of coitus induces the release of prolactin (twice per 24 h; see Chapter 8) and, together with LH, it maintains progestin secretion of the CL, whether the mating is fertile or not. By day 11, the placenta starts to secrete a luteotrophic complex which maintains the CL beyond the time of pseudopregnancy, and by day 18, the uterus secretes sufficient progesterone to maintain gestation. de Greef et al. (1977) have obtained evidence that the presence of decidual tissue in pseudopregnant rats increases the plasma progesterone concentrations in peripheral blood. The exact mechanism by which this occurs needs to be elucidated further, but the point of interest here is that progesterone can be secreted by the uterus in the absence of a placenta.

In the sheep, either LH alone or LH and prolactin maintain secretion of progestins, but the CL of pregnancy either is maintained because of an inhibition of the secretion of the uterine luteolytic factor or because the embryo secretes a luteotrophic hormone (see Chapter 8). After about 50 days, the placenta secretes sufficient progesterone itself to maintain pregnancy.

In the rhesus monkey (and possibly in the human), the regression of the CL, which would have occurred if no pregnancy had intervened, is prevented by the secretion of chorionic gonadotrophin (Neill and Knobil, 1972). This "rescue" of the CL, which occurs at about day 9–12 of pregnancy, maintains the pregnancy until about day 25, when the CL starts to secrete enough progesterone for maintenance of pregnancy (Neill et al., 1969).

The human placenta, although it produces steroid hormones, lacks the complete set of enzymes required for synthesis of all the hormones it produces. This apparent problem is solved by a complicated exchange of precursors and hormones between fetus and placenta. The so-called feto-placental unit is responsible for the final hormone production. (Some of the conversions occurring in the human feto-placental unit are illustrated in Figure 10.14). The human placenta secretes, in addition to hCG, another protein hormone, human placental lactogen (hPL), also called human chorionic somatomammotrophin (hCS). This hormone acts synergistically with hCG to regulate and promote progesterone and estrogen production by the placenta (Allen, 1975).

The equine placenta produces a gonadotrophic hormone, pregnant mare's serum gonadotrophin (PMSG), which has been considered to be the luteotrophic hormone necessary for maintenance of the pregnancy, because it presumably causes ovulations of accessory follicles and thus accessory CL, producing progesterone required for maintenance of pregnancy. Stewart and Allen (1981) have, however, obtained evidence that the follicular growth during early gestation occurs prior to PMSG secretion and is probably the result of increased FSH secretion by the AP. They also found no quantitative relationship between progesterone secretion and PMSG secretion; for example, mares carrying horse fetuses produced five times more PMSG than mares carrying mule fetuses, but maternal progesterone concentrations were the same for these two groups of mares. Also, in cases

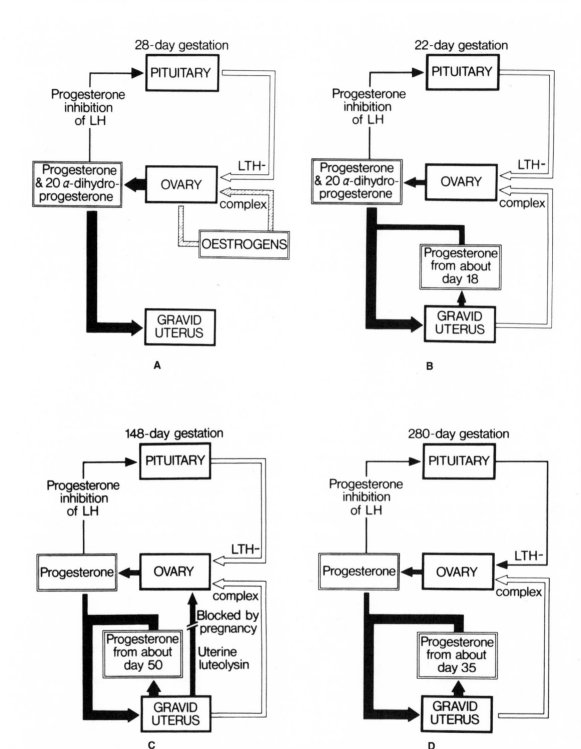

Figure 10.13 Hormonal interrelationships in the maintenance of pregnancy in the rabbit (A), rat (B), sheep (C), and human (D). (From Heap, 1972; reprinted with permission of Cambridge University Press.)

Table 10.12

Effect of fetal genotype and of dam on PMSG concentration in the dam

Dam	Sire	Fetus	PMSG in dam (IU/ml serum)
Horse (mare)	Donkey (jack)	Mule	2–10
Horse (mare)	Horse (stallion)	Horse	20–100
Donkey (jennet)	Horse (stallion)	Hinny	200–300
Donkey (jennet)	Donkey (jack)	Donkey	30–40

From Allen, 1969.

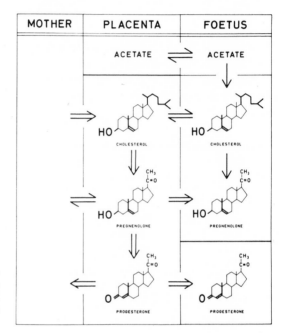

Figure 10.14 Present concept of the metabolism of cholesterol, pregnenolone, and progesterone in the human feto-placental unit at midgestation. (From Diczfalusy, 1969; reprinted with permission of E. Diczfalusy and *Acta Endocrinologica*.)

Table 10.13

Plasma proteins that bind and transport steroids during pregnancy

Plasma protein	Molecular weight	Steroids bound	Comments
Albumin (human)	69,000	Androgens, corticosteroids, estrogens, progestagens	Low affinity and high capacity for binding
α_1-Acid glycoprotein, AAG (human)	41,000	Progesterone, testosterone	Medium affinity and medium capacity for binding
Transcortin, CBG (human, monkey, guinea pig, rat and many other species)	52,000	Cortisol, corticosterone, progesterone	High affinity, low capacity for binding
Sex steroid-binding protein, SBP (human)	52,000 to 100,000	Estradiol, testosterone	High affinity, low capacity; SBP concentration increases in pregnancy
Progesterone-binding protein, PBP (guinea pig, coypu)	100,000	Progesterone	High affinity for binding; PBP concentration increases in pregnancy

From Heap, 1972; reprinted with permission of Cambridge University Press.

of pregnancy failure prior to day 40 of pregnancy, PMSG secretion continued, but the secondary CL regressed, and ovaries became small (Allen, 1975). Stewart and Allen (1981) proposed that PSMG and another factor of fetal origin regulate the secretion of progesterone during early pregnancy.

The origin of the PMSG was originally thought to be the mare, because the hormone is produced by the endometrial cups, but these cups have been shown to be of fetal origin (Allen, 1975), and Stewart and Allen (1981) have proposed that the term *PMSG* be replaced with *chorionic gonadotrophin* (CG). Not surprisingly, in view of the fetal origin of the hormone, the secretion of PMSG is strongly affected by the genotype of the fetus. Table 10.12 shows the effect of intraspecific crosses and of pure matings on the secretion of PMSG.

It should be kept in mind in any consideration of plasma concentrations of hormones that higher concentrations in pregnant than in nonpregnant animals can be the result of either a higher production of such hormones or a slower metabolic clearance rate, or both. Heap (1972) pointed out that the higher progesterone concentration during pregnancy in women is the result of a higher rate of production, but that in guinea pigs it is the result of a slower metabolic clearance rate. This slower rate of metabolic clearance can be the result of binding of the steroid to the plasma protein with high affinity for this steroid. The binding may also protect the fetus against high concentrations of hormones.

Table 10.13 lists some of the binding proteins, the steroids they bind, and the relative affinity and capacity for binding of certain steroids. Heap (1972) also pointed out that species in which there is an appreciable increase in binding proteins, such as transcortin and progesterone binding protein, have a hemochorial placenta, and higher concentrations of plasma progesterone during pregnancy than during the cycle.

11

Expulsion of the Oocyte, Embryo, or Fetus

The oocyte, the egg with the developing embryo in it, or the fetus must be expelled from the maternal reproductive tract. (In animals with external fertilization it is the oocyte, and in oviparous animals with internal fertilization it is the egg with the developing embryo or the fetus.) In pipefishes and seahorses, the embryo develops in the brood pouch of the male (as discussed in Chapter 10), and thus the embryos have to be expelled from the male's pouch. In this chapter, we will discuss mainly the endocrine control of oviposition or parturition, that is, giving birth. We will not discuss the physiology of the muscles that expel the oocyte, egg, or fetus.

CYCLOSTOMES

In this group male and female spawn with their genital openings in proximity. This means that behavioral and spawning mechanisms are closely synchronized, but this does not mean that the same hormones are necessarily involved in both processes. The endocrine mechanisms involved in spawning are not understood.

ELASMOBRANCHS

La Pointe (1977) has recently reviewed the literature on the mechanisms that control contrac-

tions of the oviduct of elasmobranches and has also reported the results of some new studies of his own. In general, the isolated oviducts of the skate (*Raja erinacea*), the spiny dogfish (*Squalus acanthias*), the smooth dogfish (*Mustelus canis*), the blue shark (*Prionace glauca*), the dogfish (*Scyliorhinus canicula*), and the smooth hound (*Mustelus californicus*) respond to low doses of acetylcholine with clearly discernible contractions, but they do not respond in the same way to various posterior pituitary hormones such as oxytocin, arginine vasotocin, aspartocin, valitocin, glumitocin, and elasmobranch neural lobe extracts. The neural lobes of sharks have been found to have aspartocin and valitocin and those of rays have glumitocin (Sawyer, 1977). The postpartum oviduct of the leopard shark (*Triakis semifasciata*) responded to aspartocin, valitocin, oxytocin, vasotocin, and glumitocin, but the immature and the gravid oviduct did not. Epinephrine in low doses caused contractions of isolated oviducts of the skate, *Raja erinacea*, the smooth dogfish, the spiny dogfish and the blue shark. Dodd (1975) has speculated that there may be "a neural or hormonal pathway between oviducts and either the hypothalamus or the ovaries or both which controls ovulation and oviposition." This hypothesis is based on observations rather than experimental evidence and thus needs further investigation.

TELEOSTS

In oviparous teleosts, ovulation and subsequent oviposition are operationally similar to parturition in aplacental viviparous teleosts (Kujala, 1978). As discussed in Chapter 7, ovulation in some oviparous teleosts, such as the rainbow trout (*Salmo gairdneri*) and the Indian catfish (*Heteropneustes fossilis*), is regulated by catecholamines, progestins, corticosteroids, and prostaglandins.

In the aplacental viviparous guppy (*Poecilia reticulata*), injection of 20 ng/g deoxycortisol induced premature parturition in 1–2 days in 2 of 7 fish. Deoxycortisol also tended to reduce the latency for parturition induced by the neurohypophyseal hormones, oxytocin and vasotocin, and by an extract of carp neurohypophyses (Kujala, 1978). The deoxycortisol may have promoted the breakdown of the follicular membrane, as it does in vitro (Kujala, 1978), and so would make the young more susceptible to expulsion by contractions of smooth muscle induced by the neurohypophyseal hormones. These hormones are implicated in the expulsion of the egg in birds and the fetus in mammals; in teleosts, they cause the spawning reflex in the killifish *Fundulus heteroclitus* (Pickford, 1952; Pickford and Strecker, 1977; Wilhelmi et al., 1955; Macy et al., 1974), oviposition in the mekada (*Oryzias latipes*) (Egami, 1959), movements related to expulsion of the young from the brood pouch of the (male) seahorse (*Hippocampus* sp.) (Fiedler, 1970), and premature parturition in *Gambusia* sp. (Ishii, 1961) and the guppy (Kujala, 1978). If corticosteroids secreted by the embryo were to affect the breakdown of the follicular wall and were to induce increased sensitivity of this structure to maternal posterior pituitary, then a mechanism would exist for the young to control its own destiny.

AMPHIBIANS

Investigations of the effect of neurohypophyseal hormones on isolated oviducts in amphibians have revealed that the oviducts of some species respond by contraction to arginine vasotocin (an amphibian pituitary hormone), oxytocin, lysine vasopressin, isotocin, mesotocin (amphibian posterior pituitary hormone), and arginine vasopressin (La Pointe, 1977). Examples of such species are the bullfrog (*Rana catesbeiana*), the toads *Bufo bufo* and *B. marinus,* the South African clawed toad (*Xenopus laevis*), the mud puppy (*Necturus maculosus*), and the salamanders *Triturus cristatus, Salamandra salamandra,* and *S. atra.* Diakow (1977, 1978) has made the interesting observation that the gravid, but not yet sexually receptive female frog (*Rana pipiens*) gives a release call in response to attempts by the male to embrace her before she is ready to spawn. This release calling is inhibited by distension of the female with water and by injection of arginine vasotocin, which acts as an antidiuretic hormone and causes retention of water, which is necessary for reproduction in frogs with external fertilization.

REPTILES

In the lizards *Zootoca vivipara* and *Sceloporus cyanogenys,* removal of the ovary or of the CL is followed by abnormal parturition, whereas in the snakes *Natrix* sp. and *Thamnophis* sp. ovariectomy late in gestation does not affect parturition (Yaron, 1972). It thus appears that progesterone or other ovarian hormones affect parturition. Unfortunately, no data seem to be available on the changes in hormone concentrations in reptiles before, during, and shortly after parturition.

Injections of neurohypophyseal extracts induce premature expulsion of the egg or the embryo in the lizards *Lacerta vivipara, Sceloporus undulatus, Uta stansburiana* (Fox, 1977), and *Zootoca vivipara* (Fitzpatrick, 1966) and in the snakes *Natrix* sp., *Thamnophis* sp., and *Storeria* sp. (Fitzpatrick, 1966). The effects of posterior pituitary hormones on isolated reptilian oviducts are discussed by La Pointe (1977).

Hypophysectomy (AP), causes abnormal parturition in the water snake *Natrix sipedon* and the garter snake *Thamnophis sirtalis.* Although AP results in a decrease in the cellular size of the CL (Fox, 1977) it seems that this effect of AP is independent of the CL since CL removal does not affect parturition.

BIRDS

Much of the information about oviposition in birds has been obtained with the domestic fowl. The chicken is well suited for such investigations because the time of expected oviposition can be accurately predicted and because hypophysectomy and neurohypophysectomy can be performed by a skilled experimenter without affecting the survival of the animal. Furthermore, serial blood samples can be obtained conveniently and in fairly large numbers from the same hen.

In Chapter 8, we pointed out that except for the last oviposition of a sequence, there was a close temporal relationship between oviposition of one egg and the ovulation of the next sequence. This phenomenon led to the discovery by Fraps (1942) that LH-induced premature ovulation was accompanied by premature expulsion of an oviductal egg. Other hormones have been implicated in the control of ovulation, i.e., posterior pituitary hormones, hormones from the ruptured follicle, and steroid hormones. The plasma vasotocin concentration increases more than 40-fold immediately prior to and during oviposition (Sturkie and Mueller, 1976). Injection of avian posterior pituitary hormones, i.e., arginine vasotocin, and oxytocin can induce premature expulsion of the egg from the shell gland. Electric stimulation of the hen's preoptic area of the brain results in premature oviposition (Opel, 1964). Presumably, this stimulation results in the secretion of the hypothalamic hormones that are stored in the posterior pituitary. It should be noted, however, that neurohypophysectomy affects neither the pattern nor the timing of oviposition (van Tienhoven, 1968; Sturkie and Mueller, 1976).

Rothchild and Fraps (1944) observed that removal of the recently ruptured follicle resulted in a delay of oviposition, sometimes for several days. In 1978 Gilbert et al., showed that removal of the granulosa cells from the ruptured follicle was as effective as removal of the follicle itself. There are two schools of thought about the nature of the hormone secreted by the recently ruptured follicle, which is a rapidly regressing structure. On the one hand, Tanaka and Goto (1976) found that the active fraction of extracts from recently ruptured

follicles was a small peptide with a disulfide bridge, which thus resembled the posterior pituitary hormones. On the other hand, evidence by other investigators strongly suggests that prostaglandins are the hormones in the ruptured follicle that cause expulsion of the egg. First, plasma concentrations of prostaglandin E (PGE) are higher when there is an egg in the shell gland than shortly after oviposition (Hertelendy and Biellier, 1978a). Second, shortly before oviposition the PGF concentration in the ruptured follicle rises sharply (Day and Nalbandov, 1977). Third, administration of indomethacin, which blocks the synthesis of PG, delays oviposition in the chicken (Hertelendy and Biellier, 1978b; Shimada and Asai, 1979), and this effect can be overcome by injection of PGE into the shell gland (Hertelendy and Biellier, 1978b). Indomethacin also inhibits the normal preoviposition increase in electrical activity of the uterine muscle (Shimada and Asai, 1979). Finally, PGE induces premature oviposition in chickens and Japanese quail (Hertelendy, 1972). Prostaglandin $F_{2\alpha}$ ($PGF_{2\alpha}$) injections into laying hens resulted in increased uterine contractions, as measured by the electrical activity of the uterus. Injection of $PGF_{2\alpha}$, 2 h prior to the expected time of oviposition was followed by premature oviposition (Shimada and Asai, 1979). The evidence is thus quite persuasive that prostaglandins are involved in oviposition, but this does not mean that the recently ruptured follicle is the exclusive source of prostaglandins. Recently, Ogasawara and Koga (1978) have presented evidence that prostaglandins are also produced in the shell gland and that these prostaglandins affect oviposition.

Among the steroid hormones, exogenous progesterone (P_4) delayed oviposition between 0 h and 6 h after ovulation, so that oviposition followed P_4 injection by about 31 h, an interval similar to the interval between the preovulatory P_4 peak and oviposition. Incidentally, this exogenous P_4 also delayed ovulation, so that the original C_3 follicle became the C_1 follicle of a new sequence. When P_4 was administered 6–15 h after ovulation, it also delayed the C_3 ovulation, so that it became the C_1 ovulation of a new sequence, and the P_4 also delayed oviposition by 10 h, so that ovulation occurred at about the same time as the delayed oviposition. Injection of P_4 just prior to the occurrence of the preovulatory P_4 peak (9–12 h prior to ovulation) resulted in premature ovulation and oviposition of the oviductal egg, so that ovulation and oviposition occurred nearly simultaneously (Wilson and Sharp, 1976). The results obtained with injection of P_4 between 6 h and 15 h after ovulation, and between 9 h and 12 h prior to ovulation provide an explanation of the finding of Fraps (1942) that LH injection induced oviposition. If we assume that the LH-induced P_4 secretion (see Chapter 8) and that the bursting of the follicle resulted in PG secretion, then P_4 and PG would be present in concentrations similar to those found at normal oviposition. No data appear to be available about the concentration of posterior pituitary hormones around the time of LH-induced ovipositions.

The sharp peak in plasma corticosterone concentration at the time of oviposition (Chapter 8) is accompanied by a number of physiological and behavioral changes. Hens that are held in floor pens and that are provided with nests will make egg-laying calls, examine nests, and assume the egg-laying posture (Wood-Gush, 1971; Bobr and Sheldon, 1977). Hens that are kept in cages will show restlessness and so-called panic behavior and may give egg-laying calls. These behavioral changes are accompanied by an increase in body temperature (Bobr and Sheldon, 1977; van Tienhoven et al., 1979) and by uterine contractions prior to oviposition. Although normally associated with oviposition, the changes in behavior, body temperature, and corticosterone concentration can be disassociated from oviposition experimentally. After removal of the recently ruptured follicle, oviposition is delayed, but laying behavior, rise in body temperature, and increase in corticosterone occur at the time of the normally expected oviposition, yet fail to occur when the egg is eventually laid (Bobr and Sheldon, 1977; Beuving and Vonder, 1981). When oviposition is advanced by the injection of vasopressin, there is an increase in corticosterone that is probably the result of the vasopressin itself, and the corticosterone peak may or may not be the result of the expulsion of the egg. The important point,

however, is that at the time that the egg would normally have been laid, the hen shows normal egg-laying behavior and an increase in corticosterone concentration (Beuving and Vonder, 1981). It appears that the laying behavior, corticosterone peak, and temperature increase are dependent on the presence of the recently ruptured follicle and that the actual oviposition contributes only slightly to the increase in corticosterone and body temperature (Beuving and Vonder, 1981; Bobr and Sheldon, 1977). Similarly, contractions of the uterus occur at the time of expected oviposition 26–28 h after ovulation, even when there is no egg in the oviduct, as for example when the yolk falls in the body cavity and is resorbed, instead of being engulfed by the infundibulum of the oviduct (Shimada and Asai, 1978).

Although not concerned directly with oviposition, uterine contractions, as determined by measurements of electrical activity, occur not only at the time of oviposition, but also when the egg arrives in the uterus and at about 1 h before ovulation of the C_1 follicle, at a time when there is no egg in the oviduct (Shimada and Asai, 1978). The C_1 follicle (Chapter 8) apparently produces a hormone, which acts either directly on the uterine masculature or acts indirectly through the neurohypophyseal hormones. Shimada et al. (1981), in a series of elegant experiments, in which they ligated the stalk of the C_1 follicle at various times before ovulation, demonstrated that this hormone was released at 1.5 h to 2 h before ovulation. The chemical nature of the hormone remains to be determined.

In the hen, oviposition appears to be independent of the adrenergic and cholinergic nervous system; neither denervation nor specific blocking agents affect the timing or the process of oviposition in a significant manner (Sturkie and Mueller, 1976).

In birds, all embryos are in about the same stage of development at the time of oviposition, because all eggs are in the oviduct for about the same length of time after fertilization in the infundibulum of the oviduct. The initiation of oviposition is probably mainly controlled by the hormones produced by the ruptured follicle. In mammals, the initiation of parturition is, at least in some species, controlled by the secretion of hormones by the fetus, thus ensuring that the young are born at about the same stage of development for the particular species.

MAMMALS

Various external factors and the genotype are known to affect the duration of gestation in mammals, and thus to affect the approximate timing of the onset of parturition. One of the external factors that has an effect on parturition is the time in the breeding season when insemination occurs. Sheep bred early in the breeding season, for example, have a longer period of gestation than sheep bred late in the season. Horses bred between September and December have an average gestation period of 328.0 d, but horses bred between March and June have a gestation period of 342.2 d, the difference being significant. In cattle, which unlike sheep and horses, do not have a breeding season, the effect of month of insemination on duration of gestation is not significant. The hour of the day is important in horses and alpacas; 86 percent of foals are born between 1900 and 0700, with the highest frequency of births occurring between 2200 and 2300; in alpacas, 79 percent of the deliveries occur between 0700 and 1100 h (Nathanielsz, 1977). George (1969) found that during the winter, 63 percent of Dorset horn ewes lambed between 0800 and 1800, but that only 37 percent of Merino ewes lambed during that period; during the summer, 23 percent of Dorset horn and 39 percent of Merino ewes lambed during those hours. Thus both season and breed affected the distribution of the time of parturition during the day. A less pronounced, but definite effect of time of day on the distribution of births is reported for pigs, mice, Chinese hamsters, and humans, but not for cattle.

The effect of the genotype of the fetus on gestation length can be illustrated most dramatically by the genotype of horse-donkey crosses. The foal (stallion × mare) is carried 340 days, the hinney foal (stallion × jennet) 345 days, the mule foal (jack × mare) 355 days, and the donkey foal (jack × jennet) 365 days (Catchpole, 1969). Less dramatic, but nevertheless significant, is the effect of

certain bulls on the length of time their offspring are carried. In dairy cattle, the heritability of the effect of the bull on gestation length as a characteristic of the fetus is estimated to be 0.42; for horses the corresponding figure is 0.36, for swine 0.30, and for sheep, in which the fetal and maternal effects were not separated, 0.40 to 0.50 (van Tienhoven, 1968).

The sex of the fetus can affect the length of gestation to some extent. For example, the duration of gestation is 1.97 days longer for bull calves than for heifer calves, and male foals are carried longer than mare foals. In sheep, however, the fetal sex has no apparent effect on the length of gestation (van Tienhoven, 1968).

In mice, the size of the litter is negatively correlated with the duration of gestation. In experiments, the weight of individual offspring was increased by crossing inbred lines of mice, suggesting that the total fetal mass (total weight of the litter) was negatively correlated with the length of the gestation period; however, the weight of the placenta played no important part in determining the length of gestation, although the placental weight, of course, contributed to the complete contents of the uterus. Similar correlations between fetal mass and the length of the gestation period have been found in rabbits, guinea pigs, cattle, goats, sheep, mink, and humans, but apparently not in pigs (van Tienhoven, 1968).

Before parturition, the fetus undergoes changes that aid its survival after birth (Liggins, 1972). Surfactants appear in the alveoli of the lungs, which aid in the maintenance of the air-distended state of the lungs after birth. Glycogen is deposited in the liver, where it acts as readily available reserve to carbohydrate, and in the heart muscle, where it serves as a source of energy during hypoxia, which may occur during parturition. Finally, white and brown fat are deposited, the brown fat being particularly important for thermogenesis.

The mother also undergoes several changes prior to parturition, including dilation of the cervix and changes in the pelvis. Fitzpatrick (1977) concluded that in sheep, prostaglandins probably are required for the activation of enzymes that degrade the collagen in the cervix. According to

Naaktgeboren and Slijper (1970), it is relaxin that causes the softening of the cervix in pigs, cattle, rats, monkeys and humans. However, Fitzpatrick (1977) has stated that studies done with respect to relaxin and cervical dilation are "inadequate and should be repeated in relation to known hormonal changes." Recently, Hollingworth et al. (1979) reported that in pregnant rats the extensibility of the cervix increased sharply at day 22. Pregnant rats, ovariectomized on day 16 and treated with estradiol benzoate (EB) and P_4 to maintain the pregnancy, showed little increase in cervical extensibility; however, a combined treatment of EB, P, and relaxin resulted in an increase in extensibility to about 50 percent of the value found in day 22 in normal pregnant rats (Hollingworth et al., 1979).

It is also possible that prostaglandins (PG) cause release of relaxin. In pigs, there is a peak in the plasma relaxin concentration at delivery. This peak occurs prematurely if labor is induced prematurely by PG infusion, and it is delayed if delivery is delayed by indomethacin treatment (Sherwood et al., 1979).

The symphysis pubis in several species (e.g., the mole, some bats, and the Pinnepedia), consists of loose connective tissue. In some other species, it consists of cartilage, which near parturition, under the influence of relaxin, is changed into a soft connective tissue that permits the passage of large young, as, for instance, in guinea pigs. Even in some animals with a calcified symphysis pubis, this structure becomes soft (Naaktgeboren and Slijper, 1970), as in the pocket gophers (*Geomys bursarius* and *Thomomys bottae*), the mole rat (*Spalax leucodon*), and mongooses. The onset of parturition is under different control mechanisms depending upon whether the CL is required throughout gestation or whether the placenta is producing its own gestational hormones (Table 10.11) (Liggins, 1972). Much of the initial breakthrough in the understanding of the regulation of the initiation of parturition has come from pathological conditions. In the CL-dependent cattle, there are some breeds, e.g., Friesians, in which an autosomal recessive gene when present in the homozygous condition in the calf, prevents the normal delivery of that calf, and at the expected time of parturition, the dam does not show the relaxation of the pelvic ligaments, the softening of the cervix, and the decrease in plasma progesterone concentration that normally occur at the time of parturition. Such calves have an adrenal insufficiency. In Guernseys, a homozygous recessive gene, when present in the calf, results in a failure to grow after the seventh month of gestation and a failure in the onset of parturition at the normally expected time; such calves lack an anterior pituitary and, as a result, also have adrenal insufficiency (van Tienhoven, 1968). Among Angora goats, some does habitually abort. This abnormality was traced to hyperplasia of the adrenal, suggesting a role of the fetal adrenal in initiation of parturition (First, 1979). In sheep, a species in which ovariectomy does not lead to abortion, the eating of skunk cabbage (*Veratrum californicum*) between day 1 and day 15 of gestation resulted in abnormal fetuses, characterized by cyclopia, fetal adrenal and thyroid hypoplasia, and fetal adrenals that in vitro produced more testosterone, but less cortisol, cortisone, and corticosterone than normal fetal adrenals (Van Kampen and Ellis, 1972). In South Africa, the eating by pregnant ewes of the shrub, *Salsola tuberculata* (Fenx ex Moq) Schinz var. *tomentosa* (A. Smith ex Allen), during the last 50 days of the normal gestation period, resulted in gestations that were as long as 213 d instead of the normal 140 d, lambs that were up to three times the normal size and that showed pituitary and adrenal atrophy (Basson et al., 1969). This information on cattle, goats, and sheep suggested that the fetal pituitary and adrenal might have a key function in the initiation of parturition. Such a key role for the adrenal is attractive also because it is known that corticosteroids induce the formation of lung surfactants, which are necessary for adequate functioning of the lungs immediately after the umbilical cord is severed. As the result of the considerable skill of animal surgeons, it became possible to test this hypothesis by performing AP and adrenalectomy on fetuses in utero.

For the sake of convenience, we will discuss in the account that follows the regulation of parturition in species in which the CL is required for the entire or nearly the entire period of gestation, i.e., CL-dependent species, and species in which the CL can be removed relatively early during gestation without affecting the duration of gestation significantly, i.e., CL-independent species.

Initiation of Parturition in CL-Dependent Species

We will discuss as examples goats, cattle, swine, rabbits, and rats.

Goats. About 48 h before the onset of parturition, the maternal plasma progesterone (P_4) concentration starts to decrease sharply and maternal plasma estradiol-17β starts to increase (Flint et al., 1979). The normal gestation period is about 148 d, and ovariectomy leads to immediate abortion (Currie and Thorburn, 1977). These data suggested that the decrease in P_4 concentration might cause the onset of uterine contractions. Such an effect might act through P_4 inhibiting oxytocin release or by inhibition of myometrial contractions by high P_4 concentrations. Roberts (1971) obtained evidence that P_4 inhibited oxytocin release by acting on the nervous system; however, Blank and DeBias (1977) have stated that progesterone infusion during pregnancy does not always inhibit oxytocin release.

The signal for regression of the CL and thus for diminished P_4 secretion is probably of fetal origin. The evidence to support this hypothesis comes from several sources. First, fetal AP prolongs gestation, and the kids, when delivered, have small adrenals. Second, fetal plasma corticosteroid concentrations increase from near zero at 15 d before parturition to about 40 ng/ml at 5 d before parturition and then further to about 300 ng/ml at parturition (Currie and Thorburn, 1977). Moreover, infusion of ACTH in physiological amounts (Nathanielsz, 1978) results in a sharp drop in maternal plasma P_4 concentration, an increase in the prostaglandin F (PGF) concentration, and premature delivery of the kids (Currie and Thorburn, 1977). Finally, administration of corticosteroids to the fetus induced premature parturition (Currie and Thorburn, 1977). There is no evidence, however, that the ACTH concentration in fetal plasma increases before parturition or before fetal plasma cortisol

concentration increases (Nathanielsz, 1978). These and other considerations led Nathanielsz (1978) to conclude that there is no proof that the signal for parturition is of fetal pituitary origin. However, much of the evidence justifies considering the fetal AP-adrenal axis as the initiator of parturition as a working hypothesis. High fetal plasma corticosteroid concentrations stimulate placental enzymes, which are needed for the synthesis of estrogens from pregnenolone (Flint et al., 1979). The estrogens probably serve two functions; (1) they promote the dissolution of the feto-maternal connections, and (2) they result in luteolysis and thus lower P_4 secretion. This may be a direct effect of the estrogens on the CL, or the estrogens may act by increasing the synthesis and secretion of PGF. Evidence that fetal corticosteroids are implicated in estrogen production is provided by the findings (1) that delivery of AP kids was not preceded by an increase in maternal plasma estrogen concentrations (Rawlings and Ward, 1978 a), and (2) that after administration of dexamethasone (a synthetic corticosteroid), the maternal plasma concentration decreased and that of estrogen increased, as they did during spontaneous and during premature delivery of kids (Flint et al. 1978). The role of estrogens in parturition is indicated by the results of estrogen injection into pregnant goats. Injection of estrogen induced premature delivery of kids, which did not survive, presumably because corticosteroid secretion was not sufficient for secretion of surfactants in the fetal lungs (Currie and Thorburn, 1977).

Infusion of $PGF_{2\alpha}$ into the tributaries of the uterine vein ipsilateral to the CL causes luteolysis, and PGF concentrations rise about 24 h prior to parturition at the time that luteolysis starts. The $PGF_{2\alpha}$ may thus also serve two purposes: (1) it may induce luteolysis and thus withdrawal of progesterone from the myometrium, and (2) it may induce contractions of the myometrium. The importance of the role of P_4 is shown by the fact that progesterone administration can prevent ACTH-induced parturition (Currie and Thorburn, 1977). Figure 11.1 summarizes in schematic form a working hypothesis for the initiation of parturition in the goat.

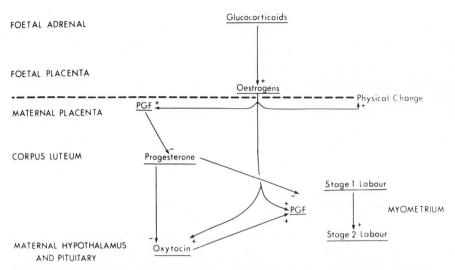

Figure 11.1 A model for initiation of parturition in the goat. (From Currie and Thorburn, 1977; reprinted with permission of W. B. Currie and *Excerpta Medica*.)

Cattle. The plasma P_4 concentration in dairy cattle decreases about 30–40 h before the onset of parturition at the end of the gestation period of about 282 d. Before day 215 ovariectomy generally results in abortion; after day 215 it generally does not result in abortion, perhaps because there may be enough P_4 secreted by the maternal adrenals (First, 1979) or released from fat stores (Hoffman, et al., 1979) to maintain pregnancy.

Maternal plasma estradiol and estrone concentrations are 3- to 10-fold higher at the time of parturition than at 26 d before parturition (Figure 11.2). Immediately before delivery, the concentration of plasma estrogens in the umbilical vein rises sharply; this has been interpreted as rapid degradation of estrogens by the fetus (First, 1979). After day 270 of gestation, there is a definite increase in fetal plasma corticosteroid concentration and an increase in the ratio of fetal cortisol to fetal corticosterone.

Parturition can be induced prematurely by four methods: First, it can be induced by fetal infusion of either ACTH, to stimulate fetal corticosteroid secretion (Comline et al., 1974), or of glucocorticosteroids, such as the naturally occurring cortisol or the synthetic dexamethasone. Such experiments suggest that, as in goats, the fetal corticosteroids induce parturition. Second, it can be induced by administering synthetic glucocorticosteroids, dexamethasone (Comline et al.,

1974; Wagner et al., 1974; First, 1979), flumethasone (Wagner et al., 1974; Hoffman et al., 1974; First, 1979), and triamcinolone acetonide (First, 1979) to the dam. Glucocorticosteroids administered to cows do not induce expulsion of the fetus if it has died (First, 1979); in vivo they do not inhibit P_4 synthesis by CL. These observations suggest that glucocorticosteroids do not induce parturition mainly as the result of their luteolytic effect. Third, parturition can be induced by administration of $PGF_{2\alpha}$ (First, 1979), and finally, by administration of estrogen (Thorburn et al., 1977), especially of diethylstilbestrol (First, 1979). This evidence, although not as convincing as that obtained with goats, suggests that the fetal pituitary, by its stimulation of fetal corticosteroid production, induces parturition in a manner similar to that suggested for the goat. Experiments by Hoffmann et al. (1979), however, have shown that the fetal cotyledons are the main source of estrogens and that estrogen production is not affected by the presence of a fetus, the precursors for estrogen production coming from the mother and not the fetus. These findings suggest that neither fetal glucocorticosteroids nor fetal precursors for estrogen production are the indirect luteolytic agents. The mechanisms for luteolysis at the end of gestation in cattle are not well enough understood to support the idea that parturition in goats

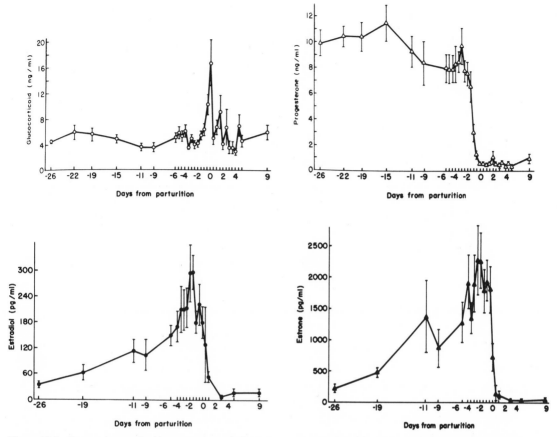

Figure 11.2 Serum glucocorticoid, serum progestin (expressed as progesterone), serum estradiol, and serum estrone in 10 cows from 26 days before to 9 days after parturition. (From Smith et al., 1973; reprinted with permission of *Journal of Animal Science*.)

and cattle is mediated by the same sequence of events.

Swine. The gestation period for swine is about 114 d; parturition is preceded by a sharp increase of fetal plasma corticosteroid concentration about 4 d before parturition, but maternal plasma corticosteroids do not increase significantly until the day of parturition (Silver et al., 1979). Maternal plasma estrogen concentrations (Figures 11.3 and 11.4), specifically those of estradiol and estrone (First, 1979), start to increase at about the same time that the fetal glucocorticosteroids start to increase. These estrogens are produced by the placenta and not by the maternal ovaries and adrenals or fetal ovaries and adrenals (Heap et al., 1977). At about 24 h before the onset of parturition, the maternal plasma P_4 concentration

starts to decrease sharply (Figure 11.3 and 11.4; Baldwin and Stabenfeldt, 1975; Silver et al., 1979; Ellendorff et al., 1979), possibly as the result of an initial rise in $PGF_{2\alpha}$ (First, 1979). At the time of delivery, there is a sharp rise in PGF concentrations and in the PGF metabolite, 13,14-dihydro-15-oxo prostaglandin in the amniotic fluid (Silver et al., 1979), and in maternal plasma $PGF_{2\alpha}$ metabolite concentrations. The maternal plasma relaxin concentration rises just before delivery (Sherwood et al., 1979) and remains high during spontaneous delivery and during $PGF_{2\alpha}$-induced delivery (Afele et al., 1979). Kertiles and Anderson (1979) found that injections of porcine relaxin from day 105 to day 110 of gestation caused premature cervical dilation in gilts. Just before delivery, maternal plasma corticosteroid concentrations increase sharply (Fig-

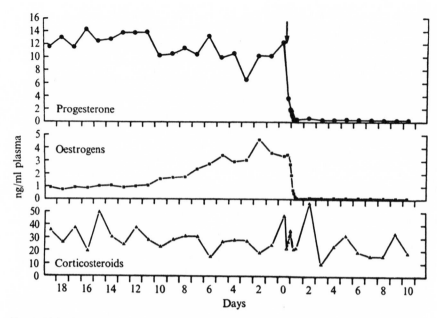

Figure 11.3 Plasma concentrations of progesterone, total unconjugated estrogens, and corticosteroids in sow (5th pregnancy) during late pregnancy, parturition (14 piglets), and early lactation. Day 0 and the arrow denote the day of parturition. (From Ash and Heap, 1975; reprinted with permission of *Journal of Endocrinology*.)

ure 11.3 and 11.4; Silver et al., 1979). Hypophysectomy and ovariectomy at any time during gestation are followed by abortion (Table 10.11). The evidence thus suggests that the decrease in maternal P_4 secretion may control the onset of parturition, and that, as in goats and cattle, a sig-

nal from the fetus may control the maternal P_4 secretion. This idea of fetal control of the onset of parturition in swine is supported by the several sources of evidence. Hypophysectomy of 50 percent or more of the fetuses in the litter prevents the decline of the CL and of maternal plasma P_4

Figure 11.4 Plasma concentrations of progesterone, total unconjugated estrogens, and corticosteroids in sow (5th pregnancy) during late pregnancy, parturition (11 piglets), and lactation and after weaning. Parturition occurred on day 0; weaning was completed by day 31, and behavioral estrus was observed on day 34. (From Ash and Heap, 1975; reprinted with permission of *Journal of Endocrinology*.)

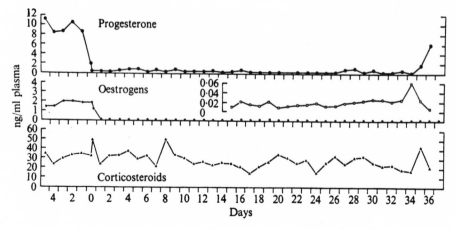

concentration and delays the onset of parturition (Bosc et al., 1974); decapitation of fetuses results in prolonged gestation (Stryker and Dziuk, 1976). Administration of ACTH to fetal pigs induces premature parturition as does the administration of dexamethasone, which is followed by a rapid decline in maternal plasma P_4 concentration (First, 1979). Administration of dexamethasone to mothers does not result in premature parturition when newly induced CL are present, suggesting that dexamethasone is itself not luteolytic (First, 1979). When PG synthesis is inhibited by administration of indomethacin, however, dexamethasone does not induce premature parturition and, of course, plasma PG concentrations do not increase. In the absence of indomethacin treatment, dexamethasone administration results in an increase in maternal plasma $PGF_{2\alpha}$ metabolites and in induction of parturition. These results suggest that dexamethasone induces parturition by stimulating the production of $PGF_{2\alpha}$ (First, 1979). It appears that in pigs, the fetal pituitary regulates the onset of parturition through adrenal glucocorticosteroid production, which, in turn, increases PG production in the uterus, with the PG causing luteolysis and a decline in P_4 production; the PGs possibly also cause uterine contractions of the now estrogen-dominated uterus. The available evidence suggests that in swine, a change in the fetal pituitary-adrenal axis may initiate the sequence of events that results in parturition.

Rabbits. Rabbits have a gestation period of 28 d, and as Table 10.11 shows, both AP and ovary are required for maintenance of pregnancy. The signal for parturition in the rabbit is probably not of fetal origin, for (1) fetal decapitation does not prolong gestation and may, in fact, shorten it (Heap et al., 1977), and (2) removal of the entire fetus, leaving the placenta in place, does not affect the length of gestation (Chiboka, 1977). These observations suggest that the control of delivery is maternal. The site of this control may be the hypothalamus, as is suggested by experiments of Lincoln (1971) with electrical stimulation of the infundibulum and of the median eminence. Such stimulation induced labor in pregnant rabbits and milk ejection in lactating does, indi-

cating that in this species, oxytocin release is involved in induction of labor. Contractions of the uterus induced by oxytocin are inhibited by progesterone (van Tienhoven, 1968). Prior to parturition, the plasma progesterone concentration decreases, and administration of progesterone or hCG, which induces new functional CL, delays the onset of labor. This suggests that oxytocin release, to be effective in causing expulsion of the fetuses, needs to be preceded by a decrease in progesterone secretion. The question then remains as to what causes the decrease in progesterone secretion. In view of the dependence of rabbit CL on the pituitary (see Chapter 8), it seems reasonable to assume that the maternal hypothalamus controls the fate of the CL and thus of parturition. The nature of the signal for the regression of the CL is, however, not known. The expansion of the uterus in relation to the myometrial progesterone concentration, may determine the onset of parturition (van Tienhoven, 1968).

Although the fetus is apparently not the source of the signal for the initiation of parturition, there is evidence that cortisol administered to either the doe or the fetus results in a decrease in plasma progesterone (Abel et al., 1973) and in early delivery (Nathanielsz and Abel, 1973), which can be prevented by administration of progesterone to the doe (Nathanielsz and Abel, 1973). Chiboka (1977) found that dexamethasone administered on day 21 did not cause expulsion of the placenta in fetectomized does, but administered on day 25, it caused premature parturition, whether a fetus was present or absent. Furthermore, he found no evidence for a local utero-ovarian pathway for the maintenance or termination of gestation in hemiovariectomized does.

Laudánski et al. (1979) reported that infusion of $PGF_{2\alpha}$ in pregnant does at day 20–21 and in pseudopregnant does on day 11–12 induced luteolysis and increased myometrial activity in pseudopregnant, but not in pregnant females. These results suggest that a factor in the conceptus other than progesterone may inhibit myometrial activity and thus control the onset of labor. These results and those of Chiboka (1977) suggest that a change in $PGF_{2\alpha}$ synthesis is not the main factor in inducing parturition in the rabbit, so that the exact mechanism for the timing of the onset of parturition in this species remains an enigma.

Rats. Gestation in the rat normally lasts 22 d. Until day 12, pregnancy is dependent upon the presence of the maternal AP to maintain the CL; after day 12, the CL are maintained by a fetal luteotrophin. During gestation, the maternal plasma progesterone concentration starts to decrease at day 17 (Figure 11.5), and the estradiol-17β concentration starts to increase sharply at day 16. In fetal plasma, the ACTH concentration reaches a peak between day 16 and 19 and then de-

creases, and the corticosterone concentration reaches a peak on day 18 of gestation and then decreases (Chatelain et al., 1980). These findings suggest that the fetal AP-adrenal axis may not be the determining factor in the initiation of parturition. This suggestion is further supported by the observations that neither fetal decapitation (Swaab et al., 1973) nor fetectomy (van Tienhoven, 1968) affects the duration of gestation, which in the case of fetectomized rats is judged by the expulsion of the placentas on day 22.

The timing of parturition is affected by photoperiod. In rats kept under 12L : 12D, with

Figure 11.5 Serum LH, prolactin, and progesterone levels during pregnancy in the rat. Number of samples accompanies each mean value. Vertical lines represent ± SEM. A = 0800–1100 h E.S.T.; vertical lines between A's on time scale represent 1430–1800 h E.S.T. Serum progesterone levels for day 12 and from day 15 to term are adapted from Pepe and Rothchild, 1972. (From Morishige et al., 1973. Copyright © 1973 by The Endocrine Society. Reprinted with permission of W. K. Morishige and Williams & Wilkins.)

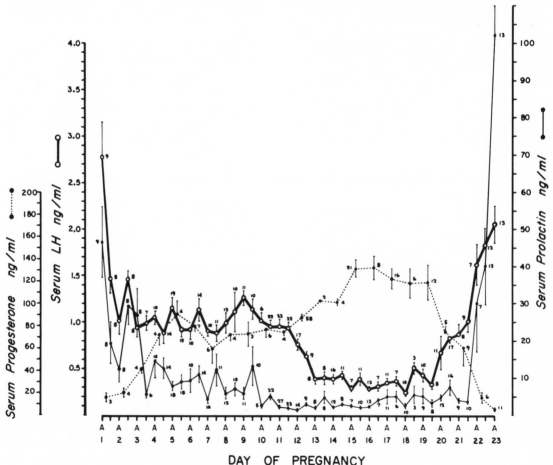

lights on from 0600 to 1800, most births occurred between 0800 and 1400.; when the light period was advanced by 8 h before midpregnancy, parturition was delayed by 12 h (Lincoln and Porter, 1979). In groups of rats kept under the control and the advanced photoperiod conditions, there was a correlation between the time of parturition on the one hand and the decline in maternal plasma P_4 concentration, the rise in maternal plasma prolactin, and the appearance of the enzyme 20α-hydroxysteroid dehydrogenase, which catalyzes the conversion of P_4 to 20α-hydroxyprogesterone, on the other hand (Lincoln and Porter, 1979). The effect of advancing the photoperiod on delay of parturition was eliminated by lesions in the maternal hypothalamus (Lincoln and Porter, 1979).

The hormone P_4 may be important in the initiation of parturition. The administration of P_4 to pregnant rats delays parturition (Lincoln and Porter, 1979); this supports the suggestion that a decrease in P_4 is essential for parturition to occur. The mechanism responsible for this decrease is not known, but since $PGF_{2\alpha}$ is luteolytic in rats and since high estrogen concentrations may increase PG production by the endometrium (Lincoln and Porter, 1979), estrogen synthesis may initiate the regression of the CL. The exact function of estrogens in parturition in the rat is not clear. Estradiolbenzoate administered on day 18 advanced parturition by 5–6 h, but administered on day 19 or 20, it delayed it by 6.5–8.5 h (Downing et al., 1981). Ovariectomy delayed the onset of labor in otherwise intact rats, an effect that was reversed by administration of estrogen. The administration of estrogen to fetectomized rats did not affect the time of onset of parturition. These data suggest that estrogen, which is of maternal ovarian origin, may be one of the important hormones in determining the onset of parturition, but that the fetus is required for the estrogen to have its effect. It is difficult to explain why administration of the anti-estrogen tamoxifen on either day 20 or 21 advanced parturition by about 10 h (Lincoln and Porter, 1979). Estrogens and P_4 interact with each other, as was shown by the delay of parturition and fetal death after administration of estrogen and P_4 (Acker, 1969).

Flower (1977) has discussed ways in which PGs may affect parturition. They may affect the uterine tone, depending on the hormonal influences on the uterus. Their effect on the response of the uterus will be discussed in more detail later in this chapter under human parturition. Prostaglandins may have an effect by vasoconstriction, thus causing local ischemia, or they may act as luteolytic agents and cause a drop in progesterone concentrations, thus making the uterus more sensitive to the contraction-inducing effects of PG. The following evidence supports this interpretation: (1) Injection of $PGF_{2\alpha}$ at days 19 and 20 of pregnancy was followed by a drop in progesterone concentration in peripheral blood and by premature delivery of the pups. (2) Injection of $PGF_{2\alpha}$ antibodies on day 17 of pregnancy delayed parturition in some rats until day 24. (3) Administration of prostaglandin synthase inhibitors, such as aspirin, indomethacin, and fenclozic acid, delayed parturition. (4) Abortions were caused by compounds that increased concentrations of PGs either by blocking their metabolism or stimulating their synthesis. (5) Administration of indomethacin and $PGF_{2\alpha}$ together did not affect the length of the gestation period. (6) Administration of estrogen (which can induce parturition) increased the activity of PG synthase and the concentration of PG in intact and in ovariectomized rats (Flower, 1977). Dukes et al. (1974) and Fuchs et al. (1976) stated that $PGF_{2\alpha}$ acts principally as a luteolysin and does not in itself cause contractions. In a clever series of experiments using unilaterally ovariectomized rats with unilaterally pregnant horns, Cavaillé and Maltier (1978) found evidence that a substance produced by one uterine horn was transmitted to the ipsilateral ovary and induced luteolysis, and subsequently parturition. The luteolytic substance may well be a PG. Further research is required both in vitro and in vivo with rat uteri under different hormonal conditions to determine whether PGs can induce contractions that are comparable to those observed at parturition.

Parturition can be induced by the release of maternal oxytocin. The periparturient uterus is sensitive to oxytocin, and although the receptors for oxytocin and PG receptors in the myometrium are separate entities (Chan, 1977), these hormones interact with each other. Oxytocin increases PG synthesis in the pregnant uterus more than tenfold (Chan, 1977) and such PG could then have a dual function: (1) causing luteolysis and so resulting in a decrease in progesterone and (2) causing contractions of the uterus. If oxytocin and PG each caused contractions, they would enhance delivery. Inhibitors of PG synthesis do not affect the response of rat uteri to oxytocin in vitro (Chan, 1977), and oxytocin inhibitors do not affect the uterine response to PG (Chan, 1977). In vivo, indomethacin does not affect the induction of parturition by oxytocin (Fuchs et al., 1976). It thus appears that the interaction between oxytocin and PG described by Chan (1977) is a redundant mechanism, which ensures that parturition will take place efficiently. It is not known, however, what signal would trigger the release of oxytocin from the posterior pituitary. As will be discussed later in this chapter, in several species, oxytocin release occurs after the fetus has distended either the cervix or vagina, that is, after parturition has been initiated and has progressed. If oxytocin release indeed initiates parturition, then the signal for its release needs to be determined.

Although it thus appears that no signal from the fetal pituitary or adrenal is necessary for induction of labor, plasma corticosterone concentrations in the fetal rat increase sharply just before delivery (Eguchi et al., 1977), an increase that is prevented by fetal decapitation (Eguchi et al., 1977). This increase in fetal plasma corticosterone coincides with the increase in the ACTH content of the fetal pituitary (Dupouy, 1976). These changes in corticosterone may be important for the fetus, increasing the surfactant secretion in the lungs for example, but they may not constitute a signal for the onset of labor.

Initiation of Parturition in CL-Independent Species

We will use sheep, rhesus monkeys, and humans as examples.

Sheep. The duration of gestation in sheep is about 140 d, and after day 50 the AP and ovary are dispensable, because the placenta produces sufficient amounts of P_4 for maintenance of preg-

nancy. The changes in the concentrations of estrone, estradiol, P_4, corticosteroids, LH, and prolactin are illustrated in Figures 11.6–11.8. Maternal corticosteroid concentrations increase 4 d prior to parturition, at about the same time that P_4 starts to decrease. The decrease in maternal P_4 is not a requirement for the initiation of parturition (First, 1979), as it is in goats. Estrone, estradiol, and prolactin concentrations in the maternal plasma increase about 2 d before parturition, whereas LH concentrations do not change significantly.

The evidence that the fetal AP and adrenal initiate parturition in sheep is persuasive. Hypophysectomy or adrenalectomy of the fetus, but not of the mother, prolongs gestation (Liggins, 1972). Infusion of synacthen ($ACTH_{1-24}$) into AP fetuses resulted in parturition, despite fetal corticosteroid concentrations being only about 50 ng/ml and in one case 20 ng/ml (Jones et al., 1978b), in contrast to the average of 100–120 ng/ml in intact control fetuses (Bassett and Thorburn, 1969). Synacthen infusion into intact fetuses at 125 d resulted also in premature par-

Figure 11.6 Changes in concentration of estrone (■–■) and estradiol-17β (□–□) in peripheral plasma [of ewes] (mean ± SEM) until just before parturition, N = 12; in subsequent collections N = 6–8. The mean hour of lambing is indicated by the arrow. (From Chamley et al, 1973; reprinted with permission of *Biology of Reproduction*.)

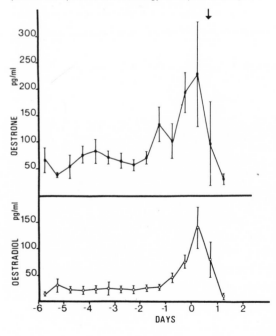

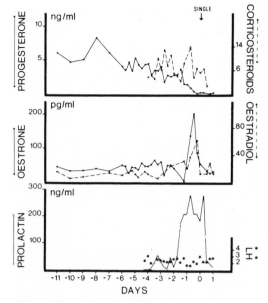

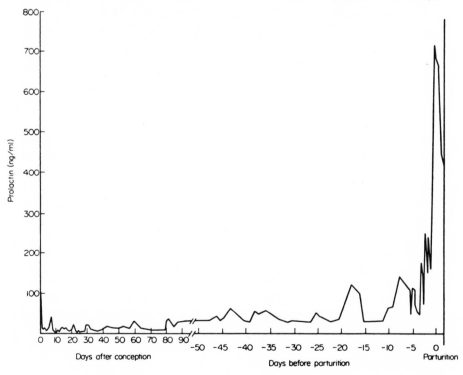

Figure 11.7 Changes in the peripheral plasma concentration of progesterone, corticosteroids, estrone, estradiol–17β, LH, and prolactin in one ewe which delivered a single lamb. (From Chamley et al., 1973; reprinted with permission of *Biology of Reproduction*.)

turition within 4 d and was accompanied by suppression of endogenous ACTH secretion during the first 24 h, followed by large fluctuations, indicating stimulation of ACTH secretion. Fetal corticosteroid concentrations fluctuated widely after the first 24 h of infusion (Jones et al., 1978a). Jones et al. (1978a) speculated that maturational changes in the central nervous system stimulate secretion of ACTH and make it less sensitive to inhibition by high corticosteroid concentrations. Robinson (1979) has proposed that in addition to ACTH, another factor of AP origin is required to mimic the increase in fetal corticosteroid concentration and in maternal plasma estrogen concentrations before parturition in intact ewes.

These observations, together with the finding that there is an increase in corticosteroids several days before parturition (Bassett and Thorburn, 1969) but an increase in ACTH that *follows* the corticosteroid increase (Rees et al., 1974), show that although the fetal pituitary is required for initiating parturition, it is not the cause for the in-

Figure 11.8 Average prolactin levels in plasma of seven ewes during gestation. (From Kann and Denamur, 1974; reprinted with permission of G. Kann and *Journal of Reproduction and Fertility*.)

crease in corticosteroid secretion. The increase in corticosteroids may, however, be caused by prostaglandins. Louis et al. (1976) found that fetal infusion of $PGF_{2\alpha}$ into the carotid artery resulted in about a threefold increase in corticosteroid concentrations. It remains to be determined whether the fetal pituitary is required for this response to $PGF_{2\alpha}$. The high corticosteroid concentrations (Liggins et al., 1977a), that are associated with the onset of parturition are apparently not a requirement for initiation of this process, as the experiments by Jones et al., (1978b) with AP fetuses, cited above, indicate.

The high cortisol concentrations about the time of parturition are correlated with a decrease in progesterone concentrations in the uterine venous blood and an increase in the concentration of estrone and estradiol-17α in fetal and maternal blood (Liggins et al., 1977a). Liggins et al. (1977a) proposed that cortisol might stimulate the activity of 17α-hydroxylase, 17-20 lyase, and possibly aromatase in the placenta. Anderson et al. (1978b) found that cortisol infusion indeed induced 17α-hydroxylase, but it did not stimulate either 17-20 lyase (Anderson et al., 1978b) or aromatase (Anderson et al., 1978a) activity.

The decrease in progesterone concentration (Scaramuzzi et al., 1977) and the increase in estrogen concentration (Liggins et al., 1977a; Scaramuzzi et al., 1977) may, together, stimulate $PGF_{2\alpha}$ release, or the decreasing progesterone may stimulate, and the increasing estrogens facilitate, $PGF_{2\alpha}$ release from the maternal placenta. A single intramuscular injection of 20 mg estradiol benzoate in ewes between day 142 and day 148 resulted in a high incidence of parturitions within 96 h. (Cahill et al., 1976).

The prostaglandins may be important in inducing and sustaining parturition. They may initiate contractions of the myometrium, or they may lower the threshold of the myometrium to oxytocin, which also induces uterine contractions (Liggins et al., 1977a). Oxytocin and PGF complement each other in sustaining uterine contractions, with oxytocin stimulating PGF synthesis by the maternal placenta (Liggins et al., 1977a), and PGF reducing the threshold for oxytocin-induced contractions. An increase in $PGF_{2\alpha}$ release (Flint et al., 1975; Mitchell et al., 1977) pre-

ceded by an oxytocin release (Flint et al., 1975) occurs upon distension of the vagina when the fetus is expelled.

Rawlings and Ward (1978b) found that in the ewe, at the onset of spontaneous increased uterine activity, progesterone concentrations were still high and that fetal and maternal estrogens rose before parturition. The maternal estrogen concentration was correlated with fetal estradiol concentration (r = 0.50) and fetal corticosteroid was negatively correlated with maternal progesterone concentration (r = −0.50). Maternal $PGF_{2\alpha}$ concentrations rose before parturition and remained high, showing a high correlation with uterine activity (r = 0.79) and with plasma estradiol (r = 0.79), and a negative correlation with plasma progesterone (r = −0.54). These correlations agree with the account presented above.

Infusion of either synthetic ACTH or of dexamethasone into either intact or AP fetuses resulted in an increase in maternal plasma progesterone and 13,14-dihydro-15-oxo $PGF_{2\alpha}$ concentrations and in parturition. Maternal estradiol concentration did not increase in the absence of the fetal AP. This suggests that the fetal AP is necessary for the increase in estradiol before parturition, but that it is not required either for the increase in maternal plasma P_4 or PG or for parturition (Kendall et al., 1977; Robinson, 1979). The initiation of parturition in the sheep and goat is similar in that in both species, the fetal pituitary initiates the secretion of fetal corticosteroids, which stimulate placental enzymes that convert progesterone to estrogens. The higher estrogen concentrations increase $PGF_{2\alpha}$, which in the goat is luteolytic and thus causes a decrease in P_4 secretion. In sheep the CL is not required for gestation, and the decrease in maternal plasma P_4 is not required for parturition to occur, whereas such a decrease in P_4 concentration is required in the goat.

Rhesus Monkeys. In this species, which has a gestation period of about 164 d, evidence for the fetus having control over its delivery is not clearcut. In one report (Hutchinson et al., 1962), 2 of 8 fetuses, in which the AP was destroyed by radioactive yttrium implantation, were carried beyond normal term, but Chez (1973) found that 4 of 5 surviving fetuses were carried beyond term (Lanman, 1977). Decapitation at 73–78 d of gestation of 18 fetuses resulted in 4 abortions within 10 d, delivery of 2 partially absorbed fetuses 30–65 d after surgery, and 2 fetuses dead and eventually delivered by Caesarean section. Of the 10 remaining fetuses, 2 were delivered by Caesarean section at 177 d and 188 d, and the other 8 were delivered between 137 d and 186 d, with an average length of gestation of 159.6 d, and controls having an average length of gestation of 167 d (Novy, 1977). These data do not support the concept that the fetal AP is required for normal delivery. Adrenalectomy of the fetus (Mueller-Heubach et al., 1972) results in too high an incidence of death to draw any valid conclusions about its effect on length of gestation.

Lanman (1977) found that fetectomy frequently resulted in retention of the placenta beyond normal term. After fetectomy, plasma estradiol decreased to about one-third of that of controls, and there was no increase in estradiol concentration prior to explusion of the placenta, whereas in normal parturition, there was an increase in maternal estradiol and estrone (Challis et al., 1977; Lanman, 1977).

In the rhesus monkey, when 90 percent of the gestation time has passed, the maternal progesterone concentration is low (2–4 ng/ml) (Challis et al., 1977; Lanman, 1977) in contrast with the chimpanzee (about 100 ng/ml), marmoset (about 150 ng/ml) and humans (about 180 ng/ml) (Lanman, 1977). Maternal progesterone concentrations do not decrease, but P_4 concentrations in the uterine vein and umbilical vein decline gradually towards the time of parturition (Lanman, 1977).

Fetectomy does not affect maternal progesterone concentrations, probably because the maternal ovary compensates for the loss of placental progesterone production (Lanman, 1977). The mother's endocrine system does not compensate, however, for the loss of estrogen production as a result of fetectomy (which deprives the placenta of precursors for estrogen synthesis) (Lanman, 1977). It would be attractive to propose that the decrease in progesterone in the uterine vein and the increase in estrogen concentration are responsible for the uterine contractions that lead to

delivery of the fetus, but that administration of estrogen does not result in parturition in the rhesus monkey because progesterone secretion does not decrease.

In primates, dexamethasone and cortisol permeate the placenta and thus inhibit the fetal adrenal glands. This inhibition results in a loss of precursors for estrogen synthesis by the maternal placenta and, therefore, a decrease in maternal plasma estrogen concentration (Challis et al., 1977; Lanman, 1977). Corticosteroid administration to either mother or fetus does not affect the length of gestation (Challis et al., 1977).

Prostaglandins are implicated in the delivery of the fetus. Exogenous $PGF_{2\alpha}$ can induce premature parturition (Challis et al., 1977), and oral administration of indomethacin results in delivery at 180 d instead of at 161 d (Novy, 1977). At parturition, but not prior to its onset, there is a sharp increase in 13, 14-dehydro-15-keto $PGF_{2\alpha}$ in the maternal plasma and in the amniotic fluid (Challis et al., 1977; Mitchell et al., 1976). During administration of ACTH to the fetus, a procedure that induces premature delivery, there is a significant increase in the concentrations of estrone, PGF, and PGE in the amniotic fluid, and in the concentrations of estrone and estradiol in the maternal and fetal plasma (Novy, 1977). It remains to be determined, however, what the mechanism is that signals the initiation of parturtion.

Humans. Much of the evidence pertaining to initiation of parturition depends on clinical data and is fragmentary, and so open to various interpretations. It is, therefore, not surprising that there are strong differences of opinion about the controlling mechanisms in parturition in humans.

Swaab et al. (1973) have reported that anencephalic and normal fetuses are born at the same fetal age and that fetal injections of ACTH plus oxytocin do not cause earlier delivery. Anencephaly results in labor being prolonged. These and other data (Liggins et al., 1977b) suggest that the fetal pituitary and the fetal adrenal do not play crucial roles in the initiation of parturition. The fact that the synthetic corticosteroid betamethasone, administered to pregnant women, caused a sharp decrease in cortisol concentrations in fetal plasma and in amniotic fluid, but no change in the time of spontaneous delivery (Gennser et al., 1977) suggests that the fetal adrenal does not provide the signal for the onset of parturition. However, Turnbull et al. (1977) interpret the sudden rise in cortisol secretion by the adrenal in normal pregnancies as the possible final trigger for parturition in women.

There seems to be reasonably good agreement that toward the end of the nine-month gestation period, maternal peripheral plasma progesterone concentrations decline and estrogen concentrations increase (Turnbull et al., 1977; Csapo 1977). Liggins et al. (1977b) state, however, that there are no "readily measurable changes in placental hormone metabolism before labor starts." They do, however, cite evidence that there is an increase in plasma estradiol-17β prior to onset of labor in women at risk of premature labor.

Csapo (1977) has proposed that placental progesterone suppresses myometrial activity and that labor is the result of a sharp drop in the P_4 concentration in the uterine vein. As a result of the decrease in P_4 concentration, prostaglandins may induce contractions of the myometrium. Csapo has named this the "see-saw theory." It is his contention that the fetus determines the maintenance of pregnancy (the death of a fetus generally induces labor). He implies that the role of the fetus in initiating labor may be the stretching of the uterus. This may result in interference with the blood supply of the placenta, leading to a decrease in uterine progesterone concentrations. This increases the amounts of prostaglandins (which are synthesized all the time in the rabbit, considered by Csapo as a model for human parturition) toward the end of gestation and so induces myometrial contractions. Oxytocin presumably maintains labor once it has been initiated (Liggins et al., 1977b).

General Considerations on Initiation of Parturition

It is apparent from the account just given that there is good evidence that in the goat, sheep, and to a lesser extent the cow and pig, parturition is initiated by a signal from the fetal pituitary. This signal, by way of the adrenals and their cor-

ticosteroid production, causes a decrease in progesterone concentrations and an increase in PGs. The combination of low progesterone and high PG starts myometrial contractions, which force the fetus through the cervix, which has become dilated and softened under the influence of PG and/or relaxin. As the fetus enters the vagina and stretches it, it initiates an increase in the concentration of PGF in the utero-ovarian vein, and, in sheep, an increase in uterine contractions (Mitchell et al., 1977). There is also a release of oxytocin in sheep (Forshing et al., 1979), as well as in goats, pigs, horses (Forshing et al., 1979), and rabbits (Beyers and Mena, 1970). Section of the pelvic nerves in lactating goats prevents the oxytocin release induced by stretching of the vagina, the so-called Ferguson reflex (Peeters et al., 1971). In rabbits, spinal cord transection between thoracic vertebrae 2 and 3 did not affect the onset of parturition, but labor was prolonged, and a high proportion of the fetuses were dead at birth (Beyer and Mena, 1970), probably because the spinal cord transection interfered with oxytocin release. Oxytocin helps to stimulate uterine contractions which help in the expulsion of the fetus, and which may also stop uterine bleeding when the young nurses immediately after it is born, since this nursing results in oxytocin release.

The hormones involved in expulsion of either the egg or the embryo or fetus in teleosts, birds, and mammals are oxytocin, PGs, progesterone and corticosteroids. The relative importance of each differs, but similarities, though interesting, cannot be used to draw general conclusions, since our knowledge is too fragmentary.

Parturition

The activity of the myometrium that expels the fetus or fetuses has received much less attention than the endocrinology of the initiation of parturition (Taverne et al., 1979). Uterine contractions are either in a tubocervical or in a cervicotubal direction, and during parturition, propagation of both types of contractions occurs in the rabbit, the rat, and the pig (Taverne et al., 1979).

During parturition in the sow, myometrial activity appears to be initiated most frequently at either the tubal or at the cervical end of the uterus and less frequently in the remainder of the uterine horn. Before and during expulsion of a piglet, cervicotubal propagation of contractions dominate tubocervical propagation. Once a uterine horn is emptied, the contractions decrease (Taverne et al., 1979). During the peak of uterine contractions in sheep, abdominal contractions begin and increase once the cervix is dilated. These abdominal contractions override the intrauterine pressure waves, and the lambs are delivered (Hindson et al., 1965).

Two effects of hormones of the myometrium are discussed: (1) the effect of steroid hormones and oxytocin on contractions of the smooth muscle and (2) the effect of steroid hormones on the ultrastructure of the myometrium. The study of the contraction of strips of rabbit myometrium in vitro has shown that: (1) there was little spontaneous motility or development of tension either after electric stimulation or after oxytocin administration when the strips were obtained from either ovariectomized or immature rabbits; (2) strips from estrogen-treated rabbits showed spontaneous motility and a response to either electric or oxytocin stimulation; and (3) progesterone administration to estrogen-primed rabbits eliminated the spontaneous motility and also the responsiveness to electric and to oxytocin stimulation (van Tienhoven, 1968). This effect of progesterone may be mediated by the binding of activator calcium (A-Ca) necessary for the actinomycin-adenosine triphosphate (ATP) interaction which causes muscle cell contraction (Csapo, 1977). Calcium ion transport in smooth muscle is cyclic-AMP-dependent (Krall and Korenman, 1977). Currie (1980) has found the strips of myometrium from pregnant rabbits can be stimulated by addition of 10^{-3} to 3^{-2} molar imidazole to the bathing medium. The imidazole probably acts by stimulating phosphodiesterase activity, thus lowering cyclic-AMP and allowing an influx of Ca^{2+} into the contractile cells.

In vivo studies with pregnant and parturient rabbits have shown that P_4 suppressed mechanical and electrical activity of the myometrium and that in the absence of P_4, action potentials were generated, with the electric activity being synchronous, so that pressure cycles of great amplitude were generated (van Tienhoven, 1968).

In rats, P_4 does not inhibit uterine contractions, but it restricts the contractions to local ones, which are not as rapid as the contraction waves found in estrogen-dominated uteri (Fuchs, 1974, 1975; Fuchs et al., 1976).

Immediately prior to, during, and immediately after parturition, gap junctions (nexuses) can be found between myometrial smooth cells in rats (Garfield et al., 1977, 1978, 1979a), guinea pigs, sheep, humans (Garfield et al., 1979b) and golden hamsters (Blaha and Niewenhuis, 1980). These gap junctions, by facilitating electric coupling, improve the synchronization of muscle contractions required for parturition. In rats, ovariectomy at day 16–17 of pregnancy results in abortion and the formation of gap junctions; administration of progesterone prevents normal parturition and the formation of gap junctions (Garfield et al., 1978). In sheep the increase in estrogen concentration and decrease in progesterone concentration just prior to parturition is correlated with the appearance of gap junctions (Garfield et al., 1979b). Old golden hamsters deliver their young at day 16–18 of pregnancy while young hamsters deliver theirs at day 15. This delayed parturition is accompanied by a delayed appearance of gap junctions (Blaha and Niewenhuis, 1980). In general there has been so far an excellent correlation between onset of parturition and the presence of gap junctions and their absence in nonparturient uteri.

This review of the initiation of parturition and of the control of uterine contractions shows that both the control of the initiation of parturition and the effect of a particular hormone on the contractions of the uterus vary considerably among species. The adaptive significance of each of these mechanisms is not clear.

12

Reproduction and Immunology

The interrelationships between the immune and the reproductive system are too many and too complex to be given extensive treatment in this book. Here we attempt only to mention briefly some of the fundamental questions, some medical and biological phenomena, and some applications of immunological techniques to reproductive physiology. Much of the discussion will be limited to mammals, because the overwhelming amount of research has been done on this class of animals.

We will discuss the use of immunological techniques that have advanced knowledge about reproductive physiology, the relationship between immunology and male reproduction, and the relationship between immunology and female reproduction.

IMMUNOLOGICAL TECHNIQUES

Three immunological techniques, radioimmunoassay (RIA), passive and active immunization, and localization of sites of hormone production will be discussed.

Radioimmunoassay

The account of the development of this technique should be read in the Nobel Prize lecture by Yallow (1978), who was one of the developers of this valuable technique. Radioimmunoassay can be defined as "Assay techniques based on competitive binding of a known amount of isotope labeled hormone and variable amounts of unlabeled hormone ('unknown') to anti-hormone antibody molecules" (Herbert and Wilkinson, 1972). For protein hormones, antibodies are raised usually in rabbits or larger laboratory or domestic animals. For the production of steroid hormone antisera, the steroid needs to be conjugated as a hapten to a protein. The position of the steroid molecule to which the protein is conjugated to a large extent determines the specificity of the antisera. Details about the techniques used for the RIA of different hormones can be found in Jaffe and Behrman (1974).

The great sensitivity of RIA has made it possible to make great strides in the understanding of reproductive endocrinology. The technique has made it possible to evaluate the dynamic changes in hormone concentrations in plasma and organs during so-called spontaneous physiological events, and before, during, and after experimental treatments. Much of the evidence used in previous chapters has been based on RIAs of various hormones.

Passive and Active Immunization against Hormones

Passive immunization (i.e., with the injection of antisera), has made it possible to study the effect of short-term inactivations of the recipient's own hormones, thus allowing the study of the requirement of an endogenous hormone for a specific physiologic function or event. We have mentioned an example of this in the study of the function of LH and FSH in the reproduction of the male (Chapter 6) and the female (Chapter 7) rat.

Active immunization, in which the animals to be studied are induced to produce antibodies against the injected antigens, makes it possible to study the long-term effects of selective hormone deprivation. Information on the effects of antibody production against gonadal steroids (conjugated to a protein) in mammals is provided in the excellent review by Nieschlag and Wickings (1978). According to these authors, active immunization against testosterone has the following repercussions: (1) plasma testosterone concentrations become very high (up to 100 times the normal concentration); (2) testosterone binding in serum is increased, as is testosterone production; and (3) testosterone clearance is decreased. Plasma luteinizing hormone concentrations are also higher in testosterone-immunized animals than in controls, but plasma FSH concentrations are similar in experimental and control animals. Several functions that are testosterone-dependent are adversely affected in testosterone-immunized animals; e.g., the accessory reproductive organs are smaller and show histologic signs of atrophy; male offspring of female rabbits actively immunized against testosterone are feminized; and male rabbits immunized against testosterone show less sexual activity than controls. The male pheromone produced by male boars in fatty tissue is reduced in testosterone-immunized boars. We thus find the apparently paradoxical situation of high testosterone titers and diminished activity of testosterone-dependent functions. This can be explained by the lack of negative feedback as a result of the antibodies inactivating androgen activity.

Localization of Sites of Hormone Production

Fluorochromes (Herbert and Wilkinson, 1972) coupled to an antigen or its antibody make it possible to localize a particular hormone in specific cells. This technique can be used for qualitative and quantitative procedures, but an unlabeled antibody-enzyme technique can be used instead (Sternberger, 1974). These techniques can be used for investigations with both light and electron microscopes (Zimmerman, 1976).

IMMUNOLOGY AND MALE REPRODUCTION

H-Y Antigen

The role of the H-Y antigen in sexual differentiation has been discussed in Chapter 1, and needs no elaboration here.

Mating Preferences and Major Histocompatibility Genes

In mice, there are strain differences among males in their preferences for females, which are similar or dissimilar with respect to the male in the major histocompatibility (H-2 complex) genes (Yamazaki et al., 1976). For example, Andrews and Boyse (1978) have shown that B-6 and B6–T1a^a males mate preferentially with B6–T1a^a females when given a choice of B6 or B6–T1a^a females. Further investigations have shown that in female mice, a recognition-of-identity gene (Ri–1) influences mating preference of males, and that in males there is another gene (Ri–2) that influences their mating preference. The approximate location of these genes within the major histocompatibility complex has been established (Andrews and Boyse, 1978; Yamaguchi et al., 1978).

Antigens of the Male Reproductive System

Bishop (1968) has used the enzyme sorbitol dehydrogenase (SDH), which is associated with spermatogenesis throughout the animal kingdom, both for monitoring spermatogenetic activity and as an antigen to induce aspermatogenesis in guinea pigs. He found that antigen to be effective when used for active immunization, but antibodies to SDH raised in rabbits did not affect spermatogenesis in guinea pigs.

The secondary reproductive organs of various mammals (e.g., dog, rabbit, rat, mouse, hamster, and gerbil) have highly organ-specific antigens (Barnes, 1972). In some cases, such antigens are shared by related genera, e.g., mice, rats, hamsters, and gerbils but they are not shared with rabbits and guinea pigs. In rats, the presence of these antigens has been shown to be androgen-dependent (Barnes, 1972). Such antigens can be used to determine the embryologic relationships among accessory organs and the phylogenetic relationships among species.

Seminal plasma, as expected, contains antigens also found in accessory reproductive organs. An example of such an antigen is the sperm-coating protein antigen (SCA) found in the seminal vesicles, seminal plasma, and on the ejaculated sperm in humans and rabbits, but not on either epididymal or testicular sperm (Barnes, 1972). Neither this nor any other antigen secreted by the accessory organs is essential for fertilization, as is evident from the fact that epididymal sperm are capable of fertilizing oocytes (Barnes, 1972). According to Beer and Billingham (1976), no autoimmune lesions have been produced in any of the accessory organs of reproduction except possibly in the seminal vesicles of the guinea pig.

The testis is an immunologically privileged site, because the blood-testis barrier prevents self-tolerance. If the sperm penetrate the blood-testis barrier, autoimmune responses may be expected. They do indeed occur and may, in turn, lead to destruction of the germ cells, azoospermia, or permanent sterility (Beer and Billingham, 1976).

An important practical aspect of autoimmune sterility involves the consequences of vasectomy. There is evidence in guinea pigs (Alexander, 1973), rabbits (Bigazzi et al., 1977), rhesus monkeys (Alexander, 1975) and humans (Alexander et al., 1974) that after vasectomy there is an increase in sperm antibodies in the blood. This increase may be associated with orchitis, e.g., in rabbits (Bigazzi et al., 1976), but in guinea pigs, Tung and Alexander (1977) found that the histopathological changes that followed vasectomy were not correlated with a humoral immune response to sperm. Tung (1978) reported that peritoneal exudate cells from long-term vasoligated guinea pigs to syngeneic recipients induced al-

lergic orchitis. It appears, therefore, that there is both a specific and a nonspecific immune response in the testicular damage after vasectomy.

These findings have clear implications for the practice of vasectomy in humans, especially if future restoration of fertility is contemplated. Alexander (1975) has found, however, that in rhesus monkeys, the sperm count after vasovasotomy (rejoining of the cut ends of the ductus deferens) was initially lower than prior to vasectomy, but that after about eight months, sperm counts were normal. A more serious potential side effect than orchitis after vasectomy is the finding by Alexander and Clarkson (1978) that the severity of atherosclerosis induced by diet is higher in vasectomized than in either sham-vasectomized or control cynomolgus monkeys (*Macaca fascicularis*). We shall discuss the effects of semen and sperm antibodies on the reproduction of the female later in this chapter.

Sperm Surface Antigens

It is known that ABO blood group antigens are expressed on sperm, but it is not known whether there is selection in the woman's genital system for sperm with a specific antigen. The possibility of such selection is indicated by the observation that antibodies to ABO antigens are found in the cervical mucus (Beer and Billingham, 1976). So far there is no evidence that ABO incompatibility is a common cause of infertility (Beer and Billingham, 1976).

There is suggestive evidence in the rat that sperm bearing determinants different from those on the oocyte may have a selective advantage over sperm bearing determinants identical to those in the oocyte. Such a phenomenon would result in a greater heterozygosity than in the absence of such selective fertilization (Johnson, 1976).

We have discussed in Chapter 10 the peculiar inheritance of the gene for short tail (brachyury) in the mouse. Recently, it has been found that the genes for this characteristic can be detected by serological methods as surface antigens on the sperm (Bennett et al., 1972). In view of the possibility of selective fertilization, this might offer a partial explanation for the unique inheritance of the *T*-locus genes; it would not, however, explain the effects of late insemination discussed in Chapter 9.

IMMUNOLOGY AND FEMALE REPRODUCTION

Antigens of the Male Reproductive System and the Female

The question has arisen whether the antigens in the seminal plasma and on the sperm might not cause immunization in the female. This aspect of the immunology of reproduction has been recently reviewed by Jones (1976), and the major part of the following account is taken from this review. Acute anaphylactic reactions to semen in women are apparently extremely rare, but they have been reported to occur. The evidence that isoimmunization against semen antigens is important in explaining long-term infertility in women without major or minor organic causes of infertility is not convincing for mammals. In chickens, Wentworth and Mellen (1964) found that after repeated artificial inseminations, the antibody titer to rooster sperm became higher than that in virgin females and that the duration of fertility was negatively correlated (r = −0.46) with sperm antibody titer. It is, of course, possible to raise antibody titers to semen products by active immunization, using classical immunological methods, such as injection of antigen with or without Freund's adjuvant.*

After such immunization, fertility in turkey hens was reduced below that in controls, the extent of the reduced fertility being dependent to some extent on the male line from which semen was used for artificial insemination (Burke et al., 1971). In laboratory mammals (Jones, 1976; Kummerfeld and Foote, 1976) and in cattle (Jones, 1976), active immunization against

---

*An adjuvant is a substance added to a drug to heighten its effect. Freund's adjuvant consists of "a water-oil emulsion adjuvant in which killed, dried, mycobacteria (usually *Mycobacterium tuberculosis*) are suspended in oil phase." Injection of this adjuvant stimulates cell-mediated immunity and antibody production (Herbert and Wilkinson, 1972). Freund's incomplete adjuvant does not contain the suspension of bacteria.

sperm has resulted in a decrease in fertility, mainly, at least in rabbits, through interference with fertilization (Kummerfeld and Foote, 1976).

Immunization with either sperm or semen sometimes introduces many antigens, and thus the antibodies produced are not very specific. Instead of sperm and semen, an isozyme specific for male germ cells, lactate dehydrogenase-C_4 (LDH-C_4, also called LDH-X), has been isolated and used to immunize mammals in attempts to affect their fertility. This isozyme may function as an autoantigen, and an antiserum against LDH-C_4 shows considerable cross-reactivity to the C_4 isozyme found in several species.

Lerum and Goldberg (1974) have reported that after active immunization with LDH-X on day 1–4 post coitus, mice showed a decrease in incidence of pregnancy. Their observations are in accord with observations on rabbits, that such immunization significantly decreased the precentage of embryos to corpora lutea to 29.3 percent from the 88.3 percent found in controls (Goldberg, 1973). These investigators (Goldberg and Lerum, 1972; Lerum and Goldberg, 1974) also found that, in mice, passive immunization with rabbit antiserum to the mouse LDH-X antigen on days 1–4, or days 4–7, or days 7–13 post coitus, reduced the incidence of pregnancy. An antiserum to a differently prepared isozyme (Goldberg and Lerum, 1972; Lerum and Goldberg, 1974) diminished the incidence of fertilization *in vitro*, but neither diminished the *in vivo* fertilization nor the survival of embryos (Erickson et al., 1975). At first thought, it seems difficult to explain the effect of antiserum to a sperm-specific antigen, when such an antiserum is injected as late as 7 d post coitus. According to Lerum and Goldberg (1974) and Goldberg (1974), the effectiveness of the immunization at 7 d post coitus may be the result of supernumerary sperm within the perivitelline space of fertilized oocytes (zygotes), which provide specific recognition sites for the LDH-X antibody. They speculate that the induced heteroantigenicity leads to immunological rejection of the blastocyst analogous to the rejection of blastocysts of another species by the uterus (see section on immunology of gestation in this chapter).

Active immunization against LDH-X in male

rabbits (Goldberg and Wheat, 1976) and guinea pigs (Wellerson et al., 1974) leads to infertility. In the guinea pig, the aspermatogenesis so induced is transient; in 50 percent of the affected males, fertility is eventually restored and normal offspring are born (Wellerson et al., 1974). Goldberg (1977) did not find aspermatogenesis after active immunization of guinea pigs with LDH-X and explained the results obtained by Wellerson et al. (1974) as due to a protein impurity in their preparation of LDH-X.

Antigens of Female Germ Cells

Mature mammalian oocytes are surrounded by a gelatinous-like layer, the zona pellucida (ZP). This layer has been assigned a large number of functions, e.g., sperm recognition, block to polyspermy (Chapter 9), osmotic regulation, and others (Shivers and Dunbar, 1977). The ZP has antigenic properties, and after treatment of oocytes with ZP antibodies, several functions of the ZP are altered, so that sperm fail to attach themselves to the oocyte surface and fertilization fails to occur (Garavagno et al., 1974).

The question whether women produce autoantibodies against the ZP and, if so, whether this affects fertility has been investigated by Shivers and Dunbar (1977). They found that the sera from 22 infertile women produced a strong antibody reaction (measured by an immunofluorescence technique) in 6 cases and a moderate response in 9 cases. These numbers are too small to draw definite conclusions about the biological significance of such autoantibodies, but the results suggest that, at least in some cases, infertility may be explained on this basis. If is unfortunate that no measurements were made on the sera of fertile women.

Blood Groups and Reproduction

In humans, the best-known instance of interaction between blood groups and reproduction is the rhesus factor. About 85 percent of Caucasians, 93 percent of all black people, and virtually all Mongolian people possess a blood antigen that is also found in the rhesus monkey. The antigen is inherited as a simple dominant factor. In a woman who lacks the rhesus factor (i.e., is

Rh negative), problems may arise if her fetus is Rh positive and the woman has been sensitized to the Rh factor and has produced antibodies to it. The antibodies may attack the antigens on the fetus' red blood cells, destroying the red blood cells, resulting in erythroblastosis foetalis. If this occurs late enough during pregnancy, the fetus may be given intra-uterine blood transfusions, or after either normal or prematurely induced delivery, it may be given blood transfusions immediately after birth. With the idea that prevention is better than cure, women at risk of immunization with the Rh factor are passively immunized by administration of anti-Rh antibody (Beer and Billingham, 1976). The antibody should be administered prior to antibody formation by the mother, so that no sensitization occurs.

Some interesting complications have arisen with the Rh factor. Sensitization can occur by either a previous blood transfusion with Rh-positive blood or by the passage of Rh-positive blood from fetus to mother at the time of delivery. Usually during the first pregnancy (if the mother has not received blood transfusion with Rh-positive blood), the chances of sensitization are small, and the chance of damage to the fetus is small. It has been reported, however, that if immunization during the first pregnancy occurs, the fetus is almost always male. The sensitization of the mother occurs either as the result of fetal bleeding, which causes red blood cells to enter the maternal circulation, or from the mother's own sensitization during *her* fetal life in the uterus of an Rh-positive mother (Beer and Billingham, 1976). Of 25 Rh-negative women who were carrying their first Rh-positive fetus and who developed antibodies prior to delivery, 21 were daughters of Rh-positive mothers. In 23 of 25 cases, the sensitization was produced by an ABO-compatible*, Rh-positive male fetus. The sensitization by the male fetus may be either the result of a synergistic effect of the Y-antigen on the immunogenicity of the erythrocytes or the result of a greater traffic of fetal and maternal red blood cells in the placentas

*ABO compatibility means that a blood transfusion between mother and fetus or vice versa can be carried out without provoking an A or B antigen-antibody reaction; ABO incompatibility means that such a transfusion would provoke a response.

of male fetuses as compared to the placentas of female fetuses (Scott and Beer, 1973).

ABO blood groups play a role in determining the effect of Rh sensitization. The risk of immunization by an ABO-incompatible, Rh-positive pregnancy is about 12.5–20 percent of that of an ABO compatible, Rh-positive pregnancy (Beer and Billingham, 1976). One explanation for this interaction is as follows: If an Rh-negative mother with blood group 0 carries an Rh-positive fetus with blood group A, then fetal red blood cells that enter the maternal circulation will be destroyed by the anti-A antibodies of the mother. The cells will be removed in the liver, probably before sensitization of the mother can occur (Beer and Billingham, 1976).

Evidence that ABO incompatibility between mother and fetus (when not complicated by differences in Rh factor) affects the success of pregnancy is inconsistent (Beer and Billingham, 1976). There may, however, be an effect of blood groups on sex ratio. Allen (1975), combining data from the literature, found that the sex ratio of babies born to mothers who had at least one A gene (AO, AA, AB) was higher (1.15) when they had a different blood group than their mothers, and lower (1.01) when they had the same blood group as their mothers. The sex ratio of babies that lacked the A gene (i.e., blood groups B and 0) was high (1.11) when their blood group was the same as their mothers and low (1.04) when their blood group was different from their mothers. The physiological and immunological differences for these differences in sex ratio have not been determined and require further investigations.

Immunology of Gestation

Probably one of the most intriguing and possibly one of the most fundamental questions in mammalian reproduction (mentioned in Chapters 10 and 11) is: Why is the fetus, which is an allograft, not rejected by the mother? We will discuss some of the explanations that have been proposed.

Immune Reponse Early during Gestation. Immediately (6–24 h) after insemination and fertilization, an early pregnancy factor can be detected by a rosette-inhibition test* in the serum of mice (Morton et al., 1974, 1976), women (Morton et al., 1977), and sheep (Morton et al., 1979). This early pregnancy factor, which in mice is present from 6 h after fertilization until 4–6 d before birth (Morton et al., 1976) has, as the above test suggests, immunorepressive properties and may thus influence or be responsible for prevention of the allograft injection. The early pregnancy factor can be inactivated at 72°C, but remains stable at 56°C; it is not dialyzable and not species specific (Morton et al., 1976). Cerini et al. (1976), using a hemagglutination test, were able to detect antigens of 14-day-old sheep embryos in the maternal blood of pregnant ewes from day 6 after mating. It is not clear, however, whether this antigen and the early pregnancy factor are identical.

The early pregnancy factor is present prior to implantation, and the question arises whether implantation involves immunological response mechanisms. As Johnson (1976) has pointed out, in highly inbred lines, fetus and mother may be considered to be not immunologically different. This argues against the requirement for an immunological mechanism in implantation. This and other evidence, reviewed critically by Johnson (1976) led him to conclude: ''There is therefore no good evidence to suggest that immunological interactions between mother and early embryo are either essential for facilitatory in the peri-implantation period in any specific manner.''

Immune Response after Implantation. After implantation proceeds, the question why the fetus is not rejected becomes considerably more interesting. Some of the explanations that have been proposed for the nonrejection of the fetus are, for example: (1) inability of the fetus to produce antigens because of immaturity, (2) implantation in an immunologically privileged site, (3) suppression of the mother's immune response by non-

*The rosette technique or immunocyte adherence technique is defined as a ''technique by which cells which carry immunoglobulin on their surfaces (either because they have formed the immunoglobulin or because it has become bound to the cell) can be detected. Red cells coated with antigen are mixed with the cells and bind to those cells bearing antibody to form a 'rosette''' (Herbert and Wilkinson, 1972).

specific factors, and (4) special properties of the trophoblast, among which are special immunological properties and the possession of a special immunological barrier.

1. The idea that the fetus is not capable of producing antigens is not an appropriate explanation for the nonrejection of the fetus, because it has been shown that major transplantation antigens are present as early as the two-cell stage of the embryo (Beer and Billingham,1978; Billington, 1979).

2. There are several immunologically privileged sites in which tissue from an immunologically incompatible donor can be transplanted without evoking a rejection, as, for example the anterior eye chamber and the cheek pouch of the hamster, which are characterized by the absence of lymph vessels. The uterus, however, is not such a privileged site, because it does have lymph vessels, and more importantly, tissue grafts made in the uterus are rejected (Beer and Billingham, 1978; Billington, 1979). However, the uterus after formation of decidual tissue, does respond differently from other sites under special conditions; for example, experimentally immunized female mice reject alien embryos transplanted to sites other than the uterus, but do not reject such embryos transplanted to the uterus (Billington, 1979). The decidual tissue that develops under the implanting trophoblast may interfere with lymphatic drainage away from the zygote.

3. There are several factors secreted only during pregnancy, and these interfere with lymphocyte functions that are important in the immune response. Examples of such factors are pig uterine protein secretions (Murray et al., 1978), human chorionic gondotrophin (hCG) (Contractor and Davies, 1973), human placental lactogen (HPL) (Billington, 1979), progesterone, estrogens, corticosteroids, and alpha feto-protein (AFP) (Beer and Billingham, 1978; Billington, 1979). The AFP binds estrogens (Chapter 2), and this AFP-estrogen complex becomes bound to lymphocytes and makes them less responsive to exogenous antigens (Beer and Billingham, 1978). Billington (1979) also mentions the possibility that the developing embryo may provide a source of prolonged exposure of the mother's immune system to relatively weak antigens and that these antigens themselves or as antigen-antibody complexes may block either lymphocyte receptors or target antigens and thus diminish the immune response of the mother.

4. The trophoblast has some special properties, which suggest that it may be an immunologically privileged tissue. For example, a choriocarcinoma, a malignant derivative of trophoblast cells, is refractory to transplantation immunity directed against the paternal antigens. Beer and Billingham (1978) state, "Small grafts of trophoblastic cells obtained from ectoplacental cones behave as if they are invulnerable to transplantation immunity." Thus nonmalignant trophoblastic tissue shows some resistance to immunity.

In virgin rats, removal of the para-aortic lymph nodes, several weeks before pregnancy, results in smaller F_1 hybrid fetoplacental units than in controls. It has also been found that the para-aortic lymph nodes that drain the uterus show a striking hypertrophy during pregnancy. Beer and Billingham (1978) proposed the hypothesis that the lymphocytes within these stimulated para-aortic lymph nodes generate a noncytopathogenic humoral immunity, which contributes to the immunologically privileged status of the trophoblast. These lines of evidence support the idea that the trophoblast may be a privileged immunological tissue. The immunology of the hydatiform mole, however, suggests otherwise. The hydatiform mole is a curious example of a trophoblast abnormality. It is the product of an abnormal pregnancy and has grossly swollen villi, but no embryo, cord, or amniotic membrane. The hydatiform mole consists of trophoblast cells only and frequently becomes an invasive malignant chorionic carcinoma (Kajii and Ohama, 1977). Kajii and Ohama (1977) determined that hydatiform moles were androgenetic in origin and might be either the result of fertilization of an oocyte by a sperm that was diploid as the result of nondivision of meiosis or the result of nondivision of a normally fertilized oocyte at blastomere mitosis. Human major histocompatibility antigens of paternal origin are present on such hydatiform moles, and the mother produces antibodies against these antigens. This evidence argues against the hypothesis that the trophoblast is an immunologically privileged tissue.

The trophoblast has paternally inherited alloantigens (Beer and Billingham, 1976), and the uterus is immunocompetent (Beer and Billingham, 1976); therefore, either the trophoblast must have some intrinsic immunological properties or the maternal immune response must be different from its response to other tissues (Mendenhall, 1976).

The possibility of an immunological barrier of the trophoblast was mentioned in Chapter 10. In humans, the trophoblast is invested by a fibrinoid layer, which is rarely if ever complete, however (Mendenhall, 1976). In mice, there is a peritrophoblastic sialomucin coating, which may make the trophoblast immunologically inert. This inertness may be the result of free carboxyl groups of sialic acid, which cause a net negative charge that repels negatively charged lymphocytes. Human chorionic gonadotrophin (hCG), which decreases lymphocyte activity, may further increase the negative charge of lymphocytes (Beer and Billingham, 1978). Uterine protein secretions in pigs (Murray et al., 1978), human chorionic somatomammotrophin and hCG in humans (Contractor and Davies, 1973) suppress the response of lymphocytes to phytohemagglutinins* (Adcock et al., 1973).

Intergeneric hybrids between goat (♀) and sheep (♂) show earlier embryonic mortality if the goat carries such a hybrid fetus for the second time than if she carries it for the first time (McGovern, 1973). This finding and the ultrastructural damage to maternal uterine blood vessels (Beer and Billingham, 1976) suggest that there is an immunological basis for this *particular* rejection. It should be kept in mind that interspecific hybrids, e.g., between horse and donkey, are carried to term normally.

The placenta provides immunological protection for the fetus possibly by soaking up immune complexes and autoantibodies (Mendenhall, 1976). The placenta also forms a selective barrier for immune globulins and antigens between mother and fetus. For instance, immunoglobulin G (IgG) (molecular weight 150,000) crosses the human placenta, but some do not, e.g., IgA (MW monomer = 160,000 and MW polymer = 400,000 in humans), IgD (MW = 150,000), IgE (MW = 200,000) and IgM (MW = 900,000). Maternal white and red blood cells may pass to the fetus (Mendenhall, 1976), and fetal red blood cells may pass to the mother.

Rejection of Allograft and Short Gestation in Marsupials. In Chapter 11 we mentioned that the short gestation period of marsupials might be related to peculiarities of the immunological response of the mother to the fetal allograft. Beer and Billingham (1976) reject the hypothesis that the onset of parturition may be the result of an immunological sensitization of the mother to the fetoplacental unit in placental mammals. Their principal arguments are that the hypothesis would predict that successive pregnancies would become shorter as the mother became sensitized to the antigens of the fetus, and initial pregnancies should be abnormally long in cases where the mother had an immunological deficiency disease. Neither of these predictions has been fulfilled.

At the end of gestation the decreased secretion of protein hormones such as hCG by the placenta removes the suppression of the immune system so that rejection of the fetus and subsequent parturition occur (Contractor and Davies, 1973). Recently Siiteri and Stites (1982) have presented persuasive evidence that progesterone that is present in high concentrations in pregnant Eutherians (except possibly the elephant; see p. 233) may be the immunosuppressive agent that prevents rejection of the fetus and that a decrease in progesterone secretion induces rejection and subsequent parturition.

After birth, limited passive immunity is transferred from the mother to the young via the immunoglobulins in the colostrum and milk. In birds, antibodies are transferred to the yolk, which is absorbed into the gut of the hatchling during the first few days after hatching.

It is beyond the scope of this book to discuss the pathological, clinical, and contraceptive or abortifacient aspects of immunological defects and applications. The interested reader is referred to the book by Beer and Billingham (1976) and to reviews in the book edited by Scott and Jones (1976).

---

*"Phytohemagglutinins are lectins extracted from seeds (beans) of *Phaseolus vulgaris* or *P. communis*. . . . Phytohemagglutinin stimulates lymphocyte transformation and causes agglutination of red cells" (Herbert and Wilkinson, 1972).

13

Endocrinology of Reproductive Behavior

In this chapter we will discuss mainly the endocrinological aspects of reproductive behavior broadly defined, so as to include the behavior that brings males and females together in their normal breeding area, territorial courtship and mating behavior, and the behavior at the time of oviposition or parturition. We will not include feeding behavior and behavior during nursing of the young.

MIGRATORY BEHAVIOR

Many species migrate to breeding areas that are inhospitable during a large part of the year, but are favorable for breeding and for rearing the young for a short period. Such breeding areas may have first been explored to escape competition and subsequently became breeding areas because they were advantageous to the populations that had explored them. The distances that are covered by some species of birds to migrate from their wintering grounds to their breeding grounds are spectacular. Some species cover 4,000–6,000 km during the round trip from the arctic breeding grounds to their wintering grounds in the southern hemisphere (Emlen, 1975). For a thorough discussion of the problems of navigation and orientation during migration the reader is referred to Emlen (1975).

Cyclostomes

Lampreys generally migrate upstream to spawn; the time of migration is dependent upon species and latitude. For instance, *Lampetra fluviatilis* in northern and western Europe starts to migrate in late summer and fall, but in Italy it starts between December and May. In northern and western Europe, the sea lamprey (*Petromyzon marinus*) migrates in late spring and early summer (Hardisty and Potter, 1971). The males of the sea lamprey arrive at the spawning sites prior to the females. Spawning and courtship behavior, in contrast to migratory behavior, occur during the day; lampreys apparently prefer lighted areas. During courtship, the female passes the posterior part of the body near the male's head, suggesting that olfactory stimuli may play a role in the final stages of mating and spawning (Hardisty and Potter, 1971).

Teleosts

Some teleost fishes migrate spectacular distances and change from fresh water to sea water and vice versa—for example, the European eel (*Anguilla anguilla*), the American eel (*A. rostrata*), the Atlantic salmon (*Salmo salar*) and all five species of the North Pacific salmons, the coho (*Oncorhynchus kisutch*), the sockeye (*O. nerka*), the chum (*O. keta*), the pink (*O. gorbuscha*), and the quinat (*O. tshawytscha*). The three-spined stickleback (*Gasterosteus aculeatus*) migrates shorter distances than the fishes named, but it migrates from the sea to fresh water and back to the sea. The European eel migrates from the rivers and ditches of western Europe, where it has grown to maturity during a period of several years, to the Sargossa sea, where it spawns (Koch, 1968; Tesch, 1978). Before migration, the eels develop a metallic sheen on their skin pigment, and the anal fin becomes slightly pink; the skin and subcutaneous tissue increase in thickness, the snout becomes narrower, the pectoral fin becomes lanceolate, the animals stop feeding, the digestive tract regresses, and thyroid activity increases (Sinha and Jones, 1975). The plasma cortisol concentration is not different between fresh water and sea water eels, although the production is higher in eels in sea water, but so is the metabolic clearance rate (Henderson et al., 1974). After spawning has occurred, the developing young eventually migrate to the fresh waters in western Europe.

The Atlantic and North Pacific salmons, after hatching in fresh water and some growth, migrate to the ocean, where they grow to sexual maturity, sometimes in one year. When sexually mature, the animals migrate to the streams where they were hatched. During this migration, none of these species feed; the intestinal tract and cardiovascular system degenerate, and the gonads increase in size. The bones in the head of the males undergo more pronounced changes than the same bones in the females, resulting in a pronounced sexual dimorphism, in the male the lower jaw forming a large hook (Koch, 1968). In young salmon, a preference for sea water is correlated with a long photoperiod and high thyroxine concentrations in the blood, whereas a preference for fresh water is correlated with short photoperiods and low thyroxine levels (Baggerman, 1978). Prior to the migration of sockeye salmon to the spawning waters, prolactin cells in the AP appear to be active (Clarke and Bern, 1980); during the migration the adrenals become greatly enlarged, and the combined concentration of cortisone and cortisol increases 5-fold to 6-fold (Idler et al.,

1959). Both the prolactin and the corticosteroids are important in the regulation of the mineral balance, which is, of course, important for fish that migrate to a medium with a lower salt concentration.

The migration of the three-spined stickleback has been studied in some detail, and because it will breed in captivity, it has been subjected to experimental conditions that could be controlled rather rigorously. In western Europe, this fish lives from late summer until about January either in the sea or in brackish coastal waters; in January it migrates to fresh water ditches and streams, where it breeds. As will be discussed in Chapter 14, photoperiod and water temperature are the important environmental cues that control this migration. Before and during this migration, the gonads increase in size, but the relationship of this increase in the size of the gonads to migration to fresh water is not a causal one, since gonadectomy does not affect migration (Baggerman, 1957). The migration is instead correlated with, and apparently caused by, changes in thyroid activity. In January, with increasing photoperiod, thyroxine secretion increases, resulting in migration to fresh water; in late summer, with decreasing photoperiod, thyroxine secretion decreases, resulting in migration to the sea. In laboratory experiments, Baggerman (1957) demonstrated that thyroxine administration resulted in a preference for fresh water and that administration of thiourea, which blocks thyroid hormone synthesis, caused a preference for sea water. Baggerman (1978) also pointed out that in young Pacific salmon, long days also resulted in a high thyroxine secretion, but in this species this led to a preference for sea water, which is in accordance with the preference for sea water by young salmon. In Pacific salmon, therefore, short photoperiods result in low thyroxine secretion and subsequent preference for fresh water.

It should be pointed out that some populations of three-spined sticklebacks in the Netherlands do not migrate to the sea and that the advantages of migration to the sea are not clear (Baggerman, 1978). In the spring, the male three-spined stickleback builds a nest with leaves and other plant material. These materials are held together by a mucus secreted by the kidney. After the nest is built, a mature female is enticed to enter the nest and to spawn. The male then spawns, guards the nest, and aerates the eggs by waving his pectoral fins (Baggerman, 1978).

The secretion of mucus by the kidney is controlled by testosterone. Under the influence of this hormone, the kidney epithelium changes; the nephron and ureters contain an extensive rough endoplasmic reticulum laden with granules, and the kidneys lose their Na^+/K^+ ATPase activity. It appears that the kidneys have become a gland and have lost their function of maintaining the water and mineral balance, a function that is apparently taken over by the intestine (de Ruiter, 1976). As expected, the kidneys play an important role in making it possible for fish to migrate back and forth between fresh and salt water. In fresh water, fish need to excrete much water and conserve salts; in sea water they need to conserve water and excrete salts. This excretion of salts occurs via kidney and gut for bivalent ions and via the gills for monovalent ions. In a series of publications Wendelaar Bonga and colleagues (Wendelaar Bonga, 1973, 1976; Wendelaar Bonga and Veenhuis, 1974a,b; Wendelaar Bonga et al., 1976) have described the changes that occur in the kidney and the effect of prolactin on kidney structure. In brief, in salt water, the nephron cells of sticklebacks have a smaller nucleus, a smaller basal labyrinth, and fewer mitochondria than in fresh water. Prolactin injections given to sticklebacks transferred from sea water to fresh water speeded up the changes in the kidney (3 d versus 6–9 d for the controls). Indirect evidence (electron microscopy of the AP) showed that the prolactin cells of sticklebacks captured in the sea were less active than those of fish caught in fresh water. It was also shown that this was not the result of a change in the reproductive system. Sticklebacks brought to sexual maturity in salt water and in fresh water (by 16L:8D at 20°C) showed a greater prolactin activity in fresh water than in sea water.

Amphibians

The American newt, *Diemictylus (Triturus) viridescens,* and the European newts, *T. cristatus*, *T. palmatius,* and *Notophthalmus vir-*

idescens, migrate to land after they have undergone their first metamorphosis in water. Then after two or three years of terrestial life, the newts undergo a second metamorphosis, which ends in sexual maturity, a change in skin pigmentation and texture, and the appearance of a keel on the tail (Koch, 1968; Clarke and Bern, 1980), after which the animals return to the water. This water-drive can be induced in intact males and females, castrated males, and AP newts by injections of (mammalian) prolactin. The injections, however, do not produce the morphological changes that occur at the second metamorphosis. A water-drive has also been induced in the rough-skinned newt (*Taricha granulosa*) and the tiger salamander (*Ambystoma tigrinum*) (Clarke and Bern, 1980). Whether prolactin is also required for the osmoregulatory processes that accompany the change from terrestrial to aquatic life, needs to be investigated further.

Reptiles

The green turtle (*Chelonia mydas*) and the hawksbill (*Eretmochelys imbricata*) nest and lay their eggs on tropical beaches. After hatching, the young migrate to the water and then to the breeding grounds, which are about 1600 km away in the ocean (Pritchard, 1979). The endocrine changes associated with this migration have apparently not been studied.

Birds

We will discuss the events during the migratory cycle of the white-crowned sparrow (*Zonotrichia leucophrys gambelii*) and the white-throated sparrow (*Z. albicollis*), because more information is available about the endocrinological changes associated with migration in these species than other avian species. While the birds are in nonbreeding condition on their wintering grounds in January and February, they successively (1) undergo a prenuptial molt, (2) exhibit hyperphagia, and as a consequence, deposit fat, and (3) show slow recrudescence of the testes. In April, the vernal migration to the north starts and lasts through early May. During migration, the testes increase further in size and reach their maximal size shortly after arrival at the breeding grounds. The ovaries do not start to increase in size very much until the females are on the breeding grounds. The breeding cycle of the white-crowned sparrow has been described in Chapter 8. After the gonads have regressed, as a result of the onset of photorefractoriness (to be discussed in Chapter 14), the birds (1) undergo a postnuptial molt, (2) become hyperphagic, but to a lesser extent than in the spring, and as a result deposit less fat than in the spring, and (3) then migrate south, in the fall, to the wintering grounds (Meyer, 1976; Wingfield and Farner, 1980). In birds kept under laboratory conditions, the hyperphagia and fat deposition can be evaluated by measuring food intake and increases in body weight; the condition of the testes can be assessed by visual inspection after laparotomy; and migratory activity can be measured by using special cages equipped with microswitches, so that the locomotor activity can be recorded. The migratory activity measured under laboratory conditions is called *Zugunruhe*, i.e., migratory restlessness or migratory activity. As will be discussed in Chapter 14, photoperiod is the principal environmental stimulus for the stimulation of gonadal growth, for hyperphagia and migratory restlessness in the laboratory, and for the migration north in the field in the spring.

The relationship between the gonads, vernal premigratory fattening, and migratory behavior has been studied experimentally. In both the white-throated and the white-crowned sparrow, castration prior to exposure to long days prevents migratory fattening, but castration performed after some photostimulation does not interfere with premigratory fat deposition (Wingfield and Farner, 1980). Castration delays and shortens the *Zugunruhe* and reduces its intensity, but does not eliminate it (Wingfield and Farner, 1980), and in white-crowned sparrows, it does not affect the autumnal fat deposition (Wingfield and Farner, 1980). In male and female white-crowned sparrows, the plasma testosterone concentrations are similar early during vernal migration; therefore, it may be that testosterone induces hyperphagia. If this is indeed the case, then testosterone has quite a different effect in these species and at this time than its usual nitrogen-

retention effect. According to Wingfield and Farner (1980) the endocrine components of migration in this species have not been elucidated.

The endocrinology of the migration of the white-throated sparrow (*Zonotrichia albicollis*) has been elucidated by Meier (1976). Figure 13.1 illustrates the interactions between photoperiod and the endocrine responses that ensue, and the subsequent effects on body fat stores, gonadal size, migratory activity, and the orientation of the birds during migration. According to this scheme, there is a circannual clock (i.e., an endogenous clock with a period of about one year; see Chapter 14 for further details) which sets the phase angle* of a circadian oscillator (i.e., an endogenous oscillator with a period of about 24 h; see Chapter 14 for further details) for the release of corticosterone and prolactin. The daily photoperiod entrains the circadian oscillator, so that it has a period of exactly 24 h, and it entrains the photoinducible phases for the stimulation of LH and FSH plus prolactin secretion in response to light. In Chapter 14 we will discuss this in more detail; it is enough here to state that the sensitivity of the central nervous system (CNS)-AP axis to stimulation by light varies during the 24-h day, and that the daily photoperiod in its entraining mode determines at what period of the day the CNS-AP system will respond to light by the secretion of the pituitary gonadotrophins. The response of the white-throated sparrow will vary when the difference between the time of occurrence of a high corticosterone secretion and a peak in prolactin secretion vary; i.e., when a peak in corticosterone concentration is followed 4 h later by a peak in prolactin concentration, hyperphagia and thus deposition of fat, and migratory activity will be high, but the gonads will not be stimulated, and the birds will orient themselves in a southerly direction. When the time difference between the peaks in corticosteroid and prolactin concentrations is 8 h, no hyperphagia and no migratory activity will take place, and the gonads will remain small; whereas when

---

*Phase angle is the difference in degrees (after the 24-h period has been converted to 360°) between the onset of light and the occurrence of the highest amplitude that the oscillator reaches.

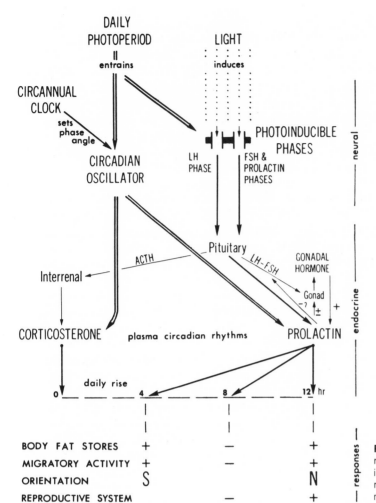

Figure 13.1 Circadian and circannual elements regulating reproductive and migratory conditions in the white-throated sparrow. (From Meier, 1976; reprinted with permission of the Australian Academy of Science.)

the time difference between the peaks is 16 h, the result is hyperphagia, fat deposition, stimulation of the gonads, increased migratory activity, and orientation in a northerly direction. So far, there seems to be no published evidence that the mechanisms proposed by Meier (1976) for white-throated sparrows also apply to white-crowned sparrows.

Mammals

The northern fur seal (*Callorhinus ursinus*) breeds on small islands in the Bering Sea, but soon after the breeding season is over, migrates about 4800 km to the Pacific Ocean, where it spends the

remainder of the year. Whales also migrate amazing distances, breeding in tropical waters but feeding in the polar waters of both hemispheres (Dawbin, 1966). Little appears to be known about the endocrinological changes during the migratory cycle. One would expect to find the changes in hypophyseal and gonadal hormone concentrations that are usually found when animals start to reproduce. It is not known, however, whether such changes stimulate migration to the breeding grounds and, if so, which hormones are the most important. Since the fur seals and whales remain in the ocean, one would not expect to find the changes in thyroid hormone and corticosteroid hormone secretion that are found in the lampreys, teleosts, and amphibia, which either migrate between fresh and sea water or between an aquatic and a terrestrial environment.

SEXUAL BEHAVIOR

The potential for appropriate sexual behavior is influenced by genetics (Siegel, 1975), as well as by the environment, especially the perinatal environment, as was discussed in Chapter 2. We will not elaborate on this further here.

The fertilization of the oocyte requires that males and females be brought in proximity and, for internal fertilization, that the male inseminate the female. Bringing males and females into proximity is greatly facilitated by the advertising displays of the presence of the males or females near the breeding grounds. The signals emitted should serve to enhance species recognition and to allow recognition of the sexual status of the emitter. In Chapter 14, several aspects of visual, olfactory, auditory, and tactile stimuli will be discussed in detail.

It is, of course, not surprising that more information about the control of sexual behavior is available for laboratory and domestic animals than for other groups. We will discuss the sexual behavior in the various classes of vertebrates and will consider as sexual behaviors those behaviors that enhance the probability of successful fertilization.

Cyclostomes

The males of the sea lamprey (*Petromyzon marinus*) arrive in the upstream, fresh-water breeding territory before the females, and may start building nests of stones in the gravel of the stream, where breeding will take place. The nest may also be built by a male and a female after the females have migrated upstream. Males may attract females by the emission of a chemical cue. The female must attach herself to a rock for copulation to succeed. The male approaches the female from downstream and then seizes the back of her head with his mouth. The two bring the anterior parts of their bodies in a parallel position, and the male curls the posterior part of his body in a right-handed spiral around the posterior part of the female, bringing the sexual openings in proximity. During the mating and spawning, the pair move their bodies quite rapidly, stirring up sand, which will cover the oocytes and afford them some protection against predation. Once spawning is over, the male and female cover the eggs with sand. The pair remains around the nest for a few days, occasionally covering the fertilized oocytes with more sand, then they die (Breder and Rosen, 1966; Hardisty and Potter, 1971).

Elasmobranchs

The male shark, *Heterodontis francisci,* starts courtship by biting the female on some part of the body. He then grasps the left pectoral fin of the female and maneuvers his tail over the female's back immediately in front of her second dorsal fin. He subsequently inserts the right clasper into the cloaca, and copulation lasts about 30 minutes (Wourms, 1977). Large sharks apparently copulate in parallel positions and swim in synchrony (Wourms, 1977).

Small-sized species of skates copulate with the ventral surfaces ventral-to-ventral close together. The male clear-nosed skate (*Raja eglanteria*) bites the caudal margin of the female's caudal fin. He bends his tail 75° beneath the female's tail, flexes one clasper medially 90°, and inserts it in the cloaca. The pair rests ventral side down on the bottom during this copulation, which lasts for more than two hours. Rays may copulate in the same manner as skates, but there are no detailed reports on their sexual behavior. Electrical stimulation of the brain (the location of the electrodes were not specified) induced clasper movements in adult bonnet-head shark (*Sphyrna tiburo*) (Demski, 1977).

Teleosts

The variation in the manner in which sperm and oocytes are brought into proximity is probably greater in the teleosts than in any other class of vertebrates. Breder and Rosen (1966) have given an excellent and exhaustive review of these various patterns, but it is not possible in this book to summarize even the principal findings. We shall, therefore, only give some examples.

The sexual behavior of the Japanese rice fish (*Oryzias latipes*) has a genetic component. As was mentioned in Chapter 1, in a competitive situation, YY males are more successful in mating with females than are XY males. That genetics does not determine the behavior of an individual in other species of teleosts is illustrated by the teleosts in which sex reversal can be induced by changing the social situation, as in the coral reef fish *Anthias squamipinnis*. This small fish may be found either in relatively small, single-male groups or in large groups consisting of as many as 350 females and 35 males (Shapiro, 1979). The males and females differ in color (see Shapiro, 1979, for description) and in sexual behavior. In groups of one male and six females kept in aquaria, the removal of the male resulted in the following behaviors: (1) The dominant, generally the largest, female changed from the female to the male color pattern, a change that was completed in about 14 d. (2) The third dorsal spine became elongated. (3) The rate of male behaviors, such as aggressive rushes or movements, U-swims (a behavior performed by males to females both in social and sexual contexts), and rapid erections of the elongated dorsal spine increased. These changes in behavior were gradual, but after about 14 d, the sex-reversed female had reached the normal male level of performance. (4) The sex-reversing female did not receive aggressive rushes or U-swims, since there was no male to perform these and none of the other females initiated such behaviors. However, the sex-reversing female became the recipient of some behaviors that females perform towards males, such as: more lateral dorsal displays and more bent approaches. (5) The gonads changed from ovary to testes, a change that took about 14 d to complete (Shapiro, 1979). The review by Shapiro (1979) gives many additional, interesting details about the social behavior and the group structure on sex reversal, but the main point made here is that in this species, it is the social situation that determines the behavior; the genetic component of the sexual behavior is not impor-

tant, except, of course, that the ability to show sex-reversal is genetically determined.

The extent to which gonadal hormones determine the sexual behavior of teleosts seems to be more limited than it is in amphibians, reptiles, birds, and mammals. Castration had no effect on courtship behavior in male jewel fish (*Hemichromis bimaculatus*), and Siamese fighting fish (*Betta splendens*), or on nest building in *Tilapia macrocephala*. Castration decreased nest-building activity in the blue acara (*Aequidens latifrons*) but not mating patterns, except for an increase in rubbing of the genital papilla over the nest. In the viviparous swordtail (*Xiphophorus maculatus*), castration reduced the frequency of copulations (van Tienhoven, 1968). In the three-spined stickleback, castration has different effects on the *trachurus* form, which breeds in fresh water but spends the winter in the sea, than on the *leiurus* form, which breeds and winters in fresh water. Males of the leiurus form, castrated either during the pre-nest-building stage and maintained under long photoperiods or after nests have been built, retain their aggressiveness. In the trachurus form, castration of males after breeding has started and the nest has been built reduces both aggressive and sexual behavior; however, castration one week before the expected time of nest building increases agonistic, but decreases sexual, behavior. This difference in response to castration during different phases of the reproductive cycle has been explained by assuming that aggressive behavior is controlled by hypophyseal hormones during the prebreeding period, but later comes under the control of gonadal hormones (Liley, 1969).

Castration of the goby (*Bathygobius soporator*) results in indiscriminate courting of males, and of gravid and nongravid females, whereas intact males court gravid females only. Castration may affect gonadal hormones, which affect the olfactory sensitivity of the male, since males "recognize" gravid females by olfactory cues Liley (1969) (Chapter 14).

Evidence of the effect of gonadal hormones on female sexual behavior is even more meager than the evidence for males. Stacey and Liley (1972) showed that in the goldfish (*Carassius auratus*), spawning behavior could be induced in females with regressed ovaries by injections of estradiol and by the placement of eggs in the oviduct; however, neither of these two treatments alone induced spawning. Ovariectomy of jewel fish and Siamese fighting fish eliminated female sexual behavior, while ovariectomy of guppies resulted in a decrease, but not a loss, of the female courting response to a displaying male. Estradiol-17β increased the level of the response and the period over which the response could be evoked in a dose-related manner. Other steroids, such as progesterone, methyltestosterone, and corticosterone, had no significant effect on this behavior of ovariectomized females. Hypophysectomy had an effect similar to ovariectomy, and administration of estradiol-17β restored the level of courting behavior to that found in controls (Liley, 1972). These results suggest that there is a basal level of female sexual behavior that is ovary-dependent, and that estrogen can increase it. Baggerman (1968) and van Tienhoven (1968) both concluded tentatively that female sex hormones may be essential for female sexual behavior in teleosts. Ovariectomized three-spined sticklebacks treated with methyltestosterone did not show male sexual behavior. Incompletely ovariectomized Siamese fighting fish, in which the regenerating gonad was testicular, showed male sexual and aggressive behavior (Baggerman, 1968; van Tienhoven, 1968). Administration of testosterone to intact or ovariectomized swordtails (*Xiphophorus helleri*) was followed by a rise in the social hierarchy and copulation attempts (in spite of a poorly developed gonopodium) by the treated females (Noble and Borne, 1940). Similarly, masculinized females of *Platypoecilus maculatus* made efforts to copulate, but in *P. variatus*, methyltestosterone treatment did not induce male precopulatory or copulatory behavior (Baggerman, 1968). Thus in some teleosts, male sex hormones can induce male sexual behavior in genetic females.

The spawning reflex, which consists of making S-shaped movements, can be induced by the injection of either oxytocin or vasopressin in intact or AP killifish (*Fundulus heteroclitus*), in intact male and female Japanese rice fish, in the Japanese bitterlings (*Rhodeus* sp.), and in the mosquito fish (*Gambusia* sp.) (van Tienhoven,

1968; Liley, 1969). It is not certain, however, whether the effect of these injections is specific, or whether the hormones act on smooth muscle and thus cause the behavioral S-shape.

Electrical stimulation of the hypothalamus of male *Tilapia heudeloti macrocephala* evokes aggressive behavior (Demski, 1973). In the male green sunfish (*Lepomis cyanellus*), electrical stimulation of the preoptic area, the dorsal hypothalamus, the thalamus, the midbrain tegmentum, the basolateral midbrain, or the medulla results in sperm discharge. This suggests that the pathway for sperm discharge is from the preoptic area via the hypothalamus, to the midbrain tegmentum, the basal midbrain, and the medulla to the spinal cord (Demski et al., 1975). It is not known, however, whether steroid hormones affect the threshold of stimulation for obtaining sperm discharge.

Among the 495 known families of teleost fishes, about 20 percent show some form of parental care (Gittleman, 1981), which may vary from guarding of the eggs to viviparity, as can be seen from the list in Table 13.1. The care may be given by the male, as is the case in 60 percent of the families that show parental care, by male and female, as in the genera *Loricaria*, *Ophicephalus*, *Amphiprion*, *Eupomacentrus*, and *Pomacentrus*, *Apistogramma*, and *Nannochromis*, and by females, as among species of the genera *Apistogramma* and *Nannochromis* (Gittleman, 1981).

Blumer (1979) proposes that parental care could evolve only in species with external fertilization, with genetic relatedness of the male to the offspring at the site of oviposition. Males can minimize the investment of care for the offspring by strategies that allow mating and care to take place simultaneously. Paternal care may consist of (1) guarding of the eggs, the most frequent form of paternal care, as in the European sheatfish *Silurus glanis;* (2) guarding and aerating the eggs, as in *Blennius sphynx;* (3) nest building and guarding of the fertilized oocytes and young, as in the three-spined stickleback; (4) carrying the eggs, e.g., on a hook on the head, as in the families of the humpheads (Kurtidae), in *Kurtus gulliveri* and *K. indicus;* (5) carrying the eggs on the lower lip, as in *Loricaria vetula, L. anus,* and

Table 13.1

Forms of parental care by males and females in teleost fishes

	Number of families		
Form of parental care	Care by males and females	Care by males	Care by females
Guarding	63	57	27
Nest building, substrate cleaning or both	39	37	18
Fanning	30	30	11
Removal	12	12	5
Oral brooding	10	7	4
Retrieval	8	7	5
Cleaning of eggs	6	6	4
External egg carrying	6	4	3
Egg burying	5	4	3
Moving	5	5	3
Coiling	4	3	2
Ectodermal feeding	3	3	3
Brood pouch	2	1	1
Splashing	2	2	—
Gestation	14	—	14

From Blumer, 1979; reprinted with permission of *The Quarterly Review of Biology*.

L. maculatus; (6) carrying the eggs on the ventral surface, as in the family of the pipefishes and seahorses, the Syngnathidae; (7) carrying the eggs in a pouch with the formation of a placenta (see Chapter 10), as in the seahorse *Hippocampus erectus;* and (8) mouth brooding, as in *Tilapia melanotheron, T. microcephala, Betta picta, B. brederi, B. anabantoides, B. pugnax,* and *Opisthognatus aurifrons* (Breder and Rosen, 1966; Ridley, 1978). The extensive reviews by Breder and Rosen (1966) and Ridley (1978) should be consulted for more complete listings of the different forms of paternal care in teleosts. Some aspects of parental behavior are under the control of prolactin or are at least influenced by it, for example, nest building, proliferation of the lining of the brood pouch of seahorses, fanning of eggs by the discus fish *Symphysodon* sp., the blue acara, *Pterophyllum scalare,* and *Crenilabris ocellatus,* secretion of mucus by the discus fish and the blue acara, and buccal incubation of eggs by mouth-brooding fishes (van Tienhoven, 1968; Nicoll and Bern, 1972; Clarke and Bern, 1980).

Amphibians

Anura. When frogs and toads migrate to their breeding ponds or streams, they may use olfactory cues as well as celestial cues. To be able to use celestial cues for compass direction, an internal clock is necessary to provide information about the local time (Salthe and Mecham, 1974). Once in the breeding territory, auditory cues become important in species recognition, in sex recognition, and in attracting females to the breeding site. Often calls are given in choruses by several to hundreds of males (Salthe and Mecham, 1974). In the species in which the male calls, the female apparently selects the male on the basis of body size, with the larger males being preferred (Fairchild, 1981; Ryan, 1980). In Fowler's toad (*Bufo woodhousei fowleri*), the larger males tend to produce calls that have a lower frequency and pulse rate and that last longer than the calls of smaller conspecifics. However, the same male at a higher temperature gives calls that are higher in frequency and pulse rate and shorter in duration than calls given at a lower temperature. Apparently, a toad at a lower temperature than a competitor of the same size sounds bigger than the one at a higher temperature. Under natural conditions most of the large toads are in the colder water of a pond and the smaller toads that have been driven out are at the edge of the pond, where the temperature is about 2° C higher. Thus females would be attracted to the larger males in the water, but the smaller males would have earlier access to the females as they come in from the fields and woods to enter the pond (Fairchild, 1981).

In the green tree frog (*Hyla cinerea*), there are noncalling satellite male frogs that associate with a calling male. These satellite males are capable of intercepting females that are attracted by the calling male, and some satellite males are successful in achieving amplexus (embracing of the female until spawning has been achieved by male and female). Such satellite males are in effect sexual parasites (Perrill et al., 1978).

Some anurans are voiceless, e.g., *Ascaphus* sp., and others have a reduced mating call, e.g., *Bufo alvarius* and *Rana sylvatica*. In these species, tactile factors are important, with the male moving around in his breeding territory and attempting to clasp any other frog, male or female, (Salthe and Mecham, 1974). Tactile cues probably give information about whether the clasped frog is a conspecific and whether it is a receptive female. For example, when a male leopard frog (*Rana pipiens*) tries to clasp an unreceptive female, she gives a release call. This is inhibited either by distending the abdomen or by the injection of either arginine vasotocin (AVT), prostaglandin E_2 (PGE_2) or $PGF_{2\alpha}$. This effect of AVT is inhibited by prior administration of indomethacin to block PG synthesis, suggesting that AVT has its effect through a stimulatory effect on PG synthesis (Diakow and Nemiroff, 1981). The development of the call-producing apparatus, the calling itself, the development of the musculature of the arm, and the thickening of the thumb pads are all under the control of testicular androgens. However, in the leopard frog, the sexual behavior, i.e., the clasping of the female, cannot be induced in castrated males by testicular implants. In intact sexually inactive males, implants of testosterone also fail to induce male sexual behavior; however, anterior pituitary implants evoke such behavior in intact, but not in castrated, sexually inactive males (Palka and Gorbman, 1973), suggesting that a testicular hormone other than testosterone is required for inducing male sexual behavior. Subsequent experiments have shown that there is no consistent correlation between high plasma testosterone concentrations and male sexual behavior (Wada et al., 1976). It appears that the failure of systemic androgens to induce male sexual behavior may be related to a failure of androgens to reach the site in the brain that regulates the amplectic behavior. There is support for this idea in the finding that testosterone implanted in the preoptic area of the brain of castrated males evokes the sexual behavior, although the latency is longer and the behavior less vigorous than in intact males treated with pituitary homogenates (Wada and Gorbman, 1977). It is also possible that androgens are not metabolized to active metabolites in the absence of a testicular factor (see discussion on Japanese quail in the section on birds). Electrical stimulation of the medial basal preoptic nucleus induces male sexual behavior in freely moving male leopard frogs (Wada and Gorbman, 1977). In frogs that mate in water, the amplexus is usually a pelvic embrace, whereas in species that mate on land it is usually a pectoral one. On land, after amplexus occurs, male and female usually move to the water and spawn except in the midwife toad (*Alytes obstetricans*). In this species, spawning occurs on land, but amplexus is first pelvic; the male's inner toes are in contact with the cloaca, and he strokes the cloaca with lateral movements. As the eggs are released from the cloaca, the male shifts his grip to just anterior to the female's forelegs, releases sperm, and, after an interval, pushes his hind leg through the double string of eggs released by the female. These strands of eggs are thus wound around his leg, and he carries them until the larvae hatch (Salthe and Mecham, 1974).

The Surinam toad (*Pipa pipa*) has to go through almost acrobatic exercises to accomplish fertilization of the eggs (oocytes) and then deposition of the fertilized eggs in the egg chambers on the back of the female. The male and female, while in amplexus, with the male on top of the female, in the water, go through a series of turnovers carried out largely by the female with the male passively participating. The female makes a sidewise upward movement, during which the male massages her abdomen; she then pauses briefly in a horizontal upside-down position and spawns. The female then descends head first to a tilted resting position at the bottom of the water. During this descent, the male probably spawns in response to the egg emission by the female and arches his body, which results in the formation of folds in his belly in which the eggs (oocytes) are caught. As the female rights herself from the tilting resting position, the eggs drop from the temporary belly folds into the egg chambers on the back of the female; the male, which has been clasping the female, increases the pressure of his grasp and thus presses the eggs into the chambers (Salthe and Mecham, 1974). The female then carries the eggs and later the embryos on her back and supplies them with oxygen through blood vessels that surround the vascular egg chambers, until the froglets are fully developed, after which they leave the egg chambers (Salthe and Mecham, 1974). The East African frog (*Nectophrynoides occidentalis*) is viviparous and thus cares for the young from fertilization of the oocyte until birth, as discussed in Chapter 10.

Many tree frogs deposit eggs in the water in foam nests, which are formed by beating the cloacal fluid with the hind legs. As the foam dries, it forms a crust that protects the developing embryos, while the central part of the nest becomes fluid, thus providing water for the embryos (Noble, 1931; Salthe and Mecham, 1974).

Paternal care consists of either guarding the nest or the male carrying the progeny, as in the midwife toad, the Chilean toad, and the Australian toad mentioned in Chapter 10. Ridley (1978) lists the occurrence of paternal care in species of the Anura. There seems to be little experimental evidence about the endocrine control of either maternal or paternal behavior in the Anura.

Urodeles. Urodeles are generally voiceless and thus auditory cues, as expected, are not important in their breeding. Visual and olfactory factors are the most important means of communicating about species, sex, and breeding. Once the animals are in the breeding area and in proximity, tactile stimuli may provide additional information about the breeding condition of the female.

In the primitive Japanese salamander, *Hynobius nebulosus,* the male develops, during the breeding season, a white gular patch, which is exposed when he lifts his head in response to the presence of a female. By pumping of the hyoid apparatus, this white patch is moved so that it becomes more conspicuous (Thorn, 1966), suggesting that a visual signal may be important during courtship. The male during courtship induces the female to deposit her eggs, which are in gelatinous sacs or egg cases, on a support of his choosing in his territory, usually a twig in stagnant water. The female clasps this support with her hind legs and presses her cloaca against it and attaches the egg sacs to the support. She is helped by the male as she attempts to free herself from the attached egg sacs (Thorn, 1963), and after she is free, the male deposits his semen on the oocytes (Thorn, 1963). The male remains near the eggs for about a month and may defend them (Thorn, 1966).

In the genus, *Triturus,* the males have bright nuptial colors and patterns of dark blotches and spots, among them dorsal crests. These patterns may serve as visual cues for recognizing the sex and breeding condition. Dorsal crests differ in morphology among species and probably serve in species recognition. The elaborate courtship pattern also functions as a visual cue. In the smooth newt (*T. vulgaris*), three aspects of the courtship behavior, wave, whip, and fan are particularly important. The wave is variable in form, and in one of these forms, the tail is held out to one side so that it makes an obtuse angle with the side of his body nearest to the female; the body is slightly inclined towards the female, and the male gently bobs up and down. The whip is a vigorous tail lash hitting the male's flank and creating a powerful blast of water towards the head of the female, so that it may push her backwards. The fan is a delicate sustained movement of the tail, so that the tip of the tail vibrates and produces a steady stream of water directed towards the snout of the female (Halliday, 1974). The rate of these displays is highly correlated with the number of spermatophores (sperm packets, Chapter 5) the male will deposit (Halliday and Houston, 1978). Halliday and Houston (1978) have suggested that the newt is "an honest salesman" in advertising how many spermatophores he will produce during his courtship. This "honesty" is presumably the result of the fact that deception serves little purpose. If only one spermatophore is produced, the chances of having it picked up may be less than 50 percent. It is assumed that the male advertises even though only one spermatophore will be deposited, since in some situations females are encountered rarely and it may be advantageous to chance fertilization.

Besides these visual cues, olfactory cues are also important in *Triturus*. The head has hedonic glands on its side that secrete olfactory substances. During courtship the male rubs his head and his cloaca over the head of the female, thus transmitting olfactory signals to her. Moreover, during the visual displays of fan and whip, he probably wafts olfactory substances from his cloaca in the direction of the female (Salthe and Mecham, 1974). After the visual displays, the male moves away and the female follows him; the male subsequently deposits one to three spermatophores. Under laboratory conditions, only 43 percent of all spermatophores are picked up by the female (Halliday and Houston, 1978), although the male deposits the spermatophores apparently in response to a tactile signal by the female (Salthe and Mecham, 1974).

In the salamandrids, the male captures the female by sliding under her and grabbing her forelegs. He then carries her around for some time, deposits his spermatophores, and then lowers her to the ground near the spermatophores. In most of the noncryptobranchoid urodele families, the male clasps the female in the pectoral region in a dorsal-to-dorsal position. He rubs the hedonic glands of his chin across her snout and his cloaca across her body. He then dismounts, deposits his spermatophores, and the female picks these up (Salthe and Mecham, 1974). In *Euproctus, Amphiuma,* and *Salamandra atra,* the cloacae of the male and female are in apposition, allowing direct transfer of spermatophores from the male to the female (Salthe and Mecham, 1974).

As in *Rana pipiens,* the courtship behavior (i.e., clasping of the female) of the male in the rough-skinned newt (*Taricha granulosa*) is not related to concentrations of testosterone, dihydrotestosterone (DHT), or 11-ketotestosterone (Moore et al., 1978). Implants of testosterone plus DHT are more effective than implants of either hormone alone in inducing sexual behavior in sexually active males that have been castrated; however, implants of testosterone plus DHT do not induce such behavior in males that were sexually inactive prior to castration (Moore et al., 1978). Moore (1978) speculates that a nontesticular hormone may be required to induce sexual behavior in such males; however, he presents no evidence that testicular implants or homogenates are more effective than testosterone plus DHT implants. It may well be, as will be discussed in the section on birds, that the conversion of androgens in the brain is better correlated with sexual behavior than plasma concentrations of hormones.

Parental care consists of guarding the eggs, which may be carried out by the female or by the male in species in which fertilization is external and occurs in the male's territory, as in *Hynobius nebulosus.* In aplacental viviparous salamanders, such as *Salamandra salamandra,* the eggs are carried by the female for as long as five months, and the young are in the late larval stage when born. In the viviparous *S. atra,* the young are born almost fully developed (Salthe and Mecham, 1974; Chapter 10) and are protected from predation by other species, but they feed in the uterus, first on degenerated embryos and later on uterine milk.

Reptiles

The neuroendocrinology and psychobiology of the American chameleon (*Anolis carolinensis*) has been studied in more detail than in any other reptile. The reproductive cycle of this lizard has been described briefly in Chapter 8, and the environmental effects on reproductive activity will be discussed in Chapter 14. Here, only the sexual behavior will be discussed.

In the spring, in response to increasing temperatures, males emerge from hibernation before the females and establish territories. During this period, the males make assertion displays to conspecific intruders. This display consists of rhythmical bobbing movements of the forebody and extension of the dewlap. If the intruder does not either leave or assume a submissive posture, the territorial male starts a challenge display characterized by lateral compression of the body and engorgement of the throat. This display may be followed by fighting (Crews and Greenberg, 1981). After the females emerge and enter the male territories (Crews and Greenberg, 1981), or if a female is introduced into a male's cage (Adkins-Regan, 1981), the male starts assertion displays that change to courtship displays, if the female makes a subordination display. The courtship behavior involves bobbing and extension of the dewlap, as in the assertion display, but as the dewlap retracts, the male advances while nodding his head rapidly. Initially, females, even when receptive, flee, but eventually, if a female is receptive, she arches her back, allowing the male to mount her. He seizes her neck with his mouth, inserts one of his hemipenes into her cloaca (Chapter 9) and ejaculates, and then dismounts (Crews, 1975, 1979; Crews and Greenberg, 1981). The entire sequence of events from the initiation of courtship displays to dismounting takes about 20 minutes (Crews, 1979). After copulation, the female is unreceptive until the largest follicle is ovulated and another follicle on the contralateral ovary develops. The exposure of a female to a sexually active male stimulates ovarian recrudescence, provided environmental temperatures are sufficiently high (Chapter 14). When a female is introduced to a group of competing males, however, ovarian recrudescence fails to occur because of the male-male aggression and interference by other males with the courting male (Crews, 1975).

The inhibition of receptivity of the female following mating is ovary-dependent; ovariectomized estrogen-treated females remain receptive for 6 h after copulation (Crews, 1979) and are receptive again 24 h after copulation (Crews and Greenberg, 1981). It is probable that a lack of progesterone after ovariectomy prevents the inhibition of receptivity (Crews, 1979). The receptivity of the female during the breeding season is eliminated by ovariectomy, but can be reinstated by estrogen administration in a dose-related manner (Crews, 1979) or by a subthreshold priming dose of estrogen followed by progesterone (McNicol and Crews, 1979). The action of estrogen can be explained by the fact that it increases the concentration of progesterone receptor in the diencephalon which, as will be described below, is an important area of the brain in regulating male and female sexual behavior.

In the ovariectomized estrogen-primed female, injections of either 1 μg LH-RH or 1 μg thyrotrophin-releasing hormone (TRH) induce sexual receptivity in less than 2 h with a maximal receptivity at 6 h after the injection of the hypothalamic hormones (Crews, 1980). This effect of LH-RH on female sexual behavior has also been noted in ringdoves and rats, as will be discussed in this chapter, when the behavior of these species is considered, but the effect of TRH has not been found in birds and mammals (Crews, 1980).

Castration eliminated the challenging behavior and sexual behavior of male American chameleons, but both types of behaviors were restored after administration of either testosterone propionate (TP) or dihydrotestosterone propionate (DHTP); in ovariectomized females, both these hormones induced aggressive behavior when tested with male sex partners, and sexual behavior when tested with female sex partners (Adkins and Schlesinger, 1979). These findings suggest that androgens do not need to be converted to estrogens in order to stimulate sexual behavior. Crews (1979) made the interesting observation that after treatment with estradiol and DHT, half of the population of castrated American chameleons responded with an immediate increase in sexual behavior, including mounting and mating; whereas after treatment with testosterone, the response was gradual and it took several days for mounting behavior to appear. It is possible that the stimulation of the nervous system by estrogen is enhanced by the stimulation of the male secondary sex organs, especially the hemipenes, which, as mentioned in Chapter 9, have a feedback loop to the testes. Crews (1979) cautions, however, against this interpretation because of the variation in the response obtained in different individuals in the population, 50 percent of which failed to respond.

Ovariectomy induced the receptivity of the females when compared with intact females, and subsequent administration of TP, but not DHTP, increased the receptivity above that of intact females. These results suggest that for induction of male behavior, testosterone need not be converted to estrogens, but that for induction of female behavior, it needs to be converted into estrogens (Adkins and Schlesinger, 1979). If, after castration, the male is returned to his home cage, he will continue to challenge intruders for at least two weeks (Crews and Greenberg, 1981), suggesting that androgens may modulate aggressive behavior in a familiar surrounding, but are not absolutely required.

Lesions of the preoptic-anterior hypothalamus (POA-AH) result in a loss of male sexual behavior that cannot be restored by androgen therapy. Lesions rostral to the POA-AH and lesions in the basal hypothalamus also result in a loss of sexual behavior, but androgen therapy can, in these cases, restore the male's sexual behavior. The importance of the POA-AH area is further revealed by the fact that implants of testosterone in this area in castrated males restore their sexual behavior, whereas implants of testosterone in other brain areas do not (Crews, 1979). The sequence of behavioral patterns following after testosterone implants in the POA-AH is the same as that in intact but inactive males after environmental stimulation that causes testicular recrudescence (Crews, 1979). Somewhat surprisingly, the implantation of either estrogen or DHT in the POA-AH induced sexual behavior in long-term castrated American chameleons and resulted in a shorter latency than testosterone implants, whereas castrated males did not respond to systemically administered estradiol or DHT.

Lesions of the septum of either intact or androgen-treated castrated American chameleons caused a decrease of 43 percent in courtship frequency and a 76 percent decrease in aggressive behavior; lesions in the rostral amygdaloid area decreased courtship frequency by 96 percent, eliminated challenge behavior towards females, and decreased the challenging towards males to low levels but did not affect assertion behavior. Bilateral implants of testosterone in long-term castrated males restored the challenge display toward females, but did not affect the frequency of challenge display toward males. Lesions of the caudal amygdaloid area did not affect courtship behavior of sexually active males, and implants of testosterone in this region in long-term castrated males did not restore sexual behavior (Crews, 1979). These observations thus suggest important functions in aggressive and sexual behavior for the POA-AH, the rostral amygdala, and the septum. Furthermore, after injections of radioactive testosterone, the radioactive label was concentrated in these areas and in the ventromedial and periventricular hypothalamus (Crews, 1979).

Compared with the rather extensive knowledge about the neuroendocrinology of reproduction of the American chameleon, the knowledge about the endocrine aspects of reproduction in other reptiles is almost anecdotal, but Bellairs (1970) and Crews (1980) give illustrated accounts of some of the courting and reproductive behaviors of several reptiles.

Different species of garter snakes of the genus, *Thamnophis,* hibernate together in northern latitudes, sometimes in enormous numbers (10,000 or more) in underground dens. In the spring, the males emerge first and remain around the opening of the den. As a female emerges, usually a few weeks after the males have emerged, as many as one hundred conspecific males may pursue and court her, so that the female becomes one within a ball of snakes, her pursuers (Crews, 1980). As soon as a male has achieved intromission and inseminated the female, a copulatory plug forms, which prevents other snakes from inseminating her and which probably emits an olfactory cue that inhibits courtship immediately (Chapter 9). The initiation of the courtship also seems to be induced by an olfactory cue, emanating from the skin of the female, and the attractiveness of the female is increased by shedding of the skin.

Estrogen treatment of nonreceptive females induces male courting behavior, i.e., chin pressing on the dorsal surface of the female. An estrogen-treated female is not courted just prior to shedding; however, after she sheds, both she and her pen mates are courted. This transfer of cues can probably be explained by the fact that the chemicals that are attractive to the male are produced after shedding and are transferred to nonshedding females (Kubie et al., 1978). After ovariectomy, sexually attractive females become unattractive to the male, but estrogen treatment will restore the attractiveness. In the red-sided garter snake (*T. sirtalis parietalis*), the male's interest in the female is not dependent upon the presence of the testes (Camazine et al., 1980).

The endocrinology of the reproductive behavior of turtles and crocodilians has not been explored. Crews (1980) has discovered that in all female, parthenogenetic species of lizards of the genus *Cnemidophorus,* females engage in sexual behavior that remarkably resembles the sexual behavior of male-female pairs of related bisexual species. In the parthenogenetic species, the courted female has ovaries with preovulatory follicles, and the courting animal has regressed ovaries. Nothing appears to be known about hormone concentrations or the response of the females to administration of estrogens, progesterone, or testosterone.

Parental behavior in most reptiles seems to be limited to guarding and sometimes defending the eggs and developing embryos or to building a nest. In Chapter 10, we mentioned that pythons may coil around their eggs and that some skinks warm their eggs by basking in the sun and transferring the heat to the incubating eggs. American alligators (*Alligator mississippiensis*) and Nile crocodiles (*Crocodylus niloticus*) apparently help their young out of the egg and out of the nest (Pooley and Gans, 1976).

Birds

For the discussion of the neuroendocrinology of avian reproductive behavior we select three species—the chicken (*Gallus domesticus*), the Japanese quail (*Coturnix coturnix japonica*), and the ringdove, also called Barbary dove (*Streptopelia risoria*), because of the breadth and depth of the knowledge of the events and hormonal interrelationships in these species as compared to others. The chicken and Japanese quail are chosen because, in addition to a considerable amount of information about the neuroendocrine system, lines have been selected for different mating ability. Comparisons among such lines allow a study of some of the mechanisms involved in mating better than studies within a more homogeneous population. The ringdove is selected because male and female sexual behaviors are distinct, and complete reproductive cycles from the initiation of ovarian growth until fledging of the young have been followed, both from a neuroendocrine and behavioral point of view.

Chicken. The aggressive and courtship displays, waltzing (dropping of a wing and advancing towards the opponent sideways or circling the opponent), wing flapping, tidbitting (pecking at the ground with or without food calls) of roosters

are androgen dependent; i.e., the displays are eliminated or diminished after castration and are restored after administration of testosterone (Wood-Gush, 1971). The copulatory behavior consists of: (1) approaching—seizing the hen's comb or feathers of the neck in the rooster's beak and starting to put a foot on her back; (2) mounting—standing on the back or outstretched humeri of the female; (3) treading—rhythmical up-and-down movement of the feet by the mounted male followed by tail depression of the male, which brings the male and female cloacae in close contact (Barfield, 1969). Balthazart et al. (1981) investigated the effects of administration of (1) testosterone; (2) 5β-dihydrotestosterone (5β-DHT), which is one of the principal conversion products of testosterone in avian brains, instead of 5α-DHT, which is not produced by chick brain tissue but which is one of the principal conversion products of testosterone in mammalian brains; and (3) a combination of 5β-DHT and estradiol, which by itself has no effect, on the copulatory behavior of 1- to 16-day-old baby chicks. In this test, the attempts of the chicks to copulate with the observer's hand were scored. Testosterone and 5β-DHT were about equally effective in inducing male copulatory behavior, but estradiol did not increase the effect of 5β-DHT. The 5β-DHT was also effective in castrated baby chicks and in chicks in which the possible effect of endogenous testosterone was blocked by the administration of cyproterone acetate.

Testosterone implanted in the medial preoptic area of the brain induced copulatory behavior but did not induce either aggressive behavior or crowing in capons (Barfield, 1969). McCollom et al. (1971) injected testosterone cypionate (TC) into three different lines of capons: a line selected for a high cumulative number of completed matings (HM line), a line selected for a low cumulative number of completed matings (LM line), and a random-bred control population. For TC, they found that the threshold for courting was lower than for mounting, treading, and completing copulation. Surprisingly, the LM line was more sensitive to TC than the HM line. In a subsequent series of experiments, Benoff et al. (1978) determined that the concentration of plasma testosterone and the corresponding number of completed matings were: 29.9 ± 2.4 ng/ml and 7.3 ± 1.0 for HM roosters, 1.4 ± 0.8 ng/ml and 4.5 ± 0.6 for LM roosters, and 6.8 ± 1.9 ng/ml and 5.9 ± 0.6 for control males.

These two studies show that in a comparison between lines, low mating frequency is correlated with plasma testosterone concentrations and not with sensitivity of the neural system to testosterone, although within a particular population such a relationship may not exist (see under Japanese quail). It is, however, the virtue of establishing such divergent lines that makes it possible to detect such relationships. We shall see later that in Japanese quail and in drakes at least, behavior may be better correlated with the plasma concentrations of testosterone than it is in chickens. We should point out, however, that if one were to start with a different population of chickens and select for high and low mating in a similar manner as was used for LM and HM lines, one might find a different relationship than was observed. Finding that plasma testosterone concentration and mating were correlated in a similar manner, as found by Benoff et al. (1978), would greatly strengthen the hypothesis that this relationship is the best explanation for low and high mating tendencies.

Female sexual behavior is generally passive, although a female may crouch to the male and initiate copulation (Wood-Gush, 1971). In ovariectomized females, crouching can be induced by estrogen injection (Barfield, 1979).

Parental behavior of chickens can be divided into incubation behavior (sitting on the eggs) and caring for the hatched chicks. The prolactin content concentration in the plasma of incubating bantam hens was 23 percent higher than in the plasma of nonincubating bantam hens (Chapter 10). Experimentally, incubation of eggs and brooding of chicks can be induced in molting hens, which have regressed ovaries, by providing them with either eggs or chicks in a warm, dark, or dimly lit room (Wood-Gush, 1971). Care of the young has been induced in roosters by a variety of treatments, such as mammalian prolactin injections, castration, or oral administration of 9 ml/kg of body weight of 33 percent ethanol. These various treatments may reduce aggressiveness in the male, although this has not been in-

vestigated specifically for ethanol administration. Since chicks can induce brooding behavior only in the absence of aggressiveness (van Tienhoven, 1968), the males treated with prolactin or ethanol, and the capons, can respond to the stimuli produced by the chicks.

Japanese Quail. The sexual behavior of the Japanese quail has been described in detail by Sachs (1969). The following steps can be distinguished: approaching by the male (which may be accompanied by a strut), crouching of the female, grabbing of the female's neck feathers by the male with his beak, mounting and treading by the male, lifting of the tail feathers by the hen and diverting them to the side, depression of the male's tail, contacting of the cloacae, and dismounting by the male, followed sometimes by fluffing by the female (Sachs, 1969). These male sexual behaviors are androgen dependent, as demonstrated by testosterone propionate injections into males with regressed testes, while they were on a short photoperiod, and by replacement therapy of castrated males (Sachs, 1969).

The effects of prenatal steroid treatment on the sexual differentiation of the brain of the Japanese quail, that is, the demasculinization of males after treatment of the embryos with either estradiol or testosterone, was discussed in Chapter 2. In adult castrated males or in males on short photoperiods with regressed testes, strutting can be induced by administration of testosterone propionate (TP) and 5α-dihydrotestosterone propionate (5α-DHTP), but not of estradiol benzoate (EB), although EB acts synergistically with 5α-DHTP (Adkins and Dniewski, 1978). Crowing can be induced by administration of TP, testosterone, or androstenedione (AE), which are converted to estrogens (aromatized) by mammalian brain tissue, and also by 6α-fluoro-testosterone propionate (FTP), 17α-methyltestosterone (MT) fluoxymesterone (FM), and 5α-DHTP. Of these steroids, 5α-DHTP, FTP, and FM are not aromatized by human placental enzyme systems, and MT is not extensively aromatized, but FTP may be aromatized by mammalian brain tissue (Adkins et al., 1980). Copulation is induced in male Japanese quail with inactive testes by administraton of testosterone, TP, AE, and EB, but not by 5α-DHTP, FTP, FM, or MT (Adkins et al., 1980), suggesting that conversion to estrogen may be required for the induction of copulation. Further evidence supporting this suggestion consists of the observation that administration of an anti-estrogen blocks the TP-induced copulations in castrated Japanese quail (Adkins and Nock, 1976), and administration of an aromatase inhibitor, 1,4,6-androstradien-3,17-dione (ATD), blocks the copulation-inducing effect of TP, but not of EB (Adkins et al., 1980). The induction of copulation in baby chicks by 5β-DHTP warrants investigation of the effect of this hormone on copulation in Japanese quail.

Balthazart et al. (1979) found that plasma testosterone concentrations in male quail exposed to long days were not correlated with either aggressive behavior (pecks and chases toward males) or strutting toward males and females, but that there was a significant ($P < 0.05$) negative correlation ($r = -0.464$) between concentrations of hypothalamic 5β reduced metabolites (5β-DHT and 5β-androstan-3α,17β-diol) and aggressive behavior, and a significant ($P < 0.05$) positive correlation between strutting and hypothalamic androstenedione concentration ($r = 0.468$). There was also a significant ($P < 0.05$) positive correlation between androstenedione concentration of the hyperstriatum, an area not known to be involved in aggressive behavior, and aggressive behavior ($r = 0.442$). However, strutting behavior was correlated with testicular weight ($r = 0.563$, $P < 0.05$) and cloacal gland area ($r = 0.614$, $P < 0.05$) on day 11 of exposure to long photoperiod. Balthazart et al. (1979) concluded that the plasma concentration of metabolites is more important than the concentration of testosterone. They also found that the cloacal gland produced androstenedione and 5α-DHT, but little 5β-DHT, the males that produced the greatest amount of androstenedione having the largest cloacal gland, after exposure to long photoperiods.

Ringdoves. The breeding cycle of the ringdove consists normally of 7–10 d of courtship and nesting, 14–15 d of incubation of the eggs, and 21 d of parental care. The sexes are not externally dimorphic, but they can be distinguished by

their behavior, and, of course, by inspection of the gonads after laparotomy. When a male and female are introduced together in a cage that has a glass nest bowl on the bottom and has nesting material, food, and water, the male starts to court the female. Initially, his most frequent displays are strutting and the aggressive courtship display of hop-charging (the male hops or steps toward the female with his head forward and he may try to peck at her). These behaviors gradually are replaced by bow-coos (the male faces the female and alternately stands erect and bows; during the bows, he utters a cooing sound), and by the nest-soliciting displays, which consist of (1) wing-flipping (the male bends forward so that his head is close to the floor and his tail and back point upward, while he vibrates his wings), and (2) nest-cooing (the male stands near the nest, gives the wing-flipping display and utters a cooing sound, which an experienced observer can distinguish from the call given during the bow-cooing display). The female may respond by: (1) preening, (2) nest-cooing (same displays as the nest-cooing by the male), (3) wing-flipping, (4) chasing (she runs towards the male with her head forward and her rump feathers fluffed while making a cackling sound), and (5) sexual crouching (she holds her body and tail close to the ground and spreads her wings). During these interactions, the ovary and oviduct increase in size in the female and androgen secretion increases in the male (Chapter 14). After a few days of courtship, the male starts to gather nesting material, which he then takes to the female, who is near the nest bowl, and she constructs a nest. During the nest-building period, which lasts about a week, the pair copulates frequently. These copulations do not result in fertilization, unless they include copulations during the immediate preovulatory period (Cheng et al., 1981). These frequent copulations may ensure the female of the male's commitment to caring for his offspring. The female in selecting a mate will presumably select one that invests heavily in his reproductive effort, i.e., courting only her and providing nesting material (which in feral populations means flying back and forth to the nest site), so that paternal care of the young is ensured.

About 7–10 d after the male and female are

placed together, a clutch of two eggs is laid, and male and female take turns incubating the eggs. After the young (squabs) hatch, both parents feed them pigeon milk, consisting of the epithelial cells of the lining of the crop glands, which are two lateral lobes of the crop sac. This pigeon milk has a high concentration of fat and protein and is regurgitated by each parent and fed to the young. The secretion of pigeon milk is under the control of hypophyseal prolactin.

After the squabs fledge, a new reproductive cycle starts and the next clutch of eggs is usually laid about 41 d after the previous one. Cheng (1977b) has investigated whether the laying or incubation of infertile eggs affected the interval between clutches. When infertile eggs were laid, the latency to the start of a new cycle had a median of 20 d (11–28 d), when there had been no copulation. After copulation with a sterile male, the latency had a median of 14 days (12–23 d). When infertile eggs were substituted for fertile eggs, recycling occurred in 55.6 percent of the females, with a median of 25 d (10–29 d). When infertile eggs were substituted for fertile eggs, recycling occurred in 50 percent of the females, but with a median latency of 38 d (30–48 d). The female thus perceives internal as well as external cues about the fertility of the eggs, and this information to a large extent determines the time it takes before a new cycle starts. If young are hatched, no new cycle will start until the squabs are fledged (Cheng, 1977b).

The male, by courting only one female and by spending time and energy in nest building (feral ringdoves need to fly back and forth in gathering and carrying nesting materials to the female), has a large investment in his reproductive effort and it would, therefore, be to his advantage if the female were not to copulate with another male whose sperm could replace his. By surveillance of the female and by guarding her, he prevents cuckoldry (Lumpkin, 1981). This guarding by the male is an advantage for the female, because it is a protection against intruders and predators. Erickson and Zenone (1976), in experiments in which inexperienced females were exposed to experienced males for 15 minutes each day, found that such an association made the females less attractive to other males. This diminished attrac-

tiveness was inferred from less frequent bow-cooing and more frequent aggressive courtship by the male towards the female. These investigators interpreted their finding as an avoidance of cuckoldry by the first male. This implies that the lack of attractiveness was imparted to the female by the male. As mentioned above, the behavioral interaction between male and female does change her hormonal status, and thus may change the female's attractiveness to other males. Rissman (1981), using ringdoves that had raised at least one squab to the fledging stage, found that exposure of the male to a female other than his mate could also change his behavior and the behavior of females towards him. For instance, in one experiment in which males and females that had been paired for 6 d were used and tested with unpaired males and females, unpaired males wing-flipped towards paired and unpaired females, but paired males did not, indicating that the males' behavior was affected by the courtship of his mate. In an experiment in which males and females were exposed for 15 minutes each day for six days either to a ringdove of the opposite sex (pre-exposed males or females) or kept in isolation (unexposed males or females), pre-exposed females nest solicited (wing-flipped) more often than unexposed females, and they did so more towards pre-exposed than toward nonexposed males; unexposed females wing-flipped more often towards unexposed than toward pre-exposed males. Pre-exposed and non-exposed males pecked more often at pre-exposed than at non-exposed females. These data can be interpreted as showing that synchronization of male and female behavior is necessary to avoid aggression by the male. Desynchronization would then make the female less attractive to another male, but also would make a new male less attractive to a female.

The neuroendocrinology of the sexual behavior of ringdoves has been studied in some detail, and we will mention the salient findings. The effects of exposure of the embryos or of the squabs to gonadal hormones have not been reported for the ringdove, but in the pigeon (*Columba livia*), which shows behavior patterns very similar to those of the ringdove, neonatal estrogen treatment of male squabs resulted in a lower incidence of bow-cooing and nest calling, and an increased incidence of female behavior (squatting) compared with controls, when the animals were adults (Orcutt, 1971). Adkins (1977) pointed out that there was no evidence concerning gonadal hormone concentrations in the plasma of the adult pigeons that had been treated neonatally with estrogens. Therefore, no conclusions can be drawn about the differentiation of the brain.

The endocrine changes during the reproductive cycle are as follows: During courtship, the plasma estradiol concentration in the female increases from an initial value of 27 pg/ml to 85 pg/ml, decreases to 67 pg/ml during nest building, and then drops steeply to undetectable values ($<$ 15 pg/ml) during the incubation phase. In the male, estradiol values remain consistently low (Korenbrot et al., 1974). During the courting phase of the cycle, progesterone concentrations in the female rise from an initial value of about 0.5 ng/ml on day 1 to about 3.0 ng/ml on day 5; during incubation and brooding, this value is about 1.1 ng/ml. In the male, there is no significant fluctuation in progesterone concentration throughout the cycle (Silver et al., 1974). The effects of gonadal hormones on the reproductive behavior have been studied with respect to aggressive courtship behavior, incubation behavior, and feeding behavior.

In adult ringdoves, the sexual behavior of the male and female are androgen and estrogen dependent, respectively. In the male, hop-charging, bow-cooing, wing-flipping, and nest-cooing were eliminated or their frequency was reduced by castration, but these behaviors were restored after administration of TP. After 5α-DHTP administration, there was some bow-cooing and hop-charging, but no wing-flipping; whereas after estradiol benzoate (EB) wing-flipping occurred frequently, but bow-cooing and hop-charging did not (Adkins-Regan, 1981). Female sexual behavior (sexual crouching) was induced in gonadectomized ringdoves of both sexes by administration of estrogens (Cheng and Lehrman, 1975). This seems to confirm the hypothesis mentioned in Chapter 2, that the female sex is the more bisexual of the two sexes. Behaviors that normally occur in both sexes (wing-flipping,

nest-cooing) are more easily induced in each sex by their homologous than by their heterologous gonadal steroids (Cheng and Lehrman, 1975). Paradoxically, in ovariectomized ringdoves, a high dose (100 μg) of EB suppressed sexual behavior, whereas a lower dose (50 μg) induced it (Cheng, 1973). This paradoxical effect can be explained on the basis of an inhibitory effect of EB on LH-RH secretion or release. Cheng (1977a) found that LH-RH injections into ovariectomized ringdoves treated with the high dose of EB restored the sexual behavior in the female, whereas LH injections failed to do so, suggesting that LH-RH acts directly on the nervous system, in the ringdove, as it does in the rat (see section on mammalian sexual behavior). She also found that exogenous LH-RH acted synergistically with low doses of EB in inducing female sexual behavior, and that an anti-ovulatory analogue of LH-RH injected into intact females reduced sexual crouching, but not nonsexual squatting (squatting preceded by a ritual display of an elaborate sequence of self-preening, begging, billing, courtship, feeding).

Hutchison and Steimer (1981) found that systemic administration of TP decreased the effectiveness of TP in inducing a male sexual behavioral display, perch calling (which is given in visual isolation from a female), as the interval between castration and TP administration increased. Perch calling was selected because when male ringdoves can see females, the male sexual behavioral response to TP is more variable than when they can't see the females. This decrease in effectiveness of TP with time elapsed since castration is correlated with an increased reduction of testosterone to 5β reduction products (5β-DHT, 5β-androstane-3α,17β-diol, 5β-androstenedione) specifically in the preoptic area (POA) of the brain, and not in adjacent areas of the brain (Steimer and Hutchison, 1981). These 5β reduced metabolites of testosterone are not effective in inducing male sexual behavior [and as will be discussed in the next paragraph, the POA is the target area for testosterone in inducing male sexual behavior in ringdoves]. Progesterone administration to TP-treated castrated males inhibits the effect of TP on male courtship behavior, with the strutting not being affected, the aggressive behavior (chasing) declining in 3–6 d, and the nest-oriented behavior disappearing in 10–20 d (Hutchison, 1976). This effect of progesterone may be exerted by blocking the testosterone receptors in the hypothalamus.

The preoptic area of the ringdove brain is of paramount importance in mediating the effect of testosterone on behavior. It is in the POA, which concentrates injected, radioactively labeled testosterone (Steimer and Hutchison, 1981), that testosterone is converted to estrogen (Steimer and Hutchison, 1980). Implants of TP in the POA of castrated male ringdoves activated bow-cooing, nest-cooing, and copulation when placed with a female, and pecking when placed with a male (Barfield, 1971). Steimer and Hutchison (1981) found that, as in systemic administration of TP, the male sexual behavioral responses after implantation of TP in the POA decreased as the interval between castration and implantation of the hormone increased. According to Hutchison (1971, 1976), implants of TP in either the POA, the anterior hypothalamus (AH), or the lateral hypothalamus are effective in restoring the displays of chasing, bowing, and nest soliciting by the male, although not to precastration levels, in terms of duration. Implants of testosterone, 5α-DHT, or 5α-DHT acetate in the POA or AH stimulated short periods of chasing and nest soliciting, whereas EB implants in these areas stimulated nest soliciting, and in a few castrated males they induced chases, but these chases were shorter in both peak duration and total duration than after TP or T implants (Hutchison, 1976).

In addition to the POA, the nucleus intercollicularis (ICo), which is involved in control of vocalization of many avian species (Phillips and Peek, 1975), has an important function in the control of reproductive behavior. Lesions of the ICo reduced nest-cooing in a otherwise intact males, whereas implants of TP, 5α-DHT, or estradiol but not 5β-DHT, in the nucleus of castrated males, increased bow-cooing by the males. In ovariectomized females, EB implanted in the ICo, increased nest-cooing by the females. The importance of vocalizations is indicated by the finding that in the female, severing of the hypo-glossal nerves, which prevents the animal from vocalizing, eliminates the growth of the ovary in response to an intact, courting male (Cohen, 1981).

The nest soliciting response to systemically administered TP in castrated male ringdoves occurs more rapidly in birds with lesions in the septum than birds without lesions, but the bow-cooing response to TP is not affected by such lesions (Cooper and Erickson, 1976), indicating that the septum is also involved in the mechanism of androgen-stimulated bow-cooing.

Incubation behavior. In both sexes, incubation behavior can readily be increased by the injection of 100 μg progesterone each day for seven days, but not by estrogen or prolactin (Lehrman, 1958). Implants of progesterone on the POA-AH or in the lateral forebrain evoke incubation behavior in both sexes and inhibit courtship behavior in the male (Komisaruk, 1967). During the reproductive cycle and also after progesterone injections, increased incubation behavior is accompanied by decreased courting behavior, as might be expected.

Silver and Feder (1973) reported, however, that progesterone injected into intact males interfered with incubation behavior. This observation seems irreconcilable, at first, with the data presented by Lehrman (1958). However, in Lehrman's experiments, the experimental birds were tested in pairs, whereas in Silver and Feder's experiments, individual birds were tested with respect to their response to a female, from the day an egg was present in the nest. It is thus possible that the testing situation affected the response. However, the fact that plasma progesterone concentrations did not fluctuate in the male during the breeding cycle casts some doubt on the importance of progesterone in inducing or maintaining incubation behavior in the male.

Cheng (1975) analyzed the effects of exogenous progesterone on intact males that were introduced into a cage with a female that was sitting on a nest with eggs and found that progesterone treatment reduced the latency of the onset of cooing in the nest bowl. In Silver and Feder's (1973) experiments, the stimulation by situational cues that were conducive to incubation was maximized, whereas in Cheng's (1975) experiments the stimulation was minimized. The results of these experiments suggest that for induction of incubation behavior, progesterone has an important function in the male under certain social conditions, and has an important function in the female under all conditions tested. For the female, this is further substantiated by experiments by Cheng and Silver (1975), in which estrogen and progesterone replacement therapy was necessary to induce nest building and incubation behavior. Intact males, which were paired with steroid-treated females, showed nest building and incubation behavior in response to the female's activities, re-emphasizing the importance of the cues from the female for the male.

Feeding behavior. The feeding of the young is done by both the male and the female. Pigeon milk is produced in the crop gland under the influence of prolactin (Lehrman, 1961). As with other aspects of the reproductive cycle of the ringdove and the pigeon, visual stimuli are important. Patel (1936) showed that an isolated male that can see an incubating female, starts to secrete pigeon milk.

The choice of the ringdove for psychoneuroendocrinological investigations has been an excellent one, for the investigations have clarified some important brain-hormone interrelationships. Much credit for the choice of this animal and for this clarification should go to Lehrman, who pioneered much of this work.

Other Avian Species. Male red-winged blackbirds (*Agelaius phoeniceus*), during aggressive encounters, had a plasma LH concentration lower than that of foraging males, but plasma corticosterone concentrations were about the same. The log of the plasma DHT concentration in aggressive males showed a positive regression on the log of LH concentration, whereas in foraging males this regression was negative (Harding and Follett, 1979). These data suggest that aggression, by lowering LH, may eventually affect reproduction. Under crowded conditions, more frequent encounters may occur and thus reduce the reproductive potential of certain males. What also needs to be determined is whether there are differences in plasma LH concentrations between winners and losers or between territorial males and intruders.

Mating patterns. The various types of mating patterns, monogamous, polygynous, monogynous, monandrous, polyandrous, and polygamous, have been reviewed and discussed by Emlen and Oring (1977) and their combinations in relation to parental and paternal care have been reviewed by Ridley (1978).

Care by the male only is often associated with polyandry, since the male's care of the eggs and young allows the female to produce multiple clutches, as long as males are available for mating and caring for the eggs and/or young (Emlen and Oring, 1977). This pattern occurs in the greater rhea (*Rhea americana*) and in some species of tinamous (Ridley, 1978).

In the phalaropes, there is a role reversal of the sexes. The females of the northern phalarope (*Lobipes lobatus*), called red-necked phalarope in Europe, are larger and more conspicuously colored than the males. The females arrive at the breeding grounds before the males, though whether they are territorial is not definitely established. They are polyandrous, and the males incubate the eggs, although the females remain near the nest until the young hatch and then leave (Ridley, 1978). The males also take care of brooding and raising the young. The behavior of Wilson's phalarope (*Steganopus tricolor*) is somewhat similar to that of the northern phalarope, but the female leaves after laying the clutch of eggs (Ridley, 1978).

In a monogamous mating system, both parents usually care for the young. This may be because potential losses to the parent that courts another potential partner and that neglects the young may be greater than the gains made by courting another bird (Emlen and Oring, 1977). Monogamy is probably the result of the inability of the species to take advantage of any potential for polygamy, and it is the prevalent mating system found among 90 percent of avian species studied (Emlen and Oring, 1977).

Polygyny exists for various possible reasons. (1) It provides a defense for resources, the male controlling females indirectly by controlling the crucial resources. (2) It also provides a defense of females, in which the male directly controls his female(s). (3) It promotes male dominance, in which the females select males from aggregations of males. (4) It also promotes breeding assemblages in which there is highly synchronized mating of populations and both sexes come together for a short time. (5) It promotes the formation of leks—an organized aggregation, which meets in a communal display area for the sole purpose of attracting and courting females and to which the females come for mating (Emlen and Oring, 1977)—in which the males remain sexually active for the duration of the females' breeding season, which is not highly synchronized. Under these mating systems, the female takes care of incubation and brooding of the young, and the contribution by the male varies according to species.

Polygamy is often a rapid multiple-clutch polygamy, in which both sexes have multiple breedings in rapid succession. Males and females each incubate separate clutches of eggs. This system is found in the red-legged partridge (*Alectoris rufa*), the mountain plover (*Charadrius montana*), and Temminck's stint (*Calidris temminckii*). All these species are ground-nesting and the precocial young are often lost as a result of predation.

The reader is referred to the papers by Emlen and Oring (1977) and Ridley (1978) for detailed discussion of the evolutionary and ecological aspects of these mating systems and of the paternal care system. For an excellent review of the mechanics of incubation, the physical aspects of warming and cooling of the eggs, and the hatching of the chicks, the review by Drent (1975) should be consulted.

Vocalizations. The male's vocalizations, which may serve to advertise his presence to friends and foes of both sexes, are controlled by androgens. In 1939, Leonard showed that injections of androgens caused female canaries to sing. This was confirmed by Nottebohm and Ar-

nold (1976), who also noted, however, that singing was not induced by androgen treatment in female zebra finches (*Poephila guttata*).

The neural centers involved in vocalizations—the caudal part of the ventral hyperstriatum (HVc), the robust nucleus of the archistriatum (RA), and the hypoglossal nucleus of the medulla (cranial nerve XII)—are strikingly larger in male canaries and zebra finches than in the females. A fourth area, area X of the lobus olfactorius, is well developed in males of both species, small in female canaries, and absent in female zebra finches (Nottebohm and Arnold, 1976). The influence of these areas in controlling vocalization has been demonstrated by lesioning (Nottebohm et al., 1976).

After injection of tritiated testosterone ($[^3H]$-T) in male and female zebra finches, less radioactivity was found to be concentrated in the HVc and in the magnocellular nucleus of the anterior neostriatum (MAN)* of females than in these same areas in males (Arnold and Saltiel, 1979). Testosterone was also concentrated in the RA, n XII, the nucleus intercollicularis—also involved in song control—and in areas of the hypothalamus (medial preoptic area, periventricular magnocellular nucleus, infundibular region) (Arnold and Saltiel, 1979; Arnold et al., 1976).

Early experience. Early experience affects sexual behavior. In poultry husbandry, rearing of turkeys in isolation may lead to abnormal orientation by males during mating (Schein and Hale, 1965). Depending on the species, birds can imprint on their parents or on any other object, shortly after hatching. The bird so imprinted may respond to this object as a conspecific. Immelman (1965a) arranged to have hatchling zebra finches reared by another finch (*Lonchura striata f. domestica*), a domesticated songbird. Such cross-fostered zebra finch males courted *Lonchura striata* females even in the presence of female zebra finches. Cross-fostered female zebra finches initially made advances towards *Lonchura striata* males, but soon courted male

*There is no evidence that the MAN is involved in song; however, there are heavy projections from MAN to HVc and RA, according to Arnold and Saltiel (1979).

zebra finches. As adults, such cross-fostered male zebra finches sang the *L. striata* song, even though the cross-fostered young were separated from the cross-foster parents before the young had started to sing (Immelman, 1965b).

Harris (1970) exchanged eggs between two species of gulls in Britain: herring gulls (*Larus argentatus*), a sedentary species in Britain, and *L. fuscus,* a species that migrates to Iberia and northwest Africa. The cross-fostered herring gulls migrated to France, Spain, and Portugal, but not as far as *L. fuscus* controls. The cross-fostered *L. fuscus* migrated normally. Cross-fostered females of each species mated with males of the foster species, whereas cross-fostered males mated with either species. This method has potential use as a biological method to control gull populations that have become or are becoming a nuisance or danger, e.g., near airfields.

Mammals

A large part of this discussion of mammalian sexual behavior will be based on the rat, because more details are known about this animal than about any other species.

Perinatal Conditions and Behavior of Adults.
The importance of the perinatal environment in determining the sexual behavior of the adult has been discussed in detail in Chapter 2. Other aspects of early environment, for instance early handling of pups and its repercussions on the animal as an adult, were also discussed in Chapter 2.

Being reared in isolation can have pronounced effects on the sexual behavior of the adult animal. Social isolation of cats (Rosenblatt, 1965), dogs (Beach, 1968), cattle (Schein and Hale, 1965), guinea pigs (Gerall, 1963), rats (Gruendel and Arnold, 1969), and rhesus monkeys (Harlow et al., 1966) leads to abnormal sexual behavior, the extent depending on the species. The deficit observed in most species is the failure of the male to orient appropriately. For instance, after semi-isolated rearing, dogs may mount towards the head or flank and, thus, have fewer intromissions than controls. Boars reared in isolation show behavioral courting deficits, have fewer copula-

tions, ejaculate faster, and have a longer latency to mount (Hemsworth et al., 1977). Rearing bulls without giving them an opportunity to mount other animals results, as in dogs, in a lack of appropriate spatial orientation, so that as adults they mount any part of the body. Rhesus monkeys, raised in isolation in a playroom that offered a stimulating situation for play and social behavior, showed severe deficits in initiation of sexual behavior (courting), of response to the hind quarters of females, of orientation and, thus, of intromission. These males were also more aggressive toward females (Harlow et al., 1966).

Sexual Behavior in Males.
The sexual behavior of males can be divided into courting behavior and copulatory behavior. Most attention has been paid to the endocrine effects on copulatory behavior and on the classification of different patterns of copulatory behavior. Dewsbury (1972) has proposed a classification of 16 patterns based on the presence or absence of locks, thrusting, multiple intromissions, and multiple ejaculations.

Locks, the retention of the penis in the vagina because the swelling of the bulbus glandis at the root of the penis during copulation prevents the withdrawal of the penis from the vagina until detumescence of the penis, are found in dogs and wolves and may last from a few minutes to an hour (Beach, 1969). Not all 16 patterns have been found, but Dewsbury (1972) has listed the patterns for 118 of about 4,000 mammalian species. Dewsbury (1975) has been able to correlate some copulatory patterns with the anatomy of the penis. For example, within the muroid rodents, the ratio of the diameter of the glans penis to glans length $\times$ 100 for rodents with a single baculum is correlated with lock and/or thrust. When this ratio is 31 or above, the species shows lock and/or thrust; when it is 28 or below, there is no lock or thrust. Locks also seem to be associated with a reduction in the components of the accessory sex glands, except for the preputial glands.

In certain species, postejaculatory intromissions (without ejaculations occurring) may be required in order to induce pregnancy. Female cactus mice (*Peromyscus eremicus*), which were separated from the males after ejaculation showed an incidence of 5 percent pseudopregnancy and 5 percent pregnancy, whereas females that were permitted to copulate to satiety (the criterion for satiety was 30 min without copulation) showed an incidence of 32 percent pseudopregnancy and 47 percent pregnancy, although after the initial ejaculation no transfer of semen took place during the subsequent intromissions (Dewsbury and Estep, 1975).

The copulatory sequence of events in the rat and the biological significance of these events for achieving pregnancy have been discussed in Chapter 8. Pfaff and Lewis (1974) have given a detailed description of the male's sexual behavior based on film analysis. Sachs and Barfield (1976) have described and discussed the arousal and inhibitory mechanisms in pacing the sequence of events in the copulatory patterns of the rat, and the interested reader should consult this excellent presentation.

The male golden hamster (*Mesocricetus auratus*) differs from the males of most other mammalian species in that castrated males treated with estrogen and progesterone consistently show female sexual behavior (Noble, 1979). In many of the other species investigated, the sexual behavior of the male is androgen-dependent, as indicated by castration and androgen replacement therapy. However, after castration, the various components of copulatory behavior do not disappear at the same rate. Castrated rats retain the mounting behavior longer than intromission, while the ejaculation reflex and intromission are often retained together; in castrated guinea pigs, ejaculation disappears first, followed by intromission and mounting (Young, 1961).

The age at which castration occurs and the extent to which the animal is experienced also affect the reduction of copulatory behavior. For instance, tomcats with sexual experience (whether gained while intact or castrated), when treated with androgens, retained their copulatory behavior, whereas inexperienced controls failed to show copulatory behavior (Rosenblatt, 1965). Male guinea pigs reared by spayed females are deficient in sexual behavior. The "learning" of the appropriate copulatory behavior does not require androgens, for castrated males reared socially but treated with androgens at the age of puberty showed normal sexual behavior (Rosenblatt, 1965). In some species, castration has little effect on copulatory behavior; e.g., castrated, experienced dogs retain virtually all aspects of copulatory behavior for as long as 5 years or more (Beach, 1969).

In rats, the concentration of androgen may be a limiting factor for sexual behavior. Craig et al. (1954) found that after castration, sexual behavior could be increased above the precastration level by androgen treatment. However, in strains of guinea pigs that showed deficiencies in copulatory behavior, no increase was obtained following castration and androgen replacement therapy (Young, 1961).

When comparisons were made among eight strains of mice for the relationship between plasma testosterone concentration and male sexual behavior, Batty (1978a), surprisingly, found *negative* significant correlations between plasma testosterone concentrations and the percentage of tests in which the males mounted, achieved intromission, and ejaculated. It is difficult to account for these correlations. Within strains, none of the above correlations were significant. In subsequent investigations, Batty (1978b) found that the plasma testosterone concentration in male mice increased upon exposure to an estrous female (see Chapter 14). Within four strains of mice, there was a significant, negative correlation between the plasma testosterone concentration, after 30 minutes of exposure to an estrous female, and latency of mounting ($r = -0.33$ to -0.73), but there was no significant correlation between plasma testosterone concentration and latency of ejaculation.

Neuroendocrinology of Male Sexual Behavior. The neuroendocrinologic aspects of male sexual behavior have been reviewed by Davidson (1977) and much of the following discussion is based on his review. The concept that androgens need to be aromatized to estrogens in order to be effective in restoring sexual behavior in castrates is probably incorrect. Yahr (1979), reviewing the available evidence, concludes that: (1) DHT is more potent than DHT propionate in restoring sexual behavior in castrates; (2) DHT can stimulate sexual behavior in rats, guinea pigs, and rhesus monkeys; (3) DHTP is less potent than TP, particularly in sexually naive castrates; and (4) DHT seems to promote longer latencies of intromission and ejaculation without affecting intromission frequency. The conclusion that androgens, in order to be effective, do not need to be converted to estrogens is further supported by the fact that 6α-fluor-testosterone is effective in eliciting male sexual behavior in castrated male rats, and, incidentally receptivity in ovariectomized rats (Yahr and Gerling, 1978), even though it is not aromatized.

There is evidence that the site of action of androgens on the nervous system can be the brain, the spinal cord, or the peripheral nervous system. In restoring male sexual behavior of castrated rats, implants of testosterone in the anterior hypothalamic-preoptic area (AH-POA), were more effective than implants in either the posterior hypothalamus or in the cortex, although less testosterone appeared in the circulation after implants in the AH-POA than after implants in the other two sites. The evidence is thus persuasive that testosterone acts at the site of the AH-POA; however, as Davidson (1977) pointed out, in castrated rats treated systemically with testosterone, 100 percent showed male sexual behavior, whereas after testosterone implants in the AH-POA, only 80 percent responded, thus suggesting that testosterone may act also at sites other than the AH-POA.

Other evidence for the involvement of the medial preoptic area (MPO) in sexual behavior is the deficiency of normal sexual behavior in animals with lesions in this area. Such lesions may especially affect the arousal mechanism, as evidenced by the observation that captive rhesus monkeys with lesions of the MPO do not attempt to copulate, but masturbate and ejaculate at the same rate as normal captive monkeys (Sachs and Barfield, 1976). Such lesions also inhibit copulatory behavior in rats, cats, and dogs (Davidson, 1977).

The involvement of androgens at the site of the spinal cord is suggested by experiments with castrated dogs and rats that have had their spinal cord severed. Animals with a severed spinal cord, but otherwise intact, show a number of sexual reflexes in response to manipulation of the genitalia. For instance, such dogs when placed on an estrous bitch have erections and make movements similar to copulatory movements (Hart and Kitchell, 1966). Rats, when placed on their back flip their penis and have erections when the preputial sheath is pushed behind the penis (Hart, 1979). Castrated animals with a severed spinal cord do not show such responses, but following testosterone treatment, the responses reappear.

The significance of these observations is not clear in view of the following considerations: In castrated animals, an implant of androgens in the hypothalamus can restore copulatory behavior in 50–80 percent of the rats. It is, of course, possible that there is some androgen leakage and that these androgens reach the spinal cord cells that are androgen dependent. A more difficult observation to explain is that testosterone and dihydrotestosterone are nearly equipotent in restoring these spinal reflexes, but that estradiol benzoate (EB) is totally ineffective (Hart, 1979); whereas either EB or testosterone is effective when given systemically in restoring copulatory behavior in castrated males. Hart (1979) concludes that "The results, with EB in particular, point out that the display of intromissive and ejaculatory patterns in rats may not involve spinal neural mechanisms that are customarily associated with these behavioral patterns."

The evidence available to determine whether androgens affect the male's sexual behavior via the peripheral nervous system is not definite. There is a possibility that the ejaculatory response is affected partly via the peripheral nervous system and that mounting behavior is not so affected. The effects of electrical stimulation of different areas of the brain on sexual behavior of squirrel monkeys were discussed in Chapter 9.

Sexual Behavior in Females. In order to show female sexual behavior in adulthood, it is necessary either that the perinatal environment be free of androgens and estrogens or that these hormones do not reach the central nervous system in an active form; for example, alpha-fetoprotein probably binds estrogens so that they are not taken up in the brain (Chapter 2).

The rearing environment is as important for females as it is for males. For example, socially deprived female rhesus monkeys showed a lower incidence of adequate support for penile intromission, thrusting, and insemination than controls (Harlow et al., 1966). Such females also showed a greater incidence of threat and aggressive behavior in addition to self-clutching and self-biting than control females. Female guinea pigs, raised in a socially deprived environment, also showed abnormal sexual behavior (Harlow et al., 1966). The remainder of this discussion relates to the sexual behavior of normal females and their interactions with normal males.

Before copulation occurs, various interactions take place between the sexual partners. Beach (1976) has divided the successive phases of the female's participation into attractiveness, proceptivity, and receptivity, and has considered the endocrine aspects of each of these phases. Sexual attractivity is inferred from the male's performance as related to different forms of stimulations provided by the female. In Chapter 14, we will discuss such forms of stimuli, e.g., visual, tactile, and olfactory. In general, the female is attractive exclusively when she is in heat, although in some primates (e.g., hamadryas and chacma baboons) estrus increases the attractiveness above that observed when she is not in estrus. Beach (1976) points out that a female's attractiveness is not entirely hormonally determined, but that there are other factors that make females preferred over others, especially in dogs (Beach, 1969). The attractiveness signals that the female emits can, in most cases, be expected to be species-specific, so that they also serve as isolating mechanisms and tend to prevent hybridization.

Proceptivity consists of appetitive behaviors of the female in response to stimuli received from the male. For instance, in pigs it is the sow that is attracted to the boar as well as the boar being attracted to the sow; the same is true in sheep (Signoret, 1976). In general, proceptivity either occurs at estrus only or it increases at estrus. It should be remembered, of course, that the male's attractiveness to the female plays a role in determining her proceptivity.

Beach (1976) defines receptivity as a stimulus-response relationship in which females exhibit behavior in response to stimuli normally pro-

vided by conspecific males. Receptive reactions are the consumatory phase of the mating sequence. These reactions can be expected to be species-specific. Several measures can be used to quantitate the receptivity of the female. Generally, a ratio of responses (e.g., completed ejaculations, lordosis response) to attempts by the male is calculated and used as an index. Receptivity is, in most mammals, a condition that occurs only at estrus or is heightened at the time of estrus. The stimuli that cause receptivity to occur may be visual, olfactory, auditory, tactile, or some combination of these. In primates, differences among individuals are apparently more important than hormonal determinants (Beach, 1976). The musk shrew (*Suncus murinus*) is an apparent exception in the induction of estrus by the ovarian hormones, estrogen and progesterone. The sexual behavior of the female is not affected by ovariectomy and, neither ovariectomy nor injection of 25 μg estradiol either intramuscularly or into the uterus, affect uterine and vaginal weight or histological appearance of the uterus (Anderson and Dryden, 1977).

Neuroendocrinology of Female Sexual Receptivity. The neural and endocrine control of sexual receptivity have been studied in most detail in the rat. Pfaff and Lewis (1974) have described their results of the analysis of films made of rats' copulatory behaviors. The main behavior studied is the lordosis behavior, consisting of elevation of the rump and head with depression of the back. Lordosis behavior in response to conspecific males is estrogen and progesterone dependent. In ovariectomized rats, the sequence of administration needs to be estrogen priming followed by an injection of progesterone. This sequence is not the same for all species, however, for in the ewe, progesterone administration must precede estrogen administration in order to obtain behavioral estrus (Young, 1961). The rat's lordosis reflex, provided the endocrine concentrations are appropriate, is induced by tactile stimulation of the flank region and pelvic thrusting by the male. The lordosis posture is maintained longer after an intromission than without an intromission (Komisaruk, 1974). Steel (1979) has reported that hamsters that have copulated successfully once not only are less receptive, but also show a decrease in proceptive behavior. This decrease in proceptivity and receptivity may be related to the secretion of LH-RH in response to intromission. Moss (1974) demonstrated that in rats that were both ovariectomized and adrenalectomized, either estrogen plus progesterone or estrogen plus LH-RH could induce the lordosis reflex in response to mounting by the male, whereas LH in combination with estrogen was ineffective.

Komisaruk (1974) found that during cervical probing a high proportion of ovariectomized rats showed the lordosis reflex in response to tactile stimulation of the flank and perineum by the experimenter. The intensity of lordosis is related to estrogen levels, and progesterone acts synergistically with estrogen. Cervical probing prior to flank and perineum stimulation also increased the response to the latter manipulation. Probing of the cervix in addition to priming the animal for the lordosis reflex in response to flank and perineum stimulation also blocked the response to general pain stimuli (pinching of the tail or ear), but not to touch; thus cervical probing acted as an analgesic but not as an anesthetic (Komisaruk, 1974). Komisaruk (1974) has proposed that lordosis is an extensor-dominated, flexor-suppressed reflex that is facilitated by estrogen and progesterone.

The site at which estrogen and progesterone act has been investigated by implanting these hormones in different areas of the brain. The most recent evidence in which female estrous behavior was investigated in ovariectomized rats in response to a male showed that small implants (30-gauge needles filled with EB) were most effective when placed in the ventral medial hypothalamus (VMH); 22 out of 30 rats responded. After similar implants in the POA-AH, 7 out of 28 rats showed estrous behavior, and after implants in the anterior hypothalamus and lateral hypothalamus, 4 out of 23 animals responded. Besides the difference in incidence of lordotic behavior, there was also a difference in the extent of estrous behavior, with animals with implants in the VMH scoring more than twice as high as the other two groups of rats. Systemic administration of progesterone to these rats did not change the incidence of animals showing estrous

behavioral response, but it did eliminate the difference in the behavioral scores between rats with an implant in the VMH and those with an implant in the POA-AH, by raising both groups to the same score (Barfield and Chen, 1977). It thus appears that the VMH is the principal site for the estradiol effect on female sexual behavior of rats. Barfield and Chen (1977) point out, on the basis of intracerebral implant studies, that this same area is implicated in the control of sexual behavior of guinea pigs, rabbits, and sheep. However, Ciaccio and Lisk (1973) found that small implants (27-gauge needles filled with estradiol) were not effective in eliciting female sexual behavior in ovariectomized hamsters; however, larger estradiol implants (23-gauge needles) were effective when placed either in the anterior hypothalamus or in the filiform nucleus, but not effective when placed in the VMH. It is not clear why such estradiol implants in the VMH are not effective in hamsters but are effective in rats, guinea pigs, and sheep. It is possible that the hamster, in which the female is aggressive except during estrus, has a different neuronal circuit that controls female sexual behavior than the other three species.

Mode of action of estrogens. How estrogen affects the brain has been investigated by intracerebral administration of antibiotics that block DNA-dependent RNA synthesis (actinomycin D) or that block protein synthesis at the ribosomal level (cycloheximide). Since actinomycin D is cytotoxic, careful controls need to be used. Ovariectomized rats, treated with EB, showed no damage of the hypothalamus after actinomycin D infusion into the third ventricle. These rats showed a significant depression in the lordosis quotient (LQ), when the actinomycin D was perfused 6 h after EB injection, and also showed full recovery 10–14 d after infusion of the antibiotic when tested again with an EB injection. After infusion into the POA, actinomycin D was effective in reducing the LQ of EB-treated ovariectomized rats when given either simultaneously or 6 h after EB. This effect was reversible even in animals with unilateral lesions in the POA (Terkel et al., 1973). Hough et al. (1974) confirmed these results, i.e., actinomycin D infused into the POA of ovariectomized rats treated with estrogen and progesterone had a reversible inhibitory effect on the LQ and on the fine structure of the nucleolus. These data argue for a necessary effect of estrogen on RNA synthesis, for estrogen to affect the sexual behavior of the female.

Subsequent experiments in which cycloheximide, instead of actinomycin D, was used have shown that: (1) infusion of cycloheximide, 6 h before or 12 h after estrodiol benzoate (EB), into either the POA or the third ventricle of ovariectomized rats treated with EB plus progesterone (progesterone was injected 36 h after EB and the animals tested 4–6 h after progesterone injection) diminished the LQ (Quadagno and Ho, 1975); (2) injections of cycloheximide into the POA were as effective as injections into the VMH in inhibiting the lordosis behavior induced by estrogen and progesterone, both in frequency and in quality (Meinkoth et al., 1979); and (3) injections of cycloheximide outside the hypothalamus did not inhibit sexual receptivity in ovariectomized rats treated with estrogen and progesterone (Leehan et al., 1979). These and the data obtained with actinomycin D injections thus strongly indicate that estrogen has its effect through the synthesis of a protein that induces female sexual behavior. It would be extremely interesting if such a protein were isolated and its effects tested.

Although in most mammalian species, female receptivity diminishes after ovariectomy, this does not seem to be true for rhesus monkeys and women (Everitt et al., 1972). The adrenal glands, however, are essential for sexual receptivity in rhesus monkeys. Everitt et al. (1972) found that after adrenalectomy and cortisol injections, androstenedione administration restored female receptivity, while dehydroepiandrosterone administration gave inconsistent results. It appears, therefore, that adrenal androgens are essential for the sexual behavior of female rhesus monkeys. The androstenedione treatment did not, however, affect the size of the clitoris, the color of the sexual skin, aggressive behavior, and grooming of the female by the male (Everitt et al., 1972). Sexual receptivity can be restored in ovariectomized, estrogen-treated, adrenalectomized rhesus monkeys by implants of testosterone propionate in the POA-AH (Everitt and Herbert, 1975).

Besides the interaction between estrogens and the hypothalamus in inducing female sexual behavior, there is evidence that other areas of the brain are important in sexual behavior. For instance, in the absence of estrogen, lesions of the septum of rats increase the intensity of lordosis in ovariectomized rats in response to cervical probing and flank and perineum stimulation, although these rats did not show lordosis in response to males. Ovariectomized, septal-lesioned rats showed a greater sensitivity after 1 μg EB/100 g body weight than ovariectomized, estrogen-treated controls with an intact septum (Komisaruk, 1974). These results suggest an inhibitory effect of the septum on sexual behavior.

Perinatal Maternal Behavior. The following account of perinatal maternal behavior is based on the reviews by Naaktgeboren (1979) and Naaktgeboren and Slijper (1970).

Nest building. Many polytocous eutherian mammals build some type of nest. Exceptions are the European hare (*Lepus europaeus*) and the guinea pig (*Cavia porcellus*), both of which give birth to well-developed young, which therefore may not need the protection provided by a nest. The nest building can consist of the digging of a hole in the soil, as in dogs (*Canis familiaris*), wolves (*Canis lupus*), red foxes (*Vulpes vulpes*), and pigs (*Sus scrofa*); or it can be an elaborately woven nest between stems of wheat, as in the little harvest mouse (*Micromys minutus*), or a nest at the end of a tunnel with the nest lined with hair pulled out of the mother's pelt, as in the wild rabbit (*Oryctolagus cuniculus*). Laboratory rabbits usually make a nest from straw and line it with hair. Other species may dig a hole in the snow, as in the wolverine (*Gulo gulo*), the ringed seal (*Phoca hispida*), and the hooded seal (*Cystophora cristata*). Hibernating bears give birth during hibernation in their hibernaculum, whereas bats give birth after hibernation, usually in the caves in which they spend the day.

In animals that live in herds, the female about to give birth may leave the herd, as in cattle; or the herd may surround the female while she is giving birth, thus giving protection against predators, as in the hartebeest (*Alacephalus* sp).

Whales and dolphins give birth under water and conspecifics may help the young to reach the surface to take its first breath.

Periparturient and parturient behavior. Just before birth, the mother may lick the vulva; this is especially true in polytocous species. In some species, the mother may pull the young out of the birth canal, as is the case in the red fox, the deermouse (*Peromyscus maniculatus*), and the rhesus monkey, whereas in other species, conspecifics may help the young out of the birth canal, as seems to be the case in dogs, mice, and spring mice (*Acomys* sp). After a young animal is born and the fetal membranes are broken, the mother, but sometimes conspecifics, will lick it to clean and dry it. The umbilical cord may be severed during the expulsion of the fetus, or it may be severed by the mother biting through it. In cats, this occurs while she eats the placenta. In cattle and horses, the umbilical cord has a preformed site of breakage, and on the side closest to the fetus, it has a strong sphincter muscle, which contracts and prevents the young from bleeding after the cord is broken. In many species, including herbivorous ones such as the rabbit, the placenta is eaten by the mother, probably to keep the nest clean.

Endocrine regulation of postparturient maternal behavior. After the young are born, they usually find the mammary glands and start to nurse. The secretion of milk is controlled by several hormones among which prolactin is one of the essential ones. Our knowledge of the endocrine regulation of maternal behavior is much less complete than our knowledge of sexual behavior. This is partly because maternal behavior has been studied less intensively than sexual behavior, but partly because different investigators have assessed the maternal behavior differently. In rats some investigators report the latency to retrieve pups only (Zarrow et al., 1971b), whereas others report an array of behaviors, e.g., the quality of the nest built, pup retrieval, licking of the pups, touching the pups, and crouching over the pups. It should be pointed out that when ovariectomized rats are daily exposed to pups, retrieval behavior starts after about 4.9 days (Zar-

row et al., 1971b). The presence of the pups is the stimulus for this maternal behavior.

Maternal behavior (pup retrieval) has a non-hormonal and a hormonal basis, as indicated by two types of evidence. First, pup retrieval, crouching over the pups, licking of the pups, and building a nest can all be induced by exposure to pups, with a latency of 6 days or more, in intact females, ovariectomized females, hypophysectomized females, and intact and castrated males, without any significant differences in the latency of these behaviors (Rosenblatt, 1967). Second, transfusion of blood from "spontaneously" retrieving females into virgin females fails to induce retrieval with short latency in the recipient virgins (Terkel and Rosenblatt, 1971). The firm basis for humoral control of maternal behavior in rats has been established by Terkel and Rosenblatt (1972) in the short-latency induction (14.5 h) of pup retrieval in virgin females after transfusion of blood from rats that had just given birth to young. The questions of which hormones are involved and what sequence of changes in hormone concentrations is required to initiate maternal behavior, have not been answered entirely (Krehbiel and LeRoy, 1979).

Efforts have been made to induce or speed up the maternal behavior of virgin rats toward pups by simulating the concentrations of estrogens, progesterone, and prolactin found at the time of parturition. The shortest latency in obtaining retrieval of pups and displaying the nursing position has been found after transfusion of blood from parturient to virgin rats. In pregnant rats, injections of hormones have produced short latencies only after hysterectomy between day 10 and 19 of pregnancy. Presumably, the hysterectomy has its effect (shortening latencies) by causing a decrease in progesterone secretion, but ovariectomy counteracts the effects of hysterectomy on maternal behavior. This counteracting effect of ovariectomy, in turn, can be overcome by a single injection of either 20 µg or 100 µg of EB, with an additional injection of 500 µg of progesterone having no effect (Siegel and Rosenblatt, 1975a). In hysterectomized, ovariectomized virgin rats, 100 µg of EB administered at the time of surgery (Siegel and Rosenblatt, 1975b) or 8 weeks later (Siegel and Rosenblatt,

1975c) induced short latency in pup retrieval. In such virgin, hysterectomized, ovariectomized rats, 20 μg of EB was not effective in inducing this response (Siegel and Rosenblatt, 1975c). A dose of 500 μg of progesterone had no effect when given 44 h after EB; it inhibited the EB response partially when given simultaneously with EB, and it completely inhibited the response when given at 24 h after EB (Siegel and Rosenblatt, 1975b). A dose of 5.0 mg of progesterone inhibited the retrieval response in these EB-treated rats completely (Siegel and Rosenblatt, 1975b). Presenting the pups immediately after the injection of EB to virgin, hysterectomized, ovariectomized rats gave a retrieval response of 46 percent at 24 h after injection; at 48 h, 92 percent of these rats showed this maternal behavior (Siegel and Rosenblatt, 1975d).

It is clear from the experiments using hysterectomy that the uterus has an important function in experimentally induced maternal behavior; however, maternal behavior normally occurs in intact rats, and given enough time and appropriate stimulation develops in virgin intact rats, albeit with a long latency. The question that has not been answered is what the role of the uterus is in delaying maternal behavior. Siegel and Rosenblatt (1975c) have speculated that the estrogen necessary for the induction of maternal behavior is bound by the uterus, thus making it unavailable to the nervous system.

Leon et al. (1973) reported that virgin rats ovariectomized 8 weeks prior to testing for pup retrieval retrieved pups with a median latency of about 24 h, whereas virgin rats ovariectomized 4 weeks before testing, had a median latency of about 120 h, which was not different from that of intact controls. Injections of an anti-estrogen (50 mg MER-25/day for 4 d) to rats ovariectomized 4 weeks earlier reduced the median latency to 48 h, whereas 12 μg EB/d for 8 d prior to testing in rats ovariectomized 8 weeks earlier increased median latency to 168 h. Males castrated 4 weeks before testing had a median latency for pup retrieval of 168 h, whereas males castrated 8 weeks before testing showed a corresponding value of 8 h. Administration of TP increased this to 96 h, while intact controls showed a median latency of 120 h. Leon et al. (1973) concluded that estrogen inhib-

ited maternal behavior, whereas Siegel and Rosenblatt (1975 a–d) concluded that estrogen induced it. In view of the difference in doses of EB used and the different durations of hormone administration it is difficult to decide between these opposing views, especially since nothing is reported about the effects of the different treatments on the hypophyseal hormone secretions and about the prolactin concentrations in the plasma, although the evidence that prolactin is probably not involved in initiation of maternal behavior of rats is persuasive (Zarrow et al., 1971b; Bridges et al., 1974; Numan et al., 1972, 1977). However, it should be realized that no experiments seem to have been reported in which the induction of maternal behavior has been investigated in hypophysectomized rats treated with purified gonadotrophic hormones, including prolactin.

Consideration should also be given to the observation that certain treatments (e.g., administration of estrogen, progesterone, and prolactin to ovariectomized rats) may have no effect on the latency with which a certain maternal behavior occurs, but that the hormones can affect the quality of the behaviors, such as the duration of licking or sniffing the pups, and the quality of the nest built (Krehbiel and LeRoy, 1979).

The relationship between maternal behavior and specific areas of the central nervous system has been studied by hormonal implants and lesion experiments. Virgin rats and rats that were ovariectomized and hysterectomized on day 16 of pregnancy have been used in these studies. The latency for pup retrieval by such ovariectomized-hysterectomized rats was shorter after estradiol benzoate (EB) was implanted in the medial pre-optic area (MPO) than after implants of EB in the ventromedial hypothalamus and after cholesterol implantation in the MPO. Lesions of the MPO of virgin rats decreased frequency of pup retrieval, nursing, licking the pups and nest-building and increased the latency in those animals that showed these behaviors (Numan et al., 1977). Lesions of the MPO and even introduction of the electrodes decreased the pup retrieval, nursing, and nest-building behavior of rats that were showing these maternal behaviors prior to the surgery (Numan, 1974). It thus appears that the MPO is required for the onset and the maintenance of several aspects of maternal behavior.

Severance of lateral connections of the POA-AH reduced the frequency of nursing by 50 percent and eliminated retrieval and nest building completely. Although these lesions and the lesions of MPO interfered severely with maintenance of maternal behavior, they did not affect sexual behavior (Numan et al., 1977). Somewhat similar results have been obtained with hamsters, in which severance of the lateral connections of the POA-AH had a deleterious effect on maternal behavior without affecting sexual receptivity, whereas severance of the posterior connections to the hypothalamus reduced sexual receptivity but did not affect maternal behavior (Marques et al., 1979).

The controlling mechanisms for maternal behavior in species other than the rat have not been studied in as much detail. In the mouse (*Mus musculus*), nest building occurs during pseudopregnancy (Gandelman et al., 1979). In intact and ovariectomized mice, progesterone stimulated maternal nest building; estrogen alone had little effect on this behavior but acted synergistically with progesterone in eliciting it (Lisk, 1971). Implants of progesterone into the hypothalamus of mice induced nest building but did not affect other maternal behaviors. Prolactin, given either systemically or as an implant in the hypothalamus induced short-latency pup retrieval and increased time spent in the nest, licking of the pups, and nest-building behavior (Voci and Carlson, 1970), thus implicating prolactin in maternal behavior in this species, while at the same time suggesting that the hypothalamus is the principal part of the circuit involved in this behavior. Besides the hypothalamus, the olfactory system apparently has a role in nest-building behavior, for removal of the olfactory bulb diminishes nest building in virgin mice, pregnant mice, and progesterone-treated virgin mice (Zarrow et al., 1971a).

Strains of rabbits (*Oryctolagus cuniculus*) differ in the frequency of nest building during pseudopregnancy and in the duration of estrogen plus progesterone administration required to induce this behavior in ovariectomized does (Zarrow et al., 1965). The effect of estrogen and progesterone on maternal behavior requires the presence of the anterior pituitary. Ergocornine administration, which inhibits prolactin secretion, inhibits nest building in pregnant does, an effect that can be overcome by prolactin injections, suggesting but not demonstrating that prolactin may be involved in maternal behavior in the rabbit.

The sheep (*Ovis aries*) is a domestic farm animal for which there is information about the endocrinology of maternal behavior. In sheep, a long-term mother-young bond is established soon after birth. As in other mammals, the ewe licks the young, and within an hour after birth the lamb attempts to suck. The mother then usually accepts the lamb, although she usually rejects alien lambs that try to suck. It appears that experienced multiparous ewes show a low incidence (about 8 percent) of aberrant maternal behavior, whereas primiparous ewes may show an incidence of aberrant behavior as high as 60 percent, mainly a failure to stand for the lamb while it tries to reach the udder (Poindron and Le Neindre, 1980). When the ewe is not allowed to have contact with her lamb shortly after parturition, her maternal behavior declines and eventually she will not accept her own lamb. The period during which she will accept her lamb is here called, the sensitive period. It has been shown experimentally that both the induction of maternal behavior and the duration of the sensitive period are controlled by estrogens and that neither parturition, lactation, nor a long estrogen treatment are required for a multiparous ewe to show the maternal behavior repertoire of licking a newborn lamb (from another ewe), emission of low-pitched bleats, acceptance at suckling, and establishment of a long-lasting bond. In primiparous ewes, the induction of maternal behavior by estrogen injection has, so far, not been successful, indicating that experience is an important factor in the induction of maternal behavior (Poindron and Le Neindre, 1980).

The induction of maternal behavior by administration of EB has not been successful for all sheep in an experiment, suggesting that other hormones may be important and that the lamb used for stimulation may have a function also. Poindron and Le Neindre (1980) obtained evidence that as the lamb ages, it loses some of its attractiveness to the ewe and that olfactory stimuli from the lamb are more important than visual and auditory stimuli in establishing the discriminatory behavior that ewes show towards the lambs they have accepted in comparison to alien lambs.

The formation of milk is necessary for feeding the young in most species, except for guinea pigs, which can survive without milk, although initially their growth is less than that of controls that are nursed by the mother. Only in a few species have hypophysectomized animals been used to establish the endocrine requirements for lactation; however, studies have been conducted on mammary gland tissue cultured in vitro (Baldwin and Plucinski, 1977). The subject of lactation and its control falls outside the scope of this book. The reader interested in comparative aspects of lactation is referred to *Comparative Aspects of Lactation,* edited by Peaker (1977).

14

Environment and Reproduction

The reproductive success of organisms depends on the interaction between their genetic endowment and the environment. As species evolve, those populations or individuals whose young are born at the optimal time for growth, development, and survival have the best chance to pass their genes to the next generation. Optimal time is used here to describe the period with the best combination of such conditions as availability of food, suitable temperature, and low incidence of predation.

Some vertebrate species, especially among fish (e.g., salmons) and birds (e.g., geese, the swallows of Capistrano) have used their rather efficient means of travel to migrate to favorable breeding grounds and to spend either their growing period and/or their nonreproductive period in environments that are most favorable for those parts of their life cycle (Chapter 13).

For migratory as well as for nonmigratory species, natural selection has favored the retention of those control systems (to be discussed) that use the most reliable information predicting the environment and that cause reproduction to occur at an optimal time for survival of the young (Farner and Follett, 1966). Baker (1938), in a classical paper, proposed two sets of causes for determining the optimal time of reproduction, i.e., (1) ultimate causes, which exert selection pressure so that *populations* that reproduce during the optimal season survive; and (2) proximate causes,

which provide *individuals* with the appropriate information, so that reproductive processes are initiated and continued in such a way that the young are born at the optimal time of survival. A priori, it appears that for aquatic animals, temperature or, depending on the depth at which they live, temperature and light would provide information that reasonably predicts future conditions. For most terrestrial organisms, day length is the most reliable source of information. We shall discuss the roles of various components of the environment but especially of light.

ENDOGENOUS RHYTHMS

In order to understand how environmental information is integrated, or in other words, how environmental stimuli exert their effect, it is necessary to consider endogenous rhythms first, because endogenous and exogenous rhythms interact with each other to stimulate or terminate reproduction. The two principal endogenous rhythms with which we will be concerned are the circadian and the circannual rhythms.*

---

*We will use the following definitions for the various rhythms:

Circadian rhythm, an endogenous free-running rhythm that persists in a cueless environment and that has a period of *about* 24 h.

Daily rhythm (Wurtman, 1967), a rhythm with a period

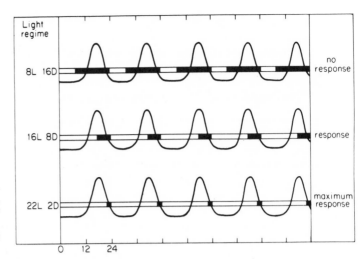

Figure 14.1 Schematic representation of an external coincidence model. The daily photoperiodic regime entrains a circadian cycle in photosensitivity into a daily cycle. The photoperiodic response is a function of the extent to which the daily photophase extends into the photo-inducible phase (above the horizontal bars). The "shape" of the photosensitivity curve is an arbitrary selection. (From Farner, 1975; reprinted with permission of *American Zoologist*.)

Light can have two functions in setting the stage for the reproductive season: (1) it can serve as a signal (Zeitgeber, entraining agent, synchronizer) that entrains an endogenous biological rhythm, or (2) it may serve as a stimulus of the neuroendocrine system, such that gametogenesis, spermiation, ovulation, endocrine stimulation of secondary sex organs and sex characters, and behavior occur, thus permitting the animal to breed. An example of light functioning as a signal is the 6-hour shift of locomotor activity of house sparrows (*Passer domesticus*) when the lights on-off schedule is changed from lights on at 600, lights off at 1800 to lights on at noon, lights off at midnight. An example of light serving as a stimulus is the increase in testicular size when photosensitive white-crowned sparrows (*Zonotrichia leucophrys gambelii*) are transferred from a short day (8L:16D) to a long day of 16L:8D.

of exactly 24 h, which is obtained in environments with a periodicity of 24 h. It is in effect a circadian rhythm that has been synchronized to exactly 24 h by the external oscillator.

Diurnal rhythm, a rhythm of biological variations or events occurring between sunrise and sunset or during the illuminated fraction of a near-daily schedule of artificial light and darkness (Halberg and Katinas, 1973).

Nocturnal rhythm, a rhythm of biological variations or events occurring between sunset and sunrise or during the dark fraction of a near-daily schedule of artificial light and darkness (Halberg and Katinas, 1973).

Circannual rhythm, an endogenous rhythm that persists in a constant environment and that has a period of about one year.

Circadian Rhythm

The role played by a circadian rhythm of photosensitivity in the photoperiodic response of organisms has been studied for a wide range of organisms and responses, such as flowering of plants, diapause in insects, emergence of pupae of *Drosophila,* and the testicular response of birds and fishes. It lies outside the scope of this book to consider the important theoretical aspects of this aspect of circadian rhythms of photosensitivity, and interested readers are referred to the following papers and reviews: Bünning (1936), Farner (1975), Farner et al. (1977), Gwinner (1975), Pittendrigh and Minis (1964), and Saunders (1976).

Two models have been proposed to explain the photoperiodic response of the neuroendocrine system that affects avian gonadal growth (which has been studied mainly in males): the external coincidence and the internal coincidence model. In the external coincidence model (Figure 14.1), it is assumed that light acts as a signal to synchronize the endogenous, circadian rhythm of photosensitivity, and that light, acting as a stimulus, falls in a period of photosensitivity and thus induces the response. This model is supported by two types of experiments: (1) by resonance experiments (Figure 14.2), in which light-dark cycles are used with a relatively short light period, e.g., 8 h in "days" of 24, 36, 48, 60, 72, 84 h; and (2) by the photoscan or interrupted-night type of experiments (Figure 14.3). After a light period

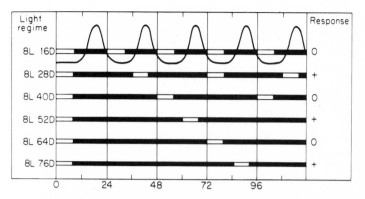

of 6 or 8 h (to signal the onset of dawn), the subsequent dark period is interrupted by a short (15 min–2 h) light period that comes at different times during the night for different experimental groups.

In the internal coincidence model (Figure 14.4), light serves as a signal and not as a stimulus, and if one can replace light by another signal, such as temperature change, one should obtain the same type of response to light as to a temperature change (Saunders, 1966). In the internal coincidence model, it is assumed that there

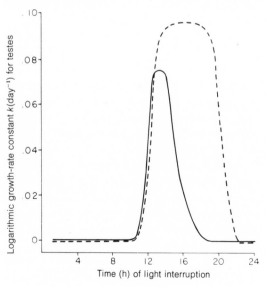

are at least two oscillators in which one is phased by dawn or by "lights on" and the other by dusk or by "lights off." When the phases of these two internal oscillators are in an optimal configuration, the maximal response, e.g., testicular growth, occurs. This model is supported by two lines of evidence. In one line of evidence, light as a signal can be replaced by a change in temperature; for example, a temperature cycle should phase the two oscillators, and the same type of response should then be found after exposure to a temperature cycle as after exposure to a similarly timed light cycle. Apparently, there has not been an experiment in which this has been tested for the avian gonadal response. In another line of evidence, the organism may be exposed to a stimulatory light cycle (long day) for a relatively short period, so that the full (gonadal) response is not yet obtained, and the organism is then transferred either to continuous darkness (DD) or to a short photoperiod, e.g., 8L:16D. If the internal coincidence model is applicable, then the response should continue under DD, i.e., the testes continue to enlarge; whereas under 8L:16D, the response should diminish, i.e., the testes should regress. Vaugien and Vaugien (1961), using English sparrows, and Wolfson (1966), using white-throated sparrows (*Zonotrichia albicollis*), have indeed found that after transfer from long days to DD, the testicular growth was sustained and after transfer to short days, the testes regressed. Farner et al. (1977) have pointed out that the number of birds used in each experiment by Vaugien and Vaugien (1966) and by Wolfson (1966) was low and that lack of measurements at critical time points in the experiment leave the data open to

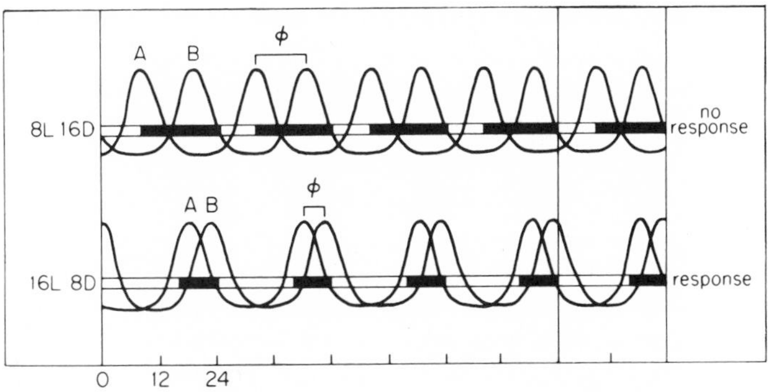

Figure 14.4 Schematic representation of internal coincidence model. A change in photoperiodic regime from 8L:16D to 16L:8D results in a change in phase angle (φ) between oscillators A and B. The smaller φ represents the necessary coincidence for the photoperiodic response. (From Farner, 1975; reprinted with permission of *American Zoologist*.)

other interpretations. From their own experiments, in which they investigated the applicability of the internal coincidence model on English sparrows, Farner et al. (1977) concluded that the internal coincidence model is too complex for the explanation of their findings. They concluded that their results could be explained on the basis of an external coincidence model that included a carryover phenomenon of the neuroendocrine events; i.e., gonadotrophins were released tonically after the stimulatory photoperiod was ended, thus sustaining gonadal growth. Rutledge (1980) investigated whether the internal coincidence model applied to the starling (*Sturnus vulgaris*). Male starlings were placed under (1) DD, after involution of their testes, following a natural photoperiod, or (2) a natural photoperiod followed by LL, or (3) under 12L:12D 45, 90, or 135 d after the testes had recrudesced and involuted. The time it took to obtain spermatogenesis was significantly shorter after exposure to natural photoperiods than after exposure to any of the other light regimes (about 60 d versus 100–130 d). These results suggest that in this species, an internal coincidence model may indeed be applicable, because it seems unlikely that a carry-over effect could explain the similar results obtained for the groups that were held 45, 90, and 135 d on 12L:12D, after testicular involution under the 12L:12D light regime.

The testicular response to variations on photoperiod has been investigated in most detail in birds. Farner et al. (1977) state that in more than 50 species in 15 avian families, it has been demonstrated that day length serves as important environmental information in the control of reproduction. In some species for which measurements are available, the logarithmic rate of testicular growth is linearly correlated with day length over the range of 9–10 to about 20 h. This has been found, e.g., for the house finch (*Carpodacus mexicanus*), the white-crowned sparrow (*Zonotrichia leucophrys gambelii* and *Z. leucophrys pugetensis,* the slate-colored junco (*Junco hyemalis*), the green finch (*Carduelis chloris*), and the Japanese quail (*Coturnix coturnix japonica*) (Farner et al., 1977). The English or house sparrow is an exception to this general relationship between day length and rate of testicular growth. In this species, in addition to the correlation between day length of above 9 h and testicular growth, there is also a testicular growth rate that is an inverse function of day length of less than 7 h. For the other species named, it appears that the external coincidence model alone can explain the observation, whereas in the English sparrow, it may involve an external coincidence model plus a "carry-over" phenomenon.

Circannual Cycle

An alternative to the idea that testicular growth is the result of light acting as either an entrainer and a stimulus (external coincidence model) or as an entrainer for circadian oscillators (internal coincidence model) is the idea that light acts as an entrainer for a circannual gonadal cycle. In order to discuss the possibility of such light-entrained circannual cycles, it is necessary to review briefly what the evidence is for circannual cycles. It is clearly, a priori, more difficult to establish the existence of an endogenous circannual rhythm than of a circadian rhythm. Moreover, there are important questions as to the definition of constant conditions, as Farner and Wingfield (1978) have pointed out. Does it mean that conditions are the same from day to day (e.g., 12L:12D throughout the year) or does it mean that conditions are constant each day and thus throughout the year? Very few experiments have been carried out under LL or DD to study gonadal cycles, except for the study in Paris, in which Pekin drakes (*Anas platyrhynchos*) were kept under LL for 54 months and under DD for 70 months, and controls were kept under natural photoperiods. (The drakes under LL had such a high mortality that the experiment could not be extended beyond 54 months). Spectral analysis of the data for the drakes under DD showed that there were two periodicities, one with a duration of 156 d and one with a duration of 319 d. Drakes kept under natural photoperiods had corresponding periods with durations of about 140–147 d and 365 d (Assenmacher, 1974). Assenmacher (1974) proposed three possible explanations for the testicular cycle in the drake: (1) a circannual clock, and light acting as the Zeitgeber; (2) a circa-semestrial clock either entrained by light or with incomplete synchronization and a biphasic profile of testicular activity (as was indeed found); (3) two endogenous rhythms fluctuating at a circa-semestrial and a circannual periodicity. Assenmacher (1974) suggests that the second explanation may explain the data better than the other two.

Gwinner (1975), in a summary of evidence for avian circannual rhythms, lists circannual testicular cycles for the garden warbler (*Sylvia borin*), the blackcap (*S. artricapilla*), the African weaver finch (*Quelea quelea*) and the starling. It should be noted that of the species named, only starlings were kept under LL, whereas the other species were kept under daily light-dark cycles. The day length under which birds are kept is an important factor in revealing the existence of a circannual rhythm. For instance, starlings exposed to 11 h or fewer of light or to 12 h or more of light do not show a circannual testicular periodicity, but starlings kept under 12L:12D show such a cycle (Schwab, 1971).

An alternative to the rhythmic variation in photosensitivity for gonadal stimulation is, that there is the possibility of an additive effect of photoperiod on the photoreceptor-hypothalamo-AP system. This effect would be comparable to an hourglass, whereas the models based on a circadian rhythm would be comparable to an oscillator. There appears to be only one example of such an hourglass model for gonadal stimulation in vertebrates, i.e., the American chameleon (*Anolis carolinensis*) (Underwood, 1979).

INFLUENCE OF LIGHT ON THE NEUROENDOCRINE-GONADAL AXIS

The onset of the reproductive season in many North Temperate Zone birds is timed by day length. The end of the reproductive period in several species (van Tienhoven and Planck, 1973) is the result of the onset of refractoriness of the neuroendocrine-gonadal system. The gonads regress in spite of the fact that the photoperiod that initiated the onset of reproduction continues. The "site" of the refractoriness is not established, and it may vary among species. We will discuss briefly the evidence for each of the components of the neuroendocrine-gonadal system that is considered responsible for photorefractoriness.

There is apparently no inherent mechanism in the testes that causes regression. Starlings (Schwab, 1971) and canaries (*Serinus canarius*) (Storey and Nicholls, 1976) maintained under 11L:13D show continuous spermatogenesis for extended periods of time. Injections of gonadotrophins in photorefractory birds will induce resumption of spermatogenesis (van Tienhoven and Planck, 1973). These types of experiments thus argue against the idea that testes become refractory. The photorefractoriness might, however, be the result of a negative feedback effect by gonadal hormones. By determining the LH concentration in the plasma of white-crowned sparrows, Mattocks et al. (1976) were able to show that photorefractoriness occurred in intact and in castrated birds. It thus appears that a negative feedback by gonadal hormones does not play a crucial role in photorefractoriness.

It is conceivable that the hypophysis becomes refractory to releasing-hormone stimulation; however, the experiments by Schwab (1971) cited above seem to argue against the idea that it is the inherent pituitary failure to respond to long-term stimulation that causes the photorefractory state. Farner and Wingfield (1978) suggest that the site of photorefractoriness is either at the hypothalamus or at a higher level in the brain.

Experiments with starlings and domestic drakes (*Anas platyrhynchos*) have shown that the thyroid influences plasma LH concentrations and the onset of testicular regression at the end of the breeding season. Wieselthier and van Tienhoven (1972) found that thyroidectomy of starlings prior to exposure to 17L:7D resulted in a failure of testicular regression at the time of regression in sham-operated controls. When thyroidectomy was carried out 4 weeks after the onset of exposure to 17L:7D, testicular regression followed, but after this a second period of testicular growth started, suggesting that thyroidectomy stimulated gonadotrophin secretion and the effect of thyroidectomy was not the result of an increased testicular sensitivity to gonadotrophins. The thyroid was not required for regression itself, and thyroidectomy prior to the refractory period did not affect the time of onset of the refractory period. The failure of refractoriness to occur after thyroidectomy before exposure to long photoperiod, is probably an indirect result of the high gonadotrophin secretion in this species after thyroid removal.

Jallageas and Assenmacher (1979) reported that domestic (Pekin) drakes, thyroidectomized in April, showed an increase in plasma LH concentrations in May; whereas in controls, plasma LH concentrations started to decline. In the thyroidectomized drakes, plasma LH concentrations reached a peak in August, then declined sharply; in the controls, plasma LH levels reached a nadir in July, then rose to a small peak in August, which was followed by a decline. Plasma testosterone concentrations were higher in thyroidectomized than in control drakes, and the postnuptial decrease in plasma testosterone concentration was delayed from June in the controls to October in the thyroidectomized birds. It is of particular interest that in intact drakes, plasma thyroxine (T_4) concentrations show a sharp peak in June–July. After castration, plasma T_4 concentrations were higher than in intact drakes, but in May they declined gradually and in

July–August steeply, to reach control concentrations in February. Jallageas and Assenmacher (1979) speculate that there is "a temporary escape of thyroxine (TSH?) secretion from the inhibitory effects of elevated testosterone levels, which would lead to an annual thyroxine peak in June and to the resulting gonadotropic depression."

It thus seems that in the starling and drake, the thyroid influences the onset of regression, but that this hormone does not affect the refractory period itself. Nevertheless, thyroid hormone affects the expression of the photorefractory state either temporarily, as in drakes, or permanently as in Indian finches, i.e., the weaver bird (*Ploceus philippinus*), the black-headed munia (*Munia malacca malacca*), the spotted munia (*Lonchura punctulata*), and the chestnut-bellied munia (*M. atricapilla*). In these finches, the gonads remain at their maximal size for an indefinite period after thyroidectomy (Thapliyal and Chandola, 1972).

As was discussed in Chapter 13, the photorefractory and photosensitive phases of the reproductive cycle of the white-throated sparrow may be controlled by different phase relationships between prolactin and corticosterone secretion (Figure 13.1). As Farner and Wingfield (1978) have pointed out, whether this mechanism applies to other avian species has not been tested.

It should be kept in mind that in species that breed early in the season in the temperate zone, young birds will be exposed to long photoperiods, which in adults cause gonadal recrudescence. Such young are photorefractory, but it is not known what mechanisms are involved in causing this refractoriness. It may be comparable to the onset of puberty in mammals, but not enough is known to make such a statement more than speculative.

The photorefractory state that is usually maintained during exposure to "long"* days can be dissipated by exposure to "short"* days (van Tienhoven and Planck, 1973). An extensive

---

*"Long" and "short" days refer to long and short days of the particular species as evidenced by, e.g., the gonadal response.

treatment of the physiology of the refractory period can be found in Murton and Westwood (1977, Chapter 12).

Water Temperature and Day Length

In the teleost fish, the three-spined stickleback (*Gasterosteus aculeatus*), certain combinations of water temperature and day length interact with internal factors in regulating the breeding season. These interactions are also important during the juvenile photorefractory period, which, of course, prevents reproduction in the young at an early age. Table 14.1 shows that long days and high water temperatures do not stimulate the gonads when they are in phase 0, the condition found in immature young. However, short days and high water temperatures, such as prevail in late summer and early fall, can cause gonadal development to phase 1a. Gonads in this phase of development respond to the short days and low temperatures that prevail in the winter, and gonadal development proceeds similarly in adults and juveniles.

Juvenile sticklebacks exposed to 16L:8D at 20° C from July failed to show sexual maturation during a year of exposure, whereas sticklebacks kept on 16L:8D from July and exposed to 20° C water from July until August, to 4° C for 24 days and then to gradually increasing water temperature (from 4° C to 20° C in 18 days) matured in December (Baggerman, 1957). Thus, in this species, the temperature of the water is important in dissipating the juvenile refractory period when the fish are exposed to 16L:8D.

At the end of the breeding season (which lasts until about July in the Netherlands), the gonads regress. Under laboratory conditions, male three spined sticklebacks kept under 16L:8D and 20° C showed repeated alternate breeding and non-breeding cycles, with a successive decrease in the duration of the nonbreeding cycles from an average of 108 d for the first nonbreeding period to 46.7 d for the second and 25 d for the third nonbreeding period. Thus there seems to be a refractory period, but unlike conditions for the birds discussed, there is no requirement for exposure to short days to dissipate the photorefractory period (Baggerman, 1957).

The testicular response of adult sticklebacks to 16L:8D at 20° C in the fall is not substantially affected by prior exposure for 45 days to 16L:8D at 6° C, 8L:16D at 20° C or 8L:16D at 6° C. However, both the stage of development of the gonads and the prior light and temperature regimes affect the response to 8L:16D at 20° C. None of the fish exposed first to 8L:16D at 20° C responded, thus demonstrating a refractory condition. Three of four fish with testes initially in phase 1a and exposed first to 8L:16D at 6° C responded, but only one of four with testes in phase 1b responded to 8L:16D at 20° C (Baggerman, 1957). Thus, the refractoriness of the testes to a change in temperature from 6° C to 20° C depends on the stage of testicular development.

None of the males with testes in phase 1a and exposed first to 16L:8D at 6° C responded to 8L:16D at 20° C, whereas six of eight males initially in phase 1b responded (Baggerman, 1957).

Table 14.1

Interaction of a photoperiod and water temperature on gonads of three-spined sticklebacks at various phases of gonadal development

Phase	Day length: 16 h Temp.: 20° C	Day length: 8 h Temp.: 20° C	Day length: 16 h Temp.: 4°–6° C	Day length: 8 h Temp.: 4°–6° C
0	−	+	+	+ (?)
1 < 1a	+	−	−	+
1b	+	−	+	−
2	+	+	+	+

Gonadal development proceeds from phase 0 to 2 via 1a and 1b. Plus sign indicates that the fish responded and the gonadal development proceeded to the next phase. Stage 0 is found in juvenile fish only.

From Baggerman, 1957; reprinted with permission of *Netherlands Journal of Zoology*.

Therefore, in this species, interactions between temperature, photoperiod, and the internal status of the animal prior to exposure to a stimulatory combination of temperature and photoperiod are important.

ENVIRONMENTAL STIMULI AND REPRODUCTION

Photoperiod and temperature or a combination of them are probably the principal proximate causes that supply information to animals of the temperate zones for prediction of the optimal period for mating.

For animals that live and breed in the tropics, one would expect factors other than day length and temperature to time the breeding season, if there is one. Similarly, for opportunistic breeders, such as the budgerigar, other factors would be expected to be important. Moreover, the exact timing of breeding of animals in the temperate zones might be expected to be affected directly or indirectly by factors other than photoperiod and temperature.

We will discuss, by giving a few examples, the particular factors that are important, and in selected cases, the pathways that are involved. It is useful to refer also to Chapter 8, in which we mentioned the principal proximate causes for a number of species.

Cyclostomes

Temperature of the water is the principal factor that times the migration upstream of *Lampetra fluviatilis* and *Petromyzon marinus* (Chapter 13). Temperature is also the main factor for the timing of spawning. Hardisty and Potter (1971) list the various species of lampreys and the temperatures at which they spawn.

Spawning and courtship behavior, in contrast to migratory behavior, occur during the day and apparently with a preference for lighted areas. During courtship, the female passes the posterior part of the body near the male's head, suggesting that olfactory stimuli may play a role in the final stages of mating and spawning (Hardisty and Potter, 1971).

Elasmobranchs

In elasmobranchs, the main proximate cause is probably the temperature of the water.

Teleosts

For nontropical teleost fishes, photoperiod and temperature are probably the main proximate causes in timing the reproductive season (Chapter 8). In the three-spined stickleback, there is evidence that the threshold value for gonadal stimulation by photoperiod varies throughout the year, it being high in June–July and low in March–April (Baggerman, 1972, 1978, 1980). In the rainbow trout (*Salmo gairdneri*) which spawns in the fall, gonadotrophin secretion and spermatogenesis are advanced by exposure to decreasing photoperiods, with these responses being greater at 16° C than at 8° C (Peter and Crim, 1979). In female rainbow trout, the change in gonadotrophin secretion during exposure to decreasing photoperiods and high temperature is similar to that in the male (Peter, 1981). Most freshwater teleost fishes spawn in the spring, and usually a combination of increasing photoperiods or of long photoperiods in combination with an optimal water temperature, which may be different for different species, brings about an increase in gonadotrophin secretion and recrudescence of the gonads. For the goldfish (*Carassius auratus*), the optimal temperature range for gonadal growth is between 17° C and 24° C. At 30° C gonadotrophin secretion is stimulated, but the testes remain small (Peter and Hontela, 1978), and ovarian follicles become atretic (Peter and Crim, 1979; Peter, 1981).

The Pineal and Photostimulation of Reproduction. The way the photoperiod has its effect on gonadal recrudescence may vary according to the species. Several lines of evidence suggest that the pineal is involved in the photoperiodic response of some teleost fishes. In the eastern brook trout (*Salvelinus fontinalis*), the pineal organ is involved in (1) the entrainment of circadian activity by light-dark cycles, (2) the diurnal variation of plasma and liver metabolites, and (3) hypothalamic and endocrine cycles (de Vlaming and Olcese, 1981).

In certain fishes, e.g., rainbow trout (*Salmo gairdneri irideus*) and in *Pterophyllum scalare,* the pineal has been demonstrated to have a photosensory function (Dodt, 1963; Morita and Bergmann, 1971). In other species, there is morphological evidence for a secretory function of the pineal, e.g., the mosquito fish (*Gambusia affinis*) and *Symphodus melops* (Chèze, 1969; Chèze and Lahaye, 1969). In the golden shiner (*Notemigonus crysoleucas*), kept under 15.5L:8.5D, pinealectomy results in regression of the testes at 24° C and of the ovaries at 24° and 15° C. However, when this fish was kept under 9L:15D at 25° C, gonadal activity was increased by pinealectomy during the prespawning season (de Vlaming and Vodicnik, 1977). Pinealectomy caused disappearance of the daily cycle in GTH content of the AP, lowered the hypothalamic gonadotrophin-releasing hormone (GnRH) content of fish kept under 15.5L:8.5D at 25° C, and increased the hypothalamic GnRH content of fish kept under 9L:15D at 25° C. Thus the effect of pinealectomy on reproduction of this species depends on the photoperiod, water temperature, and the internal condition of the animal at the time of pinealectomy.

Pinealectomy of goldfish in either the fall or spring did not affect the gonadosomatic index (GSI), i.e., the gonadal weight per unit of body weight or the serum estradiol concentration. However, sectioning of the optic tract resulted in a lower serum estradiol concentration in the spring, but not in the fall (Delahunty et al., 1979). These results suggest that the effect of light, at least on estrogen secretion, may be mediated by a retinal pathway and that in the goldfish the pineal does not have a crucial influence on the secretion of estrogen. On the basis of the data available in the literature, de Vlaming and Olcese (1981) suggest that, although the results are not always consistent, the pineal may be a component in the pathway by which light affects the gonadal activity in goldfish.

In the Japanese rice fish (*Oryzias latipes*), the opposite response was obtained. In fish kept under LL, pinealectomy reduced the GSI in females, whereas blinding, which is presumably equivalent to DD, did not affect the GSI, and blinded pinealectomized females had a similar GSI as pinealectomized females (Urasaki, 1972). When the GSI was measured regularly during the year, blinding was found to reduce the maximal GSI in controls kept under natural photoperiods almost one-half. Blinded pinealectomized females had an even smaller GSI than blinded ones. Unfortunately, no data for pinealectomized fish kept under similar conditions as the blinded and blinded pinealectomized ones were given (Urasaki, 1976). Urasaki (1972, 1976) concluded that the effect of light on reproduction of the Japanese rice fish is mediated through the eye and the pineal, with the slightly greater contribution made by the pineal, which he considers to be an endocrine as well as a photosensitive organ.

The pineal of fishes, as well as of other vertebrates, contains melatonin. Using the killifish (*Fundulus similis*), de Vlaming et al. (1974) have shown that melatonin injections cause reduction of the GSI in males and females kept under a long photoperiod in January. Male killifish, kept under long photoperiod (13L:11D) at the start of the spawning season (May), showed some reduction in the GSI after melatonin treatment. Males kept under a short photoperiod (10L:14D) showed almost a 25 percent increase in the GSI, a response that was reduced almost one-half by melatonin treatment. Thus melatonin must be considered a possible mediator in the photoperiodic response in this species. As we shall see in the discussion of mammals, in hamsters the role of melatonin can be either stimulatory or inhibitory, depending on the time of administration during the day, whereas in the killifish, de Vlaming et al. (1974) found no difference in results following melatonin injections made at 2 h and 8 h after the onset of the photoperiod.

Other Influences. The reproductive cycle of the Indian catfish (*Heteropneustes fossilis*) consists of four periods: (1) a preparatory period (February–April), (2) a prespawning period (May–June), (3) a spawning period (July–August), and (4) a postspawning period (September–January) (Sundararaj, 1978). A threshold temperature of 25° C is required for the formation of yolky follicles, and the rate of vitellogenesis is proportional to increasing temperature, once the threshold temperature has been reached. The preferred temperature, i.e., the temperature that the fish selects when given a choice, is between 28.6° C and 32.0° C. During the postspawning period, the follicles regress more slowly at 30° C than at 25° C, but eventually, regardless of photoperiod and temperature, the ovaries regress. Experiments in which this fish was kept under LL and under DD, indicate that there is an endogenous circannual cycle of ovarian follicular activity. It thus appears that daylight may act as a synchronizer of this endogenous rhythm and that water temperature regulates the exact timing of the reproductive cycle.

Visual stimuli play a role in determining the rate of reproduction and possibly the onset of breeding. By the use of colored models, it has been shown that visual signals are important in eliciting courtship behavior in three-spined sticklebacks (Marler and Hamilton, 1966). Polder (1971) found that isolated *Aequidens portalegrensis* males and females showed no evidence of quivering, skimming, or spawning. However, all females spawned within four weeks when they were provided with a stone, on which to lay their eggs, and a mirror placed behind the stone in the aquarium. These females thus apparently were stimulated to reproduce by the visual stimuli of their own image. *Sarotherodon (Tilapia) macrocephala,* kept in isolation so that they cannot see other members of their species, spawn less frequently than when they can see either a male, female, or gonadectomized conspecific in another aquarium (Aronson, 1957). In these two examples, stimuli other than visual ones have thus been excluded. Similar observations have been made for *Sarotherodon (Tilapia) mossambicus* by Silverman (1978a), who also observed that the first spawning by young *S. mossambicus* was delayed by visual isolation, compared with young females that could see conspecifics of the same age in another aquarium (Silverman, 1978b).

The European bitterling (*Rhodeus amarus*) shows courtship behavior, nuptial coloration in the male, and ovipositor growth in the female only if certain fresh water mussels (e.g., *Unio, Anodonta*) are present (Banarescu, 1975). The female lays her eggs in such a mussel, as discussed in Chapter 10. In the absence of a male, she will still lay her eggs appropriately; therefore the interaction with a male is not required for spawning, although it is, of course, for fertilization of the eggs. We postulate here that visual stimuli are involved, but that other, e.g., olfactory stimuli cannot be excluded.

Olfactory signals apparently can be important in the courtship of the gobiid fish (*Bathygobius soporator*). Males of this species attack conspecific males and nongravid females that enter their territory, but gravid (ready to spawn) females are courted. Tavolga (1955) discovered that water in which gravid females had been present induced courtship behavior in the male in a few seconds, but when the nostrils were plugged or cauterized this effect was not observed (Tavolga, 1955, 1956). The olfactory signal is apparently present in the ovarian fluid (Tavolga, 1955). In this same species, auditory signals also affect courtship behavior. The male can make grunting sounds, to which the female responds by darting movements. However, these darting movements show a definite orientation only in the presence of another goby, suggesting that visual and auditory stimuli are required for this behavior to be meaningful.

Tidal effects apparently time the breeding of the grunion (*Leuresthes tenuis*) and the New Zealand white bait. The grunion's breeding season lasts from March to September. The fish land on the beaches during the high tide with one wave, and the males and females, with their genital openings in proximity, spawn in the sand, in which they have made a hole. With the next wave the fish swim back to the sea. The fertilized eggs develop in the warm wet sand and with the next high tide the embryos are returned to the sea, where further development occurs. However, although the spawning seems to be timed by the tides, the factors that time the maturation of oocytes and sperm have not been determined.

Tropical fishes do not experience a sufficient change in the length of photoperiod to have photoperiod function as a signal for the timing of reproductive activity. However, the great differences in rainfall during different times of the year could be used as signals for timing of reproduction, either directly by changing the salinity of the water, or the water level, or indirectly, by increasing the food supply for the fish. For the tambaquí (*Colossoma bidens*), the jaraquí (*Prochilodus* sp), branquínha (*Anodus* sp), matrinchà (*Brycon* sp), and peito de aço (*Patamorhina pristigaster*) of the Amazon basin, the rising water level in the lakes in which these fishes live apparently regulates their time of breeding (Schwassmann, 1978) by increasing the amount of living space and the food supply. In general, one would expect availability of food, water temperature, and rainfall to be the most significant factors in the timing of reproduction of fishes in the tropics.

Amphibians

Anura. For anurans that breed in the temperate zone, temperature and rainfall appear to be the principal proximate causes (Salthe and Mecham, 1974). According to Jørgensen et al. (1978), this effect of temperature may be partly an indirect one, in that it allows the animals (e.g., the toad, *Bufo bufo bufo*) to feed. This is an important consideration, since the nutritional status of the female is of the utmost importance in determining the growth of the follicles. In the fall and during hibernation, temperatures probably act directly on the animal, low temperatures preventing the follicles from becoming atretic. For anurans of the tropics, subtropics and deserts, rainfall, and thus probably also indirectly the availability of food may time the breeding period.

For the edible frog (*Rana esculenta*), photoperiod as well as temperature are important in stimulating gonadotrophin secretion and testicular activity. Both the eye and the pineal, which in anurans has a photosensory function (de Vlaming and Olcese, 1981), are required for these responses to long photoperiods (Figure 6.3).

In courtship and the establishment of a territory, vocal communications are very important in many anurans, although there are exceptions, for example, the ascaphids, which do not have vocal signals (Madison, 1977). The vocalizations of the male green tree frog (*Hyla cinerea*) (Garton and

Brandon, 1975) can be divided into three parts: (1) the so-called A calls, uttered when the male moves out to establish calling stations, (2) the B calls, which function in agonistic encounters and in the dispersion of the males, and (3) choruses of C calls that begin at about 2100 and last until about midnight. These choruses apparently attract the females, because by the time these choruses stop, all females found are in amplexus. Garton and Brandon (1975) speculate that in this species, it is the vocalizations rather than amplexus that induce ovulation. Oplinger (1966) has made a similar suggestion for the tree frog (*Hyla crucifer crucifer*), in which he found that ovulation preceded amplexus. He also suggested that humidity and vocalizations were the factors that timed ovulation.

The central neural efferent pathway for the control of calling in anurans involves (1) the ventral magnocellular preoptic (VMP) nucleus, which contains androgen receptors (Schmidt, 1973), although testosterone needs to be aromatized to estrogen in order to affect calling behavior in the South African clawed frog (*Xenopus laevis*) (Kelley, 1980); (2) the ventral infundibulum, which like the VMP contains androgen receptors, and where testosterone is aromatized to estrogen in the South African clawed frog (Kelley, 1980); (3) the dorsal tegmentum of the medulla, which concentrates dihydrotestosterone (DHT), and (4) the area of the hypoglossal and vagus nerve motor neurons, which also concentrate DHT in the South African clawed frog (Kelley, 1980). Moreover, in this frog a nucleus of the sensory auditory pathway, the laminar nucleus of the torus semicircularis, concentrates DHT and estradiol (Kelley, 1980), suggesting that testosterone is aromatized in this nucleus. These findings provide persuasive evidence that the androgen-induced calling is mediated by androgen affecting the efferent pathways at multiple sites and that the sensory pathway is also affected by the hormonal status of the recipient of the cells.

Narins and Capranica (1976) have investigated the neurophysiological aspect of the two-note "co-qui" call of the neotropical frog (*Eleutherodactylus coqui*). In this species the resident male starts with a co-qui call. If he is approached by an intruding male, he drops the "qui" note and utters a "co" call. If the intruder approaches to within 0.6 m, he attacks the intruder. Females are attracted by either the two-note co-qui call or the "qui" call, but apparently not by the "co" call. Narins and Capranica (1976) found that this "dichotomy of the two notes in the male call reflects a difference in the distribution of the best excitatory frequencies of primary auditory neurons for males and females"; in other words, the male's and female's auditory system are "tuned in" on different notes of the two-note call.

In Chapter 9, we indicated that the release call by the female depends on the distension of the abdomen. Olfactory cues may have a function in attracting males to females. In the Surinam toad (*Pipa pipa*), the water in which a receptive female has been held, causes the male to become agitated and to attempt to clasp nonreceptive females (Madison, 1977).

Urodeles. In the temperate zone, temperature, humidity, and photoperiod are the most likely proximate causes for the breeding cycles of urodeles; in the tropics, rainfall is probably the principal factor (Chapter 8). Olfactory signals are important in the courtship of the Plethodontidae, Desmognathidae, Salamandridae and Ambystomatidae, as discussed by Madison (1977). For the salamander, *Taricha rivularis,* there is evidence that odors secreted by the female can attract males, and the same may be true for *Triturus vulgaris* (Madison, 1977). Vocal communication is not pertinent in salamanders as they are avocal. For an extensive review of reproductive and courtship patterns in amphibia, the reader is referred to Salthe and Mecham (1974).

Reptiles

Tropical geckos and the tropical aquatic oviparous snake, *Hopolapsis buccatta,* have been reported to be continuous breeders. It is important to distinguish, however, between species in which breeding individuals can be found every month and between continuously breeding individuals, although, in each case there is no breeding season. Not all tropical reptiles lack breeding seasons; for example, the equatorial lizard

(*Agama agama lionotus*) in Kenya lays eggs in July and August only. This cycle is timed by the availability of insects for food (Lofts, 1978).

We have discussed some of the breeding cycles found in reptiles in Chapter 8. The proximate causes that time the breeding cycle have been reviewed by Lofts (1978), and the following summary is largely from his review article. Endogenous circannual rhythms may exist and time the reproductive cycle, but incontrovertible evidence for this is lacking (Lofts, 1978). The single most important timing factor for most reptiles (Licht, 1972) is temperature. For the Chinese cobra (*Naja naja*), Lofts (1978) speculates that there may be a thermorefractory period; i.e., gonads will not recrudesce after involution when the cobras are kept at the temperature that stimulates the recrudescence of the gonads. This is analogous to the photorefractory period in many avian species in which the gonads remain involuted as long as the birds are kept under long photoperiods. Day length affects the duration of the thermorefractory phase, with increasing photoperiods resulting in an increase in the duration of the photorefractory phase in the lizard, *Lacerta sicula.*

The American chameleon (*Anolis carolinensis*) has been studied in the greatest detail for the temperature-photoperiod interaction. The photosexual response, i.e., the increase in gonadal size in response to long photoperiods can be obtained in males at constant body temperatures between 28° and 32° C, but not at 25° C (Licht, 1972). When body temperature is not constant, the photosexual response can be obtained when the body temperature is 32° C for about 8 h each day. As mentioned earlier in this chapter, the effect of photoperiod is additive (i.e., having hourglass clock) (Underwood, 1979).

Anolis carolinensis breeds in the spring when the neuroendocrine-gonadal system is not sensitive to photoperiod, and the activation of the gonads is the result of a response between July and October when day lengths are decreasing; however, when the animals are exposed to long days experimentally, the testes are maintained. In November, the sensitivity to photostimulation is lost, and temperature becomes the single most important signal for the timing of the reproductive phase (Licht, 1972). According to Licht (1972), the pathway by which temperature has this effect may involve: (1) a direct effect on the central nervous system, which then controls the secretion of gonadotrophin-releasing hormones, and (2) an effect on the response of the gonads to gonadotrophins (GTH). Licht (1972) has presented evidence for a strong effect of ambient temperature on the response of the gonads to exogenous GTH in the lizards, *A. carolinensis* and *Xantusia vigilis*. For example, at 21° C the ovarian weight of *A. carolinensis* after 12 daily injections of 10 μg ovine FSH was 9.7 ± 0.4 mg; for the females kept at 29° C, however, the ovarian weight after 10 daily injections of 10 μg ovine FSH was 161.7 ± 22.0 mg. There were no differences between the ovarian weights of control-injected animals kept at 21° C and at 29° C (Licht, 1972). Licht (1972) mentions two other possible pathways for the effect of temperature: (1) an effect on the AP, and (2) an effect on the half-life of the gonadotrophic and gonadal hormones, but neither of these possibilities was investigated experimentally. In this lizard, the pineal is apparently not required for ovarian recrudescence in the spring; however, in females kept on short days, 6L:18D at 31° C, pinealectomy stimulates ovarian development, suggesting that on short days the pineal may secrete a substance that inhibits ovarian development either by acting on the hypothalamus, the AP, or the ovary. The substance may be melatonin (de Vlaming and Olcese, 1981), but the site of action has not been elucidated.

When a female is housed with one male (isolated pair), ovarian recrudescence occurs in nearly 100 percent of the females, provided that the photoperiod is 14L:10D, the temperature 32° C during the light and 23° C during the dark phase, the relative humidity 70–80 percent. When several pairs of these lizards are housed in one cage, the incidence of ovarian recrudescence drops to about 30 percent, if there is male-male aggression, but it is nearly 100 percent if the group is stable, that is, there is little male-male aggression. Females exposed to castrated males showed some delay in the growth of the follicles. It thus appears that courtship facilitates the ovarian response to a stimulatory environment and that

male-male aggression inhibits this response. Long photoperiods have been shown to stimulate gonadal growth in the following species: MacCall's horned lizard (*Phrynosoma m'calli*), the Texas horned lizard (*P. cornutum*), the desert night lizard (*Xantusia vigilis*), the wall lizard (*Lacerta sicula*), the fringe-toed lizards (*Uma notata, U. inornata,* and *U. scoparia*) and the red-eared pond turtle (*Pseudemys scripta elegans*) (van Tienhoven, 1968). In contrast to findings in the American chameleon, the response of the desert night lizard to long photoperiod is not affected by temperatures between 8° C and 19° C. This difference is explainable on the basis of ecological differences; however, the physiological basis for this difference has not been elucidated.

Thapliyal and Haldar (1979) have reported that in the Indian garden lizard (*Calotes versicolor*), pinealectomy early in the summer partially prevents the regression of the gonads that normally occurs upon exposure to 6L:18D. Pinealectomized males exposed to either natural daylight and temperature (10 h, 15 min light at 22 ± 2° C) or 15L:5D at 30° C, showed accelerated testicular growth and androgen secretion compared with the controls. This indicates that the pineal is inhibitory during long as well as during short days. On the basis of the information available, it is not possible to state what functions of the pineal of this lizard are essential for its effects on reproduction, photosensory function, and/or endocrine function.

Madison (1977), in an extensive review, states that in the lizard families, Iguanidae, Agamidae, and Chamaeleontidae, visual signals are the main releasers in courtship, whereas in the Scincidae, Lacertidae, and Teiidae, tactile, olfactory, and vocal signals are important. In the Iguanidae, the visual signals are of primary importance, but olfactory cues are also used during courtship. In snakes, intraspecific communication relies mainly on visual and olfactory signals. For species and sex recognition, olfactory cues are of primary importance; visual signals play a minor role (Madison, 1977). Garstka and Crews (1981) found that yolk, or lipid extract of skin and serum of estrogen-treated female red-sided garter snakes (*Thamnophis sirtalis parietalis*) when applied to the skin of males, elicited courtship behavior by other males. The substance (a pheromone) is probably produced in the liver under the influence of estrogen and may be associated with vitellogenin. According to Madison (1977), chemical cues are in all probability the main way of communication in turtles; however, visual cues cannot be excluded. The Crocodilia use visual, auditory, and chemical signals for courtship and territorial signaling (Madison, 1977).

Rainfall may time the breeding cycle of the females of several tropical species, e.g., *Anolis grahami, A. lineatopus,* and *A. sagrei.* However, the relationship between rainfall and ovarian activity seems to be based principally on correlations between rainfall and the onset of the breeding season, but no controlled experiments seem to have been carried out. Rainfall, of course, can act directly or indirectly by increasing the food supply.

Birds

Some of the breeding cycles encountered among birds are (Chapter 8): (1) continuous breeding, (2) breeding intervals of less than a year, (3) six-month cycles, (4) opportunistic breeders, and (5) yearly cycles. We also mentioned in this chapter that there may be endogenous circannual testicular cycles in birds that breed in the temperate zone and that changes in day length may synchronize such cycles to yearly cycles. We will now discuss some of the environmental factors that time the onset and cessation of the breeding period.

Photoperiod. Photoperiod is the factor most investigated, especially for birds that breed in the North Temperate Zone. Generalizations that can be made (Murton and Westwood, 1977) about species for which photoperiod is the main proximate factor are:

1. The rate of gametogenesis in photosensitive individuals within a species is determined by the length of the photoperiod; thus, the start of the breeding period is timed precisely.
2. The length of the photoperiod determines the increase in testicular size, probably because it determines the level of LH secreted in response to long days (Farner and Wingfield, 1978).
3. The duration of the response or the latency between the onset of exposure to long days and gonadal regression is determined by day length; the longer the day length, the shorter the latency. Storey and Nicholls (1976) have speculated that the longer days stimulate a higher level of secretory activity by the neuroendocrine system and that this system may become exhausted sooner than when the days are shorter, but still stimulatory.
4. The duration of the refractory period is proportionally related to the day length; the shorter the day length, the shorter the duration of the photorefractory period.

Storey and Nicholls (1976) have suggested that the duration of the photosexual response may be directly related to the duration of the period of pretreatment with nonstimulatory day lengths.

All but the first generalization are illustrated by observations made on English sparrows: In populations of these sparrows at 34° N (maximum day length about 15 h), the testes reach a size of about 400 mm^3, spermatogenesis lasts about 138 d, and the refractory period lasts about 13 d. Populations at 52° N (maximum day length about 17 h) have a testicular size of about 500 mm^3, spermatogenesis lasts 106 d, and the refractory period is about 100 d (Murton and Westwood, 1977).

Among closely related species, those species exposed to the longer day lengths have a larger ratio of testes to body weight than species exposed to shorter day lengths (Murton and Westwood, 1977). Within a species, a more northern population may thus start to breed later because the day length threshold is reached later than for a more southern population; however, the more northern population would show more rapid testicular growth than the southern population. This rapid growth is an adaptive response, because the duration of a favorable period to breed and raise young is shorter for the northern population than for the southern one. This variation in testicular growth with latitude has in fact been found for chaffinches (*Fringilla coelebs*) in Russia (Murton and Westwood, 1977).

Transequatorial migrants encounter special problems, because they will experience long days during their stay on the wintering grounds. Apparently the bobolink (*Dolichonyx oryzivorus*), which is a transequatorial migrant, has a slow response to 14-15-h light periods, and its refractory period can be dissipated by 12-h photoperiods, a photoperiod not effective in dissipating the refractory period of white-throated sparrows and juncos (*Junco hyemalis*) (Murton and Westwood, 1977).

Changes in day length are too small in tropical species to play a role in timing of the breeding cycle; nevertheless, the African weaver finch (*Quelea quelea*) can respond to exposure to different photoperiods, although the response is not the same as in most species in the North Temperate Zone. *Quelea quelea* kept under 12L:12D had enlarged testes for about 7 months, then had a 42-d refractory period, and subsequently showed testicular recrudescence. When the birds were transferred from 12L:12D to 17L:7D, the testes initially increased in size, but the testes remained enlarged for a shorter period than those of birds kept on 12L:12D. When the birds were transferred from 17L:7D to 8L:16D, the testes remained regressed. After the birds were transferred from three weeks on 8L:16D to 17L:7D, the testes showed recrudescence, but at the same time as birds that had been kept on 17L:7D throughout. This and other evidence indicate that the photorefractory period in this species is not shortened by exposure to short days. Exposure of *Quelea quelea* to the natural photoperiod to which English sparrows are exposed in the North Temperate Zone resulted in a gonadal cycle quite similar to that of English sparrows exposed to the same day lengths. Conversely, English sparrows subjected to weekly stimulatory photoperiods had gonadal cycles that resembled those of *Quelea quelea* (Murton and Westwood, 1977). Readers interested in other types of breeding cycles are referred to Murton and Westwood (1977), who discuss many different types of breeding cycles more extensively than can be done here.

Mention should be made of the lack of an inhibitory effect of continuous darkness (DD) on reproduction of Australian grass parakeets or budgerigars (*Melopsittacus undulatus*). Females provided with a nest box and exposed to male vocal-izations start laying after a latency that is the same for birds exposed to 6L:18D, 2L:22D, or DD, but that is longer than for birds on 14L:10D (Putnam and Hinde, 1973; Hinde and Steel, 1978). In the absence of a nest box, however, this latency is shorter for birds in DD than for birds under 14L:10D (Hinde and Putnam, 1973). Apparently the DD to some extent simulates the darkness of the nest box, and such darkness may be a requirement for the female to reproduce. When given skeleton photoperiods consisting of an initial 6L followed by 2L after various periods of darkness, 6L:6D:2L:10D proved to be the most effective for stimulation of the gonads of male and female budgerigars. This is similar to the most effective skeleton photoperiod of this basic pattern for temperate zone birds (Hinde and Steel, 1978).

The pathway by which photoperiod induces gonadal recrudescence has been extensively investigated in drakes, Japanese quail, English sparrows, and white-crowned sparrows. The evidence demonstrates that blinded birds can show this photosexual response, and so suggests the presence of an extra-retinal receptor, as proposed originally by Benoit in 1935 (Benoit, 1970, for review). Further investigations with different techniques by which light was directed at exactly localized places in the brain (e.g., by quartz rods, optic fibers, phosphorescent dyes, diodes) have revealed that extra-retinal receptors are located in the hypothalamus. The most precise localization has been accomplished by Yokoyama et al. (1978) in the white-crowned sparrow. These investigators found that receptors for the photosexual response were located in the ventromedial hypothalamus and/or in the tuberal complex, and perhaps also in or near the nucleus rotundus in the thalamus. Oliver (1979), using Japanese quail, found that phosphorescent pellets of 0.6×0.2 mm induced the photosexual response when implanted in the infundibular region, but not when placed in either the preoptic region or in any region of the hypothalamus.

The fact that there is no anatomical evidence for the presence of photoreceptors or photopigment in these photosensitive regions of the hypothalamus (Yokoyama et al., 1978) makes the interpretation of the mediation of the effect of light difficult. A thermal effect does not seem to be likely, but it cannot be excluded. However, the fact that infrared stimulation of sites that respond to visible radiation does not provoke a photosexual response argues against a thermal effect (Homma et al., 1977).

The relative roles of the retinal and extra-retinal receptors in the photosexual response have been investigated by the use of blinded birds and birds in which access of light to the extra-retinal receptors was blocked, e.g., by the injection of India ink under the skin on top of the skull or by hooding of the birds. In English sparrows (Menaker, 1971), white-crowned sparrows (Gwinner et al., 1971), and golden-crowned sparrows (*Zonotrichia atricapilla*) (Gwinner et al., 1971), blocking of the extra-retinal receptors decreased the testicular growth, migratory restlessness, and fat deposition, compared to controls. However, some increase in these three measurements occurred in the experimental birds after transfer from short to long days. These data suggest that with extra-retinal receptors blocked, the information about the photoperiod that is perceived by the eyes is ignored. There is, however, evidence that information through the retina may have an inhibitory effect. In Japanese quail, blinded males and females do not show gonadal regression when transferred from a long day to a short day, if the birds have experienced long days prior to blinding (Homma et al., 1972). In female white-crowned sparrows, information through the retina may inhibit secretion of gonadotrophic hormone (LH) and the final stages of development of ovarian follicles (Yokoyama and Farner, 1976). This information may, of course, consist of either visual or photic stimuli.

On the basis of the evidence presently available it seems that the pineal is not involved in the photosexual response of either intact or blinded birds (Homma et al., 1972; Turek, 1978), although the pineal clearly plays a role in circadian rhythmicity of locomotor activity, body temperature, and migratory restlessness of several species, e.g., the house sparrow, white-crowned sparrow, whitethroated sparrow, and house finch (*Carpodacus mexicanus*). It is intriguing that in the adult chicken, the pineal seems to have no effect on circadian rhythmicity but that in the baby chick, the pineal in vitro retains a circadian rhythm of N-acetyl

transferase activity and that such an explanted pineal is photosensitive (Takahashi and Menaker, 1979; Turek, 1978).

Oliver (1979) has shown that in the Japanese quail, hemispherectomy did not affect the photosexual response, that the area in which light could provoke this response was limited to the infundibular complex, and that red light (620 nm) was slightly more effective than green light (530 nm) in inducing the response. Electrolytic lesions of the posterior part of the basal infundibular nucleus or of the dorsal part of the infundibular complex (IC) or of the preoptic-anterior hypothalamus (POA) block the photosexual response (Follett and Davies, 1975; Oliver, 1979). Deafferentation of the basal medial hypothalamus prevents the photosexual response (Follett and Davies, 1975; Oliver, 1979).

Oliver (1979) has explored, by electrophysiologic techniques, the extent to which the POA and the IC areas differ. He found: (1) shorter latency of evoked potentials in the POA than in the IC, (2) a greater reduction in spontaneous multiunit activity (MUA) in the POA than in the IC of quail on short days (8L:16D) by severance of the optic nerve, and (3) direct retinal connections to the POA, but not to the IC. In both areas, long days (18L:6D) diminished spontaneous MUA after light flashes. However, in the IC a decrease in MUA could be obtained either by exposure to 16L:8D or by implantation of radioluminous pellets, indicating that a retinal pathway to the IC was not required for this response to long days. Oliver (1979) suggests that there are two hypothalamic areas, one with a direct responsiveness to light, the other requiring retinal input, but both have the capacity to regulate gonadotrophin secretion.

Pinealectomy of Indian weaver birds (*Ploceus philippinus*) indicates that the pineal may have an antigonadal function; for example: (1) after pinealectomy of young birds, sexual maturity occurs precociously; (2) pinealectomy accelerates gonadal recrudescence in response to long photoperiods; and (3) the gonads recrudesce in pinealectomized birds exposed to short photoperiods (8L:16D) that normally do not stimulate gonadal growth. Pinealectomy has no apparent effect on the gonadal response to long photoperiods in the white-crowned sparrows (*Zonotrichia leucophrys gambelii*), Harris' sparrows (*Z. querula*), house finches (*Carpodacus mexicanus*), and Japanese quail (*Coturnix coturnix japonica*). Evidence that pinealectomy decreased the ovarian response of blinded mallard ducks, *Anas platyrhynchos,* to continuous light, and the testicular response of mallard drakes to natural long daylight in the spring is not convincing (Ralph, 1981).

Photoperiod affects not only intact birds and gonadal activity, but it can also affect the response to exogenous hormones, as in the following examples:

1. Canaries (*Serinus canarius*) on long days and treated with PMSG showed greater nest-building activity than PMSG-treated females under a short photoperiod.

2. The nest-building activity of estrogen-treated ovariectomized canaries, intact, photorefractory, estrogen-treated canaries, and castrated estrogen-treated males is considerably higher under 20L:4D than under 8L:16D (Hinde and Steel, 1978). This difference between the results obtained under short and long days is, according to Hinde and Steel (1978), probably not the result of the greater amount of time available for nest building under long days than under short days.

3. A higher portion of estrogen-treated ovariectomized females enters the nest box under 14L:10D than under 8L:16D (Hinde and Steel, 1978).

Temperature. Other factors than photoperiod influence the time of breeding. Temperature probably has a modifying effect on the photoperiodic gonadal response (Murton and Westwood, 1977). Correlations between the onset of breeding and temperature may be the result of an indirect effect of the temperature on food supply, and availability of food is of particular importance for ovarian growth because of the extensive metabolic demands made for yolk deposition. Murton and Westwood (1977) state, "In no temperate-zone photoperiodic species have any of these other factors been shown to cause gonadal recrudescence in the absence of appropriate photostimulation." Under experimental conditions, testicular growth of white-crowned sparrows (*Z. leucophrys gambelii*) is somewhat greater at higher temperatures (in the range of 5.2–34.1° C) under long days (Murton and Westwood, 1977).

Visual Stimuli. The effects of visual stimuli have been investigated elegantly, both under laboratory conditions and under field conditions. In the ringdove (*Streptopelia risoria*), it was shown first that visual stimuli, produced by castrated males were less effective in inducing ovarian and oviductal growth than stimuli from intact males (Erickson and Lehrman, 1964). Subsequently it was demonstrated that ovarian and oviductal recrudescence was nearly linearly related to the dose of androgen administered to the males to which the females were exposed (Erickson, 1970). The dose of androgen determined the courtship behavior of the male, and the female's reproductive organs showed a corresponding response.

The visual stimulation provided by the female in turn affects the testicular activity of the male; however, this can be demonstrated only under short day length. Compensatory growth of the left testis in unilaterally castrated male ringdoves was greater than in isolated males, but this occurred under a short-day and not under a long-day regime, presumably because under the long-day regime the testis was already maximally stimulated (Cheng, 1976).

A second example of visual stimuli on avian reproduction is found in the experiments by Smith (1966), in which the main features used by four sympatric gull species for "recognition" of their own species were the yellow eye ring and yellow iris in the herring gull (*Larus argentatus*), the reddish-purple eye ring and dark iris in Thayer's gull (*L. thayeri*), and the reddish-purple eye ring and irises varying between very light and dark in Kumlien's gull (*L. glaucoides kumlieni*). In these gulls, the female selects the male, and she chooses males with an eye-head contrast similar to her own. In mated pairs, the female's eye-head contrast serves as a "releaser" for the male to mount. By capturing males and females and painting eye rings of different color, Smith was able to deceive females into selecting males of a species other than their own. By subsequently changing the eye-head contrast of the female, so that it resembled that of the male's own species, the males were induced to mate with these females. When

Smith changed the eye-head contrast of the female in male-female pairs of the same species, the testicular growth of the testes observed in controls was prevented. The pathways by which the visual stimuli affect the neuroendocrine-gonad unit need to be elucidated.

Olfactory Stimuli. In general, the olfactory system of birds is not well developed; however, Balthazart and Schoffeniels (1979) showed that after severance of the olfactory nerve in domesticated Rouen ducks (*Anas platyrhynchos*), sexual displays and social behavior decreased. These investigators suggest that the difference in chemical composition between secretions of the preen gland during the reproductive season and nonreproductive season may serve as a cue for the male. Sexual displays have an important role in courtship, and, as discussed for ringdoves above, such displays may affect the development of the females' reproductive system. It is thus conceivable that the olfactory system affects the interplay between male and female necessary not only for mating but also for development of the reproductive organs to their fullest potential.

Auditory Stimuli. Receiving auditory stimuli, or being able to give them, is important for full development of testicular or ovarian size in at least two species, the budgerigar (*Melopsittacus undulatus*) and the ringdove. Under experimental conditions, isolated pairs of budgerigars will not reproduce unless they can hear the vocalizations of another male, or even specific isolated parts of a male's vocalization, such as the tusk and soft warble (both of which are normally given during precopulatory, behavioral displays) when played on a tape recorder (Brockway, 1965). The male of the isolated pair, however, needs himself to be able to vocalize in response to the vocalizations received from another male, in order to show full testicular development (Brockway, 1967). The time of the day that vocalizations are received affects the testicular response. Gosney and Hinde (1976) have shown that during a 14-h day the female's reproductive development is stimulated to a greater degree, if the male song is presented during the first half of the day than if it is presented during the second half of the day. Subsequently, Steel et al.

(1977) demonstrated that in ovariectomized birds, treated with estrogen and prolactin male vocalizations tended to increase the proportion of birds that entered the nest box and the amount of time spent in the nest box. They also showed that male vocalizations were more effective in the morning than in the afternoon of a 14-h day.

Although visual stimuli are important in inducing ovarian and oviductal growth in ringdoves, auditory stimuli also have a stimulatory function. Auditory stimuli from a breeding colony increased the response in a female exposed only to the visual signals given by a courting male (Lott et al., 1967). In the absence of visual stimuli by a courting male, auditory stimuli from a breeding colony could induce ovarian and oviductal growth, albeit not to the same extent as the visual stimuli given by a courting male (Lehrman and Friedman, 1969). When vocalizations of female ringdoves were altered, so that the cooing sound was changed by either severing the hypoglossal motor nerves to the syrinx (Cohen and Cheng, 1979a) or by lesioning the nucleus intercollicularis (Cohen and Cheng, 1979b), the ovarian and oviductal response to a courting male were inhibited. Thus in the ringdove and in the budgerigar, the gonadal response to male stimuli requires either that the recipient be able to make or hear his or her own normal auditory responses, respectively.

In canaries under a short daylight regime (11L:13D), songs by the male accelerated ovarian growth and increased plasma LH concentrations, but this effect of auditory stimulation was not found under 14L:10D, presumably because the gonadal system was maximally stimulated by this photoperiod (Hinde and Steel, 1978).

Rainfall. Rainfall is correlated with the reproduction of budgerigars, probably because after the rain, grass seeds can germinate and the grass will then form seeds, which then become a source of food for the birds. The reproduction of the African weaver bird (*Quelea quelea*) is also correlated with rainfall. These birds migrate from areas where the rainfall has started, which makes seeds unavailable because they start to germinate, to areas where rain has not started. After about a month, they return to the original area from which they migrated to harvest the seeds of the grasses that have bloomed after the rainfall. The availability of green grass to build nests probably is also important in timing reproduction. In both the African weaver bird (Murton and Westwood, 1977) and the budgerigar (W. C. Dilger, personal communication), neither the testes nor the ovaries regress completely between breeding periods, but they appear to be ready to develop as soon as the opportunity arises.

Captivity. Captivity can severely inhibit reproduction, especially among birds. In many species the male in captivity will show normal testicular development, but in females, ovarian development is not completed to the stage in which large yolky follicles are formed, as for example in the wild mallard duck (*Anas platyrhynchos*). In ducklings obtained from eggs collected from wild mallards, it has been possible to increase ovarian development by imprinting ducklings on the experimenter. Correlated with the increased ovarian development was a diminished fear response of the ducks when they were adult (Phillips, 1964). The pathway by which captivity has its effect on fear behavior and on the inhibition of ovarian development probably involves the archistriatum, which has projections to the hypothalamus. Lesions of the archistriatum diminished the fear response and increased ovarian development in wild ducks not imprinted on humans (Phillips, 1964).

Nutritional Factors. The Piñon jay (*Gymnorhinus cyanocephalus*) is apparently dependent for its reproduction on the availability of green pine cones of *Pinus edulis*. Feeding on such pine cones in the fall induces testicular recrudescence in Piñon jays (Ligon, 1974).

Subtle Factors. Brockway (1962) found that budgerigars did not visit the nest box and did not breed unless the vertical distance between the roosting perch and the nest hole was at least 10 cm. Within minutes after an adjustment was made to increase the distance from less than 10 cm to 10 cm, birds started to visit the nest box, and eventually breeding occurred. This requirement for a distance between a roosting perch and

a nest hole may have evolved as a protection against predators.

Mammals

Photoperiod. Evidence that photoperiod is the proximate factor for the onset of breeding of mammals is limited to few species. Among mammals that respond reproductively to photoperiod, two groups can be distinguished: one group that breeds with the onset of long days and another group that breeds with the onset of short days. To the first group belong, among others, the golden hamster (*Mesocricetus auratus*), the voles *Microtus montanus* and *M. arvalis,* the Djungarian hamster (*Phodopus sungorus*), the raccoon (*Procyon lotor*), the ferret (*Mustela furo*), the mink (*M. vison*), the sable (*Martes zibellina*), the hare (*Lepus timidus*), and the horse (*Equus caballus*). To the second group, the short-day breeders, belong, among others, the sheep (*Ovis aries*), the goat (*Capra hircus*), the white-tailed deer (*Odocoileus virginianus*), and the silver fox (*Vulpes fulva*).

The pathway of the photoperiodic response involves: (1) the retina, (2) retino-hypothalamic and retino-hypothalamic-pineal connections, and (3) the pineal. The retina is required for the photoperiod to affect reproduction. Blinding of hamsters has the same effect on the gonadal regression of hamsters as exposure to short days. As discussed at the beginning of this chapter, the external coincidence hypothesis implies a time-measuring device in the organism. Turek and Campbell (1979) state, "the mammalian eye is clearly involved in the perception of light used to measure time," but they point out that the identity of the retinal photoreceptors is not unequivocally established.

The retino-hypothalamic tract, a tract from the retina which projects mainly to the contralateral suprachiasmatic nucleus (SCN), but with a few projections to the ipsilateral SCN, is required for the photoperiodic response. Experimental interruption of the retino-hypothalamic tract, while sparing the primary optic tract, also destroys the SCN because of its location (Turek and Campbell, 1979). This poses a serious difficulty for interpretation of the data, because the SCN is probably the location of the biological clock, or it is such a crucial part of the biological clock that the clock cannot function after the SCN is destroyed. Rusak states, ''No circadian rhythm has survived SCN ablation in rodents, but a variety of noncircadian cycles can be generated by lesioned animals.'' Thus lesions of the SCN interrupt not only the transmission of photic stimulation, but also the time-measuring mechanism.

After the SCN is destroyed, exposure to short days does not result in gonadal regression (Stetson and Watson-Whitmyre, 1976). Since blinding causes gonadal regression, it is clear that lesions of the SCN not only interrupt the transmission of photic stimuli to the hypothalamic-AP-gonadal system, but also have the effect of interrupting or destroying the circadian clock.

The pineal in mammals receives information about the photoperiod by a pathway from the retina → the retino-hypothalamic tract (an unmyelinated tract in the optic nerve) → the SCN → connections to the periventricular and ventral tuberal areas of the hypothalamus → connections to the lateral hypothalamus → a connection probably through the medial forebrain bundle and the midbrain reticular formation to the upper thoracic intermediolateral cell column → preganglionic fibers to the superior cervical ganglion → postganglionic fibers which form the nerve conarii → pineal (Ueck, 1979).

The physiology of the photosexual response has been studied most extensively in the golden and the Djungarian hamster, the ferret, and sheep. We will discuss these species largely on the basis of the reviews by Hoffman (1981), Reiter (1978 and 1981), Turek and Campbell (1979), and Legan and Karsch (1979).

Golden hamster. The seasonal reproductive cycle of the golden hamster is ecologically explained as follows. In the fall, when day length becomes shorter, the testes regress, and as temperatures drop, the animal will hide in a burrow and begin hibernation. During the hibernation in the dark, the gonads and secondary sex organs regress; presumably, this conserves energy. As spring approaches, the gonads recrudesce pre-

paratory to the emergence from hibernation, so that when the weather becomes favorable, those animals that have fully functional gonads can start to reproduce immediately. The evolutionary advantage of such preparation is obvious.

The regression of the ovaries and uteri and of the testes and secondary sex organs that normally occurs in the fall can be induced by exposure of the animals to short days (1L:23D). In spite of continued exposure to short days, however, gonadal recrudescence takes place in about 20 weeks, a period equivalent to that spent by hamsters in burrows. After such gonadal recrudescence, the reproductive system is refractory to the effect of short days unless the animals are first exposed to about 11 weeks of long days (Turek and Campbell, 1979).

Some physiological aspects of this refractoriness have been elucidated:

1. The amount of testosterone required to suppress gonadotrophin secretion in castrated golden hamsters is higher in hamsters under long days than under short days. This higher sensitivity decreases in hamsters kept on 6L:18D (Elles et al., 1979) and this correlates with the dissipation of the refractoriness in intact control hamsters.

2. Melatonin, a hormone secreted by the pineal, administered to hamsters with recrudescing testes under short photoperiods, does not inhibit the growth of the testes (Bittman, 1978), indicating a lack of sensitivity of the target organs for melatonin.

3. The induction of photorefractoriness is, according to Turek and Losee (1979), not the result of exposure to short days in themselves, but the result of an inhibition of the neuroendocrine-gonadal system. In melatonin-treated hamsters transferred to short days, the gonads did not regress as long as the melatonin was administered (24 weeks), but regressed after melatonin was no longer administered, indicating that the neuroendocrine-gonadal system still responded to short days. In hamsters receiving melatonin from week 2–24, testes were regressed at 12 weeks, active at 24 weeks, and active at 32 weeks, so that these animals were refractory, indicating that melatonin treatment did not prevent refractoriness.

4. Exposure to long photoperiods restores the involution response of the testes when hamsters are subsequently exposed to short photoperiods. This dissipation of photorefractoriness is accompanied by a restoration of the sensitivity of the target organs to the inhibitory effect of melatonin. This restoration by long days of the inhibitory response to short days and to the inhibitory effect of melatonin on the reproductive system requires the presence of the pineal (Bittman and Zucker, 1981).

5. For photoperiodic stimulation of the testicular activity, the golden hamster has to receive at least 12.5 h of light per day if uninterrupted light-dark schedules are used (Elliott, 1976).

The effect of pinealectomy on reproduction has been studied extensively in the golden hamster. The salient findings for the male are: (1) After blinding, by removal of the eye (enucleation) or after exposure to short photoperiods, such as 1L:23D, the testes and secondary sex organs regress and are atrophied after 10–12 weeks; however, after an additional 14–16 weeks, they recrudesce in spite of the blinding or the exposure to 1L:23D. (2) Either pinealectomy or superior cervical ganglionectomy (SCGX) prevents this response if carried out prior to blinding or exposure to 1L:23D, or reverses the regression if performed after the regression has started. (3) In general, FSH, LH, and prolactin (PRL) concentrations in the plasma are lower in hamsters with regressed testes due to enucleation or exposure to 1L:23D than in comparably treated hamsters with active testes which have had a pinealectomy or a SCGX. (4) After implanting two pituitaries from donor hamsters under the renal capsule, in order to obtain high plasma PRL concentrations, and after twice-daily injections of LH-RH, the testes of hamsters enucleated or exposed to 1L:23D were similar in size to those of controls exposed to long photoperiods, whereas either transplanting two pituitaries alone or twice daily LH-RH injections alone failed to restore testicular size. (5) The inhibitory effect of the pineal on testicular size of golden hamsters exposed to 1L:23D is, in all probability, the result of melatonin secretion by the pineal (Reiter, 1981). (6) The amount of testosterone required to

suppress LH and FSH concentrations in plasma of castrated hamsters is considerably lower under short-day periods than under long-day periods. In other words, the neuroendocrine unit (hypothalamus-AP) is more sensitive to the negative feedback by androgen under short-day than under long-day periods (Ellis and Turek, 1979; Turek and Campbell, 1979). This effect is at least partly mediated by the pineal (Turek, 1979).

In female golden hamsters, the following findings are important: (1) Enuclation or exposure to 1L:23D results, within several weeks, in acyclicity of the vaginal morphology as shown by vaginal smears, whereas in control females there is a rather rigid 4-day cyclicity. The ovaries become devoid of follicles but become heavier than those of controls because the interstitial tissue proliferates, and the uteri become atrophic in about 10–12 weeks. (2) Pinealectomy or SCGX prevents the effects of either blinding or exposure to 1L:23D, as it does in males. (3) The FSH and LH concentrations in hamsters blinded or exposed to 1L:23D are not different from control females under long photoperiods. The surges of FSH and LH which occur daily in the afternoon in controls, also occur in hamsters that are enucleated or exposed to 1L:23D. (4) The data on PRL concentrations in the plasma are not sufficient to draw valid conclusions, although from the available data, it appears that plasma PRL concentrations are lower in the acyclic hamsters than in the cyclic controls (Reiter, 1981).

The identity of the pineal hormone that inhibits the gonads is not completely established, although most of the evidence indicates that it probably is melatonin. According to Turek and Campbell (1979), melatonin is of most interest because: (1) the pineal is the principal source of melatonin; (2) synthesis and release of melatonin are influenced by photoperiod, with high secretion occurring during darkness; (3) melatonin, if injected into hamsters under a long photoperiod 6.5 h or later after lights on, results in involution of the testes and a cyclicity of the vagina shown in vaginal smears. However, when injected in the morning, melatonin has no such effect, and continuous administration of melatonin mixed in a pellet of beeswax implanted under the skin to provide continuous delivery of the melatonin resulted in maintenance of the gonads of male and female hamsters exposed to 1L:23D (Reiter, 1981). This effect is exactly the opposite of the one expected, if melatonin secretion is to function as an inhibitor of the gonads. Reiter (1981) explains this finding as due to a decrease in melatonin receptors as the result of continuously high melatonin concentrations. The evidence presented by Reiter (1981) is persuasive for the experimental conditions under which melatonin is continuously supplied. It does not explain, however, why under 1L:23D, when there should be a continuously high melatonin secretion, such a decrease in melatonin receptors, and thus a stimulatory effect on the gonads or at least maintenance of the gonads, does not occur. Turek et al. (1976), using Silastic capsules to deliver melatonin continuously, found that melatonin suppressed testicular activity of golden hamsters kept under 14L:10D. Reiter (1981) states that he has been unable to repeat these experimental results. The reason for the difference between the results obtained by Reiter (1981) and by Turek et al. (1976) is not clear.

Turek and Campbell (1979) suggest that the following generalizations can be made about the function of melatonin in stimulating testicular growth of golden hamsters: (1) Melatonin is stimulatory for gonadal activity during exposure to short days, but is inhibitory to gonadal function when the animals are exposed to long days. (2) Melatonin inhibits testicular growth that is dependent upon long photoperiods, but it cannot inhibit testicular growth that occurs independent of long photoperiods, such as testicular growth during the first nine weeks of life or the spontaneous recrudescence that occurs when the golden hamsters are maintained under 1L:23D for about 28–30 weeks.

Arginine vasotocin (AVT) has been tentatively identified in the pineal of golden hamsters, but its effect on testicular regression has been mainly investigated in rats and mice; however, its effect on reproduction in the hamster is not determined (Reiter, 1978). In bovine pineals, a polypeptide other than AVT, which also suppresses testicular activity of mice, has been isolated (Benson et al., 1976). It has not been determined whether such a peptide is present in the pineals of hamsters and whether this bovine polypeptide affects the gonadal activity of hamsters.

The discussion of reproduction of the hamster to this point has been limited to the proximate causes which regulate the seasonal breeding. Within the breeding season, there are communication systems between males and females. Vaginal secretions from diestrous and estrous females are attractive to males and these odors have sexual excitant effects (Johnston, 1977). Singer et al. (1976) isolated dimethyl disulfide from hamster vaginal secretions and found it to be an attractant for male hamsters.

In addition to olfactory signals, hamsters have vocal signals for communication. Female and male hamsters emit ultrasounds, which in each sex depend on the endocrine status of the female. The rate of ultrasound emission by the female is dependent on estrogen, and the emission by the male depends on signals (acoustic and olfactory?) emitted by the female. For example, either testosterone administration to ovariectomized females or hypophysectomy of adult females eliminates the emission of ultrasounds by the male. These auditory signals may function to advertise the presence of a female and her endocrine status to the male. The signals by the male also indicate the presence of a conspecific animal, and they have a facilitating effect on the female's reproductive behavior, but the female does not give ultrasounds during lordosis. It thus appears probable that these signals are not important during heterosexual contact. The utterance of ultrasound signals by the male is androgen-dependent, for after castration the rate of ultrasound production is reduced, and androgen therapy restores the ability to produce ultrasounds at a rate similar to that of intact males.

Djungarian hamster. The Djungarian or hairy-footed hamster (*Phodopus sungorus*) lives in the steppes of Mongolia and western Siberia. This animal does not hibernate, but the changes in testicular weight and weight of the secondary accessory organs are similar to those described for golden hamsters. The size of the testes can be manipulated by changes in photoperiod. Exposure to short days causes regression of fully developed testes and exposure to long photoperiods causes a recrudescence of involuted testes. In contrast to golden hamsters, there are daily cycles of plasma testosterone and prolactin concentrations in the plasma, so that the time at which blood samples are taken becomes an important factor in making comparisons between Djungarian male hamsters with regressed testes on short photoperiods and males with large testes on long photoperiods. In Djungarian hamsters, pinealectomized males do not show regression of the testes upon exposure to short photoperiods, but in contrast to the findings in golden hamsters, pinealectomy with regression of the testes and secondary sex organs diminishes the increase in weight of the testes and secondary sex organs upon exposure to long photoperiods, i.e., photoperiods of 13 h or longer. In pinealectomized Djungarian hamsters exposed to 8L:16D until 45 d of age, the testes do not reach the same size as the testes of either pinealectomized or intact males exposed to 16L:8D for the same length of time. This suggests that in the Djungarian hamster, not all effects of long and short photoperiod are mediated through the pineal. Melatonin administration to Djungarian hamsters has yielded contradictory findings, as it has in golden hamsters, and at present it is difficult to give a satisfactory interpretation of these results. Hoffmann (1981) concludes that the pineal in Djungarian hamsters is not specifically concerned with regulation of gonadal functions (the pineal also affects the pellage of these animals), but that it plays an integral role in the transduction of photoperiodic stimuli upon the neuroendocrine axis and thereby may influence many functions.

Ferret. The breeding season of the ferret (*Mustela putorius*) starts in the spring and lasts through August. Estrus starts in the spring and, unless ovulation is induced by coitus, lasts for 20–25 weeks. Exposure of estrous females to short photoperiods induces anestrus and exposure of anestrous females to long photoperiods induces estrus. However, animals kept on long days eventually end their estrus and then remain in anestrus, permanently, as do several avian species, as long as they are receiving long photoperiods, suggesting that they have become photorefractory. After transfer to short days for 7 or 8 weeks, the animals are capable of responding again to long photoperiods by going into estrus. By exposing ferrets to 6-, 4- or 2-month cycles, of long and short photoperiods, 1, 1 ½, and 3 estrous periods can be obtained in one year; however, alternating cycles of long and short days of 1 month fail to induce regular cycles, but yield, instead, irregular intervals between estrous periods (Herbert and Klinowska, 1978). The length of the photoperiod does not affect the induction of estrus in ovariectomized ferrets treated with gonadal hormones (Baum and Schretlen, 1978), whereas in ovariectomized canaries, the nest-building activity induced by estrogen administration is greater under long than under short photoperiods (Hinde and Steel, 1978).

When ferrets were blinded neonatally, they came into estrus in the spring, as in intact controls, but in subsequent years the controls came into estrus in the spring, whereas the blinded ferrets had recurring estrous cycles (for as long as five years), which were not synchronized to photoperiod or other factors (Herbert et al., 1978). The fact that blinded and control ferrets came into estrus at the same time may be the result of a preprogrammed onset of puberty, which may be related either to age or to body weight. The recurrent breeding cycles of the blind animals being out of synchrony with the controls suggests that in the controls, light may act as a signal for synchronization of an endogenous circannual rhythm. If this were the case, then the blinded ferrets should show circannual rhythms of estrus. Herbert et al. (1978) interpret their data as showing no evidence for such a circannual rhythm in the blinded ferrets. It appears, however, that no statistical analysis was carried out to ascertain whether there was a periodicity in the data. It would be worthwhile to analyze the data by appropriate statistical methods.

Pinealectomized ferrets resemble blinded ones in having recurrent unsynchronized breeding cycles (Herbert et al., 1978), suggesting that both the pineal and the eye are required for the transmission of information about photoperiod to the neuroendocrine axis. The effect of the pineal on ferret reproduction does not seem to be mediated by melatonin (Herbert and Klinowska, 1978). However, as discussed for the golden hamster, the time and method of administration of melatonin affect the response; therefore, no premature conclusions about the function of melatonin in the reproductive cycle of the ferret should be drawn. Herbert and Klinowska (1978) suggest that serotonin, which is present in high concentrations in the pineal of the ferret, and the SCN may be involved either in transmitting or modulating the effects of photostimulation by long photoperiods.

Mink. Under natural conditions, mink mate from the middle of January until the middle of March in the North Temperate Zone. The young are born near the end of May, independent of the time of mating because implantation does not occur until mid-March, when prolactin concentrations apparently increase sufficiently to induce implantation. Exposure to long photoperiods in late fall advances the time of estrus, and exposure to long photoperiods of mated females advances the time of implantation (Canivenc, 1970). A similar response of advanced implantation has been observed in the sable (*Martes zibellina*), the pine marten (*M. americana*), and the European marten (*M. martes*) (van Tienhoven, 1968; Canivenc, 1970).

An unexpected effect of additional illumination (17L:7D) has been reported by Belyaev and Zhelezova (1978) for mink carrying the autosomal mutation shadow. This is a dominant mutation for coat color with a recessive lethal effect. All homozygotes die, mostly prior to implantation. By lengthening the photoperiod, Belyaev and Zhelezova (1978) were able to reduce embryonic mortality of shadow kits and to increase the ratio of shadow to nonshadow kits (from a shadow × shadow mating) significantly by additional illumination. These authors also state that under long photoperiods, embryos heterozygous for shadow were less viable than normal nonmutants.

Belyaev et al. (1975) have reported that in the Georgian white fox, there is an incompletely dominant autosomal coat mutation with a recessive lethal effect on the embryos, which die before implantation. As in the mink, the mortality was reduced by providing long days. Belyaev et al. (1975) also found that as litter size increased, the survival of homozygous embryos increased.

They ascribed the effect of long days and of large litter size on survival of these embryos to an increase of CL activity; however, they gave no quantitative data on progesterone secretion. Long days did not favorably affect the survival of two other mutations, silver- and white-faced, in which homozygous embryos also die early.

Manipulation of the photoperiod is used in a few domesticated species either to improve reproduction or to obtain young at a specific time. In horse breeding, it is important to have the young born early in the calendar year because of the criteria used in defining 2-year-olds for competitive purposes. By using a lighting and temperature program that was phase shifted several months, Sharp and Ginther (1975) were able to induce estrus in 7 out of 7 mares by day 112 after the start of the increasing light and modified temperature program. Oxender et al. (1977) found that exposure of mares to 16L:8D induced estrus and ovulation, which was accompanied by increases in LH and progesterone. In the control group, kept outdoors, estradiol (E_2) peaks were higher than in the 16L:8D group, but these were not associated with estrus, whereas the smaller E_2 peaks in the 16L:8D group were associated with behavioral estrus; this suggests that factors other than E_2 concentrations are important in the induction of estrus in the horse. Supplemental lighting is used for mink and foxes to induce earlier breeding and, in the case of mink, to induce earlier implantation.

Sheep. The reproductive season of sheep in the temperate zones starts in the late summer or early fall and seems to be initiated by decreasing photoperiods. By manipulation of the photoperiod, the onset of estrus can be induced at will (Thibault et al., 1966); however, there is increasing evidence that there is an endogenous rhythm. Under constant light regimes of 6L:18D, 12L:12D, 18L:6D or LL, first estrus occurred in ewes at about the same time as in ewes that had been exposed to a decreasing or to an increasing photoperiod. During 2 ½ years on such constant photoperiods, all groups experienced three breeding seasons. This suggests the presence of an inherent rhythm, which controls estrus in the absence of changes in photoperiod (Thibault et al.,

1966; Ducker et al., 1973) but which will respond to changes in photoperiod.

Other evidence for an inherent circannual rhythm in sheep is found in the changes in plasma prolactin concentrations. In France, in ewes and rams under a natural photoperiod, plasma prolactin concentrations are high ($>$ 200 ng/ml) in June and low ($<$ 25 ng/ml) in December, and generally parallel those of the duration of daylight during the year (Ortavant et al., 1978). By the use of skeleton photoperiods, e.g., 7L:6D:1L:10D, it was demonstrated that there was a variation in photosensitivity, and that prolactin concentrations could be increased by manipulation of the photoperiod. However, in ewes maintained on natural daylight, 8L:16D, 7L:3D:1L:13D, 7L:6D:1L:10D, 7L:9D:1L:7D, or 7L:12D:1L:4D, the variations in prolactin concentration during the year were parallel for the different photoperiods (Ortavant et al., 1978).

After transfer of rams from 16L:8D to 8L:16D, there is an initial rise in the plasma concentration of FSH and LH in about four weeks with peak concentrations at 5–9 weeks for FSH and 3–8 weeks for LH, followed by an increase in plasma testosterone concentration, peaking at 10–18 weeks and an increase in testicular diameter at 11–17 weeks (Lincoln and Davidson, 1977), such a change in photoperiod also results in a decrease in plasma prolactin concentration (Lincoln et al., 1978). Lincoln and Peet (1977) also found that transfer from 16L:8D to 8L:16D increased the frequency of episodic gonadotrophin release (most LH peaks occur during the dark period). It thus appears that a combination of changes in the secretion of FSH and LH, which both increase in concentration and in frequency of episodic release, and prolactin, which decreases in concentration in response to short photoperiod, stimulates testicular activity. The sudden transfer from short to long photoperiods resulted in progressive increase in plasma prolactin and a decrease in plasma FSH concentration (Lincoln et al., 1978).

In Chapter 8, we mentioned the hormonal changes that occur at the transition from the breeding season to the period of anestrus (Figure 8.11). The main difference between the breeding and nonbreeding season appears to be the effect of estradiol on LH secretion. When the negative feedback effect of estrogen is high during the nonbreeding season, LH is suppressed and estrogen does not have a stimulatory effect on LH secretion. This concept has been discussed in greater detail by Legan and Karsch (1979), and more supporting evidence is presented in their review paper, which should be consulted by the interested reader.

The effect of photoperiod on the reproductive system may be mediated through the hypothalamus, the AP, or the pineal or any combination of those. Investigations by Pelletier and Ortavant (1970) suggest that the LH-RH content of the hypothalamus of rams, as in hamsters, increases during photoinhibitory regimes (long days in sheep, short days in hamsters), whereas FSH and LH concentrations decrease. This suggests that long days inhibit release of LH-RH. Land et al. (1979) found that the response of ovariectomized ewes to LH-RH varied with the time of the year. This response would have to be investigated under carefully controlled photoperiods, before this variation can be ascribed to photoperiodic variation. Pinealectomy, either before or during the breeding season, did not affect the number of estrous cycles, the duration of estrus, the incidence of ovulations during anestrus, the LH concentration between peaks of the cycle, or LH concentrations during anestrus (Roche et al., 1970). In rams, pinealectomy resulted in a lower mean plasma prolactin concentration in the summer (14 $\pm$ 3 ng/ml versus 29 $\pm$ 2 ng/ml for controls), a higher concentration in the fall (53 $\pm$ 8 ng/ml versus 26 $\pm$ 3 ng/ml for controls), but no effect on spring and winter plasma prolactin concentrations. Control rams showed seasonal differences in LH peak frequencies, with higher frequencies in the summer, whereas pinealectomized rams did not show such a seasonal change. The peak plasma concentration of FSH found during the fall in controls was absent in pinealectomized rams, but there was no significant effect of pinealectomy on plasma testosterone concentrations during any of the seasons (Kennaway et al., 1981). After SCGX of rams in February, testicular recrudescence started and continued as in control rams. During exposure to sequential 16-week periods of short days (8L:16D) and long days (16L:8D), SCGX rams had a continuously high testicular weight, secreted high amounts of testosterone, FSH, and LH, and showed less variation in plasma prolactin concentrations between long and short days than controls. Control rams had smaller testes, lower plasma testosterone, FSH and LH concentrations, and higher prolactin concentrations, during long days than during short days (Lincoln, 1979). These data suggest that a functional pineal is not required for the normal seasonal change in testicular size, but that it is necessary for the response to abrupt changes in light regimes.

Goats. The endocrine profile of male goats under natural daylight conditions resembles that of rams, i.e., LH and testosterone start to rise in August, reach a peak in October, and decrease to basal concentrations in November to December. FSH shows a small, but not statistically significant peak in September and a large peak in April. Prolactin concentrations parallel the photoperiod (Muduuli et al., 1979). Exposure of female goats to 19L:5D, for 70 d starting at the end of January and subsequent exposure to natural photoperiods (at about 37° N) induced estrus, LH release, and ovulation 60–80 d in advance of control goats kept under the natural photoperiod (BonDurant et al., 1981).

Deer. The changes in the plasma concentrations of LH, testosterone, and prolactin in captive, male white-tailed deer, show a pattern similar to that found in sheep and goats; FSH concentrations are parallel with those of LH, and no peaks are found in the spring (Mirarchi et al., 1978).

Swine. In wild pigs (*Sus scrofa*), reproduction is seasonal, with most estrous periods between December and February, although a sow may have another estrus in April or May if she loses her litter or if she has her young early and they are weaned before April (Mauget, 1978). According to Mauget (1978), the reproductive efficiency of domesticated pigs in France decreases in the summer, the season in which wild pigs cease to reproduce. It appears, therefore, that the tendency for seasonal reproduction may still be present in domesticated pigs. In domesticated boars that were exposed to 15L:9D, puberty occurred earlier than in controls on natural daylight

(9.3–12.7 h light) (Mahone et al., 1979), suggesting that photoperiod may be one of the proximate causes in the seasonal reproduction of this species. Once puberty is reached, olfactory and auditory signals are important in reproduction of swine. It appears that the olfactory signals emitted by the male are of primary importance in bringing male and female together, whereas in sheep, male and female each is attracted by the signals emitted by the sexually active member of the opposite sex (Signoret, 1976). The chemical signals emitted by the boar are metabolites of testosterone, i.e., 5 α-androst-16-en-3-one and 3 α-hydroxy-5-androst-16-ene (Signoret, 1976). Testosterone is metabolized to these sex attractants in the salivary glands, and the material is excreted in the saliva. Females treated with testosterone became as attractive to an estrous female as to a boar. A sow or gilt in estrus will stand almost rigidly when touched on the back. To obtain this so-called standing response, the sow has to be exposed both to olfactory signals, of which 5 α-androst-16-en-3-one is the most effective, and to auditory signals, such as the grunting by the boar. Additional visual (seeing the boar) and tactile stimuli (being mounted by the boar) add relatively little compared to the effect obtained by the olfactory and auditory stimuli. When artificial insemination is used commercially, it is sometimes difficult to determine which gilts or sows are in estrus and it is a common practice to spray 5 α-androst-16-en-3-one (which is available in spray cans) in the vicinity of the sow and then touch her back to determine whether she will stand rigidly and thus is in estrus.

Dietary Factors. In the mammalian species discussed so far in this chapter, photoperiod was the most important proximate cause in timing of the breeding season. In at least three species of voles, *Microtus agrestis*, *M. arvalis*, and *M. montanus*, which show seasonal reproduction, the timing of the breeding season is related to the availability of green feeds, such as grass, alfalfa, wheat. The growing season for these plants is, of course, in turn modulated by photoperiod and temperature. In *M. montanus*, the reproductive activity seems to be triggered mainly through the availability of certain plant foods. Negus and Berger (1977) showed that supplying a natural population of this vole with fresh green wheat grass for two weeks resulted in a high incidence of pregnant females, whereas no pregnant females were detected in the control population. Berger et al. (1981) subsequently showed that a cyclic carbamate, 6-methoxybenzoxazolinone (6-MBOA), extracted from green wheat of about 10 cm in height (Sanders et al., 1981) induced breeding in a natural winter population of *M. montanus*. It appears that the end of the breeding season is not only the result of a lack of 6-MBOA, but also the result of inhibitors of reproduction, such as cinnamic acids, which are relatively abundant in grasses after they have flowered, fruited, and become dehydrated, which usually occurs in late summer and fall (Berger et al., 1977). Under laboratory conditions, the presence of green feeds, such as wheat, grass, or alfalfa, in the diet is essential for breeding of *M. montanus* (Negus and Pinter, 1966; Pinter and Negus, 1965) and *M. arvalis* (Delost, 1972, 1973). Photoperiod does have a modulating effect on reproduction in *M. agrestis* and *M. arvalis*, to the extent that 15–20 h light per day is optimal, although *M. arvalis* will reproduce, but at a lower rate, under a 5L:19D regime (Martinet, 1970). In *M. arvalis*, testicular weights are less under 6L:18D than under 18L:6D, but some spermatogenesis occurs under the short photoperiod (Grocock and Clarke, 1974).

As in golden hamsters, the pineal exerts an inhibitory effect on reproduction of *M. agrestis* kept under short photoperiods. For example, destruction of the sympathetic nervous system by 6-hydroxydopamine, which blocks the transmission of information of photoperiod to the pineal, and pinealectomy each stimulated gonadal activity of both sexes kept under short photoperiods (Farrar and Clarke, 1976). Negus and Berger (1971) reported that the feeding of green foods reduced the pineal weight of *M. montanus* kept under 8L:16D, 12L:12D, and 16L:8D. It is not clear, however, what is cause and what is effect. It is conceivable that pineal weight decreases as a result of the higher GTH and steroid hormone concentrations in the animals fed greens, and it is also possible that the feeding of greens inhibits pineal activity and thus permits gonadal activity even on short days. Presumably, on long days,

the pineal has little inhibitory effect on reproduction of this species.

In laboratory strains of *M. agrestis*, long photoperiods, 16L:8D, in comparison to 6L:18D, stimulated testicular and ovarian activity independent of the ambient temperature (6.5 ± 1.5° C and 20.0 ± 2.0° C). In wild-trapped *M. agrestis*, long photoperiods maintained testicular activity of sexually mature animals, but with 6L:18D at 20 ± 2° C testicular weights were also maintained, which did not occur in the laboratory strain (Clarke and Kennedy (1967). Roth (1974), using red-back voles (*Clethrionomys gapperi athabascae*) caught in the wild, found no effect of photoperiod or temperature and no interaction between these environmental variables in a factorial experiment with 20L:4D, 4L:20D, 20° C, and 5° C as the variables. In all four environments, gonadal size increased to about the same extent.

Temperature. It is, of course, of interest to know to what extent ambient temperature determines the start or the end of the reproductive period or to what extent ambient temperature affects reproductive performance. As Wimsatt (1969) pointed out in his review, there appears to be a mutual antagonism between hibernation and reproduction. For example, when female thirteen-lined ground squirrels (*Citellus tridecemlineatus*) are maintained at room temperature without hibernation, reproductive development is inhibited compared to controls that are kept in cold and darkness for several months, during which time they hibernate. Such inhibition is not found in males. During the breeding season, exposure to low temperature does not induce either testicular regression or hibernation and, as long as the males are maintained at these low temperatures, testicular regression is prevented. In the spring, males and females may hibernate, but the tendency to do so decreases as gonadal activity increases. During this spring phase, gonadectomy of either males or females increases the tendency to hibernate; however, this effect is restricted to the spring.

Prolonging the winter environment in the spring by keeping the animals in a cold and dark environment delays the onset of reproduction of chipmunks (*Tamias striatus*) (Kayser, 1970). In this case the effects of temperature and photoperiod cannot be separated. The length of the gestation period of the pippistrelle bat (*Pippistrellus pippistrellus*) is extended by exposure to a temperature between 5° C and 14° C, and it is shortened by exposure to temperatures of 35° C (Racey, 1973).

Ambient temperature can affect mammalian reproduction of non-seasonally reproducing species. House mice can be bred for many generations at 3° C; however, comparison with mice bred at 21° C has shown that at 3° C there are more barren pairs, and fertile pairs rear fewer young, although litters are larger at birth and the young are heavier, so that the smaller number of young reared is the result of greater postnatal mortality. Mice that had been bred for 10 generations at 3° C and were then transferred to 21° C were more fertile than controls bred at 21° C for 10 generations (Barnett et al., 1975).

High ambient temperatures seem, in general, to have more deleterious effects on reproductive processes than low ambient temperatures. We have discussed the effects of high ambient temperatures on spermatogenesis in Chapters 4 and 6. Effects of high ambient temperature on reproduction have been studied in some detail in sheep, when sheep were exposed during different stages of the reproductive cycle to high ambient temperatures (40° C), it was found that (1) the unfertilized egg was not damaged (Ulberg and Sheean, 1973); (2) spermatozoa in ewes at the time of capacitation were capable of fertilization, but there was a high pre-implantation mortality (Ulberg, 1966); and (3) exposure when the fertilized egg was undergoing its first cell division resulted in post-implantation mortality.

Olfactory Signals. The mammalian species discussed so far in this chapter have seasonal breeding, and we have discussed the effect of photoperiod, photoperiod and temperature, and the availability of certain feeds as proximate causes for the timing of reproduction. In the species to be discussed in this next section, there may be no breeding season, and the environmental factors to be discussed are factors that are created within the population itself. Olfactory signals in the form of chemicals, of which the identity is not always known, may be excreted in the saliva, the urine, in the feces, or secreted from specialized skin glands. They announce the presence of conspecifics to each other, and they may also serve to announce the sex and sexual status of the emitter of the signal.

Prairie voles. As discussed in Chapter 8, estrus in the prairie vole (*Microtus ochrogaster*) is induced, and the induction occurs through the male urine. Under laboratory conditions, olfactory signals have several important effects on the reproductive physiology of the female.

Olfactory urinary signals emitted by the male affect female reproduction by advancing puberty, inducing estrus in sexually mature but reproductively inactive females, and interrupting pregnancy induced by another male. Advancement of puberty has been found when female prairie voles were exposed to an unfamiliar adult male or to a male of the same age but from a different litter. It remains to be determined whether this advancement of puberty is influenced by prepuberal exposure to males (Richmond and Stehn, 1976).

The induction of estrus is clearly an advantage for the male, since it allows him to copulate and thus transmit the alleles of the genes he carries to the next generation. This advantage is based on the assumption that induction of estrus also occurs in feral populations. In the laboratory, induction of estrus occurs only when male and female are unfamiliar with each other. For example, an adult male in a cage with his female partner and one or more of the litters of the pair does not induce estrus in his daughters, unless he is removed for 8 d or more and then returned to his cage. Upon return, he induces estrus in most of his daughters of 36 d of age and all his daughters of over 45 d of age. If a female in a pair has become anestrus, removal of the male and return after 8 days or more, brings her into estrus (Richmond and Stehn, 1976). Evidence that the signal is an olfactory one and that it is present in the urine is provided by different experiments. First, anosmic females will not come into estrus. Second, the placement of one drop of urine on the upper lip of reproductively inactive females causes a rapid increase in LH-RH in the olfactory bulb, an increase in serum LH concentration, and a decrease in norepinephrine concentration in the posterior olfactory bulb (Dluzen et al., 1981). Once the female is in estrus, she will remain so for as long as a month if she does not copulate. After copulation, which induces ovulation, the presence of the male is required for at least 24 h for pregnancy to continue. If the male is removed within 24 h post coitus, only about 35 percent of the females remain pregnant; if the male is removed 5 d post coitus, 95 percent of the females remain pregnant. This effect of the presence of a male in maintaining pregnancy is probably not the result of olfactory signals, but the result of additional copulations that are necessary to maintain the CL of pregnancy (Richmond and Stehn, 1976). Finally interruption of pregnancy by a strange male occurs if the stud male is removed, then when a strange male is placed with the female, the pregnancy induced by the stud male is interrupted and the female returns to estrus. This interruption of pregnancy can occur as late as day 17 of pregnancy, with a normal gestation length of 21.4 d in this species (Stehn and Richmond, 1975).

Males not only emit olfactory signals, they also receive them from females and other males. Male prairie voles show a preference for estrous vaginal and estrous urine odors, a preference that is androgen-dependent (Richmond and Stehn, 1976). In a population of male prairie voles that are behaviorally compatible, either the introduction of female estrous odors or making the males anosmic, results in increased threatening and increased aggression, suggesting that odors serve as important signals among males for recognition and avoidance of aggressive encounters.

House mice. In feral populations, which are usually of low density, reproduction of the house mouse (*Mus musculus*) is seasonal (Bronson, 1979b). The proximate causes, which time the breeding in these populations have not been determined, although availability of food to meet the caloric intake required for reproduction is the most likely single candidate; however, factors such as photoperiod and temperature may modulate the requirement for calories or may modulate the neuroendocrine-axis. Commensal populations, closely associated with humans, usually have high density and usually show no seasonality (Bronson, 1979b). With commensal and laboratory populations of mice, olfactory signals have been found to be of primary importance in the regulation of reproduction. Figure 14.5 illustrates many of the interactions that have been found, and we will discuss these in sequence. Mouse populations typically live in a territory that is defended by a single male, who marks the boundaries of his territory by urinary markings (Bronson, 1979b). Within the territory live the male, ten or fewer breeding females, some of their offspring, and some subordinate males. Females and males use urinary markings to indicate their presence and to affect the reproduction of other conspecifics. The marking behavior by the male is androgen-dependent and status-dependent; that is, subordinate males urine-mark infrequently and the territorial dominant male may mark as often as 3,000 times per night. The female's urinary marking is not dependent upon her endocrine status, but is increased by the presence of an intact male but not by a castrated male (Bronson, 1979b). A male exposed to the urine of a female will show an increase in plasma LH and testosterone concentrations (Macrides et al., 1975; Maruniak et al., 1979). The male habituates to the female cue and his response diminishes upon continued exposure to the same female or her urine. The function of this elevation of LH and testosterone concentrations is not immediately obvious, since spermatogenesis is a continuous process, and under conditions in which the male is not exposed to female urinary odors, LH and testosterone are released episodically. Bronson (1979b) has proposed that the elevated testosterone secretion in response to female urine may increase the synthesis of the male's own urinary olfactory signals. Male urine has several effects on female mouse reproduction: (1) acceleration of puberty, (2) dissipation of the estrus-suppressing effect of grouping of female mice, (3) synchronization of estrous cycles, and (4) interruption of pregnancy induced by another male.

1. *Acceleration of puberty.* The induction of puberty in females at an earlier age than in females not receiving the olfactory male urine

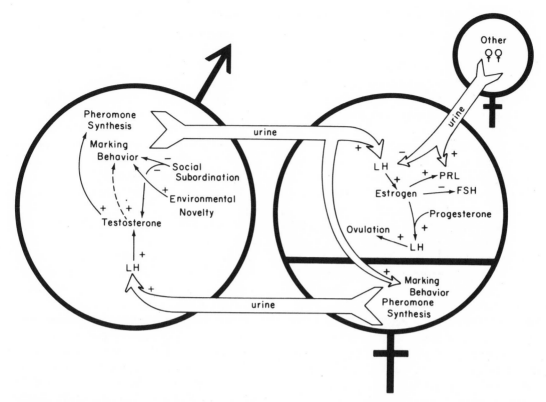

Figure 14.5 The pheromonal cueing system of the house mouse. LH = luteinizing hormone, PRL = prolactin, FSH = follicle-stimulating hormone, + = stimulation, − = inhibition. (From Bronson, 1979; reprinted with permission of *Quarterly Review of Biology*.)

signal is called the Vandenbergh effect after its discoverer (Vandenbergh, 1967). The puberty-accelerating effect is greatly enhanced if it is combined with tactile stimuli (Bronson, 1976). Kirchhof-Glazier (1979), using mice pups nursed by prairie deer mice (*Peromyscus maniculatus bairdii*) mothers, showed that this effect was genetically determined and not acquired. Massey and Vandenbergh (1981) found that male urine collected from a feral population at its lowest density in June and at its highest density in January accelerated puberty in laboratory mice. The potency of the puberty-accelerating factor was similar in the samples collected in June and January, suggesting that population density did not affect the excretion of this factor.

The hormonal changes in the female prepubertal mice in response to the puberty-accelerating urine consist of an increase in plasma LH concentration followed by an increase in plasma es-

tradiol concentration. On day 3, the estradiol concentration returns to its base level, and, provided that the mechanism for the stimulatory effect to release LH has matured, is followed one day later by a preovulatory FSH and LH release followed by ovulation. The chemical that causes the acceleration of puberty is probably a protein. It is present in bladder urine and is excreted with the urine, and its production is androgen-dependent (Lombardi et al., 1976; Marchlewska-Koj, 1977).

2. *Dissipation of the estrus-suppressing effect of grouping of female mice without males.* Female mice grouped together show a high incidence of pseudopregnancies in the absence of mating (van der Lee-Boot effect). The presence of a male or as little as 0.1 ml of male urine introduced into the cage with grouped mice results in resumption of normal 4- and 5-d cycles (Bronson, 1979b).

3. *Synchronization of estrus*. In a population of mice with 4-d cycles, one expects 25 percent of the population to be in estrus on any given day, as is generally found unless the van der Lee-Boot effect interferes. If a male or male urine is introduced in a population of regularly cyclic female mice, a peak of estrous females occurs on day 3 after introduction of the male (Whitten, 1965). This is called the Whitten effect, and it is not restricted to mice, but also occurs in sheep and goats (van Tienhoven, 1968). The material excreted in the urine that induces the Whitten effect in mice is probably lipoidal (Monder et al., 1978) and is present in the urine in the bladder. The excretion of the chemical is androgen-dependent but does not involve the secondary sex organs or their secretions (Bronson and Whitten, 1968).

The effect of the male olfactory signals in synchronization of the estrous cycles of mice seems to be the result of a depression of progesterone secretion within 36 h after exposure to the male. This is followed by an increase in plasma estradiol and by a preovulatory peak in plasma LH concentration (Ryan and Schwartz, 1976).

4. *Interruption of pregnancy by a strange male*. If the male of a mated pair of mice is replaced by another male, i.e., a strange male within the first five days post coitus, pregnancy is interrupted. This pregnancy-blocking effect or Bruce effect (Bruce, 1959) is followed by the induction of estrus. The greater the genetic difference between the stud male and the strange male, the higher the incidence of blocked pregnancies (van Tienhoven, 1968). The effect of the strange male can be simulated by exposure of the females to the urine of a strange male. The production of the chemical(s), a peptide, that cause(s) the Bruce effect is androgen-dependent, but the site of its production has not been ascertained (Marchlewska-Koj, 1981).

The endocrinological changes in the pregnant female in response to a strange male are probably interruption of prolactin secretion and release of LH, which are necessary to induce the new estrous cycle and ovulation. Chapman et al. (1970) suggest that increased LH secretion preceding a decrease in prolactin secretion interrupts progesterone secretion by the CL of pregnancy and thus causes the pregnancy block. This hypothesis is based on changes in AP content, not on changes in plasma concentrations of LH and prolactin; therefore, it needs further substantiation.

The Bruce effect has also been observed in *Microtus ochrogaster, M. agrestis, M. pennsylvanicus,* and *Peromyscus maniculatus bairdii* (Richmond and Stehn, 1976; Terman, 1969). The Bruce effect probably is rarely, if ever, encountered in feral populations and may be a result of artificial selection (Bronson, 1979b).

Suppression of estrus by females. Female mice emit urinary signals that suppress estrus in prepubertal females as they approach puberty and in adult female mice when adults are grouped together. Exposure of young female mice to urine from adult or juvenile grouped females for 7 d during the first two weeks after weaning delayed the onset of puberty by 3–5 d in comparison with controls that were not exposed to urine. Tactile stimuli by the grouped mice had no effect on the urinary cue effect, in contrast to the enhancement by tactile stimuli of the puberty-accelerating effect of male urine (Drickamer, 1977). Massey and Vandenbergh (1980) have shown that urine from grouped female laboratory house mice was effective in delaying puberty in a feral mouse population.

When female mice are grouped without a male or without exposure to odors from male urine, estrous cycles become irregular; there may be a high incidence of pseudopregnancies in the absence of any copulations (van der Lee-Boot effect); or there is a high incidence of anestrus but without any CL formation. The occurrence of either of these two forms of interrupted estrous cycles seems to depend on the number of mice housed together; with about 4 mice there is a high incidence of pseudopregnancies; with 30 mice there is a high incidence of anestrus (Whitten, 1966). As mentioned in the discussion of olfactory cues emitted by male mice, exposure of such grouped mice to male urine eliminates the inhibitory effect of female olfactory signals on the estrous cycle.

The pathway by which odors affect the neuroendocrine unit probably includes the vomeronasal organ → the vomeronasal nerve → the accessory olfactory bulb → the medial and posteromedial amygdaloid nuclei → the stria terminalis → the medial preoptic area, ventromedial nucleus, nucleus tuberis lateralis, and premammilary nucleus (Scalia and Winans, 1976). There is persuasive evidence that sectioning of the vomeronasal nerve of hamsters interrupts copulatory behavior, whereas destruction of the nasal mucosa does not (Scalia and Winans, 1976). The importance of this accessory olfactory system may explain the significance of the Flehmen behavior in bulls, stallions, rams, and other mammals. This Flehmen is the moving up of the head and the curling of the upper lip (Signoret, 1976). This behavior may allow the chemical signals easier access to the vomeronasal organ.

Wild mice are nocturnal, and in spite of many generations of laboratory breeding, laboratory mice (CF-1 strain) showed a preovulatory LH peak that was 2.4 times greater under DD than under 14L:10D. A similar peak was observed in ovariectomized estrogen-treated mice in which LH release was induced by progesterone (Bronson and Vom Saal, 1979). The number of oocytes ovulated, however, was not affected by the light treatment, and a tenfold increase in light intensity did not affect the peak LH concentration. However, wild mice caught and exposed to 14L:10D at 1000 lx showed a depression in body weight of males, and an over 50 percent reduction in litter size, whereas a strain of laboratory mice did not show these effects. Under DD, both stocks reproduced normally (Bronson, 1979a).

Rats. Two types of populations of wild brown rats (*Rattus norvegicus*), from which most strains of laboratory albino rats are derived, exist; these are the feral populations that exist independent of human activities and the populations closely associated with human activities, such as city rat populations. Much of the present knowledge about this species is based on the laboratory strains, which may be quite different from the wild populations. For example, McClintock and Adler (1979) found that the estrous cycle of wild rats caught in a city and then studied in the laboratory had an estrous period of one or two weeks instead of the 4 or 5 days found in laboratory rats.

Steniger (1950), in a study of a feral rat population in North Germany, found no pregnant rats

between October and March. He attributed this seasonal reproduction to the influence of temperature, but provided no data to substantiate this claim. In a city population, Davis (1951) found evidence for some seasonal fluctuation in the incidence of pregnancy, but pregnant rats were found throughout the year. Calhoun (1962), in an extensive study of rats caught in a city and studied in an outside pen with a natural environment, found clear evidence of seasonal reproduction by examination of the nipples of females and by diagnosis of conceptions. In this population, no pregnancies occurred between September and January. Temperature was not the major factor controlling reproduction, because rats conceived in February when the mean minimum nightly temperature, $-3°$,C was below the $15°$ C of September. Calhoun (1962) suggested that photoperiod might be a controlling factor. Food was always available in abundance, thus an indirect effect of season through an effect on food availability can be eliminated as one of the proximate causes for the seasonal reproduction. At present, too little information is available to state what the proximate causes of the seasonal reproduction of wild rats are.

The estrous cycle of laboratory rats is affected by photoperiod; under 12L:12D rats showed exclusively 4-d cycles, whereas under 14L:10D and 16L:8D, there was a high incidence of 5-d cycles (Hoffmann, 1973). In most strains of laboratory rats, exposure to LL induces a permanent vaginal estrus syndrome (PVE), which is characterized by a failure of ovulations to occur, cystic ovaries, and persistent estrus shown in vaginal smears. After a few days of exposure to LL, the preovulatory LH surge disappears (Turek and Campbell, 1979). Recent investigations have shown that in rats under LL, taking smears at 24-h intervals instead of at random times each day delayed the onset of the PVE (Weber and Adler, 1979). This suggests that the daily light-dark periods may act as synchronizers and that LL causes PVE partly by the removal of such a signal and partly by an inherent effect on the neuroendocrine unit.

For LL to have the effect of inducing PVE, the light needs to be bright (Turek and Campbell, 1979). Weber and Adler (1979) subjected rats to either LL (700 lx), LL (10 lx) or a light regime providing a contrast in light intensities, 12L (700 lx):12L (10 lx). On the 12L (700 lx):12L (10 lx) regime, fewer animals came into constant estrus than under either of the other two LL regimes, again pointing to light functioning as a cue instead of as a stimulator or inhibitor.

During solicitation and mounting, males emit ultrasounds of 50 kHz. The number of such calls are higher before successful than before unsuccessful matings and are higher before mounts with intromissions than before mounts without intromissions (Barfield et al., 1979). During the postejaculatory refractory period, the male rat emits ultrasounds of 22 kHz, which are correlated with a sleeplike electroencephalogram pattern. Barfield et al. (1979) suggest that the 22-kHz vocalizations may serve to maintain contact between partners, at the same time preventing too early recopulation (see Chapter 8). Adler and Anisko (1979) submit that the 22-kHz ultrasounds are uttered when the rat is in a submissive or helpless state. The emission of postejaculatory ultrasounds by male rats does not appear to depend on the presence of testicular hormones, but estradiol dipropionate has an inhibitory effect (Parrott and Barfield, 1975).

The production of ultrasounds may involve the central gray area (Floody, 1979). The pathway by which the ultrasounds affect the neuroendocrine unit involved in presexual, sexual, and postsexual behavior needs to be elucidated.

Olfactory communication occur among rats as they do in hamsters and mice, but the effects of female olfactory signals on other females are different in rats than in mice. In grouped rats and in rats housed singly but exposed to odors from grouped rats, the variation in the length of the estrous cycle was shorter than in isolated rats housed singly. Furthermore, the estrous cycles of grouped rats and of single rats exposed to odors from grouped rats became synchronized, i.e., a majority of females came into estrus on the same day (McClintock, 1981). Some ingenious experiments have provided evidence that the odors from rats with the LL-induced PVE syndrome impart the acyclicity on other rats downwind, but not in rats upwind with respect to the PVE rats. Odors from proestrous rats desynchronized the synchronous estrous cycles of rats housed together. Odors from metestrous rats had a syn-

chronizing effect on rats in different stages of the cycle. It thus appears that proestrous rats are the leaders in synchronizing the estrous cycle in grouped rats. The proestrous rat first gives the signals that tend to synchronize the rats and then when the proestrus rat goes into estrous, she emits an olfactory signal that brings the group into synchrony (McClintock, 1981).

In addition to the male-female olfactory communication, there is olfactory communication between the lactating mother and her offspring. The extensive review and discussion by Leon (1978) should be consulted for these interrelationships in rats. In brief, it appears that there are: (1) maternal odors that inhibit gross motor activity between age 2–12 d, (2) olfactory cues necessary for orientation of the pup towards the nipple (anosmic pups fail to orient towards the nipple), and (3) home odor cues that allow the pup to orient toward the nest. This cue may be imparted by the mother to the bedding. That the mother emits olfactory cues attractive to the pups has been shown by a number of experiments. The maternal odor of lactating females is derived from her excreta, and is attractive to pups between the ages of 12 d and 27 d. It is apparently present in the caecotrophe (a semisolid, light-colored material formed in the caecum and excreted through the anus). Lactating rats excrete more caecotrophe than nonlactating females. The difference between the attractiveness of excreta from virgins and lactating rats may be the result of the greater food consumption by the lactating females and the greater amount of caecotrophe excreted in conjunction with this. The amount of food ingested is related to the number and age of pups being nursed. The emission of the odor (but not its synthesis) is correlated with high prolactin secretion, which, in turn, increases food intake.

The odor is apparently synthesized by caecal microorganisms. The weakening of the pheromonal bond is probably the result of ingestion of maternal fecal material and the subsequent production of their own caecotrophe and their own odor, which resembles that of their mother. At the same time, this odor production by the mother diminishes and the attraction of the young to the feces of the mother is transferred to their own feces. As rats normally eat their own feces, such a transfer makes evolutionary sense.

The stimulus provided to a male rat by an estrous female may well consist of a complex mixture of auditory, olfactory, and tactile signals. Even when the exact nature of the stimulus is not known, it is possible to obtain the hormonal response in a male rat by classical conditioning of the male. Exposure of sexually naive males first to a neutral stimulus, which did not result in LH release and testosterone secretion (e.g., vapors of methylsalycilate), and then to an estrous female once daily for 14 trials, resulted in LH release and testosterone secretion after exposure to the neutral stimulus alone (Graham and Desjardins, 1980).

Rhesus monkeys. Captive rhesus monkeys (*Macaca mulatta*) in an outdoor environment show a definite seasonal pattern of reproduction. The onset of mating activity occurs in late summer to early fall and continues for 4 to 5 months. Individual females are receptive for 5-11 d during the follicular and periovulatory part of the cycle. Rhesus monkeys kept in a laboratory environment show no seasonal pattern of reproduction and females are receptive every day of the menstrual cycle, although there is a peak of copulatory activity around the time of ovulation (Gordon, 1981). The proximate cause for the seasonal reproduction of free-ranging rhesus monkeys has not been determined. Within laboratory populations of rhesus monkeys, much research has been done on the role of olfactory cues emitted by the female in the initiation of mating behavior by the male. Goldfoot (1981) has reviewed both the evidence and the interpretation of the evidence. It appears that for sexually experienced males, odors from midcycle products of the vagina make a female more attractive; however, such olfactory cues appear neither necessary nor, by themselves, sufficient to induce matings.

Nutrition. Certain foods can act as proximate factors in timing of the breeding season, as was discussed for *Microtus montanus, M. agrestis,* and *M. arvalis.* Nutrition can also affect reproduction without serving as a proximate cause. In general, little is known about the nutritional requirements of wild populations, whereas the nutritional requirements of some laboratory and farm animals are known in considerable detail. It

lies outside the scope of this book to review the effects of specific nutrients on reproduction, but in most cases the deficiency of a nutrient leads to inanition (except for vitamin A and vitamin E deficiency), so that it becomes difficult to assess whether a specific nutrient deficiency affects reproduction. Wild populations may occasionally face deficiencies of water, calories, and proteins, and we will discuss these briefly. Water deprivation, even in the gerbil (*Meriones unguiculatus*), a desert animal, leads to severe depression of reproduction, without affecting ovarian or testicular weights. Gerbils do not need water for survival, because they can use the water present in food and metabolic water; for reproduction, however, water is a requirement (Yahr and Kessler, 1975).

Deficiencies of caloric intake may range from intermittent deficiencies to total starvation. On theoretical grounds, one would expect caloric deficiency, especially for a protracted period, to have the severest repercussions on females in which the oocyte contains large amounts of yolk, such as the elasmobranchs, reptiles, and birds, whereas the effect on males might be expected not to be as severe as the effect on the females. In species that incorporate little yolk in the oocyte, one might expect that caloric deficiency would not affect ovulation and fertilization, but might result in abnormal embryonic development or rejection and resorption of the embryo.

Experiments with mice have shown that fasting of grouped males and females for as little as 18–30 h reduced fertility without affecting estrous behavior; however, starvation for 36 h diminished estrous behavior. That the effects were the result of a deficient caloric intake was shown by the fact that feeding a 75 percent glucose solution prevented the effects of starvation (McClure, 1966).

Withdrawal of food from mice for the 48-h period from day 3 to day 5 after copulatory plugs were found, resulted in failure to give birth to young. Withdrawal of food at other times between day 1 and day 7 after copulatory plugs were found had less effect on the birth rate. Apparently, at the time of implantation, the embryo-uterine unit is particularly sensitive to caloric deficiency. This effect is mediated through the CL, since injections of either hCG or progesterone can prevent this effect of inanition of the mother (McClure, 1962).

Many experimental procedures have been used to study the effects of malnutrition (e.g., feeding of protein-free diets, total starvation, vitamin-deficient diets) on reproduction of female rats. In many of these cases, the embryos are resorbed or aborted, and these effects can be prevented by progesterone administration (van Tienhoven, 1968). Srebnik et al. (1978) have presented evidence that malnutrition depresses the activity of the neuroendocrine unit but that it is the hypothalamus that is supersensitive in malnourished rats. They speculate that steroidogenesis is affected first and that failure of gonadotrophin secretion (primarily prolactin and LH, with little effect on FSH) is an effect of the failure of steroidogenesis instead of its cause.

Starvation of laying chickens results in atresia of the large follicles which, of course, results in cessation of egg production. Egg production can, however, be maintained for about 10 days by the injection of either mammalian FSH and LH or by chicken AP extracts. The eggs become smaller as starvation progresses in such gonadotrophin-injected hens until the reserves that can be mobilized are exhausted (Morris and Nalbandov, 1961).

Deficiency of calcium in the diet of laying hens leads to inanition and to a severe reduction in egg production and an increase in the proportion of soft-shelled eggs. Injection of chicken hypothalamic extracts does not increase the number of eggs laid during a 21-day period of calcium deficiency, but it does increase the ratio of hard-shelled to soft-shelled eggs compared to calcium-deficient hens injected with saline or mammalian LH, mammalian FSH, or chicken pituitary extracts (Roland et al., 1974). It would be of interest to follow the concentrations of steroid and gonadotrophic hormones during the feeding of such a calcium-deficient diet and to determine whether the stimulatory effect of progesterone on LH secretion and vice versa is maintained in such hens.

Several food sources have been identified as inhibitors of reproduction, two of which we mentioned in Chapter 11, i.e., skunk cabbage and the bush *Salsola tuberulata.* Both caused the failure

of labor, because they damaged pituitary development of the fetus. In Australia, it has been found that sheep grazing on subterranean clover, *Trifolium subterraneum,* become sterile largely because of estrogenic compounds present in this clover (Bennetts et al., 1945).

It is beyond the scope of this book to review or discuss the effects of pesticides, fungicides and herbicides or of drugs, alcohol, caffeine, nicotine, etc. on the reproduction of laboratory animals, wild animals, or humans.

Population Cycles. The number of animals in populations of small rodents, such as mice, voles, and lemmings, fluctuates with a periodicity of about three to four years. One explanation of this periodicity is that in high-density populations, more agonistic encounters occur that lead to an increase in the secretion of ACTH corticosterone. As the population density increases, high concentrations of corticosterone become more or less permanent and start to have deleterious effects on reproductive performance and survival of the young (Christian, 1971). Laboratory experiments with mice tend to support this hypothesis, but whether it explains fluctuations in feral populations is still a matter of controversy (Bronson, 1979b). Myers and Krebs (1974) suggest that, within populations of rodents showing these fluctuations in density, there may be two genotypes, one intolerant of high population density but reproductively superior, and another genotype with a low rate of reproduction but adapted to high density. They suggest that as the population density increases, the individuals intolerant of crowding, migrate, thus decreasing the population density.

The relationship between stress by agonistic encounters, reproduction, and population density is illustrated in a unique manner in the genus *Antechinus* of dasyurid marsupials. In *A. stuarti, A. swainsoni, A. minimus, A. flavipes, A. bellus,* and *A. bilarni,* the males die soon after mating, in *A. stuarti* as early as three weeks post coitus (Lee et al. 1977). In this species, mortality is associated with several symptoms considered as reliable indicators of stress, such as high plasma corticosteroid concentrations, high liver glycogen content, high adrenal weight, and a wide fasciculate zone (Barnett, 1973). Not only does plasma corticosteroid concentration in males rise from 2.9 ng/ml prior to mating to 5.2 ng/ml after mating, but there is also a decrease in the corticosteroid-binding capacity of the serum (Lee et al., 1977), thus making more free corticosteroid available. Females, in contrast to males, are capable of reproducing during two or three consecutive years.

Injections of cortisol acetate into captive males of *A. stuarti,* caused mortality in 9 of 27 animals, but no death among 11 controls. Mortality of males captured prior to mating and kept isolated in captivity survived beyond the normal life expectancy, but males captured during the last week prior to the expected time of death did not survive in isolation. The apparent stress which causes the syndrome, may consist of agonistic encounters and copulations. The phenomenon of once-in-a-life-time reproduction, which has been described for lampreys and salmon previously in this book, has been called *semelparity* (Lee et al., 1977). Lee et al. (1977) speculate that the high male mortality in brown lemmings (*Lemmus trimucronatus*) and in *Microtus pennsylvanicus* during highly synchronized spring matings may be similar to the semelparity of *A. stuarti.*

One researcher (von Holst, 1969) noted that when the tree shrew *Tupaia belangeri* is disturbed, the sympathetic nervous system is stimulated, which results, among other events, in activation of the arrectores pilorum muscles which raise the hairs on the tail. By expressing the time that the animal has a bushy tail as a percentage of the time that the animal was observed, an indicator for the amount of disturbance was obtained. Most (90 percent) of the bushy tail resulted from confrontations with a dominant male, and von Holst (1969) found a strong relationship between the percentage of time an animal had a bushy tail and reproductive disturbances. When the percentage of time of bushy tail was between 20 and 40 percent, female tree shrews showed male instead of female sexual behavior and a high frequency of eating of their young, presumably because they had failed to mark them with their sternal scent gland. When the percentage of time that the animal had a bushy tail was 50 percent or

more, females were sterile and had small ovaries without mature follicles. In lactating females, when this percentage was above 20, milk secretion decreased. In tree shrew males, the testes were in the scrotum but could be retracted into the abdomen when the percentage of time that the animal had a bushy tail was between 50 and 70 percent; above 70 percent the testes remained in the abdomen, and the germinal epithelium degenerated; whereas below 50 percent the testes remained in the scrotum and were not retracted into the abdomen. These data von Holst (1969) interpreted as indicating that confrontations with a dominant male stimulate the sympathetic nervous system, resulting in raising of the hairs on the tail and also affecting the hypothalamic neuroendocrine unit, so that the secretion of FSH, LH, TSH, and growth hormone are decreased, but ACTH secretion is increased. This unusual method of evaluating stress should be used in conjunction with determinations of corticosteroids, gonadal, and the pituitary hormones mentioned in order to verify von Holst's interpretation (1969).

SUMMARY

In summary, it appears that for many species of nontropical zones, light acts as a timer of the breeding season, either by acting as a signal that synchronizes an endogenous circannual rhythm or by acting as a stimulus for the neuroendocrine-gonadal axis. Temperature, rainfall, and humidity may act also as signals, but they may interact with photoperiod to stimulate reproduction through the neuroendocrine unit, or they may act by making food available and thus indirectly time the breeding season.

Once the stage is set and ovaries and gonads have reached a certain degree of development, intraspecific communication through visual, auditory, and olfactory cues may fine-tune the further progress of reproductive events. The information is integrated by male and female in such a way that the interplay between the male and female can result in insemination, fertilization of oocytes, development of the embryos, and birth of the young.

References

1 | Sex and Its Determination

ADLER, D. A., J. D. WEST, and V. N. CHAPMAN. Expression of α-galactosidase in preimplantation mouse embryos. *Nature* 267:838–839, 1977.

BAHR, G. F. Separation of X- and Y-bearing spermatozoa by gravity: a reconsideration. In *Sex Ratio at Birth—Prospects for Control,* eds. C. A. Kiddy and H. D. Hafs, pp. 28–36. American Society of Animal Science, 1971.

BEATTY, R. A. Chromosome deviations and sex in vertebrates. In *Intersexuality in Vertebrates including Man,* eds. C. N. Armstrong and A. J. Marshall, pp. 17–143. New York: Academic Press, 1964.

BEÇAK, W., and M. L. BEÇAK. W-sex chromatin fluorescence in snakes. *Experientia (Basel)* 28:228–229, 1972.

BENIRSCHKE, K., and T. C. HSU. *Chromosome Atlas: Fish, Amphibians, Reptiles, and Birds,* Vols. 1–2. New York: Springer, 1971, 1973.

BENNETT, D., and E. A. BOYSE. Sex ratio in progeny of mice inseminated with sperm treated with H-Y antiserum. *Nature* 246:308–309, 1973.

BLOOM, S. E. Current knowledge about the avian W chromosome. *BioScience* 24:340–344, 1974.

BROWN, S. W., and H. S. CHANDRA. Inactivation system of the mammalian X chromosome. *Proc. Nat. Acad. Sci. USA* 70:195–199, 1973.

CASPERSSON, T., L. ZECH, C. JOHANSSON, J. LINDSTEN, and M. HULTÉN. Fluorescent staining of heteropycnotic chromosome regions in human interphase nuclei. *Exp. Cell Res.* 61:472–474, 1970.

CATTANACH, B. M. Sex reversal in the mouse and other mammals. In *The Early Development of Mammals,* eds. M. Balls and A. E. Wild, pp. 305–317. Cambridge: Cambridge University Press, 1975.

CLARK, E. Functional hermaphroditism and self-fertilization in a serranid fish. *Science* 129:215–216, 1959.

COHEN, J. *Reproduction.* London: Butterworth, 1977.

COLE, C. J., C. H. LOWE, and J. W. WRIGHT. Sex chromosomes in lizards. *Science* 155:1028–1029, 1967.

———. Chromosomes in teiid whiptail lizards (genus *Cnemidophorus*). *Am. Mus. Novit.* 2395:1–14, 1969.

DREW, J. S., W. T. LONDON, E. D. LUSTBADER, J. E. HESSER, and B. S. BLUMBERG. Hepatitis B virus and sex ratio of offspring. *Science* 201(4357):687–692, 1978.

FECHHEIMER, N. S., and R. A. BEATTY. Chromosome abnormalities and sex ratio in rabbit blastocysts. *J. Reprod. Fertil.* 37:331–341, 1974.

FINEMAN, R., J. HAMILTON, and G. CHASE. Reproductive performance of male and female phenotypes in three sex chromosomal genotypes (XX, XY, YY) in the killifish *Oryzias latipes*. *J. Exp. Zool.* 192:349–354, 1975.

FINEMAN, R., J. HAMILTON, and W. SILVER. Duration of life and mortality rates in male and female phenotypes in three sex chromosomal genotypes (XX, XY, YY) in the killifish *Oryzias latipes*. *J. Exp. Zool.* 188:35–40, 1974.

FISHER, R. A. *The Genetical Theory of Natural Selection.* Oxford: Clarendon Press, 1930.

FOOTE, W. D., and M. M. QUEVEDO. Sex ratio following subjection of semen to reduced atmospheric pressure. In *Sex Ratio at Birth—Prospects for Control,* eds. C. A. Kiddy and H. D. Hafs, pp. 55–57. American Society of Animal Science, 1971.

FREDGA, K. Unusual sex chromosome inheritance in mammals. *Philos. Trans. R. Soc. Lond. B Biol. Sci.* 259:15–36, 1970.

FREDGA, K., A. GROPP, H. WINKING, and F. FRANK. Fertile XX- and XY-type females in the wood lemming *Myopus schisticolor. Nature* 261:225–227, 1976.

GALLIEN, L. Genetic control of sexual differentiation in vertebrates. In *Organogenesis,* eds. R. L. DeHaan and H. Ursprung, pp. 583–610, New York: Holt, Rinehart, and Winston, 1965.

GERMAN, J., J. L. Simpson, R. S. K. CHAGANTI, R. L. SUMMITT, L. B. REID, and I. R. MERKATZ. Genetically determined sex-reversal in 46,XY humans. *Science* 202:53–56, 1978.

GORMAN, J., GORMAN, G. C., and L. ATKINS. Chromosomal heteromorphism in some male lizards of the genus *Anolis. Am. Nat.* 100:579–583, 1966.

GORMAN, G. C., L. ATKINS, and T. HOLZINGER. New karyotypic data on 15 genera of lizards in the family Iguanidae, with a discussion of taxonomic and cytological implications. *Cytogenetics (Basel)* 6:286–299, 1967.

GORMAN, G. C., and F. GRESS. Sex chromosomes of a pygopodid lizard, *Lialis burtonis. Experientia (Basel)* 26:206–207, 1970.

GROPP, A., K. FREDGA, H. WINKING, and F. FRANK. Regulation of sex chromosome constitution of somatic and germ cells in the wood lemming. *Ann. Biol. Anim. Biochim. Biophys.* 18(2B):367–375, 1978.

HAFS, H. D., and L. J. BOYD. Galvanic separation of X- and Y-chromosome–bearing sperm. In *Sex Ratio at Birth—Prospects for Control,* eds. C. A. Kiddy and H. D. Hafs, pp. 85–96. American Society of Animal Science, 1971.

HAMERTON, J. L. Significance of sex chromosome derived heterochromatin in mammals. *Nature* 219:910–914, 1968.

HAMILTON, J. B., and R. O. WALTER. Supermale (YY sex chromosomes) and ordinary male (XY): studies in killifish of competition for mating with females and evidence that a second Y chromosome results in effects during adulthood. *Anat. Rec.* 160:359, 1968 (Abstract).

HAYS, F. A. The primary sex ratio in domestic chickens. *Am. Nat.* 79:184–186, 1945.

HOOK, E. B. Behavioral implications of the human XYY genotype. *Science* 179:139–150, 1973.

HOWE, H. F. Sex-ratio adjustment in the common grackle. *Science* 198:744–746, 1977.

KAUFMAN, M. H. Analysis of the first cleavage division to determine the sex-ratio and incidence of chromosome anomalies at conception in the mouse. *J. Reprod. Fertil.* 35:67–72, 1973.

KIDDY, C. A. and J. D. HAFS, eds. *Sex Ratio at Birth—Prospects for Control.* American Society of Animal Science, 1971.

KING, M. The evolution of sex chromosomes in lizards.

In *Proc. Fourth Symp. Comparative Biology of Reproduction, Canberra, 1976,* pp. 55–60. Canberra: Australian Academy of Science, 1977.

LAVON, U., R. VOLCANI, and D. DANON. An attempt to separate ejaculated bull spermatozoa into groups in order to improve fertility. In *Sex Ratio at Birth—Prospects for Control,* eds. C. A. Kiddy and H. D. Hafs, pp. 19–26. American Society of Animal Science, 1971.

LIFSCHYTZ, E. X chromosome inactivation: an essential feature of normal spermiogenesis in male heterogametic organisms. In *Proc. Int. Symp. The Genetics of the Spermatozoon,* eds. R. A. Beatty and S. Gluecksohn-Waelsch, pp. 223–232. Edinburgh: Department of Genetics, University of Edinburgh, 1972.

LIFSCHYTZ, E., and D. L. LINDSEY. The role of X-chromosome inactivation during spermatogenesis. Proc. Nat. Acad. Sci. USA 69:182–186, 1972.

LINDAHL, P. E. Centrifugation as a means of separating X- and Y-chromosome–bearing spermatozoa. In *Sex Ratio at Birth—Prospects for Control,* eds. C. A. Kiddy and J. D. Hafs, pp. 69–74. American Society of Animal Science, 1971.

LOWE, T. P., and J. R. LARKIN. Sex reversal in *Betta splendens* Regan with emphasis on the problem of sex determination. *J. Exp. Zool.* 191:25–32, 1975.

LUTZ, H., and Y. LUTZ-OSTERTAG. Etude d'un freemartin chez les oiseaux. *Arch. Anat. Microsc. Morphol. Exp.* 47:205–210, 1958.

LYON, M. F. X-chromosome inactivation and developmental patterns in mammals. *Biol. Rev. Camb. Philos. Soc.* 47:1–35, 1972.

————. Mechanism and the evolutionary origins of variable X-chromosome activity in mammals. *Proc. R. Soc. Lond. B Biol. Sci.* 187:243–268, 1974a.

————. Sex chromosome activity in germ cells. In *Physiology and Genetics of Reproduction,* eds. E. M. Coutinho and F. Fuchs, Vol. 4A, pp. 63–71. New York: Plenum Press, 1974b.

MARCUM, J. B. The freemartin syndrome. *Anim. Breed. Abstr.* 42:227–242, 1974.

MARSHALL, A. J. Introduction to *Intersexuality,* eds. C. N. Armstrong and A. J. Marshall, pp. 1–16. New York: Academic Press, 1964.

MAYNARD SMITH, J. *The Evolution of Sex.* Cambridge: Cambridge University Press, 1978.

McMILLEN, M. M. Differential mortality by sex in fetal and neonatal deaths. *Science* 204:89–91, 1979.

MIGEON, B. R., and K. JELALIAN. Evidence for two active X chromosomes in germ cells of female before meiotic entry. *Nature* 269:242–243, 1977.

MITTWOCH, U. *Genetics of Sex Differentiation.* New York: Academic Press, 1973.

MOORE, K. L. Sex chromatin patterns in various animals. In *The Sex Chromatin,* ed. K. L. Moore, pp. 16–73. Philadelphia: W. B. Saunders, 1966.

MÜLLER, M., M. T. TENZES, M. WOLF, W. ENGEL, and J. P. WENIGER. Appearance of H-W (H-Y) antigen in gonads of oestradiol sex-reversed male chicken embryos. *Nature* 280:142–144, 1979.

OHNO, S. *Sex Chromosomes and Sex-Linked Genes.* New York: Springer, 1967.

———. *Major Sex-Determining Genes.* New York: Springer, 1979.

OHNO, S., L. C. CHRISTIAN, S. S. WACHTEL, and G. C. KOO. Hormone-like role of H-Y antigen in bovine freemartin gonad. *Nature* 261:597–599, 1976.

PENNOCK, L. A., D. W. TINKLE, and M. W. SHAW. Minute Y chromosome in the lizard genus *Uta* (family Iguanidae). *Cytogenetics (Basel)* 8:9–19, 1969.

PRICE, W. H., and P. B. WHATMORE. Criminal behavior and the XYY male. *Nature* 213:815–816, 1967.

ROBERTSON, D. R. Social control of sex reversal in a coral-reef fish. *Science* 177:1007–1009, 1972.

SCHILLING, E. Sedimentation as an approach to the problem of separating X- and Y-chromosome-bearing spermatozoa. In *Sex Ratio at Birth—Prospects for Control,* eds. C. A. Kiddy and H. D. Hafs, pp. 76–83. American Society of Animal Science, 1971.

SCHULTZ, R. J. Special adaptive problems associated with unisexual fishes. *Am. Zool.* 11:351–360, 1971.

SELDEN, J. R., S. S. WACHTEL, G. C. KOO, M. E. HASKINS, and D. F. PATTERSON. Genetic basis of XX male syndrome and XX true hermaphroditism: evidence in the dog. *Science* 201:644–646, 1978.

SELLMANOFF, M. K., J. A. JUMONVILLE, S. C. MAXON, and B. E. GINSBURG. Evidence for a Y chromosomal contribution to an aggressive phentotype in inbred mice. *Nature* 253:529–530, 1975.

SHARMA, M., H. F. L. MEYER-BAHLBURG, D. A. BOON, W. R. SLAUNWHITE, Jr., and J. A. EDWARDS. Testosterone production by XXY subjects. *Steroids* 26:175–180, 1975.

SHORT, R. V. Sex determination and differentiation. In *Reproduction in Mammals,* eds. C. R. Austin and R. V. Short, Vol. 2, pp. 43–71. Cambridge: Cambridge University Press, 1972.

———. New thoughts on sex determination and differentiation. *Glaxo Vol.* 39:5–20, 1974.

SILVERS, W. K., and S. S. WACHTEL. H-Y antigen: behavior and function. *Science* 195:956–960, 1977.

SPECTOR, W. S., ed. *Handbook of Biological Data.* Philadelphia: W. B. Saunders, 1956.

STANLEY, S. N. Clades versus clones in evolution: why we have sex. *Science* 190:382–383, 1975.

SUMNER, A. T., and J. A. Robinson. A difference in dry mass between the heads of X- and Y-bearing human spermatozoa. *J. Reprod. Fertil.* 48:9–15, 1976.

TAKAGI, N., and M. SASAKI. Preferential inactivation of the paternally derived X chromosome on the extraembryonic membranes of the mouse. *Nature* 256:640–642, 1975.

WACHTEL, S. S. H-Y antigen and the genetics of sex determination. *Science* 198:797–799, 1977.

WACHTEL, S. S., P. BASRUR, and G. C. KOO. Recessive male-determining genes. *Cell* 15:279–281, 1978.

WACHTEL, S. S., G. C. KOO, and E. A. BOYSE. Evolutionary conservation of H-Y ('male') antigen. *Nature* 254:270–272, 1975.

WACHTEL, S. S., G. C. KOO, S. OHNO, A. GROPP, V. G. DEV, R. TANTRAVAHI, D. A. MILLER, and O. J. MILLER. H-Y antigen and the origin of XY female wood lemmings (*Myopus schisticolor*). *Nature* 264:638–639, 1976.

WEIR, J. A. Genetic control of sex ratio in mice. In *Sex Ratio at Birth—Prospects for Control,* eds. C. A. Kiddy and H. D. Hafs, pp. 43–52. American Society of Animal Science, 1971.

WERREN, J. H., and E. L. CHARNOV. Facultative sex ratios and population dynamics. *Nature* 272:349–350, 1978.

WILLIAMS, G. C. *Sex and Evolution.* Princeton: Princeton University Press, 1975.

2 | Sexual Development

ADER, R. Early experience and hormones: emotional behavior and adrenocortical function. In *Hormonal Correlates of Behavior,* eds. B. E. Eleftheriou and R. L. Sprott, Vol. 1, pp. 7–33. New York: Plenum Press, 1975.

ADKINS, E. A. Sex steroids and the differentiation of avian reproductive behavior. *Am. Zool.* 18:501–509, 1978.

ADKINS, E. K. Hormonal basis of sexual differentiation in the Japanese quail. *J. Comp. Physiol. Psychol.* 89:61–71, 1975.

ANDERSON, C. O., M. X. ZARROW, and V. H. DENENBERG. Maternal behavior in the rabbit. Effects of androgen treatment during gestation upon the nest-building behavior of the mother and her offspring. *Horm. Behav.* 1:337–345, 1970.

ARAI, Y. Advancement of masculine differentiation of the brain by synthetic LH-releasing hormone (LH-RH) in the male rat. *Endocrinol. Jpn.* 21:121–123, 1974.

ARAI, Y., and K. SERISAWA. Effect of gonadotropins on neonatal testicular activity and sexual differentiation of the brain in the rat. *Proc. Soc. Exp. Biol. Med.* 143:656–660, 1973.

AREY, L. B. *Developmental Anatomy,* fifth ed. Philadelphia: W. B. Saunders, 1946.

ARNOLD, A. P. Sexual differences in the brain. *Am. Sci.* 68:165–173, 1980.

AZIZI, F., A. G. VAGENAKIS, J. BOLLINGER, S. REICHLIN, L. E. BRAVERMAN, and S. H. INGBAR. Persistent abnormalities in pituitary function following neonatal thyrotoxicosis in the rat. *Endocrinology* 94:1681–1688, 1974.

BAKER, T. G. Primordial germ cells. In *Reproduction in Mammals,* eds. C. R. Austin and R. V. Short, Vol. 1, pp. 1–13, Cambridge: Cambridge University Press, 1972.

BAKKE, J. L., N. L. LAWRENCE, J. BENNETT, and S. ROBINSON. The late effects of neonatal hyperthyroidism upon the feedback regulation of TSH secretion in rats. *Endocrinology* 97:659–664, 1975.

BALLARD, W. W. *Comparative Anatomy and Embryology.* New York: Ronald Press, 1964.

BARNEA, A., and H. R. LINDNER. Short-term inhibition of macromolecular synthesis and androgen-induced sexual differentiation of the rat brain. *Brain Res.* 45:479–487, 1972.

BARNETT, S. A., and J. BURN. Early stimulation and maternal behaviour. *Nature* 213:150–152, 1967.

BARRACLOUGH, C. A. Modifications in the CNS regulation of reproduction after exposure of prepubertal rats to steroid hormones. *Recent Prog. Horm. Res.* 22:503–529, 1966.

———. Sex steroid regulation of reproductive neuroendocrine processes. In *Handbook of Physiology,* Sec. 7, Vol. 2, Part 1, pp. 29–56. Washington, D.C.: American Physiological Society, 1973.

BEACH, F. A. Effects of gonadal hormones on urinary behavior in dogs. *Physiol. Behav.* 12:1005–1013, 1974.

BEACH, F. A., and R. E. KUEHN. Coital behavior in dogs. X. Effects of androgenic stimulation during development on feminine mating responses in females and males. *Horm. Behav.* 1:346–367, 1970.

BEACH, F. A., R. E. KUEHN, R. H. SPRAGUE, and J. J. ANISKO. Coital behavior in dogs. XI. Effects of androgen stimulation during development on masculine mating responses in females. *Horm. Behav.* 3:143–168, 1972.

BEATTY, W. W., and P. A. BEATTY. Hormonal determinants of sex differences in avoidance behavior and reactivity to electric shock in the rat. *J. Comp. Physiol. Psychol.* 73:446–455, 1970.

BELLEC, A. and J. STOLKOWSKI. Influence du rapport potassium/calcium (K^+/Ca^{++}) du milieu d'élevage sur la distribution des sexes chez *Discoglossus pictus* OTTH—nouvelles observations. *Ann. Endocrinol.* 26:51–64, 1965.

BLACKER, A. W. Germ-cell transfer and sex ratio in *Xenopus laevis. J. Embryol. Exp. Morphol.* 13:51–61, 1965.

BLACKLER, A. W., and C. A. GECKING. Transmission of sex cells of one species through the body of a second species in the genus *Xenopus.* I. Intraspecific matings. *Dev. Biol.* 27:376–384, 1972.

BLANCHARD, M. G., and N. JOSSO. Source of anti-Müllerian hormone synthesized by the fetal testis: Müllerian-inhibiting activity of fetal bovine Sertoli cells in tissue culture. *Pediatr. Res.* 8:968–971, 1974.

BLIZARD, D., and C. DENEF. Neonatal androgen effects on open-field activity and sexual behavior in the female rat: the modifying influence of ovarian secretions during development. *Physiol. Behav.* 11:65–69, 1973.

BRAYSHAW, J. S., and H. H. SWANSON. The effects of implanting testosterone propionate into the brain of neonatal female hamsters on their sexual behaviour when adult. *J. Endocrinol.* 57: XI, 1973.

BREUER, J., and G. Köster. Interaction between estrogens and neurotransmitters. Biochemical mechanism. *Adv. Biosci.* 15:287–298, 1975.

BRONSON, F. H., and C. DESJARDINS. Steroid hormones and aggressive behavior in mammals. In *The Physiology of Aggression and Defeat,* eds. B. Eleftheriou and J. P. Scott, pp. 43–63, New York: Plenum Press, 1971.

BRONSON, F. H., J. M. WHITSETT, and T. H. HAMILTON. Responsiveness of accessory glands of adult mice to testosterone: priming with neonatal injections. *Endocrinology* 90:10–16, 1972.

BROWN-GRANT, K. A re-examination of the lordosis response in female rats given high doses of testosterone propionate or estradiol benzoate in the neonatal period. *Horm. Behav.* 6:351–378, 1975.

BROWN-GRANT, K., and M. R. SHERWOOD. The 'early androgen syndrome' in the guinea-pig. *J. Endocrinol.* 49:277–291, 1971.

BULL, J. J., and R. C. VOGT. Temperature-dependent sex determination in turtles. *Science* 206:1186–1188, 1979.

BURNS, R. K. Role of hormones in the differentiation of sex. In *Sex and Internal Secretions,* ed. W. C. Young, Vol. 1, pp. 76–158. Baltimore: Williams and Wilkins, 1961.

CALLARD, G. V., Z. PETRO, and K. J. RYAN. Conversion of androgen to estrogen and other steroids in the vertebrate brain. *Am. Zool.* 18:511–523, 1978.

CHUNG, L. W. K., and G. FERLAND-RAYMOND. Differences among rat accessory glands in their neonatal androgen dependency. *Endocrinology* 97:145–153, 1975.

CLARKE, I. J. The sexual behaviour of prenatally androgenized ewes observed in the field. *J. Reprod. Fertil.* 49:311–315, 1977.

CLAYTON, R. B., J. KOGURA, and H. C. KRAEMER. Sexual differentiation of the brain: effects of testosterone on brain RNA metabolism in newborn female rats. *Nature* 226:810–812, 1970.

CLEMENS, L. G. Neurohormonal control of male sexual behavior. *Adv. Behav. Biol.* 11:23–53, 1974.

CUMINGE, D., and R. DUBOIS. Méchanisme de pénétration des gonocytes primordiaux dans les épithéliums germinatifs attractifs, chez l'embryon de poulet. *C. R. Hebd. Séances Acad. Sci. Ser. D Sci. Nat. (Paris)* 268:1200–1202, 1969.

CUNNINGHAM, N. F., N. SABA, and C. D. BOARER. The acute effects of oestradiol-17β and synthetic LH-RH on plasma LH levels in freemartin heifers. *J. Reprod. Fertil.* 51:29–33, 1977.

DAVENPORT, J. W., and L. M. GONZALEZ. Neonatal thyroxine stimulation in rats. *J. Comp. Physiol. Psychol.* 85:397–408, 1973.

DEBOLD, J. F., and R. E. WHALEN. Differential sensitivity of mounting and lordosis control systems to early androgen treatment in male and female hamsters. *Horm. Behav.* 6:197–209, 1975.

DENEF, C., and P. DEMOOR. Sexual differentiation of steroid metabolizing enzymes in the rat liver. Further studies on predetermination by testosterone at birth. *Endocrinology* 91:374–384, 1972.

DENENBERG, V. H., J. T. BRUMAGHIM, G. C. HALTMEYER, and M. X. ZARROW. Increased adrenocortical activity in the neonatal rat following handling. *Endocrinology* 81:1047–1052, 1967.

DENENBERG, V. H., D. DESANTIS, S. WAITE, and E. B. THOMAN. The effects of handling in infancy on behavioral states in the rabbit. *Physiol. Behav.* 18:553–557, 1977.

DENENBERG, V. H., and K. M. ROSENBERG. Nongenetic transmission of information. *Nature* 216:549–550, 1967.

DENENBERG, V. H., K. M. ROSENBERG, and M. X. ZARROW. Mice reared with rat aunts: effects in adulthood upon plasma corticosterone and open field activity. *Physiol. Behav.* 4:705–707, 1969.

DENENBERG, V. H., M. V. WYLY, J. K. BURNS, and M. X. ZARROW. Behavioral effects of handling rabbits in infancy. *Physiol. Behav.* 10:1001–1004, 1973.

DIXIT, V. P., and M. NIEMI. Action of testosterone administered neonatally on the rat perineal complex. *J. Endocrinol.* 59:379–380, 1973.

DÖRNER, G., and G. HINZ. Androgen dependent brain differentiation and life span. *Endokrinologie* 65:378–380, 1975.

DÖRNER, G., G. HINZ, F. DÖCKE, and R. TÖNJES. Effects of psychotrophic drugs on brain differentiation in female rats. *Endokrinologie* 70:113–123, 1977.

DUBOIS, R. Sur l'attraction exercée par le jeune épithélium germinatif sur les gonocytes primaires de l'embryon de poulet en culture *in vitro:* démonstration à l'aide de la thymidine tritiée. *C. R. Hebd. Séances Acad. Sci. Ser. D Sci. Nat. (Paris)* 260:5885–5887, 1965.

———. Localisation et migration des cellules germinales du blastoderme non incubés de poulet d'après les résultats de culture *in vitro. Arch. Anat. Microsc. Morphol. Exp.* 56:245–264, 1967.

———. Le mécanisme d'entrée des cellules germinales primordiales dans le réseau vasculaire, chez l'embryon de poulet. *J. Embryol. Exp. Morphol.* 21:255–270, 1969.

DUBOIS, R., and Y. CROISILLE. Germ-cell line and sexual differentiation in birds. *Philos. Trans. R. Soc. Lond. B Biol. Sci.* 259:73–89, 1970.

DUBOIS, R., and D. CUMINGE. Chimotactisme et organisation biologique: étude de l'installation de la lignée germinale dans les ébauches gônadiques chez l'embryon de poulet. *Année Biol.* 13:241–269, 1974.

DUFAURE, J. P. Recherches descriptives et expérimentales sur les modalités et facteurs du développement de l'appareil génital chez le lézard vivipare (*Lacerta vivipara* Jacquin). *Arch. Anat. Microsc. Morphol. Exp.* 55:437–537, 1966.

DYER, R. G., F. ELLENDORFF, and N. K. MACLEOD. Electrophysiology of endocrine sex. *J. Physiol. (Lond.)* 254:18P–19P, 1976.

EDWARDS, D. A. Neonatal administration of androstenedione, tetosterone or testosterone propionate: effects on ovulation, sexual receptivity and aggressive behavior in female mice. *Physiol. Behav.* 6:223–228, 1971.

EVANS, E. P., C. E. FORD, and M. F. LYON. Direct evidence of the capacity of the XY germ cell in the mouse to become an oocyte. *Nature* 267:430–431, 1977.

FARGEIX, N. La colonisation des gonades par les cellules germinales dans les cas de gémellité expérimentale chez l'embryon de caille (*Coturnix coturnix japonica*). *C. R. Hebd. Séances Acad. Sci. Ser. D Sci. Nat. (Paris)* 285:725–728, 1977.

FEDER, H. H., and G. N. WADE. Integrative actions of perinatal hormones on neural tissues mediating adult sexual behavior. In *The Neurosciences, Third Study Program,* eds. F. O. Schmitt and F. G. Worden, pp. 583–586. Cambridge: M.I.T. Press, 1974.

FORD, F. C., E. P. EVANS, M. D. BURTENSHAW, H. M. CLEGG, M. TUFFREY, and R. D. BARNES. A functional 'sex reversed' oocyte in the mouse. *Proc. R. Soc. Lond. B Biol. Sci.* 190:189–197, 1975.

FUGO, N. W. Effects of hypophysectomy in the chick embryo. *J. Exp. Zool.* 85:271–297, 1940.

FUJIMOTO, T., A. UKESHIMA, and R. KIYOFUJI. The origin, migration and morphology of the primordial germ cells in the chick embryo. *Anat. Rec.* 185:139–154, 1976.

FULLER, G. B., M. X. ZARROW, C. O. ANDERSON, and V. H. DENENBERG. Testosterone propionate during gestation in the rabbit. Effect on subsequent maternal behaviour. *J. Reprod. Fertil.* 23:285–290, 1970.

GALLI, F. E., and G. F. WASSERMANN. Steroid biosynthesis by gonads of 7- and 10-day-old chick embryos. *Gen. Comp. Endocrinol.* 21:77–83, 1973.

GERALL, A. A., M. M. MCMURRAY, and A. FARRELL. Suppression of the development of the female hamster behaviour by implants of testosterone and nonaromatizable androgens administered neonatally. *J. Endocrinol.* 67:439–445, 1975.

GOLDFOOT, D. A., and J. J. VAN DER WERFF TEN BOSCH. Mounting behavior of female guinea pigs after prenatal and adult administration of the propionates of testosterone, dihydrotestosterone and androstanediol. *Horm. Behav.* 6:139-148, 1975.

GORSKI, R. A. Gonadal hormones and the perinatal development of neuroendocrine function. In *Frontiers in Neuroendocrinoloy 1971,* eds. L. Martini and W. F. Ganong, pp. 237-290. New York: Academic Press, 1971.

GORSKI, R. A., and B. D. GOLDMAN. Antigonadotropin treatment and sexual differentiation of the brain in the male rat. *Physiologist* 14:153, 1971 (Abstract).

GORSKI, R. A., and J. SHRYNE. Intracerebral antibiotics and androgenization of the neonatal female rat. *Neuroendocrinology* 10:109-120, 1972.

GOY, R. W., and D. A. GOLDFOOT. Hormonal influences on sexual dimorphic behavior. In *Handbook of Physiology* Sec. 7, Vol. 2, Part 1, pp. 169-186, Washington, D.C.: American Physiological Society, 1973.

_____. Neuroendocrinology: animal models and problems of human sexuality. *Archiv. Sex. Behav.* 4:405-420, 1975.

GOY, R. W., and J. A. RESKO. Gonadal hormones and behavior of normal and pseudohermaphroditic nonhuman female primates. *Recent Prog. Horm. Res.* 28:707-733, 1972.

GREENE, W. A., L. MOGIL, and R. H. FOOTE. Behavioral characteristics of freemartins administered estradiol, estrone, testosterone and dihydrotestosterone. *Horm. Behav.* 10:71-84, 1978.

GROPP, A., and S. OHNO. The presence of a common embryonic blastema for ovarian and testicular parenchymal (follicular, interstitial and tubular) cells in cattle, *Bos Taurus. Z. Zellforsch. Mikrosk. Anat.* 74:505-528, 1966.

GRIFFITHS, E. C., and K. C. HOOPER. The effect of neonatal androgen on the activity of certain enzymes in the rat hypothalamus. *Acta Endocrinol.* 70:767-774, 1972.

_____. The effects of orchidectomy and testosterone propionate injection on peptidase activity in the male rat hypothalamus. *Acta Endocrinol.* 72:1-8, 1973.

GUSTAFSSON, J. Å., M. INGELMAN-SUNDBERG, and A. STERNBERG. Neonatal androgenic programming of hepatic steroid metabolism in rats. *J. Steroid Biochem.* 6:643-649, 1975.

HAFFEN, K. Intersexualité chez la caille (*Coturnix coturnix*). Obtention d'un cas de ponte ovulaire par un mâle génétique. *C. R. Hebd. Séances Acad. Sci. Ser. D Sci. Nat.* (Paris) 261:3876-3879, 1965.

HAMERTON, J. L. Significance of sex chromosome derived heterochromatin in mammals. *Nature* 219:910-914, 1968.

HAMERTON, J. L., J. M. DICKSON, C. E. POLLARD, S. A. GRIEVES, and R. V. SHORT. Genetic intersexuality in goats. *J. Reprod. Fertil. Suppl.* 7, pp. 25-51, 1969.

HARDIN, C. M. Sex differences and the effects of testosterone injections on biogenic amine levels of neonatal rat brain. *Brain Res.* 62:286-290, 1973.

HARDISTY, M. W. Sex differentiation and gonadogenesis in lampreys. I. *J. Zool. (Lond.)* 146:305-345, 1965a.

_____. Sex differentiation and gonadogenesis in lampreys. II. *J. Zool. (Lond.)* 146:346-387, 1965b.

_____. Gonadogenesis, sex differentiation and gametogenesis. In *The Biology of Lampreys,* eds. M. W. Hardisty and I. C. Potter, Vol. 1, pp. 295-357. New York: Academic Press, 1971.

HARLAN, R. E., and R. A. GORSKI. Steroid regulation of luteinizing hormone secretion in normal and androgenized rats at different ages. *Endocrinology* 101:741-749, 1977a.

_____. Correlations between ovarian sensitivity, vaginal cyclicity and luteinizing hormone and prolactin secretion in lightly androgenized rats. *Endocrinology* 101:750-759, 1977b.

HARRINGTON, R. W., JR. Sex determination and differentiation in fishes. In *Control of Sex in Fishes,* ed. C. B. Schrek, pp. 4-12. Blacksburg: Virginia Polytechnic Institute and State University, 1974.

_____. Sex determination and differentiation among uniparental homozygotes of the hermaphroditic fish *Rivulus marmoratus* (Cyprinodontidae: Atheriformes). In *Intersexuality in the Animal Kingdom,* ed. R. Reinboth, pp. 249-262. New York: Springer, 1975.

HART, B. L. Neonatal castration: influence on neural organization of sexual reflexes in male rats. *Science* 160:1135-1136, 1968.

_____. Manipulation of neonatal androgen: effects on sexual responses and penile development in male rats. *Physiol. Behav.* 8:841-845, 1972.

HAYASHI, S., and R. A. GORSKI. Critical exposure time for androgenization by intracranial crystals of testosterone propionate in neonatal female rats. *Endocrinology* 94:1161-1167, 1974.

HEFFNER, L. J., and A. van TIENHOVEN. Effects of neonatal ovariectomy upon ^{3}H-estradiol uptake by target tissues of androgen-sterilized female rats. *Neuroendocrinology* 29:237-246, 1979.

HENRY, J. L., and F. R. CALARESU. Topography and numerical distribution of neurons of the thoracolumbar intermedio-lateral nucleus in the cat. *J. Comp. Neurol.* 144:205-214, 1972.

HINZ, G., G. SCHLENKER, and G. DÖRNER. Pränatale Behandlung von Schweinen mit Testosteronpropionat. *Endokrinologie* 63:161-165, 1974.

HOAR, W. S. Reproduction. In *Fish Physiology,* eds. W. S. Hoar and D. J. Randall, Vol. 3, pp. 1-72. New York: Academic Press, 1969.

IWASAWA, H. Effects of thyroidectomy, hypophysectomy and thiourea treatment on the development of gonads in frog larvae. *Jpn. J. Zool.* 13:69-77, 1961.

JOHNSON, W. A. Neonatal androgenic stimulation and adult sexual behavior in male and female golden hamsters. *J. Comp. Physiol. Psychol.* 89:433–441, 1975.

JOST, A. Problems of fetal endocrinology: the gonadal and hypophyseal hormones. *Recent Prog. Horm. Res.* 8:379–413, 1953.

KARSCH, F. J., D. J. DIERSCHKE, and E. KNOBIL. Sexual differentiation of pituitary function; apparent difference between primates and rodents. *Science* 179:484–486, 1973.

KARSCH, F. J., and D. L. FOSTER. Sexual differentiation of the mechanism controlling the preovulatory discharge of luteinizing hormone in sheep. *Endocrinology* 97:373–379, 1975.

KAWAKAMI, M., and E. TERASAWA. A possible role of the hippocampus and the amygdala in the androgenized rat: effect of electrical or electrochemical stimulation of the brain on gonadotropin secretion. *Endocrinol. Jpn.* 19:349–358, 1972.

KELLEY, D. B., and D. W. PFAFF. Hormone effects on male sex behavior in adult South African clawed frogs, *Xenopus laevis. Horm. Behav.* 7:159–182, 1976.

KINCL, F. A., and M. MAQUEO. Prevention by progesterone of steroid-induced sterility in neonatal male and female rats. *Endocrinology* 77:859–862, 1965.

KING, J. A., and B. ELEFTHERIOU. Effects of early handling upon adult behavior in two subspecies of deer mice, *Peromyscus maniculatus. J. Comp. Physiol. Psychol.* 52:82–88, 1959.

KORENBROT, C. C., D. C. PAUP, and R. A. GORSKI. Effects of testosterone propionate or dihydrotestosterone propionate on plasma FSH and LH levels in neonatal rats and on sexual differentiation of the brain. *Endocrinology* 97:709–717, 1975.

KRAMEN, M. A., and D. C. JOHNSON. Uterine decidualization in rats given testosterone propionate neonatally. *J. Reprod. Fertil.* 42:559–562, 1975.

KUBO, K., S. P. MENNIN, and R. A. GORSKI. Similarity of plasma LH release in androgenized and normal rats following electrochemical stimulation of the basal forebrain. *Endocrinology* 96:492–500, 1975.

LEE, C. T., and W. GRIFFO. Early androgenization and aggression pheromone in inbred mice. *Horm. Behav.* 4:181–189, 1973.

LESHNER, A. I., and S. M. SCHWARTZ. Neonatal corticosterone treatment increases submissiveness in adulthood in mice. *Physiol. Behav.* 19:163–165, 1977.

LEVINE, S. Plasma-free corticosteroid response to electric shock in rats stimulated in infancy. *Science* 135:795–796, 1962.

———. Neuroendocrine factors and the ontogeny of behaviour. In *Normal and Abnormal Development of Brain and Behaviour,* eds. G. B. A. Stoelinga and J. J. van der Werff ten Bosch, pp. 284–294. Baltimore: Williams and Wilkins, 1971.

LEVINE, S., M. ALPERT, and G. W. LEWIS. Infantile experience and the maturation of the pituitary adrenal axis. *Science* 126:1347, 1957.

LEVINE, S., and C. COHEN. Differential survival to leukemia as a function of infantile stimulation in DBA/2 mice. *Proc. Soc. Exp. Biol. Med.* 102:53–54, 1959.

LITTERIA, M., and G. S. O'BRIEN. Long-lasting inhibitory effect of neonatal estrogenization on incorporation of ³H-lysine in specific hypothalamic nuclei of adult rats. *Fed. Proc.* 34:340, 1975 (Abstract).

LOBL, R. T., and R. A. GORSKI. Neonatal intrahypothalamic androgen administration: the influence of dose and age on androgenization of female rats. *Endocrinology* 94:1325–1330, 1974.

LOBL, R. T., and R. M. MAENZA. Androgenization: alterations in uterine growth and morphology. *Biol. Reprod.* 13:255–268, 1975.

———. The ontogeny of uterine pathology and pathophysiology following neonatal androgen administration. *Biol. Reprod.* 16:182–189, 1977.

LOFTS, B. Reproduction. In *Physiology of the Amphibia,* ed. B. Lofts, Vol. 2, pp. 107–218. New York: Academic Press, 1974.

LUTZ-OSTERTAG, Y. Action de la chaleur sur le développement de l'appareil génital de l'embryon de caille *Coturnix coturnix japonica. C. R. Hebd. Seánces Acad. Sci. Ser. D Sci. Nat. (Paris)* 262:133–135, 1965.

MALLAMPATI, R. S., and D. C. JOHNSON. Gonadotropins in female rats androgenized by various treatments: prolactin as an index to hypothalamic damage. *Neuroendocrinology* 15:255–266, 1974.

MANNING, A., and T. E. McGILL. Neonatal androgen and sexual behavior in female house mice. *Horm. Behav.* 5:19–31, 1974.

McCULLOUGH, J., D. M. QUADAGNO, and B. D. GOLDMAN. Neonatal gonadal hormones: effect on maternal and sexual behavior in the male rat. *Physiol. Behav.* 12:183–188, 1974.

McEWEN, B. S. Neural gonadal steroid actions. *Science* 211:1303–1311, 1981.

McEWEN, B. S., C. J. DENEF, J. L. GERLACH, and L. PLAPINGER. Chemical studies of the brain as a steroid hormone target tissue. In *The Neurosciences, Third Study Program,* eds. F. O. Schmitt and F. G. Worden, pp. 599–620. Cambridge: M.I.T. Press, 1974.

McEWEN, B. L., L. PLAPINGER, C. CHAPTAL, J GERLACH, and G. WALLACH. Role of fetoneonatal estrogen binding proteins in the associations of estrogen with neonatal brain cell nuclear receptors. *Brain Res.* 96:400–406, 1975.

MENNIN, S. P., and R. A. GORSKI. Effects of ovarian steroids on plasma LH in normal and persistent estrous adult female rats. *Endocrinology* 96:486–491, 1975.

MENNIN, S. P., K. KUBO, and R. A. GORSKI. Pituitary responsiveness to luteinizing hormone-releasing factor in normal and androgenized female rats. *Endocrinology* 95:412–416, 1974.

MITTWOCH, U. *Genetics of Sex Differentiation.* New York: Academic Press, 1973.

———. Chromosomes and sex differentiation. In *Intersexuality in the Animal Kingdom,* ed. R. Reinboth, pp. 438–446. New York: Springer, 1975.

MONEY, J., and A. A. EHRHARDT. Prenatal hormonal exposure: possible effects on behaviour in man. In *Endocrinology and Human Behaviour,* ed. R. P. Michael, pp. 32–48. London: Oxford University Press, 1968.

MYERSON, B. J. Drugs and sexual motivation in the female rat. In *Sexual Behavior: Pharmacology and Biochemistry,* eds. M. Sandler and G. L. Gessa, pp. 21–31. New York: Raven Press, 1975.

NADLER, R. D. Intrahypothalamic exploration of androgen-sensitive brain loci in neonatal female rats. *Trans. N.Y. Acad. Sci.* 34:572–581, 1972.

NAFTOLIN, F., K. BROWN-GRANT, and C. S. CORKER. Plasma and pituitary luteinizing hormone and peripheral plasma oestradiol concentrations in the normal oestrous cycle of the rat and after experimental manipulation of the cycle. *J. Endocrinol.* 53:17–30, 1972.

NAFTOLIN, F., and K. J. RYAN. The metabolism of androgens in central neuroendocrine tissues. *J. Steroid Biochem.* 6:993–997, 1975.

NAFTOLIN, F., K. J. RYAN, I. J. DAVIES, V. V. REDDY, F. FLORES, Z. PETRO, M. KUHN, R.J. WHITE, Y. TAKAOKA, and L. WOLIN. The formation of estrogens by central neuroendocrine tissues. *Recent Prog. Horm. Res.* 31:295–315, 1975.

NAMIKI, H., W. RUCH, and A. GORBMAN. Further studies of qualitative changes in RNA transcription in brains of female rats given testosterone soon after birth. *Comp. Biochem. Physiol.* 42(B):563–568, 1972.

NEUMANN, F., W. ELGER, and H. STEINBECK. Drug-induced intersexuality in mammals. *J. Reprod. Fertil. Suppl.* 7:9–24, 1969.

NISHIZUKA, M. Neuropharmacological study on the induction of hypothalamic masculinization in female mice. *Neuroendocrinology* 20:157–165, 1976.

NOBLE, R. G. Mounting in female hamsters: effects of different hormone regimens. *Physiol. Behav.* 19:519–526, 1977.

O, W. S., and T. G. BAKER. Germinal and somatic cell interrelationships in gonadal sex differentiation. *Ann. Biol. Anim. Biochim. Biophys.* 18:351–357, 1978.

OHNO, S. *Sex Chromosomes and Sex-linked Genes.* New York: Springer, 1967.

———. Homology of X-linked genes in mammals and evolution of sex determining mechanism. In *Proc. Fourth Symp. Comparative Physiology of Reproduction, Canberra, 1976,* pp. 19–53. Canberra: Australian Academy of Science, 1977.

OHNO, S., Y. NAGAI, and S. CICCARESE. Testicular

cells lysostripped of H-Y antigen organize ovarian follicle-like aggregates. *Cytogenet. Cell Genet.* 20:351–364, 1978.

PAYNE, A. P., and H. H. SWANSON. The effects of neonatal androgen administration on the aggression and related behaviour of male golden hamsters during interactions with females. *J. Endocrinol.* 58:627–636, 1973.

PFEIFFER, C. A. Sexual differences of the hypophyses and their determination by the gonads. *Am. J. Anat.* 58:195–225, 1936.

PHOENIX, C. H., R. W. GOY, A. A. GERALL, and W. C. YOUNG. Organizing action of prenatally administered testosterone propionate on the tissues mediating mating behavior in the female guinea pig. *Endocrinology* 65:369–382, 1959.

PIEAU, C. Sur la proportion sexuelle chez les embryons de deux chéloniens (*Testudo graeca* L. et *Emys orbicularis* L.) issus d'oeufs incubés artificiellement. *C. R. Hebd. Séances Acad. Sci. Ser. D Sci. Nat. (Paris)* 272:3071–3074, 1971.

———. Nouvelles données expérimentales concernant les effets de la température sur la différenciation sexuelle chez les embryons de chéloniens. *C. R. Hebd. Séances Acad. Sci. Ser. D Sci. Nat. (Paris)* 277:2789–2792, 1973.

POLLAK, E. I., and B. D. SACHS. Masculine sexual behavior and morphology: paradoxical effects of perinatal androgen treatment in male and female rats. *Behav. Biol.* 13:401–411, 1975.

PRICE, D., J. J. P. ZAAIJER, and E. ORTIZ. Prenatal development of the oviduct in vivo and in vitro. In *The Mammalian Oviduct,* eds. E. S. E. Hafez and R. J. Blandau, pp. 29–46. Chicago: University of Chicago Press, 1969.

PRICE, D., J. J. P. ZAAIJER, E. ORTIZ, and A. O. BRINKMANN. Current views on embryonic sex differentiation in reptiles, birds, and mammals. *Am. Zool.* 15 (*Suppl.* 1):173–195, 1975.

RAISMAN, G., and P. M. FIELD. Sexual dimorphism in the neuropil of the preoptic area of the rat and its dependence on neonatal androgen. *Brain Res.* 54:1–29, 1973.

REDDY, V. V. R., F. NAFTOLIN, and K. J. RYAN. Conversion of androstenedione to estrone by neural tissues from fetal and neonatal rats. *Endocrinology* 94:117–121, 1974.

REMACLE, C., P. DELAERE, and P. JACQUET. Actions hormonales sur les cellules germinales femelles de *Carassius auratus* L. en culture organotypique. Renversement sexuel et ovogenèse in vitro. *Gen. Comp. Endocrinol.* 29:212–224, 1976.

RESKO, J. A. Fetal hormones and their effect on the differentiation of the central nervous system in primates. *Fed. Proc.* 34:1650–1655, 1975.

RESKO, J. A., H. H. FEDER, and R. W. GOY, Androgen concentrations in plasma and testes of developing rats. *J. Endocrinol.* 40:485–491, 1968.

ROSENBERG, K. M., V. H. DENENBERG, M. X. ZAR-

ROW, and B. L. FRANK. Effects of neonatal castration and testosterone on the rat's pup-killing behavior and activity. *Physiol. Behav.* 7:363–368, 1971.

SALAMAN, D. F., and S. BIRKETT. Androgen-induced sexual differentiation of the brain is blocked by inhibitors of DNA and RNA synthesis. *Nature* 247:109–112, 1974.

SHERIDAN, P. J., M. X. ZARROW, and V. H. DENENBERG. The role of gonadotropins in the development of cyclicity in the rat. *Endocrinology* 92: 500–508, 1973.

SHORT, R. V. New thoughts on sex determination and differentiation. *Glaxo Vol.* 39:5–20, 1974.

SIMON, E. D. Contribution à l'étude de la circulation et du transport, des gonocytes primaires dans les blastodermes d'oiseau cultivés in vitro. *Arch. Anat. Microsc. Morphol. Exp.* 49:93–176, 1960.

STANLEY, A. J., and E. WITSCHI. Germ cell migration in relation to asymmetry in the sex glands of hawks. *Anat. Rec.* 76:329–342, 1940.

STEINER, R. A., D. K. CLIFTON, H. G. SPIES, and J. A. RESKO. Sexual differentiation and feedback control of luteinizing hormone secretion in the rhesus monkey. *Biol. Reprod.* 15:206–212, 1976.

SUTHERLAND, S. D., and R. A. GORSKI. An evaluation of the inhibition of androgenization of the neonatal female rat brain by barbiturate. *Neuroendocrinology* 10:94–108, 1972.

SWANSON, H. H. Effects of castration at birth in hamsters of both sexes on luteinization of ovarian implants, oestrous cycles and sexual behaviour. *J. Reprod. Fertil.* 21:183–186, 1970.

_____. Determination of the sex role in hamsters by the action of sex hormones in infancy. In *Influence of Hormones on the Nervous System. Proc. Int. Soc. Psychoneuroendocrinology, Brooklyn, 1970,* pp. 424–440. Basel: Karger, 1971.

SWARTZ, W. J. Effect of steroids on definitive localization of primordial germ cells in the chick. *Am. J. Anat.* 142:499–514, 1975.

_____. Effect of cyproterone acetate on primordial germ cell colonization of gonads in the chick embryo. *Gen. Comp. Endocrinol.* 32:474–480, 1977.

TABEI, T., and W. L. HEINRICHS. Enzymatic oxidation and reduction of C_{19}-Δ^5-3β-hydroxysteroids by hepatic microsomes. III. Critical period for the neonatal differentiation of certain mixed-function oxidases. *Endocrinology* 94:97–103, 1974.

_____. Enzymatic oxidation and reduction of C_{19}-Δ^5-3β-hydroxysteroids by hepatic microsomes. V. Testosterone as a neonatal determinant in rats of the 7- and 16α-hydroxylation and reduction of 3β-hydroxyandrost-5-en-17-one (DHA). *Endocrinology* 97:418–424, 1975.

TACHINANTE, F. Sur les échanges interspécifiques de cellules germinales entre le poulet et la caille, en culture organotypique et en greffes coelomiques. *C. R. Hebd. Séances Acad. Sci. Ser. D Sci. Nat. (Paris)* 278:1895–1898, 1973.

_____. Sur l'activité migratrice des cellules germinales de souris soumises à l'attraction de l'épithélium germinatif de poulet en culture in vitro. *C. R. Hebd. Séances Acad. Sci. Ser. D Sci. Nat. (Paris)* 278:3135–3138, 1974.

TAKASUGI, N. Cytological basis for permanent vaginal changes in mice treated neonatally with steroid hormones. *Int. Rev. Cytol.* 44:193–224, 1976.

THIEBOLD, J. Some effects of embryonic gonad and non-gonadal grafts in the development of primary sexual characteristics in the chick. In *Intersexuality in the Animal Kingdom,* ed. R. Reinboth, pp. 375–381. New York: Springer, 1975.

TORAN-ALLERAND, C. D. Sex steroids and the development of the newborn mouse hypothalamus and preoptic area *in vitro:* implications for sexual differentiation. *Brain Res.* 106:407–412, 1976.

_____. Gonadal hormones and brain development: cellular aspects of sexual differentiation. *Am. Zool.* 18:553–565, 1978.

TORREY, T. W. Intraocular grafts of embryonic gonads of the rat. *J. Exp. Zool.* 115:37–57, 1950.

TRAN, D., and N. JOSSO. Relationship between avian and mammalian anti-Müllerian hormones. *Biol. Reprod.* 16:267–273, 1977.

TRAN, D., N. MEUSY-DESSOLLE, and N. JOSSO. Anti-Müllerian hormone is a functional marker of foetal Sertoli cells. *Nature* 269:411–412, 1977.

TURNER, C. D. *General Endocrinology.* Philadelphia: W. B. Saunders, 1966.

TURNER, J. W., JR. Influence of neonatal androgen on the display of territorial marking behavior in the gerbil. *Physiol. Behav.* 15:265–270, 1975.

ULRICH, R., A. YUWILER, and E. GELLER. Neonatal hydrocortisone: effect of the development of the stress response and diurnal rhythm of corticosterone. *Neuroendocrinology* 21:49–57, 1976.

VAN LIMBORGH, J. La répartition numérique des cellules germinales chez des embryons de canard génétiquement femelles et mâles aux premiers stades postsomitiques. *Arch. Anat. Microsc. Morphol. Exp.* 55:423–436, 1966.

_____. Le premier indice de la différenciation sexuelle des gonades chez l'embryon de poulet. *Arch. Anat. Microsc. Morphol. Exp.* 53:79–90, 1968.

VAN TIENHOVEN, A. A method of "controlling sex" by dipping of eggs in hormone solutions. *Poult. Sci.* 36:628–632, 1957.

_____. *Reproductive Physiology of Vertebrates.* Philadelphia: W. B. Saunders, 1968.

VOM SAAL, F. S., and F. H. BRONSON. In utero proximity of female mouse fetuses to males: effect on reproductive performance during later life. *Biol. Reprod.* 19:842–853, 1978.

VREEBURG, J. T. M., P. D. M. VAN DER VAART, and P. VAN DER SCHOOT. Prevention of central defeminization but not masculinization in male rats by inhibition neonatally of oestrogen biosynthesis. *J. Endocrinol.* 74:375–382, 1977.

WADE, G. N., and I. ZUCKER. Taste preferences of female rats: modification by neonatal hormones, food deprivation and prior experience. *Physiol. Behav.* 4:935–943, 1969.

WARD, I. L. Prenatal stress feminizes and demasculinizes the behavior of males. *Science* 175:82–84, 1972.

———. Sexual behavior differentiation: prenatal hormone and environmental control. In *Sex Differences in Behavior,* eds. R. C. Friedman, R. M. Richart, and R. L. Vande Wiele, pp. 3–17. New York: Wiley, 1975.

WEATHERSBEE, P. S., R. L. AX, and J. R. LODGE. Caffeine-mediated changes of sex ratio in Chinese hamsters, *Cricetulus griseus. J. Reprod. Fertil.* 43:141–143, 1975.

WESTLEY, B. R., and D. F. SALAMAN. Role of oestrogen receptor in androgen-induced sexual differentiation of the brain. *Nature* 262:407–408, 1976.

WHITNEY, J. B., and L. R. HERRENKOHL. Effects of anterior hypothalamic lesions on the sexual behavior of prenatally-stressed male rats. *Physiol. Behav.* 19:167–169, 1977.

WHITSETT, J. M., E. W. IRWIN, F. W. EDENS, and J. P. THAXTON. Demasculinization of male Japanese quail by prenatal estrogen treatment. *Horm. Behav.* 8:254–263, 1977.

WITSCHI, E. Studies on sex differentiation and sex determination in amphibians. III. Rudimentary hermaphroditism and Y chromosomes in *Rana temporaria. J. Exp. Zool.* 54:157–223, 1929.

———. Studies on sex differentiation and sex determination in amphibians. IV. The geographical distribution of sex races of the European grass frog (*Rana temporaria*). A contribution to the problem of evolution of sex. *J. Exp. Zool.* 56:149–165, 1930.

———. Overripeness of the egg as a cause of twinning and teratogenesis: a review. *Cancer Res.* 12:763–786, 1952.

———. Biochemistry of sex differentiation in vertebrate embryos. In *Biochemistry of Animal Development,* ed. R. Weber, Vol. 2, pp. 193–225. New York: Academic Press, 1968.

WOODS, J. E., E. S. PODCZASKI, L. H. ERTON, J. E. RUTHERFORD, and C. F. McCARTER. Establishment of the adenohypophyseal-testicular axis in the chick embryo. *Gen. Comp. Endocrinol.* 32:390–394, 1977.

YNTEMA, C. L. Effects of incubation temperatures on sexual differentiation of the turtle, *Chelydra serpentina. J. Morphol.* 150:453–462, 1976.

ZENZES, M. T., U. WOLF, E. GÜNTHER, and W. ENGEL. Studies on the function of H-Y antigen: dissociation and reorganization experiments on rat gonadal tissue. *Cytogenet. Cell Genet.* 20:365–372, 1978.

3 | Intersexes

ARMSTRONG, C. N. Intersexuality in man. In *Intersexuality in Vertebrates including Man,* eds. C. N. Armstrong and A. J. Marshall, pp. 349–393. New York: Academic Press, 1964.

ATZ, J. W. Intersexuality in fishes. In *Intersexuality in Vertebrates including Man,* eds. C. N. Armstrong and A. J. Marshall, pp. 145–232. New York: Academic Press, 1964.

BECKER, P., H. ROLAND, and R. REINBOTH. An unusual approach to experimental sex inversion in the teleost fish, *Betta* and *Macropodus.* In *Intersexuality in the Animal Kingdom,* ed. R. Reinboth, pp. 236–242. New York: Springer, 1975.

BRUSLÉ, J., and S. BRUSLÉ. Ovarian and testicular intersexuality in two protogymous Mediterranean groupers, *Epinephelus aeneus* and *Epinephelus guaza.* In *Intersexuality in the Animal Kingdom,* ed. R. Reinboth, pp. 222–227. New York: Springer, 1975.

CHAN, S. T. H., WAI-SUM O, and S. W. B. HUI. The gonadal and adenohypophysial functions of natural sex reversal. In *Intersexuality in the Animal Kingdom,* ed. R. Reinboth, pp. 201–221. New York: Springer, 1975.

CHARNOV, E. L., J. M. SMITH, and J. J. BULL. Why be a hermaphrodite? *Nature* 263:125–126, 1976.

CHOAT, J. H., and D. R. ROBERTSON. Protogynous hermaphroditism in fishes of the family Scaridae. In *Intersexuality in the Animal Kingdom,* ed. R. Reinboth, pp. 263–283. New York: Springer, 1975.

COLE, CH. J. Evolution of parthenogenetic species of reptiles. In *Intersexuality in the Animal Kingdom,* ed. R. Reinboth, pp. 340–355. New York: Springer, 1975.

CREW, F. A. E. Studies in intersexuality. II. Sex reversal in the fowl. *Proc. R. Soc. Lond. B Biol. Sci.* 95:256–278, 1923.

FOOTE, C. L. Intersexuality in amphibians. In *Intersexuality in Vertebrates including Man,* eds. C. N. Armstrong and A. J. Marshall, pp. 233–272. New York: Academic Press, 1964.

FORBES, T. R. Intersexuality in reptiles. In *Intersexuality in Vertebrates including Man,* eds. C. N. Armstrong and A. J. Marshall, pp. 274–283. New York: Academic Press, 1964.

FORD, C. E., and E. P. EVANS. Cytogenetic observations on XX/XY chimaeras and a reassessment of the evidence for germ cell chimaerism in heterosexual twin cattle and marmosets. *J. Reprod. Fertil.* 49:25–33, 1977.

FRANKENHUIS, M. T. Een Poging tot Autofertilisatie bij

Gallus domesticus. Ph.D. Diss., University of Utrecht, 1974.

FRICKE, H., and S. FRICKE. Monogamy and sex change by aggressive dominance in coral reef fish. *Nature* 266:830-832, 1977.

GARDNER, W. A., Jr., A. H. WOOD, and E. TABER. Demonstration of a nonestrogenic gonadal inhibitor produced by the ovary of the brown Leghorn. *Gen. Comp. Endocrinol.* 4:673-683, 1964.

GERMAN, J., J. L. SIMPSON, R. S. K. CHAGANTI, R. L. SUMMITT, L. B. REID, and I. R. MERKATZ. Genetically determined sex-reversal in 46, XY humans. *Science* 202:53-56, 1978.

GOMOT, L. Intersexuality in birds. Study of the effects of hybridization and post-embryonic ovariectomy. In *Intersexuality in the Animal Kingdom*, ed. R. Reinboth, pp. 356-374. New York: Springer, 1975.

HARRINGTON, R. W., Jr. Sex determination and differentiation among uniparental homozygotes of the hermaphroditic fish *Rivulus marmoratus* (Cyprinodontidae: Atheriniformes). In *Intersexuality in the Animal Kingdom*, ed. R. Reinboth, pp. 249-262. New York: Springer, 1975.

JACOBS, P. A. Abnormalities of the sex chromosomes in man. *Adv. Reprod. Physiol.* 1:61-91, 1966.

JOST, A., M. CHODKIEWICZ, and P. MAULÉON. Intersexualité du foetus de veau produite par des androgènes. Comparaison entre l'hormone foetale responsable du free-martinisme et l'hormone testiculaire adulte. *C. R. Hebd. Séances Acad. Sci. Ser. D Sci. Nat. (Paris)* 256:274-276, 1963.

KANNANKERIL, J. V., and L. V. DOMM. The influence of gonadectomy on sexual characters in the Japanese quail. *J. Morphol.* 126:395-412, 1968.

LILLIE, F. R. The free-martin: a study of the action of sex hormones in the foetal life of cattle. *J. Exp. Zool.* 23:371-452, 1917.

LUTZ, H., and Y. LUTZ-OSTERTAG. Intersexuality of the genital system and "free-martinism" in birds. In *Intersexuality in the Animal Kingdom*, ed. R. Reinboth, pp. 382-391. New York: Springer, 1975.

OHNO, S., L. C. CHRISTIAN, S. S. WACHTEL, and G. C. KOO. Hormone-like role of H-Y antigen in bovine freemartin gonad. *Nature* 261:597-599, 1976.

PIEAU, C. Temperature and sex differentiation in embryos of two chelonians, *Emys orbicularis* L., and *Testudo graeca* L. In *Intersexuality in the Animal Kingdom*, ed. R. Reinboth, pp. 332-339. New York: Springer, 1975.

RIDDLE, O., H. H. DUNHAM, and J. P. SCHOOLEY. Genetic hermaphroditism in a strain of pigeons. *Genetics* 27:165, 1942. (Abstract).

SMITH, C. L. The evolution of hermaphroditism in fishes. In *Intersexuality in the Animal Kingdom*, ed. R. Reinboth, pp. 295-310. New York: Springer, 1975.

TABER, E. Intersexuality in birds. In *Intersexuality in Vertebrates including Man*, eds. C. N. Armstrong and A. J. Marshall, pp. 285-310. New York: Academic Press, 1964.

TANDLER, J., and K. KELLER. Ueber das Verhalten des Chorions bei verschiedengeschlechtlicher Zwillings— graviditat des Rindes und über die Morphologie des Genitales der weiblichen Tiere, welche einer solchen Gravidität entstammen. *Deut. Tieraerztl. Woch enschr.* 19:148-149, 1911.

TANG, F., S. T. H. CHAN, and B. LOFTS. Effect of mammalian luteinizing hormone on the natural sex reversal in the rice-field eel, *Monopterus albus* (Zuiew). *Gen. Comp. Endocrinol.* 24:242-248, 1974.

———. A study on the 3β- and 17β-hydroxysteroid dehydrogenase activities in the gonads of *Monopterus albus* (Pisces: Teleostei) at various sexual phases during natural sex reversal. *J. Zool. (Lond.)* 175:571-580, 1975.

VAGUE, J., J. GUIDON, J. F. MATTEI, J. M. LUCIANI, and S. ANGELETTI. Les hommes à caryotype 46, XX. *Ann. Endocrinol.* 38:311-321, 1977.

VAN TIENHOVEN, A. Endocrinology of reproduction in birds. In *Sex and Internal Secretions*, ed. W. C. Young, Vol. 2, pp. 1088-1169. Baltimore: Williams and Wilkins, 1961.

———. *Reproductive Physiology of Vertebrates.* Philadelphia: W. B. Saunders, 1968.

VIGIER, B., A. LOCATELLI, J. PRÉPIN, F. DU MESNIL DU BUISSON, and A. JOST. Les premières manifestations du "freemartinisme" chez le foetus de veau ne dépendent pas du chimérisme chromosomique XX/XY. *C. R. Hebd. Séances Acad. Sci. Ser. D Sci. Nat. (Paris)* 282:1355-1358, 1976.

VIGIER, B., J. PRÉPIN, and A. JOST. Absence de corrélation entre le chimérisme XX/XY dans le foie et les premiers signes de free-martinisme chez le foetus de veau. *Cytogenetics* 11:81-101, 1972.

WARREN, R. R. The adaptive significance of sequential hermaphroditism in animals. *Am. Nat.* 109:61-82, 1975.

4 | Puberty

ABRAHAMS, G. E. Wall chart. *Research in Reproduction* 3 (5), 1961.

ADVIS, J. P., and E. O. ALVAREZ. Changes in uterine responsiveness to estradiol in maturing female rats with precocious puberty induced by hypothalamic lesions. *Biol. Reprod.* 17:321-326, 1977.

ADVIS, J. P., C. BERBECKEN, E. O. ALVAREZ, and S. RODRIGUEZ. Ovulatory response to gonadotropin administration in maturing rats with lesions in the anterior hypothalamic area. *Biol. Reprod.* 17:327–332, 1977.

ADVIS, J. P., and S. R. OJEDA. Hyperprolactinemia induced precocious puberty in the female rat: ovarian site of action. *Endocrinology* 103:924–935, 1978.

ADVIS, J. P., and V. D. RAMIREZ. Plasma levels of LH and FSH in female rats with precocious puberty induced by hypothalamic lesions. *Biol. Reprod.* 17:313–320, 1977.

ADVIS, J. P., J. W. SIMPKINS, H. T. CHEN, and J. MEITES. Relation of biogenic amines to onset of puberty in the female rat. *Endocrinology* 103:11–16, 1978.

ADVIS, J. P., S. SMITH WHITE, and S. R. OJEDA. Activation of growth hormone short loop negative feedback delays puberty in the female rat. *Endocrinology* 108:1343–1352, 1981.

ALVAREZ, E. O., J. L. HANCKE, and J. P. ADVIS. Indirect evidence of prolactin involvement in precocious puberty induced by hypothalamic lesions in female rats. *Acta Endocrinol.* 85:11–17, 1977.

ANDREWS, W. W., and S. R. OJEDA. On the feedback actions of estrogen on gonadotropin and prolactin release in infantile female rats. *Endocrinology* 101:1517–1523, 1977.

BARTKE, A. Influence of luteotrophin on fertility of dwarf mice. *J. Reprod. Fertil.* 10:93–103, 1965.

———. Effects of prolactin on spermatogenesis in hypophysectomized mice. *J. Endocrinol.* 49:311–316, 1971.

BARTKE, A., B. T. CROFT, and S. DALTERIO. Prolactin restores testosterone levels and stimulates testicular growth in hamsters exposed to short day-lengths. *Endocrinology* 97:1601–1604, 1975.

BARTKE, A., and S. DALTERIO. Effects of prolactin on the sensitivity of the testis to LH. *Biol. Reprod.* 15:90–93, 1976.

BARTKE, A., B. D. GOLDMAN, F. BEX, and S. DALTERIO. Effects of prolactin (PRL) on pituitary and testicular function in mice with hereditary PRL deficiency. *Endocrinology* 101:1760–1766, 1977a.

BARTKE, A., M. S. SMITH, S. D. MICHAEL, F. G. PERON, and S. DALTERIO. Effects of experimentally-induced hyperprolactinemia on testosterone and gonadotropin levels in male rats and mice. *Endocrinology* 100:182–186, 1977b.

BARTKE, A., J. A. WEIR, P. MATHISON, C. ROBERTSON, and S. DALTERIO. Testicular function in mouse strains with different age of sexual maturation. *J. Hered.* 65:204–208, 1974.

BAUM, M. J., and D. A. GOLDFOOT. Effect of hypothalamic lesions on maturation and annual cyclicity of the ferret testis. *J. Endocrinol.* 62:59–73, 1974.

———. Effect of amygdaloid lesions on gonadal maturation in male and female ferrets. *Am. J. Physiol.* 228:1646–1651, 1975.

BECK, W., and W. WUTTKE. Desensitization of the

dopaminergic inhibition of pituitary luteinizing hormone release by prolactin in ovariectomized rats. *J. Endocrinol.* 74:67–74, 1977.

BECK, W., S. ENGELBART, M. GELATO, and W. WUTTKE. Antigonadotrophic effect of prolactin in adult castrated and in immature female rats. *Acta Endocrinol.* 84:62–71, 1977.

BESEDOVSKY, H. O., and E. SORKIN. Thymus involvement in female sexual maturation. *Nature* 249:356–358, 1974.

BICKNELL, R. J., and B. K. FOLLETT. Quantitative bioassay of luteinizing hormone releasing activity in the quail hypothalamus during photostimulated sexual development. *Gen. Comp. Endocrinol.* 31:466–474, 1977.

BLOCH, G. J., J. MASKEN, C. L. KRAGT, and W. F. GANONG. Effect of testosterone on plasma LH in male rats at various ages. *Endocrinology* 94:947–951, 1974.

BRONSON, F. H. The regulation of luteinizing hormone secretion by estrogen: relationships among negative feedback, surge potential, and male stimulation in juvenile, peripubertal, and adult female mice. *Endocrinology* 108:506–516, 1981.

CARMEL, P. W., S. ARAKI, and M. FERIN. Pituitary stalk portal blood collection in rhesus monkeys: evidence for pulsatile release of gonadotropin-releasing hormone (GnRH). *Endocrinology* 99:243–248, 1976.

COGBURN, L. A., and P. C. HARRISON. Retardation of sexual development in pinealectomized single comb white Leghorn cockerels. *Poult. Sci.* 56:876–882, 1977.

COLLU, R., M. MOTTA, R. MASSA, and L. MARTINI. Effect of hypothalamic deafferentations on puberty in the male rat. *Endocrinology* 94:1496–1501, 1974.

DAVIDSON, J. M. Hypothalamic-pituitary regulation of puberty, evidence from animal experimentation. In *Control of the Onset of Puberty,* eds. M. M. Grumbach, G. D. Grave, and F. E. Mayer, pp. 79–101. New York: Wiley, 1974.

DEBELJUK, L., A. ARIMURA, and A. V. SCHALLY. Studies on the pituitary responsiveness to luteinizing hormone-releasing hormone (LH-RH) in intact male rats at different ages. *Endocrinology* 90:585–588, 1972.

DE JONG, F. H., and R. M. SHARPE. The onset and establishment of spermatogenesis in rats in relation to gonadotrophin and testosterone levels. *J. Endocrinol.* 75:197–207, 1977.

DIERSCHKE, D. J., F. J. KARSCH, R. F. WEICK, G. WEISS, J. HOTCHKISS, and E. KNOBIL. Hypothalamic pituitary regulation of puberty: feed-back control of gonadotropin secretion in the rhesus monkey. In *Control of Onset of Puberty,* eds. M. M. Grumbach, G. D. Grave, and F. E. Mayer, pp. 104–114. New York: Wiley, 1974b.

DIERSCHKE, D. J., G. WEISS, and E. KNOBIL. Sexual maturation in the female rhesus monkey and the development of estrogen-induced gonadotropic hormone release. *Endocrinology* 94:198–206, 1974a.

DÖCKE, F. Differential effects of amygdaloid and hippocampal lesions on female puberty. *Neuroendocrinology* 14:345–350, 1974.

————. Neuro-hormonal control of puberty in female rats. In *Endocrinology of Sex*, ed. G. Dörner, pp. 344–351. Leipzig: Barth, 1975.

DÖCKE, F., M. LEMKE, and R. OKRASA. Studies on the puberty-controlling function of the mediocortical amygdala in the immature female rat. *Neuroendocrinology* 20:166–175, 1976.

DÖHLER, K. D., and W. WUTTKE. Serum LH, FSH, prolactin and progesterone from birth to puberty in female and male rats. *Endocrinology* 94:1003–1008, 1974.

————. Changes with age in levels of serum gonadotropins, prolactin, and gonadal steroids in prepubertal male and female rats. *Endocrinology* 97:898–907, 1975.

DONOVAN, B. T., and J. J. VAN DER WERFF TEN BOSCH. *Physiology of Puberty*. London: Edward Arnold, 1965.

DÖRNER, G., G. HINZ, F. DÖCKE, and R. TÖNJES. Effects of psychotrophic drugs on brain differentiation in female rats. *Endokrinologie* 70:113–123, 1977.

ELDRIDGE, J. C., W. P. DMOWSKI, and V. B. MAHESH. Effects of castration of immature rats on serum FSH and LH and of various steroid treatments after castration. *Biol. Reprod.* 10:438–446, 1974.

FITZGERALD, J. A. Serum concentration of LH, FSH, progesterone, and prolactin in ewe lambs during growth and development toward sexual maturity. Masters Thesis, Cornell University, 1978.

FOLLETT, B. K. Plasma follicle-stimulating hormone during photoperiodically induced sexual maturation in male Japanese quail. *J. Endocrinol.* 69:117–126, 1976.

FOSTER, D. L., and F. J. KARSCH. Development of the mechanism regulating the preovulatory surge of luteinizing hormone in sheep. *Endocrinology* 97:1205–1209, 1975.

FRISCH, R. E. Critical weight at menarche, initiation of the adolescent growth spurt, and control of puberty. In *Control of the Onset of Puberty*, eds. M. M. Grumbach, G. D. Grave, and F. E. Mayer, pp. 403–423. New York: Wiley, 1974.

GELATO, M., J. DIBBET, S. MARSHALL, J. MEITES, and W. WUTTKE. Prolactin-adrenal interactions in the immature female rat. *Ann. Biol. Anim. Biochem. Biophys.* 16:395–397, 1976.

GLEDHILL, B., and B. K. FOLLETT. Diurnal variations and episodic release of plasma gonadotrophins in Japanese quail during a photoperiodically induced gonadal cycle. *J. Endocrinol.* 71:245–257, 1976.

GORSKI, R. A. Extrahypothalamic influences on gonadotropin regulation. In *Control of the Onset of Puberty*, eds. M. M. Grumbach, G. D. Grave, and F. E. Mayer, pp. 182–207. New York: Wiley, 1974.

GREENSTEIN, B. D. The role of hormone receptors in development and puberty. *J. Reprod. Fertil.* 52:419–426, 1978.

GRUMBACH, M. M., J. C. ROTH, S. L. KAPLAN, and R. P. KELCH. Hypothalamic-pituitary regulation of puberty in man: evidence and concepts derived from clinical research. In *Control of the Onset of Puberty*, eds. M. M. Grumbach, G. D. Grave, and F. E. Mayer, pp. 115–166. New York: Wiley, 1974.

GUPTA, D., K. RAGER, J. ZARZYCKI, and M. EICHNER. Levels of luteinizing hormone, follicle-stimulating hormone, testosterone and dihydrotestosterone in the circulation of sexually maturing intact male rats and after orchidectomy and experimental bilateral cryptorchidism. *J. Endocrinol.* 66:183–193, 1975.

JOHANSON, A. Fluctuations of gonadotropin levels in children. *J. Clin. Endocrinol. Metab.* 39:154–159, 1974.

KATONGOLE, C. B., F. NAFTOLIN, and R. V. SHORT. Relationship between blood levels of luteinizing hormone and testosterone in bulls and the effect of sexual stimulation. *J. Endocrinol.* 50:457–466, 1971.

KULIN, H. E., and R. J. SANTEN. Endocrinology of puberty in man. In *Regulatory Mechanisms of Male Reproductive Physiology*, eds. C. H. Spilman, T. J. Lobl, and K. T. Kirton, pp. 175–190. Amsterdam: Excerpta Medica, 1976.

LARSEN, L. O. Hormonal control of sexual maturation in lampreys. In *Comparative Endocrinology*, eds. P. J. Gaillard and H. H. Boer, pp. 105–108. Amsterdam: Elsevier-Holland Biomedical, 1978.

LINCOLN, G. A., and M. J. PEET. Photoperiodic control of gonadotrophin secretion in the ram: a detailed study of the temporal changes in plasma levels of follicle-stimulating hormone, luteinizing hormone and testosterone following an abrupt switch from long to short days. *J. Endocrinol.* 74:355–367, 1977.

LINTERN-MOORE, S. Effect of athymia on the initiation of follicular growth in the rat ovary. *Biol. Reprod.* 17:155–161, 1977.

MACKINNON, P. C. B., E. PUIG-DURAN, and R. LAYNES. Reflections on the attainment of puberty in the rat: have circadian signals a role to play in its onset? *J. Reprod. Fertil.* 52:401–412, 1978.

MARSHALL, W. A. The relationship of puberty to other maturity indicators and body composition in man. *J. Reprod. Fertil.* 52:437–443, 1978.

McCANN, S. M., S. OJEDA, and A. NEGRO-VILLAR. Sex steroid, pituitary and hypothalamic hormones during puberty in experimental animals. In *Control of the Onset of Puberty*, eds. M. M. Grumbach, G. D. Grave, and F. E. Mayer, pp. 1–19. New York: Wiley, 1974.

MEYS-ROELOFS, H. M. A., W. J. DEGREEF, and J. TH. J. UILENBROEK. Plasma progesterone and its relationship to serum gonadotrophins in immature female rats. *J. Endocrinol.* 64:329–336, 1975a.

MEYS-ROELOFS, H. M. A., and J. MOLL. Sexual maturation and the adrenal glands. *J. Reprod. Fertil.* 52:413–418, 1978.

MEYS-ROELOFS, H. M. A., J. TH. J. UILENBROEK, W. J. DEGREEF, F. H. DEJONG, and P. KRAMER. Gonadotrophin and steroid levels around the time of first ovulation in the rat. *J. Endocrinol.* 67:275–282, 1975b.

MOGER, W. H. Serum 5 α-androstane-3α, 17β-diol, androsterone, and testosterone concentrations in the male rat. Influence of age and gonadotropin stimulation. *Endocrinology* 100:1027–1032, 1977.

MOGER, W. H., and D. T. ARMSTRONG. Changes in serum testosterone levels following acute LH treatment in immature and mature rats. *Biol. Reprod.* 11:1–6, 1974.

NANKIN, H. R., and P. TROEN. Repetitive luteinizing hormone elevations in serum of normal men. *J. Clin. Endocrinol. Metab.* 33:558–560, 1971.

————. Overnight patterns of serum luteinizing hormone in normal men. *J. Clin. Endocrinol. Metab.* 35:705–710, 1972.

NEGRI, A., and V. L. GAY. Differing effects of comparable serum testosterone concentration and gonadotropin secretion in pre- and postpubertal orchidectomized rats. *Biol. Reprod.* 15:375–380, 1976.

ODELL, W. D., and R. S. SWERDLOFF. The role of testicular sensitivity to gonadotropins in sexual maturation of the male rat. In *Control of Onset of Puberty*, eds. M. M. Grumbach, G. D. Graves, and F. E. Mayer, pp. 313–332. New York: Wiley, 1974.

OJEDA, S. R., J. E. WHEATON, H. E. JAMESON, and S. M. MCCANN. The onset of puberty in the female rat: changes in plasma prolactin, gonadotropins, luteinizing hormone-releasing hormone (LHRH) and hypothalamic LHRH content. *Endocrinology* 98:630–638, 1976.

PAHNKE, V. G., F. A. LEIDENBERGER, and H. J. KÜNZIG. Correlation between HCG (LH)-binding capacity, Leydig cell number and secretory activity of rat testis throughout pubescence. *Acta Endocrinol.* 79:610–618, 1975.

PARKER, C. R., JR., and V. B. MAHESH. Hormonal events surrounding the natural onset of puberty in female rats. *Biol. Reprod.* 14:347–353, 1976.

PHILLIPS, R. E. Endocrine mechanisms of the failure of pintails (*Anas acuta*) to reproduce in captivity. Ph.D. Diss., Cornell University, 1959.

RAMALEY, J. A. Adrenal-gonadal interactions at puberty. *Life Sci.* 14:1623–1633, 1974.

————. Development of gonadotropin regulation in the prepubertal mammal. *Biol. Reprod.* 20:1–31, 1979.

RAMALEY, J. A., and D. BARTOSIK. Effect of adrenalectomy on light-induced precocious puberty in rats. *Biol. Reprod.* 13:347–352, 1975.

RAMALEY, J. A., and G. T. CAMPBELL. Serum prolactin concentrations in the adrenalectomized rat: relationships to puberty onset. *Endocrinology* 101:890–897, 1977.

RAMIREZ, V. D. Endocrinology of puberty. In *Handbook of Physiology*, Sec. 7, Vol. 2, Part 1, pp. 1–28. Washington, D.C.: American Physiological Society, 1973.

ROWE, P. H., P. A. RACEY, G. A. LINCOLN, M. ELLWOOD, J. LEHANE, and J. C. SHENTON. The temporal relationship between the secretion of luteinizing hormone and testosterone in man. *J. Endocrinol.* 64:17–26, 1975.

RYAN, K. D., and D. L. FOSTER. Neuroendocrine mechanisms involved in onset of puberty in the female: concepts derived from the lamb. *Fed. Proc.* 39:2372–2377, 1980.

SCHREIBMAN, M. P., and K. D. KALLMAN. The genetic control of the pituitary-gonadal axis in the platyfish, *Xiphophorus maculatus. J. Exp. Zool.* 200:277–294, 1977.

SHARP, P. J. A comparison of variations in plasma luteinizing hormone concentrations in male and female domestic chickens (*Gallus domesticus*) from hatch to sexual maturity. *J. Endocrinol.* 67:211–223, 1975.

SHARP, P. J., J. CULBERT, and J. W. WELLS. Variations in stored and plasma concentrations of androgens and luteinizing hormone during sexual development in the cockerel. *J. Endocrinol.* 74:467–476, 1977.

SMITH, E. R., D. A. DAMASSA, and J. M. DAVIDSON. Feedback mechanisms and male puberty in the rat. *Physiologist* 18:395, 1975.

SMITH, W. R. A mathematical model of the hypothalamic-pituitary-gonadal axis. II. Feedback control of gonadotropin secretion. Unpublished manuscript, 1978.

TROUNSON, A. O., S. M. WILLADSEN, and R. M. MOOR. Reproductive function in prepubertal lambs: ovulation, embryo development and ovarian steroidogenesis. *J. Reprod. Fertil.* 49:69–75, 1977.

VAN TIENHOVEN, A. Endocrinology of reproduction in birds. In *Sex and Internal Secretions*, ed. W. C. Young, Vol. 2, pp. 1088–1169. Baltimore: Williams and Wilkins, 1961.

WALTON, J. S., J. R. MCNEILLY, A. S. MCNEILLY, and F. J. CUNNINGHAM. Changes in concentration of follicle-stimulating hormone, luteinizing hormone, prolactin and progesterone in the plasma of ewes during the transition from anoestrus to breeding activity. *J. Endocrinol.* 75:127–136, 1977.

WILDT, L., G. MARSHALL, and E. KNOBIL. Experimental induction of puberty in the infantile female rhesus monkey. *Science* 207:1373–1375, 1980.

WILLIAMS, J. B., and P. J. SHARP. A comparison of plasma progesterone and luteinizing hormone in growing hens from eight weeks of age to sexual maturity. *J. Endocrinol.* 75:447–448, 1977.

WILSON, S. C., and P. J. SHARP. Episodic release of luteinizing hormone in the domestic fowl. *J. Endocrinol.* 64:77–86, 1975.

WUTTKE, W., K. D. DÖHLER, and M. GELATO. Oestrogens and prolactin as possible regulations of puberty. *J. Endocrinol.* 68:391–396, 1976.

WUTTKE, W., and M. GELATO. Maturation of positive feedback action of estradiol and its inhibition by pro-lactin in female rats. *Ann. Biol. Anim. Biochim. Biophys.* 16:349–362, 1976.

5 | Anatomy of the Reproductive System

ADAMS, A. E., and E. E. RAE. An experimental study of the fat bodies in triturus (Diemyctylus) viridescens. *Anat. Rec.* 41:181–203, 1929.

AITKEN, R. N. C. The oviduct. In *Physiology and Biochemistry of the Domestic Fowl*, eds. D. J. Bell and B. M. Freeman, Vol. 3, pp. 1237–1289. New York: Academic Press, 1971.

AREY, L. B. *Developmental Anatomy*. Philadelphia: W. B. Saunders, 1946.

ARNOULT, J. Comportement et reproduction en captivité de *Polypterus senegalus* Cuvier. *Acta Zool.* [Stockholm] 45:191–199, 1964.

BAER, J. G. *Comparative Anatomy of Vertebrates*. Washington: Butterworths, 1964.

BAILEY, R. J. The osteology and relationships of the phallostethid fishes. *J. Morphol.* 59:453–483, 1936.

BANARESCU, P. Carp. In *Grzimek's Animal Life Encyclopedia*, ed. B. Grzimek, Vol. 4, pp. 305–360. New York: Van Nostrand Reinhold, 1973.

BECK, L. R., and L. R. BOOTS. The comparative anatomy, histology and morphology of the mammalian oviduct. In *The Oviduct and Its Functions*, eds. A. D. JOHNSON and C. W. FOLEY, pp. 1–51. New York: Academic Press, 1974.

BEDFORD, J. M. Anatomical evidence for the epididymis as the prime mover in the evolution of the scrotum. *Am. J. Anat.* 152:483–508, 1978a.

———. Influence of abdominal temperature on epididymal function in the rat and rabbit. *Am. J. Anat.* 152:509–522, 1978b.

BELLAIRS, A. *The Life of Reptiles*, Vol. 2. New York: Universe Books, 1970.

BÖHLKE, J. E., and V. G. SPRINGER. A review of the Atlantic species of the clinid fish genus *Starksia*. *Proc. Acad. Nat. Sci. Phila.* 113:29–60, 1961.

BREDER, C. M., JR., and D. E. ROSEN. *Modes of Reproduction in Fishes*. Garden City, New York: Natural History Press, 1966.

BRENNER, R. M., and R. G. W. ANDERSON. Endocrine control of ciliogenesis in the primate oviduct. In *Handbook of Physiology*, Sec. 7, Vol. 2, Part 2, pp. 123–139. Washington, D.C.: American Physiological Society, 1973.

BRETSCHNEIDER, L. H., and J. DUYVENÉ DE WIT. *Reproductive Physiology of Nonmammalian Vertebrates*. Amsterdam: Elsevier, 1947.

BRYDEN, M. M. Testicular temperature in the Southern elephant seal, *Mirounga leonina* (Linn). *J. Reprod. Fertil.* 13:583–584, 1967.

CHIEFFI, G. Endocrine aspects of reproduction in elasmobranch fishes. *Gen. Comp. Endocrinol. Suppl.* 1:275–285, 1962.

CHIEFFI, G., R. K. RASTOGI, L. IELA, and M. MILONE. The function of fat bodies in relation to the hypothalamo-hypophyseal-gonadal axis in the frog, *Rana esculenta*. *Cell Tissue Res.* 161:157–165, 1975.

COE, W. R., and B. W. KUNKEL. The female urogenital organs of the limbless lizard *Anniella*. *Anat. Anz.* 26:219–222, 1905.

CREWS, D., and P. LICHT. Inhibition by corpora atretica of ovarian sensitivity to environmental and hormonal stimulation in the lizard, *Anolis carolinensis*. *Endocrinology* 95:102–106, 1975.

CUELLAR, H. S. Relationships among hormones, food consumption, fat reserves and testes growth in male lizards, *Anolis carolinensis* (Reptilia: Iguanidae). *Dissert. Abstr. Int. B Sci. Eng.* 33:3476, 1973.

DEVINE, M. C. Copulatory plugs in snakes: enforced chastity. *Science* 187:844–845, 1975.

———. Copulatory plugs, restricted mating opportunities and reproductive competition among male garter snakes. *Nature* 267:345–346, 1977.

DODD, J. M. Ovarian control in cyclostomes and elasmobranchs. *Am. Zool.* 12:325–339, 1972.

DRYDEN, G. L., and J. N. ANDERSON. Ovarian hormone: lack of effect on reproductive structures of female Asian musk shrews. *Science* 197:782–784, 1977.

ECKSTEIN, P., and S. ZUCKERMAN. Morphology of the reproductive tract. In *Marshall's Physiology of Reproduction*, ed. A. S. Parkes, Vol. 1, Part 1, pp. 43–155. London: Longmans, Green, and Co., 1956.

ERPINO, M. H. Histogenesis of atretic ovarian follicles in a seasonally breeding bird. *J. Morphol.* 139:239–250, 1973.

EVANS, H. E. Introduction and anatomy. In *Zoo and Wild Animal Medicine*, ed. M. E. Fowler, pp. 91–113. Philadelphia: W. B. Saunders, 1978.

FOX, H. The urogenital system of reptiles. In *Biology of the Reptilia*, ed. C. Gans, Vol. 6, pp. 1–157. New York: Academic Press, 1977.

FOX, W. Seminal receptacles of snakes. *Anat. Rec.* 124:519–539, 1956.

———. Special tubules for sperm storage in female lizards. *Nature* 198:500–501, 1963.

GÉRARD, P. Organes reproducteurs. In *Traité de Zoologie*, Vol. 13, Part 2, pp. 1565–1583. Paris: Masson et Cie, 1957.

GIERSBERG, H., and P. RIETSCHEL. *Vergleichende Anatomie der Wirbeltiere*, Vol. 2. Jena: Gustav Fischer, 1968.

GILBERT, A. B. The female reproductive effort. In *Physiology and Biochemistry of the Domestic Fowl*, eds. D. J. Bell and B. M. Freeman, Vol. 3, pp. 1153–1208. New York: Academic Press, 1971.

GILBERT, TH. Das Os priapi der Säugethiere. *Morphol. Jb.* 18:805-831, 1892.

GREENSTEIN, J. S., and R. G. HART. The effects of removal of the accessory glands separately or in paired combinations on the reproductive performance of the male rat. In *Fifth Intern. Congr. Animal Reprod.* [Proc.] (*Trento*) Vol. 3, pp. 414-420, Trento, 1964.

GRIER, H. J. Cellular organization of the testis and spermatogenesis in fishes. *Am. Zool.* 21:345-357, 1981.

GRIER, H. J., and J. R. LINTON. Ultrastructural identification of the Sertoli cells in the testis of the Northern pike, *Esox lucius*. *Am. J. Anat.* 149:283-288, 1977.

GRIER, H. J., J. R. LINTON, J. F. LEATERLAND, and V. L. DE VLAMING. Structural evidence for two different testicular types in teleost fishes. *Am. J. Anat.* 159:331-345, 1980.

GUIBÉ, J. L'appareil uro-génital. In *Traité de Zoologie,* Vol. 14, Part 3, pp. 801-828. Paris: Masson et Cie, 1970.

HAFEZ, E. S. E. Female reproductive organs. In *Reproduction and Breeding Techniques for Laboratory Animals,* ed. E. S. E. Hafez, pp. 74-106. Philadelphia: Lea and Febiger, 1970.

————. Anatomy and physiology of the mammalian uterotubal junction. In *Handbook of Physiology,* Sec. 7, Vol. 2, Part 2, pp. 87-95. Washington, D.C.: American Physiological Society, 1973a.

————. Endocrine control of the structure and function of the mammalian oviduct. In *Handbook of Physiology,* Sec. 7, Vol. 2, Part 2, pp. 97-122. Washington, D.C.: American Physiological Society, 1973b.

HARDER, W. *Anatomy of Fishes.* Stuttgart: E. Sweizerbar 'tsche Verlagsbuchhandlung, 1975.

HOAR, W. S. Reproduction. In *Fish Physiology,* eds. W. S. Hoar and D. J. Randall, Vol. 3, pp. 1-72. New York: Academic Press, 1969.

HOFFMAN, R. A. Gonads, spermatic ducts, and spermatogenesis in the reproductive system of male toad fish, *Opsanus tau*. *Chesapeake Sci.* 4:21-29, 1963.

HOLDEN, M. J. Significance of sexual dimorphism of the anal fin of *Polypteridae*. *Nature* 232:135-136, 1971.

HOLTZ, W. H. Structure, function and secretions of reproductive organs in the male rabbit. Ph.D. Diss., Cornell University, 1972.

HUBBS, C. L. A revision of the viviparous perches. *Proc. Biol. Soc. Wash.* 31:9-14, 1918.

————. Fishes from the caves of Yucatan. *Carnegie Inst. Wash. Publ.* 491:261-295, 1938.

IWASAWA, H., and M. MICHIBATA. Comparative morphology of sperm storage portion of Wolffian duct in Japanese anurans. *Annot. Zool. Jpn.* 45:218-233, 1972.

KING, A. S., and J. MCLELLAND. *Outlines of Avian Anatomy.* London: Baillière Tindall, 1975.

KINSKY, F. C. The consistent presence of paired ovaries in the kiwi (*Apteryx*) with some discussion of this condition in other birds. *J. Ornithol.* 112:334-357, 1971.

KNIGHT, C. E., and R. D. KLEMM. Anatomy of the foam glands in the *Coturnix coturnix japonica*. *Poult. Sci.* 51:1825, 1972 (Abstract).

LAMBERT, J. G. D. The ovary of the guppy *Poecilia reticulata*. *Gen. Comp. Endocrinol.* 15:464-476, 1970.

LANCE, V., and I. P. CALLARD. A histochemical study of ovarian function in the ovoviviparous elasmobranch, *Squalus acanthias*. *Gen. Comp. Endocrinol.* 13:255-267, 1969.

LOFTS, B. Reproduction. In *Physiology of the Amphibia,* ed. B. Lofts, Vol. 2 pp. 107-218. New York: Academic Press, 1974.

LOWE, C. H., and S. R. GOLDBERG. Variation in the circumtesticular Leydig cell tunic of Teiid lizards (*Cnemidophorus* and *Ameiva*). *J. Morphol.* 119:277-282, 1966.

MANN, T. *The Biochemistry of Semen and the Male Reproductive Tract.* New York: Wiley, 1964.

MARSHALL, A. J. Reproduction in male bony fish. *Symp. Zool. Soc. Lond.* 1:137-151, 1960.

MATTHEWS, L. H. Reproduction in the basking shark (*Cetorhinus maximus*) Gunner. *Trans. Roy. Soc. London* 234:247-316, 1950.

MCKEEVER, S. Male reproductive organs. In *Reproduction and Breeding Techniques for Laboratory Animals,* ed. E. S. E. Hafez, pp. 28-55. Philadelphia: Lea and Febiger, 1970.

MCKENZIE, F. F., J. C. MILLER, and L. C. BAUGUESS. *The Reproductive Organs and Semen of the Boar.* Mo. Agric. Exp. Stn. Res. Bull. 279, 1938.

MIZUE, K. Studies on the Scorpaneous fish *Sebasticus marmoratus* Cuvier et Valenciennes. IV. On the copulatory organ of the marine ovoviviparous teleost. *Bull. Fac. Fish. Nagasaki Univ.* 8:80-83, 1959.

MOHSEN, T. Sur la présence d'un organe copulateur interne, très évolué chez *Skiffia lermae*, Cyprinodonte *Goodeidae*. *C. R. Hebd. Séances Acad. Sci. Ser. D Sci. Nat. (Paris)* 252:3327-3329, 1961.

MORGAN, W., and W. KOHLMEYER. Hens with bilateral oviducts. *Nature* 180:98, 1957.

MORRIS, R. W. Clasping mechanism of the cottid fish, *Oligocottus synderi*, Greely. *Pac. Sci.* 10:314-317, 1956.

MOSSMAN, H. W., and K. L. DUKE. *Comparative Morphology of the Mammalian Ovary.* Madison: University of Wisconsin Press, 1973a.

————. Some comparative aspects of the mammalian ovary. In *Handbook of Physiology,* Sec. 7, Vol. 2, Part 1, pp. 389-402. Washington, D.C.: American Physiological Society, 1973b.

NALBANDOV, A. V., and M. F. JAMES. The bloodvascular system of the chicken ovary. *Am. J. Anat.* 85:347-377, 1949.

NISHIYAMA, H. Studies on the accessory reproductive organs in the cock. *J. Fac. Agric. Kyushu Univ.* 10:277-305, 1955.

NISHIYAMA, H., N. NAKASHIMA, and N. FUJIHARA.

Studies on the accessory reproductive organs in the drake. I. Addition to semen of the fluid from the ejaculatory groove region. *Poult. Sci.* 55:234–242, 1976.

NOBLE, G. K. *The Biology of the Amphibia.* New York: Dover Publications, 1954.

PEDERSEN, T., and H. PETERS. Proposal for a classification of oocytes and follicles in the mouse ovary. *J. Reprod. Fertil.* 17:555–557, 1968.

PRASAD, M. R. N. Männliche Geschlechtsorgane. In *Handbuch der Zoologie,* eds. J.-G. Helmcke, D. Stark, and H. Wermuth, Vol. 8, No. 51, pp. 1–150. Berlin: Walter de Gruyter, 1974.

RAMASWAMI, L. S., and D. JACOB. Effect of testosterone propionate on the urogenital organs of immature crocodile *Crocodylus palustris* Lesson. *Experientia* 21:206–207, 1965.

RAYNAUD, A. Les organes génitaux des mammifères. In *Traité de Zoologie,* Vol. 16, Part 6, pp. 149–636. Paris: Masson et Cie, 1959.

REGAN, C. T. The morphology of the Cyprinodont fishes of the subfamily Phallostethinae, with descriptions of a new genus and two new species. In *Proc. Zool. Soc. Lond.,* pp. 1–26, 1916.

ROMANOFF, A. L., and A. J. ROMANOFF. *The Avian Egg.* New York: Wiley and Sons, 1949.

ROSEN, D. E., and R. M. BAILEY. The poeciliid fishes (Cyprinodontiformes), their structure, zoogeography, and systematics. *Bull. Amer. Mus. Nat. Hist.* 126:1–176, 1963.

ROSEN, D. E., and M. GORDON. Functional anatomy and evolution of male genitalia in poeciliid fishes. *Zoologica* 38:1–52, 1953.

ROSS, P., JR., and D. CREWS. Influence of seminal plug on mating behaviour in the garter snake. *Nature* 267:344–345, 1977.

SHARMAN, G. B. Evolution of viviparity in mammals. In *Reproduction in Mammals,* eds. C. R. Austin and R. V. Short, Vol. 6, pp. 32–70. Cambridge: Cambridge University Press, 1976.

SHORT, R. V. Species differences. In *Reproduction in Mammals,* eds. C. R. Austin and R. V. Short, Vol. 4, pp. 1–33. Cambridge: Cambridge University Press, 1972.

SNEED, K. E., and H. P. CLEMENS. The morphology of the testes and accessory reproductive glands of the catfishes (Ictaluridae). *Copeia:*606–611, 1963.

STANLEY, A. J., and E. WITSCHI. Germ cell migration in relation to asymmetry in the sex glands of hawks. *Anat. Rec.* 76:329–342, 1940.

STRAUSS, F. Weibliche Geschlechtsorgane. In *Hand-buch der Zoologie,* eds. J.-G. Helmcke, D. Starck, and H. Wermuth, Vol. 8, No. 36, pp. 1–96, Berlin: Walter de Gruyter, 1964.

_____. Weibliche Geschlechtsorgane. 2 Teil. In *Handbuch der Zoologie,* Vol. 8, No. 40, pp. 97–202, Berlin: Walter de Gruyter, 1966.

STURKIE, P. D. *Avian Physiology,* third ed. New York: Springer, 1976.

SUNDARARAJ, B. I., and S. K. NAYYAR. Effect of extirpation of "seminal vesicles" on the reproductive performance of the male catfish, Heteropneustes fossilis (Bloch). *Physiol. Zoöl.* 42:429–437, 1969.

TAVOLGA, W. N. Effects of gonadectomy and hypophysectomy on prespawning behavior in males of the gobiid fish, Bathygobius soporator. *Physiol. Zoöl.* 28:218–233, 1955.

TINGARI, M. D. On the structure of the epididymal region and ductus deferens of the domestic fowl (*Gallus domesticus*). *J. Anat.* 109:423–435, 1971.

TURNER, C. D. *General Endocrinology,* fourth ed. Philadelphia: W. B. Saunders, 1966.

TURNER, C. L. Male secondary sexual characters of *Dinematichthys iluocoeteoides. Copeia:*92–96,

_____. The gonopodium of the viviparous Jenynsia lineata. *Anat. Rec.* 101:675–676, 1948. (Abstract).

VAN DEN HURK, R. Morphological and functional aspects of the testis of the black molly (Mollienisia latipinna). Ph.D. Dis., University of Utrecht, 1975.

VAN TIENHOVEN, A. *Reproductive Physiology of Vertebrates.* Philadelphia: W. B. Saunders, 1968.

VON IHERING, R. Oviducal fertilization in the South American catfish *Trachycorystes. Copeia:*201–205, 1937.

WEBSTER, D., and M. WEBSTER. *Comparative Vertebrate Morphology.* New York: Academic Press, d1974.

WEISEL, G. F. The seminal vesicles and testes of *Gillichthys,* a marine teleost. *Copeia:*201–205, 1937.

WIMSATT, W. A. Some comparative aspects of implantation. *Biol. Reprod.* 12:1–40, 1975.

WOLFSON, A. Sperm storage at lower-than-body temperature outside the body cavity in some passerine birds. *Science* 120:68–71, 1954.

WOURMS, J. P. Reproduction and development in chondrichthyan fishes. *Am. Zool.* 17:379–410, 1977.

WUNDER. W. Brutpflege und Nestbau bei Fischen. *Ergeb. Biol.* 7:118–192, 1931.

XAVIER, F. La pseudogestation chez *Nectophrynoides occidentalis* Angel. Gen. Comp. Endocrinol. 22: 98–115, 1974.

6 | The Testis

ABRAHAMS, G. E. Wall chart. *Research in Reproduction* 3(5), 1971.

AHMAD, N., G. C. HALTMEYER, and K. B. EIK-NES. Maintenance of spermatogenesis with testosterone or dihydrotestosterone in hypophysectomized rats. *J. Reprod. Fertil.* 44:103–107, 1975.

AHSAN, S. N. Effects of gonadotropic hormones on male hypophysectomized lake chub, Couesius plumbeus. *Can. J. Zool.* 44:703–717, 1966.

AMANN, R. P., and V. K. GANJAM. Steroid production

by the bovine testis and steroid transfer across the pampiniform plexus. *Biol. Reprod.* 15:695–703, 1976.

ANDRIEUX, B., A. COLLENOT, G. COLLENOT, and C. PERGRALE. Aspects morphologiques de l'action d'hormones gonadotropes mammaliennes sur l'activité testiculaire du triton Pleurodèles, mature hypophysectomisé. *Ann. Endocrinol.* 34:711–712, 1973.

AUDY, M. C. Etude ultrastructurale des cellules de Leydig et de Sertoli au cours du cycle sexuel saisonnier de la fouine (*Martes foina erx*). *Gen. Comp. Endocrinol.* 36:462–476, 1978.

BARTKE, A. A. HAFIEZ, F. J. BEX, and S. DALTERIO. Hormonal interactions in regulation of androgen secretion. *Biol. Reprod.* 18:44–54, 1978.

BASU, S. L., and J. NANDI. Effects of testosterone and gonadotropins on spermatogenesis in *Rana pipiens* Schreber. *J. Exp. Zool.* 159:93–112, 1965.

BASU, S. L., J. NANDI, and S. NANDI. Effects of hormones on adult frog (*Rana pipiens*) testes in organ culture. *J. Exp. Zool.* 162:245–256, 1966.

BAYLÉ, J. D., M. KRAUS, and A. VAN TIENHOVEN. The effects of hypophysectomy and testosterone propionate on the testes of Japanese quail, *Coturnix coturnix japonica*. *J. Endocrinol.* 46:403–404, 1970.

BILLARD, R., E. BURZAWA-GERARD, and B. BRETON. Régéneration de la spermatogenèse du Cyprin hypophysectomisé *Carassius auratus* L. par un facteur gonadotrope hautement purifié de Carpe. *C. R. Hebd. Séances Acad. Sci. Ser. D Sci. Nat. (Paris)* 271:1896–1899, 1970.

BLACKSHAW, A. W. Temperature and seasonal influences. *In The Testis,* eds. A. D. Johnson and W. R. Gomes, Vol. 4, pp. 517–545. New York: Academic Press, 1977.

BLANC, M. R., M. TH. HOCHEREAU-DE REVIERS, C. CAHOREAU, M. COUROT, and J. L. DACHEUX. Inhibin: effects on gonadotropin secretion and testis function in the ram and the rat. In *Intragonadal Regulation of reproduction,* eds. P. Franchimont and C. P. Channing, pp. 299–326. New York: Academic Press, 1981.

BOGDANOVE, E. M., J. M. NOLIN, and G. T. CAMPBELL. Qualitative and quantitative gonad-pituitary feedback. *Recent Prog. Horm. Res.* 31:567–619, 1975.

BRETON, B., R. BILLARD, and B. JALABERT. Spécificité d'action et relations immunologiques des hormones gonadotropes de quelques téléostéens. *Ann. Biol. Anim. Biochim. Biophys.* 13:347–362, 1973.

BROWN, N. L., J. D. BAYLÉ, C. G. SCANES, and B. K. FOLLETT. Chicken gonadotrophins. Their effects on the testes of immature and hypophysectomized Japanese quail. *Cell Tissue Res.* 156:499–520, 1975.

BROWN, N. L., and B. K. FOLLETT. Effects of androgens on the testes of intact and hypophysectomized

Japanese quail. *Gen. Comp. Endocrinol.* 33:267–277, 1977.

CALLARD, I. P., G. V. CALLARD, V. LANCE, J. F. BOLAFFI, and J. S. ROSSET. Testicular regulation in nonmammalian vertebrates. *Biol. Reprod.* 18:16–43, 1978.

CALLARD, I. P., G. V. CALLARD, V. LANCE, and S. ECCLES. Seasonal changes in testicular structure and function and the effects of gonadotropins in the fresh water turtle, *Chrysemys picta. Gen. Comp. Endocrinol.* 30:347–356, 1976.

CHIEFFI, G. Comparative endocrinology of the vertebrate testis. *Am. Zool.* 12:207–211, 1972.

CHIEFFI, G., R. K. RASTOGI, L. IELA, and M. MILONE. The function of fat bodies in relation to the hypothalamo-hypophyseal-gonadal axis in the frog, *Rana esculenta. Cell Tissue Res.* 161:157–165, 1975.

CHU, J. P. The effects of oestrone and testosterone and of pituitary extracts on the gonads of hypophysectomized pigeons. *J. Endocrinol.* 2:21–37, 1940.

CHU, J. P., and S. S. YOU. Gonad stimulation by androgens in hypophysectomized pigeons. *J. Endocrinol.* 4:431–435, 1946.

CLERMONT, Y. Structure de l'épithélium séminal et mode de renouvellement des spermatogonies chez le canard. *Arch. Anat. Microsc. Morphol. Exp.* 47:47–66, 1958.

―――. Quantitative analysis of spermatogenesis of the rat: a revised model for the renewal of spermatogonia. *Am. J. Anat.* 111:111–129, 1962.

―――. Kinetics of spermatogenesis in mammals: seminiferous epithelium cycle and spermatogonial renewal. *Physiol. Rev.* 52:198–236, 1972.

COUROT, M., M. T. HOCHEREAU-DE REVIERS, and R. ORTAVANT. Spermatogenesis. In *The Testis,* eds. A. D. Johnson, W. R. Gomes, and N. L. VanDemark, Vol. 1, pp. 339–432. New York: Academic Press, 1970.

DE JONG, F. H., E. H. J. M. JANSEN, and H. J. VAN DER MOLEN. Purification and characterization of inhibin. In *Intragonadal Regulation of Reproduction,* eds. P. Franchimont and C. P. Channing, pp. 229–250. New York: Academic Press, 1981.

DEMOULIN, A., J. HUSTIN, R. LAMBOTTE, and P. FRANCHIMONT. Effect of inhibin on testicular function. In *Intragonadal Regulation of Reproduction,* eds. P. Franchimont and C. P. Channing, pp. 327–342. New York: Academic Press, 1981.

DESJARDINS, C., and F. W. TUREK. Effects of testosterone on spermatogenesis and luteinizing hormone release in Japanese quail. *Gen. Comp. Endocrinol.* 33:293–303, 1977.

DEVLAMING, V. L. Environmental and endocrine control of teleost reproduction. In *Control of Sex in Fishes,* ed. C. B. Schreck, pp. 13–83. Blacksburg: Virginia Polytechnic Institute and State University, 1974.

D'ISTRIA, M., G. DELRIO, V. BOTTE, and G. CHIEFFI.

Radioimmunoassay of testosterone, 17β oestradiol, oestrone in the male and female plasma in *Rana esculenta* during the sexual cycle. *Steroids Lipids Res.* 5:42–48, 1972.

DOBSON, S., and J. M. DODD. Endocrine control of the testis in the dogfish *Scyliorhinus canicula* L. I. Effects of partial hypophysectomy on gravimetric, hormonal and biochemical aspects of testis function. *Gen. Comp. Endocrinol.* 32:41–52, 1977a.

_____. Endocrine control of the testis in the dogfish, *Scyliorhinus canicula* L. II. Histological and ultrastructural changes in the testis after partial hypophysectomy (ventral lobectomy). *Gen. Comp. Endocrinol.* 32:53–71, 1977b.

_____. The roles of temperature and photoperiod in the response of the testis of the dogfish, *Scyliorhinus canicula* L. to partial hypophysectomy (ventral lobectomy). *Gen. Comp. Endocrinol.* 32:114–115, 1977c.

DODD, J. M. The hormones of sex and reproduction and their effects in fish and lower chordates: twenty years on. *Am. Zool.* 15 (*Suppl.* 1):137–171, 1975.

DONALDSON, E. M. Reproductive endocrinology of fishes. *Am. Zool.* 13:909–927, 1973.

DORRINGTON, J. M., I. B. FRITZ, and D. T. ARMSTRONG. Control of testicular estrogen synthesis. *Biol. Reprod.* 18:55–64, 1978.

EWING, L., and B. L. BROWN. Testicular steroidogenesis. In *The Testis*, eds. A. D. Johnson and W. R. Gomes, Vol. 4, pp. 239–287. New York: Academic Press, 1977.

FRANCHIMONT, P., K. HENDERSON, G. VERHOEVEN, M.-T. HAZEE-HAGELSTEIN, C. CHARLET-RENARD, A. DEMOULIN, J.-P. BOURGIGNON, and M.-J. LECOMTE-YERNA. Inhibin: mechanisms and action and secretion. In *Intragonadal Regulation of Reproduction*, eds. P. Franchimont and C. P. Channing, pp. 167–191. New York: Academic Press, 1981.

FRENCH, F. S., S. N. NAYFEH, E. M. RITZEN, and V. HANSSON. FSH and a testicular androgen-binding protein in the maintenance of spermatogenesis. *Res. Reprod.* 6(4):2–3, 1974.

GARNIER, D. Etude de la fonction endocrine du testicule chez le canard pékin au cours du cycle saisonnier— aspects biochimiques et cytologiques. Thesis, University of Paris, 1972.

GIGON-DEPEIGES, A., and J. P. DUFAURE. Secretory activity of the lizard epididymis and its control by testosterone. *Gen. Comp. Endocrinol.* 33:473–479, 1977.

GORBMAN, A., and K. TSUNEKI. A technique for hypophysectomy of the Pacific hag fish: first observations. *Gen. Comp. Endocrinol.* 26:420–422, 1975.

GRIER, H. J. Aspects of germinal cyst and sperm development in *Poecilia latipinna* (Teleoster: Poecillidae). *J. Morphol.* 146:229–250, 1975.

GUHA, K. K., and C. B. JØRGENSEN. Effects of hypophysectomy on structure and function of testis in

adult toads, *Bufo bufo bufo* L. *Gen. Comp. Endocrinol.* 34:201–210, 1978.

GUSTAFSON, A. W., and M. SCHEMESH. Changes in plasma testosterone levels during the annual reproductive cycle of the hibernating bat, Myotis lucifugus lucifugus with a survey of plasma testosterone levels in adult male vertebrates. *Biol. Reprod.* 15:9–24, 1976.

HAFIEZ, A. A., C. W. LLOYD, and A. BARTKE. The role of prolactin in the regulation of testis function: the effects of prolactin and luteinizing hormone on the plasma levels of testosterone and androstenedione in hypophysectomized rats. *J. Endocrinol.* 52:327–332, 1972.

HAGENOS, L., and E. M. RITZEN. Impaired Sertoli cell function in experimental cryptorchidism in the rat. *Mol. Cell. Endocrinol.* 4:25–34, 1976.

HANSSON, V., S. C. WEDDINGTON, W. S. MCLEAN, A. A. SMITH, S. N. NAYFEH, F. S. FRENCH, and E. M. RITZÉN. Regulation of seminiferous tubular function by FSH and androgen. *J. Reprod. Fertil.* 44:363–375, 1975.

HERIN, R. A., N. H. BOOTH, and R. M. JOHNSON. Thermoregulatory effects of abdominal air sacs on spermatogenesis in domestic fowl. *Am. J. Physiol.* 198:1343–1345, 1960.

HOAR, W. S. Reproduction. In *Fish Physiology*, eds. W. S. Hoar and D. J. Randall, Vol. 3, pp. 1–72. New York: Academic Press, 1969.

HODSON, N. The nerves of the testis, epididymis, and scrotum. In *The Testis*, eds. A. D. Johnson, W. R. Gomes, and N. L. VanDemark, Vol. 1, pp. 47–99. New York: Academic Press, 1970.

HOLSTEIN, A.-F. Zur Frage der lokalen Steuerung der Spermatogenese beim Dornhai (*Squalus acanthias* L.). *Z. Zellforsch. Mikrosk. Anat.* 93:265–281, 1969.

KIME, D. E., and E. A. HEWS. *In vitro* biosynthesis of 11β-hydroxy and 11-oxotestosterone by testes of the pike (*Esox lucius*) and the perch (*Perca fluviatilis*). *Gen. Comp. Endocrinol.* 36:604–608, 1978.

KOTITE, N. J., S. N. NAYFEH, and F. S. FRENCH. FSH and androgen regulation of Sertoli cell function in the immature rat. *Biol. Reprod.* 18:65–73, 1978.

LEBLOND, C. P., and Y. CLERMONT. Spermiogenesis of rat, mouse, hamster and guineapig as revealed by the "Periodic Acid–Fuchsin Sulfurous Acid" technique. *Am. J. Anat.* 90:167–216, 1952.

LEHRMAN, D. S. Hormonal regulation of parental behavior in birds and infrahuman mammals. In *Sex and Internal Secretions*, ed. W. C. Young, Vol. 2, pp. 1268–1382. Baltimore: Williams and Wilkins, 1961.

LICHT, P. The relation between preferred body temperatures and testicular heat sensitivity in lizards. *Copeia*: 428–436, 1965.

_____. Testicular function after partial and total adenohypophysectomy and gonadotropin-treatment in the lizard *Anolis carolinensis*. *Am. Zool.* 8:759, 1968.

_____. Actions of mammalian pituitary gonadotropins (FSH and LH) in reptiles. I. Male snakes. *Gen. Comp. Endocrinol.* 19:273–281, 1972a.

_____. Actions of mammalian pituitary gonadotropins (FSH and LH) in reptiles. II. Turtles. *Gen. Comp. Endocrinol.* 19:282–289, 1972b.

_____. Induction of spermiation in anurans by mammalian pituitary gonadotropins and their subunits. *Gen. Comp. Endocrinol.* 20:522–529, 1973.

_____. Luteinizing hormone (LH) in the reptilian pituitary gland. *Gen. Comp. Endocrinol.* 22:463–469, 1974.

LICHT, P., and E. M. DONALDSON. Gonadotropic activity of salmon pituitary extract in the male lizard (*Anolis carolinensis*). *Biol. Reprod.* 1:307–314, 1969.

LICHT, P., S. W. FARMER, and H. PAPKOFF. Further studies on the chemical nature of reptilian gonadotropins: FSH and LH in the American alligator and green sea turtle. *Biol. Reprod.* 14:222–232, 1976.

LICHT, P., and A. R. MIDGLEY, JR. In vitro binding of radioiodinated human follicle-stimulating hormone to reptilian and avian gonads: radioligand studies with mammalian hormones. *Biol. Reprod.* 15:195–205, 1976a.

_____. Competition for the *in vitro* binding of radioiodinated human follicle-stimulating hormone in reptilian, avian, and mammalian gonads by nonmammalian gonadotropins. *Gen. Comp. Endocrinol.* 30:364–371, 1976b.

_____. Autoradiographic localization of binding sites for human follicle-stimulating hormone in reptilian testes and ovaries. *Biol. Reprod.* 16:117–121, 1977.

LICHT, P., and H. PAPKOFF. Gonadotropic activities of the subunits of ovine FSH and LH in the lizard *Anolis carolinensis. Gen. Comp. Endocrinol.* 16:586–593, 1971.

_____. Evidence for an intrinsic gonadotropic activity of ovine LH in the lizard. *Gen. Comp. Endocrinol.* 20:172–176, 1973.

_____. Separation of two distinct gonadotropins from the pituitary gland of the snapping turtle (*Chelydra serpertina*). *Gen. Comp. Endocrinol.* 22:218–237, 1974.

LICHT, P., H. PAPKOFF, S. W. FARMER, C. MULLER, H. W. TSUI, and D. CREWS. Evolution in gonadotropin structure and function. *Recent Prog. Horm. Res.* 33:169–243, 1977.

LICHT, P., and A. K. PEARSONS. Effects of mammalian gonadotropin (FSH and LH) on the testes of the lizard *Anolis carolinensis. Gen. Comp. Endocrinol.* 13:367–381, 1969.

LICHT, P., and H. W. TSUI. Evidence for intrinsic activity of ovine FSH on spermatogenesis, ovarian growth, steroidogenesis and ovulation in lizards. *Biol. Reprod.* 12:346–350, 1975.

LOFTS, B. The effects of follicle-stimulating hormone and luteinizing hormone on the testis of hypophysec-tomized frogs (*Rana temporaria*). *Gen. Comp. Endocrinol.* 1:179–189, 1961.

_____. Reproduction. In *Physiology of the Amphibia,* ed. B. Lofts, Vol. 2, pp. 107–218. New York: Academic Press, 1974.

LOFTS, B., and A. J. MARSHALL. The post-nuptial occurrence of progestins in the seminiferous tubules of birds. *J. Endocrinol.* 19:16–21, 1959.

LOFTS, B., G. E. PICKFORD, and J. W. ATZ. Effects of methyl testosterone on the testes of a hypophysec-tomized cyprinodont fish, *Fundulus heteroclitus. Gen. Comp. Endocrinol.* 6:74–88, 1966.

LOUIS, B. G., and I. B. FRITZ. Follicle-stimulating hormone and testosterone independently increase the production of androgen-binding protein by Sertoli cells in culture. *Endocrinology* 104:454–461, 1979.

MATTY, A. J., K. TSUNEKI, W. W. DICKHOFF, and A. GORBMAN. Thyroid and gonadal function in hypophysectomized hagfish *Eptatretus stouti. Gen. Comp. Endocrinol.* 30:500–516, 1976.

MEANS, A. R. Mechanisms of action of follicle-stimulating hormone (FSH). In *The Testis,* eds. A. D. Johnson and W. R. Gomes, Vol. 4, pp. 163–188. New York: Academic Press, 1977.

MONET-KUNTZ, C., M. TERQUI, A. LOCATELLI, M. T. HOCHEREAU DE REVIERS, and M. COUROT. Effets de la supplémentation en testostérone sur la sper-matogenèse de béliers hypophysectomisés. *C. R. Hebd. Séances Acad. Sci. Ser. D Sci. Nat. (Paris)* 283:1763–1766, 1976.

MOORE, F. L. Spermatogenesis in larval *Ambystoma tigrinum:* positive and negative interactions of FSH and testosterone. *Gen. Comp. Endocrinol.* 26:523–533, 1975.

MULLER, C. H. Gonadotropin regulation of the bullfrog testis. *Am. Zool.* 16:259, 1976.

_____. *In vitro* stimulation of 5α-dihydro-testosterone and testosterone secretion from bullfrog testis by nonmammalian and mammalian gonadotropins. *Gen. Comp. Endocrinol.* 33:109–121, 1977a.

_____. Plasma 5α-dihydrotestosterone and testosterone in the bullfrog, *Rana catesbeiana:* stimulation by bullfrog LH. *Gen. Comp. Endocrinol.* 33:122–132, 1977b.

NALBANDOV, A. V., R. K. MEYER, and W. H. McSHAN. The role of a third gonadotrophic hormone in the mechanism of androgen secretion in chicken testes. *Anat. Rec.* 110:475–493, 1951.

NANDI, J. Comparative endocrinology of steroid hormones in vertebrates. *Am. Zool.* 7:115–133, 1967.

NAYYAR, S. K., P. KESHAVANATH, B. I. SUNDARARAJ, and E. M. DONALDSON. Maintenance of spermatogenesis and seminal vesicles in the hypophysec-tomized catfish, *Heteropneustes fossilis* (Bloch): effects of ovine and salmon gonadotropin, and testosterone. *Can. J. Zool.* 54:285–292, 1976.

OREGEBIN-CRIST, M. C. Studies on the function of the epididymis. *Biol. Reprod. Suppl.* 1:155–175, 1969.

ORGEBIN-CRIST, M. C., B. J. DANZO, and J. DAVIES. Endocrine control of the development and maintenance of sperm fertilizing ability in the epididymis. In *Handbook of Physiology,* Sec. 7, Vol. 5, pp. 319-338. Washington, D.C.: American Physiological Society, 1975.

PICKFORD, G. E., B. LOFTS, G. BARA, and J. W. ATZ. Testis stimulation in hypophysectomized male killifish, *Fundulus heteroclitus,* treated with mammalian growth hormone and/or luteinizing hormone. *Biol. Reprod.* 7:370-386, 1972.

RAJ, H. G. M., and M. DYM. The effects of selective withdrawal of FSH or LH on spermatogenesis in the immature rat. *Biol. Reprod.* 14:489-494, 1976.

RASTOGI, R. K., and G. CHIEFFI. Cytological changes in the pars distalis of the pituitary of the green frog, *Rana esculenta* during the reproductive cycle. *Z. Zellforsch.* 111:505-618, 1970.

RASTOGI, R. K., L. IELA, P. K. SAXENA, and G. CHIEFFI. The control of spermatogenesis in the green frog, *Rana esculenta. J. Exp. Zool.* 196:151-166, 1976.

RAVONA, H., N. SNAPIR, and M. PEREK. The effect on the gonadal axis in cockerels of electrolytic lesions in various regions of the basal hypothalamus. *Gen. Comp. Endocrinol.* 20:112-124, 1973.

ROMMERTS, F. F. G., J. A. GROOTEGOED, and H. J. VAN DER MOLEN. Physiological role for androgen-binding protein-steroid complex in testis. *Steroids* 28:43-49, 1976.

ROOSEN-RUNGE, E. C. *The Process of Spermatogenesis in Animals.* London: Cambridge University Press, 1977.

RUSSELL, L. D., and Y. CLERMONT. Degeneration of germ cells in normal, hypophysectomized and hormone treated hypophysectomized rats. *Anat. Rec.* 187:347-366, 1977.

SCHAEFER, W. H. Hypophysectomy and thyroidectomy of snakes. *Proc. Soc. Exp. Biol. Med.* 30:1363-1365, 1933.

SCHANBACHER, B. D. Testosterone secretion in cryptorchid and intact bulls injected with gonadotropin-releasing hormone and luteinizing hormone. *Endocrinology* 104:360-364, 1979.

SETCHELL, B. P. *The Mammalian Testis.* Ithaca: Cornell University Press, 1978.

SETCHELL, B. P., R. V. DAVIES, and S. J. MAIN. Inhibin. In *The Testis,* eds. A. D. Johnson and W. R. Gomes, Vol. 4, pp. 189-238. New York: Academic Press, 1977.

SILVER, R., C. REBOULLEAU, D. S. LEHRMAN, and H. H. FEDER. Radioimmunoassay of plasma progesterone during the reproductive cycle of male and female ring doves (*Streptopelia risoria*). *Endocrinology* 94:1547-1554, 1974.

SNAPIR, N., I. NIR, F. FURUTA, and S. LEPKOVSKY. Effect of administered testosterone propionate on cocks functionally castrated by hypothalamic lesions. *Endocrinology* 84:611-618, 1969.

STANLEY, H. P. The structure and development of the seminiferous follicle in *Scyliorhinus caniculus* and *Torpedo marmorata* (Elasmobranchii) *Z. Zellforsch. Mikrosk. Anat.* 75:453-468, 1966.

STEINBERGER, A., and E. STEINBERGER. The Sertoli cells. In *The Testis,* eds. A. D. Johnson and W. R. Gomes, Vol. 4, pp. 371-399. New York: Academic Press, 1977.

STEINBERGER, E., and A. STEINBERGER. Hormonal control of testicular function in mammals. In *Handbook of Physiology,* Sec. 7, Vol. 4, pp. 325-347. Washington, D.C.: American Physiological Society, 1974.

STEINBERGER, E., A. STEINBERGER, and B. SANBORN. Endocrine control of spermatogenesis. In *Physiology and Genetics of Reproduction,* eds. E. M. Coutinho and F. Fuchs, Part A, pp. 163-181. New York: Plenum Press, 1974.

SUNDARARAJ, B. I., and S. K. NAYYAR. Efects of exogenous gonadotrophins and gonadal hormones on the testes and seminal vesicles of hypophysectomized catfish, *Heteropneustes fossilis* (Bloch). *Gen. Comp. Endocrinol.* 8:403-416, 1967.

SUNDARARAJ, B. I., S. K. NAYYAR, T. C. ANAND, and E. M. DONALDSON. Effects of salmon pituitary gonadotropin, ovine luteinizing hormone, and testosterone on the testes and seminal vesicles of hypophysectomized catfish, *Heteropneustes fossilis* (Bloch). *Gen. Comp. Endocrinol.* 17:73-82, 1971.

VANDEMARK, N. L., and M. J. FREE. Temperature effects. In *The Testis,* eds. A. D. Johnson, W. R. Gomes, and N. L. VanDemark, Vol. 3, pp. 233-312. New York: Academic Press, 1970.

VAN TIENHOVEN, A. Endocrinology of reproduction in birds. In *Sex and Internal Secretions,* ed. W. C. Young, Vol. 2, pp. 1088-1169. Baltimore: Williams and Wilkins, 1961.

————. *Reproductive Physiology of Vertebrates.* Philadelphia: W. B. Saunders, 1968.

WAITES, G. M. H. Temperature regulation and the testis. In *The Testis,* eds. A. D. Johnson, W. R. Gomes, and N. L. VanDemark, Vol. 1, pp. 241-279. New York: Academic Press, 1970.

WAITES, G. M. H., and G. R. MOULE. Relation of vascular heat exchange to temperature regulation in the testis of the ram. *J. Reprod. Fertil.* 2:213-224, 1961.

WEISBART, M., J. H. YOUSON, and J. P. WIEBE. Biochemical, histochemical, and ultrastructural analysis of presumed steroid-producing tissues in sexually mature sea lamprey, *Petromyzon marinus* L. *Gen. Comp. Endocrinol.* 34:25-37, 1978.

WILLIAMS, D. D. A histological study of the effects of subnormal temperature on the testis of the fowl. *Anat. Rec.* 130:225-241, 1958a.

_____. Effect of heat on transplanted testis material of the fowl. *Transplant. Bull.* 5:32–35, 1958b.

WOLFSON, A., and B. K. HARRIS. Maintenance of gonadal activity with testosterone following inhibition of the pituitary with short days. *Anat. Rec.* 134:656, 1959.

YAMAZAKI, F., and E. M. DONALDSON. The effects of partially purified salmon pituitary gonadotropin on spermatogenesis, vitellogenesis, and ovulation in hypophysectomized goldfish. (*Carassius auratus*). *Gen. Comp. Endocrinol.* 11:292–299, 1968.

_____. Involvement of gonadotropin and steroid hormones in the spermiation of the goldfish (*Carassius auratus*). *Gen. Comp. Endocrinol.* 12:491–497, 1969.

7 | The Ovary

ANAND, T. C, and B. I. SUNDARARAJ. Ovarian maintenance in the hypophysectomized catfish, *Heteropneustes fossilis* (Bloch), with mammalian hypophyseal and placental hormones and gonadal and adrenocortical steroids. *Gen. Comp. Endocrinol.* 22:154–168, 1974.

ANDERSON, L. D., F. W. SCHAERF, and C. P. CHANNING. Effects of follicular development on the ability of cultured porcine granulosa cells to convert androgens to estrogens. *Adv. Exp. Med. Biol.* 112:187–195, 1979.

ANDERSON, L. L., J. D. BAST, and R. M. MELAMPY. Relaxin in ovarian tissue during different reproductive stages in the rat. *J. Endocrinol.* 59:371–372, 1973.

ANDERSON, M. L., and J. A. LONG. Localization of relaxin in the pregnant rat. Bioassay of tissue extracts and cell fractionation studies. *Biol. Reprod.* 18:110–117, 1978.

ANDERSON, M. L., J. A. LONG, and T. HAYASHIDA. Immunofluorescence studies on the localization of relaxin in the corpus luteum of the pregnant rat. *Biol. Reprod.* 13:499–504, 1975.

ARMSTRONG, D. T., and J. H. DORRINGTON. Androgens augment FSH-induced progesterone secretion by cultured rat granulosa cells. *Endocrinology* 99:1411–1414, 1976.

BAHR, J., L. KAO, and A. V. NALBANDOV. The role of catecholamines and nerves in ovulation. *Biol. Reprod.* 10:273–290, 1974.

BAKER, T. G. Oogenesis and ovulation. In *Reproduction in Mammals*, eds. C. R. Austin and R. V. Short, Vol. 1, pp. 14–45. Cambridge: Cambridge University Press, 1972.

BARKER-JØRGENSEN, C. Mechanisms regulating ovarian cycle in the toad *Bufo bufo bufo* (L): role of presence of second growth phase oocytes in controlling recruitment from pool of first growth phase oocytes. *Gen. Comp. Endocrinol.* 23:170–177, 1974.

_____. Factors controlling the annual ovarian cycle in the toad. *Bufo bufo bufo* (L). *Gen. Comp. Endocrinol.* 25:264–273, 1975.

BEHRMAN, H. R. Prostaglandins in hypothalamo-pituitary and ovarian function. *Annu. Rev. Physiol.* 41:685–700, 1979.

BENTLEY, P. J. *Comparative Vertebrate Endocrinology*. Cambridge: Cambridge University Press, 1977.

BJERSING, L., and S. CAJANDER. Ovulation and the mechanism of follicle rupture. I. Light microscopic changes in rabbit ovarian follicles prior to induced ovulation. *Cell Tissue Res.* 149:287–300, 1974a.

_____. Ovulation and the mechanism of follicle rupture. II. Scanning electron microscopy of rabbit germinal epithelium prior to induced ovulation. *Cell Tissue Res.* 149:301–312, 1974b.

_____. Ovulation and the mechanism of follicle rupture. III. Transmission electron microscopy of rabbit germinal epithelium prior to induced ovulation. *Cell Tissue Res.* 149:313–317, 1974c.

_____. Ovulation and the mechanism of follicle rupture. IV. Ultrastructure of membrana granulosa of rabbit Graafian follicles prior to induced ovulation. *Cell Tissue Res.* 153:1–14, 1974d.

_____. Ovulation and the mechanism of follicle rupture. V. Ultrastructure of tunica albuginea and theca externa of rabbit Graafian follicles prior to induced ovulation. *Cell Tissue Res.* 153:15–30, 1974e.

_____. Ovulation and the mechanism of follicle rupture. VI. Ultrastructure of theca interna and the inner vascular network surrounding rabbit Graafian follicles prior to induced ovulation. *Cell Tissue Res.* 153:31–44, 1974f.

BROWNE, C. L., H. S. WILEY, and J. N. DUMONT. Oocyte-follicle cell gap junctions in *Xenopus laevis* and the effects of gonadotropin on their permeability. *Science* 203:182–183, 1979.

BRYANT, G. D., and W. A. CHAMLEY. Plasma relaxin and prolactin immunoactivities in pregnancy and at parturition in the ewe. *J. Reprod. Fertil.* 48:201–204, 1976.

BULLOCK, D. W., and B. H. KAPPAUF. Dissociation of gonadotropin-induced ovulation and steroidogenesis in immature rats. *Endocrinology* 92:1625–1628, 1973.

BYSKOV, A. G. The role of the rete ovarii in meiosis and follicle formation in the cat, mink and ferret. *J. Reprod. Fertil.* 45:201–209, 1975.

_____. The anatomy and ultrastructure of the rete system in the fetal mouse ovary. *Biol. Reprod.* 19:720–735, 1978.

CALLARD, I. P., S. W. C. CHAN, and M. A. POTTS. The control of the reptilian gonad. *Am. Zool.* 12:273–287, 1972.

CAMPBELL, C. M., and D. R. IDLER. Hormonal control of vitellogenesis in hypophysectomized winter flounder (*Pseudopleuronectes americanus* Walbaum). *Gen. Comp. Endocrinol.* 28:143–150, 1976.

CAMPBELL, C. M., J. M. WALSH, and D. R. IDLER. Steroids in the plasma of the winter flounder (*Pseudopleuronectes americanus* Walbaum). A seasonal study and investigation of steroid involvement in oocyte maturation. *Gen. Comp. Endocrinol.* 2:14–20, 1976.

CHAMLEY, W. A., T. STELMASIAK, and G. D. BRYANT. Plasma relaxin immunoactivity during the oestrous cycle of the ewe. *J. Reprod. Fertil.* 45:455–461, 1975.

CHANNING, C. P. General discussion—steroidogenesis. *Adv. Exp. Med. Biol.* 112:209–210, 1979a.

_____. Follicular non-steroidal regulators. *Adv. Exp. Med. Biol.* 112:327–343, 1979b.

CHANNING, C. P., S. L. STONE, A. S. KRIPNER, and S. H. POMERANTZ. Studies on an oocyte maturation inhibitor present in porcine follicular fluid. In *Novel Aspects of Reproductive Physiology,* eds. C. H. Spilman and J. W. Wilks, pp. 37–54. New York: Spectrum Publications, 1978.

COLOMBO, L., P. C. BELVEDERE, P. PRANDO, P. SCAFFAI, and T. CISOTTO. Biosynthesis of 11-deoxycorticosteroids and androgens by the ovary of the newt *Triturus alpestris alpestris* Laur. *Gen. Comp. Endocrinol.* 33:480–495, 1977.

COLOMBO, L., H. A. BERN, J. PIEPRZYK, and D. W. JOHNSON. Biosynthesis of 11-deoxycorticosteroids by teleost ovaries and discussion of their possible role in oocyte maturation and ovulation. *Gen. Comp. Endocrinol.* 21:168–178, 1973.

CRAIK, J. C. A. Effects of hypophysectomy on vitellogenesis in the elasmobranch *Scyliorhinus canicula* L. *Gen. Comp. Endocrinol.* 36:63–67, 1978.

CREWS, D., and P. LICHT. Inhibition by corpora atretica of ovarian sensitivity to environmental and hormonal stimulation of the lizard, *Anolis carolinensis. Endocrinology* 95:102–106, 1974.

CROZE, F., and R. J. ETCHES. The ovulation-inducing properties of various androgens in the domestic fowl. *Poult. Sci.* 56:1706–1707, 1977 (Abstract).

DAY, S. L., and A. V. NALBANDOV. Presence of prostaglandin F (PGF) in hen follicles and its physiological role in ovulation and oviposition. *Biol. Reprod.* 16:486–494, 1977.

deVLAMING, V. L. Environmental and endocrine control of teleost reproduction. In *Control of Sex in Fishes,* ed. C. B. Schreck, pp. 13–83. Blacksburg; Virginia Polytechnic Institute and State University, 1974.

DICK, H. R., J. CULBERT, J. W. WELLS, A. B. GILBERT, and M. F. DAVIDSON. Steroid hormones in the postovulatory follicle of the domestic fowl (*Gallus domesticus*). *J. Reprod. Fertil.* 53:103–107, 1978.

DODD, J. M. Ovarian control in cyclostomes and elasmobranchs. *Am. Zool.* 12:325–339, 1972.

_____. The hormones of sex and reproduction and their effects in fish and lower chordates: twenty years on. *Am. Zool.* 15 (*Suppl.* 1):137–171, 1975.

DODSON, K. S., and J. WATSON. Effects of oestradiol-17β and gonadotrophin on prostaglandin production by pre-ovulatory pig follicles superfused *in vitro. Adv. Exp. Med. Biol.* 112:95–103, 1979.

DORRINGTON, J. H., and R. KILPATRICK. Effect of pituitary hormones on progestational hormone production by the rabbit ovary *in vivo* and *in vitro. J. Endocrinol.* 35:53–63, 1966.

EAYERS, J. T., E. GLASS, and H. H. SWANSON. The ovary and nervous system in relation to behavior. In *The Ovary,* eds. Lord Zuckerman and B. J. Weir, Vol. 2, pp. 399–457. New York: Academic Press, 1977.

ECKSTEIN, P. Endocrine activities of the ovary. In *The Ovary,* eds. Lord Zuckerman and B. J. Weir, Vol. 2, 275–313. New York: Academic Press, 1977.

EDWARDS, R. G., R. E. FOWLER, R. E. GORE-LANGTON, R. G. GOSDEN, E. C. JONES, C. READHEAD, and P. C. STEPTOE. Normal and abnormal follicular growth in mouse, rat and human ovaries. *J. Reprod. Fertil.* 51:237–263, 1977.

ELBAUM, D. J., and P. L. KEYES. Synthesis of 17β-estradiol by isolated ovarian tissues of the pregnant rat: aromatization in the corpus luteum. *Endocrinology* 99:573–579, 1976.

EL-FOULY, M. A., B. COOK, M. NEKOLA, and A. V. NALBANDOV. Role of the ovum in follicular luteinization. *Endocrinology* 87:288–293, 1970.

ERICKSON, G. F., and K. J. RYAN. Spontaneous maturation of oocytes isolated from ovaries of immature hypophysectomized rats. *J. Exp. Zool.* 195:153–158, 1976.

ESPEY, L. L. Ovarian proteolytic enzymes and ovulation. *Biol. Reprod.* 10:216–235, 1974.

ESPEY, L. L. Ovulation as an inflammatory reaction—a hypothesis. *Biol. Reprod.* 22:73–106, 1980.

ESPEY, L. L., and P. J. Coons. Factors which influence ovulatory degradation of rabbit ovarian follicles. *Biol. Reprod.* 14:233–245, 1976.

FOLLETT, B. K., and M. R. REDSHAW. The physiology of vitellogenesis. In *Physiology of the Amphibia,* ed. B. Lofts, Vol. 2, pp. 219–308. New York: Academic Press, 1974.

FORTUNE, J. E., and D. T. ARMSTRONG. Androgen production by theca and granulosa isolated from proestrous rat follicles. *Endocrinology* 100:1341–1347, 1977.

_____. Hormonal control of 17β-estradiol biosynthesis in proestrous rat follicles: estradiol production by isolated theca *versus* granulosa. *Endocrinology* 102:227–235, 1978.

FORTUNE, J. E., P. W. CONCANNON, and W. HANSEL. Ovarian progesterone levels during *in vitro* oocyte maturation and ovulation in *Xenopus laevis*. *Biol. Reprod.* 13:561-567, 1975.

FORTUNE, J. E., and W. HANSEL. Modulation of thecal progesterone secretion by estradiol-17β. *Adv. Exp. Med. Biol.* 112:203-208, 1979.

GOETZ, F. W., and H. L. BERGMAN. The effects of steroids on final maturation and ovulation of oocytes from brook trout (Salvelinus fontinalis) and yellow perch (Perca flavescens). *Biol. Reprod.* 18:293-298, 1978.

GOSPODAROWICZ, D. La production "in vitro" d'androgènes par des corps jaunes de lapine. *Biochim. Biophys. Acta* 100:618-620, 1965.

GOSWAMI, S. V., and B. I. SUNDARARAJ. *In vitro* maturation and ovulation of oocytes of the catfish, *Heteropneustes fossilis* (Bloch): effects of mammalian hypophyseal hormones, catfish pituitary homogenate, steroid precursors and metabolites, and gonadal and adrenocortical steroids. *J. Exp. Zool:* 178:467-478, 1971.

GREENWALD, G. S. Preovulatory changes in ovulating hormone in the cyclic hamster. *Endocrinology* 88:671-677, 1971.

————. Role of follicle-stimulating hormone and luteinizing hormone in follicular development and ovulation. In *Handbook of Physiology*, Sec. 7, Vol. 4, Part 2, pp. 293-323. Washington, D.C.: American Physiological Society, 1974.

GUERRIER, P., M. MOREAU, and M. DORÉE. Inhibition de la réinitiation de la méiose des ovocytes de *Xenopus laevis* par trois antiprotéases naturelles, l'antipaïne, la chymostatine et la leupeptine. *C. R. Hebd. Séances Acad. Sci. Ser. D Sci. Nat. (Paris)* 284:317-319, 1977.

HILLENSJÖ, T., A. S. KRIPNER, S. H. POMERANTZ, and C. P. CHANNING. Action of porcine follicular fluid oocyte maturation in vitro: possible role of the cumulus cells. *Adv. Exp. Med. Biol.* 112:283-290, 1979.

HILLIARD, J. Corpus luteum function in guinea pigs, hamsters, rats, mice and rabbits. *Biol. Reprod.* 8:203-221, 1973.

HILLIARD, J., D. ARCHIBALD, and C. H. SAWYER. Gonadotropic activation of preovulatory synthesis and release of progestin in the rabbit. *Endocrinology* 72:59-66, 1963.

HILLIER, S. G., R. A. KNAZEK, and G. T. ROSS. Androgenic stimulation of progesterone production by granulosa cells from preantral ovarian follicles: further *in vitro* studies using replicate cell cultures. *Endocrinology* 100:1539-1549, 1977.

HIROSE, K. Endocrine control of ovulation in medaka (*Oryzias latipes*) and ayu (*Plecoglossus altivelis*). *J. Fish. Res. Board Can.* 33:989-994, 1976.

HIRSCHFIELD, A. N., and A. R. MIDGLEY, JR. The role of FSH in the selection of large ovarian follicles in the rat. *Biol. Reprod.* 19:606-611, 1978.

HOFFMAN, J. C., and N. B. SCHWARTZ. Attenuation of the proestrous primary FSH surge, but not the LH surge by porcine follicular fluid (PPF). *Fed. Proc.* 37:724, 1978 (Abstract 2686).

HUANG, E. S-R. General discussion—steroidogenesis. *Adv. Exp. Med. Biol.* 112:210-211, 1979.

HUANG, E. S-R., K. J. KAO, and A. V. NALBANDOV. Synthesis of sex steroids by cellular components of chicken follicles. *Biol. Reprod.* 20:454-461, 1979.

HUANG, E. S-R., and A. V. NALBANDOV. Testosterone synthesis by chicken follicular cells. *Adv. Exp. Med. Biol.* 112:197-202, 1979a.

————. Steroidogenesis of chicken granulosa and theca cells: in vitro incubation system. *Biol. Reprod.* 20:442-453, 1979b.

HUNTER, R. H. F., B. COOK, and T. G. BAKER. Dissociation of response to injected gonadotropin between the Graafian follicle and oocyte in pigs. *Nature* 260:156-157, 1976.

JALABERT, B. In vitro oocyte maturation and ovulation in rainbow trout (*Salmo gairdneri*), Northern pike (*Esox lucius*), and goldfish (*Carassius auratus*). *J. Fish. Res. Board Can.* 33:974-988, 1976.

JONES, R. E., K. T. FITZGERALD, and D. DUVALL. Quantitative analysis of the ovarian cycle of the lizard *Lepidodactylus lugubris*. *Gen. Comp. Endocrinol.* 35:70-76, 1978.

JONES, R. E., R. R. TOKARZ, F. T. LAGREEK, and K. T. FITZGERALD. Endocrine control of clutch size in reptiles. VI. Patterns of FSH-induced ovarian stimulation in adult *Anolis carolinensis*. *Gen. Comp. Endocrinol.* 30:101-116, 1976.

KAO, L. W. L., and A. V. NALBANDOV. The effect of antiadrenergic drugs on ovulation in hens. *Endocrinology* 90:1343-1349, 1972.

KATZ, Y., and B. ECKSTEIN. Changes in steroid concentration in the blood of female *Tilapia aurea* (Teleostei, Cichlidae) during initiation of spawning. *Endocrinology* 95:963-967, 1974.

KENDALL, J. Z., C. G. PLOPPER, and G. D. BRYANT-GREENWOOD. Ultrastructural immunoperoxidase demonstration of relaxin in corpora lutea from a pregnant rat. *Biol. Reprod.* 18:94-98, 1978.

KENNELLY, J. J., and R. H. FOOTE. Oocytogenesis in rabbits. The role of neogenesis in the formation of the definitive ova and the stability of oocyte DNA measured with tritiated thymidine. *Am. J. Anat.* 118:573-590, 1966.

LANCE, V., and I. P. CALLARD. Steroidogenesis by enzyme-dispersed turtle (*Chrysemys picta*) ovarian cells in response to ovine gonadotropins (FSH and LH). *Gen. Comp. Endocrinol.* 34:304-311, 1978.

LARKIN, L. H., P. A. FIELDS, and R. M. OLIVER. Production of antisera against electrophoretically separated relaxin and immunofluorescent localization of relaxin in the porcine corpus luteum. *Endocrinology* 101:679-685, 1977.

LARSEN, L. O. Effects of hypophysectomy in cyclostome, *Lampetra fluviatilis* (L) Gray. *Gen. Comp. Endocrinol.* 5:16-30, 1965.

————. Development in adult, freshwater river lampreys and its hormonal control. Thesis, University of Copenhagen, 1973.

LEDWITZ-RIGBY, F., and B. W. RIGBY. The influence of follicular fluid on progesterone secretion by porcine granulosa cells in vitro. *Adv. Exp. Med. Biol.* 112:347–359, 1979.

LICHT, P., H. PAPKOFF, S. W. FARMER, C. H. MULLER, H. W. TSUI, and D. CREWS. Evolution of gonadotropin structure and function. *Recent Prog. Horm. Res.* 33:169–243, 1977.

LINDNER, H. R., A. AMSTERDAM, Y. SALOMON, A. TSAFRIRI, A. NIMROD, S. A. LAMPRECHT, U. ZOR, and Y. KOCH. Intraovarian factors in ovulation: determinants of follicular response to gonadotrophins. *J. Reprod. Fertil.* 51:215–235, 1977.

LIPNER, H. Mechanism of mammalian ovulation. In *Handbook of Physiology*, Sec. 7, Vol. 2, Part 1, pp. 409–437. Washington, D.C.: American Physiological Society, 1973.

LOUVET, J. P., S. MITCHELL HARMAN, J. R. SCHREIBER, and G. T. ROSS. Evidence for a role of androgens in follicular maturation. *Endocrinology* 97:366–372, 1975.

LUCKY, A. W., J. R. SCHREIBER, S. G. HILLIER, J. D. SCHULMAN, and G. T. ROSS. Progesterone stimulation by cultured preantral rat granulosa cells: stimulation by androgens. *Endocrinology* 100:128–133, 1977.

LUNENFELD, B., Z KRAIEM, and A. ESHKOL. The function of the growing of follicle. *J. Reprod. Fertil.* 45:567–574, 1975.

MACDONALD, G. J., D. T. ARMSTRONG, and R. O. GREEP. Stimulation of estrogen secretion from normal rat corpora lutea by luteinizing hormone. *Endocrinology* 79:289–293, 1966.

————. Initiation of blastocyst implantation by luteinizing hormone. *Endocrinology* 80:172–176, 1967.

MACDONALD, G. J., P. L. KEYES, and R. O. GREEP. Steroid secreting capacity of X-irradiated rat ovaries. *Endocrinology* 84:1004–1008, 1969.

MALLER, J. L., and E. G. KREBS. Progesterone-stimulated meiotic cell division in *Xenopus* oocytes. *J. Biol. Chem.* 252:1712–1718, 1977.

MARDER, M. L., C. P. CHANNING, and N. B. SCHWARTZ. Suppression of serum follicle stimulating hormone in intact and acutely ovariectomized rats by porcine follicular fluid. *Endocrinology* 101:1639–1642, 1977.

MATTY, A. J., K. TSUNEKI, W. W. DICKHOFF, and A. GORBMAN. Thyroid and gonadal function in hypophysectomized hagfish, *Eptatretus stouti. Gen. Comp. Endocrinol.* 30:500–516, 1976.

MORRILL, G. A., F. SCHATZ, A. B. KOSTELLOW, and J. M. POUPKO. Changes in cyclic AMP levels in the amphibian ovarian follicle following progesterone induction of meiotic maturation. *Differentiation* 8:97–104, 1977.

MOSSMAN, H. W., and K. L. DUKE. *Comparative Morphology of the Mammalian Ovary.* Madison: University of Wisconsin Press, 1973.

MULNER, O., C. THIBIER, and R. OZON. Steroid biosynthesis by ovarian follicles of *Xenopus laevis, in vitro* during oogenesis. *Gen. Comp. Endocrinol.* 34:287–295, 1978.

NAKAJO, S., A. H. ZAKARIA, and K. IMAI. Effect of local administration of proteolytic enzymes on the rupture of the ovarian follicle in the domestic fowl, *Gallus domesticus. J. Reprod. Fertil.* 34:235–240, 1973.

NAKANO, R., T. MIZUNO, K. KATAYAMA, and S. TOJO. Growth of ovarian follicles in the absence of gonadotrophins. *J. Reprod. Fertil.* 45:545–546, 1975.

NALBANDOV, A. V. *Reproductive Physiology of Mammals and Birds,* third ed. San Francisco: W. H. Freeman, 1976.

NALBANDOV, A. V., and M. F. JAMES. The blood-vascular system of the chicken ovary. *Am. J. Anat.* 85:347–377, 1949.

NEKOLA, M. V., and A. V. NALBANDOV. Morphological changes of rat follicular cells as influenced by oocytes. *Biol. Reprod.* 4:154–160, 1971.

NG, T. B., and D. R. IDLER. "Big" and "little" forms of plaice vitellogenic and maturational hormones. *Gen. Comp. Endocrinol.* 34:408–420, 1978.

O'BYRNE, E. M., W. K. SAWYER, M. C. BUTLER, and B. G. STEINETZ. Serum immuno reactive relaxin and softening of the uterine cervix in pregnant hamsters. *Endocrinology* 99:1333–1335, 1976.

O'BYRNE, E. M., and B. G. STEINETZ. Radioimmunoassay (RIA) of relaxin in sera of various species using an antiserum to porcine relaxin. *Proc. Soc. Exp. Biol. Med.* 152:272–276, 1976.

O'MALLEY, B. W., S. L. WOO, S. E. HARRIS, J. M. ROSEN, J. P. COMSTOCK, L. CHAN, C. B. BORDELON, J. W. HOLDER, P. SPERRY, and A. R. MEANS. Steroid hormone action in animal cells. *Am. Zool.* 15 (*Suppl.* 1):215–225, 1975.

OPEL, H., and A. V. NALBANDOV. Onset of follicular atresia following hypophysectomy of the laying hen. *Proc. Soc. Exp. Biol. Med.* 107:233–235, 1961.

————. Ovulability of ovarian follicles in the hypophysectomized hen. *Endocrinology* 69:1029–1035, 1962.

PARR, E. L. Rupture of ovarian follicles at ovulation. *J. Reprod. Fertil. Suppl.* 22:1–17, 1975.

PEARSON, O. P. Reproduction in the shrew (*Blarina previcauda* Say). *Am. J. Anat.* 75:39–93, 1944.

PETERS, H., A. G. BYSKOV, and M. FABER. Intraovarian regulation of follicle growth in the immature mouse. In *The Development and Maturation of the Ovary and Its Functions*, ed. H. Peters, pp. 20–23. Amsterdam: Excerpta Medica, 1973.

PETERS, H., A. G. BYSKOV, R. HIMELSTEIN-BRAW, and M. FABER. Follicular growth: the basic event in the mouse and human ovary. *J. Reprod. Fertil.* 45:559–566, 1975.

PHILLIPS, R. E., and A. VAN TIENHOVEN. Endocrine

factors involved in the failure of pintail ducks *Anas acuta* to reproduce in captivity. *J. Endocrinol.* 21:253–261, 1960.

REDSHAW, M. R. The hormonal control of the amphibian ovary. *Am. Zool.* 12:289–306, 1972.

REINBOTH, R. Hormonal control of the teleost ovary. *Am. Zool.* 12:307–324, 1972.

REMACLE, C., P. DELAERE, and P. JACQUET. Actions hormonales sur les cellules germinals femelles de *Carassius auratus* L. en culture organotypique. Renversement sexual en ovogenèse *in vitro. Gen. Comp. Endocrinol.* 29:212–224, 1976.

RICHARDS, JOANNE S. Estradiol receptor content in rat granulosa cells during follicular development: modification by estradiol and gonadotropins. *Endocrinology* 97:1174–1184, 1975.

ROTHCHILD, I. Summing up. *Adv. Exp. Med. Biol.* 112:767–789, 1979.

ROWLANDS, I. W., and B. J. WEIR. The ovarian cycle in vertebrates. In *The Ovary,* eds. Lord Zuckerman and B. J. Weir, Vol. 2, pp. 217–274. New York: Academic Press, 1977.

SAMSONOVITCH, M., and P. C. LAGUË. Effects of prostaglandin E_1, E_2 and indomethacin on ovulation in the domestic fowl. *Poult. Sci.* 56:1754, 1977 (Abstract).

SAVARD, K. The biochemistry of the corpus luteum. *Biol. Reprod.* 8:183–202, 1973.

SCHATZ, F., and G. A. MORRILL. Effects of dibutyryl 3', 5'-AMP, caffeine and theophylline on meiotic maturation and ovulation in *R. pipiens. Fifth Intern. Congr. Pharmacol.,* San Francisco, 1972.

SCHIMKE, R. T., G. S. MCKNIGHT, D. J. SHAPIRO, D. SULLIVAN, and R. PALACIOS. Hormonal regulation of ovalbumin synthesis in the chick oviduct. *Recent Prog. Horm. Res.* 31:175–208, 1975.

SCHOMBERG, D. W. Steroidal modulation of steroid secretion *in vitro*: an experimental approach to intrafollicular regulatory mechanisms. *Adv. Exp. Med. Biol.* 112:155–168, 1979.

SCHREIBER, J. R., R. REID, and G. T. ROSS. A receptor-like testosterone-binding protein in ovaries from estrogen-stimulated hypophysectomized immature female rats. *Endocrinology* 98:1206–1213, 1976.

SCHREIBER, J. R., and G. T. ROSS. Further characterization of a rat ovarian testosterone receptor with evidence for nuclear translocation. *Endocrinology* 99:590–596, 1976.

SCHUETZ, A. W. Role of hormones in oocyte maturation. *Biol. Reprod.* 10:150–178, 1974.

———. Induction of oocyte maturation and differentiation: mode of progesterone action. *Ann. N.Y. Acad. Sci.* 286:408–419, 1977.

———. The somatic-germ cell complex: interactions and transformations in relation to oocyte maturation. *Adv. Exp. Med. Biol.* 112:307–314, 1979.

SCHWARTZ, N. B. The role of FSH and LH and of their antibodies on follicle growth and on ovulation. *Biol. Reprod.* 10:236–272, 1974.

SENIOR, B. E. Oestradiol concentration in the peripheral plasma of the domestic hen from 7 weeks of age until the time of sexual maturity. *J. Reprod. Fertil.* 41:107–112, 1974.

SHAFFNER, C. S. Feather papilla stimulation by progesterone. *Science* 120:345, 1954.

———. Progesterone induced molt. *Poult. Sci.* 34:840–842, 1955.

SHAHABI, N. A. H., W. NORTON, and A. V. NALBANDOV. Steroid levels in follicles and the plasma of hens during the ovulatory cycle. *Endocrinology* 96:962–968, 1975.

SHERWOOD, O. D., and E. M. O'BYRNE. Purification and characterization of porcine relaxin. *Arch. Biochem. Biophys.* 160:185–196, 1976.

SHERWOOD, O. D., K. R. ROSENTRETER, and M. L. BIRKHIMER. Development of a radio immunoassay for porcine relaxin using ^{125}I-labeled polytyrosyl-relaxin. *Endocrinology* 96:1106–1113, 1975.

SLIJKHUIS, H. Ultrastructural evidence for two types of gonadotropic cells in the pituitary gland of the male three-spined stickleback, *Gasterosteus aculeatus. Gen. Comp. Endocrinol.* 36:639–641, 1978.

SMITH, H. M., G. SINELNIK, J. D. FAWCETT, and R. E. JONES. A survey of the chronology of ovulation in anoline lizard genera. *Trans. Kans. Acad. Sci.* 75:107–120, 1972.

SPEAKER, M. G., and F. R. BUTCHER. Cyclic nucleotide fluctuations during steroid induced meiotic maturation of frog oocytes. *Nature* 267:848–850, 1977.

STACEY, N. E., A. F. COOK, and R. E. PETER. Ovulatory surge of gonadotropin in the goldfish. *Carassius auratus. Gen. Comp. Endocrinol.* 37:246–249, 1979.

STEINETZ, B. G. Secretion and function of ovarian estrogens. In *Handbook of Physiology*, Sec. 7, Vol. 2, Part 1, pp. 439–466. Washington, D.C.: American Physiological Society, 1973.

STOCKELL-HARTREE, A., and F. J. CUNNINGHAM. The pituitary gland. In *Physiology and Biochemistry of the Domestic Fowl,* eds. D. J. Bell and B. M. Freeman, pp. 427–457. New York: Academic Press, 1971.

STOKLOSOWA, S., and A. V. NALBANDOV. Luteinization and steroidogenic activity of rat ovarian follicles cultured *in vitro. Endocrinology* 91:25–32, 1972.

SUMPTER, J. P., B. K. FOLLETT, N. JENKINS, and J. M. DODD. Studies on the purification and properties of gonadotrophin from ventral lobes of the pituitary gland of the dogfish (*Scyliorhinus canicula L.*). *Gen. Comp. Endocrinol.* 36:264–274, 1978.

SUNDARARAJ, B. I., T. C. ANAND, and E. M. DONALDSON. Effects of partially purified salmon pituitary gonadotropin on ovarian maintenance, ovulation, and vitellogenesis in the hypophysectomized catfish, *Heteropneustes fossilis* (Bloch). *Gen. Comp. Endocrinol.* 18:102–114, 1972a.

SUNDARARAJ, B. I., T. C. ANAND, and V. R. P. SINHA.

Effects of carp pituitary fractions on vitellogenesis, ovarian maintenance, and ovulation in hypophysectomized catfish, *Heteropneustes fossilis* (Bloch). *J. Endocrinol.* 54:87–98, 1972b.

SUNDARARAJ, B. I., and S. V. GOSWAMI. Role of interrenal in luteinizing hormone-induced ovulation and spawning of the catfish, *Heteropneustes fossilis* (Bloch). *Gen. Comp. Endocrinol. Suppl.* 2:374–384, 1969.

SUNDARARAJ, B. I., S. K. NAYYAR, E. BURZAWA-GÉRARD, and Y. A. FONTAINE. Effects of carp gonadotropin on ovarian maintenance, maturation, and ovulation in hypophysectomized catfish, *Heteropneustes fossilis* (Bloch). *Gen. Comp. Endocrinol.* 30: 472–476, 1976.

TAKAHASHI, M. J., J. J. FORD, K. YOSHINAGA, and R. O. GREEP. Induction of ovulation in hypophysectomized rats by progesterone. *Endocrinology* 95:1322–1326, 1974.

TERRANOVA, P. F., J. S. CONNOR, and G. S. GREENWALD. In vitro steroidogenesis in corpora lutea and nonluteal ovarian tissues of the cyclic hamster. *Biol. Reprod.* 19:249–255, 1978.

THIBAULT, C. Are follicular maturation and oocyte maturation independent processes? *J. Reprod. Fertil.* 51:1–15, 1977.

TOKARZ, R. R. An autoradiographic study of the effects of mammalian gonadotropins (Follicle-Stimulating hormone and Luteinizing hormone) and estradiol-17β on [^{3}H] thymidine labeling of surface epithelial cells, prefollicular cells, and oogonia in the ovary of the lizard *Anolis carolinensis. Gen. Comp. Endocrinol.* 35:179–188, 1978.

TRUSCOTT, B., D. R. IDLER, B. I. SUNDARARAJ, and S. V. GOSWAMI. Effects of gonadotropins and adrenocorticotropin on plasmatic steroids of the catfish *Heteropneustes fossilis* (Bloch). *Gen. Comp. Endocrinol.* 34:149–157, 1978.

TSAFRIRI, A. Mammalian oocyte maturation: model systems and their physiological relevance. *Adv. Exp. Med. Biol.* 112:269–281, 1979.

TSAFRIRI, A., S. H. POMERANTZ, and C. P. CHANNING. Inhibition of oocyte maturation by porcine follicular fluid: partial characterization of the inhibitor. *Biol. Reprod.* 14:511–516, 1976.

UNGAR, F., R. GUNVILLE, B. I. SUNDARARAJ, and S. V. GOSWAMI. Formation of 3α-hydroxy-5 β pregnan-20-one in the ovaries of catfish *Heteropneustes fossilis* (Bloch). *Gen. Comp. Endocrinol.* 31:53–59, 1977.

URIST, M. R., and A. O. SCHJEIDE. The partition of calcium and protein in the blood of oviparous vertebrates during estrus. *J. Gen. Physiol.* 44:743–756, 1961.

VAN TIENHOVEN, A. Endocrinology of reproduction in birds. In *Sex and Internal Secretions,* ed. W. C. Young, Vol. 2, pp. 1088–1169. Baltimore: Williams and Wilkins, 1961.

———. *Reproductive Physiology of Vertebrates.* Philadelphia: W. B. Saunders, 1968.

VEVERS, N. G. The influence of the ovaries on secondary sexual characters. In *The Ovary,* eds. Lord Zuckerman and B. J. Weir, Vol. 1, pp. 447–473. New York: Academic Press, 1977.

WASSARMAN, P. M., R. M. SCHULTZ, G. E. LETOURNEAU, M. J. LaMARCA, W. J. JOSEFOWICZ, and J. D. BLEIL. Meiotic maturation of mouse oocytes *in vitro. Adv. Exp. Med. Biol.* 112:251–267, 1979.

WAYNFORTH, H. B., and D. M. ROBERTSON. Oestradiol content of ovarian venous blood and ovarian tissue in hypophysectomized rats during late pregnancy. *J. Endocrinol.* 54:79–85, 1972.

WEISS, G., E. M. O'BYRNE, and B. G. STEINETZ. Relaxin: a product of the human corpus luteum of pregnancy. *Science* 194:948–949, 1976.

WILSON, S. C., and P. J. SHARP. Changes in plasma concentrations of luteinizing hormone after injection of progesterone at various times during the ovulatory cycle of the domestic hen (*Gallus domesticus*). *J. Endocrinol.* 67:59–70, 1975.

———. Induction of luteinizing hormone release by gonadal steroids in the ovariectomized domestic hen. *J. Endocrinol.* 71:87–98, 1976.

WOODS, J. E., and L. V. DOMM. A histochemical identification of the androgen-producing cells in the gonads of the domestic fowl and albino rat. *Gen. Comp. Endocrinol.* 7:559–570, 1966.

ZUCKERMAN, S., and T. G. BAKER. The development of the ovary and the process of oogenesis. In *The Ovary,* eds. Lord Zuckerman and B. J. Weir, Vol. 1, pp. 41–67. New York: Academic Press, 1977.

8 | Reproductive Cycles

ADLER, N. T. The behavioral control of reproductive physiology. *Adv. Behav. Physiol.* 11:259–286, 1974.

ADLER, N. T., P. G. DAVIS, and B. R. KOMISARUK. Variation in the size and sensitivity of a genital sensory field in relation to the estrous cycle in rats. *Horm. Behav.* 9:334–344, 1977.

AKAKA, J., E. O'LAUGHLIN-PHILLIPS, E. ANTCZAK,

and I. ROTHCHILD. The relation between the age of the corpus luteum (CL) and the luteolytic effect of an LH-antiserum (LH-AS): comparison of hysterectomized pseudopregnant rats with intact pregnant rats for their response to LH-AS treatment at four stages of CL activity. *Endocrinology* 100:1334–1340, 1977.

ANDERSON, L. L. Effects of hysterectomy and other

factors of luteal function. In *Handbook of Physiology*, Sec. 7, Vol. 2, Part 2, pp. 69–86. Washington, D.C.: American Physiological Society, 1973.

ANDERSON, L. L., G. W. DYCK, H. MORI, D. M. HENRICKS, and R. M. MELAMPY. Ovarian function in pigs following hypophysial stalk transsection or hypophysectomy. *Am. J. Physiol.* 212:1188–1194, 1967.

ANDERSON, L. L., P. C. LÉGLISE, F. DU MESNIL DU BUISSON, and P. ROMBAUTS. Interaction des hormones gonadotropes et de l'utérus dans le maintien du tissu lutéal ovarien chez la truie. *C. R. Hebd. Séances Acad. Sci.* (Paris) 261:3675–3678, 1965.

ANDERSON, L. L., and R. M. MELAMPY. Hypophysial and uterine influences on pig luteal function. In *Reproduction in the Female Mammal,* eds. G. E. Lamming and E. C. Amoroso, pp. 285–313. London: Butterworths, 1967.

ASDELL, S. A. *Patterns of Mammalian Reproduction,* second ed. Ithaca: Cornell University Press, 1964.

AULETTA, F. J., L. SPEROFF, and B. V. CALDWELL. Prostaglandin $F_{2\alpha}$ induced steroidogenesis and luteolysis in the primate corpus luteum. *J. Clin. Endocrinol. Metab.* 36:405–407, 1973.

AX, R. L., JR. Development and validation of a bioassay for chicken LH and application to chicken serum. Ph.D. Diss., University of Illinois, 1978.

BAGGERMAN, B. Het endocriene systeem en aanpassing van stekelbaarzen aan hun milieu. In *Endocrinologie,* eds. J. Lever and J. de Wilde, pp. 64–92. Wageningen, The Netherlands: Centrum voor Landbouwpublikaties and Landbouwdocumentatie, 1978.

BAGWELL, J. N., D. L. DAVIES, and J. R. RUBY. The effects of prostaglandin $F_{2\alpha}$ on the fine structure of the corpus luteum of the hysterectomized guinea pig. *Anat. Rec.* 183:229–242, 1975.

BAKER, J. R. The evolution of breeding seasons. In *Evolution,* ed. G. R. de Beer, pp. 161–177. Oxford: Clarendon Press, 1938.

BANKS, J. A., and M. E. FREEMAN. The temporal requirement of progesterone on proestrus for extinction of the estrogen-induced daily signal controlling luteinizing hormone release in the rat. *Endocrinology* 102:426–432, 1978.

BARRY, T. W. Effect of late seasons on Atlantic brant reproduction. *J. Wildl. Manage.* 26:19–26, 1962.

BEAL, W. E., J. H. LUKASZEWSKA, and W. HANSEL. Luteotropic effects of bovine blastocysts. *J. Anim. Sci.* 52:567–574, 1981.

BEAL, W. E., R. A. MILVAE, and W. HANSEL. Oestrous length and plasma progesterone concentrations following administration of prostaglandin F-2α early in the bovine oestrous cycle. *J. Reprod. Fertil.* 59:393–396, 1980.

BENTLEY, P. J. *Comparative Endocrinology.* Cambridge: Cambridge University Press, 1976.

BETTERIDGE, K. J., and D. MITCHELL. Direct evidence of retention of unfertilized ova in the oviduct of the mare. *J. Reprod. Fertil.* 39:145–148, 1974.

BEUVING, G., and G. M. A. VONDER. Daily rhythm of corticosterone in laying hens and the influence of egg laying. *J. Reprod. Fertil.* 51:169–173, 1977.

——. The influence of ovulation and oviposition on corticosterone levels in the plasma of laying hens. *Gen. Comp. Endocrinol.* 44:382–388, 1981.

BILLARD, R., and B. BRETON. Rhythms of reproduction in teleost fish. In *Rhythmic Activity of Fishes,* ed. J. E. Thorpe, pp. 31–53. New York: Academic Press, 1978.

BLATCHLEY, F. R., B. T. DONOVAN, E. W. HORTON, and N. L. POYSER. The release of prostaglandins and progestin into the utero-ovarian venous blood of guinea pigs during the oestrous cycle and following oestrogen treatment. *J. Physiol.* 223:69–88, 1972.

BOHNET, H. G., R. P. C. SHIU, D. GRINWICH, and H. G. FRIESEN. *In vivo* effects of antisera to prolactin receptors in female rats. *Endocrinology* 102:1657–1661, 1978.

BONNEY, R. C., F. J. CUNNINGHAM, and B. J. A. FURR. Effect of synthetic luteinizing hormone releasing hormone on plasma luteinizing hormone in the female domestic fowl (*Gallus domesticus*). *J. Endocrinol.* 63:539–547, 1974.

BREDER, C. M., JR., and D. E. ROSEN. *Modes of Reproduction in Fishes.* Garden City, New York: Natural History Press, 1966.

BROWN, K. M., and O. J. SEXTON. Stimulation of reproductive activity of female Anolis sagrei by moisture. *Physiol. Zool.* 46:168–172, 1973.

BURGER, J. W. On the relation of day length to the phases of testicular involution and inactivity of the spermatogenetic cycle of the starling. *J. Exp. Zool.* 105:259–268.

CALLARD, I. P., G. V. CALLARD, V. LANCE, and S. ECCLES. Seasonal changes in testicular structure and function and the effects of gonadotropins in the freshwater turtle, *Chrysemys picta. Gen. Comp. Endocrinol.* 30:347–356, 1976.

CALLARD, I. P., J. DOOLITTLE, W. L. BANKS, JR., and S. W. C. CHAN. Recent studies on the control of the reptilian ovarian cycle. *Gen. Comp. Endocrinol. Suppl.* 3:65–75, 1972.

CAMPBELL, C. M., J. M. WALSH, and D. R. IDLER. Steroids in the plasma of the winter flounder (*Pseudopleuronectes americanus* Walbaum). A seasonal study and investigation of steroid involvement in oocyte maturation. *Gen. Comp. Endocrinol.* 29:14–20, 1976.

CAMPER, P. M., and W. H. BURKE. The effects of prolactin on the gonadotropin induced rise in serum estradiol and progesterone of the laying turkey. *Gen. Comp. Endocrinol.* 32:72–77, 1977.

CHARLTON, H. M., F. NAFTOLIN, M. C. SOOD, and R. W. WORTH. The effect of mating upon LH release in male and female voles in the species *Microtus agrestis. J. Reprod. Fertil.* 42:167–170, 1975.

CHURCH, G. The reproductive cycles of the Javanese geckos, *Cosymbotus platyurus, Hemidactylus fre-*

natus and *Peropus mutilatus. Copeia:*262-269, 1962.

CONAWAY, C. H. Ecological adaptation and mammalian reproduction. *Biol. Reprod.* 4:239-247, 1971.

CONCANNON, P., R. COWAN, and W. HANSEL. LH release in ovariectomized dogs in response to estrogen withdrawal and its facilitation by progesterone. *Biol. Reprod.* 20:523-531, 1979a.

CONCANNON, P. W., W. HANSEL, and K. MCENTEE. Changes in LH, progesterone and sexual behavior associated with preovulatory luteinization in the bitch. *Biol. Reprod.* 17:604-613, 1977.

CONCANNON, P. W., W. HANSEL, and W. J. VISEK. The ovarian cycle of the bitch: plasma estrogen, LH and progesterone. *Biol. Reprod.* 13 : 112-121, 1975.

CONCANNON, P. W., N. WEIGAND, S. WILSON, and W. HANSEL. Sexual behavior in ovariectomized bitches in response to estrogen and progesterone treatments. *Biol. Reprod.* 20:799-809, 1979b.

CREWS, D. Psychobiology of reptilian reproduction. *Science* 189:1059-1065, 1975.

CRIM, L. W., E. G. WATTS, and D. M. EVANS. The plasma gonadotropin profile during sexual maturation in a variety of salmonid fishes. *Gen. Comp. Endocrinol.* 27:62-70, 1975.

CROZE, F., and R. J. ETCHES. The ovulation-inducing properties of various androgens in the domestic fowl. *Poult. Sci.* 56:1706-1707, 1977.

_____. The physiological significance of androgen-induced ovulation in the hen. *J. Endocrinol.* 84:163-171, 1980.

CUNNINGHAM, F. J., and B. J. A. FURR. Plasma levels of luteinizing hormone and progesterone during the ovulatory cycle of the hen. In *Egg Formation and Production,* eds. B. M. Freeman and P. E. Lake, pp. 51-64. Edinburgh: British Poultry Science, 1972.

DAY, D. L., and A. V. NALBANDOV. Presence of prostaglandin F (PGF) in hen follicles and its physiological role in ovulation and oviposition. *Biol. Reprod.* 16:486-494, 1977.

DEANESLY, R. Experimental observations on the ferret corpus luteum of pregnancy. *J. Reprod. Fertil.* 13:183-185, 1967.

DE GREEF, W. J., J. DULLAART, and G. H. ZEILMAKER. Serum concentrations of progesterone, luteinizing hormone, follicle stimulating hormone and prolactin in pseudopregnant rats: effect of decidualization. *Endocrinology* 101:1054-1063, 1977.

DE GREEF, W. J., and G. H. ZEILMAKER. Regulation of prolactin secretion during the luteal phase in the rat. *Endocrinology* 102:1190-1198, 1978.

DENAMUR, R. Luteotrophic factors in the sheep. *J. Reprod. Fertil.* 38:251-259, 1974.

DODD, J. M. Ovarian control in cyclostomes and elasmobranchs. *Am. Zool.* 12:325-339, 1972.

_____. The hormones of sex and reproduction and their effects in fish and lower chordates: twenty years on. *Am. Zool.* 15 (*Suppl.* 1):137-171, 1975.

DONOVAN, B. T. The effect of pituitary stalk section of luteal function in the ferret. *J. Endocrinol.* 27:201-211, 1963.

_____. The hypophysis and the control of corpus luteum function. In *Symposium on Reproduction. Congress of the Hungarian Society for Endocrinology and Metabolism,* ed. K. Lissák, pp. 81-98. Budapest: Akadémiai Kiadó, 1967.

DZIUK, P. J. Reproduction in pigs. In *Reproduction in Domestic Animals,* third ed., eds. H. H. Cole and P. T. Cupps, pp. 455-474. New York: Academic Press, 1977.

ETCHES, R. J. A radioimmunoassay for corticosterone and its application to the measurement of stress in poultry. *Steroids* 28:763-773, 1976.

_____. Plasma concentrations of progesterone and corticosterone during the ovulation cycle of the hen (*Gallus domesticus*). *Poult. Sci.* 58:211-216, 1979.

ETCHES, R. J., and F. J. CUNNINGHAM. The effect of pregnenolone, progesterone, deoxycorticosterone or corticosterone on the time of ovulation and oviposition in the hen. *Br. Poult. Sci.* 17:637-642, 1976a.

_____. The interrelationship between progesterone and luteinizing hormone during the ovulatory cycle of the hen (*Gallus domesticus*). *J. Endocrinol.* 71:51-58, 1976b.

_____. The plasma concentrations of testosterone and LH during the ovulation cycle of the hen (*Gallus domesticus*). *Acta Endocrinol.* 84:357-366, 1977.

ETCHES, R. J., A. S. MCNEILLY, and C. E. DUKE, Plasma concentrations of prolactin during the reproductive cycle of the domestic turkey (*Meleagris gallopavo*). *Poult. Sci.* 58:963-970, 1979.

FERNANDEZ-BACA, S., W. HANSEL, and C. NOVOA. Embryonic mortality in the alpaca. *Biol. Reprod.* 3:243-251, 1970a.

_____. Corpus luteum function in the alpaca. *Biol. Reprod.* 3:252-261, 1970b.

FERNANDEZ-BACA, S., W. HANSEL, R. SAATMAN, J. SUMAR, and C. NOVOA. Differential luteolytic effect of right and left uterine horns in the alpaca. *Biol. Reprod.* 20:586-595, 1979.

FERNANDEZ-BACA, S., D. H. L. MADDEN, and C. NOVOA. Effect of different mating stimuli on induction of ovulation in the alpaca. *J. Reprod. Fertil.* 22:261-267, 1970c.

FIRTH, H. J. Discussion. In *Breeding Biology of Birds,* ed. D. S. Farner, pp. 147-154. Washington, D.C.: National Academy of Sciences, 1973.

FLYNN, T. T., and J. P. HILL. The development of the monotremata. IV. Growth of the ovarian ovum, maturation, fertilization and early cleavage. *Trans. Zool. Soc. Lond.* 24:445-622, 1939.

FOLLETT, B. K., and D. T. DAVIES. The endocrine control of ovulation in birds. In *Beltsville Symposium, Agricultural Research III, Animal Reproduction,* ed. H. W. Hawk, pp. 323-344. Montclair, New Jersey: Alanheld, Osmun, and Co., 1979.

FRAPS, R. M. Egg production and fertility in poultry. In *Progress in the Physiology of Farm Animals,* ed. J. Hammond, Vol. 2, pp. 661–740. London: Butterworths, 1955.

FRASER, H. M., and P. J. SHARP. Prevention of positive feedback in the hen (*Gallus domesticus*) by antibodies to luteinizing hormone releasing hormone. *J. Endocrinol.* 76:181–182, 1978.

FREEMAN, M. E., M. S. SMITH, S. J. NAZIAN, and J. D. NEILL. Ovarian and hypothalamic control of the daily surges of prolactin secretion during pseudopregnancy in the rat. *Endocrinology* 94:875–882, 1974.

FREEMAN, M. E., and J. R. STERMAN. Ovarian steroid modulation of prolactin surges in cervically stimulated ovariectomized rats. *Endocrinology* 102:1915–1920, 1978.

FURR, B. J. A., R. C. BONNEY, R. J. ENGLAND, and F. J. CUNNINGHAM. Luteinizing hormone and progesterone in peripheral blood during the ovulatory cycle of the hen *Gallus domesticus. J. Endocrinol.* 57:159–169, 1973.

FURR, B. J. A., and G. K. SMITH. Effects of antisera against gonadal steroids on ovulation in the hen *Gallus domesticus. J. Endocrinol.* 66:303–304, 1975.

GENGENBACH, D. R., J. E. HIXON, and W. HANSEL. A luteolytic interaction between estradiol and prostaglandin $F_{2\alpha}$ in hysterectomized ewes. *Biol. Reprod.* 16:571–579, 1977.

GIBORI, G., R. RODWAY, and I. ROTHCHILD. The luteotrophic effect of estrogen in the rat: prevention by estradiol of the luteolytic effect of an antiserum to luteinizing hormone in the pregnant rat. *Endocrinology* 101:1683–1689, 1977.

GILMAN, D. P., L. F. MERCER, and J. C. HITT. Influence of female copulatory behavior on the induction of pseudopregnancy in the female rat. *Physiol. Behav.* 22:675–678, 1979.

GINTHER, O. J. Internal regulation of physiological processes through local venoarterial pathways: a review. *J. Anim. Sci.* 39:550–564, 1974.

GODING, J. R. Demonstration that $PGF_{2\alpha}$ is the uterine luteolysin in the ewe. *J. Reprod. Fertil.* 38:261–271, 1974.

GOODMAN, A. L., and J. D. Neill. Ovarian regulation of postcoital gonadotropin release in the rabbit: reexamination of a functional role for 20α dihydroprogesterone. *Endocrinology* 99:852–860, 1976.

GORBMAN, A., and H. A. BERN. *Textbook of Comparative Endocrinology.* New York: Wiley and Sons, 1962.

GRÄF, K. J. Serum oestrogen, progesterone and prolactin concentrations in cyclic, pregnant and lactating beagle dogs. *J. Reprod. Fertil.* 52:9–14, 1978.

GRAHAM, J. B. Low-temperature acclimation and the seasonal temperature sensitivity of some tropical marine fishes. *Physiol. Zool.* 45:1–13, 1972.

GREENWALD, G. S. Proestrous hormone surges dissociated from ovulation in the estrogen-treated hamster. *Endocrinology* 97:878–884, 1975.

GREEP, R. O. Regulation of luteal cell function. *Proc. Third Cong. Hormonal Steroids, Hamburg, 1970,* eds. V. H. T. James and L. Martini, pp. 670–677. Amsterdam: Excerpta Medica, 1971.

GRIFFITHS, M. *The Biology of the Monotremes.* New York: Academic Press, 1978.

GUERRO, R., T. ASO, P. F. BRENNER, Z. CEKAN, B. M. LANDGREN, K. HAGENFELDT, and E. DICZFALUSY. Studies on the patterns of circulating steroids in the normal menstrual cycle. I. Simultaneous assays of progesterone, pregnenolone, dehydroepiandrosterone, testosterone, dihydrotestosterone, androstenedione, oestradiol and oestrone. *Acta Endocrinol.* 81:133–149, 1976.

GWINNER, E. Circadian and circannual rhythms in birds. In *Avian Biology,* eds. D. S. Farner and J. R. King, Vol. 5, pp. 221–285. New York: Academic Press, 1975.

HAFEZ, E. S. E. Rabbits. In *Reproduction and Breeding Techniques for Laboratory Animals,* ed. E. S. E. Hafez, pp. 273–298. Philadelphia: Lea and Febiger, 1970.

HALL, T. R., and A. CHADWICK. Hypothalamic control of prolactin and growth hormone secretion in different vertebrate species. *Gen. Comp. Endocrinol.* 37:333–342, 1979.

HANKS, J. Comparative aspects of reproduction in the male hyrax and elephant. *Proc. Fourth Symp. Comparative Biology of Reproduction, Canberra, 1976,* pp. 155–164. Canberra: Australian Academy of Science, 1977.

HANSEL, W., P. W. CONCANNON, and J. H. LUKASZEWSKA. Corpora lutea of the large domestic animals. *Biol. Reprod.* 8:222–245, 1973.

HANSEL, W., and K. H. SIEFART. Maintenance of luteal function in the cow. *J. Dairy Sci.* 50:1948–1958, 1967.

HAUGER, R. L., F. J. KARSCH, and D. L. FOSTER. A new concept of the control of the estrous cycle of the ewe based on the temporal relationships between luteinizing hormone, estradiol and progesterone in peripheral serum and evidence that progesterone inhibits tonic LH secretion. *Endocrinology* 101:807–817, 1977.

HAYNES, N. B., K. J. COOPER, and M. J. KAY. Plasma progesterone concentration in the hen in relation to the ovulatory cycle. *Br. Poult. Sci.* 14: 349–357, 1973.

HEAP, R. B., and J. HAMMOND, JR. Plasma progesterone levels in pregnant and pseudopregnant ferrets. *J. Reprod. Fertil.* 39:149–152, 1974.

HEAP, R. B., J. S. PERRY, and I. W. ROWLANDS. Corpus luteum function in the guinea pig; arterial and luteal progesterone levels, and the effects of hysterectomy and hypophysectomy. *J. Reprod. Fertil.* 13:537–553, 1967.

HEARN, J. P. The pituitary gland and implantation in the tammar wallaby, *Macropus eugenii. J. Reprod. Fertil.* 39:235–241, 1974.

HEARN, J. P., R. V. SHORT, and D. T. BAIRD. Evolution of the luteotrophic control of the mammalian corpus luteum. *Proc. Fourth Symp. Comparative Biology of Reproduction, Canberra, 1976,* pp. 255–263. Canberra: Australian Academy of Science, 1977.

HERTELENDY, F., and H. V. BIELLIER. Prostaglandin levels in avian blood and reproductive organs. *Biol. Reprod.* 18:204–211, 1978.

HERTELENDY, F., M. YEH, and H. V. BIELLIER. Induction of oviposition in the domestic hen by prostaglandins. *Gen. Comp. Endocrinol.* 22:529–531, 1974.

HILL, C. J. The development of the monotremata. I. The histology of the oviduct during gestation. *Trans. Zool. Soc. Lond.* 21:413–443, 1933.

HILL, J. P., and J. B. GATENBY. The corpus luteum of the monotremata. *Proc. Zool. Soc. Lond.* 45:715–763, 1926.

HILLIARD, J. Corpus luteum function in guinea pigs, hamsters, rats, mice and rabbits. *Biol. Reprod.* 8:203–221, 1973.

HILLIARD, J., R. PENARDI, and C. H. SAWYER. A functional role for 20α-hydroxypregn-4-en-3-one in the rabbit. *Endocrinology* 80:901–909, 1967.

HILLIARD, J., H. G. SPIES, and C. H. SAWYER. Cholesterol storage and progestin secretion during pregnancy and pseudopregnancy in the rabbit. *Endocrinology* 82:157–165, 1968.

HOLTAN, D. W., T. M. NETT, and V. L. ESTEGREEN. Plasma progestins in pregnant postpartum and cycling mares. *J. Anim. Sci.* 40:251–260, 1975.

HORTON, E. W., and N. L. POYSER. Uterine luteolytic hormone: a physiological role for prostaglandin $F_{2\alpha}$. *Physiol. Rev.* 56:595–651, 1976.

HSUEH, S. C., and J. L. VOOGT. Effect of L-dopa on nocturnal prolactin surges during pseudopregnancy. *Biol. Reprod.* 13:223–227, 1975.

HUANG, E. S.-R., K. J. KAO, and A. V. NALBANDOV. Synthesis of sex steroids by cellular components of chicken follicles. *Biol. Reprod.* 20:454–461, 1979.

HUANG, E. S.-R., and A. V. NALBANDOV. Steroidogenesis of chicken granulosa and theca cells: in vitro incubation system. *Biol. Reprod.* 20:442–453, 1979.

ILLINGWORTH, D. V., and J. S. PERRY. The effect of hypophysial stalk section on the corpus luteum of the guinea-pig. *J. Endocrinol.* 50:625–635, 1971.

———. Effects of oestrogen administered early or late in the oestrous cycle, upon the survival and regression of the corpus luteum of the guinea-pig. *J. Reprod. Fertil.* 33:457–467, 1973.

IMAI, K., and A. V. NALBANDOV. Changes in FSH activity of anterior pituitary glands and of blood plasma during the laying cycle of the hen. *Endocrinology* 88:1465–1470, 1971.

IMMELMANN, K. Role of the environment in reproduction as source of "predictive" information. In *Breeding Biology of Birds,* ed. D. S. Farner, pp. 121–147. Washington, D.C.: National Academy of Sciences, 1973.

JÖCHLE, W. Coitus-induced ovulation. *Contraception* 7:523–564, 1973.

———. Current research in coitus-induced ovulation: a review. *J. Reprod. Fertil. Suppl.* 22:165–202, 1975.

JOHNSON, A. L., and A. VAN TIENHOVEN. Plasma concentration of six steroids and LH during the ovulatory cycle of the hen, Gallus domesticus. *Biol. Reprod.* 23:386–393, 1980a.

———. Hypothalamo-hypophyseal sensitivity to hormones in the hen. I. Plasma concentrations of LH, progesterone, and testosterone in response to central injections of progesterone and R5020. *Biol. Reprod.* 23:910–917, 1980b.

———. Hypothalamo-hypophyseal sensitivity to hormones in the hen. II. Plasma concentrations of LH, progesterone, and testosterone in response to peripheral and central injections of LHRH or testosterone. *Biol. Reprod.* 25:153–161, 1981.

JOSHI, H. S., D. J. WATSON, and A. P. LABHSETWAR. Ovarian secretion of oestradiol, oestrone and 20-dihydroprogesterone and progesterone during the oestrous cycle of the guinea-pig. *J. Reprod. Fertil.* 35:177–181, 1973.

KALRA, P. S., and S. P. KALRA. Temporal changes in the hypothalamic and serum luteinizing hormone - releasing hormone (LH-RH) levels and the circulating ovarian steroids during the rat oestrous cycle. *Acta Endocrinol.* 85:449–455, 1977.

KAMEL, F., E. J. MOCK, W. W. WRIGHT, and A. I. FRANKEL. Alterations in plasma concentrations of testerone, LH, and prolactin associated with mating in the male rat. *Horm. Behav.* 6:277–288, 1975.

KAPPAUF, B., and A. VAN TIENHOVEN. Progesterone concentrations in the peripheral plasma of laying hens in relation to the time of ovulation. *Endocrinology* 90:1350–1355, 1972.

KAWAKAMI, M., E. YOSHIOKA, N. KONDA, J. ARITA, and S. VISESSUVAN. Data on the sites of stimulatory feedback action of gonadal steroids indispensable for luteinizing hormone release in the rat. *Endocrinology* 102:791–798, 1978.

KAWASHIMA, M., M. INAGAMI, M. KAMIYOSHI, and K. TANAKA. Effect of testosterone on the cock pituitary *in vitro* leading to the release of gonadotropins. *Jpn. Poult. Sci.* 15:170–175, 1978b.

KAWASHIMA, M., M. KAMIYOSHI, and K. TANAKA. A cytoplasmic progesterone receptor in hen pituitary and hypothalamic tissues. *Endocrinology* 102:1207–1213, 1978a.

———. Cytoplasmic progesterone receptor concentrations in the hen hypothalamus and pituitary: difference between laying and nonlaying hens and changes during the ovulatory cycle. *Biol. Reprod.* 20:581–585, 1979.

KELLY, R. W. Prostaglandin synthesis in the male and female reproductive tract. *J. Reprod. Fertil.* 62:293–304, 1981.

KEYES, P. L., and D. W. BULLOCK. Effects of prostaglandin $F_{2\alpha}$ on ectopic and ovarian corpora lutea of the rabbit. *Biol. Reprod.* 10:519-525, 1974.

KNOBIL, E. On the regulation of the primate corpus luteum. *Biol. Reprod.* 8:246-258, 1973.

——. On the control of gonadotropin secretion in the rhesus monkey. *Recent Prog. Horm. Res.* 30:1-36, 1974.

KNOBIL, E., and T. M. PLANT. Neuroendocrine control of gonadotropin secretion in the female rhesus monkey. In *Frontiers in Neuroendocrinology,* eds. W. F. Ganong and L. Martini, Vol. 5, pp. 249-264. New York: Raven Press, 1978.

KOERING, M. J., and S. A. SHOLL. Ovarian interstitial gland tissue and serum progesterone levels during the first periovulatory period in the mated rabbit. *Biol. Reprod.* 19:936-948, 1978.

KOMISARUK, B. R. Neural and hormonal interactions in the reproductive behavior of female rats. *Adv. Behav. Biol.* 11:97-129, 1974.

LABHSETWAR, A. P. Pituitary gonadotrophic function (FSH and LH) in various reproductive states. *Adv. Reprod. Physiol.* 6:97-183, 1973.

LAGUË, P. C., A. van TIENHOVEN, and F. J. CUNNINGHAM. Concentrations of estrogens, progesterone and LH during the ovulatory cycle of the laying hen (*Gallus domesticus*). *Biol. Reprod.* 12:590-598, 1975.

LAU, I. F., S. K. SAKSENA, and M. C. CHANG. Serum concentrations of progestins, estrogens, testosterone and gonadotropins in pseudopregnant rats with special reference to the effect of prostaglandin $F_{2\alpha}$. *Biol. Reprod.* 20:575-580, 1979.

LEE, C., P. L. KEYES, and H. I. JACOBSON. Estrogen receptor in the rabbit corpus luteum. *Science* 173:1032-1033, 1971.

LEGAN, S. J., G. A. COON, and F. J. KARSCH. Role of estrogen as initiator of daily LH surges in the ovariectomized rat. *Endocrinology* 96:50-56, 1975.

LEGAN, S. J., and F. J. KARSCH. A daily signal for the LH surge in the rat. *Endocrinology* 96:57-62, 1975.

LEGAN, S. J., F. J. KARSCH, and D. L. FOSTER. The endocrine control of seasonal reproductive function in the ewe: a marked change in response to the negative feedback action of estradiol on luteinizing hormone secretion. *Endocrinology* 101:818-824, 1977.

LEHRMAN, D. S. Hormonal regulation of parental behavior in birds and infrahuman mammals. In *Sex and Internal Secretions,* ed. W. C. Young, Vol. 2, pp. 1268-1382. Philadelphia: Williams and Wilkins, 1961.

L'HERMITE, M., G. D. NISWENDER, L. E. REICHERT, JR., and A. R. MIDGLEY, JR. Serum follicle-stimulating hormone in sheep as measured by radioimmunoassay. *Biol. Reprod.* 6:325-332, 1972.

LICHT, P. Environmental control of annual testicular cycles in the lizard *Anolis carolinensis*. I. Interaction of light and temperature in the initiation of testicular recrudescence. *J. Exp. Zool.* 165:505-516, 1967a.

——. Environmental control of annual testicular cycles in the lizard *Anolis carolinensis*. II. Seasonal variations in the effects of photoperiod and temperature on testicular recrudescence. *J. Exp. Zool.* 166:243-254, 1967b.

——. Problems in experimentation of timing mechanisms for annual physiological cycles in reptiles. In *Hibernation and Hypothermia, Perspectives and Challenges,* eds. F. E. South, J. P. Hannon, J. R. Willis, E. T. Pengeley, and N. R. Alpert, pp. 681-711. New York: Elsevier, 1972.

LOFTS, B. Reproduction. In *Physiology of the Amphibia,* ed. B. Lofts, Vol. 2, pp. 107-218. New York: Academic Press, 1974.

——. Reptilian reproductive cycles and environmental regulators. In *Environmental Endocrinology,* eds. I. Assenmacher and D. Farner, pp. 37-43. New York: Springer, 1978.

LUKASZEWSKA, J., and W. HANSEL. Corpus luteum maintenance during early pregnancy in the cow. *J. Reprod. Fertil.* 59:485-493, 1980.

MALVEN, P. V. Luteotrophic and luteolytic responses to prolactin in hypophysectomized rats. *Endocrinology* 84:1224-1229, 1969.

MANSUETI, A. J., and J. D. HARDY, JR. *Development of Fishes of the Chesapeake Bay Region,* In *Part 1,* ed. E. E. Deubler, Jr. College Park, Maryland: Natural Resources Institute, University of Maryland, 1967.

MAPLETOFT, R. J., D. R. LAPIN, and O. J. GINTHER. The ovarian artery as the final component of the local luteotropic pathway between a gravid uterine horn and ovary in ewes. *Biol. Reprod.* 15:414-421, 1976.

MARTAL, J., M.-C. LACROIX, C. LOUDES, M. SAUNIER, and S. WINTERBERGER-TORRÈS. Trophoblastin, an antiluteolytic protein present in early pregnancy in sheep. *J. Reprod. Fertil.* 56:63-73, 1979.

MASLAR, I. A. Life span of the corpus luteum following complete or partial hypophysectomy and pituitary autotransplantation. Ph.D. Diss. Duke University, 1977.

MATTHEWS, M. K., JR., and N. T. ADLER. Systematic interrelationship of mating, vaginal plug position and sperm transport in the rat. *Physiol. Behav.* 20:303-309, 1978.

McCRACKEN, J. A., D. T. BAIRD, and J. R. GODING. Factors affecting the secretion of steroids from the transplanted ovary in the sheep. *Recent Prog. Horm. Res.* 27:537-582, 1971.

McLEAN, B. K., and M. B. NIKITOVITCH-WINER. Cholinergic control of the nocturnal prolactin surge in the pseudopregnant rat. *Endocrinology* 97:763-770, 1975.

McNATTY, K. P., W. M. HUNTER, A. S. McNEILLY, and R. S. SAWERS. Changes in the concentration of pituitary and steroid hormones in the follicular fluid of human Graafian follicles throughout the menstrual cycle. *J. Endocrinol.* 64:555-571, 1975.

MEIER, A. H. Temporal synergism of prolactin and adrenal steroids. *Gen. Comp. Endocrinol. Suppl.* 3:499–508, 1972.

MELLIN, T. N., G. P. ORCZYK, M. HICHENS, and H. R. BEHRMAN. Serum profiles of luteinizing hormone, progesterone and total estrogens during the canine estrous cycle. *Theriogenology* 5:175–185, 1976.

MENDOZA, G. The reproductive cycles of three viviparous teleosts Allophorus robustus, Goodea luitpoldi and Neoophorus diazi. *Biol. Bull. (Woods Hole)* 123:351–365, 1962.

MICHAEL, R. P., and J. WELEGALLA. Ovarian hormones and the sexual behaviour of the female rhesus monkey (*Macaca mulatta*) under laboratory conditions. *J. Endocrinol.* 41:407–420, 1968.

MILLER, A. H. Capacity for photoperiodic response and endoginous factors in the reproductive cycles of an equatorial sparrow. *Proc. Nat. Acad. Sci. USA* 54:97–101, 1965.

MILLER, J. B., and P. L. KEYES. Transition of the rabbit corpus luteum to estrogen dependence during early luteal development. *Endocrinology* 102:31–38, 1978.

MILLER, K. F., S. L. BERG, D. C. SHARP, and O. J. GINTHER. Concentrations of circulating gonadotropins during various reproductive states in mares. *Biol. Reprod.* 22:744–750, 1980.

MILVAE, R. A., and W. HANSEL. The effects of prostaglandin (PGI$_2$) and 6-keto-PGF$_{1\alpha}$ on bovine plasma progesterone and LH concentrations. *Prostaglandins* 20:641–647, 1980.

MØLLER, O. M. The progesterone concentrations in the peripheral plasma of the mink (*Mustela vison*) during pregnancy. *J. Endocrinol.* 56:121–132, 1973.

MOORE, F. L. Effects of progesterone on sexual attractivity in female rough-skinned newts, *Taricha granulosa*. *Copeia*: 530–532, 1978.

MOORE, F. L., C. McCORMACK, and L. SWANSON. Induced ovulation: effects of sexual behavior and insemination on ovulation and progesterone levels in *Taricha granulosa*. *Gen. Comp. Endocrinol.* 39:262–269, 1979.

MORRIS, T. R. The effects of ahemeral light and dark cycles on egg production in the fowl. *Poult. Sci.* 52:423–445, 1973.

MOSSMAN, H. W., and K. L. DUKE. *Comparative Morphology of the Mammalian Ovary*. Madison: University of Wisconsin Press, 1973.

NALBANDOV, A. V. Control of luteal function in mammals. In *Handbook of Physiology*, Sec. 7, Vol. 2, Part *1*, pp. 153–167. Washington, D.C.: American Physiological Society, 1973.

NEILL, J. D., M. E. FREEMAN, and S. A. TILLSON. Control · of the proestrus surge of prolactin and luteinizing hormone secretion by estrogens in the rat. *Endocrinology* 89:1448–1453, 1971.

NEQUIN, L. G., J. ALVAREZ, and N. B. SCHWARTZ. Measurement of serum steroid and gonadotropin levels and uterine and ovarian variables throughout 4 day and 5 day estrous cycles in the rat. *Biol. Reprod.* 20:659–670, 1979.

NOBLE, G. K. *The Biology of the Amphibia*. New York: Dover, 1954.

OGAWA, K., H. TOJO, and Y. KAJIMOTO. Inhibition of multiple ovulations in hypophysectomized hens treated with gonadotrophins. *J. Reprod. Fertil.* 47:373–375, 1976.

———. Inhibitory effect of FSH administration prior to LH injection on the multiple ovulations induced by exogenous LH in the hypophysectomized laying-hens. *Mem. Fac. Agric. Kagoshima Univ.* 13:101–108, 1977.

OPEL, H., and P. D. LEPORE. *In vivo* studies of luteinizing hormone-releasing factor in the chicken hypothalamus. *Poult. Sci.* 51:1004–1014, 1972.

PETER, R. E., and L. W. CRIM. Reproductive endocrinology of fishes: gonadal cycles and gonadotropin in teleosts. *Annu. Rev. Physiol.* 41:323–335, 1979.

PETERSON, A. J., and R. H. COMMON. Progesterone concentration in peripheral plasma of laying hens as determined by competitive protein-binding assay. *Can. J. Zool.* 49:599–604, 1971.

———. Estrone and estradiol concentrations in peripheral plasma of laying hens as determined by radioimmunoassay. *Can. J. Zool.* 50:395–404, 1972.

PETERSON, A. J., G. O., HENNEBERRY, and R. H. COMMON. Androgen concentrations in the peripheral plasma of laying hens. *Can. J. Zool.* 50:395–404, 1973.

PLANCK, R. J., and H. J. JOHNSON. Oviposition patterns and photoresponsivity in Japanese quail (*Coturnix coturnix japonica*). *J. Interdiscipl. Cycle Res.* 6:131–140, 1975.

POLDER, J. J. W. On gonads and reproductive behaviour in the cichlid fish Aequidens portalegrensis (Hensel). *Neth. J. Zool.* 21:265–365, 1971.

POSTON, H. A. Neuroendocrine mediation of photoperiod and other environmental influences on physiological responses in Salmonids: a review. *US Fish Wildl. Serv. Tech. Pap.* 96, 1978.

RADFORD, H. M., C. D. NANCARROW, and P. E. MATTNER. Ovarian function in suckling and nonsuckling beef cows *post partum*. *J. Reprod. Fertil.* 54:49–56, 1978.

RALPH, C. L. Some effects of hypothalamic lesions on gonadotrophin release in the hen. *Anat. Rec.* 134:411–431, 1959.

RALPH, C. L., and R. M. FRAPS. Effects of hypothalamic lesions on progesterone-induced ovulation in the hen. *Endocrinology* 65:819–824, 1959.

———. Induction of ovulation in the hen by injection of progesterone in the brain. *Endocrinology* 66:269–272, 1960.

RIBEIRO, W. O., D. R. MISHELL, JR., and I. H. THORNEYCROFT. Comparison of the patterns of androstenedione, progesterone and estradiol during the human menstrual cycle. *Am. J. Obstet. Gynecol.* 119(8): 1026–1032, 1974.

RICHMOND, M., and R. STEHN. Olfaction and reproductive behavior in microtine rodents. In *Mammalian Olfaction, Reproductive Processes and Behavior,* ed. R. L. Doty, pp. 197–216. New York: Academic Press, 1976.

ROBERTSON, H. A. Reproduction in the ewe and the goat. In *Reproduction in Domestic Animals,* third ed., eds. H. H. Cole and P. T. Cupps, pp. 475–498. New York: Academic Press, 1977.

ROBINSON, T. J. Reproduction in cattle. In *Reproduction in Domestic Animals,* third ed., eds. H. H. Cole and P. T. Cupps, pp. 433–454. New York: Academic Press, 1977.

ROTHCHILD, I., G. J. PEPE, and W. K. MORISHIGE. Factors affecting the dependency on LH in the regulation of corpus luteum progesterone secretion in the rat. *Endocrinology* 95:280–288, 1974.

ROWLANDS, I. W., and B. J. WEIR. The ovarian cycle in vertebrates. In *The Ovary,* eds. Lord Zuckerman and B. J. Weir, Vol. 2, pp. 217–273. New York: Academic Press, 1977.

SAIDAPUR, S. K., and G. S. GREENWALD. Ovarian steroidogenesis in the proestrous hamster. *Biol. Reprod.* 20:226–234, 1979.

SALTHE, S. N., and J. S. MECHAM. Reproductive and courtship patterns. In *Physiology of the Amphibia,* ed. B. Lofts, Vol. 2, pp. 309–521. New York: Academic Press, 1974.

SCANES, C. G., A. CHADWICK, P. J. SHARP, and N. J. BOLTON. Diurnal variation in plasma luteinizing hormone levels in the domestic fowl (*Gallus domesticus*). *Gen. Comp. Endocrinol.* 34:45–49, 1978.

SCANES, C. G., P. M. M. GODDEN, and P. J. SHARP. An homologous radioimmunoassay for chicken follicle-stimulating hormone: observations on the ovulatory cycle. *J. Endocrinol.* 73:473–481, 1977a.

SCANES, C. G., P. J. SHARP, and A. CHADWICK. Changes in plasma prolactin concentrations during the ovulatory cycle of the chicken. *J. Endocrinol.* 72:401–402, 1977b.

SCHAMS, D., and E. SCHALLENBERGER. Heterologous radioimmunoassay for bovine follicle-stimulating hormone and its application during the oestrous cycle in cattle. *Acta Endocrinol.* 81:461–473, 1976.

SCHWARTZ, N. B. The role of FSH and LH and their antibodies on follicle growth and ovulation. *Biol. Reprod.* 10:236–272, 1974.

SELANDER, R. K., and R. J. HAUSER. Gonadal and behavioral cycles in the great-tailed grackle. *Condor* 67:157–182, 1965.

SENIOR, B. E. Changes in the concentrations of oestrone and oestradiol in the peripheral plasma of the domestic hen during the ovulatory cycle. *Acta Endocrinol.* 77:588–596, 1974.

SENIOR, B. E., and F. J. CUNNINGHAM. Oestradiol and luteinizing hormone during the ovulatory cycle of the hen. *J. Endocrinol.* 60:201–202, 1974.

SHAHABI, N. A., J. M. BAHR, and A. V. NALBANDOV. Effect of LH injection on plasma and follicular steroids in the chicken. *Endocrinology* 96:969–972, 1975b.

SHAHABI, N. A., H. W. NORTON, and A. V. NALBANDOV. Steroid levels in follicles and the plasma of hens during the ovulatory cycle. *Endocrinology* 96:962–968, 1975a.

SHARMAN, G. B. Studies on marsupial reproduction. II. The oestrous cycle of *Setonix brachyurus. Aust. J. Zool.* 3:44–45, 1955a.

————. Studies on marsupial reproduction. III. Normal and delayed pregnancy in *Setonix brachyurus. Aust. J. Zool.* 3:56–70, 1955b.

————. Reproductive physiology of marsupials. *Science* 167(3922):1221–1228, 1970.

SHARP, P. J. and G. BEUVING. The role of corticosterone in the ovulatory cycle of the hen. *J. Endocrinol.* 78:195–200, 1978.

SHARP, P. J., C. J. SCANES, and A. B. GILBERT. *In vivo* effects of an antiserum to partially purified chicken luteinizing hormone (CM 2) in laying hens. *Gen. Comp. Endocrinol.* 34:296–299, 1978.

SHIKHSHABEKOV, M. M. Features of the sexual cycle of some semi-diadromous fishes in the lower reaches of the Terek. *J. Ichtyol. Engl. Transl. Vopr. Ikhtiol.* 14:79–87, 1974.

SHODONO, M., T. NAKAMURA, Y. TANABE, and K. WAKABAYASHI. Simultaneous determinations of oestradiol-17β, progesterone and luteinizing hormone in the plasma during the ovulatory cycle of the hen. *Acta Endocrinol.* 78:565–573, 1975.

SHORT, R. V. Species differences. In *Reproduction in Mammals,* eds. C. R. Austin and R. V. Short, Vol. 4, pp. 1–32. Cambridge: Cambridge University Press, 1972.

SHUTT, D. A., A. H. CLARKE, I. S. FRASER, P. GOH, G. R. McMAHON, D. M. SAUNDERS, and R. P. SHEARMAN. Changes in concentrations of prostaglandin F and steroids in human corpora lutea in relation to growth of the corpus luteum and luteolysis. *J. Endocrinol.* 71:453–454, 1976.

SILBERZAHN, P., D. QUINCY, C. ROSIER, and P. LEYMARIE. Testosterone and progesterone in peripheral plasma during the oestrous cycle of the mare. *J. Reprod. Fertil.* 53:1–5, 1978.

SMITH, M. S., and L. E. McDONALD. Serum levels of luteinizing hormone and progesterone during the estrous cycle, pseudopregnancy and pregnancy in the dog. *Endocrinology* 94:404–412, 1974.

SOLIMAN, K. F. A., and T. M. HUSTON. Involvement of the adrenal gland in ovulation of the fowl. *Poult. Sci.* 53: 1664–1667, 1974.

SPECKER, J. L., and F. L. MOORE. Annual cycle of plasma androgens and testicular composition in the rough-skinned newt, *Taricha granulosa. Gen. Comp. Endocrinol.* 42:297–303, 1980.

Squires, E. L., R. H. Douglas, W. P. Steffenha-
gen, and O. J. Ginther. Ovarian changes during the
estrous cycle and pregnancy in mares. *J. Animal Sci.*
38:330-338, 1974.

Stabenfeldt, G. H., and J. P. Hughes. Reproduction
in horses. In *Reproduction in Domestic Animals,*
third ed., eds. H. H. Cole and P. T. Cupps, pp.
401-431. New York: Academic Press, 1977.

Stabenfeldt, G. H., and V. M. Shille. Reproduction in the dog and cat. In *Reproduction in Domestic
Animals,* third ed., eds. H. H. Cole and P. T. Cupps,
pp. 499-527. New York: Academic Press, 1977.

Stacey, N. E., A. F. Cook, and R. E. Peter. Ovulatory surge of gonadotropin in the gold fish, *Carassius
auratus. Gen. Comp. Endocrinol.* 37:246-249,
1979.

Stavy, M., J. Terkel, and F. Kohen. Plasma progesterone levels during pregnancy and pseudopregnancy
in the hare (*Lepus europaeus syriacus*). *J. Reprod.
Fertil* 54:9-14, 1978.

Sturkie, P. D., and Y. C. Lin. Further studies on
oviposition and vasotocin release in the hen. *Poult.
Sci.* 46:1591-1592, 1967.

Sundararaj, B. I., and S. Vasal. Photoperiod and
temperature control in the regulation of reproduction
in the female catfish *Heteropneustes fossilis. J. Fish.
Res. Board Can.* 33:959-973, 1976.

Takahashi, M., K. Shiota, and Y. Suzuki. Prepro-
gramming mechanism of luteinizing hormone in the
determination of the life span of the corpus luteum.
Endocrinology 102:494-498, 1978.

Tanabe, Y., and T. Nakamura. Endocrine mechanism of ovulation in chickens (*Gallus domesticus*),
quail (*Coturnix coturnix japonica*), and duck (*Anas
platyrhynchos domestica*). In *Biological Rhythms
in Birds: Neural and Endocrine Aspects,* eds. Y.
Tanabe, K. Tanaka, and T. Ookawa, pp. 179-188.
New York: Springer, 1980.

Tanaka, K., M. Kamiyoshi, and Y. Tanabe. Inhibition of premature ovulation by prolactin in the hen.
Poult. Sci. 50:63-66, 1971.

Temple, S. A. Plasma testosterone titers during the
annual reproductive cycle of starlings (*Sturnus vulgaris*). *Gen. Comp. Endocrinol.* 22:470-479, 1974.

Terasawa, E., M. K. King, S. J. Wiegand, W. E.
Bridson, and R. W. Goy. Barbiturate anesthesia
blocks the positive feedback effect of progesterone,
but not of estrogen on luteinizing hormone release in
ovariectomized   guinea   pigs.   *Endocrinology*
104:687-692, 1979b.

Terasawa, E., J. S. Rodriguez, W. E. Bridson, and
S. J. Wiegand. Factors influencing the positive
feedback action of estrogen upon the luteinizing hormone surge in the ovariectomized guinea pig. *Endocrinology* 104:680-686, 1979a.

Terranova, P. F., and G. S. Greenwald. Steroid and
gonadotropin responses to ergocryptine-suppressed
prolactin secretion in hysterectomized, pseudopregnant hamsters. *Endocrinology* 103:845-853, 1978.

Tyndale-Biscoe, C. H., and J. Hawkins. The corpora
lutea of marsupials: aspects of function and control.
*Proc. Fourth Symp. Comparative Biology of Re-
production Canberra, 1976,* pp. 245-252. Canberra:
Australian Academy of Science, 1977.

Udry, J. R., and N. M. Morris. The distribution of
events in the human menstrual cycle. *J. Reprod.
Fertil.* 51:419-425, 1977.

van Oordt, P. G. W. J. Het binnenspel van vissen en
derzelver voortplanting. In *Endocrinologie*, eds. J.
Lever and J. de Wilde, pp. 42-63. Wageningen, The
Netherlands: Centrum voor Landbouwpublikaties en
Landbouwdocumentatie, 1978.

van Tienhoven, A. The effect of massive doses of
corticotrophin and of corticosterone on ovulation in
the chicken (*Gallus domesticus*). *Acta Endocrinol.*
38:407-412, 1961.

————. *Reproductive Physiology of Vertebrates.*
Philadelphia: W. B. Saunders, 1968.

van Tienhoven, A., and R. J. Planck. The effect of
light on avian reproductive activity. In *Handbook of
Physiology,* Sec. 7, Vol. 2, Part 1, pp. 79-107.
Washington, D.C.: American Physiological Society,
1973.

van Tienhoven, A., and A. V. Schally. Mammalian luteinizing hormone-releasing hormone induces
ovulation in the domestic fowl. *Gen. Comp. Endocrinol.* 19:594-595, 1973.

Verhage, H. G., N. B. Beamer, and R. M. Brenner.
Plasma levels of estradiol and progesterone in the cat
during polyestrus, pregnancy and pseudopregnancy.
Biol. Reprod. 14:579-585, 1976.

Ward, K. E., L. C. Longwell, J. L. Kreider, and R.
A. Godke. Effect of unilateral hysterectomy on cy-
cling beef heifers. *J. Anim. Sci.* 43:309, 1976
(Abstract).

Warren, D.C., and H. M. Scott. The time factor in
egg formation. *Poult. Sci.* 14:195-207, 1935.

Wiebe, J. P. The reproductive cycle of the viviparous
sea perch *Cymatogaster aggregata* gibbons. *Can. J.
Zool.* 46:1221-1234, 1968.

Wildt, D. E., W. B. Panko, P. K. Chakraborty,
and S. W. J. Seager. Relationship of serum estrone,
estradiol-17β and progesterone to LH, sexual behavior and time of ovulation in the bitch. *Biol. Reprod.* 20:648-658, 1979.

Williams, J. B., and P. J. Sharp. Control of the
preovulatory surge of luteinizing hormone in the hen
(*Gallus domesticus*): the role of progesterone and androgens. *J. Endocrinol.* 77:57-65, 1978.

Wilson, S. C. Relationship between plasma concentration of luteinizing hormone and intensity of lay in the
domestic hen. *Br. Poult. Sci.* 19:643-650, 1978.

Wilson, S. C., and F. J. Cunningham. Modification
by metyrapone of the "open period" for preovulatory
LH release in the hen. *Br. Poult. Sci.* 21:351-361,
1980.

————. Effects of an anti-oestrogen, tamoxifen (ICI

46, 474) on luteinizing hormone release and ovulation in the hen. *J. Endocrinol.* 88:309-316, 1981.

WILSON, S. C., and P. J. SHARP. Variations in plasma LH levels during the ovulatory cycle of the hen, *Gallus domesticus*. *J. Reprod. Fertil.* 35:561-564, 1973.

_____. Changes in plasma concentrations of luteinizing hormone after injection of progesterone at various times during the ovulatory cycle of the domestic hen (*Gallus domesticus*). *J. Endocrinol.* 67:59-70, 1975.

_____. Effects of androgens, oestrogens and deoxycorticosterone acetate on plasma concentrations of luteinizing hormone in laying hens. *J. Endocrinol.* 69:93-102, 1976a.

_____. Induction of luteinizing hormone release by gonadal steroids in the ovariectomized domestic hen. *J. Endocrinol.* 71:87-98, 1976b.

_____. Effects of androgens, oestrogens and deoxycorticosterone acetate on plasma concentrations of luteinizing hormone in laying hens. *J. Endocrinol.* 68:93-102, 1976.

WINGFIELD, J. C., and D. S. FARNER. The endocrinology of a natural breeding population of the white-crowned sparrow (*Zonotrichia leucophrys pugetensis*). *Physiol. Zoöl.* 51:188-205, 1978a.

_____. The annual cycle of plasma ir LH and steroid hormones in feral populations of the white-crowned sparrow, *Zonotrichia leucophrys gambelii*. *Biol. Reprod.* 19:1046-1056, 1978b.

WITSCHI, E. Vertebrate gonadotrophins. *Mem. Soc. Endocrinol.* 4:149-163, 1955.

WOURMS, J. P. Reproduction and development in chondrichthyan fishes. *Am. Zool.* 17:379-410, 1977.

XAVIER, F. La pseudogestation chez *Nectophrynoides occidentalis* Angel. *Gen. Comp. Endocrinol.* 22:98-115, 1974.

XAVIER, F., and R. OZON. Recherches sur l'activité endocrine de l'ovaire de *Nectophrynoides occidentalis* Angel (Amphibien Anoure vivipare) II. Synthèse *in vitro* des steroïdes. *Gen. Comp. Endocrinol.* 16:30-40, 1971.

YEN, S. S. C., B. L. LASLEY, C. F. WANG, H. LE-BLANC, and T. M. SILER. The operating characteristics of the hypothalamic-pituitary system during the menstrual cycle and observations of biological action of somatostatin. *Recent Prog. Horm. Res.* 31:321-357, 1975.

YOUNG, W. C. The hormones and mating behavior. In *Sex and Internal Secretions*, ed. W. C. Young, Vol. 2, pp. 1173-1239. Baltimore: Williams and Wilkins, 1961.

ZUBER-VOGELI, M., and F. XAVIER. Les modifications cytologiques de l'hypophyse distale des femelles de Nectophrynoïdes occidentalis Angel après ovariectomie. *Gen. Comp. Endocrinol.* 20:199-213, 1973.

9 | Insemination and Fertilization

AONUMA, S., T. MAYINA, K. SAZUKI, T. NAGUDU, M. IWAI, and M. OKABE. Studies of sperm capacitation. I. The relationship between the guinea pig sperm-coating antigen and a sperm capacitation phenomenon. *J. Reprod. Fertil.* 35:425-432, 1973.

ASDELL, S. A. *Patterns of Mammalian Reproduction.* Ithaca: Cornell University Press, 1964.

AUSTIN, C. R. Fertilization. Englewood Cliffs, New Jersey: Prentice-Hall, 1965.

_____. Chromosome deterioration of aging eggs of the rabbit. *Nature* 213:1018-1019, 1967.

_____. Fertilization. In *Reproduction in Mammals,* eds. C. R. Austin and R. V. Short, Vol. 1, pp. 103-133. Cambridge: Cambridge University Press, 1972.

_____. Fertilization. In *Concepts of Development,* eds. J. Lash and J. R. Whittaker, pp. 48-75. Stamford, Connecticut: Sinauer, 1974a.

_____. Recent progress in the study of eggs and spermatozoa. Insemination and ovulation to implantation. In *Reproductive Physiology,* ed. R. O. Greep, M.T.P. International Review of Science Vol. 8, pp. 95-131. London: Butterworths, 1974b.

_____. Patterns in metazoan fertilization. *Curr. Top. Dev. Biol.* 12:1-9, 1978.

AUSTIN, C. R., and E. C. AMOROSO. The mammalian egg. *Endeavour* 18:130-143, 1959.

AUSTIN, C. R., and M. W. H. BISHOP. Fertilization in mammals. *Biol. Rev.* 32:296-349, 1957.

BAKST, M. R., and B. HOWARTH, JR. Hydrolysis of the hen's perivitelline layer by cock sperm *in vitro*. *Biol. Reprod.* 17:370-379, 1977.

BEDFORD, J. M. Sperm capacitation and fertilization in mammals. *Biol. Reprod. Suppl.* 2:128-158, 1970.

_____. Maturation, transport, and fate of spermatozoa in the epididymis. *Handbook of Physiology,* Sec. 7, Vol. 5, pp. 303-318. Washington, D.C.: American Physiological Society, 1975.

_____. Evolution of the sperm maturation and sperm storage functions of the epididymis. In *The Spermatozoon,* eds. D. W. Fawcett and J. M. Bedford, pp. 7-21. Baltimore: Urban and Schwartzenberg, 1979.

BEDFORD, J. M., and N. L. CROSS. Normal penetration of rabbit spermatozoa through a trypsin and acrosin resistant zona pellucida. *J. Reprod. Fertil.* 54:385-392, 1978.

BENNETT, D. The T-locus of the mouse. *Cell* 6:441-454, 1975.

BERTIN, L. Sexualité et fécondation. In *Traité de Zoologie*, Vol. 13, pp. 1584-1652, Paris: Masson, 1958.

BLANDAU, R. J. Gamete transport in the female mammal. In *Handbook of Physiology,* Sec. 7, Vol. 2, Part

2, pp. 153–183. Washington, D.C.: American Physiological Society, 1973.

BOBR, L. W., F. X. OGASAWARA, and F. W. LORENZ. Distribution of spermatozoa in the oviduct and fertility in domestic birds. I. Residence sites of spermatozoa in fowl oviducts. *J. Reprod. Fertil.* 8:39–47, 1964.

BRADEN, A. W. H. T-locus in mice: segregation distortion and sterility in the male. In *Proc. Int. Symp., The Genetics of the Spermatozoon*, eds. R. A. Beatty and S. Gluecksohn-Waelsch, pp. 289–305. Edinburgh: Department of Genetics, University of Edinburgh, 1972.

BRADFORD, M. M., R. A. MCRORIE, and W. L. WILLIAMS. A role for esterases in the fertilization process. *J. Exp. Zool.* 197:297–301, 1976.

BRANDT, H., T. S. SCOTT, D. J. JOHNSON, and D. D. HOSKINS. Evidence for an epididymal origin of bovine sperm forward mobility protein. *Biol. Reprod.* 19:830–835, 1978.

BREDER, C. M., JR., and D. E. ROSEN. *Modes of Reproduction in Fishes*. Garden City, New York: Natural History Press, 1966.

BURKE, W. H., and F. X. OGASAWARA. Presence of spermatozoa in uterovaginal fluids of the hen at various stages of the ovulatory cycle. *Poult. Sci.* 48:408–413, 1969.

BUTCHER, R. L., and N. W. FUGO. Overripeness and the mammalian ova. II. Delayed ovulation and chromosome anomalies. *Fertil. Steril.* 18:297–302, 1967.

BUTCHER, R. L., J. D. BLUE, and N. W. FUGO. Overripeness and the mammalian ova. III. Fetal development at midgestation and at term. *Fertil. Steril.* 20:223–231, 1969.

COHEN, J. The comparative physiology of gamete populations. *Adv. Comp. Physiol. Biochem.* 4:267–380, 1971.

———. *Reproduction*. London: Butterworths, 1977.

COMPTON, M. M., and H. P. VAN KREY. Emptying of the uterovaginal sperm storage glands in the absence of ovulation and oviposition in the domestic hen. *Poult. Sci.* 58:187–190, 1979.

CREWS, D. Hemipenile preference: stimulus control of male mounting behavior in the lizard *Anolis carolinensis*. *Science* 197:195–196, 1978.

CUELLAR, O. On the origin of parthenogenesis in vertebrates: the cytogenic factors. *Am. Nat.* 108:625–648, 1974.

———. Genetic homogeneity and speciation in the parthenogenetic lizards *Cnemidophorus velox* and *C. neomexicanus:* evidence from intraspecific histocompatibility. *Evolution* 31:24–31, 1977a.

———. Animal parthenogenesis. *Science* 197:837–843, 1977b.

DARCEY, K. M., E. G. BUSS, S. E. BLOOM, and M. W. OLSEN. A cytological study of early cell population in developing parthenogenetic blastodiscs of the turkey. *Genetics* 69:479–489, 1971.

DEVINE, M. C. Copulatory plugs in snakes: enforced chastity. *Science* 187:844–845, 1975.

———. Copulatory plugs, restricted mating opportunities and reproductive competition among male garter snakes. *Nature* 267:345–346, 1977.

DHARMARAJAN, M. Effect on the embryo of staleness of the sperm at the time of fertilization in the domestic hen. *Nature* 165:398, 1950.

DIAKOW, C. Initiation and inhibition of the release croak of *Rana pipiens*. *Physiol. Behav.* 19:607–610, 1977.

———. Hormonal basis for breeding behavior in female frogs; vasotocin inhibits the release call of *Rana pipiens*. *Science* 199:1456–1457, 1978.

DUKELOW, W. R., and G. D. RIEGLE. Transport of gametes and survival of ovum as functions of the oviduct. In *The Oviduct and Its Functions*, eds. A. D. Johnson and C. W. Foley, pp. 193–220. New York: Academic Press, 1974.

EPEL, D. Mechanisms of activation of sperm and egg during fertilization of sea urchin gametes. *Curr. Top. Dev. Biol.* 12:185–246, 1978.

EPEL, D., and E. J. CARROLL, JR. Molecular mechanisms for prevention of polyspermy. *Res. Reprod.* 7 (2):2–3, 1975.

FISCHER, B., and C. E. ADAMS. Fertilization following mixed insemination with 'cervix-selected' and 'unselected' spermatozoa in the rabbit. *J. Reprod. Fertil.* 62:337–343, 1981.

FUCHS, A. R. Uterine activity during and after mating in the rabbit. *Fertil. Steril.* 23:915–923, 1972.

GERALL, A. A. Role of the nervous system in reproductive behavior. In *Comparative Reproduction in Nonhuman Primates*, ed. E. S. E. Hafez, pp. 58–82. Springfield, Illinois: Thomas, 1971.

GILBERT, P. W., and G. W. HEATH. The clasper-siphon sac mechanism in *Squalus acanthias* and *Mustelus canis*. *Comp. Biochem. Physiol. A Comp. Physiol.* 42:97–119, 1972.

GINSBURG, A. S. Fertilization in the sturgeon. I. The fusion of the gametes. *Cytologia* (Russian) 1:510, 1959.

GRIGG, G. W. The structure of stored sperm in the hen and the nature of the release mechanism. *Poult. Sci.* 36:450–451, 1957.

GWATKIN, R. B. L. *Fertilization Mechanisms in Man and Mammals*. New York: Plenum Press, 1977.

HALE, E. B. Duration of fertility and hatchability following natural matings in turkeys. *Poult. Sci.* 34:228–233, 1955.

HARPER, M. J., and M. C. CHANG. Some aspects of the biology of mammalian eggs and spermatozoa. *Adv. Reprod. Physiol.* 5:167–218, 1971.

HEALEY, W. V., P. S. RUSSELL, H. K. POOLE, and M. M. OLSEN. A skin-grafting analysis of fowl parthenogens: evidence for a new type of genetic histocompatibility. *Ann. N. Y. Acad. Sci.* 99(3):698–705, 1962.

HEFFNER, L. J. Separation of calcium effects on

motility and binding to zona pellucida in mouse sperm. *Biol. Reprod.* 20 (*Suppl.* 1):78A, 1979.

HOWARTH, B., JR., and M. B. PALMER. An examination of the need for sperm capacitation in the turkey *Meleagris gallopavo. J. Reprod. Fertil.* 28:443–445, 1972.

JANSEN, R. P. S. Fallopian tube isthmic mucus and ovum transport. *Science* 201:349–351, 1978.

JOHNSON, W. L., and A. G. HUNTER. Seminal antigens: their alteration in the genital tract of female rabbits and during *in vitro* capacitation with beta amylase and beta glucuronidase. *Biol. Reprod.* 7:322–340, 1972.

KAUFMAN, M. H. The experimental production of mammalian parthenogenetic embryos. In *Methods in Mammalian Reproduction*, ed. J. C. Daniel, Jr., pp. 21–47. New York: Academic Press, 1978.

KILLE, R. A. Fertilization of the lamprey egg. *Exp. Cell Res.* 20:12–27, 1960.

KOEHLER, J. K. Changes in antigenic cell distribution on rabbit spermatozoa after incubation in "capacitating" media. *Biol. Reprod.* 15:444–456, 1976.

LOFTS, B. Reproduction. In *Physiology of the Amphibia*, ed. B. Lofts, Vol. 2, pp. 107–218. New York: Academic Press, 1974.

LONGO, F. J., and M. KUNKLE. Transformations of sperm nuclei upon insemination. *Curr. Top. Dev. Biol.* 12:149–184, 1978.

MUNRO, S. S. Functional changes in fowl sperm during their passage through the excurrent ducts of the male. *J. Exp. Zool.* 79:71–92, 1938.

NALBANDOV, A. V., and L. E. CARD. Effect of stale sperm on fertility and hatchability of chicken eggs. *Poult. Sci.* 22:218–226, 1943.

OLDS-CLARKE, P., and A. BECKER. The effect of the T/t locus on sperm penetration in vivo in the house mouse. *Biol. Reprod.* 18:132–140, 1978.

OLIPHANT, G., and B. G. BRACKETT. Immunological assessment of surface change of rabbit sperm undergoing capacitation. *Biol. Reprod.* 9:404–414, 1973.

OLSEN, M. W. Fowl pox vaccine associated with parthenogenesis in chicken and turkey eggs. *Science* 124:1078–1079, 1956.

———. Nine year summary of parthenogenesis in turkeys. *Proc. Soc. Exp. Biol. Med.* 105:279–281, 1960.

———. Parthenogenesis in eggs of White Leghorn chickens following an outbreak of visceral lymphomatosis. *Proc. Soc. Exp. Biol. Med.* 122:977–980, 1966.

OLSEN, M. W., and E. G. BUSS. Segregation of two alleles for color of down in parthenogenetic and normal turkey embryos and poults. *Genetics* 72:69–75, 1972.

OVERSTREET, J. W., and D. F. KATZ. Sperm transport and selection in the female genital tract. In *Development in Mammals,* ed. M. H. Johnson, Vol. 2, pp.

31–65. Amsterdam: North Holland Publishing, 1977.

POOLE, H. K., and M. W. OLSEN. Incidence of parthenogenetic development in eggs laid by three strains of dark Cornish chickens. *Proc. Soc. Exp. Biol. Med.* 97:477–478, 1958.

RACEY, P. A. The prolonged storage and survival of spermatozoa in Chiroptera. *J. Reprod. Fertil.* 56:391–402, 1979.

ROMANOFF, A. L. *The Avian Embryo.* New York: Macmillan, 1960.

ROSS, P., JR., and D. CREWS. Influence of seminal plug on mating behaviour in the garter snake. *Nature* 267:344–345, 1977.

SALISBURY, G. W., and R. G. HART. Gamete aging and its consequences. *Biol. Reprod. Suppl.* 2:1–13, 1970.

SALISBURY, G. W., R. G. HART, and J. R. LODGE. The fertile life of spermatozoa. *Perspect. Biol. Med.* 19:213–230, 1976.

SRIVASTAVA, P. N., and W. L. WILLIAMS. Sperm capacitation and decapacitation factor. *Proc. 15th Nobel Symposium, Södergarn, Lidingö, Sweden, 1970*, eds. E. Diczfalusy and U. Borell, pp. 73–83. New York: Wiley Interscience, 1971.

TYNDALE-BISCOE, C. H., and J. C. RODGER. Differential transport of spermatozoa into the two sides of the genital tract of a monovular marsupial, the tammar wallaby (*Macropus eugenii*). *J. Reprod. Fertil.* 52:37–43, 1978.

VAN TIENHOVEN, A. *Reproductive Physiology of Vertebrates.* Philadelphia: W. B. Saunders, 1968.

WALLACE, H. Chiasmata have no effect on fertility. *Heredity* 33:423–428, 1974.

WALTON, A. Copulation and natural insemination. In *Marshall's Physiology of Reproduction*, ed. A. S. Parkes, Vol. 1, Part 2, pp. 130–160, London: Longmans, Green, and Co., 1960.

WEIR, B. J. The reproductive organs of the female plains viscacha, *Lagostomus maximus. J. Reprod. Fertil.* 25:365–373, 1971.

WHITELAW, G. P., and R. H. SMITHWICK. Some secondary effects of sympathectomy. *New Engl. J. Med.* 245:121–130, 1951.

WICKLER, W. 'Egg-dummies' as natural releasers in mouth breeding cichlids. *Nature* 194:1092–1093, 1962.

———. Signal value of the genital tossel in the male *Tilapia macrochir* Blgr (Pisces: Cichlidae). *Nature* 208:595–596, 1965.

WITSCHI, E., and R. LAGUENS. Chromosomal aberrations in embryos from overripe eggs. *Dev. Biol.* 7:605–616, 1963.

YANAGIMACHI, R. Some properties of the sperm-activating factor in the micropyle area of the herring egg. *Ann. Zool. Jpn.* 30:114–119, 1957.

YANAGIMACHI, R. Sperm-egg associations in mammals. *Curr. Top. Dev. Biol.* 12:83–105, 1978a.

_____. Calcium requirement for sperm-egg fusion in mammals. _Biol. Reprod._ 19:949–958, 1978b.

YANAGISAWA, K., D. BENNETT, E. A. BOYSE, L. C. DUNN, and A. DIMEO. Serological identification of sperm antigens specified by lethal _(t)_-alleles in the mouse. _Immunogenetics_ 1:57–67, 1974.

10 | Care of the Embryo and Fetus

AITKEN, R. J. Aspects of delayed implantation in the roe deer (_Capreolus capreolus_). _J. Reprod. Fertil. Suppl._ 29:83–95, 1981.

ALLEN, W. R. Factors influencing pregnant mare serum gonadotrophin production. _Nature_ 223:64–66, 1969.

_____. Endocrine functions of the placenta. In _Comparative Placentation,_ ed. D. H. Steven, pp. 214–267. New York: Academic Press, 1975.

AMOROSO, E. C. Placentation. In _Marshall's Physiology of Reproduction._ ed. A. S. Parkes, Vol. 2, pp. 127–311. London: Longmans Green and Co., 1952.

_____. Viviparity in fishes. _Symposia Zool. Soc., London._ 1:153–181, 1960.

AMOROSO, E. C., and J. S. PERRY. Ovarian activity during gestation. In _The Ovary,_ eds. Lord Zuckerman and B. J. Weir, Vol. 2, pp. 315–398. New York: Academic Press, 1977.

ASH, R. W., and R. B. HEAP. Oestrogen, progesterone and corticosteroid concentrations in peripheral plasma of sows during pregnancy, parturition, lactation and after weaning. _J. Endocrinol._ 64:141–154, 1975.

AUSTAD, R., A. LUNDE, and Ø. V. SJAASTAD. Peripheral plasma levels of oestradiol-17β and progesterone in the bitch during the oestrous cycle, in normal pregnancy and after dexamethasone treatment. _J. Reprod. Fertil._ 46:129–136, 1976.

BAKER, E. C. S. _Cuckoo Problems. H. F. and G._ Witherby, 1942.

BARANCZUK, R. and G. S. GREENWALD. Plasma levels of oestrogen and progesterone in pregnant and lactating hamsters. _J. Endocrinol._ 63:125–135, 1974.

BEATO, M. Hormonal control of uteroglobin biosynthesis. In _Development in Mammals,_ ed. M. H. Johnson, Vol. 1, pp. 361–384. Amsterdam: North-Holland Publishing, 1977.

BEIER, H. M. Physiology of uteroglobin. In _Novel Aspects of Reproductive Physiology,_ eds. C. H. Spilman and J. W. Wilks, pp. 219–245. New York: Spectrum Publications, 1978.

BERGSTRÖM, S. _Surface Ultrastructure of Mouse Blastocysts before and at Implantation._ Uppsala, Sweden: Department of Anatomy, University of Uppsala, 1971(?).

BERTIN, L. _Nidification._ In _Traité de Zoologie,_ Vol. 13, pp. 1653–1684. Paris: Masson, 1958.

BLACKBURN, D. G. An evolutionary analysis of vertebrate viviparity. _Am. Zool._ 21:936, 1981.

BLATCHLEY, F. R., F. M. MAULE WALKER, and N. L. POYSER. Progesterone, prostaglandin F$_{2\alpha}$ and oestradiol in the utero-ovarian venous plasma of non-pregnant and early, unilaterally pregnant guinea-pigs. _J. Endocrinol._ 67:225–229, 1975.

BOISSEAU, J.-P. Effets de la castration et de l'hypophysectomie sur l'incubation de L'Hippocampe mâle (_Hippocampus hippocampus_ L.) _C.R. Hebd. Séances Acad. Sci. (Paris)_ 259:4839–4840, 1964.

BONNIN, M., R. CANIVENC, and C. RIBES. Plasma progesterone levels during delayed implantation in the European badger (_Meles meles_). _J. Reprod. Fertil._ 52:55–58, 1978.

BORLAND, R. Placenta as an allograft. In _Comparative Placentation,_ ed. D. H. Steven, pp. 268–281. New York: Academic Press, 1975.

BREDER, C. M., JR., and D. E. ROSEN. _Modes of Reproduction in Fishes._ Garden City, New York: Natural History Press, 1966.

BUDKER, P. _The Life of Sharks._ New York: Columbia University Press, 1971.

BULLOCK, D. W. Steroids from the pre-implantation blastocyst. In _Development in Mammals,_ ed. M. H. Johnson, Vol. 2, pp. 199–208. Amsterdam: North-Holland Publishing, 1977.

BURKE, W. H., and P. T. DENNISON. Prolactin and luteinizing hormone levels in female turkeys (_Meleagris gallopavo_) during a photoinduced reproductive cycle and broodiness. _Gen. Comp. Endocrinol._ 41:92–100, 1980.

BURKE, W. H., and H. PAPKOFF. Purification of turkey prolactin and the development of a homologous radioimmunoassay for its measurement. _Gen. Comp. Endocrinol._ 40:297–307, 1980.

CANIVENC, R. A study of progestation in the European Badger, _Meles meles_ L. _J. Reprod. Fertil._ 9:364, 1965.

CANIVENC, R. and M. BONNIN. Environmental control of delayed implantation in the European badger (_Meles meles_). _J. Reprod. Fert. Suppl._ 29:25–33, 1981.

CANIVENC, R., M. BONNIN-LAFFARGUE, and M. LAJUS-BOUÉ. Réalisation expérimentale précoce de l'ovo-implantation chez le blaireau européen (_Meles meles_ L.) pendant la période de latence blastocytaire. _C.R. Hebd. Séances Acad. Sci. Ser. D Sci. Nat. (Paris)_ 273:1855–1857, 1971.

CATCHPOLE, H. R. Hormonal mechanisms in pregnancy and parturition. In _Reproduction and Domestic Animals,_ eds. H. H. Cole and P. T. Cupps, pp. 341–368. New York: Academic Press, 1977.

CHALLIS, JOHN R. G., I. JOHN DAVIES, and KENNETH J. RYAN. The concentrations of progesterone, estrone and estradiol-17β in the plasma of pregnant rabbits. *Endocrinology* 93:971–976, 1973.

CHAMLEY, W. A., J. M. BUCKMASTER, M. E. CERINI, I. A. CUMMINGS, J. R. GODING, J. M. OBST, A. WILLIAMS, and C. WINFIELD. Changes in the levels of progesterone, corticosteroids, estrone, estradiol-17β, luteinizing hormone and prolactin in the peripheral plasma of the ewe during late pregnancy and at parturition. *Biol. Reprod.* 9:30–35, 1973.

CHENG, M. F. Induction of incubation behaviour in male ring doves (*Streptopelia risoria*): a behavioural analysis. *J. Reprod. Fertil.* 42:267–276, 1975.

CLARKE, W. C., and H. A. BERN. Comparative endocrinology of prolactin. In *Hormonal Proteins and Peptides*, ed. C. H. Li, Vol. 8, pp. 106–197. New York: Academic press, 1980.

COHEN, J. Reproduction. London: Butterworths, 1977.

CORBEN, C. J., G. J. INGRAM, and M. J. TYLER. Gastric breeding: unique form of parental care in an Australian frog. *Science* 186:946–947, 1974.

CRAIGHEAD, J. J., M. G. HORNOCKER, and F. C. CRAIGHEAD, JR. Reproductive biology of young female grizzly bears. *J. Reprod. Fertil. Suppl.* 6:447–475, 1969.

DANIEL, J. C., JR. Delayed implantation in the northern fur seal (*Callorhinus ursinus*) and other pinnipeds. *J. Reprod. Fertil. Suppl.* 29:35–50, 1981.

de GREEF, W. J., J. DULLAART, and G. H. ZEILMAKER. Serum concentrations of progesterone, luteinizing hormone, follicle stimulating hormone and prolactin in pseudopregnant rats: effect of decidualization. *Endocrinology* 101:1054–1063, 1977.

DICKMANN, Z. Hormonal requirements for the survival of blastocysts in the uterus of the rat. *J. Endocrinol.* 37:455–461, 1967.

DICKMANN, Z., S. K. DEY, and J. S. GUPTA. A new concept: control of early pregnancy by steroid hormones originating in the preimplantation embryo. *Vitam. Horm.* 34:215–242, 1976.

DICZFALUSY, E. Steroid metabolism in the human foeto-placental unit. *Acta Endocrinol.* 61:649–664, 1969.

DITTRICH, L., and H. KRONBERGER. Biologische-Anatomische Untersuchungen über die Fortpflanzungs-biologie des Braunbären (*Ursus arctos* L.) und anderer Ursiden in Gefangenschaft. *Z. Säugetierkd.* 28 (3):129–155, 1962.

DRENT, R. Incubation. In *Avian Biology,* eds. D. S. Farner and J. R. King, Vol. 5, pp. 333–420. New York: Academic Press, 1975.

DUKELOW, W. R., and G. D. RIEGLE. Transport of gametes and survival of the ovum as functions of the oviduct. In *The Oviduct and its Functions*, eds. A. D. Johnson and C. W. Foley, pp. 193–220. New York: Academic Press, 1974.

EL HALAWANI, M. E., W. H. BURKE, and P. T. DENNISON. Effect of nest-deprivation on serum prolactin level in nesting female turkeys. *Biol. Reprod.* 23:118–123, 1980.

ENDERS, A. C., and R. L. GIVEN. The endometrium of delayed and early implantation. In *Biology of the Uterus*, second ed., ed. R. M. Wynn, pp. 203–243. New York: Plenum Press, 1977.

ETCHES, R. J., A. S. McNEILLY, and C. E. DUKE. Plasma concentrations of prolactin during the reproductive cycle of the domestic turkey (*Meleagris gallopavo*). *Poult. Sci.* 58:963–970, 1979.

FINN, C. A. The implantation reaction. In *Biology of the Uterus,* second ed., ed. R. M. Wynn, pp. 245–308. New York: Plenum Press, 1977.

GIDLEY-BAIRD, A. A. Endocrine control of implantation and delayed implantation in rats and mice. *J. Reprod. Fertil. Suppl.* 29:97–109, 1981.

GOIN, C. J., and O. B. GOIN. *Introduction to Herpetology.* San Francisco: W. H. Freeman, 1962.

GREENWALD, G. S. Species differences in egg transport in response to exogenous estrogen. *Anat. Rec.* 157:163–172, 1967.

GREENWALD, G. S., and J. D. BAST. Hormone patterns in pregnant or pseudopregnant hamsters after unilateral ovariectomy or hysterectomy. *Biol. Reprod.* 18:658–662, 1978.

GULAMHUSEIN, A. P., and A. R. THAWLEY. Plasma progesterone levels in the stoat. *J. Reprod. Fertil.* 36:405–408, 1974.

HAFEZ, E. S. E. Differentiation of mammalian blastocysts. In *Biology of Mammalian Fertilization and Implantation,* eds. K. S. Moghisi and E. S. E. Hafez, pp. 296–342. Springfield, Illinois: C. C. Thomas, 1972.

HEAP, R. B. Role of hormones in pregnancy. In *Reproduction in Mammals,* eds. C. R. Austin and R. V. Short, Vol. 3, pp. 73–105. Cambridge: Cambridge University Press, 1972.

HEAP, R. B., J. S. Perry, and J. R. G. CHALLIS. Hormonal maintenance of pregnancy. In *Handbook of Physiology,* Sec. 7, Vol. 2, Part 2, pp. 217–260. Washington, D.C.: American Physiological Society, 1973.

HINDE, R. A. Behavior. In *Avian Biology*, eds. D. S. Farner and J. R. King, Vol. 3, pp. 479–535. New York: Academic Press, 1973.

HOAR, W. S. Reproduction. In *Fish Physiology*, eds. W. S. Hoar and D. J. Randall, Vol. 3, pp. 1–72. New York: Academic Press, 1969.

HOFFMAN, L. H. Placentation in the garter snake, *Thamnophis sirtalis. J. Morphol.* 131:57–88, 1970.

HONEGGER, R. E. Breeding and maintaining reptiles in captivity. In *Breeding Endangered Species in Captivity,* ed. R. D. Martin, pp. 1–12. New York: Academic Press, 1975.

KANE, M. T., and R. H. FOOTE. Culture of two- and four-cell rabbit embryos to the expanding blastocyst stage in synthetic media. *Proc. Soc. Exp. Biol. Med.* 133:921–925, 1970.

KANN, G., and R. DENAMUR. Possible rôle of prolactin during the oestrous cycle and gestation in the ewe. *J. Reprod. Fertil.* 39:473–483, 1974.

KERN, M. D., and A. BUSHRA. Is the incubation patch required for the construction of a normal nest? *Condor* 82:328–334, 1980.

KOMISARUK, B. R. Effects of local brain implants of progesterone on reproductive behavior in ring doves. *J. Comp. Physiol. Psychol.* 64:219–224, 1967.

LEHRMAN, D. Hormonal regulation of parental behavior in birds and infrahuman mammals. In *Sex and Internal Secretions,* third ed., ed. W. C. Young. Vol. 2, pp. 1268–1382. Baltimore: Williams and Wilkins, 1961.

LEHRMAN, D. S. Interaction between internal and external environments in the regulation of the reproductive cycle of the ring dove. In *Sex and Behavior,* ed. F. A. Beach, pp. 355–380. New York: Wiley and Sons, 1965.

LINTON, J. R., and B. L. SOLOFF. The physiology of the brood pouch of the male sea horse *Hippocampus erectus. Bull. Mar. Sci. Gulf Caribb.* 14:45–61, 1964.

LUCKETT, W. P. Comparative development and evolution of the placenta in primates. *Contributions to Primatology,* Vol. 3, pp. 142–234. Basel: Karger, 1974.

MACDONALD, G. J., D. T. ARMSTRONG, and R. O. GREEP. Initiation of blastocyst implantation in luteinizing hormone. *Endocrinology* 80:172–176, 1967.

MARTEL, D., and A. PSYCHOYOS. Estrogen receptors in the nidatory sites of the rat endometrium. *Science* 211:1454–1455, 1981.

MARTINET, L., C. ALLAIS, and D. ALLAIN. The role of prolactin and LH in luteal function and blastocyst growth in mink (*Mustela vison*). *J. Reprod. Fertil. Suppl.* 29:119–130, 1981.

McCORMACK, J. T., and G. S. GREENWALD. Progesterone and oestradiol-17β concentrations in the peripheral plasma during pregnancy in the mouse. *J. Endocrinol.* 62:101–107, 1974.

MEAD, R. A. Effects of light and blinding upon delayed implantion in the spotted skunk. *Biol. Reprod.* 5:214–220, 1971.

———. Effects of hypophysectomy on blastocyst survival, progesterone secretion and nidation in the spotted skunk. *Biol. Reprod.* 12:526–533, 1975.

———. Delayed implantation in mustelids with special emphasis on the spotted skunk. *J. Reprod. Fertil. Suppl.* 29:11–24, 1981.

MICHEL, G. F. Experience and progesterone in ring dove incubation. *Anim. Behav.* 25:281–285, 1977.

MØLLER, O. M. The progesterone concentrations in the peripheral plasma of the mink (*Mustela vison*) during pregnancy. *J. Endocrinol.* 56:121–132, 1973.

———. Effects of ovariectomy on the plasma progesterone and maintenance of gestation in the blue fox. *Alopex lagopus. J. Reprod. Fertil.* 37:141–143, 1974.

MORISHIGE, W. K., G. J. PEPE, and I. ROTHCHILD. Serum luteinizing hormone, prolactin and progesterone levels during pregnancy in the rat. *Endocrinology* 92:1527–1530, 1973.

MOSER, H. G. Reproduction and development of *Sebastodes paucispinis* and comparison with other rock fishes off Southern California. *Copeia*:773–797, 1967.

MURPHY, B. D., P. W. CONCANNON, H. F. TRAVIS, and W. HANSEL. Prolactin: the hypophyseal factor that terminates embryonic diapause in mink. *Biol. Reprod.* 25:487–491, 1981.

MURR, S. M., G. E. BRADFORD, and I. I. GESCHWIND. Plasma luteinizing hormone, follicle-stimulating hormone and prolactin during pregnancy in the mouse. *Endocrinology* 94:112–116, 1974.

NEILL, J. D., E. D. B. JOHANSSON, and E. KNOBIL. Patterns of circulating progesterone concentrations during the fertile menstrual cycle and the remainder of gestation in the rhesus monkey. *Endocrinology* 84:45–48, 1969.

NEILL, J. D., and E. KNOBIL. On the nature of the initial luteotropic stimulus of pregnancy in the rhesus monkey. *Endocrinology* 90:34–38, 1972.

NETT, T. M., D. W. HOLTAN, and V. L. ESTERGREEN. Plasma estrogens in pregnant and post partum mares. *J. Anim. Sci.* 37:962–970, 1973.

NILSSON, O., I. LINDQVIST, and G. RONQUIST. Decreased surface charge of mouse blastocysts at implantation. *Exp. Cell. Res.* 83:421–423, 1974.

———. Blastocyst surface charge and implantation in the mouse. *Contraception* 11:441–450, 1975.

NOBLE, G. K. *The Biology of the Amphibia.* New York: Dover Publications, 1954.

PAPKE, R. L., P. W. CONCANNON, H. F. TRAVIS, and W. HANSEL. Control of luteal function and implantation in the mink by prolactin. *J. Anim. Sci.* 50:1102–1107, 1980.

PARKER, P. An ecological comparison of marsupial and placental patterns of reproduction. In *The Biology of Marsupials*, eds. B. Stonehouse and D. Gilmore, pp. 273–286. Baltimore: University Park Press, 1977.

PEPE, G., and I. ROTHCHILD. The effect of hypophysectomy on day 12 of pregnancy on the serum progesterone level and time of parturition in the rat. *Endocrinology* 91:1380–1385, 1972.

PEPPLER, R. D., and S. C. STONE. Plasma progesterone level in the female armadillo during delayed implantation and gestation: preliminary report. *Lab. Anim. Sci.* 26:501–504, 1976.

PROUDMAN, J. A., and H. OPEL. Turkey prolactin: validation of a radioimmunoassay and measurement of changes associated with broodiness. *Biol. Reprod.* 25:573–580, 1981.

PSYCHOYOS, A. Hormonal control of ovoimplantation. *Vitam. Horm.* 31:201–256, 1973.

RAMSEY, E. M. *The Placenta of Laboratory Animals and Man*. New York: Holt, Rinehart, and Winston, 1975.

RASWEILER, J. J., IV. Preimplantation development, fate of the zona pellucida, and observations on the glycogen-rich oviduct of the little bulldog bat *Noctilio albiventris*. *Am. J. Anat.* 150:269-300, 1977.

_____. Early embryonic development and implantation in bats. *J. Reprod. Fertil.* 56:403-416, 1979.

RENFREE, M. B. Ovariectomy during gestation in the American opposum *Didelphis marsupialis virginiana*. *J. Reprod. Fertil.* 39:127-130, 1974.

RENFREE, M. B., and J. M. CALABY. Background to delayed implantation and embryonic diapause. *J. Reprod. Fertil. Suppl.* 29:1-9, 1981.

RIBBINK, A. J. Cuckoo among Lake Malawi cichlid fish. *Nature* 267:243-244, 1977.

ROMER, A. S. *Vertebrate Paleontology*, third ed. Chicago: University of Chicago Press, 1966.

RYAN, K. J. Steroid hormones in mammalian pregnancy. In *Handbook of Physiology*, Sec. 7, Vol. 2, Part 2, pp. 285-293. Washington, D.C.: American Physiological Society, 1973.

SAUER, M. J. Hormone involvement in the establishment of pregnancy. *J. Reprod. Fertil.* 56:725-743, 1979.

SCHLAFKE, S., and A. C. ENDERS. Cellular basis of interaction between trophoblast and uterus at implantation. *Biol. Reprod.* 12:41-65, 1975.

SHARP, P. J., C. G. SCANES, J. B. WILLIAMS, S. HARVEY, and A. CHADWICK. Variations in concentrations of prolactin, luteinizing hormone, growth hormone and progesterone in the plasma of broody bantams (*Gallus domesticus*). *J. Endocrinol.* 80:51-57, 1979.

SHEMESH, M., F. MILAGUIR, N. AYALON, and W. HANSEL. Steroidogenesis and prostaglandin synthesis by cultured bovine blastocysts. *J. Reprod. Fertil.* 56:181-185, 1979.

SMITH, V. G., L. A. EDGERTON, H. D. HAFS, and E. M. CONVEY. Bovine serum estrogens, progestins and glucocorticoids during late pregnancy, parturition and early lactation. *J. Anim. Sci.* 36:391-396, 1973.

STEVEN, D. Anatomy of the placental barrier. In *Comparative Placentation*, ed. D. H. Steven, pp. 25-57. New York: Academic Press, 1975.

STEVEN, D., and G. MORRISS. Development of foetal membranes. In *Comparative Placentation*, ed. D. H. STEVEN, pp. 58-86. New York: Academic Press, 1975.

STEWART, F., and W. R. ALLEN. Biological functions and receptor-binding activities of equine chorionic gonadotrophins. *J. Reprod. Fertil.* 62:527-536, 1981.

THORBURN, G. D., J. R. G. CHALLIS, and J. S. ROBINSON. Endocrine control of parturition. In *Biology of the Uterus*, second ed., R. M. Wynn, pp. 653-732. New York: Plenum Press, 1977.

TYNDALE-BISCOE, H. *Life of Marsupials*. Contemporary Biology Series. London: Edward Arnold, 1973.

TYNDALE-BISCOE, C. H. Hormonal control of embryonic diapause and reactivation in the tammar wallaby. In *Maternal Recognition of Pregnancy*, CIBA Foundation Symposia 64 (new series), pp. 173-190, Amsterdam: Elsevier, 1979.

VAN TIENHOVEN, A. *Reproductive Physiology of Vertebrates*. Philadelphia: W. B. Saunders, 1968.

VEHRENCAMP, S. L. Relative fecundity and parental effort in communally nesting anis, *Crotophaga sulcirostris*. *Science* 197: 403-405, 1977.

VOGEL, P. Occurrence and interpretation of delayed implantation in insectivores. *J. Reprod. Fertil. Suppl.* 29:51-60, 1981.

WALKER, M. T., and R. L. HUGHES. Ultrastructural changes after diapause in the uterine glands, corpus luteum and blastocyst of the red-necked wallaby, *Macropus rufogriseus banksianus*. *J. Reprod. Fertil. Suppl.* 29:151-158, 1981.

WELTY, J. C. *The Life of Birds*. Philadelphia: W. B. Saunders, 1962.

WICKLER, W. *Mimicry in Plants and Animals*. New York: World University Library, McGraw Hill, 1968.

WIMSATT, W. A. Some comparative aspects of implantation. *Biol. Reprod.* 12:1-40, 1975.

WOOD-GUSH, D. G. M. *The Behaviour of the Domestic Fowl*. London: Heinemann, 1971.

WOURMS, J. P. Reproduction and development in chondrichthyan fishes. *Am. Zool.* 17:379-410, 1977.

YARON, Z. Endocrine aspects of gestation in viviparous reptiles. *Gen. Comp. Endocrinol. Suppl.* 3:663-674, 1972.

11 | Expulsion of the Oocyte, Embryo, or Fetus

ABEL, M., J. TAUROG, and P. W. NATHANIELSZ. A comparison of the luteolytic effect of PGF_2 and cortisol in pregnant rat. *Prostaglandins* 4:431-440, 1973.

ACKER, G. Etude de déterminisme hormonal de la mise en bas chez la ratte. Importance de l'oestradiol, *Gen. Comp. Endocrinol.* 13:489, 1969.

AFELE, S., G. D. BRYANT-GREENWOOD, W. A.

CHAMLEY and E. M. DAX. Plasma relaxin immunoactivity in the pig at parturition and during nuzzling and suckling. *J. Reprod. Fertil.* 56:451-457, 1979.

ANDERSON, N. G., L. B. CURET, and A. E. COLÁS. Changes in C_{19}-steroid metabolism by ovine placentas during cortisol administration. *Biol. Reprod.* 18:643-651, 1978a.

_____. Changes in C_{21}-steroid metabolism by ovine placentas during cortisol administration. *Biol. Reprod.* 18:652-657, 1978b.

ASH, R. W., and R. B. HEAP. Oestrogen, progesterone and corticosteroid concentrations in peripheral plasma of sows during pregnancy, lactation and after weaning. *J. Endocrinol.* 64:141-154, 1975.

BALDWIN, D. M., and G. H. STABENFELDT. Endocrine changes in the pig during late pregnancy, parturition and lactation. *Biol. Reprod.* 12:508-515, 1975.

BASSETT, J. M., and G. D. THORBURN. Foetal plasma corticosteroids and the initiation of parturition in sheep. *J. Endocrinol.* 44:285-286, 1969.

BASSON, P. A., J. D. Morgenthal, R. B. Bilbrough, J. L. MARAIS, S. P. KRUGER, and J. L. DE B VANDER MERWE. "Grootlamsziekte" a specific syndrome of prolonged gestation in sheep caused by a shrub *Salsola tuberculata* (Fenzel ex Moq) Schinz Var *tomentosa* C. A. Smith ex Aellen. Onderstepoort. *J. Vet. Res.* 36:59-104, 1969.

BEUVING, G., and G. M. A. VONDER. The influence of ovulation and oviposition on corticosterone levels in the plasma of laying hens. *Gen. Comp. Endocrinol.* 44:382-388, 1981.

BEYER, C., and F. MENA. Parturition and lactogenesis in rabbits with high spinal cord transection. *Endocrinology* 87:195-197, 1970.

BLAHA, G. C., and R. J. NIEUWENHUIS. Myometrial gap junctions in golden hamsters: absence in cases of delayed parturition. *Anat. Rec.* 196:19A-20A, 1980.

BLANK, M. S., and D. A. DeBIAS. Oxytocin release during vaginal distension in the goat. *Biol. Reprod.* 17:213-223, 1977.

BOBR, L. W., and B. L. SHELDON. Analysis of ovulation-oviposition patterns in the domestic fowl by telemetry measurement of deep body temperature. *Aust. J. Biol. Sci.* 30:243-257, 1977.

BOER, K., D. W. LINCOLN, and D. F. SWAAB. Effects of electrical stimulation of the neurohypophysis on labour in the rat. *J. Endocrinol.* 65:163-175, 1975.

BOSC, M. F. DU MESNIL DU BUISSON, and A. LOCATELLI. Mise en évidence d un contrôle foetal de la parturition chez la truie. Interactions avec la fonction lutéale. *C.R. Hebd. Séances Acad. Sci. Ser. D Sci. Nat.* (Paris) 287:1507-1510, 1974.

CAHILL, L. P., B. W. KNEE, and R. A. S. LAWSON. Induction of parturition in ewes with a single injection of oestradiol benzoate. *Theriogenology* 5:289-294, 1976.

CATCHPOLE, M. R. Hormonal mechanisms during pregnancy and parturition. In *Reproduction in Domestic Animals,* second ed., eds. H. H. Cole and P. T. Cupps, pp. 415-440. New York: Academic Press, 1969.

CAVAILLÉ, F., and J. P. MALTIER. A local signal from the gravid horn to the ovary for the onset of parturition in rats. *J. Reprod. Fertil.* 54:227-231, 1978.

CHALLIS, J. R. G., J. S. ROBINSON, and G. D. THORBURN. Fetal and maternal endocrine changes during pregnancy and parturition in the rhesus monkey. In *The Fetus and Birth,* CIBA Foundation Symposium 47 (new series), pp. 211-227. Amsterdam: Elsevier, 1977.

CHAMLEY, W. A., J. M. BUCKMASTER, M. E. CERENI, I. A. CUMMINGS, J. R. GODING, J. M. OBST, A. WILLIAMS, and C. WINFIELD. Changes in the levels of progesterone, corticosteroids, estrone, estradiol-17 β, luteinizing hormone and prolactin in the peripheral plasma of the ewe during late pregnancy and at parturition. *Biol. Reprod.* 9:30-35, 1973.

CHAN, W. Y. Relationship between the uterotonic action of oxytocin and prostaglandins: oxytocin action and release of PG-activity in isolated nonpregnant and pregnant uteri. *Biol. Reprod.* 17:541-548, 1977.

CHATELAIN, A., J.-P. DUPOUY, and P. ALLAUME. Fetal-maternal adrenocorticotropin and corticosterone relationships in the rat: effects of maternal adrenalectomy. *Endocrinology* 106:1297-1303, 1980.

CHEZ, R. A. Fetal factors relating to labor. In *Uterine Contractions,* ed. J. B. Josimovich, pp. 219-221. New York: Wiley, 1973.

CHIBOKA, O. Role of fetuses, placentas, uterus and ovaries in maintenance and termination of late pregnancy in swine and rabbits. *Dissert. Abstr. Int. B Sci. Eng.* 38:83-84, 1977.

COMLINE, R. S., L. W. HALL, R. B. LAVELLE, P. W. NATHANIELSZ, and M. SILVER. Parturition in the cow: endocrine changes in animals with chronically implanted catheters in the foetal and maternal circulations. *J. Endocrinol.* 63:451-472, 1974.

CONNER, M. H., and R. M. FRAPS. Premature oviposition following subtotal excision of the hen's ruptured follicle. *Poult. Sci.* 33:1051, 1954 (Abstract).

CSAPO, A. I. The 'see-saw' theory of parturition. In *The Fetus and Birth,* CIBA Foundation Symposium 47 (new series), pp. 159-195. Amsterdam: Elsevier, 1977.

CURRIE, W. B. Enhanced excitability of the uterus of the pregnant rabbit by imidazole stimulation of cyclic AMP phosphodiesterase. *J. Reprod. Fertil.* 60:369-375, 1980.

CURRIE, W. B., and G. D. THORBURN. The fetal role in timing the initiation of parturition in the goat. In *The Fetus and Birth,* CIBA Foundation Symposium 47 (new series), pp. 49-60. Amsterdam: Elsevier, 1977.

DAY, S. L., and A. V. NALBANDOV. Presence of prostaglandin F (PGF) in hen follicles and its physiological role in ovulation and oviposition. *Biol. Reprod.* 16:486-494, 1977.

DIAKOW, C. Initiation and inhibition of release croak of *Rana pipiens. Physiol. Behav.* 19:607-610, 1977.

_____. Hormonal basis for breeding behavior in female frogs: vasotocin inhibits the release call of *Rana pipiens. Science* 199:1456-1457, 1978.

DODD, J. M. The hormones of sex and reproduction and their effects in fish and lower chordates: twenty years on. *Am. Zool.* 15 (*Suppl.* 1):137-171, 1975.

DUKES, M., R. CHESTER, and P. ATKINSON. Effects of

oestradiol and prostaglandin $F_{2\alpha}$ on the timing of parturition in the rat. *J. Reprod. Fertil.* 38:325–346, 1974.

DUPOUY, J.-P. Evolution de la teneur de l'hypophyse foetale en ACTH. Etude chez le rat, en fin de gestation. *C.R. Hebd. Séances Acad. Sci. Ser. D Sci. Nat.* (Paris) 282:211–214, 1976.

EGAMI, N. Preliminary note on the induction of the spawning reflex and oviposition in *Oryzias latipes* by the administration of neurohypophyseal substances. *Annot. Zool. Jpn.* 32:13–17, 1959.

EGUCHI, Y., K. ARISHIMA, Y. MORIKAWA, and Y. HASHIMOTO. Rise of plasma corticosterone concentrations in rats immediately before and after birth and in fetal rats after the ligation of maternal uterine blood vessels or of the umbillical cord. *Endocrinology* 100:1443–1447, 1977.

ELLENDORFF, F., M. TAVERNE, F. ELSAESSER, M. FORSLING, N. PARVIZI, C. NAAKTGEBOREN, and D. SMIDT. Endocrinology of parturition in the pig. *Anim. Reprod. Sci.* 2:323–334, 1979.

FÈVRE, J. Corticosteroïdes maternels et foeteaux chez la truie en fin de gestation. *C.R. Hebd. Séances Acad. Sci. Ser. D Sci. Nat.* (Paris) 281:2009–2012, 1976.

FIEDLER, K. Hormonale Auslösung der Geburtsbewegungen beim Seepferdchen (*Hippocampus*, Syngatridae, Teleostei) *Z. Tierpsychol.* 27:679–686, 1970.

FIRST, N. L. Mechanisms controlling parturition in farm animals. In *Animal Reproduction*, ed. H. Hawk, pp. 215–257. Montclair, N.J.: Alanheld, Osmun, 1979.

FITZPATRICK, R. J. The posterior pituitary gland and the female reproductive tract. In *The Pituitary Gland*, eds. G. W. Harris and B. T. Donovan, Vol. 3, pp. 453–504. Berkeley: University of California Press, 1966.

———. Dilatation of the uterine cervix. In *The Fetus and Birth*, Ciba Foundation Symposium 47 (new series), pp. 31–39. Amsterdam: Elsevier, 1977.

FLINT, A. P. F., M. L. FORSLING, M. D. MITCHELL, and A. C. TURNBULL. Temporal relationship between changes in oxytocin and prostaglandin F levels in response to vaginal distension in the pregnant and puerperal ewe. *J. Reprod. Fertil.* 43:551–554, 1975.

FLINT, A. P. F., E. J. KINGSTON, J. S. Robinson, and G. D. THORBURN. Initiation of parturition in the goat: evidence for control by foetal glucocorticoid through activation of placental C_{21}-steroid 17 α-hydroxylase. *J. Endocrinol.* 78:367–378, 1978.

FLINT, A. P. F., A. P. RICKETTS, and V. A. CRAIG. The control of placental steroid synthesis at parturition in domestic animals. *Anim. Reprod. Sci.* 2:239–251, 1979.

FLOWER, R. J. The role of prostaglandins in parturition, with special reference to the rat. In *The Fetus and Birth*, CIBA Foundation Symposium 47 (new series), pp. 297–312. Amsterdam: Elsevier, 1977.

FORSLING, M. L., A. A. MACDONALD, and F. ELLENDORFF. The neurohypophysial hormones. *Anim. Reprod. Sci.* 2:43–56, 1979.

FOX, H. The urogenital system of reptiles. In *Biology of the Reptilia*, ed. C. Gans, Vol. 6, pp. 1–157. New York: Academic Press, 1977.

FRAPS, R. M. Synchronized induction of ovulation and premature oviposition in the domestic fowl. *Anat. Rec.* 84:521, 1942.

FUCHS, A.-R. Myometrial response to prostaglandin enhanced by progesterone. *Am. J. Obstet. Gynecol.* 118:1093–1098, 1974.

———. Regulation of uterine activity during gestation and parturition in rabbits and rats. In *Physiology and Genetics of Reproduction*, eds. F. Fuchs and E. M. Coutonho, Part B, pp. 403–422. New York: Plenum Press, 1975.

FUCHS, A.-R., Y. SMITASIRI, and U. CHANTHARAKSRI. The effect of indomethacin on uterine contractility and luteal regression in pregnant rats at term. *J. Reprod. Fertil.* 48:331–340, 1976.

GARFIELD, R. E., S. RABIDEAU, J. R. G. CHALLIS, and E. E. DANIEL. Hormonal control of GAP junctions in sheep myometrium during parturition. *Biol. Reprod.* 21:999–1007, 1979a.

———. Ultrastructural basis for maintenance and termination of pregnancy. *Am. J. Obstet. Gynecol.* 133:308–315, 1979b.

GARFIELD, R. E., S. M. SIMS, and E. E. DANIEL. Gap junctions: their presence and necessity in myometrium during parturition. *Science* 198:958–960, 1977.

GARFIELD, R. E., S. M. SIMS, S. M. KANNAN, and E. E. DANIEL. The possible role of gap junctions in activation of the myometrium during parturition. *Am. J. Physiol.* 235:C168–C179, 1978.

GENNSER, G., S. OHRLANDER, and P. ENEROTH. Fetal cortisol and the initiation of labour in the human. In *Fetus and Birth*, CIBA Foundation Symposium 47 (new series), pp. 401–420. Amsterdam: Elsevier, 1977.

GEORGE, J. M. Variation in the time of parturition of Merino and Dorset horn ewes. *J. Agric. Sci.* 73:295–299, 1969.

GILBERT, A. B., M. F. DAVIDSON, and J. W. WELLS. Role of the granulosa cells of the postovulatory follicle of the domestic fowl in oviposition. *J. Reprod. Fertil.* 52:227–229, 1978.

HEAP, R. B., A. K. A. GALIL, F. A. HARRISON, G. JENKIN, and J. S. PERRY. Progesterone and oestrogen in pregnancy and parturition: comparative aspects and hierarchical control. In *The Fetus and Birth*, CIBA Foundation Symposium 47 (new series), pp. 127–150. Amsterdam: Elsevier, 1977.

HERTELENDY, F. Prostaglandin-induced premature oviposition in the corturnix quail. *Prostaglandins* 2:269–279, 1972.

HERTELENDY, F., and H. V. BIELLIER. Prostaglandin levels in avian blood and reproductive organs. *Biol. Reprod.* 18:204–211, 1978a.

———. Evidence for a physiological role of prostaglandins in oviposition by the hen. *J. Reprod. Fertil.* 53:71–74, 1978b.

HERTELENDY R. F., M. YEH, and H. V. BIELLIER. Induction of oviposition in the domestic hen by prostaglandins. *Gen. Comp. Physiol.* 22:529-531, 1974.

HINDSON, J. C., B. M. SCHOFIELD, C. B. TURNER, and H. S. WOLFF. Parturition in sheep. *J. Physiol. (Lond.)* 181:560-567, 1965.

HOFFMANN, B., W. C. WAGNER, J. E. HIXON, and J. BAHR. Observations concerning the functional status of the corpus luteum and the placenta around parturition in the cow. *Anim. Reprod. Sci.* 2:253-266, 1979.

HOFFMAN, R. B., W. C. WAGNER, E. RATTENBERGER, and J. SCHMIDT. Endocrine relationships during late gestation and parturition in the cow. In *The Fetus and Birth*, CIBA Foundation Symposium 47 (new series), pp. 107-118. Amsterdam: Elsevier, 1977.

HOLLINGWORTH, M., C. N. M. ISHERWOOD, and R. W. FOSTER. The effect of oestradiol benzoate, progesterone, relaxin and ovariectomy on cervical extensibility in the late pregnant rat. *J. Reprod. Fertil.* 56:471-477, 1979.

HUTCHINSON, D. L., J. L. WESTOVER, and D. W. WILL. The destruction of the maternal and fetal pituitary glands in subhuman primates. *Am. J. Obstet. Gynecol.* 83:857-865, 1962.

ISHII, S. Artificial induction of parturition in the top minnow *Gambusia* sp. *Zool. Mag. (Tokyo)* 70:3-4, 1961 (as cited by Kujala 1978).

JONES, C. T., P. JOHNSON, J. Z. KENDALL, J. W. K. RITCHIE, and G. D. Thorburn. Induction of premature parturition in sheep: adrenocorticotrophin and corticosteroid changes during infusion of synacthen into the foetus. *Acta Endocrinol.* 87:192-202, 1978a.

———. Adrenocorticotrophin and corticosteroid changes during dexamethasone infusion to intact and synacthen infusion to hypophysectomized foetuses. *Acta Endocrinol.* 87:203-211, 1978b.

KANN, G., and R. DENAMUR. Possible rôle of prolactin during the oestrus cycle and gestation in the ewe. *J. Reprod. Fertil.* 39:473-483, 1974.

KENDALL, J. Z., J. R. G. CHALLIS, I. C. HART, C. T. JONES, M. D. MITCHELL, J. W. K. RITCHIE, J. S. ROBINSON, and G. D. THORBURN. Steroid and prostaglandin concentrations in the plasma of pregnant ewes during infusion of adrenocorticotrophin or dexamethasone to intact or hypophysectomized foetuses. *J. Endocrinol.* 75:59-71, 1977.

KERTILIS, L. P., and L. L. ANDERSON. Effect of relaxin on cervical dilatation, parturition and lactation in the pig. *Biol. Reprod.* 21:57-68, 1979.

KRALL, J. F., and S. G. KORENMAN. Control of uterine contractility via cyclic AMP-dependent protein kinase. In *The Fetus and Birth*, CIBA Foundation Symposium 47 (new series), pp. 319-338. Amsterdam: Elsevier, 1977.

KUJALA, G. A. Corticosteroid and neurophypopyseal hormone control of parturition in the guppy *Poecilia*

reticulata. Gen. Comp. Endocrinol. 36:286-296, 1978.

LANMAN, J. T. Parturition in nonhuman primates. *Biol. Reprod.* 16:28-38, 1977.

LA POINTE, J. Comparative physiology of neurohypophysial hormone action on the vertebrate oviduct-uterus. *Am. Zool.* 17:763-773, 1977.

LAUDANSKI, T., S. BATRA, and M. ÅKERLUND. Prostaglandin-induced luteolysis in pregnant and pseudopregnant rabbits and the resultant effects on the myometrial activity. *J. Reprod. Fertil.* 56:141-148, 1979.

LIGGINS, G. C. The fetus and birth. In *Reproduction in Mammals*, eds. C. R. Austin and R. V. Short, Vol. 2, pp. 72-109. Cambridge: Cambridge University Press, 1972.

LIGGINS, G. C., R. J. FAIRCLOUGH, S. A. GRIEVES, C. S. FORSTER, and B. S. KNOX. Parturition in the sheep. In *The Fetus and Birth*, CIBA Foundation Symposium 47 (new series), pp. 5-25. Amsterdam: Elsevier, 1977a.

LIGGINS, G. C., C. S. FORSTER, S. A. GRIEVES, and A. L. SCHWARTZ. Control of parturition in man. *Biol. Reprod.* 16:39-56, 1977b.

LINCOLN, D. W. Labour in the rabbit: effect of electrical stimulation applied to the infundibulum and median eminence. *J. Endocrinol.* 50:607-618, 1971.

LINCOLN, D. W., and D. G. PORTER. Photoperiodic dissection of endocrine events at parturition. *Anim. Reprod. Sci.* 2:97-115, 1979.

LOUIS, T. M., J. R. G. CHALLIS, J. S. ROBINSON, and G. D. THORBURN. Rapid increase of foetal corticosteroids after prostaglandin E_2. *Nature* 264:797-799, 1976.

MACEY, M. J., G. E. PICKFORD, and R. E. PETER. Forebrain localization of the spawning reflex response to exogenous neurohypophysial hormones in the killifish *Fundulus heteroclitus*. *J. Exp. Zool.* 190:269-280, 1974.

MITCHELL, M. D., A. P. A. FLINT, and A. C. TURNBULL. Increasing uterine response to vaginal distension during late pregnancy in sheep. *J. Reprod. Fertil.* 49:35-40, 1977.

MITCHELL, M. D., J. E. PATRICK, J. S. ROBINSON, G. D. THORBURN, and J. R. G. CHALLIS. Prostaglandins in the plasma and amniotic fluid of rhesus monkeys during pregnancy and after intra-uterine foetal death. *J. Endocrinol.* 71:67-76, 1976.

MORISHIGE, W. K., A. P. PEPE, and I. ROTHCHILD. Serum luteinizing hormone, prolactin and progesterone levels during pregnancy in the rat. *Endocrinology* 92:1527-1530, 1973.

MUELLER-HEUBACH, E., R. E. MYERS, and K. ADAMSONS. Effects of adrenalectomy on pregnancy length in the rhesus monkey. *Am. J. Obstet. Gynecol.* 112:221-226, 1972.

NAAKTGEBOREN, C., and E. J. SLIJPER. *Biologie der Geburt*. Berlin: Paul Parey, 1970.

NATHANIELSZ, P. W. Discussion of paper by Heap *et al.*,

1977. In *The Fetus and Birth,* CIBA Foundation Symposium 47 (new series), p. 151. Amsterdam: Elsevier, 1977.

———. Endocrine mechanisms of parturition. *Annu. Rev. Physiol.* 40:411–445, 1978.

NATHANIELSZ, P. W., and M. ABEL. Initiation of parturition in the rabbit by maternal and foetal administration of cortisol: effect of rate and duration of administration: suppression of delivery by progesterone. *J. Endocrinol.* 57:47–54, 1973.

NOVY, M. J. Endocrine and pharmacological factors which influence the onset of labour in rhesus monkeys. In *The Fetus and Birth,* CIBA Foundation Symposium 47 (new series), pp. 259–288. Amsterdam: Elsevier 1977.

OGASAWARA, T., and O. KOGA. Prostaglandin production by the uterus of the hen in relation to spontaneous and phosphate-induced ovipositions. *Jpn. J. Zootech. Sci.* 49:523–528, 1978.

OPEL, H. Premature oviposition following operative interference with the brain of the chicken. *Endocrinology* 74:193–200, 1964.

PEETERS, G., N. DE VOS, and A. HOUVENAGHEL. Elimination of the Ferguson reflex by section of the pelvic nerves in the lactating goat. *J. Endocrinol.* 49:125–130, 1971.

PICKFORD, G. E. Induction of a spawning reflex in hypophysectomized killifish. *Nature* 170: 807–808, 1952.

PICKFORD, G. E., and E. L. STRECKER. The spawning reflex response of the killifish, *Fundulus heteroclitus.* Isotocin is relatively inactive in comparison with arginine vasotocin. *Gen. Comp. Endocrinol.* 32:132–137, 1977.

RAWLINGS, N. C., and W. R. WARD. Effect of fetal hypophysectomy on the initiation of parturition in the goat. *J. Reprod. Fertil.* 52:249–254, 1978a.

———. Correlations of maternal and fetal endocrine events with uterine pressure changes around parturition in the ewe. *J. Reprod. Fertil.* 54:1–8, 1978b.

REES, L. H., P. M. B. JACK, A. L. THOMAS, and P. W. NATHANIELSZ. Role of foetal adrenocorticotrophin during parturition in sheep. *Nature* 253:274–275, 1974.

ROBERTS, J. S. Progesterone-inhibition of oxytocin release during vaginal distension: evidence for a central site of action. *Endocrinology* 89:1137–1141, 1971.

ROBINSON, J. S. The sheep fetus at parturition. *Anim. Reprod. Sci.* 2:167–178, 1979.

ROTHCHILD, I., and R. M. FRAPS. On the function of the ruptured ovarian follicle of the domestic fowl. *Proc. Soc. Exp. Biol. Med.* 56:79–82, 1944.

SAWYER, W. H. Evolution of active neurohypophysial principles among the vertebrates. *Am. Zool.* 17:727–737, 1977.

SCARAMUZZI, R. J., D. T. BAIRD, H. P. BOYLE, R. B. LAND, and A. G. WHEELER. The secretion of prostaglandin F from the autotransplanted uterus of the ewe. *J. Reprod. Fertil.* 49:157–160, 1977.

SHERWOOD, O. D., B. S. NARA, V. E. CRNEKOVIC, and N. L. FIRST. Relaxin concentrations in pig plasma after administration of indomethacin and prostaglandin $F_{2\alpha}$ during late pregnancy. *Endocrinology* 104:1716–1721, 1979.

SHIMADA, K., and I. ASAI. Uterine contractions during the ovulatory cycle of the hen. *Biol. Reprod.* 19:1057–1062, 1978.

———. Effects of prostaglandin $F_{2\alpha}$ and indomethacin on uterine contractions in the hen. *Biol. Reprod.* 21:523–527, 1979.

SHIMADA, K., M. NISHIO, and T. HATTORI. Release of uterine-contraction-inducing factor from preovulatory follicle. *Biol. Reprod.* 24:114–118, 1981.

SILVER, M., R. J. BARNES, R. S. COMLINE, A. L. FOWDEN, L. CLOVER, and M. D. MITCHELL. Prostaglandins in the foetal pig and prepartum endocrine changes in mother and foetus. *Anim. Reprod. Sci.* 2:305–322, 1979.

SMITH, V. G., L. A. EDGERTON, H. D. HAFS, and E. M. CONVEY. Bovine serum estrogens, progestins and glucocorticoids during late pregnancy, parturition and early lactation. *J. Anim. Sci.* 36:391–396, 1973.

STRYKER, J. L., and P. L. DZIUK. Effects of fetal decapitation on fetal development, parturition and lactation in pigs. *J. Anim. Sci.* 40:282–287, 1976.

STURKIE, P. D. Hypophysis. In *Avian Physiology,* ed. P. D. Sturkie, pp. 286–301. New York: Springer, 1976.

STURKIE, P. D., and W. J. MUELLER. Reproduction in the female and egg production. In *Avian Physiology,* ed. P. D. Sturkie, pp. 302–330. New York, Springer, 1976.

SWAAB D. F., K. BOER, and W. J. HONNEBIER. The influence of the fetal hypothalamus and pituitary on the onset and course of parturition. In *The Fetus and Birth,* CIBA Foundation Symposium 47 (new series), pp. 379–393. Amsterdam: Elsevier, 1977.

SWAAB, D. F., W. J. HONNEBIER, and K. BOER. The role of the foetal rat brain in labour. *J. Endocrinol.* 57: 31–32, 1973.

TANAKA, K. Oviposition-inducing activity in the ovarian follicles of different sizes in the laying hen. *Poult. Sci.* 55:714–716, 1976.

TANAKA, K., and K. GOTO. Partial purification of the ovarian oviposition-inducing factor and estimation of its chemical nature. *Poult. Sci.* 55:1774–1778, 1976.

TAVERNE, M. A. M., C. NAAKTGEBOREN, and G. C. VAN DER WEYDEN. Myometrial activity and expulsion of fetuses. *Anim. Reprod. Sci.* 2:117–131, 1979.

THORBURN, G. D., J. R. C. CHALLIS, and W. B. CURRIE. Control of parturition in domestic animals. *Biol. Reprod.* 16:18–27, 1977.

TURNBULL, A. C., A. B. M. ANDERSON, A. P. F. FLINT, J. Y. JEREMY, M. J. N. C. KEIRSE, and M. D. MITCHELL. Human parturition. In *The Fetus and*

Birth, CIBA Foundation Symposium 47 (new series), pp. 427–452. Amsterdam: Elsevier, 1977.

VAN KAMPEN, K. R., and L. C. ELLIS. Prolonged gestation in ewes ingesting *Veratrum californicum:* morphological changes and steroid biosynthesis in the endocrine organs of cyclopic lambs. *J. Endocrinol.* 52:549–560, 1972.

VAN TIENHOVEN, A. *Reproductive Physiology of Vertebrates.* Philadelphia: W. B. Saunders, 1968.

VAN TIENHOVEN, A., N. R. SCOTT, and P. E. HILLMAN. The hypothalamus and thermoregulation: a review. *Poult. Sci.* 58:1633–1639, 1979.

WAGNER, W. C., R. L. WILLHAM, and L. E. EVANS. Controlled parturition in cattle. *J. Anim. Sci.* 38:485–489, 1974.

WILHELMI, A. E., G. E. PICKFORD, and W. H. SAWYER. Initiation of the spawning reflex response in Fundulus by the administration of fish and mammalian neurohypophysial preparations and synthetic oxytocin. *Endocrinology* 57:243–252, 1955.

WILSON, S. C., and P. J. SHARP. The effects of progesterone on oviposition and ovulation in the domestic fowl (*Gallus domesticus*). *Br. Poult. Sci.* 17:163–173, 1976.

WOOD-GUSH, D. G. M. *The Behaviour of the Domestic Fowl.* London: Heinemann, 1971.

YARON, Z. Endocrine aspects of gestation in viviparous reptiles. *Gen. Comp. Endocrinol. Suppl.* 3:663–674, 1972.

12 | Reproduction and Immunology

ADCOCK, E. W., III F. TEASDALE, C. S. AUGUST, S. COX, G. MESCHIA, F. C. BATTAGLIA, and M. A. NAUGHTON. Human chorionic gonadotropin: its possible role in maternal lymphocyte suppression. *Science* 181:845–847, 1973.

ALEXANDER, N. J. Autoimmune hypospermatogenesis in vasectomized guinea pigs. *Contraception* 8:147–164, 1973.

––––––. Immunologic and morphologic effects of vasectomy in the rhesus monkey. *Fed. Proc.* 34:1692–1697, 1975.

ALEXANDER, N. J., and T. B. CLARKSON. Vasectomy increases the severity of diet-induced atherosclerosis in *Macaca fascicularis. Science* 201:538–541, 1978.

ALEXANDER, N. J., B. J. WILSON, and G. D. PATTERSON. Vasectomy: immunologic effects in rhesus monkeys and men. *Fertil. Steril.* 25:149–156, 1974.

ALLAN, T. M. ABO blood groups and human sex ratio at birth. *J. Reprod. Fertil.* 43:209–219, 1975.

ANDREWS, P. W., and E. A. BOYSE. Mapping of an H-2 linked gene that influences mating preference in mice. *Immunogenetics* 6:265–268, 1978.

BARNES, G. W. The antigenic nature of male accessory glands of reproduction. *Biol. Reprod.* 6:384–421, 1972.

BEER, A. E., and B. E. BILLINGHAM. *The Immunology of Mammalian Reproduction.* Englewood Cliffs, New Jersey: Prentice-Hall, 1976.

––––––. Immunoregulatory aspects of pregnancy. *Fed. Proc.* 37:2374–2378, 1978.

BENNETT, D., E. GOLDBERG, L. C. DUNN, and E. A. BOYSE. Serological detection of a cell surface antigen specified by the T (brachyury) mutant gene in the house mouse. *Proc. Nat. Acad. Sci. USA* 69:2076–2080, 1972.

BIGAZZI, P. E., L. L. KOSUDA, and L. L. HARNICK. Sperm autoantibodies in vasectomized rats of different inbred strains. *Science* 197:1282–1283, 1977.

BIGAZZI, P. E., L. L. KOSUDA, K. C. HU, and G. A. ANDRES. Immune complex orchitis in vasectomized rabbits. *J. Exp. Med.* 143:382–404, 1976.

BILLINGTON, W. D. The placenta and the tumour: variations on an immunological enigma. In *Placenta: A Neglected Experimental Animal.* New York: Pergamon Press, 1979.

BISHOP, D. W. Testicular enzymes as fingerprints in the study of spermatogenesis. In *Reproduction and Sexual Behavior,* ed. M. Diamond, pp. 261–286. Bloomington: Indiana University Press, 1968.

BURKE, W. H., J. W. RIESER, and R. N. SHOFFNER. The effect of isoimmunization with semen on fertility in the turkey hen. *Poult. Sci.* 50:1841–1847, 1971.

CERINI, M., J. K. FINDLAY, and R. A. S. LAWSON. Pregnancy-specific antigens in sheep: application to the diagnosis of pregnancy. *J. Reprod. Fertil.* 46:65–69, 1976.

CONTRACTOR, S. F., and H. DAVIES. Effect of human chorionic somatomammotrophin and human chorionic gonadotrophin on phytohaemagglutinin-induced lymphocyte transformation. *Nature New Biol.* 243:284–286, 1973.

ERICKSON, R. P., P. C. HOPPE, D. TENNEBAUM, H. SPIELMANN, and C. J. EPSTEIN. Lactate dehydrogenase X: effects of antibody on mouse gametes, but not on early development *in vitro* and *in vivo. Science* 188:261–263, 1975.

GARAVAGNO, A., J. POSADA, C. BARROS, and C. A. SHIVERS. Some characteristics of the zona pellucida antigen in the hamster. *J. Exp. Zool.* 189:37–50, 1974.

GOLDBERG, E. Infertility in female rabbits immunized with lactate dehydrogenase X. *Science* 181:458–459, 1973.

––––––. Isozymes in testes and spermatozoa. In *Isozymes: Current Topics in Biological and Medical Research,* Vol. 1, pp. 79–124. New York: A. R. Liss, 1977.

GOLDBERG, E., and J. LERUM. Pregnancy suppression

by antiserum to the sperm specific lactate dehydrogenase. *Science* 176: 686–687, 1972.

GOLDBERG, E., and T. E. WHEAT. Induction of infertility in male rabbits by immunization with LDH-X. In *Regulatory Mechanisms of Male Reproductive Physiology*, eds. C. H. Spilman, J. Lobl, and K. T. Kirton, pp. 133–139. Amsterdam: Excerpta Medica, 1976.

HERBERT, W. J., and P. C. WILKINSON. *A Dictionary of Immunology*, second printing. Oxford: Blackwell Scientific, 1972.

JAFFE, B. M., and H. R. Behrman. *Methods of Hormone Radioimmunoassay*. New York: Academic Press, 1974.

JOHNSON, M. H. Fertilisation and implantation. In *Immunology of Human Reproduction*, eds. J. S. Scott and W. R. Jones, pp. 32–60. New York: Academic Press, 1976.

JONES, W. R. Immunological aspects of infertility. In *Immunology of Human Reproduction*, eds. J. S. Scott and W. R. Jones, pp. 375–413. New York: Academic Press, 1976.

KAJII, T., and K. OHAMA. Androgenetic origin of hydatiform mole. *Nature* 268:633–634, 1977.

KUMMERFELD, H. L., and R. H. FOOTE. Infertility and embryonic mortality in female rabbits immunized with different sperm preparations. *Biol. Reprod.* 14:300–305, 1976.

LERUM, J. E., and E. GOLDBERG. Immunological impairment of pregnancy in mice by lactate dehydrogenase-X. *Biol. Reprod.* 11:108–115, 1974.

McGOVERN, P. T. The effect of maternal immunity on the survival of goat × sheep embryos. *J. Reprod. Fertil.* 34:215–220, 1973.

MENDENHALL, H. W. The immunology of the fetal-maternal relationship. In *Immunology of Human Reproduction*, eds. J. S. Scott and W. R. Jones, pp. 61–80. New York: Academic Press, 1976.

MORTON, H., V. HEGH, and G. J. A. CLUNIE. Immunosuppression detected in pregnant mice by rosette inhibition test. *Nature* 249:459–460, 1974.

————. Studies of the rosette inhibition test in pregnant mice: evidence of immunosuppression? *Proc. Roy. Soc. Lond. B Biol. Sci.* 193:413–419, 1976.

MORTON, H., C. D. NANCARROW, R. J. SCARAMUZZI, B. M. EVISON, and G. J. A. CLUNIE. Detection of early pregnancy in sheep by the rosette inhibition test. *J. Reprod. Fertil.* 56: 75–80, 1979.

MORTON, H., B. ROLFE, G. J. A. CLUNIE, M. J. ANDERSON, and J. MORRISON. An early pregnancy factor detected in human serum by the rosette inhibition test. *Lancet* 1:394–397, 1977.

MURRAY, F. A., E. C. SEGERSON, and F. T. BROWN. Suppression of lymphocytes by porcine uterine secretory protein. *Biol. Reprod.* 19:15–25, 1978.

NIESCHLAG, E., and E. J. WICKINGS. Biological effects of antibodies to gonadal steroids. *Vitam. Horm.* 36:165–202, 1978.

SCOTT, J. R., and A. E. BEER. Immunological factors in first pregnancy Rh immunization. *Lancet* 1:717–718, 1973.

SCOTT, J. S., and W. R. Jones, eds. *Immunology of Human Reproduction*. New York: Academic Press, 1976.

SHIVERS, C. A., and B. S. DUNBAR. Autoantibodies to zona pellucida: a possible cause for infertility of women. *Science* 197:1084–1086, 1977.

SIITERI, P. K., and D. P. STITES. Immunologic and endocrine interrelationships in pregnancy. *Biol. Reprod.* 26:1–14, 1982.

STERNBERGER, L. A. *Immunocytochemistry*. Englewood Cliffs, New Jersey: Prentice-Hall, 1974.

TUNG, K. S. K. Allergic orchitis lesions are adoptively transferred from vasoligated guinea pigs to syngeneic recipients. *Science* 201:833–835, 1978.

TUNG, K. S. K., and N. J. ALEXANDER. Immunopathologic studies on vasectomized guinea pigs. *Biol. Reprod.* 17:241–254, 1977.

WELLERSON, R., P. WAGSTAFF, S. ASCULAI, M. HUDSON, and A. B. KUPFERBERG. Induction of aspermatogenesis in guinea pig through immunization with lactate dehydrogenase-X isozyme. *Int. J. Fertil.* 19: 65–72, 1974.

WENTWORTH, B. C., and W. J. MELLEN. Effects of spermatozoal antibodies and method of insemination on the fecundity of domestic hens. *Br. Poult. Sci.* 5:59–65, 1964.

YALLOW, R. S. Radioimmunoassay: a probe for the fine structure of biologic systems. *Science* 200:1236–1245, 1978.

YAMAGUCHI, M., K. YAMAZAKI, and E. A. BOYSE. Mating preference tests with recombinant congenic strain BALB.HTG. *Immunogenetics* 6:261–264, 1978.

YAMAZAKI, K., E. A. BOYSE, V. MIKÉ, M. T. THALER, B. J. MATHIESON, J. ABBOTT, J. BOYSE, Z. A. ZAYAS, and L. THOMAS. Control of mating preferences in mice by genes in the major histocompatibility complex. *J. Exp. Med.* 144:1324–1335, 1976.

ZIMMERMAN, E. A. Localization of hypothalamic hormones by immunocytochemical techniques. In *Frontiers in Neuroendocrinology*, eds. L. Martini and W. F. Ganong, Vol. 4, pp. 25–62. New York: Raven Press, 1976.

13 | Endocrinology of Reproductive Behavior

ADKINS, E. K. Effects of diverse androgens on the sexual behavior and morphology of castrated male quail. *Horm. Behav.* 8:201–207, 1977.

————. Sex steroids and the differentiation of avian reproductive behavior. *Am. Zool.* 18:501–509, 1978.

ADKINS, E. K., J. J. BOOP, D. K. KOUTNIK, J. B. MORRIS, and E. E. PNIEWSKI. Further evidence that androgen aromatization is essential for the activation of copulation in male quail. *Physiol. Behav.* 24:441–446, 1980.

ADKINS, E. K., and B. L. NOCK. The effects of the antiestrogen CI-628 on sexual behavior activated by androgen or estrogen in quail. *Horm. Behav.* 7:417–430, 1976.

ADKINS, E. K., and E. E. PNIEWSKI. Control of reproductive behavior by sex steroids in male quail. *J. Comp. Physiol. Psych.* 92:1169–1178, 1978.

ADKINS, E., and L. SCHLESINGER. Androgens and the social behavior of male and female lizards (*Anolis carolinensis*). *Horm. Behav.* 13:139–152, 1979.

ADKINS-REGAN, E. Hormone specificity, androgen metabolism, and social behavior. *Am. Zool.* 21:257–271, 1981.

ARNOLD, A. P., F. NOTTEBOHM, and D. W. PFAFF. Hormone concentrating cells in vocal control and other areas of the brain of the zebra finch (*Poephila guttata*). *J. Comp. Neurol.* 165:487–512, 1976.

ARNOLD, A. P., and A. SALTIEL. Sexual difference in pattern of hormone accumulation in the brain of a song bird. *Science* 205:702–705, 1979.

BAGGERMAN, B. An experimental study on the timing of breeding and migration of the three spined stickle back. *Arch. Néerl. Zool.* 12:105–317, 1957.

––––––Hormonal control of reproductive and parental behaviour in fishes. In *Perspectives in Endocrinology*, eds. E. J\ W. Barrington and C. B. Jørgensen, pp. 351–404. New York: Academic Press, 1968.

––––––. Het endocriene systeem en aanpassingen van stekelbaarzen aan hun milieu. In *Endocrinologie*, eds. J. Lever en J. de Wilde, pp. 64–92. Wageningen, The Netherlands: Centrum voor Landbouwpublicaties en Landbouwdocumentatie, 1978.

BALDWIN, R. L., and T. PLUCINSKI. Mammary gland development and lactation. In *Reproduction in Domestic Animals*, third ed., eds. H. H. Cole and P. T. Cupps, pp. 369–400. New York: Academic Press, 1977.

BALTHAZART, J., G. MALACARNE, and P. DEVICHE. Stimulatory effects of 5 β-dihydrotestosterone on the sexual behavior in the domestic chick. *Horm. Behav.* 15:246–258, 1981.

BALTHAZART, J., R. MASSA, and P. NEGRI-CESI. Photoperiodic control of testosterone metabolism, plasma gonadotrophins, cloacal gland growth, and reproductive behavior in the Japanese quail. *Gen. Comp. Endocrinol.* 39:222–235, 1979.

BARFIELD, R. J. Activation of copulatory behavior by androgen implanted into the preoptic area of the male fowl. *Horm. Behav.* 1:37–52, 1969.

––––––. Activation of sexual and aggressive behavior by androgen implanted into the male ring dove brain. *Endocrinology* 89:1470–1476, 1971.

––––––. The hypothalamus and social behavior with special reference to the hormonal control of sexual behavior. *Poult. Sci.* 58:1625–1632, 1979.

BARFIELD, R. J., and J. J. CHEN. Activation of estrous behavior in ovariectomized rats by intracerebral implants of estradiol benzoate. *Endocrinology* 101:1716–1725, 1977.

BATTY, J. Plasma levels of testosterone and male sexual behaviour in strains of the house mouse (*Mus musculus*). *Anim. Behav.* 26:339–348, 1978a.

––––––. Acute changes in plasma testosterone levels and their relation to measures of sexual behaviour in the male house mouse (*Mus musculus*). *Anim. Behav.* 26:349–357, 1978b.

BEACH, F. A. Coital behavior in dogs. III. Effects of early isolation on mating in males. *Behaviour* 30:217–238, 1968.

––––––. Locks and beagles. *Am. Psychol.* 24:971–989, 1969.

––––––. Sexual attractivity, proceptivity, and receptivity in female mammals. *Horm. Behav.* 7:105–138, 1976.

BELLAIRS, A. *The Life of Reptiles*, Vol. 2. New York: Universe Books, 1970.

BENOFF, F. H., P. B. Siegel, and H. P. Van Krey. Testosterone determinations in lines of chickens selected for differential mating frequency. *Horm. Behav.* 10:246–250, 1978.

BENTLEY, P. J. *Comparative Vertebrate Endocrinology*. New York: Cambridge University Press, 1976.

BERTHOLD, P. Migration: control and metabolic physiology. In *Avian Biology,* eds D. S. Farner and J. R. King, Vol. 5, pp. 77–128. New York: Academic Press, 1975.

BLUMER, L. S. Male parental care in the bony fishes. *Q. Rev. Biol.* 54:149–161, 1979.

BREDER, C. M., JR., and D. E. ROSEN. *Modes of Reproduction in Fishes*. Garden City, New York: Natural History Press, 1966.

BRIDGES, R. S., B. D. GOLDMAN, and L. P. BRYANT. Serum prolactin concentrations and the initiation of maternal behavior in the rat. *Horm. Behav.* 5:219–226, 1974.

CAMAZINE, B., W. GARSTKA, R. TOKARZ, and D. CREWS. Effects of castration and androgen replacement on male courtship behavior in the red-sided garter snake (*Thamnophis sirtalis parietalis*). *Horm. Behav.* 14:358–372, 1980.

CHENG, M.-F. Effect of estrogen on behavior of ovariectomized ring doves (*Streptopelia risoria*). *J. Comp. Physiol. Psychol.* 83:234–239, 1973.

––––––. Induction of incubation behaviour in male ring doves (*Streptopelia risoria*): a behavioural analysis. *J. Reprod. Fertil.* 42:267–276, 1975.

––––––. Role of gonadotrophin releasing hormones in the reproductive behaviour of female ring doves (*Streptopelia risoria*). *J. Endocrinol.* 74:37–45, 1977a.

––––––. Egg fertility and prolactin as determinants of reproductive cycling in doves. *Horm. Behav.* 9:85–98, 1977b.

CHENG, M.-F., and D. S. LEHRMAN. Gonadal hormone

specificity in the sexual behavior of ring doves. *Psychoneuroendocrinology* 1:95-102, 1975.

CHENG, M.-F., M. PORTER, and G. BALL. Do ring doves copulate more than necessary for fertilization. *Physiol. Behav.* 27:659-662, 1981.

CHENG, M.-F., and R. SILVER. Estrogen-progesterone regulation of nest-building and incubation behavior in ovariectomized ring doves (*Streptopelia risoria*). *J. Comp. Physiol.* 88:256-263, 1975.

CIACCIO, L. A., and R. D. LISK. Central control of estrous behavior in the female golden hamster. *Neuroendocrinology* 13:21-28, 1973.

CLARKE, W. C., and H. A. BERN. Comparative endocrinology of prolactin. In *Hormonal Proteins and Peptides*, ed. C. H. Li, Vol. 8, pp. 105-197. New York: Academic Press, 1980.

COHEN, J. Midbrain and motor control of courtship behavior in male and female ring doves (*Streptopelia risoria*). *Diss. Abst. Int. Sci. Eng.*, 41 Ref. 4301B, 1981.

COOPER, R. L., and C. J. ERICKSON. Effects of septal lesions on the courtship behavior of male ring doves (*Streptopelia risoria*). *Horm. Behav.* 7:441-450, 1976.

CRAIG, J. V., L. E. CASIDA, and A. B. CHAPMAN. Male infertility associated with lack of libido in the rat. *Am. Nat.* 88:365-372, 1954.

CREWS, D. Psychobiology of reptilian reproduction. *Science* 189:1059-1065, 1975.

_____. Hormonal control of male courtship behavior and female attractivity in the garter snake (*Thamnophis sirtalis sirtalis*). *Horm. Behav.* 7:451-460, 1976.

_____. Neuroendocrinology of lizard reproduction. *Biol. Reprod.* 20:51-73, 1979.

_____. Interrelationships among ecological, behavioral, and neuroendocrine processes in the reproductive cycle of *Anolis carolinensis* and other reptiles. *Adv. Study Behav.* 11:1-74, 1980.

CREWS, D., and N. GREENBERG. Function and causation of social signals in lizards. *Am. Zool.* 21:273-294, 1981.

CUNNINGHAM, D. L., P. B. SIEGEL, and H. P. VAN KREY. Androgen influence on mating behavior in selected lines of Japanese quail. *Horm. Behav.* 8:166-174, 1977.

DAVIDSON, J. M. Neurohormonal bases of male sexual behavior. In *Int. Rev. Physiol. Reproductive Physiology* II, Vol. 13, pp. 225-254. Baltimore: University Park Press, 1977.

DAWBIN, W. H. The seasonal migratory cycle of humpback whales. In *Whales, Dolphins and Porpoises*, ed. K. S. Norris, pp. 145-169. Berkeley and Los Angeles: University of California Press, 1966.

DEMSKI, L. S. Feeding and aggressive behavior evoked by hypothalamic stimulation in a cichlid fish. *Comp. Biochem. Physiol.* 44A: 685-692, 1973.

_____. Electrical stimulation of the shark brain. *Am. Zool.* 17:487-500, 1977.

DEMSKI, L. S., D. H. BAUER, and J. W. GERALD. Sperm release evoked by electrical stimulation of the fish brain: a functional-anatomical study. *J. Exp. Zool.* 191:215-232, 1975.

DE RUITER, A. J. H. A combined structural and biochemical analysis of the effect of testosterone on kidney function of the three-spined stickle back. *Gen. Comp. Endocrinol.* 29:283-284, 1976.

DEWSBURY, D. A. Patterns of copulatory behavior in male mammals. *Q. Rev. Biol.* 47:1-33, 1972.

_____. Diversity and adaptation in rodent copulatory behavior. *Science* 190:947-954, 1975.

DEWSBURY, D. A., and D. Q. ESTEP. Pregnancy in cactus mice: effects of prolonged copulation. *Science* 187:552-553, 1975.

DIAKOW, C., and A. NEMIROFF. Vasotocin, prostaglandin and female reproductive behavior in the frog, *Rana pipiens. Horm. Behav.* 15:86-93, 1981.

DRENT, R. Incubation. In *Avian Biology*, eds. D. S. Farner and J. R. King, Vol. 5, pp. 333-420. New York: Academic Press, 1975.

DRYDEN, G. L., and J. N. ANDERSON. Ovarian hormone: lack of effect on reproductive structures of female Asian musk shrews. *Science* 197:782-784, 1977.

EMLEN, S. T. Migration: orientation and navigation. In *Avian Biology*, eds. D. S. Farner and J. R. King, Vol. 5, pp. 129-219. New York: Academic Press, 1975.

EMLEN, S. T., and L. W. ORING. Ecology, sexual selection, and the evolution of mating systems. *Science* 197:215-223, 1977.

ERICKSON, C. J., and D. S. LEHRMAN. Effect of castration on male ring doves upon ovarian activity of females. *J. Comp. Physiol. Psychol.* 58:164-166, 1964.

ERICKSON, C. J., and P. G. ZENONE. Courtship differences in male ring doves: avoidance of cuckoldry? *Science* 192:1353-1354, 1976.

EVERITT, B. J., and J. Herbert. The effects of implanting testosterone propionate into the central nervous system on the sexual behaviour of adrenalectomized female rhesus monkeys. *Brain Res.* 86:109-120, 1975.

EVERITT, B. J., J. HERBERT, and J. D. HAMER. Sexual receptivity of bilaterally adrenalectomized female rhesus monkeys. *Physiol. Behav.* 8:409-415, 1972.

FAIRCHILD, L. Mate selection and behavioral thermoregulation in Fowler's toads. *Science* 212:950-951, 1981.

GANDELMAN, R., N. J. MCDERMOTT, M. KLEINMAN, and D. DEJIANNE. Maternal nest building by pseudopregnant mice. *J. Reprod. Fertil.* 56:697-699, 1979.

GERALL, A. A. An exploratory study of the effect of social isolation variables on the sexual behaviour of male guinea pigs. *Anim. Behav.* 11:274-282, 1963.

GITTLEMAN, J. L. The phylogeny of parental care in fishes. *Anim. Behav.* 29:936-941, 1981.

GRUENDEL, A. D., and W. J. ARNOLD. Effects of early

social deprivation on reproductive behavior of male rats. *J. Comp. Physiol. Psychol.* 67:123–128, 1969.

GWINNER, E. G., F. W. TUREK, and S. D. SMITH. Extraocular light perception in photoperiodic responses of the white-crowned sparrow (*Zonotrichia leucophrys*) and of the golden-crowned sparrow (*Z. atricapilla*). *Z. Vgl. Physiol.* 75:323–331, 1971.

HALLIDAY, T. T. Sexual behaviour of the smooth newt *Triturus vulgaris* (Urodela, Salamandridae). *J. Herpetol.* 8:277–292, 1974.

HALLIDAY, T., and A. HOUSTON. The newt as an honest salesman. *Anim. Behav.* 26:1273–1274, 1978.

HARDING, C. F., and B. K. FOLLETT. Hormone changes triggered by aggression in a natural population of black birds. *Science* 203:918–920, 1979.

HARDISTY, H. W., and I. C. POTTER. The general biology of adult lampreys. In *The Biology of Lampreys,* eds. M. W. Hardisty and I. C. Potter, Vol. 1, pp. 127–206. New York: Academic Press, 1971.

HARLOW, H. F., W. D. JOSLYN, M. G. SENKO, and A. DOPP. Behavioral aspects of reproduction in primates. *J. Anim. Sci.* 25 (*Suppl.*):49–67, 1966.

HARRIS, M. P. Abnormal migration and hybridization of *Larus argentatus* and *L. fuscus* after interspecies fostering experiments. *Ibis* 112:488–498, 1970.

HART, B. L. Activation of sexual reflexes of male rats by dihydrotestosterone but not estrogen. *Physiol. Behav.* 23:107–109, 1979.

HART, B. L., and R. L. KITCHELL. Penile erection and contraction of penile muscles in the spinal and intact dog. *Am. J. Physiol.* 210:257–262, 1966.

HEMSWORTH, P. H., R. G. BEILHARZ, and D. B. GALLOWAY. Influence of social conditions during rearing on the sexual behaviour of the domestic boar. *Anim. Prod.* 24:245–251, 1977.

HENDERSON, I. W., N. SA'DI, and G. HARGREAVES. Studies on the production and metabolic clearance rates of cortisol in the European eel, *Anguilla anguilla* (L). *J. Steroid. Biochem.* 5:701–707, 1974.

HOUGH, J. C., JR., G. K.-W.HO, P. H. COOKE, and D. M. QUADAGNO. Actinomycin D: reversible inhibition of lordosis behavior and correlated changes in nuclear morphology. *Horm. Behav.* 5:367–375, 1974.

HUTCHINSON, J. B. Effects of hypothalamic implants of gonadal steroids on courtship behaviour in Barbary doves (*Streptopelia risoria*). *J. Endocrinol.* 50:97–113, 1971.

⸻. Hormones and brain mechanisms of sexual behaviour: a possible relationship between cellular and behavioural events in doves. *Persp. Exp. Biol.* 1:417–436, 1976.

⸻. Hypothalamic regulation of male sexual responsiveness to androgen. In *Biological Determinants of Sexual Behaviour*, ed. J. B. Hutchison, pp. 277–317. New York: Wiley, 1978.

HUTCHISON, J. B., and T. STEIMER. Brain 5 β-reductase: a correlate of behavioral sensitivity to androgen. *Science* 213:244–246, 1981.

IDLER, D. R., A. P. RONALD, and P. J. SCHMIDT. Biochemical studies on sockeye salmon during spawning migration. VII Steroid hormones in plasma. *Can. J. Biochem. Physiol.* 37:1227–1238, 1959.

IMMELMANN, K. Objektfixierung geschlechtlicher Triebhandlungen bei Prachtfinken. *Naturwissenschaften* 52:169, 1965a.

⸻. Prägungserscheinungen in der Gesangentwicklung junger Zebrafinken. *Naturwissenschaften* 52:169–170, 1965b.

KOCH, H. J. A. Migration. In *Perspectives in Endocrinology*, eds. E. J. W. Barrington and C. B. Jørgensen, pp. 305–349. New York: Academic Press, 1968.

KOMISARUK, B. R. Effects of local brain implants of progesterone on reproductive behavior in ring doves. *J. Comp. Physiol. Psychol.* 64:219–224, 1967.

⸻. Neural and hormonal interactions in the reproductive behavior of female rats. *Adv. Behav. Biol.* 11:97–129, 1974.

KORENBROT, C. C., D. W. SCHOMBERG, and C. J. ERICKSON. Radioimmunoassay of plasma estradiol during the breeding cycle of ring doves (*Streptopelia risoria*). *Endocrinology* 94:1126–1132, 1974.

KREHBIEL, D. A., and L. M. LEROY. The quality of hormonally stimulated maternal behavior in ovariectomized rats. *Horm. Behav.* 12:243–252, 1979.

KUBIE, J. L., J. COHEN, and M. HALPERN. Shedding enhances the sexual attractiveness of oestradiol treated garter snakes and their untreated penmates. *Anim. Behav.* 26:562–570, 1978.

LEEHAN, S. W., D. M. QUADAGNO, and J. D. BAST. The effects of extrahypothalamic cycloheximide on sexual receptivity in the rat. *Horm. Behav.* 12:264–268, 1979.

LEHRMAN, D. S. Effect of female sex hormones on incubation behavior in the ring dove (*Streptopelia risoria*). *J. Comp. Physiol. Psychol.* 51:142–145, 1958.

⸻. Hormonal regulation of parental behavior in birds and infrahuman mammals. In *Sex and Internal Secretions*, ed. W. C. Young, Vol. 2, pp. 1268–1382. Baltimore: Williams and Wilkins, 1961.

⸻. The reproductive behavior of ring doves. *Sci. Am.* 211(5):48–54, 1964.

LEON, M., M. NUMAN, and H. MOLTZ. Maternal behavior in the rat: facilitation through gonadectomy. *Science* 179:1018–1019, 1973.

LEONARD, S. L. Induction of singing in female canaries by injections of male hormone. *Proc. Soc. Exp. Biol. Med.* 41:229–230, 1939.

LILEY, N. R. Hormones and reproductive behavior in fishes. In *Fish Physiology,* eds. W. S. Hoar and D. J. Randall, Vol. 3, pp. 73–116. New York: Academic Press, 1969.

⸻. The effects of estrogens and other steroids on the sexual behavior of the female guppy, *Poecilia reticulata. Gen. Comp. Endocrinol. Suppl.* 3:542–552, 1972.

LISK, R. D. Oestrogen and progesterone synergism and elicitation of maternal nest building in the mouse (*Mus musculus*). *Anim. Behav.* 19:606–610, 1971.

LUMPKIN, S. Sexual solicitation behavior in the female ring dove (*Streptopelia risoria*). *Diss. Abst. Int. Sci. Eng.* 41 Ref. 4302B, 1981.

MARQUES, D. M., C. W. MALSBURY, and J. DAOOD. Hypothalamic knife cuts dissociate maternal behaviors, sexual receptivity, and estrous cyclicity in female hamsters. *Physiol. Behav.* 23:347–355, 1979.

McCOLLOM, R. E., P. B. SIEGEL, and H. P. VAN KREY. Responses to androgen in lines of chickens selected for mating behavior. *Horm. Behav.* 2:31–42, 1971.

McNICOL, D., JR., and D. CREWS. Estrogen/progesterone synergy in the control of female sexual receptivity in the lizard, *Anolis carolinensis*. *Gen. Comp. Endocrinol.* 38:68–74, 1979.

MEIER, A. M. Chronoendocrinology of the white-throated sparrow. *Proc. 16th Int. Ornithol. Congress, Canberra, 12–17 August 1974*, pp. 355–368. Canberra: Australian Academy of Science, 1976.

MEINKOTH, J., D. M. QUADAGNO, and J. D. BAST. Depression of steroid-induced sex behavior in ovariectomized rat by intracranial injection of cycloheximide: preoptic area compared to the ventromedial hypothalamus. *Horm. Behav.* 12:199–204, 1979.

MOLTZ, H. Some mechanisms governing the induction, maintenance, and synchrony of maternal behavior in the laboratory rat. *Adv. Behav. Biol.* 11:77–96, 1974.

MOORE, F. L. Differential effects of testosterone plus dihydrotestosterone on male courtship of castrated newts, *Taricha granulosa*. *Horm. Behav.* 11:202–208, 1978.

MOORE, F. L., C. McCORMACK, and L. SWANSON. Induced ovulation: effects of sexual behavior and insemination on ovulation and progesterone levels in *Taricha granulosa*. *Gen. Comp. Endocrinol.* 39:262–269, 1979.

MOORE, F. L., J. L. SPECKER, and L. SWANSON. Effects of testosterone and methallibure on courtship and plasma androgen concentrations of male newts, *Taricha granulosa*. *Gen. Comp. Endocrinol.* 34:259–264, 1978.

MOSS, R. L. Relationship between the central regulation of gonadotropins and mating behavior in female rats. *Adv. Behav. Biol.* 11:55–76, 1974.

NAAKTGEBOREN, C. Behavioural aspects of parturition. *Anim. Reprod. Sci.* 2:155–166, 1979.

NAAKTGEBOREN, C., and E. J. SLIJPER. *Biologie der Geburt*. Berlin: Paul Parey, 1970.

NALBANDOV, A. V. Retrospects and prospects in reproductive physiology. In *Novel Aspects of Reproductive Physiology*, eds. C. H. Spilman and J. W. Wilks, pp. 3–10. Jamaica, New York: Spectrum Publications, 1978.

NICOLL, C. S., and H. A. BERN. On the actions of prolactin among the vertebrates: is there a common denominator? In *Lactogenic Hormones*, eds. G. E. W. Wolstenholme and J. Knight, pp. 299–317. London: Churchill Livingstone, 1972.

NOBLE, G. K. *The Biology of the Amphibia*. New York: McGraw-Hill, 1931.

NOBLE, G. K., and R. BORNE. The effect of sex hormones on the social hierarchy of *Xiphophorus helleri*. *Anat. Rec. Suppl.* 78:147, 1940.

NOBLE, R. G. Male hamsters display female sexual responses. *Horm. Behav.* 12:293–298, 1979.

NOTTEBOHM, F., and A. P. ARNOLD. Sexual dimorphism in vocal control areas of the songbird brain. *Science* 194:211–213, 1976.

NOTTEBOHM, F., T. M. STOKES, and C. M. LEONARD. Central control of song in the canary, *Serinus canarius*. *J. Comp. Neurol.* 165:457–486, 1976.

NUMAN, M. Medial preoptic area and maternal behavior in the female rat. *J. Comp. Physiol. Psychol.* 87:746–759, 1974.

NUMAN, M., M. LEON, and H. MOLTZ. Interference with prolactin release and maternal behavior of female rats. *Horm. Behav.* 3:29–38, 1972.

NUMAN, M., J. S. ROSENBLATT, and B. R. KOMISARUK. Medial preoptic area and onset of maternal behavior in the rat. *J. Comp. Physiol. Psychol.* 91:146–164, 1977.

ORCUTT, F. S., JR. Effects of oestrogen on the differentiation of some reproductive behaviours in male pigeons (*Columba livia*). *Anim. Behav.* 19:277–286, 1971.

ORCUTT, F. S., JR., and A. B. ORCUTT. Nesting and parental behavior in domestic common quail. *Auk* 93:135–141, 1976.

PALKA, Y. S., and A. GORBMAN. Pituitary and testicular influenced sexual behavior in male frogs, *Rana pipiens*. *Gen. Comp. Endocrinol.* 21:148–151, 1973.

PATEL, M. D. The physiology of the formation of "pigeon's milk." *Physiol. Zool.* 9:129–152, 1936.

PEAKER, M. (ed.). *Comparative Aspects of Lactation*. Symp. Zool. Soc. Lond. No. 41, 1977.

PERRILL, S. A., H. C. GERHARDT, and R. DANIEL. Sexual parasitism in the green tree frog (*Hyla cinerea*). *Science* 200:1179–1180, 1978.

PFAFF, D. W., and C. LEWIS. Film analysis of lordosis in female rats. *Horm. Behav.* 5:317–335, 1974.

PHILLIPS, R. E., and F. W. PEEK. Brain organization and neuromuscular control of vocalization in birds. In *Neural and Endocrine Aspects of Behavior in Birds*, eds. P. Wright, P. G. Caryl, and D. M. Vowles, pp. 243–274. Amsterdam: Elsevier, 1975.

POINDRON, P., and P. LE NEINDRE. Endocrine and sensory regulation of maternal behavior in the ewe. *Adv. Study Behav.* 11:75–119, 1980.

POOLEY, A. C., and C. GANS. The Nile crocodile. *Sci. Am.* 234(4):114–124, 1976.

PRITCHARD, P. C. H. Taxonomy, evolution and zoogeography. In *Turtles: Perspectives and Re-*

search, eds. M. Harless and H. Morlock, pp. 1–42. New York: Wiley, 1979.

QUADAGNO, D. M., and G. K. W. HO. The reversible inhibition of steroid-induced sexual behavior by intracranial cycloheximide. *Horm. Behav.* 6:19–26, 1975.

RIDLEY, M. Paternal care. *Anim. Behav.* 26:904–932, 1978.

RISSMAN, E. F. Effects of experience and pairing on detection of cuckoldry by male ring doves. Manuscript submitted to *Anim. Behav.,* 1981.

ROSENBLATT, J. S. Effects of experience on sexual behavior in male cats. In *Sex and Behavior,* ed. F. A. Beach, pp. 416–439. New York: Wiley, 1965.

———. Nonhormonal basis of maternal behavior in the rat. *Science* 156:1512–1514, 1967.

RYAN, M. J. Female mate choice in a neotropical frog. *Science* 209:523–525, 1980.

SACHS, B. D. Photoperiodic control of reproductive behavior and physiology of the male Japanese quail (*Coturnix coturnix japonica*). *Horm. Behav.* 1:7–24, 1969.

SACHS, B. D., and R. J. BARFIELD. Functional analysis of masculine copulatory behavior in the rat. *Adv. Study Behav.* 7:91–154, 1976.

SALTHE, S. N., and J. S. MECHAM. Reproductive and courtship patterns. In *Physiology of the Amphibia,* ed. B. Lofts, Vol. 2, pp. 309–521. New York: Academic Press, 1974.

SCHEIN, M. W., and E. B. HALE. Stimuli eliciting sexual behavior. In *Sex and Behavior,* ed. F. A. Beach, pp. 440–482. New York: Wiley, 1965.

SEFTON, A. E., and P. B. SIEGEL. Selection for mating ability in Japanese quail. *Poult. Sci.* 54:788–794, 1975.

SHAPIRO, D. Y. Social behavior, group structure, and the control of sex reversal in hermaphroditic fish. *Adv. Study Behav.* 10:43–102, 1979.

SIEGEL, H. I., and J. S. ROSENBLATT. Hormonal basis of hysterectomy-induced maternal behavior during pregnancy in the rat. *Horm. Behav.* 6:211–222, 1975a.

———. Progesterone inhibition of estrogen-induced maternal behavior in hysterectomized-ovariectomized virgin rats. *Horm. Behav.* 6:223–230, 1975b.

———. Estrogen-induced maternal behavior in hysterectomized-ovariectomized virgin rats. *Physiol. Behav.* 14:465–471, 1975c.

———. Latency and duration of estrogen induction of maternal behavior in hysterectomized-ovariectomized virgin rats: effects of pup stimulation. *Physiol. Behav.* 14:473–476. 1975d.

SIEGEL, P. B. Behavioural genetics. In *The Behaviour of Domestic Animals,* ed. E. S. E. Hafez, pp. 20–42. London: Baillière Tindall, 1975.

SIGNORET, J. P. Chemical communication and reproduction in domestic mammals. In *Mammalian Olfaction, Reproductive Processes, and Behavior,* ed. R.

L. Doty, pp. 243–256. New York: Academic Press, 1976.

SILVER, R., and H. H. FEDER. Role of gonadal hormones in incubation behavior of male ring doves (*Streptopelia risoria*). *J. Comp. Physiol. Psychol.* 84:464–471, 1973.

SILVER, R., C. REBOULLEAU, D. S. LEHRMAN, and H. H. FEDER. Radioimmunoassay of plasma progesterone during the reproductive cycle of male and female ring doves (*Streptopelia risoria*). *Endocrinology* 94:1547–1554, 1974.

SINHA, V. R. P., and J. W. Jones. *The European Freshwater Eel.* Liverpool: Liverpool University Press, 1975.

STACEY, N. E., and N. R. LILEY. Regulation of spawning behaviour in the female goldfish. *Nature* 247:71–72, 1972.

STEEL, E. Short-term, postcopulatory changes in receptive and proreceptive behavior in the female Syrian hamster. *Horm. Behav.* 12:280–292, 1979.

STEIMER, T., and J. B. HUTCHISON. Aromatization of testosterone within a discrete hypothalamic area associated with the behavioral action of androgens in the male dove. *Brain Res.* 192:586–591, 1980.

TERKEL, A. S., J. SHRYNE, and R. A. GORSKI. Inhibition of estrogen facilitation of sexual behavior by the intracerebral infusion of actinomycin-D. *Horm. Behav.* 4:377–386, 1973.

TERKEL, J., and J. S. ROSENBLATT. Aspects of nonhormonal maternal behavior in the rat. *Horm. Behav.* 2:161–171, 1971.

———. Humoral factors underlying maternal behavior at parturition: crosstransfusion between freely moving rats. *J. Comp. Physiol. Psychol.* 80:365–371, 1972.

TESCH, F.-W. Horizontal and vertical swimming of eels during the spawning migration at the edge of the continental shelf. In *Animal Migration, Navigation, and Homing,* eds. K. Schmidt-Koenig and W. T. Keeton, pp. 378–391. New York: Springer, 1978.

THORN, R. Contribution à l'étude de'une salamandre japonaise L'*Hynobius nebulosus* (Schlegel). Comportement et reproduction en captivité. *Instit. Gr-Duc. Luxem. Archiv.* N.S. 29:201–215, 1963.

———. Nouvelles observations sur l'éthologie sexuelle de l'*Hynobius nebulosus* (Temminck et Schlegel) (Caudata, Hynobiidae). *Instit. Gr-Duc. Luxem. Archiv.* (N.S.) 32:267–271, 1967.

VAN TIENHOVEN, A. *Reproductive Physiology of Vertebrates* Philadelphia: W. B. Saunders, 1968.

VOCI, V. E., and N. R. CARLSON. Enhancement of maternal behavior and nest building following systemic and diencephalic administration of prolactin and progesterone in the mouse. *J. Comp. Physiol. Psychol.* 83:388–393, 1970.

WADA, M., and A. GORBMAN. Relation of mode of administration of testosterone to evocation of male sex behavior in frogs. *Horm. Behav.* 8:310–319, 1977.

WADA, M., J. C. WINGFIELD, and A. GORBMAN. Correlation between blood levels of androgen and sexual behavior in male leopard frogs, *Rana pipiens*. *Gen. Comp. Endocrinol.* 29:72-77, 1976.

WENDELAAR BONGA, S. E. Morphometrical analysis with the light and electron microscope of the kidney of the anadromous three-spined stickle back *Gasterosteus aculeatus*, form *trachus* from freshwater and from sea water. *Z. Zellforsch. Mikrosk. Anat.* 137:563-588, 1973.

————. The effect of prolactin on kidney structure of the euryhaline teleost, *Gasterosteus aculeatus* during adaptation to fresh water. *Cell Tissue Res.* 166:319-338, 1976.

WENDELAAR BONGA, S. E., J. A. A. GREVEN, and M. VEENHUIS. The relationship between the ionic composition of the environment and the secretory activity of the endocrine cell types of Stannius corpuscles in the teleost *Gasterosteus aculeatus*. *Cell Tissue Res.* 175:297-312, 1976.

WENDELAAR BONGA, S. E., and M. VEENHUIS. The membranes of the basal labyrinth in kidney cells of the stickleback, *Gasterosteus aculeatus*, studied in ultrathin section and freeze-etch replicas. *J. Cell Sci.* 14:587-609, 1974a.

————. The effect of prolactin on the number of membrane-associated particles in kidney cells of euryhaline teleost *Gasterosteus aculeatus* during transfer from seawater to fresh water: a freeze-etch study. *J. Cell Sci.* 16:687-701, 1974b.

WINGFIELD, J. C., and D. S. FARNER. Control of seasonal reproduction in temperate-zone birds. In *Progress in Reproductive Biology,* ed. P. O. Hubinont, Vol. 5, pp. 62-101. Basel: Karger, 1980.

WOLFSON, A. Environmental and neuroendocrine regulation of annual gonadal cycles and migratory behavior in birds. *Recent Prog. Horm. Res.* 239:177-239, 1966.

WOOD-GUSH, D. G. M. *The Behaviour of the Domestic Fowl.* London: Heinemann, 1971.

WOURMS, J. P. Reproduction and development in Chondrichthyan fishes. *Am. Zool.* 17:379-410, 1977.

YAHR, P. Data and hypotheses in tales of dihydrotestosterone. *Horm. Behav.* 13:92-96, 1979.

YAHR, P., and S. A. GERLING. Aromatization and androgen stimulation of sexual behavior in male and female rats. *Horm. Behav.* 10: 128-142, 1978.

YOUNG, W. C. The hormones and mating behavior. In *Sex and Internal Secretions,* ed. W. C. Young, Vol. 2, pp. 1173-1239. Baltimore: Williams and Wilkins, 1961.

ZARROW, M. X., V. H. DENENBERG, and W. D. KALBERER. Strain differences in the endocrine basis of maternal nest-building in the rabbit. *J. Reprod. Fertil.* 10:397-401, 1965.

ZARROW, M. X., R. GANDELMAN, and V. H. DENENBERG. Lack of nest building and maternal behavior in the mouse following olfactory bulb removal. *Horm. Behav.* 2:227-238, 1971a.

————. Prolactin: is it an essential hormone for maternal behavior in the mammal? *Horm. Behav.* 2:343-354, 1971b.

14 | Environment and Reproduction

ADLER, N., and J. ANISKO. The behavior of communicating: an analysis of the 22 kHz call of rats (*Rattus norvegicus*). *Am. Zool.* 19:493-508, 1979.

ARONSON, L. R. Reproductive and parental behavior. In *The Physiology of Fishes*, ed. M. E. Brown, Vol. 2, pp. 271-304. New York: Academic Press, 1957.

ASDELL, S. A. *Patterns of Mammalian Reproduction,* second ed. Ithaca: Cornell University Press, 1946.

ASSENMACHER, I. External and internal components of the mechanism controlling reproductive cycles in drakes. In *Circannual Clocks,* ed. E. T. Pengelley, pp. 197-248. New York: Academic Press, 1974.

BAGGERMAN, B. An experimental study on the timing of breeding and migration of the three spined stickleback. *Arch. Néerl. Zool.* 12:105-317, 1957.

————. Hormonal control of reproductive and parental behaviour in fishes. In *Perspectives in Endocrinology,* eds. E. J. W. Barrington and C. Barker Jørgensen, pp. 351-404. New York: Academic Press, 1968.

————. Photoperiodic responses in the stickleback and their control by a daily rhythm of photosensitivity. *Gen. Comp. Endocrinol. Suppl.* 3:466-475, 1972.

————. Het endocriene systeem en aanpassingen van stekelbaarzen aan hun milieu. In *Endocrinologie,* eds. J. Lever and J. de Wilde, pp. 64-92. Wageningen, The Netherlands: Centrum voor Landbouwpublikaties en Landbouwdocumentatie, 1978.

————. Photoperiodic and endogenous control of the annual reproductive cycle in teleost fishes. In *Environmental Physiology of Fishes,* ed. M. S. Ali, pp. 533-567. New York: Plenum Press, 1980.

BAKER, J. B. The evolution of breeding seasons. In *Evolution: Essays on Aspects of Evolutionary Biology,* ed. G. R. de Beer (ed.), pp. 161-177. Oxford: Clarendon Press, 1938.

BALTHAZART, J., and E. SCHOFFENIELS. Pheromones are involved in the control of sexual behaviour in birds. *Naturwissenschaften* 66:55-56, 1979.

BANARESCU, P. Carp. In *Grzimek's Animal Encyclopedia,* Vol. 4, pp. 305-360. New York: Van Nostrand Reinold, 1975.

BARFIELD, R. J., P. AUERBACH, L. A. GEYER, and T. K. McINTOSH. Ultrasonic vocalizations in rat sexual behavior. *Am. Zool.* 19:469-480, 1979.

BARNETT, J. L. A stress response in *Antechinus stuartii* (Macleay). *Aust. J. Zool.* 21:501-513, 1973.

BARNETT, S. A., K. M. H. MUNRO, J. L. SMART, and R. C. STODDART. House mice bred for many generations in two environments. *J. Zool.* (Lond.) 177:153–169, 1975.

BAUM, M. J., and P. J. M. SCHRETLEN. Oestrogenic induction of sexual behaviour in ovariectomized ferrets housed under short or long photoperiods. *J. Endocrinol.* 78:295–296, 1978.

BELL, R. W. Ultrasonic control of maternal developmental implications. *Am. Zool.* 19:413–418, 1979.

BELYAEV, D. K., L. N. TRUT, and A. O. RUVINSKY. Genetics of the *W* locus in foxes and expression of its lethal effects. *J. Hered.* 66:331–338, 1975.

BELYAEV, D. K., and A. I. ZHELEZOVA. Effect of day length on lethality in mink homozygous for the shadow gene. *J. Hered.* 69:366–368, 1978.

BENNETTS, H. W., E. J. UNDERWOOD, and F. L. SHIER. A specific breeding problem of sheep on subterranean clover pastures in Western Australia. *Aust. Vet. J.* 22:2–12, 1946.

BENOIT, J. Etude de l'action des radiations visibles sur la gonado-stimulation et de leur pénétration intracranienne chez les oiseaux et les mammifères. In *La Photorégulation de la Reproduction chez les Oiseaux et les Mammifères*, eds. J. Benoit and I. Assenmacher, pp. 121–149. Paris: Centre National de la Recherche Scientifique, 1970.

BENSON, B., M. J. MATTHEWS, and V. J. HRUBY. Characterization and effects of a bovine pineal antigonadotropic peptide. *Am. Zool.* 16: 17–24, 1976.

BERGER, P. J., N. C. NEGUS, E. H. SANDERS, and P. D. GARDNER. Chemical triggering of reproduction in *Microtus montanus. Science* 214:69–70, 1981.

BERGER, P. J., E. H. SANDERS, P. D. GARDNER, and N. C. NEGUS. Phenolic plant compounds functioning as reproductive inhibitors in *Microtus montanus. Science* 195:575–577, 1977.

BERTHOLD, P. Migration: control and metabolic physiology. In *Avian Biology*, eds. D. S. Farner and J. R. King, Vol. 5, pp. 77–128. New York: Academic Press, 1975.

BEX, F., A. BARTKE, B. D. GOLDMAN, and S. DALTERIO. Prolactin, growth hormone, luteinizing hormone receptors, and seasonal changes in testicular activity in the golden hamster. *Endocrinology* 103: 2069–2080, 1978.

BITTMAN, E. L. Hamster refractoriness: the role of insensitivity of pineal target tissues. *Science* 202:648–650, 1978.

BITTMAN, E. L., and I. ZUCKER. Photoperiodic termination of hamster refractoriness: participation of the pineal gland. *Biol. Reprod.* 24:568–572, 1981.

BONDURANT, R. H., B. J. DARIEN, C. J. MUNRO, G. H. STABENFELDT, and P. WANG. Photoperiod induction of fertile oestrus and changes in LH and progesterone concentrations in yearling dairy goats (*Capra hircus*). *J. Reprod. Fertil.* 63:1–9, 1981.

BROCKWAY, B. F. The effects of nest-entrance positions and male vocalizations on reproduction in budgerigars. *Living Bird* 1:93–101, 1962.

———. Stimulation of ovarian development and egg laying by male courtship vocalizations in budgerigars (*Melopsittacus undulatus*). *Anim. Behav.* 13:575–578, 1965.

———. The influence of vocal behavior on the performer's testicular activity in budgerigars (*Melopsittacus undulatus*). *Wilson Bull.* 79:328–334, 1967.

BRONSON, F. H. Urine marking in mice: causes and effects. In *Mammalian Olfaction, Reproduction Processes and Behavior* ed. R. L. Doty, pp. 119–141. New York: Academic Press, 1976.

———. Light intensity and reproduction in wild and domestic house mice. *Biol. Reprod.* 21:235–239, 1979a.

———. The reproductive ecology of the house mouse. *Q. Rev. Biol.* 54:265–299, 1979b.

BRONSON, F. H., and F. S. VOM SAAL. The preovulatory surge of luteinizing hormone secretion in mice: variation in magnitude due to ambient light intensity. *Biol. Reprod.* 20:1005–1008, 1979.

BRONSON, F. H., and W. K. WHITTEN. Oestrus-accelerating pheromone of mice: assay, androgen-dependency, and presence in bladder urine. *J. Reprod. Fertil.* 15:131–134, 1968.

BRUCE, H. M. An exteroceptive block to pregnancy in the mouse. *Nature* 184:105, 1959.

BÜNNING, E. Die endonome Tagesrhythmik als Grundlage der photoperiodischen Reaktion. *Ber. Dtsch Bot. Ges.* 54:590–607, 1936.

CALHOUN, J. B. *The Ecology and Sociology of the Norway Rat.* U.S. Public Health Service Publication No. 1008. Washington, D.C.: U.S. Government Printing Office, 1962.

CANIVENC, R. Photopériodisme chez quelques mammifères à nidation différée. In *La Photorégulation de la Reproduction chez les Oiseaux et les Mammifères,* eds. J. Benoit and I. Assenmacher, pp. 453–466. Paris: Centre National de la Recherche Scientifique, 1970.

CHAMLEY, W. A., I. A. CUMMING, and J. R. GODING. Studies on oestrogen binding to uterine cytosol receptor prepared from normal and clover-infertile ewes. *J. Reprod. Fertil.* 36:491, 1974 (Abstract).

CHAPMAN, V. M., C. DESJARDINS, and W. K. WHITTEN. Pregnancy block in mice: changes in pituitary LH and LTH and plasma progestin levels. *J. Reprod. Fertil.* 21:333–337, 1970.

CHENG, M.-F. Interaction of lighting and other environmental variables on the activity of hypothalamo-hypophyseal-gonadal system. *Nature* 263:148–149, 1976.

CHÈZE, G. Etude morphologique, histologique et expérimentale de l'épiphyse de *Symphodus melops* (Poisson, Labridé). *Bull. Soc. Zool. Fr.* 94:697–704, 1969.

CHÈZE, G., and J. LAHAYE. Etude morphologique de la région épiphysaire de *Gambusia affinis holbrookig.*

Incidences histologiques de certains facteurs externes sur le toit diencéphalique. *Ann. Endocrinol.* 30:45–53, 1969.

CHRISTIAN, J. J. Population density and reproductive efficiency. *Biol. Reprod.* 4:248–294, 1971.

CLAESSON, A., and R. M. SILVERSTEIN. Chemical methodology in the study of mammalian communications. In *Symposium on Chemical Signals in Vertebrates, Saratoga Springs, N.Y. 1976,* eds. D. Müller-Schwarze and M. M. Mozell, pp. 71–93. New York: Plenum Press, 1977.

CLARKE, J. R., and J. P. KENNEDY. Effect of light and temperature upon gonad activity in the vole (*Microtus agrestis*). *Gen. Comp. Endocrinol.* 8:474–488, 1967.

COHEN, J. Midbrain and motor control of courtship behavior in male and female ring doves (*Streptopelia risoria*). *Diss. Abst. Int. Sci. Eng.* 41 Ref. 4301B, 1981.

COHEN, J., and M.-F. Cheng. The role of the midbrain in courtship behavior of the female ring dove (*Streptopelia risoria*): evidence from radiofrequency lesions and hormone implant studies. *Brain Res.* 207:279–301, 1981.

————. Role of vocalizations in the reproductive cycle of ring doves (*Streptopelia risoria*): effects of hypoglossal nerve section on the reproductive behavior and physiology of the female. *Horm. Behav.* 13:113–127, 1979.

COQUELIN, A., and F. H. BRONSON. Release of luteinizing hormone in male mice during exposure to females: habituation of the response. *Science* 206:1099–1101, 1979.

CREWS, D. Effects of group stability, male-male aggression, and male courtship behaviour on environmentally-induced ovarian recrudescence in the lizard, *Anolis carolinensis. J. Zool. (Lond.)* 172:419–441, 1974.

DAVIS, D. E. A comparison of reproductive potential of two rat populations. *Ecology* 32:469–475, 1951.

DELAHUNTY, G., C. SCHRECK, J. SPECKER, J. OLCESE, J. J. VODICNIK, and V. DE VLAMING. The effects of light reception on circulating estrogen levels in female goldfish, *Carassius auratus:* importance of retinal pathways versus the pineal. *Gen. Comp. Endocrinol.* 38:148–152, 1979.

DELOST, P. Etude expérimentale des causes de la reprise printanière de l'activité sexuelle chez les mammifères à cycle reproducteur saisonier. *C. R. Séances Soc. Biol. Fil.* 166:879–884, 1972.

————. Analyse du rôle de la ration alimentaire saisonnière sur le développement sexuel des mammifères sauvages à cycle reproducteur par l'étude expérimentale comparée des effets des herbages d'hiver et de printemps. *C.R. Séances Soc. Biol. Fil.* 167:977–983, 1973.

DE RUITER, A. J. H. A combined structural and biochemical analysis of the effect of testosterone on kidney function of the three-spined stickleback. *Gen. Comp. Endocrinol.* 29:283–284, 1976.

DE VLAMING, V., and J. OLCESE. The pineal and reproduction in fish, amphibians, and reptiles. In *The Pineal Gland,* ed. R. J. Reiter. Vol. 2, pp. 1–29. Boca Raton, Florida: CRC Press, 1981.

DE VLAMING, V. L., M. SAGE, and C. B. CHARLTON. The effects of melatonin treatment on the gonadosomatic index in the teleost, *Fundulus similis* and the tree frog. *Hyla cinerea. Gen. Comp. Endocrinol.* 22:433–438, 1974.

DE VLAMING, V. L., and M. J. VODICNIK. Effects of pinealectomy on pituitary gonadotrophs, pituitary gonadotropin potency and hypothalamic gonadotropin releasing activity in *Notemigonus crysoleucas. J. Fish Biol.* 10:73–86, 1977.

DE VUYST, A., G. THINÈS, L. HENRIET, and M. SOFFIÉ. Influence des stimulations auditives sur le comportement sexuel du taureau. *Experientia* 20:648–650, 1964.

DLUZEN, D. E., V. D. RAMIREZ, C. S. CARTER, and L. L. GETZ. Male vole urine changes luteinizing hormone-releasing hormone and norepinephrine in female olfactory bulb. *Science* 212:573–575, 1981.

DODT, E. Photosensitivity of the pineal organ in the teleost, *Salmo irideus,* (Gibbons). *Experientia* 19:642–643, 1963.

DRICKAMER, L. C. Delay of sexual maturation in female house mice by exposure to grouped females or urine of grouped females. *J. Reprod. Fertil.* 51:77–81, 1977.

DUCKER, M. J., J. C. BOWMAN, and A. TEMPLE. The effect of constant photoperiod on the expression of oestrus in the ewe. *J. Reprod. Fertil. Suppl.* 19:143–150, 1973.

ELLIOTT, J. A. Circadian rhythms and photoperiodic time measurement in mammals. *Fed. Proc.* 35:2339–2346, 1976.

ELLIS, G. B., S. H. LOSEE, and F. W. TUREK. Prolonged exposure of castrated male hamsters to a nonstimulatory photoperiod: spontaneous change in sensitivity of the hypothalamic-pituitary axis to testosterone feedback. *Endocrinology* 104:631–635, 1979.

ELLIS, G. B., and F. W. TUREK. Time course of the photoperiod-induced change in sensitivity of the hypothalamic-pituitary axis to testosterone feedback in castrated male hamster. *Endocrinology* 104:625–630, 1979.

ERICKSON, C. J. Induction of ovarian activity in female ring doves by an androgen treatment of castrated males. *J. Comp. Physiol. Psychol.* 71:210–215, 1970.

ERICKSON, C. J., and D. S. LEHRMAN. Effect of castration on male ring doves upon ovarian activity of females. *J. Comp. Physiol. Psychol.* 58:164–166, 1964.

FARNER, D. S. Photoperiodic controls in the secretion

of gonadotropins in birds. *Am. Zool.* 15 (*Suppl.*): 117–135, 1975.

FARNER, D. S., R. S. DONHAM, R. A. LEWIS, P. W. MATTOCKS, JR., T. R. DARDEN, and J. P. SMITH. The circadian component in the photoperiodic mechanism of the house sparrow, Passer domesticus. *Physiol. Zool.* 50:247–268, 1977.

FARNER, D. S., and B. K. FOLLETT. Light and other environmental factors affecting avian reproduction. *J. Anim. Sci.* 25 (*Suppl.*): 90–115, 1966.

FARNER, D. S., and J. C. WINGFIELD. Environmental endocrinology and the control of annual reproductive cycles in passerine birds. In *Environmental Endocrinology*, eds. I. Assenmacher and D. S. Farner, pp. 44–51. New York: Springer, 1978.

FARRAR, G. M., and J. R. CLARKE. Effect of chemical sympathectomy and pinealectomy upon gonads of voles (*Microtus agrestis*) exposed to short photoperiod. *Neuroendocrinology* 22:134–143, 1976.

FLOODY, O. R. Behavioral and physiological analyses of ultrasound production by female hamsters (*Mesocricetus auratus*). *Am. Zool.* 19:443–455, 1979.

FOLLETT, B. K., and D. T. DAVIES. Photoperiodicity and the neuroendocrine control of reproduction in birds. *Symp. Zool. Soc. Lond.* 35:199–224, 1975.

FOLLETT, B. K., and P. J. SHARP. Circadian rhythmicity in photoperiodically induced gonadotropin release and gonadal growth in the quail. *Nature* 223:968–971, 1969.

GARSTKA, W. R., and D. CREWS. Female sex pheromone in the skin and circulation of a garter snake. *Science* 214: 681–683, 1981.

GARTON, J. S., and R. A. BRANDON. Reproductive ecology of the green tree frog, *Hyla cinerea,* in Southern Illinois (Anura: Hylidae). *Herpetologica* 31:150–161, 1975.

GOLDFOOT, D. A. Olfaction, sexual behavior and the pheromone hypothesis in rhesus monkeys: a critique. *Am. Zool.* 21:153–164, 1981.

GORDON, T. P. Reproductive behavior in the rhesus monkey: social and endocrine variables. *Am. Zool.* 21:185–195, 1981.

GOSNEY, S., and R. A. HINDE. Changes in the sensitivity of female budgerigars to male vocalizations. *J. Zool. (Lond.)* 179:407–410, 1976.

GRAHAM, J. M., and C. DESJARDINS. Classical conditioning: induction of luteinizing hormone and testosterone secretion in anticipation of sexual activity. *Science* 210:1039–1041, 1980.

GROCOCK, C. A., and J. R. CLARKE. Photoperiodic control of testis activity in the vole, *Microtus agrestis. J. Reprod. Fertil.* 39:337–347, 1974.

GWINNER, E. Circadian and circannual rhythms in birds. In *Avian Biology,* eds. D. S. Farner and J. R. King, Vol. 5, pp. 221–285. New York: Academic Press, 1975.

GWINNER, E., and V. DORKA. Endogenous control of annual reproductive rhythms in birds. In *Proc. 16th Int. Ornithol. Congress, Canberra 12–17 August, 1974,* pp. 223–234, 1976.

GWINNER, E. G., F. W. TUREK, and S. D. SMITH. Extraocular light perception in photoperiodic responses of the white-crowned sparrow (*Zonotrichia leucophrys*) and of the golden-crowned sparrow (*Z. atricapilla*). *Z. Vgl. Physiol.* 75:323–331, 1971.

HALBERG, F., and G. S. KATINAS. Chronobiologic glossary. *Int. J. Chronobiol.* 1:31–63, 1973.

HAMNER, W. M. Diurnal rhythm and photoperiodism in testicular recrudescence of the house finch. *Science* 142:1294–1295, 1963.

———. Circadian control of photoperiodism in the house finch demonstrated by interrupted night experiments. *Nature* 203:1400–1401, 1964.

———. Photoperiodic control of the annual testicular cycle in the house finch (*Carpodacus mexicanus*). *Gen. Comp. Endocrinol.* 7:224–233, 1966.

HARDISTY, M. W., and I. C. POTTER. The general biology of adult lampreys. In *The Biology of Lampreys,* eds. M. W. Hardisty and I. C. Potter, Vol. 1, pp. 127–206. New York: Academic Press, 1971.

HERBERT, J., and M. KLINOWSKA. Day length and the annual reproductive cycle in the ferret (*Mustela furo*): the role of the pineal body. In *Environmental Endocrinology,* eds. I. Assenmacher and D. S. Farner, pp. 87–93. New York: Springer, 1978.

HERBERT, J., P. M. STACEY, and D. H. THORPE. Recurrent breeding seasons in pinealectomized or optic-nerve-sectioned ferrets. *J. Endocrinol.* 78:389–397, 1978.

HINDE, R. A., and R. J. PUTNAM. Why budgerigars breed in continuous darkness. *J. Zool. (Lond.)* 170:485–491, 1973.

HINDE, R. A., and E. STEEL. The influence of daylength and male vocalizations on the estrogen-dependent behavior of female canaries and budgerigars, with discussion of data from other species. *Adv. Study Behav.* 8:39–73, 1978.

HOFFMANN, J. C. The influence of photoperiods on reproductive functions in female mammals. *Handbook of Physiology*, Sec. 7, Vol. 2, Part 1, pp. 57–77. Washington, D.C.: American Physiological Society, 1973.

HOFFMANN, K. Pineal involvement in the photoperiodic control of reproduction and other functions in the Djungarian hamster *Phodopus sungorus*. In *The Pineal Gland*, ed. R. J. Reiter, Vol. 2, pp. 83–102. Boca Raton, Florida: CRC Press, 1981.

HOMMA, K., Y. SAKAKIBARA, and Y. OHTA. Potential sites and action spectra for encephalic photoreception in the Japanese quail. In *First Int. Symp. Avian Endocrinology, Calcutta, 1977,* pp. 25–26. Bangor, U. K.: University College of North Wales, 1977.

HOMMA, K., W. O. WILSON, and T. D. SIOPES. Eyes have a role in photoperiodic control of sexual activity of Coturnix. *Science* 178:421–423, 1972.

IMMELMAN, K. Ecological aspects of periodic reproduction. In *Avian Biology,* eds. D. S. Farner and J. R. King, Vol. 1, pp. 341-389. New York: Academic Press, 1971.

JALLAGEAS, M., and I. ASSENMACHER. Further evidence for reciprocal interactions between the annual sexual and thyroid cycles in male Peking ducks. *Gen. Comp. Endocrinol.* 37:44-51, 1979.

JOHNSTON, R. E. Sex pheromones in golden hamsters. In *Symposium on Chemical Signals in Vertebrates, Saratoga Springs, N.Y., 1976,* eds. D. Müller-Schwarze and M. M. Mozell, pp. 225-249. New York: Plenum Press, 1977.

JØRGENSEN, C. B., K.-E. HEDE, and L. O. LARSEN. Environmental control of annual ovarian cycle in the toad *Bufo bufo bufo* L.: role of temperature. In *Environmental Endocrinology,* eds. I. Assenmacher and D. S. Farner, pp. 29-36. New York: Springer, 1978.

KAYSER, CH. Photopériode, reproduction et hibernation des mammifères. In *La Photorégulation de la Reproduction chez les Oiseaux et les Mammifères,* eds. J. Benoit and I. Assenmacher, pp. 409-433. Paris: Centre National de la Recherche Scientifique, 1970.

KELLEY, D. B. Auditory and vocal nuclei in the frog brain concentrate sex hormones. *Science* 207:553-555, 1980.

KENNAWAY, D. J., J. M. OBST, E. A. DUNSTAN, and H. G. FRIESEN. Ultradian and seasonal rhythms in plasma gonadotropins, prolactin, cortisol, and testosterone in pinealectomized rams. *Endocrinology* 108:639-646, 1981.

KEVERNE, E. B. Sexual receptivity and attractiveness in the female rhesus monkey. *Adv. Study Behav.* 7:155-200, 1976.

KIRCHHOF-GLAZIER, D. A. Absence of sexual imprinting in house mice cross-fostered to deermice. *Physiol. Behav.* 23:1073-1080, 1979.

LAND, R. B., W. R. CARR, and R. THOMPSON. Genetic and environmental variation in the LH response of ovariectomized sheep to LH-RH. *J. Reprod. Fertil.* 56:243-248, 1979.

LEE, A. K., A. J. BRADLEY, and R. W. BRAITHWAITE. Corticosteroid levels and male mortality in Antechinus stuartii. In *The Biology of Marsupials,* eds. B. Stonehouse and D. Gilmore, pp. 209-220. Baltimore: University Park Press, 1977.

LEGAN, S. J., and F. J. KARSCH. Photoperiodic regulation of the estrous cycle and seasonal breeding of the ewe. *Biol. Reprod.* 20: 74-85, 1979.

LEHRMAN, D. S., and M. FRIEDMAN. Auditory stimulation of ovarian activity in the ring dove (*Streptopelia risoria*). *Anim. Behav.* 17:494-497, 1969.

LEON, M. Filial responsiveness to olfactory cues in the laboratory rat. *Adv. Study Behav.* 8:117-153, 1978.

LICHT, P. Problems in experimentation on timing mechanisms for annual physiological cycles in reptiles. In *Hibernation and Hypothermia, Perspectives and Challenges,* eds. F. E. South, Jr., J. P. Hannon, J. R. Willis, E. T. Pengelley, and N. R. Alpert, pp. 681-711. New York: Elsevier, 1971.

————. Environmental physiology of reptilian breeding cycles: role of temperature. *Gen. Comp. Endocrinol. Suppl.* 3:477-488, 1972.

LIGON, J. D. Green cones of the piñon pine stimulate late summer breeding in the piñon jay. *Nature* 250:80-82, 1974.

LINCOLN, G. A. Photoperiodic control of seasonal breeding in the ram: participation of the cranial sympathetic nervous system. *J. Endocrinol.* 82:144-146, 1979.

LINCOLN, G. A., and W. DAVIDSON. The relationship between sexual and aggressive behaviour, and pituitary and testicular activity during the seasonal sexual cycle of rams, and the influence of photoperiod. *J. Reprod. Fertil.* 49:267-276, 1977.

LINCOLN, G. A., A. S. MCNEILLY, and C. L. CAMERON. The effects of a sudden decrease in daylength on prolactin secretion in the ram. *J. Reprod. Fertil.* 52:305-311, 1978.

LINCOLN, G. A., and M. J. PEET. Photoperiodic control of gonadotrophin secretion in the ram: a detailed study of the temporal changes in plasma levels of follicle-stimulating hormone, luteinizing hormone and testosterone following an abrupt switch from long to short days. *J. Endocrinol.* 74:355-367, 1977.

LINCOLN, G. A., M. J. PEET, and R. A. CUNNINGHAM. Seasonal and circadian changes in the episodic release of follicle-stimulating hormone, luteinizing hormone and testosterone in rams exposed to artificial photoperiods. *J. Endocrinol.* 72:337-349, 1977.

LOFTS, B. Reptilian reproductive cycles and environmental regulators. In *Environmental Endocrinology,* eds. I. Assenmacher and D. S. Farner, pp. 37-43. New York: Springer, 1978.

LOMBARDI, J. R., J. G. VANDENBERGH, and J. M. WHITSETT. Androgen control of sexual maturation pheromone in house mouse urine. *Biol. Reprod.* 15:179-186, 1976.

LOTT, D., S. D. SCHOLZ, and D. S. LEHRMAN. Exteroceptive stimulation of the reproductive system of the female ring dove (*Streptopelia risoria*) by the mate and by the colony milieu. *Anim. Behav.* 15:433-437, 1967.

MACRIDES, F., A. BARTKE, and S. DALTERIO. Strange females increase plasma testosterone levels in male mice. *Science* 189: 1104-1106, 1975.

MADISON, D. M. Chemical communication in amphibians and reptiles. In *Symposium on Chemical Signals in Vertebrates, Saratoga Springs, N.Y., 1976,* eds. D. Müller-Schwarze and M. M. Mozell, pp. 135-168. New York: Plenum Press, 1977.

MAHONE, J. P., T. BERGER, E. D. CLEGG, and W. L. SINGLETON. Photoinduction of puberty in boars during naturally occurring short day lengths. *J. Anim. Sci.* 48:1159-1164, 1979.

MARCHLEWSKA-KOJ, A. Pregnancy block elicited by urinary proteins of male mice. *Biol. Reprod.* 17:729–732, 1977.

_____. Pregnancy block elicited by male urinary peptides in mice. *J. Reprod. Fertil.* 61:221–224, 1981.

MARLER, P., and W. J. HAMILTON, III. *Mechanisms of Animal Behavior.* New York: Wiley, 1966.

MARTINET, L. Role du photopériodisme sur la biologie sexuelle du campagnol des champs (*Microtus arvalis*). In *La Photorégulation de la Reproduction chez les Oiseaux et les Mammifères,* eds. J. Benoit and I. Assenmacher, pp. 435–450. Paris: Centre National de la Recherche Scientifique, 1970.

MARUNIAK, J. A., A. COQUELIN, and F. H. BRONSON. The release of LH in male mice in response to female urinary odors: characteristics of the response in young males. *Biol. Reprod.* 18:251–255, 1978.

MASSEY, A., and J. G. VANDENBERGH. Puberty delay by a urinary cue from female house mice in feral populations. *Science* 209:821–822, 1980.

_____. Puberty acceleration by a urinary cue from male mice in feral populations. *Biol. Reprod.* 24:523–527, 1981.

MATTOCKS, P. W., JR., D. S. FARNER, and B. K. FOLLETT. The annual cycle in luteinizing hormone in the plasma of intact and castrated white-crowned sparrows, *Zonotrichia leucophrys gambelii. Gen. Comp. Endocrinol.* 30:156–161, 1976.

MAUGET, R. Seasonal reproductive activity in the European wild boar. Comparison with the domestic sow. In *Environmental Endocrinology,* eds. I. Assenmacher and D. S. Farner, pp. 79–80. New York: Springer, 1978.

McCLINTOCK, M. K. Social control of the ovarian cycle and the function of estrous synchrony. *Am. Zool.* 21:243–256, 1981.

McCLINTOCK, M. K., and N. T. ADLER. The role of the female during copulation in wild and domestic Norway rats (*Rattus norvegicus*). *Behaviour* 67:67–96, 1978.

McCLURE, T. J. Infertility in female rodents caused by temporary inanition at or about the time of implantation. *J. Reprod. Fertil.* 4:241, 1962.

_____. Infertility in mice caused by fasting at about the time of mating. I. Mating behaviour and littering rates. *J. Reprod. Fertil.* 12:243–248, 1966.

McMILLAN, J. P., H. A. UNDERWOOD, J. A. ELLIOTT, M. H. STETSON, and M. MENAKER. Extraretinal light perception in the sparrow. IV. Further evidence that the eyes do not participate in photoperiodic photoreception. *J. Comp. Physiol.* 97:205–213, 1975.

MEIER, A. H. Chronoendocrinology of the white-throated sparrow. *Proc. 16th Int. Ornithol. Congress, Canberra, 12–17 August, 1974,* pp. 355–368, 1976.

MENAKER, M. Synchronization with the photic environment via extraretinal receptors in the avian brain. In *Biochronometry,* ed. M. Menaker, pp. 315–332. Washington, D.C.: National Academy of Sciences, 1971.

MICHAEL, R. P., and R. W. BONSALL. Chemical signals and primate behavior. In *Symposium on Chemical Signals in Vertebrates, Saratoga Springs, N.Y. 1976,* eds. D. Müller-Schwarze and M. M. Mozell, pp. 251–271. New York: Plenum Press, 1977.

MIRARCHI, R. E., B. E. HOWLAND, P. F. SCANLON, R. L. KIRKPATRICK, and L. M. SANFORD. Seasonal variation in plasma LH, FSH, prolactin, and testosterone concentrations in adult male white-tailed deer. *Can. J. Zool.* 56:121–127, 1978.

MONDER, H., C.-T. LEE, P. J. DONOVICK, and R. G. BURRIGHT. Male mouse urine extract effects on pheromonally mediated reproductive functions of female mice. *Physiol. Behav.* 20:447–452, 1978.

MORITA, Y., and G. BERGMANN. Physiologische Untersuchungen und weitere Bemerkungen zur Struktur des lichtempfindlichen Pinealorgans von *Pterophyllum scalare* Cuv. et Val. (Cichlidae, Teleostei). *Z. Zellforsch. Mikrosk. Anat.* 119:289–294, 1971.

MORRIS, T. R., and A. V. NALBANDOV. The induction of ovulation in starving pullets using mammalian and avian gonadotropins. *Endocrinology* 68:687–697, 1961.

MUDUULI, D. S., L. M. SANFORD, W. M. PALMER, and B. E. HOWLAND. Secretory patterns and circadian and seasonal changes in luteinizing hormone, follicle-stimulating hormone, prolactin and testosterone in the male pygmy goat. *J. Anim. Sci.* 49:543–553, 1979.

MURTON, R. K., and N. J. WESTWOOD. *Avian Breeding Cycles.* Oxford: Clarendon Press. 1977.

MYERS, J. M. and C. L. KREBS. Population cycles in rodents. *Scient. Am.* 230(6):38–46, 1974.

NARINS, P. M., and R. R. CAPRANICA. Sexual differences in the auditory system of the tree frog *Eleutherodactylus coqui. Science* 192:378–380, 1976.

NEGUS, N. C., and P. J. BERGER. Pineal weight response to a dietary variable in *Microtus montanus. Experientia* 27:215–216, 1971.

NEGUS, N. C., and P. J. BERGER. Experimental triggering of reproduction in a natural population of *Microtus montanus. Science* 196:1230–1231, 1977.

NEGUS, N. C., and A. J. PINTER. Reproductive responses of *Microtus montanus* to plants and plant extracts in the diet. *J. Mammal.* 47:596–601, 1966.

OLIVER, J. Photoréception et mécanismes régulateurs du reflexe photosexuel: Étude chez la Caille. Thesis, Université des Sciences et Techniques du Languedoc, Montpellier, France, 1979.

OPLINGER, C. S. Sex ratio, reproductive cycles, and time of ovulation in *Hyla crucifer crucifer* Wied. *Herpetologica* 22:276–283, 1966.

ORTAVANT, R., J. PELLETIER, J. P. RAVAULT, and J. THIMONIER. Annual cyclic variations in prolactin in sheep. In *Environmental Endocrinology*, eds. I. Assenmacher and D. S. Farner, pp. 75–78. New York: Springer, 1978.

OXENDER, W. D., P. A. NODEN, and H. O. HAFS. Estrus, ovulation, and serum progesterone, estradiol, and LH concentrations in mares after an increased photoperiod during winter. *Am. J. Vet. Res.* 38:203–207, 1977.

PARROTT, R. F., and R. J. BARFIELD. Post-ejaculatory vocalization in castrated rats treated with various steroids. *Physiol. Behav.* 15:159–163, 1975.

PELLETIER, J., and R. ORTAVANT. Influence du photopériodisme sur les activités sexuelle, hypophysaire et hypothalamique de bélier Ile-de-France. In *La Photorégulation de la Reproduction chez les Oiseaux et les Mammifères,* eds. J. Benoit and I. Assenmacher, pp. 483–493. Paris: Centre National de la Recherche Scientifique, 1970.

PETER, R. E. Gonadotropin secretion during reproductive cycles in teleosts: influences of environmental factors. *Gen. Comp. Endocrinol.* 42:294–305, 1981.

PETER, R. E., and L. W. CRIM. Reproductive endocrinology of fishes: gonadal cycles and gonadotropin in teleosts. *Ann. Rev. Physiol.* 41:323–335, 1979.

PETER, R. E., and H. HONTELA. Annual gonadal cycles in teleosts: environmental factors and gonadotropin levels in blood. In *Environmental Endocrinology* eds. I. Assenmacher and D. S. Farner, pp. 20–25. New York: Springer, 1978.

PHILLIPS, R. E. "Wildness" in the mallard duck: effects of brain lesions and stimulation on "escape behavior" and reproduction. *J. Comp. Neurol.* 122:139–155, 1964.

PINTER, A. J., and N. C. NEGUS. Effects of nutrition and photoperiod on reproductive physiology of *Microtus montanus*. *Am. J. Physiol.* 208:633–638, 1965.

PITTENDRIGH, C. S., and D. H. MINIS. The entrainment of circadian oscillations by light and their role as photoperiodic clocks. *Am. Nat.* 98:261–294, 1964.

POLDER, J. J. W. On gonads and reproductive behaviour in the cichlid fish Aequidens portalegrensis (Hensel). *Neth. J. Zool.* 21:265–365, 1971.

POSTON, H. A. Neuroendocrine mediation of photoperiod and other environmental influences on physiological responses in salmonids: a review. *U.S. Fish Wildl. Serv. Tech. Pap.* 96, 1978.

PUTMAN, R. J. and R. A. HINDE. Effect of the light regime and breeding experience on budgerigar reproduction. *J. Zool. (Lond.)* 170:475–484, 1973.

RACEY, P. A. Environmental factors affecting the length of gestation in heterothermic bats. *J. Reprod. Fertil. Suppl.* 19:175–189, 1973.

RALPH, C. L. The pineal and reproduction in birds. In *The Pineal Gland,* ed. R. J. Reiter, Vol. 2, pp. 31–43. Boca Raton, Florida: CRC Press, 1981.

REITER, R. J. Pineal-mediated reproductive event. In *Novel Aspects of Reproductive Physiology,* eds. C.

H. Spilman and J. W. Wilks, pp. 369–388. Jamaica, New York: Spectrum Publications, 1978.

RICHMOND, M., and R. STEHN. Olfaction and reproductive behavior in microtine rodents. In *Mammalian Olfaction, Reproductive Processes, and Behavior*, ed. R. L. Doty, pp. 197–217. New York: Academic Press, 1976.

ROCHE, J. F., F. J. KARSCH, F. J. FOSTER, S. TAKAGI, and P. J. DZIUK. Effects of pinealectomy on estrus, ovulation and luteinizing hormone in the ewe. *Biol. Reprod.* 2:251–254, 1970.

ROLAND, D. A., SR., D. R. SLOAN, H. R. WILSON, and R. H. HARMS. Relationship of calcium to reproductive abnormalities in the laying hen (*Gallus domesticus*). *J. Nutr.* 104: 1079–1085, 1974.

ROTH, R. R. The effect of temperature and light combinations upon the gonads of male red-back voles. *Biol. Reprod.* 10:309–314, 1974.

RUSAK, B. Neural mechanisms for entrainment and generation of mammalian circadian rhythms. *Fed. Proc.* 38:2589–2595, 1979.

RUTLEDGE J. T. The non-photic induction of spermatogenic development in the European starling. *Int. J. Biometeor.* 24:77–81, 1980.

RYAN, K. D., and N. B. SCHWARTZ. Male-induced estrus in group-housed female mice. *Fed. Proc.* 35:686, 1976.

——. Grouped female mice: demonstration of pseudopregnancy. *Biol. Reprod.* 17:578–583, 1977.

SALTHE, S. N., and J. S. MECHAM. Reproductive and courtship patterns. In *Physiology of the Amphibia,* ed. B. Lofts, Vol. 2, pp. 309–521. New York: Academic Press, 1974.

SANDERS, E. G., P. D. GARDNER, P. J. BERGER, and N. C. NEGUS. 6-Methoxybenzoxazolinine: a plant derivative that stimulates reproduction in *Microtus montanus*. *Science* 214:67–69, 1981.

SAUNDERS, D. S. *Insect Clocks.* Oxford: Pergamon, 1976.

SCALIA, F., and S. S. WINANS. New perspectives on the morphology of the olfactory system: olfactory and vomeronasal pathways in mammals. In *Mammalian Olfaction, Reproductive Processes and Behavior,* ed. R. L. Doty, pp. 7–28. New York: Academic Press, 1976.

SCHMIDT, R. S. Central mechanism of frog calling. *Am. Zool.* 13:1169–1177, 1973.

SCHWAB, R. G. Circannian testicular periodicity in the European starling in the absence of photoperiodic change. In *Biochronometry,* ed. M. Menaker, pp. 428–445. Washington, D.C.: National Academy of Sciences, 1971.

SCHWASSMANN, J. O. Times of annual spawning and reproductive strategies in Amazonian fishes. In *Rhythmic Activity of Fishes,* ed. J. E. Thorpe, pp. 187–200. New York: Academic Press, 1978.

SHARP, D. C., III, and O. J. GINTHER. Stimulation of follicular activity and estrous behavior in anestrous mares with light and temperature. *J. Anim. Sci.* 41:1368-1372, 1975.

SIGNORET, J. P. Chemical communication and reproduction in domestic mammals. In *Mammalian Olfaction, Reproductive Processes and Behavior,* ed. R. L. Doty, pp. 243-256. New York: Academic Press, 1976.

SILVERMAN, H. I. Effects of different levels of sensory contact upon reproductive activity of adult male and female *Sarotherodon (Tilapia) mossambicus* (Peters); Pisces: Cichlidae. *Anim. Behav.* 26:1081-1090, 1978a.

––––––. The effects of visual social stimulation upon age at first spawning in the mouth-brooding cichlid fish *Sarotherodon (Tilapia) mossambicus. Anim. Behav.* 26:1120-1125, 1978b.

SINGER, A. G., W. C. AGOSTA, R. J. O'CONNELL, C. PFAFFMANN, D. V. BOWEN, and F. H. FIELD. Dimethyl disulfide: an attractant pheromone in hamster vaginal secretions. *Science* 191:948-950, 1976.

SMITH, N. G. Evolution of some arctic gulls (Larus): an experimental study of isolating mechanisms. Ornithological Monographs No. 4, Washington, D.C.: American Ornithologists' Union, Smithsonian Institution, 1966.

SREBNIK, H. H., W. H. FLETCHER, and G. A. CAMPBELL. Neuroendocrine aspects of reproduction in experimental malnutrition. In *Environmental Endocrinology,* eds. I. Assenmacher and D. S. Farner, pp. 306-312. New York: Springer, 1978.

STEEL, E., S. GOSNEY, and R. A. HINDE. Effect of male vocalizations on the nest-occupation response of female budgerigars to oestrogen and prolactin. *J. Reprod. Fertil.* 49:123-125, 1977.

STEHN, R. A., and M. E. RICHMOND. Male-induced pregnancy termination in the prairie vole, Microtus ochrogaster. *Science* 187: 1211-1213, 1975.

STEINIGER, F. Beitrage zur Soziologie und sonstigen Biologie der Wanderratte *Z. f. Tierpsychol.* 7:356-379, 1950.

STETSON, M. H., and M. WATSON-WHITMYRE. Nucleus suprachiasmaticus: the biological clock in hamsters? *Science* 191:197-199, 1976.

STOREY, C. R., and T. J. NICHOLLS. Some effects of manipulation of daily photoperiod on the rate of onset of a photorefractory state in canaries (*Serinus canarius*). *Gen. Comp. Endocrinol.* 30:204-208, 1976.

SUNDARARAJ, B. I. Environmental regulation of annual reproductive cycles in the catfish *Heteropneustes fossilis.* In *Environmental Endocrinology,* eds. I. Assenmacher and D. S. Farner, pp. 26-27. New York: Springer, 1978.

TAKAHASHI, J. S., and M. MENAKER. Physiology of avian circadian pacemakers. *Fed. Proc.* 38:2583-2588, 1979.

TAVOLGA, W. N. Ovarian fluids as stimuli to courtship behavior in the gobiid fish Bathygobius soporator (C and V). *Anat. Rec.* 122:425, 1955 (Abstract).

––––––. Visual, chemical and sound stimuli as cues in the sex discriminatory behavior of the gobiid fish, *Bathygobius soporator. Zoologica* 41(2):49-64, 1956.

TERMAN, C. R. Pregnancy failure in female prairie deermice related to parity and social environment. *Anim. Behav.* 17:104-108, 1969.

THAPLIYAL, J. P., and A. CHANDOLA. Thyroid in wild finches. *Proc. Nat. Acad. Sci. India Sect. B (Biol. Sci.)* 42 Part I:76-90, 1972.

THAPLIYAL, J. P., and D. C. MISRA NÉE HALDAR. Effect of pinealectomy on the photoperiodic gonadal response of the Indian garden lizard, *Calotes versicolor. Gen. Comp. Endocrinol.* 39:79-86, 1979.

THIBAULT, C., M. COUROT, L. MARTINET, P. MAULEON, F. DU MESNIL DU BUISSON, R. ORTAVANT, J. PELLETIER, and J. P. SIGNORET. Regulation of breeding season and estrous cycles by light and external stimuli in some mammals. *J. Anim. Sci.* 25 (*Suppl.*):119-130, 1966.

TUREK, F. Diurnal rhythms and the seasonal reproductive cycle in birds. In *Environmental Endocrinology,* eds. I. Assenmacher and D. S. Farner, pp. 144-152. New York: Springer, 1978.

TUREK, F. W. Circadian involvement in the termination of the refractory period in two sparrows. *Science* 178:1112-1113, 1972.

––––––. Role of the pineal gland in photoperiod-induced changes in hypothalamic-pituitary sensitivity to testosterone feedback in castrated male hamsters. *Endocrinology* 104:636-640, 1979.

TUREK, F. W., and C. S. CAMPBELL. Photoperiodic regulation of neuroendocrine-gonadal activity. *Biol. Reprod.* 20:32-50, 1979.

TUREK, F. W., C. DESJARDIN, and M. MENAKER. Differential effects of melatonin on the testes of photoperiodic and nonphotoperiodic rodents. *Biol. Reprod.* 15:94-97, 1976.

TUREK, F. W., and S. H. LOSEE. Photoperiodic inhibition of the reproductive system: a prerequisite for the induction of the refractory period in hamsters. *Biol. Reprod.* 20:611-616, 1979.

UECK, M. Innervation of the vertebrate pineal. *Prog. Brain Res.* 52:45-87, 1979.

ULBERG, L. C. Introduction to discussion. *J. Anim. Sci.* 26 (*Suppl.*): 16-18, 1966.

ULBERG, L. C., and L. A. SHEEAN. Early development of mammalian embryos in elevated ambient temperatures. *J. Reprod. Fertil. Suppl.* 19:155-161, 1973.

UNDERWOOD, H. Photoperiodic time measurement in the male lizard, *Anolis carolinensis. J. Comp. Physiol.* 125:143-150, 1979.

URASAKI, H. Role of the pineal gland in gonadal development in the fish, *Oryzias latipes. Annot. Zool. Jpn.* 45:152-158, 1972.

––––––. The role of the pineal and eyes in the photoperiodic effect on the gonad of the medaka, *Oryzias latipes. Chronobiologia* 3:228-234, 1976.

VANDENBERGH, J. G. Effect of the presence of a male

on the sexual maturation of female mice. *Endocrinology* 81:345–349, 1967.

VAN DER LEE, S., and L. M. BOOT. Spontaneous pseudopregnancy in mice. II. *Acta Physiol. Pharmacol. Neerl.* 5:213–214, 1956.

VAN TIENHOVEN, A. *Reproductive Physiology of Vertebrates*. Philadelphia: W. B. Saunders, 1968.

VAN TIENHOVEN, A., and R. J. PLANCK. The effect of light on avian reproductive activity. In *Handbook of Physiology*, Sect. 7, Vol. 2, Part 1, pp. 79–107. Washington, D.C.: American Physiological Society, 1973.

VAUGIEN, M., and L. VAUGIEN. Le moineau domestique peut développer son activité sexuelle et la maintenir dans l'obscurité complète. *C. R. Hebd. Séances Acad. Sci. (Paris)* 253:2762–2764, 1961.

VON HOLST, D. Sozialer Stress bei Tupajas (*Tupaia belangeri*). *Z. f. vergl. Physiol.* 63:1–58, 1969.

WEBER, A. L., and N. T. ADLER. Delay of constant light-induced persistent vaginal estrus by 24-hour time cues in rats. *Science* 204:323–325, 1979.

WENDELAAR BONGA, S. E. Morphometrical analysis with the light and electron microscope of the kidney of the anadromous three-spined stickle back *Gasterosteus aculeatus,* form *trachus* from freshwater and from sea water. *Z. Zellforsch. Mikrosk. Anat.* 137:563–588, 1973.

————. The effect of prolactin on kidney structure of the euryhaline teleost, *Gasterosteus aculeatus* during adaptation to fresh water. *Cell Tissue Res.* 166:319–338, 1976.

WENDELAAR BONGA, S. E., J. A. A. GREVEN, and M. VEENHUIS. The relationship between the ionic composition of the environment and the secretory activity of the endocrine cell types of Stannius corpuscles in the teleost *Gasterosteus aculeatus. Cell Tissue Res.* 175:297–312, 1976.

WENDELAAR BONGA, S. E., and M. VEENHUIS. The membranes of the basal labyrinth in kidney cells of the stickleback, *Gasterosteus aculeatus*, studied in ultrathin section and freeze-etch replicas. *J. Cell Sci.* 14: 587–609, 1974a.

————. The effect of prolactin on the number of membrane-associated particles in kidney cells of euryhaline teleost *Gasterosteus aculeatus* during transfer from seawater to fresh water: a freeze-etch study. *J. Cell Sci.* 16:687–701, 1974b.

WHITNEY, G., and J. NYBY. Cues that elicit ultrasounds from adult male mice. *Am. Zool.* 19:457–463, 1979.

WHITTEN, W. K. Modifications of the oestrous cycle of the mouse by external stimuli associated with the male. *J. Endocrinol.* 13:399–404, 1956.

————. Pheromones and mammalian reproduction. *Adv. Reprod. Physiol.* 1:155–177, 1966.

WIESELTHIER, A. S., and A. VAN TIENHOVEN. The effect of thyroidectomy on testicular size and on photorefractory period in the starling (*Sturnus vulgaris* L.) *J. Exp. Zool.* 179:331–338, 1972.

WIMSATT, W. A. Some interrelations of reproduction and hibernation in mammals. *Symp. Soc. Exp. Biol.* 23:511–549, 1969.

WOLFSON, A. Environmental and neuroendocrine regulation of annual gonadal cycles and migratory behavior in birds. *Recent Progr. Horm. Res.* 22: 177–239, 1966.

WURTMAN, R. J. Ambiguities in the use of the term circadian. *Science* 156:104, 1967.

WURTMAN, R. J., J. AXELROD, and D. E. KELLY. *The Pineal.* New York: Academic Press, 1968.

YAHR, P., and S. KESSLER. Suppression of reproduction in water-deprived Mongolian gerbils (*Meriones unguiculatus*). *Biol. Reprod.* 12:249–255, 1975.

YOKOYAMA, K., and D. S. FARNER. Photoperiodic responses in bilaterally enucleated female white-crowned sparrows, *Zonotrichia leucophrys gambelii. Gen. Comp. Endocrinol.* 30:528–533, 1976.

YOKOYAMA, K., A. OKSCHE, T. R. DARDEN, and D. S. FARNER. The sites of encephalic photoreception in photoperiodic induction of the growth of the testes in the white-crowned sparrow, *Zonotrichia leucophrys gambelii. Cell Tissue Res.* 189:441–467, 1978.

Animal Species Index

This index lists the pages at which the scientific name of the species is used. For further information see also the subject index under the common (vernacular) name.

Dasypus sp., 133, 134
Dasypus hybridus, 296
Dasypus novemcinctus, 228, 292, 293, 296
Dasyurus sp., 116, 264, 283
Dasyurus viverrinus, 279, 280, 289
Denisonia sp., 276
Denisonia sata, 276
Denisonia superba, 276
Desmodus sp., 283, 292, 293
Desmognathus fuscus, 108, 203
Dicentrarchus labrax, 198
Diceros bicornis, 300
Dicrostonyx groenlandicus, 221
Didelphis sp., 116, 133, 264, 283
Didelphis azarae, 221
Didelphis marsupialis, 279, 282
Didelphis virginiana, 116, 217
Diemictylus sp., 107
Diemictylus (Triturus) viridescens, 334
Dinematichthys sp., 103
Dipodillus simoni, 297
Dipsosaurus dorsalis, 179, 190
Dipulus caecus, 103
Discoglossus sp., 107
Discoglossus pictus, 48, 141
Dolichonyx oryzivorus, 378
Dorosoma cepedianum, 199
Drosophila sp., 365
Drymarchon corais corais, 31
Drymarchon corais couperi, 251

Echinops telfairi, 24
Eclectus roratus, 122
Eidolon helvum, 296
Elephantulus myurus, 283
Elephantulus myurus jamesoni, 253
Elephas maximus, 163
Eleutherodactylus coqui, 142, 374
Ellobius lutescens, 24, 27
Emys leprosa, 46
Emys orbicularis, 47
Enhydra lutris, 296
Enhydrina sp., 276
Epinephelus sp., 73, 75, 198
Eptatretus burgeri (B. burgeri), 70
Eptatretus stouti, 139, 157, 159, 171, 180
Eptesicus fuscus, 22
Eptesicus fuscus fuscus, 251
Equus asinus, 22
Equus caballus, 22, 117, 228, 282, 284, 382
Eretmochelys imbricata, 334
Erignathus barbatus, 296
Erimyzon oblongus, 199
Erinaceus europaeus, 113, 135, 221
Erithizon dorsatum, 131
Erythrocebus patas, 230
Esox americanus americanus, 199
Esox lucius, 102, 159, 174, 199
Esox niger, 199

Etrumeus teres, 200
Eumeces sp., 275
Eumetopias jubata, 296
Eunectes murinus, 31
Eupomacentrus sp., 339
Euproctus sp., 343
Euproctus asper, 204
Eurycea sp., 126
Eurycea bislineata, 108, 203
Eurycea tynerensis, 204

Falco columbarius, 122
Falco tinnunculus, 122
Falco vespertinus, 122
Farancia abacura, 275
Felis catus, 78, 284
Felis domesticus, 222, 237
Fringilla coelebs, 377
Fulmarus glacialis, 122
Fundulus heteroclitus, 105, 140, 306, 339
Fundulus similis, 372
Furcipenis sp., 99

Gadus callarias, 188
Galeus sp., 119, 122
Gallus sp., 161
Gallus domesticus 27, 41, 46, 76, 190, 278, 346
Gambusia sp., 99, 201, 274, 306, 339
Gambusia affinis, 104, 372
Gambusia holbrookii, 23, 151
Gasterosteus aculeatus, 32, 159, 173, 198, 199, 252, 273, 333, 370
Gastrotheca marsupiata, 125, 126, 275
Gastrotheca pygmaea, 125, 126
Gazella gazella, 34
Geomys bursarius, 310
Gerbillus gerbillus, 24
Gerrhonotus sp., 127
Gillichthys mirabilis, 102, 140, 181
Girardinus sp., 99
Glaridichthys uninotatus, 104
Glossophaga sp., 283, 292, 293
Glossophaga soricina, 282
Glyptosternum sp., 100
Gobius minutus, 102
Gobius niger, 102
Gobius paganellus, 102, 159
Graptemys geographica, 47
Graptemys ouachitensis, 47
Graptemys pseudogeographica, 47
Gulo gulo, 296, 299, 359
Gymnophthalmus underwoodi, 269
Gymnorhinus cyanocephalus, 382
Gymnura micrura, 272

Halichoeres sp., 74
Halichoeres grypus, 296
Haplochromis chrynosotus, 273
Haplochromis macrostoma, 273

Haplochromis polystigma, 273
Haplochromis wingatii, 252
Helarctos malayanus, 296
Helminthophis sp., 175
Hemicentetes sp., 172
Hemicentetes semispinosus, 135
Hemichromis bimaculatus, 338
Hemidactylus sp., 127
Hemidactylus frenatus, 205
Herpestes auropunctatus (auropunctato), 24, 25, 222
Herpestes brachyurus, 24
Herpestes edwardsi, 24
Herpestes fuscus, 24
Herpestes ichneumon, 24
Herpestes javanicus, 24
Herpestes sanguineus, 24
Heterandria formosa, 202, 251, 274
Heterodontis sp., 197, 273
Heterodontis francisci, 337
Heteropneustes spp., 101
Heteropneustes fossilis, 102, 140, 172, 189, 200, 306, 372
Hexanchus, 70
Hippocampus sp., 306
Hippocampus erectus, 274, 340
Hippocampus hippocampus, 274
Hippoglossoides platessoides, 172, 199
Homo sp., 283
Homo sapiens, 27, 117, 135, 161, 190, 222, 230
Hopolopsis bucatta, 205, 375
Horaichthys setnai, 101
Hyaena crocuta, 136
Hybognathus nuchalis, 199
Hydrolagus colliei, 123
Hydrophis sp., 276
Hyla arborea japonica, 105
Hyla cinerea, 340, 374
Hyla crucifer crucifer, 374
Hyla regilla, 142, 146
Hynobius sp., 105
Hynobius nebulosus, 32, 204, 342, 343
Hynobius retardatus, 47
Hynobius tokyoensis, 32
Hypseleotris galii, 198

Ichthyosaurus sp., 275
Ictalurus catus, 102, 199
Ictalurus furcatus, 102
Ictalurus natalis, 199
Ictalurus nebulosus, 102, 199
Ictalurus punctatus, 188, 199
Isoodon macrourus, 279

Jenynsia sp., 201
Jenynsia lineata, 103
Junco hyemalis, 111, 149, 378

Kinosternon spp., 164
Kinosternon subrubrum, 145, 146

Subject Index

Chilko sockeye salmon, GTH concentration in, 197–198
Chipmunk
  ambient temperature and reproduction of, 391
  haemodichorial placenta of, 299
  photoperiod and reproduction of, 391
Chlorpromazin, and ovulation, 182
Chorio-allantoic placenta, 276–277
Circadian rhythm, 92, 196, 364–367
  and migration, 335–336
  suprachiasmatic nucleus and, 383
Circannual rhythm, 365–368, 386
  and migration, 196, 335–336
  and reproduction, 375
Circhoral LH-RH infusions, 229
Circhoral LH release, 229
Clasper
  anatomy of, 97–98
  and copulation, 254
Clitoris, 136
Cloacal glands, 126
  and spermatophores, 106
Coitus. See Copulation
Collagenase, and ovulation, 187
Communal nesting, 277
Copulation, 254
  and ovulation, 221–222, 235–237
  and pseudopregnancy, 241, 288
Copulatory organs, 97–117
Copulatory plug, 109, 254
Coqui
  spermiation in, 142
  vocalizations of, 374–375
Coral reef fish
  sex reversal in, 20, 75–76, 338
  sexual behavior of, 338
Corona radiata, 130
Corpus atreticum, 122–126
  and ovarian response, 126, 206
Corpus luteum (CL)
  accessory, 131, 229
  in androgenized rats, 54
  aromatization in, 193
  and blastocyst, 282
  coitus and, 288
  estrogen receptors in, 236
  formation, 131
  function of, 222–249
  and gestation, 206, 300
  or luteal gland, 131
  maintenance of, 224–249
  and parturition, 306, 311–321
  and pelvic nerve, 244
  and pouch young removal, 218–220, 280
  and progestin secretion, 192–193, 222–249
  and relaxin secretion, 193
  rescue of, 301

"Corpus luteum," 124, 197
  defined, 122
  and egg retention, 206
Corticosteroids
  and brain differentiation, 67–68
  and oocyte maturation, 173–174
  and ovulation, 180–181, 306
  and parturition, 311–322
  and puberty, 92–93
  and reproductive cycle, 198
Corticosterone
  and oviposition, 181, 211, 308
  and ovulation, 211, 214–216
Cortisol
  ovarian secretion of, 174
  and sexual behavior, 359
Corticin, 42
Cow. See Cattle
Cowper's gland, 113
Cremaster muscle, 111–112
Crepuscular LH peak, 211, 215–216
Crocodile
  androgen and, 109
  parental behavior of, 346
  signaling by, 377
Cross-fostering
  and adult behavior, 353–354
  and migration, 354
Cryptorchidism, and spermatogenesis, 167
Cumulus oophorus, 130
  and fertilization, 263
  and oocyte maturation, 185
  and ZP, 185
Cyclic 3′,5′adenosinemonophosphate (cAMP)
  and FMP and oocyte maturation, 182
  and ovulation, 183, 186
  and smooth muscle, 323
  and steroidogenesis, 186–187
Cycloheximide
  and androgenization, 64
  and sexual behavior, 358–359
Cyclostomes
  anatomy of, 96–117
  androgen secretion by, 157
  hermaphrodites in, 70
  nest building by, 270, 337
  ovulation in, 180, 197
  reproductive cycles of, 197
  spawning by, 305
  spermatogenesis in, 139
  vitellogenesis in, 171
Cynomolgus monkey, vasectomy, 326
Cyproterone acetate
  and distribution of PGCs between left and right gonad, 43–44
  and duct system differentiation, 49
Cytochalasin B, and ovulation, 182

Daily rhythm, 92, 364
  of LH release, 240
Decapacitation factor, 262
Deer
  antlers and androgen in, 169
  hormone concentrations in, 389
Defeminization, 52–67
Delayed anovulatory syndrome, 54–55
Delayed implantation
  facultative, 297–298
  obligatory, 294–297
  and photoperiod, 294–295, 387
Deoxycorticosterone
  and oocyte maturation, 173
  synthesis by ovary, 187
  and ovulation, 181
11-Deoxycortisol
  and ovulation, 181
  and parturition, 306
  synthesis by ovary, 184
Deoxyribonucleic acid (DNA)
  and brain differentiation, 63–64
  during spermatogenesis, 137
Dexamethasone
  and ovulation, 215
  and parturition, 311–315, 321
Diandry, definition of, 72
Dietary factors
  as proximate causes, 382, 397–398
  and reproduction, 382, 389–390, 397–398
  and sex ratio, 36
Diethylstilbestrol (DES), 92, 178
17α,20α-Dihydro-pregn-4-en-3-one
  in follicles, 187
  and oocyte maturation, 174
  and ovulation, 181
Dihydrotestosterone
  and androgenization, 62
  and defeminization, 63
  concentration by laminar nucleus of torus semicircularis, 374
  concentrations in blood, 208–209, 211, 214
  secretion of, 164
  and sexual behavior, 344, 346–347, 350, 355–356
  and spermatogenesis, 154
  and vocalizations, 374
Dog
  anatomy of, 120
  false pregnancy in, 234
  hormone concentrations in, 233–234
  nest building by, 359
  ovulation in, 172, 179, 234
  parturient behavior of, 360
  relaxin in, 193
  reproductive cycle of, 233–234
  sexual behavior of, 57–59, 355

Dihydrotestosterone propionate. *See*
    Dihydrotestosterone
Donkey
    cross with horse and length of ges-
        tation, 309
    diffuse placenta of, 299
    secretion of PMSG by, 303
Dopamine
    and androgenization, 64–65
    and prolactin, 90–91
    and puberty, 90–91
Duck
    captivity of, 382
    circannual rhythm in, 368
    continuous breeding in, 206
    copulatory organ of, 110
    gonadal differentiation in, 46
    olfactory cues in, 381
    reproductive cycle of, 216
    vitellogenesis in, 179
Duck-billed platypus
    breeding cycle of, 216
    estradiol concentration in blood of,
        217
    incubation of eggs by, 279
    testes of, 111
Ductus deferens, 100, 102, 104, 105

East African viviparous toad
    CL in, 125
    embryos of, 202–203, 274–275
    reproductive cycle of, 202–203
Eastern chipmunk, medullary folli-
    cles and CL, 131
Edible frog, 374
Eel
    gonadotrophin cells of, 199
    migration of, 332–333
Efferent ducts, 97
Egg
    care of, 275
    clutch size, 209
    expulsion of, 306
    formation of, 127–129, 211
    incubation of, 277–279
    See also Oocyte; Oogonia; Ovum
        transport
Ejaculation, 254–259
Elasmobranchs
    anatomy of, 96–98, 122–123
    androgen secretion by, 157
    copulation by, 254
    estrogen secretion by, 187
    gestation period of, 197
    gonadotrophin, 171
    hermaphrodites in, 70
    oviductal contractions in, 305
    oviparity in, 270–273
    ovulation in, 180
    placental anatomy of, 272
    reproductive cycles of, 197

Elasmobranchs (*cont.*)
    spermatogenesis in, 139–140
    vitellogenesis in, 171, 178
    viviparity in, 270–273
Electrodes, 55
Elephant
    estrous cycle of, 233
    progesterone secretion in, 233
Embryo
    ambient temperature and, 391
    care of, 270–304
    and CL, 224–228
    development of, 272–273
    diapause of, 217–221, 280–282,
        294–298
    losses of, 289, 391
    nutrition of, 271, 274, 286
    oviparity versus viviparity of,
        271–272
    oxygen supply for, 274
    protection of, 273
    transuterine migration of, 237
    See also Blastocyst; Zygote
Endometrium
    and gonadal hormones, 288
    and pouch young removal, 280
    and trophoblast, 287, 290–294
    and zygote, 288
English sparrow (house sparrow)
    external coincidence model for,
        366–367
    incubation patch in, 278
    internal coincidence model for,
        366–367
    photosexual response of, 367, 377
    spermatogenesis in, 149
Epididymis, 97, 108
    acrosome changes in, 250
    function of, 112
    homologies of, 111
    sperm maturation in, 250
Epinephelus, gonadal evolution of, 73
Epinephrine
    and oviductal contractions, 305
    and ovulation, 181–183
Erection, 254–259
Estradiol
    and axonal growth, 66
    blood concentrations of, 208–209,
        211, 214, 217, 223–249
    and cervix, 310
    concentration by laminar nucleus
        of torus semicircularis, 374
    hypothalamic concentration of,
        66–67
    and maternal behavior, 339, 350,
        357–358
    and oocyte maturation, 173
    and ovarian weight, 173
    and ovulation, 214
    and ovum transport, 285

Estradiol (*cont.*)
    and puberty, 88–92
    Sertoli cell secretion of, 166
    and sex reversal, 27
    and sexual behavior, 339, 350,
        357–358
Estradiolbenzoate. *See* Estradiol
Estradiolcyclopentyl propionate. *See*
    Estradiol
Estriol
    and oocyte maturation, 173
    and ovarian weight, 173
Estrogen receptors, 62, 236
Estrogens
    AFP and binding of, 92
    and blastocyst, 286–290
    and brain differentiation, 52–67
    and cancer of prostate, 114
    CL and, 194
    and endometrium, 288
    and follicle, 176
    and food intake, 194
    functions of, 194–195
    and gonadal differentiation, 43
    and gonoduct differentiation, 49
    and implantation, 288, 297
    and incubation patch, 278
    luteolytic effect of, 228
    luteotrophic effect of, 228, 232,
        236
    and nest building, 278
    and parturition, 311–321
    and peripheral nervous system,
        244, 278
    and puberty, 88, 94
    secretion of, 187–193
    and sexual behavior, 339, 345,
        350, 356–358
    in subterranean clover, 398
    and vitellogenesis, 87
    See also Estradiol; Estriol; Estrone
Estrone
    blastocyst secretion of, 286
    blood concentrations of, 208–209,
        211, 214
    luteotrophic effect of, 289
    and oocyte maturation, 173
    and ovarian maintenance, 173
Estrus
    induction of, 234, 237
    long, 233
    silent, 224
    split, 229
European badger
    delayed implantation in, 295–297
    zonary placenta of, 299
European bitterling
    fertilization in, 252
    and freshwater mussel, 125, 252,
        373
    ovipositor of, 125

European hare
  nest of, 359
  progesterone concentrations in,
    236–237
  reproductive cycle of, 236–237
  sperm storage in, 261
Eutheria
  gestation in, 282–304
  reproductive cycles of, 221–249
Extra-retinal receptors, 378–379

Factor *x*
  and Müllerian duct regression,
    49–51
  secretion by Sertoli cells, 50
Fat bodies
  reciprocal relation with ovary, 126
  and spermatogenesis, 104,
    142–143
  and vitellogenesis, 178–179
  and yolk precursors, 126, 178–179
Feedback
  and breeding season, 224
  negative, 231
  photorefractoriness and, 368
  positive, in rabbit, 235
  positive, versus stimulatory effect,
    55, 231
Female external genitalia, 135–136
Ferguson reflex, 322
Ferret
  blinding, and reproduction of,
    386
  circannual rhythm in, 386
  hormone concentrations in blood
    of, 239–240
  hypothalamic lesion in, 86
  photoperiod and reproduction of,
    382, 386–387
  pineal, and reproduction in,
    386–387
  puberty in, 86
  reproductive cycle of, 239–240,
    386
  zonary placenta of, 299
Fertilization, 204
  and acrosin, 263
  delayed, 221
  external, 252–253
  and hyaluronidase, 263
  internal, 253–264
  and polyspermy, 264–265
  proximity of gametes and, 252
  selective, and antigens, 326
  and sperm membrane plasma fu-
    sion, 264
  sperm penetration and, 263–264
Fetoplacental unit, function of,
    301–303
Fetus
  antigen production by, 329–330
  and length of gestation, 309

Fetus (*cont.*)
  and maternal preparturient
    changes, 309–310
  preparturient changes of, 309
  removal and parturition, 316, 321
Fibroblast, and ovulation, 187
Field vole, LH concentration in, 237
Fighting fish, sex reversal in, 74
Fluoxymesterone, and brain androge-
    nization, 62
Foam glands, 111
Follicle, 119, 122, 127, 170
  atretic, 130
  classification of, 130
  fate of, 124
  growth of, 175, 177–178, 205
  steroidogenesis in, 191–193
  stigma, of birds, 129
Follicle-stimulating hormone (FSH)
  ambient temperature and, 376
  and androgen secretion, 164
  in chicken, 81
  concentration in blood, 212–213,
    224–249
  in cryptorchid rats, 167
  and delayed implantation, 298
  and growth of follicles, 178
  and hierarchy of follicles, 176
  after immunization against testos-
    terone, 325
  inhibition by inhibin, 166–167
  and LH binding, 185
  luteotrophic effect of, 248
  and meiosis, 184
  and ovulation, 183
  as ovulation-inducing hormone,
    185
  photoperiod and, 384
  and puberty, 89–93
  purified chicken, 147, 176
  in quail, 81
  in rats, 81–83, 89–92
  and Sertoli cell, 154
  and spermatogenesis, 141–156
  and spermiation, 142
  starvation and, 398
  steroidogenesis and, 191–193
  stimulation of follicles by, 213
  stress and, 399
  and [^{3}H]-thymidine incorporation,
    176
  and vitellogenesis, 179
Food intake
  corticosterone and, 335–336
  estrogens and, 194
  prolactin and, 335–336
Forward mobility protein (FMP), 250
Fowlpox, and parthenogenesis, 269
Foxes
  nest building by, 359
  ovulation of primary oocyte in,
    172, 179

Foxes (*cont.*)
  parturient behavior in, 360
  photoperiod and embryo mortality,
    387
Freemartin
  androgen and, 78
  in cattle, 26–28, 77–78
  in chicken, 28, 76
  chimerism in, 26, 78
  H-Y antigen in, 27, 78
  LH release in, 59–60
  sexual behavior of, 59
Freshwater mussel, and European
    bitterling, 125
Frogs
  differentiated races of, 44
  gonadal differentiation in, 46
  semidifferentiated races of, 44
  undifferentiated races of, 44
  See also species names
Funnel (infundibulum) of oviduct,
    123, 126–129, 132

Gametes. See Oocyte; Sperm
Gametogenesis. See Oogenesis;
    Spermatogenesis
Gap junctions, 175, 323
Garter snake
  CL and parturition in, 306
  copulatory plug of, 109
  hypophysectomy and parturition
    in, 306
  hypophysectomy and sper-
    matogenesis in, 146
  olfactory cues in, 375–376
  sexual behavior of, 345
  sperm storage in, 261
  yolk sac placenta of, 275–276
Gen, and nest parasitism, 277
Genetic selection for modified sex
    ratio, 36
Genitofemoral nerve, and lordosis re-
    flex, 244
Gerbil
  male reproductive system of, 120
  water requirement of, 397
Germinal epithelium, 122, 129
Germinal vesicle breakdown
    (GVBD), 181, 184
Gestation period
  and allograft rejection, 279
  duration of, 300
  and genotype, 309
  heritability of effect of sire on, 309
  and immune response, 331
  litter size and, 309
  and sex of fetus, 309
  and time of breeding, 309
β-Glucuronidase, sperm capacitation,
    262
Goat
  adrenal and abortion in, 310

Pregnant mare's serum
    gonadotrophin, 229
    effect of fetus on secretion of,
        301–303
    and hierarchy of follicles, 176
    photoperiod and, 380
    pregnancy and, 301
    and puberty, 88
    and spermatogenesis, 139, 142
    stimulation of chicken ovary, 87
Pregnenolone
    and oocyte maturation, 173
    secretion by CL, 192–193
Preoptic-anterior hypothalamus
    and cyclic GTH release, 88
    and incubation behavior, 351–352
    and maternal behavior, 362
    and sexual behavior, 345, 356,
        358–359
Preoptic area, 240
    aromatase in, 62
    culture in vitro of, 66
    effect of androgenization on preop-
        tic-suprachiasmatic area,
        54–55
    electrochemical stimulation of, and
        LH release, 55
    latency of evoked potentials from,
        379
    preoptic-suprachiasmatic area and
        cyclic LH release, 54–55
    and sexual behavior, 351
    sexual dimorphism of, 64–65
    stimulation of, and oviposition,
        307
    and vocalizations, 374
Preputial glands, 114
Primates, perinatal androgens and
    cyclic GTH release, 56. See
    also Humans; Rhesus monkey
Primordial germ cells (PGCs)
    ameboid movement of, 38
    biochemical changes of, 41
    and condensed X chromosome, 34
    distribution between left and right
        gonad primordia, 40–41
    gonadal steroids and distribution
        of, 43–44
    location of, 38–39
    migration of, 38
    and sex differentiation, 40
    site of formation of, 41
    survival of XX PGC in XY soma,
        41
Procollagenase, and ovulation, 187
Progesterone, 187–193, 204–249,
        281–286, 311–322
    and avidin secretion, 195
    and blastocyst, 281–286
    and cAMP, 182
    and cervix, 310

Progesterone (cont.)
    concentration during reproductive
        cycle, 204–249
    and cumulus oophorus, 185
    and delayed implantation,
        295–298
    and embryonic development, 203
    and endometrium, 288
    functions of, 195
    and immune response, 331
    and incubation behavior, 278,
        351–352
    and meiosis, 182
    metabolic clearance and production
        of, 304
    and morula, 286
    and oviposition, 307–308
    and ovulation, 186
    and parental behavior, 351–352,
        360–362
    and parturition, 311–322
    and PMSG, 301–304
    secretion by CL, 192–193
    secretion by ovary, 187–193
    secretion by testis, 164
    and sexual behavior, 344
    and skin receptors, 278
    and sperm capacitation, 262
Progesterone receptor, in hypo-
    thalamus, 212
Progestins
    and oocyte maturation, 174
    and ovulation, 181–182, 306
    secretion by different vertebrates,
        188–190
Prolactin
    and androgen secretion, 164–166
    binding by interstitial cells, 165
    concentration during reproductive
        cycle, 212–213
    concentration in rat, 242, 245–247
    and delayed implantation,
        294–295
    in dwarf mice, 165
    and GTH secretion, 85
    and increase in cholesterol con-
        centration of testis,
        165–166
    and incubation behavior, 278–279
    and incubation patch, 278
    and kidney, 334
    and LH binding by testis, 165
    luteolytic effect of, 246
    luteotrophic effect of, 246
    and maintenance of CL, 226
    and maternal behavior, 360–362
    and neonatal androgenization,
        55–56
    olfactory cues and secretion of,
        394
    and parental behavior, 339, 347

Prolactin (cont.)
    photoperiod and secretion of, 384,
        388–389
    and pigeon milk, 349, 353
    and pouch breeding, 274
    and puberty, 84–88, 89–93
    in rats, 242–247
    and removal of pouch young, 280
    and skin secretion, 274
    and spermatogenesis, 156
    stimulation of FSH secretion, 166
    synergism with LH, 165
    and teleost migration, 333
    twice daily surges in rat, 245–247
    and uterine sensitivity, 90–91
    and waterdrive in newts, 334
Prolactin inhibiting factor, suckling
    and, 219. See also Dopamine
Pronghorn, embryonic losses in, 289
Proprostate glands, 112
Prostacyclin and CL, 227
Prostaglandins, 181–187
    administration of, 227
    and cervix, 309
    and implantation, 288–289
    luteolytic effect of, 225–236, 248
    and oviposition, 307
    and ovulation, 181–183, 186–187,
        306
    and parturition, 311–322
    production in ovary, 194
    and sperm transport, 260
    stimulation of progesterone pro-
        duction, 230
Prostate, 48
    cancer of, 114
    in female, 135
    in male, 113–115
Prostatic utricle. See Uterus
    masculinus
Protandrous hermaphroditism, 72,
    75
Protein synthesis
    and brain differentiation, 63–64
    and GVBD, 181
    and sexual behavior, 358
Protogynous hermaphroditism, 72–75
Proximate cause (of reproduction)
    ambient temperature as, 198,
        200–205, 371, 375–376,
        395
    food availability as, 200, 205,
        217, 233, 382, 389–390
    humidity as, 205
    photoperiod as, 196, 198,
        207–209, 217, 223, 228,
        239, 371, 377–380,
        382–389
    rainfall as, 200, 202, 205, 207,
        217, 233, 373–376,
        381–382

Scrotum, 111–112
  function and evolution of, 112,
    167
Seahorse
  deposition of oocytes in, 252
  expulsion of young by, 306
  hypophysectomy, effect of, in,
    274
  male pouch of, 252, 274
  prolactin and brood pouch of, 274
  sperm release by, 252
Seals
  nestbuilding by, 359
  penis of, 117
Seaperch
  nutrition of embryo in, 274
  reproductive cycle of, 201
  spermatogenesis in, 140, 198
Secondary sex characteristics, 52,
    95, 169
Semelparity, 399
Seminal plug, function of, 241–244
Seminal receptacles, 127
Seminal vesicles, 102, 111–114
  of mammals, 112–116
  sperm storage below body tem-
    perature in birds, 111
  sperm storage in, in anurans, 106
  of teleosts, 102
"Seminal vesicles," for sperm stor-
    age in females, 127
Septum, aromatase in, 62
Serine proteases, and ovulation, 187
Serotonin
  and androgenization of the brain,
    64
  and ovulation, 187
  and reproduction of ferret, 387
  and sperm transport, 260
Serranidae, hermaphroditism in, 71
Sertoli cells
  and estradiol secretion, 166
  FSH binding by, 146
  functions of, 101, 167
  homology in teleosts, 101
  inhibin secretion by, 166–167
  and interstitial gland, 143
  phagocytosis by, 139
  secretion of anti-Müllerian duct
    hormone by, 50
  two-cell-two-GTH hypothesis for
    estradiol synthesis by,
    166
Sex, 19–37
  diagnosis of genetic, 33
  genotypic, 22
  heterogametic, 23–32
  homogametic, 23
  phenotypic, 22
Sex chromatin, 33–35. See also
    Lyon hypothesis
Sex chromosomes, 23–36

Sex-determining mechanism
  consequences of genetic, 31–32
  environmental, 46–47
  genetic (XX-XY; ZW-ZZ), 23, 27,
    30–32
Sex inversion, 72
Sex ratio
  blood groups and, 329
  hepatitis virus and, 36–37
  modification of, 35–37
  primary, 21
  secondary, 21–23
  tertiary, 21–23
Sex reversal
  genetic, 24–29
  sequential, 71–76
  social situation, 20, 338
Sexual behavior, 207, 337–359
  attractiveness and, 357
  bisexual, 52
  classification of, 354
  effect of perinatal sex hormones
    on, 56–60
  of freemartin, 59
  genetic component of, 338,
    346–347
  hypothalamus and, 60
  limbic system and, 255–258
  long and short duration tests for,
    56
  olfactory cues and, 373
  and orientation, 354
  perinatal condition and, 354
  peripheral sex organs and, 60–61
  proceptivity and, 357
  receptivity and, 357
  spinal cord and, 60
  testosterone and, 57, 341
Sexual development, 38–69, 80–95
  development of secondary sex
    characteristics, 52, 169
  differentiation of the brain, 53–62
  differentiation of the gonad, 38–52
  differentiation of the gonoducts,
    49–52
  differentiation of the spinal cord,
    60
Sexual parasites, 340
Sexual skin, 136
Sharks
  AP and gonadal differentiation in,
    46
  aromatase and preoptic hypothala-
    mic tissue in, 62
  contractions of oviduct in, 305
  oophagy and cannibalism of em-
    bryo in, 271
  oviparity and viviparity in,
    270–273
  sexual behavior of, 337
  spermatogenesis in, 138–140
  trophonemata of, 271–272

Sheep
  anatomy of, 133
  attachment of embryo in, 287
  body weight and puberty in, 94
  CL maintenance in, 224–226
  cotyledonary placenta of, 299
  effect of sire on length of gestation
    in, 309
  endocrinology of parturition in,
    318–320
  endocrinology of pregnancy of,
    301–304
  estrogen synthesis in, 192
  fetal mass and length of gestation
    in, 309
  fetal sex and length of gestation
    in, 309
  gonadotrophins at puberty in, 84
  hormone concentration in,
    223–225
  maternal behavior of, 362–363
  migration of primordial germ cells
    in, 38
  photoperiod and reproduction of,
    223–226, 382, 387–389
  pineal and reproduction of, 388
  plants and length of gestation in,
    310
  progestins in, 193
  relaxin in, 193
  reproductive cycle of, 223–226
  sexual maturation of, 94
  superior cervical ganglion and re-
    production of, 388–389
  trophoblast and endometrium in,
    287
  uterine gap junctions in, 323
Shell deposition, 217
Shell gland
  of birds, 129
  of elasmobranchs, 123
Short-tailed weasel, progesterone and
    delayed implantation in, 295
Shrew
  fertilization in follicle in, 172
  male reproductive system of, 120
Siphon sac and copulation, 98, 254
Skates
  breeding cycle of, 197
  contractions of oviduct, 305
  oviparity and viviparity in, 271
  sexual behavior of, 337
  sperm motility in, 250
Skinks
  care of eggs by, 275
  chorio-allantoic placentas,
    276–277
Skunk cabbage
  and abnormal embryos, 310
  and length of gestation, 310, 398
Slate-colored junco
  photorefractoriness in, 378

Reproductive Physiology of Vertebrates

Designed by G. T. Whipple, Jr.
Composed by The Composing Room of Michigan, Inc.
in 10 point Times Roman, 2 points leaded,
with display lines in Times Roman.
Printed offset by Vail-Ballou Press, Inc.
on Warren's Olde Style, 60 pound basis.
Bound by Vail-Ballou Press, Inc.
in Kivar 5 cover material
and stamped in All Purpose foil.

Library of Congress Cataloging in Publication Data

Van Tienhoven, Ari.
    Reproductive physiology of vertebrates.

    Includes index.
    1. Reproduction. 2. Vertebrates—Physiology.
I. Title. [DNLM: 1. Reproduction. 2. Physiology,
Comparative. 3. Vertebrates—Physiology. QP 251 V282r]
QP251.V35   1983   596'.016   82-71595
ISBN 0-8014-1281-1